International Commission for Optics
Commission Internationale d'Optique
VOLUME VI

Advances in Information Optics and Photonics

Ari T. Friberg, René Dändliker
Editors

Bellingham, Washington USA

Library of Congress Cataloging-in-Publication Data

Advances in information optics and photonics / Ari T. Friberg and René Dändliker.
    p. cm. -- (Press monograph ; PM183)
 Includes bibliographical references and index.
 ISBN 978-0-8194-7234-2
 1. Optical communications. 2. Photonics. I. Friberg, Ari T., 1951- II. Dändliker, René.
 TK5103.59.A35 2008
 621.382'7--dc22

                                                                       2008022953

Published by

SPIE
P.O. Box 10
Bellingham, Washington 98227-0010 USA
Phone: +1 360 676 3290
Fax: +1 360 647 1445
Email: spie@spie.org
Web: http://spie.org

Copyright © 2008 Society of Photo-Optical Instrumentation Engineers

All rights reserved. No part of this publication may be reproduced or distributed
in any form or by any means without written permission of the publisher.

The content of this book reflects the work and thought of the author(s).
Every effort has been made to publish reliable and accurate information herein,
but the publisher is not responsible for the validity of the information or for any
outcomes resulting from reliance thereon.

Printed in the United States of America.

Contents

List of Contributors   ix
Preface   xv
ICO International Trends in Optics Series History   xix

I. Beam Optics

1. First-Order Optical Systems for Information Processing   1
   Tatiana Alieva

2. Applications of the Wigner Distribution to Partially Coherent Light Beams   27
   Martin J. Bastiaans

3. Characterization of Elliptic Dark Hollow Beams   57
   Julio C. Gutiérrez-Vega

4. Transfer of Information Using Helically Phased Modes   77
   Miles Padgett, Graham Gibson, and Johannes Courtial

II. Laser Photonics and Components

5. Microoptical Components for Information Optics and Photonics   89
   Christof Debaes, Heidi Ottevaere, and Hugo Thienpont

6. Intracavity Coherent Addition of Lasers   117
   Vardit Eckhouse, Amiel A. Ishaaya, Liran Shimshi, Nir Davidson, and Asher A. Friesem

7. Light Confinement in Photonic Crystal Microcavities   137
   Philippe Lalanne and Christophe Sauvan

8. Limits to Optical Components 153
   David A. B. Miller

III. Electromagnetic Coherence

9. An Overview of Coherence and Polarization Properties for Multicomponent Electromagnetic Waves 171
   Alfredo Luis

10. Intrinsic Degrees of Coherence for Electromagnetic Fields 189
    Philippe Réfrégier and Antoine Roueff

IV. Imaging, Microscopy, Holography, and Materials

11. Digital Computational Imaging 209
    Leonid Yaroslavsky

12. Superresolution Processing of the Response in Scanning Differential Heterodyne Microscopy 229
    Dmitry V. Baranov and Evgeny M. Zolotov

13. Fourier Holography Techniques for Artificial Intelligence 251
    Alexander V. Pavlov

14. Division of Recording Plane for Multiple Recording and Its Digital Reconstruction Based on Fourier Optics 271
    Guoguang Mu and Hongchen Zhai

15. Fundamentals and Advances in Holographic Materials for Optical Data Storage 285
    Maria L. Calvo and Pavel Cheben

16. Holographic Data Storage in Low-Shrinkage Doped Photopolymer 317
    Shiuan Huei Lin, Matthias Gruber, Yi-Nan Hsiao, and Ken Y. Hsu

V. Photonic Processing

17. Temporal Optical Processing Based on Talbot's Effects ... 343
    Jürgen Jahns, Adolf W. Lohmann, and Hans Knuppertz

18. Spectral Line-by-Line Shaping ... 359
    Andrew M. Weiner, Chen-Bin Huang, Zhi Jiang, Daniel E. Leaird, and Jose Caraquitena

19. Optical Processing with Longitudinally Decomposed Ultrashort Optical Pulses ... 381
    Robert Saperstein and Yeshaiahu Fainman

20. Ultrafast Information Transmission by Quasi-Discrete Spectral Supercontinuum ... 405
    Mikhail A. Bakhtin, Victor G. Bespalov, Vitali N. Krylov, Yuri A. Shpolyanskiy, and Sergei A. Kozlov

VI. Quantum Information and Matter

21. Noise in Classical and Quantum Photon-Correlation Imaging ... 423
    Bahaa E. A. Saleh and Malvin Carl Teich

22. Spectral and Correlation Properties of Two-Photon Light ... 437
    Maria V. Chekhova

23. Entanglement-Based Quantum Communication ... 457
    Alexios Beveratos and Sébastien Tanzilli

24. Exploiting Optomechanical Interactions in Quantum Information ... 489
    Claudiu Genes, David Vitali, and Paolo Tombesi

25. Optimal Approximation of Non-Physical Maps via Maximum Likelihood Estimation ... 513
    Vladimír Bužek, Mário Ziman, and Martin Plesch

26. Quantum Processing Photonic States in Optical Lattices ... 533
    Christine A. Muschik, Inés de Vega, Diego Porras, and J. Ignacio Cirac

27. Strongly Correlated Quantum Phases of Ultracold Atoms in
    Optical Lattices   555
    Immanuel Bloch

VII. Communications and Networks

28. The Intimate Integration of Photonics and Electronics   581
    Ashok V. Krishnamoorthy

29. Echelle and Arrayed Waveguide Gratings for WDM and
    Spectral Analysis   599
    Pavel Cheben, André Delâge, Siegfried Janz, and Dan-Xia Xu

30. Silicon Photonics—Recent Advances in Device Development   633
    Andrew P. Knights and J. K. Doylend

31. Toward Photonic Integrated Circuit All-Optical Signal
    Processing Base on Kerr Nonlinearities   657
    David J. Moss and Benjamin J. Eggleton

32. Ultrafast Photonic Processing Applied to Photonic Networks   687
    Hideyuki Sotobayashi

Index   713

List of Contributors

Tatiana Alieva
Universidad Complutense de Madrid, Spain

Mikhail A. Bakhtin
State University of Information Technologies, Russia

Dmitry V. Baranov
General Physics Institute of Russian Academy of Sciences, Russia

Martin J. Bastiaans
Technische Universiteit Eindhoven, The Netherlands

Victor G. Bespalov
State University of Information Technologies, Russia

Alexios Beveratos
Alcatel de Marcoussis, France

Immanuel Bloch
Johannes Gutenberg-Universität, Germany

Vladimír Bužek
*Research Center for Quantum Information
and
QUNIVERSE, Slovakia*

María L. Calvo
Universidad Complutense de Madrid, Spain

Jose Caraquitena
Purdue University, USA

Pavel Cheben
National Research Council Canada

Pavel Cheben
National Research Council Canada

Maria V. Chekhova
M.V. Lomonosov Moscow State University, Russia

J. Ignacio Cirac
Max-Planck-Institut für Quantenoptik, Germany

Johannes Courtial
University of Glasgow, Scotland

Nir Davidson
Weizmann Institute of Science, Israel

Christof Debaes
Vrije Universiteit Brussel, Belgium

André Delâge
National Research Council Canada

J. K. Doylend
McMaster University, Canada

Vardit Eckhouse
Weizmann Institute of Science, Israel

Benjamin J. Eggleton
University of Sydney, Australia

Yeshaiahu Fainman
University of California San Diego, USA

Asher A. Friesem
Weizmann Institute of Science, Israel

Claudiu Genes
Università di Camerino, Italy

Graham Gibson
University of Glasgow, Scotland

Matthias Gruber
Fern Universität Hagen, Germany

Julio C. Gutiérrez-Vega
Optics Center Tecnológico de Monterrey, México

Yi-Nan Hsiao
National Chiao Tung University, Taiwan

Ken Y. Hsu
National Chiao Tung University, Taiwan

Chen-Bin Huang
Purdue University, USA

Amiel A. Ishaaya
Weizmann Institute of Science, Israel

Jürgen Jahns
FernUniversität Hagen, Germany

Siegfried Janz
National Research Council Canada

Zhi Jiang
Purdue University, USA

Andrew P. Knights
McMaster University, Canada

Hans Knuppertz
FernUniversität Hagen, Germany

Sergei A. Kozlov
State University of Information Technologies, Russia

Ashok V. Krishnamoorthy
Sun Microsystems, USA

Vitali N. Krylov
State University of Information Technologies, Russia

Philippe Lalanne
Université Paris-Sud, France

Daniel E. Leaird
Purdue University, USA

Shiuan Huei Lin
National Chiao Tung University, Taiwan

Adolf W Lohmann
University of Erlangen, Germany

Alfredo Luis
Universidad Complutense, Spain

David A. B. Miller
Stanford University, USA

David J. Moss
University of Sydney, Australia

Guoguang Mu
Nankai University, China

Christine A. Muschik
Max-Planck-Institut für Quantenoptik, Germany

Heidi Ottevaere
Vrije Universiteit Brussel, Belgium

Miles Padgett
University of Glasgow, Scotland

Alexander V. Pavlov
St. Petersburg State University for Information Technologies, Russia

Martin Plesch
Research Center for Quantum Information
and
QUNIVERSE, Slovakia

Diego Porras
Max-Planck-Institut für Quantenoptik, Germany

Philippe Réfrégier
Institut Fresnel, Aix-Marseille Université, France

Antoine Roueff
Institut Fresnel, Aix-Marseille Université, France

Bahaa E. A. Saleh
Boston University, USA

Robert Saperstein
University of California San Diego, USA

Christophe Sauvan
Université Paris-Sud, France

Liran Shimshi
Weizmann Institute of Science, Israel

Yuri A. Shpolyanskiy
State University of Information Technologies, Russia

Hideyuki Sotobayashi
Aoyama Gakuin University and National Institute of Information and
Communications Technology, Japan

Sébastien Tanzilli
Université de Nice Sophia-Antipolis, France

Malvin Carl Teich
Boston University, USA

Hugo Thienpont
Vrije Universiteit Brussel, Belgium

Paolo Tombesi
Università di Camerino, Italy

Inés de Vega
Max-Planck-Institut für Quantenoptik, Germany

David Vitali
Università di Camerino, Italy

Andrew M. Weiner
Purdue University, USA

Dan-Xia Xu
National Research Council Canada

Leonid Yaroslavsky
Tel Aviv University, Israel

Hongchen Zhai
Nankai University, China

Mário Ziman
Research Center for Quantum Information
and
QUNIVERSE, Slovakia

Evgeny M. Zolotov
General Physics Institute of Russian Academy of Sciences, Russia

Preface

This volume is the sixth in a series of books that the International Commission for Optics (ICO) edits for publication at the time of its triennial congresses. The earlier volumes have covered a broad scope of interests in optics at the time and have dealt with fundamental subjects, while the later editions have increasingly addressed advances in applied optics and photonics. The books previously published in the series are

- *International Trends in Optics*, ed. J. W. Goodman, USA (Academic Press, 1991)
- *Current Trends in Optics*, ed. J. C. Dainty, UK (Academic Press, 1994)
- *Trends in Optics – Research, Developments and Applications*, ed. A. Consortini, Italy (Academic Press, 1996)
- *International Trends in Optics and Photonics*, ed. T. Asakura, Japan (Springer, 1999)
- *International Trends in Applied Optics*, ed. A. H. Guenther, USA (SPIE Press, 2002)

The complete history of the ICO Book series, including the Tables of Contents of the previous volumes, can be found on p. xix of this book.

Besides highlighting the main developments of international optics and photonics, the aim of this book series is to promote the general awareness of the ICO and raise funds for its global activities, in particular the travelling lecturer program, which is aimed at enhancing optics in developing nations. Therefore all royalties will go to the ICO for that purpose.

In today's 'age of light,' optical information science and technology play a central role. The ICO has a long tradition in the subjects of information optics, dating back to the ICO topical meetings in Kyoto, Japan 1994 (Frontiers in Information Optics) and Tianjin, China 1998 (Optics for Information Infrastructure). The ICO has also been a permanent sponsor of the Optical Computing/Optics in Computing conferences, a series of meetings spanning well over a decade. In 2006, the ICO organized two key events on information optics: the ICO topical meeting on Optoinformatics / Information Photonics in St. Petersburg, Russia (Chairs A. V. Pavlov, M. L. Calvo, and J. Jahns) and the ICO/ICTP Winter College on Optics in Trieste, Italy, with title "Quantum and Classical Aspects of Information Optics" (Directors P. Tombesi, M. L. Calvo, and P. Knight). Additionally, the recent ICO Prizes – most notably those in 2003 (B. J. Eggleton), 2004 (A. V. Krishnamoorthy), 2005 (I. Bloch), and 2006 (H. Sotobayashi) – have dealt with various basic and applied aspects of optical information. Hence it was quite natural to take advantage of these developments and focus the current volume of the ICO Book series on Advances in Information Optics and Photonics.

The present volume VI differs from the previous ones in at least three respects: it concentrates on a specific, though extremely important, topic within the broad field of optics and photonics, it does not contain the words 'International Trends' explicitly in the title, and it is published as a paperback. We hope that with these changes the book will find its way as a standard reading and reference material on the topic. The volume consists of 32 invited contributions from scientists or research groups working throughout the world on optical information science, technology, and applications. Many of the authors have actively participated in the ICO conferences and other activities and all of them are internationally recognized leaders in their respective subjects.

Many new concepts in classical and quantum-entangled light, coherent interaction with matter, novel materials and processes have led to remarkable breakthroughs in information science and technology. While it is difficult, and sometimes even dangerous, to group the contributions under separate headings, we have divided the chapters of this book into 7 sections:

1. Beam Optics
2. Laser Photonics and Components
3. Electromagnetic Coherence
4. Imaging, Microscopy, Holography, and Materials
5. Photonic Processing
6. Quantum Information and Matter
7. Communications and Networks

The sections contain chapters that address optical information sciences broadly in the linear, nonlinear, classical, and quantum regimes and describe the foundations, state-of-the-art devices and technologies, as well as the diverse applications of information optics and photonics. It is hoped that the reader will find chapters that are directly relevant to his/her own work or otherwise will create interest in this fascinating, rapidly advancing, and highly potential subject.

We would like to express our sincere appreciation to all of the authors who have devoted their time, effort, and expertise to write the superb and timely contributions for this volume. We would also like to thank the staff of SPIE Press, and especially Merry Schnell, Gwen Weerts, and Eric Pepper, for their professional work to produce this high-quality publication for the benefit of the global optics and photonics community.

Ari T. Friberg
President, International Commission for Optics
Royal Institute of Technology (KTH), Stockholm, Sweden
Helsinki University of Technology (TKK), Espoo, Finland
University of Joensuu, Finland

René Dändliker
Past President, International Commission for Optics
President of the Swiss Academy of Engineering Sciences
University of Neuchâtel, Switzerland

ICO *International Trends in Optics* Series History

The first book in the series appeared in 1991 under the title "International Trends in Optics" in the *Lasers and Optical Engineering* series of Academic Press. The Editor was 1987–1990 ICO President, Prof. J.W. Goodman of Stanford University, USA. It includes the following chapters:

- Integrated Optics, OEICs, or PICs? *H. Kogelnik*
- Quantum Optoelectronics for Optical Processing, *D.A.B. Miller*
- Optics in Telecommunications: Beyond Transmission, *P.W.E. Smith*
- Microoptics, *Kenichi Iga*
- Holographic Optical Elements for Use with Semiconductor Lasers, *H.P. Herzig and R. Dändliker*
- Fibre-Optic Signal Processing, *B. Culshaw, I. Andonovic*
- Optical Memories, *Yoshito Tsunoda*
- How Can Photorefractives Be Used? *H. Rajbenbach and J.-P. Huignard*
- Adaptice Interferometry: A New Area of Applications of Photorefractive Crystals, *S.I. Stepanov*
- Water Wave Optics, *J. J. Stamnes*
- About the Philosophies of Diffraction, *A.W. Lohmann*
- The Essential Journals of Optics, *J.N. Howard*
- Optics in China: Ancient and Modern Accomplishments, *Z.-M. Zhang*
- Unusual Optics: Optical Interconnects as Learned from the Eyes of Nocturnal Insects, Crayfish, Shellfish, and Similar Creatures, *P. Greguss*
- The Opposition Effect in Volume and Surface Scattering, *J.C. Dainty*

- Influence of Source Correlations on Spectra of Radiated Fields, *E. Wolf*
- Quantum Statistics and Coherence of Nonlinear Optical Processes, *J. Peřina*
- One-Photon Light Pulses versus Attenuated Classical Light Pulses, *A. Aspect and P. Grangier*
- Optical Propagation through the Atmosphere, *A. Consortini*
- Are the Fundamental Principles of Holography Sufficient for the Creation of New Types of 3-D Cinematography and Artificial Intelligence? *Y. Denisyuk*
- Medical Applications of Holographic 3-D Display, *J. Tsujiuchi*
- Moiré Fringes and Their Applications, *O. Bryngdahl*
- Breaking the Boundaries of Optical System Design and Construction, *C.H.F. Velzel*
- Interferometry: What's New Since Michelson? *P. Hariharan*
- Current Trends in Optical Testing, *D. Malacara*
- Adaptive Optics, *F. Merkle*
- Triple Correlations and Bispectra in High-Resolution Astronomical Imaging, *G. Weigelt*
- Phase-Retrieval Imaging Problems, *J.R. Fienup*
- Blind Deconvolution—Recovering the Seemingly Irrecoverable! *R.H.T Bates and H. Jiang*
- Pattern Recognition, Similarity, Neural Nets, and Optics, *H.H. Arsenault, Y. Sheng*
- Towards Nonlinear Optical Processing, *T. Szoplik and K. Chalasinska-Macukow*
- New Aspects of Optics for Optical Computing, *V. Morozov*
- Digital Optical Computing, *S.D. Smith and E.W. Martin*
- Computing: A Joint Venture for Light and Electricity? *P. Chavel*

The second book in the series appeared under the title "Current Trends in Optics" in the *Lasers and Optical Engineering* series of Academic Press Limited, London, 1994 (ISBN 0-12-20720-4). The Editor is ICO Past President, Prof. J.C. Dainty of Imperial College, London. It includes the following chapters:

- Atomic Optics, *S.M. Tan and D.F. Walls*
- Single Atoms in Cavities and Traps, *H. Walther*

- Meet a Squeezed State and Interfere in Phase Space, *D. Krähmer, E. Mayr, K. Vogel and W.P. Schleich*
- Can Light Be Localized? *A. Lagendijk*
- Time-resolved Laser-induced Breakdown Spectrometry, *G. Lupkovics, B. Nemet and L. Kozma*
- Fractal Optics, *J. Uozumi and T. Asakura*
- On the Spatial Parametric Characterization of General Light Beams, *R. Martinez-Herrero and P.M. Mejias*
- To See the Unseen: Vision in Scattering Media, *E.P. Zege and I.L. Katsev*
- Backscattering Through Turbulence, *A.S. Gurvich and A.N. Bogaturov*
- Why is the Fresnel Transform So Little Known? *F. Gori*
- Fourier Curios, *A.W. Lohmann*
- The Future of Optical Correlators, *D. Casasent*
- Spectral Hole Burning and Optical Information Processing, *K.K. Rebane*
- Holographic Storage Revisited, *G.T. Sincerbox*
- Colour Information in Optical Pattern Recognition, *M.J. Yzuel and J. Campos*
- The Optics of Confocal Microscopy, *C.J.R. Sheppard*
- Diffraction Unlimited Optics, *A. Lewis*
- Super-resolution in Microscopy, *V.P. Tychinsky and C.H.F. Velzel*
- Fringe Analysis: Anything New? *M. Kujawinska*
- Diagnosing the Aberrations of the Hubble Space Telescope, *J.R. Fienup*
- Laser Beacon Adaptive Optics:Boom or Bust? *R.Q. Fugate*

The third book in the series appeared in August 1996 under the title *Trends in Optics—Research, Developments and Applications,* ISBN 0-12-186030-2. Like its two predecessors, it was published by Academic Press. The Editor is ICO Past President, Prof. Anna Consortini of Universita degli Studi di Firenze, Italy. The Museo ed Istituto della Scienza in Florence deserves thanks for its permission to use the photography of one of its Galileo Galilei lenses as cover illustration. The book includes the following chapters:

- A Short History of the Optics Group of the Willow Run Laboratories, *E.N. Leith*
- Bio-speckles, *Y. Aizu and T. Asakura*

- Photon Migration and Imaging of Biological Tissues, *G. Zaccanti and D. Contini*
- Direct Image Processing Using Artificial Retina Chips, *E. Lange, Y. Nitta and K. Kyuma*
- Principles and Development of Diffraction Tomography, *E. Wolf*
- Diffractive Optics: From Promise to Fruition, *J. Turunen and F. Wyrowski*
- Planar Diffractive Elements for Compact Optics, *A.A. Friesem and Y. Amitai*
- Resonant Light Scattering from Weakly Rough Metal Surfaces, *K.A. O'Donnell*
- Femtosecond Time-and-Space-Domain Holography, *A. Rebane*
- Holographic 3D disks Using Shift Multiplexing, *D. Psaltis, G. Barbastathis and M. Levene*
- Dense Optical Interconnections for Silicon Electronics, *D.A.B. Miller*
- Fan-in Loss for Electrical and Optical Interconnections, *J.W. Goodman and J.C. Lain*
- Signal Processing and Storage Using Hybrid Electro-Optical Procedures, *J. Shamir*
- Young's Experiment in Signal Synthesis, *J. Ojeda-Castaeda and A.W. Lohmann*
- Resolution Enhancement by Data Inversion Techniques, *C. de Mol*
- Electronic Speckle Pattern Interferometry: An Aid in Cultural Heritage Protection, *G. Schirripa Spagnolo*
- Numerical Simulation of Irradiance Fluctuations for Optical Waves Through Atmospheric Turbulence, *S.M. Flatte*
- Optical Scintillation Methods of Measuring Atmospheric Surface Fluxes of Heat and Momentum, *R.J. Hill*
- Coherent Doppler Lidar Measurements of Winds, *R. Frehlich*
- Doing Coherent Optics with Soft X-Ray Sources, *D. Joyeux, P. Jaegle and A. l'Huillier*
- Axially Symmetric Multiple Mirror Optics for Soft X-Ray Projection Microlithography, *S.S. Lee, C.S. Rim, Y.M. Cho, D.E. Kim and C.H. Nam*
- Olmec Mirrors: An Example of Archaeological American Mirrors, *J.J. Lunazzi*
- Galileo Galilei: Research and Development of the Telescope, *G. Molesini and V. Greco*

- GRIN Optics: Practical Elements, *C. Gomez-Reino and J. Linares-Beiras*
- Photorefractive Fibers: Fabrication and Hologram Construction, *F.T.S. Yu and S. Yin*
- Optical Morphogenesis: Dynamics of Patterns in Passive Optical Systems, *F.T. Arecchi, S. Boccaletti, E. Pampaloni, P.L. Ramazza and S. Residori*
- High Sensitivity Molecular Spectroscopy with Diode Lasers, *K. Ernst*
- Sub-micrometre Optical Metrology Using Laser Diodes and Polychromatic Light Sources, *C. Gorecki and P. Sandoz*
- A Physical Method for Colour Photography, *G.G. Mu, Z.L. Fang, F.L. Liu and H.C. Zhai*
- Multiwavelength Vertical Cavity Laser Arrays by Molecular Beam Epitaxy, *C.J. Chang-Hasnain, W. Yeun, G.S. Li and L.E. Eng*
- Compact Blue-Green Laser Sources, *W.J. Kozlovsy*

The fourth book in the series appeared in August 1999 under the title *International Trends in Optics and Photonics*. It was edited by Prof. T. Asakura and published by Springer-Verlag as Volume 74 of the Springer Series in Optical Sciences. The book includes the following chapters:

- Optical Twist, *A.T. Friberg*
- Principles and Fundamentals of Near Field Optics, *M. Nieto-Vesperinas*
- Spin-Orbit Interaction of a Photon: Theory and Experiment on the Mutual Influence of Polarization and Propagation, *N.D. Kundikova and B.Ya. Zel'dovich*
- Atoms and Cavities: the Birth of a Schroedinger Cat of the Radiation Field, *J.-M. Raimond and S. Haroche*
- Quantum Tomography of Wigner Functions from Incomplete Data, *V. Buzek, JG. Drobny and H. Wiedemann*
- Some New Aspects on the Resolution in Gaussian Pupil Optics, *S.S. Lee, M.H. Lee and Y.R. Song*
- Multichannel Photography with Digital Fourier Optics, *G.-G. Mu, L. Lin and Z.-Q. Wang*
- Holographic Optics for Beamsplitting and Image Multiplication, *A.L. Mikaelian, A.N. Palagushkin and S.A. Prokopenko*
- Image Restoration, Enhancement and Target Location with Local Adaptive Linear Filters, *L. Yaroslavsky*

- Fuzzy Problem for Correlation Recognition in Optical Digital Image Processing, *G. Cheng, G. Jin, M. Wu and Y. Yan*
- All-Optical Regeneration for Global-Distance Fiber-Optic Communications, *E. Desurvire and O. Leclerc*
- Non Quantum Cryptography for Secure Optical Communications, *J.P. Goedgebuer*
- Pulsed Laser Deposition: An Overview, *I.N. Mihailescu and E. Gyorgy*
- Absolute Scale of Quadratic Nonlinear-Optical Susceptibilities, *I. Shoji, T. Kondo and R. Ito*
- Femtosecond Fourier Optics: Shaping and Processing of Ultrashort, Optical Pulses, *A.M. Weiner*
- Aperture Modulated Diffusers (AMDs), *H.P. Herzig and P. Kipfer*
- Optical Properties of Quasi-Periodic Structures: Linear and Nonlinear Analysis, *M. Bertolotti and C. Sibilia*
- Diffractive Optical Elements in Materials Inspection, *R. Silvennoinen, K.-E. Peiponen and T. Asakura*
- Multiple-Wavelength Interferometry for Absolute Distance Measurement, *R. Dandliker and Y. Salvade*
- Speckle Metrology—Some Newer Techniques and Applications, *R.S. Sirohi*
- Limits of Optical Range Sensors and How to Exploit Them, *G. Hausler, P. Ettl, M. Schenk, G. Bohn and I. Laszlo*
- Imaging Spectroscopy for the Non-Invasive Investigations of Paintings, *A. Casini, F. Lotti and M. Picollo*
- Optical Coherence Tomography in Medicine, *A.F. Fercher and C.K. Hitzenberger*
- The Spectral Optimization of Human Vision: Some Paradoxes, Errors and Resolutions, *B.H. Soffer and D.K. Lynch*
- Optical Methods for Reproducing Sounds from Old Photograph Records, *J. Uozumi and T. Asakura*

The fifth book in the series appeared in August 2002 under the title *International Trends in Applied Optics*. It was edited by Past President of ICO, Arthur H. Guenther and published by SPIE. The book includes the following chapters:
- Ultrashort-Pulse Laser-Matter Interaction and Fast Instabilities, *M. N. Libenson*

- Ultrafast Mode-locked Lasers for the Measurement of Laser Frequencies and as Optical Clockworks, *R. Holzwarth, T. Udem, and T. W. Hänsch*
- Ablation of Metals with Femtosecond Laser Pulses, *S. I. Anisimov*
- Laser Microprocessing and Applications in Microelectronics and Electronics, *Y. Feng Lu*
- There are No Fundamental Limits to Optical Lithography, *S. R. J. Brueck*
- Laser-produced Rapid Prototyping in Manufacturing, *Y. P. Kathuria*
- Computer Numerically Controlled Optics Fabrication, *H. Pollicove and D. Golini*
- Interference Coatings for the Ultraviolet Spectral Region, *N. Kaiser*
- Standardization in Optics Characterization, *D. Ristau*
- Advances in Thin Films, *K. Lewis*
- Micro-Optics for Spectroscopy, *R. Dändliker, H. P. Herzig, O. Manzardo, T. Scharf, and G. Boer*
- Defense Optics and Electro-Optics, *G. J. Simonis, G. Wood, Z. G. Sztankay, A. Goldberg, and J. Pellegrino*
- Recent Progress in System and Component Technologies for Fiber Optic Communication, *N. Shibata*
- Short-Distance Optical Interconnections with VCSELs, *H. Thienpont and V. Baukens*
- Spontaneous Emission Manipulation, *M. O. Scully, S. Zhu, and M. S. Zubairy*
- Progress in Fiber Optics and Optical Telecommunication, *A. K. Ghatak and B. P. Pal*
- Nano- and Atom Photonics, *M. Ohtsu*
- Binary Image Decompositions for Nonlinear Optical Correlations, *H.H. Arsenault and P. García-Martínez*
- Optical Pattern Recognition of Partially Occluded Images, *K. Chałasińska-Macukow*
- Optical Sensing by Fiber and Integrated Optics Devices, *G. C. Righini*
- Wave-optical Engineering, *F. Wyrowski and J. Turunen*
- Neutron Optics, Neutron Waveguides, and Applications, *M. L. Calvo and R. F. Alvarez-Estrada*
- Polarimetric Imaging, *P. Réfrégier, F. Goudail, and P. Chavel*
- Atmospheric Compensation, *R. Q. Fugate*

Chapter 1
First-Order Optical Systems for Information Processing

Tatiana Alieva
Universidad Complutense de Madrid, Spain

1.1 Introduction
1.2 Canonical Integral Transforms: Definition and Classification
    1.2.1 Definition
    1.2.2 Generalized imaging transforms
    1.2.3 Orthosymplectic canonical transforms
    1.2.4 Canonical transforms for the case $\det \mathbf{B} = 0$
1.3 Main Properties of the Canonical Integral Transforms
    1.3.1 Parseval theorem
    1.3.2 Shift theorem
    1.3.3 Convolution theorem
    1.3.4 Scaling theorem
    1.3.5 Coordinates multiplication and derivation theorems
1.4 Canonical Integral Transforms of Selected Functions
    1.4.1 Plane wave, chirp, and Gaussian functions
    1.4.2 Periodic functions
    1.4.3 Eigenfunctions for the canonical integral transforms
1.5 Generalized Convolution for Analog Optical Information Processing
    1.5.1 Analog optical information processing
    1.5.2 Generalized convolution: Definition
    1.5.3 Filtering in fractional Fourier domains
    1.5.4 Pattern recognition
    1.5.5 Localization of the generalized chirp signals
    1.5.6 Security applications
1.6 Other Optical Computing Approaches via Orthosymplectic Transforms
    1.6.1 Mode presentation on orbital Poincaré sphere

1.6.2 Orbital angular momentum manipulation
1.6.3 Geometric phase accumulation
References

1.1 Introduction

During the last decades, optics is playing an increasingly important role in acquisition, processing, transmission, and archiving of information. In order to underline the contribution of optics in the information acquisition process, let us mention such optical modalities as microscopy, tomography, speckle imaging, spectroscopy, metrology, velocimetry, particle manipulation, etc. Data transmission through optical fibers and optical data storage (CD, DVD, as well as current advances of holographic memories) make us everyday users of optical information technology. In the area of information processing, optics also has certain advantages with respect to electronic computing, thanks to its massive parallelism, operating with continuous data, possibility of direct penetration into the data acquisition process, implementation of fuzzy logic, etc.

The basis of the analog coherent optical information processing is the ability of a thin convergent lens to perform the Fourier transform (FT). More than 40 years ago, Van der Lugt introduced an optical scheme for convolution/correlation operation, based on a cascade of two optical systems performing the Fourier transform with filter mask between them, initiating an era of Fourier optics.[1] This simple scheme realizes the most important shift-invariant operations in signal/image processing, such as filtering and pattern recognition. Nowadays, the Fourier optics area has been expanded with more sophisticated signal processing tools such as wavelets, bilinear distributions, fractional transformations, etc. Nevertheless, the paraxial optical systems (also called first-order or Gaussian ones, which consist for example from several aligned lenses, or mirrors) remain the basic elements for analog optical information processing.

In paraxial approximation of the scalar diffraction theory, a coherent light propagation through such a system is described by a canonical integral transform (CT). Thus starting from the complex field amplitude at the input plane of the system, we have its CT at the output plane. The two-dimensional CTs include, among others, such well-known transformations as image rotation, scaling, fractional Fourier[2] and Fresnel transforms. We can say that the CTs represent a two-dimensional signal in different phase space domains, where the phase space is defined by the position and momentum (spatial frequency) coordinates. The signal manipulation in different phase space domains opens new perspectives for information processing. Indeed, several useful applications of the first-order optical systems for information processing have been proposed in the past decade. In particular first-order optical systems performing fractional Fourier transform have been used for shift-variant filtering, noise reduction, chirp localization, encryption, etc.[2-5] Others have served as mode converters, which transform the

Hermite−Gaussian modes into helicoidal vortex Laguerre−Gaussian ones or other structurally stable modes.[6,7] These modes, in particular, are interesting for new types of information encoding in orbital angular momentum of beam[8] or in the geometric phase accumulated when it undergoes the cyclic transformation.[9]

Moreover the beam evolution in the first-order optical systems is a good model for the analysis of two-dimensional harmonic oscillator.[10]

In this chapter, we briefly summarize the main properties of the two-dimensional CTs,[11] used for the description of the first-order systems, consider their applications to traditional analog optical signal processing tasks, such as filtering, pattern recognition, encryption, etc., and then discuss new methods of information encoding related to the orbital angular momentum transfer and geometric phase accumulation.

1.2 Canonical Integral Transforms: Definition and Classification

1.2.1 Definition

The evolution of the complex field amplitude $f(r)$ during its propagation through a first-order optical system is described by the linear integral transform

$$f_o(\boldsymbol{r}_o) = \int_{-\infty}^{\infty} f_i(\boldsymbol{r}_i) K^{\mathbf{t}}(\boldsymbol{r}_i, \boldsymbol{r}_o)\, d\boldsymbol{r}_i,$$

where subindices i and o stand for input and output planes of the system. The kernel $K^{\mathbf{t}}(\boldsymbol{r}_i, \boldsymbol{r}_o)$ is parametrized by the wavelength λ and the real symplectic ray transformation 4×4 matrix $\mathbf{t}$ that relates the position $\boldsymbol{r}_i$ and direction $\boldsymbol{q}_i$ of an incoming ray to the position $\boldsymbol{r}_o$ and direction $\boldsymbol{q}_o$ of the outgoing ray,

$$\begin{bmatrix} \boldsymbol{r}_o \\ \boldsymbol{q}_o \end{bmatrix} = \begin{bmatrix} \mathbf{a} & \mathbf{b} \\ \mathbf{c} & \mathbf{d} \end{bmatrix} \begin{bmatrix} \boldsymbol{r}_i \\ \boldsymbol{q}_i \end{bmatrix} = \mathbf{t} \begin{bmatrix} \boldsymbol{r}_i \\ \boldsymbol{q}_i \end{bmatrix}.$$

Proper normalization of the variables and the matrix parameters to some length factor w and λ leads to the dimensionless variables: $\mathbf{r} = \boldsymbol{r}/\sqrt{\lambda w}$, $\mathbf{q} = \boldsymbol{q}\sqrt{w/\lambda}$, $\mathbf{A} = \mathbf{a}$, $\mathbf{B} = \mathbf{b}/w$, $\mathbf{C} = \mathbf{c}w$, $\mathbf{D} = \mathbf{d}$, which will be used further in this chapter,

$$\begin{bmatrix} \mathbf{r}_o \\ \mathbf{q}_o \end{bmatrix} = \begin{bmatrix} \mathbf{A} & \mathbf{B} \\ \mathbf{C} & \mathbf{D} \end{bmatrix} \begin{bmatrix} \mathbf{r}_i \\ \mathbf{q}_i \end{bmatrix} = \mathbf{T} \begin{bmatrix} \mathbf{r}_i \\ \mathbf{q}_i \end{bmatrix}, \tag{1.1}$$

where $\mathbf{r} = (x, y)^t$ and $\mathbf{q} = (q_x, q_y)^t$. As usual, the superscript ^{t} denotes transposition. The normalized variable $\mathbf{q}$ can also be interpreted as spatial frequency or ray momentum. The canonical integral transform associated with matrix $\mathbf{T}$ will be represented by the operator $\mathcal{R}^{\mathbf{T}}$

$$f_o(\mathbf{r}_o) = \mathcal{R}^{\mathbf{T}}\left[f_i(\mathbf{r}_i)\right](\mathbf{r}_o) = F_{\mathbf{T}}(\mathbf{r}_o) = \int_{-\infty}^{\infty} f_i(\mathbf{r}_i) K^{\mathbf{T}}(\mathbf{r}_i, \mathbf{r}_o)\, d\mathbf{r}_i. \tag{1.2}$$

The CT is a linear transform: $\mathcal{R}^\mathbf{T}\left[f(\mathbf{r}_i) + g(\mathbf{r}_i)\right](\mathbf{r}) = \mathcal{R}^\mathbf{T}\left[f(\mathbf{r}_i)\right](\mathbf{r}) + \mathcal{R}^\mathbf{T}\left[g(\mathbf{r}_i)\right](\mathbf{r})$. It is additive in the sense that $\mathcal{R}^{\mathbf{T}_2}\mathcal{R}^{\mathbf{T}_1} = \mathcal{R}^{\mathbf{T}_2 \times \mathbf{T}_1}$. The inverse transformation is parametrized by the matrix $\mathbf{T}^{-1}$, which, because $\mathbf{T}$ is symplectic, is given by

$$\mathbf{T}^{-1} = \begin{bmatrix} \mathbf{D}^t & -\mathbf{B}^t \\ -\mathbf{C}^t & \mathbf{A}^t \end{bmatrix}. \tag{1.3}$$

Any proper normalized symplectic ray transformation matrix can be decomposed in the modified Iwasawa form as[12]

$$\mathbf{T} = \begin{bmatrix} \mathbf{A} & \mathbf{B} \\ \mathbf{C} & \mathbf{D} \end{bmatrix} = \begin{bmatrix} \mathbf{I} & 0 \\ -\mathbf{G} & \mathbf{I} \end{bmatrix} \begin{bmatrix} \mathbf{S} & 0 \\ 0 & \mathbf{S}^{-1} \end{bmatrix} \begin{bmatrix} \mathbf{X} & \mathbf{Y} \\ -\mathbf{Y} & \mathbf{X} \end{bmatrix} = \mathbf{T}_L \mathbf{T}_S \mathbf{T}_O, \tag{1.4}$$

with $\mathbf{I}$ throughout denoting the identity matrix, in which the first matrix represents a lens transform described by the symmetric matrix

$$\mathbf{G} = -(\mathbf{C}\mathbf{A}^t + \mathbf{D}\mathbf{B}^t)(\mathbf{A}\mathbf{A}^t + \mathbf{B}\mathbf{B}^t)^{-1} = \mathbf{G}^t. \tag{1.5}$$

The second matrix corresponds to a scaler described by the positive definite symmetric matrix

$$\mathbf{S} = (\mathbf{A}\mathbf{A}^t + \mathbf{B}\mathbf{B}^t)^{1/2} = \mathbf{S}^t \tag{1.6}$$

and the third is an orthosymplectic[12,13] (i.e., both orthogonal and symplectic) matrix, which can be shortly represented by the unitary matrix

$$\mathbf{U} = \mathbf{X} + i\mathbf{Y} = (\mathbf{A}\mathbf{A}^t + \mathbf{B}\mathbf{B}^t)^{-1/2}(\mathbf{A} + i\mathbf{B}). \tag{1.7}$$

Note that because $\mathbf{A} = \mathbf{SX}$ and $\mathbf{B} = \mathbf{SY}$, the products $\mathbf{B}^{-1}\mathbf{A} = \mathbf{Y}^{-1}\mathbf{X}$ and $\mathbf{A}^{-1}\mathbf{B} = \mathbf{X}^{-1}\mathbf{Y}$ used further in different relations are defined by the orthogonal matrix $\mathbf{T}_O$.

Because the ray transformation matrix $\mathbf{T}$ is symplectic and therefore

$$\begin{aligned} \mathbf{AB}^t &= \mathbf{BA}^t, \quad \mathbf{CD}^t = \mathbf{DC}^t, \quad \mathbf{AD}^t - \mathbf{BC}^t = \mathbf{I}, \\ \mathbf{A}^t\mathbf{C} &= \mathbf{C}^t\mathbf{A}, \quad \mathbf{B}^t\mathbf{D} = \mathbf{D}^t\mathbf{B}, \quad \mathbf{A}^t\mathbf{D} - \mathbf{C}^t\mathbf{B} = \mathbf{I}, \end{aligned} \tag{1.8}$$

it has only ten free parameters. We call the transform associated with $\mathbf{T}$ *separable* if the block matrices $\mathbf{A}$, $\mathbf{B}$, $\mathbf{C}$, and $\mathbf{D}$ and $\mathbf{G}$, $\mathbf{S}$, $\mathbf{X}$, and $\mathbf{Y}$ correspondingly are diagonal. A separable transform has six degrees of freedom which reduce to three for rotational symmetric case corresponding to scalar block matrices.

In the often-used case $\det \mathbf{B} \neq 0$, the CT takes the form of Collins' integral[14]

$$f_o(\mathbf{r}_o) = \mathcal{R}^\mathbf{T}\left[f_i(\mathbf{r}_i)\right](\mathbf{r}_o) = (\det i\mathbf{B})^{-1/2} \int_{-\infty}^{\infty} f_i(\mathbf{r}_i)$$
$$\times \exp\left[i\pi\left(\mathbf{r}_i^t \mathbf{B}^{-1}\mathbf{A}\mathbf{r}_i - 2\mathbf{r}_i^t\mathbf{B}^{-1}\mathbf{r}_o + \mathbf{r}_o^t\mathbf{D}\mathbf{B}^{-1}\mathbf{r}_o\right)\right] d\mathbf{r}_i. \tag{1.9}$$

The kernel corresponds to two-dimensional generalized chirp function because its phase is a polynomial of second degree of variables $\mathbf{r}_i$ and $\mathbf{r}_o$. In particular for $\mathbf{A} = \mathbf{0}$, the kernel as a function of $\mathbf{r}_i$ has a form of plane wave. Thus for $\mathbf{A} = \mathbf{D} = \mathbf{0}$ and $\mathbf{B} = -\mathbf{C} = \mathbf{I}$, we obtain apart from a constant phase factor $\exp(-i\pi/2)$ the Fourier transform $\mathcal{F}[f(\mathbf{r}_i)](\mathbf{r}_o)$

$$\mathcal{F}[f(\mathbf{r}_i)](\mathbf{r}_o) = \int_{-\infty}^{\infty} f(\mathbf{r}_i) \exp(-i2\pi \mathbf{r}_o^t \mathbf{r}_i) \, d\mathbf{r}_i, \qquad (1.10)$$

known in optics as an angular spectrum of the complex field amplitude f. Moreover from the analysis of the matrix $\mathbf{Q} = \mathbf{B}^{-1}\mathbf{A} = \mathbf{Y}^{-1}\mathbf{X}$, it follows that the kernel as a function of $\mathbf{r}_i$ corresponds to the elliptic, hyperbolic or parabolic waves if $Q = 4Q_{11}Q_{22} - (Q_{12} + Q_{21})^2$ is positive, negative, or 0, relatively.

The well-known Fresnel transform that describes the evolution of the complex field amplitude during light propagation in an isotropic homogeneous medium at distance z, which in this chapter is a normalized dimensionless variable, is associated with matrix $\mathbf{T}_F(z) : \mathbf{A} = \mathbf{D} = \mathbf{I}, \mathbf{C} = \mathbf{0}, \mathbf{B} = z\mathbf{I}$.

1.2.2 Generalized imaging transforms

The case $\mathbf{B} = \mathbf{0}$ corresponds to the generalized imaging condition

$$f_o(\mathbf{r}) = (|\det \mathbf{A}|)^{-1/2} \exp\left(i\pi \mathbf{r}^t \mathbf{C}\mathbf{A}^{-1}\mathbf{r}\right) f_i(\mathbf{A}^{-1}\mathbf{r}), \qquad (1.11)$$

which includes a possible scaling and rotation of the input function accompanied by an additional phase modulation.

In the case $\mathbf{C} = \mathbf{B} = \mathbf{0}$, we have a family of the imaging transforms without phase modulation, which includes image rotation, scaling, and shearing.

The rotator transform, associated with $\mathbf{T}_r(\alpha)$: $\mathbf{C}_r = \mathbf{B}_r = \mathbf{0}$ and

$$\mathbf{A}_r = \mathbf{D}_r = \mathbf{X}_r = \begin{bmatrix} \cos\alpha & \sin\alpha \\ -\sin\alpha & \cos\alpha \end{bmatrix},$$

produces a clockwise rotation of f_i in $x - y$ plane and, correspondingly, its FT (the angular spectrum) in $q_x - q_y$ plane at angle α

$$f_o(x, y) = f_i\left(x\cos\alpha - y\sin\alpha, x\sin\alpha + y\cos\alpha\right).$$

A flexible optical scheme performing a rotation at angle α by only the appropriate rotating of cylindrical lenses composing the setup has been recently proposed.[15] Alternatively, Dove prisms can be used for optical rotator realization.

The separable scaling transform associated with the block matrices $\mathbf{A} = \mathbf{D}^{-1} = \mathbf{S}_s$, $\mathbf{C} = \mathbf{B} = \mathbf{0}$, where

$$\mathbf{S}_s = \begin{bmatrix} s_x & 0 \\ 0 & s_y \end{bmatrix},$$

defines the separable fractional FT $\mathcal{R}^{\mathbf{T}_f(\gamma_x,\gamma_y)}$ at angles $\gamma_x = \varphi + \gamma$ and $\gamma_y = \varphi - \gamma$. More information about the fractional FT can be found in Refs. 2, 3, 17–19 and references there in.

The gyrator transform (GT), associated with $\mathbf{T}_g(\vartheta)$, corresponds to twisting, i.e., rotations in the (x, q_y) and (y, q_x) planes of phase space, and is described by unitary matrix

$$\mathbf{U}_g(\vartheta) = \begin{bmatrix} \cos\vartheta & i\sin\vartheta \\ i\sin\vartheta & \cos\vartheta \end{bmatrix}. \tag{1.18}$$

The kernel of the GT has a form of the hyperbolic wave

$$K^{\mathbf{T}_g(\vartheta)}(\mathbf{r}_i, \mathbf{r}_o) = \frac{1}{|\sin\vartheta|}\exp\left[i2\pi\frac{(x_o y_o + x_i y_i)\cos\vartheta - (x_i y_o + x_o y_i)}{\sin\vartheta}\right], \tag{1.19}$$

which reduces to $\delta(\mathbf{r}_i - \mathbf{r}_o)$ for $\vartheta = 0$, to $\delta(\mathbf{r}_i + \mathbf{r}_o)$ for $\vartheta = \pi$, and to the twisted FT kernel $\exp\left[\mp i2\pi(x_i y_o + x_o y_i)\right]$ for $\vartheta = \pm\pi/2$. A detailed analysis of the GT can be found in Refs. 20–22.

Based on the matrix formalism flexible optical setups, which perform the antisymmetric fractional FT $\mathcal{R}^{\mathbf{T}_f(\gamma,-\gamma)}$, and the GT $\mathcal{R}^{\mathbf{T}_g(\vartheta)}$ have been designed.[15,21] These optical schemes contain three generalized lenses, $\mathcal{L}_1$, $\mathcal{L}_2$, and $\mathcal{L}_3 = \mathcal{L}_1$, with fixed equal distances between them denoted by z. The transformation angle is changed by rotation of the cylindrical lenses which form the generalized lenses.

In the case of the antisymmetric fractional FT setup every generalized lens $\mathcal{L}_j$ ($j = 1, 2$) consists from a convergent spherical lens with focal distance $p_j^{-1} = z/j$ and convergent and divergent cylindrical lenses with focal distances $\pm z$ rotated at angle $\phi_1^{(j)} = \phi^{(j)}$, $\phi_2^{(j)} = \pi/2 - \phi^{(j)}$ with respect to OX axis. The antisymmetric fractional FT at angle $(\gamma, -\gamma)$ is achieved if $\cos(2\phi^{(1)}) = \cot(\gamma/2)$ and $2\phi^{(2)} = \pi/2 - \gamma$. It is easy to see from the last relation that this setup is able to perform the antisymmetric fractional FT for the angles $[\pi/2, 3\pi/2]$ that cover a π interval needed for the different applications.

In the case of the GT, every generalized lens $\mathcal{L}_j$ ($j = 1, 2$) is a combination of two convergent cylindrical lenses of equal focus distance z/j rotated at angle $\phi_1^{(j)} = -\phi^{(j)}$, $\phi_2^{(j)} = \phi^{(j)} - \pi/2$. The GT at angle ϑ is achieved if $\cos(2\phi^{(1)}) = \cot(\vartheta/2)$ and $\sin(2\phi^{(2)}) = (\sin\vartheta)/2$. We again observe that this setup is able to perform the GT for the angles from π interval $[\pi/2, 3\pi/2]$.

It has been shown[23] that any orthosymplectic matrix can be decomposed in the form

$$\mathbf{T}_O = \mathbf{T}_r(\beta)\,\mathbf{T}_f(\gamma_x, \gamma_y)\,\mathbf{T}_r(\alpha). \tag{1.20}$$

It means that $R^{\mathbf{T}_O}$ is a separable fractional Fourier transformer $R^{\mathbf{T}_f}$ embedded between two rotators $R^{\mathbf{T}_r}$. In particular for the gyrator matrix, we obtain $\mathbf{T}_g(\vartheta) = \mathbf{T}_r(-\pi/4)\,\mathbf{T}_f(\vartheta, -\vartheta)\,\mathbf{T}_r(\pi/4)$. Therefore, based on the optical setups, performing the fractional FT and rotator a system for arbitrary orthosymplectic transformation can be constructed.

Moreover, because the generalized lens transform and scaler from the decomposition (1.4) also can be presented as separable ones embedded into two rotators, any CT can be written in the form

$$\mathbf{T} = \mathbf{T}_r(\alpha_4)\mathbf{T}_{L_s}(g_x, g_y)\mathbf{T}_r(\alpha_3)\mathbf{T}_{S_s}(s_x, s_y)\,\mathbf{T}_r(\alpha_2)\mathbf{T}_f(\gamma_x, \gamma_y)\,\mathbf{T}_r(\alpha_1), \quad (1.21)$$

where we used the angle additivity of the rotator transform. The 10 parameters of the CT are defined then as 4 rotation angles α_j ($j = 1, \ldots, 4$) and 2 parameters for three separable transforms: lens (g_x, g_y), scaler (s_x, s_y), and fractional FT (γ_x, γ_y).

In general, the ray matrix decomposition into a cascade of the others is very useful for the analysis of a particular optical system. Thus, the Fresnel transform is a combination of the symmetric fractional FT at angle $\arctan z$, scaling with $s = \sqrt{1 + z^2}$ and spherical lens transform with $g = z/(1 + z^2)$. On the other side, the Fresnel transform itself can be considered as a basic element for the system design, because the above-mentioned fractional Fourier and gyrator transformers are constructed as a cascade of lenses and homogeneous medium intervals, described by the Fresnel transform.

1.2.4 Canonical transforms for the case $\det \mathbf{B} = 0$

In order to deal with the singular case $\det \mathbf{B} = 0$, but $\mathbf{B} \neq \mathbf{0}$, we can use the Iwasawa decomposition (1.4), from which we note that $\mathbf{B} = \mathbf{SY}$. Because the scaling matrix $\mathbf{S}$ is nonsingular, a singularity of $\mathbf{B}$ is only due to the orthosymplectic part described by $\mathbf{T}_O$. Moreover from the analysis of Eq. (1.20), we conclude that the separable fractional Fourier transformer is responsible for a singularity of the submatrix $\mathbf{B}$. Thus, $\det \mathbf{B} = 0$ if for at least one coordinate the fractional Fourier transformer acts as an identity system, i.e., $\gamma_x = 0$ and/or $\gamma_y = 0$. Then, based on the modified Iwasawa decomposition (1.4) and Eq. (1.20), we can write a general representation of the CT, which is valid for any ray transformation matrix, including a singular submatrix $\mathbf{B}$[23]

$$\begin{aligned}f_o(\mathbf{r}_o) &= \mathcal{R}^{\mathbf{T}}\left[f_i(\mathbf{r}_i)\right](\mathbf{r}_o) = (\det \mathbf{S})^{-1/2} \exp(-i\pi \mathbf{r}_o^t \mathbf{G} \mathbf{r}_o) \\ &\quad \times \mathcal{R}^{\mathbf{T}_f(\gamma_x, \gamma_y)}\left[f_i\left(\mathbf{X}_r(\alpha)\,\mathbf{r}_i\right)\right]\left(\mathbf{X}_r(-\beta)\mathbf{S}^{-1}\mathbf{r}_o\right).\end{aligned} \quad (1.22)$$

1.3 Main Properties of the Canonical Integral Transforms

In order to properly use the first-order optical systems for information processing tasks, we need to know the basic properties of the CTs describing these systems. In this section, the well-known theorems for the FT, such as shift, convolution, scaling, etc., are generalized to the case of the CTs.[11]

In information processing, a signal $f(\mathbf{r})$ is often described by its Fourier transform $\mathcal{F}[f(\mathbf{r})](\mathbf{q})$ [see Eq. (1.10)], which corresponds in coherent optics to the angular spectrum. If a complex field amplitude $f(\mathbf{r})$ is canonically transformed with the matrix $\mathbf{T}$, then its angular spectrum is canonically transformed with $(\mathbf{T}^t)^{-1}$

$$(\mathbf{T}^t)^{-1} = \begin{bmatrix} \mathbf{D} & -\mathbf{C} \\ -\mathbf{B} & \mathbf{A} \end{bmatrix}. \tag{1.23}$$

In particular, for the orthosymplectic CTs, $\mathbf{T} = (\mathbf{T}^t)^{-1}$.

1.3.1 Parseval theorem

The complex conjugated version of the CT related to matrix $\mathbf{T}$ of $f(\mathbf{r}_i)$ is equal to the CT described by matrix $\tilde{\mathbf{T}}$ of the complex conjugated version of that function $f^*(\mathbf{r}_i)$, i.e., $\{\mathcal{R}^\mathbf{T}[f(\mathbf{r}_i)](\mathbf{r})\}^* = \mathcal{R}^{\tilde{\mathbf{T}}}[f^*(\mathbf{r}_i)](\mathbf{r})$, where

$$\tilde{\mathbf{T}} = \begin{bmatrix} \mathbf{A} & -\mathbf{B} \\ -\mathbf{C} & \mathbf{D} \end{bmatrix}. \tag{1.24}$$

The well-known Parseval theorem holds for the entire class of the CTs

$$\int_{-\infty}^{\infty} f(\mathbf{r}_i) \, g^*(\mathbf{r}_i) \, d\mathbf{r}_i = \int_{-\infty}^{\infty} \mathcal{R}^\mathbf{T}[f(\mathbf{r}_i)](\mathbf{r}_o) \, \{\mathcal{R}^\mathbf{T}[g(\mathbf{r}_i)](\mathbf{r}_o)\}^* \, d\mathbf{r}_o, \tag{1.25}$$

which, in particular, yields the energy preservation law

$$\int_{-\infty}^{\infty} |f(\mathbf{r}_i)|^2 d\mathbf{r}_i = \int_{-\infty}^{\infty} |\mathcal{R}^\mathbf{T}[f(\mathbf{r}_i)](\mathbf{r}_o)|^2 \, d\mathbf{r}_o. \tag{1.26}$$

1.3.2 Shift theorem

A shift of the input field in the transversal with respect to the optical axis plane by a vector $\mathbf{v}$, $f_i(\mathbf{r}) \to f_i(\mathbf{r} - \mathbf{v})$, leads to a shift of the output signal by the vector $\mathbf{Av}$ and to an additional quadratic phase factor

$$\mathcal{R}^\mathbf{T}[f_i(\mathbf{r}_i - \mathbf{v})](\mathbf{r}_o) = \exp[i\pi(2\mathbf{r}_o - \mathbf{Av})^t \mathbf{Cv}] \, \mathcal{R}^\mathbf{T}[f_i(\mathbf{r}_i)](\mathbf{r}_o - \mathbf{Av}), \tag{1.27}$$

where we have used the symplecticity conditions (1.8) and the fact that $\mathbf{v}^t \mathbf{Zq} = \mathbf{q}^t \mathbf{Z}^t \mathbf{v}$. This implies that the intensity distribution does not change due to a displacement by $\mathbf{v}$, but is merely shifted by $\mathbf{Av}$:

$$\left| \mathcal{R}^\mathbf{T}[f(\mathbf{r}_i - \mathbf{v})](\mathbf{r}_o) \right| = |F_\mathbf{T}(\mathbf{r}_o - \mathbf{Av})|.$$

In particular, Eq. (1.27) reduces to $\mathcal{R}^\mathbf{T}[f(\mathbf{r}_i - \mathbf{v})](\mathbf{r}_o) = F_\mathbf{T}(\mathbf{r}_o - \mathbf{Av})$ and to $\mathcal{R}^\mathbf{T}[f(\mathbf{r}_i - \mathbf{v})](\mathbf{r}_o) = \exp(i\pi 2\mathbf{r}_o^t \mathbf{Cv}) \, F_\mathbf{T}(\mathbf{r}_o)$ for $\mathbf{C} = \mathbf{0}$, and for $\mathbf{A} = \mathbf{0}$, respectively. The shift theorem explains the position-variant nature of signal processing in the phase space domains, which are different from the Fourier ones. It also describes the evolution of the coherent state wave function of the two-dimensional harmonic oscillator.[10]

1.3.3 Convolution theorem

Using the shift theorem, the CT of the convolution between f and h

$$(f * h)(\mathbf{r}) = \int_{-\infty}^{\infty} f(\mathbf{r} - \mathbf{v}) h(\mathbf{v}) d\mathbf{v} = \int_{-\infty}^{\infty} h(\mathbf{r} - \mathbf{v}) f(\mathbf{v}) d\mathbf{v} \quad (1.28)$$

can be written in the form

$$\mathcal{R}^{\mathbf{T}}\left[(f * h)(\mathbf{r}_i)\right](\mathbf{r}_o) = \int_{-\infty}^{\infty} \exp[i\pi(2\mathbf{r}_o - \mathbf{A}\mathbf{v})^t \mathbf{C}\mathbf{v}] F_{\mathbf{T}}(\mathbf{r}_o - \mathbf{A}\mathbf{v}) h(\mathbf{v}) d\mathbf{v}. \quad (1.29)$$

In the case $\mathbf{A} = \mathbf{0}$ (and, thus, also $\mathbf{C}^t = -\mathbf{B}^{-1}$), we get

$$\mathcal{R}^{\mathbf{T}}\left[(f * h)(\mathbf{r}_i)\right](\mathbf{r}_o) = (\det i\mathbf{B})^{1/2} \exp(-i\pi \mathbf{r}_o^t \mathbf{D}\mathbf{B}^{-1}\mathbf{r}_o) F_{\mathbf{T}}(\mathbf{r}_o) H_{\mathbf{T}}(\mathbf{r}_o),$$

which reduces to the product $(\det i\mathbf{B})^{1/2} F_{\mathbf{T}}(\mathbf{r}_o) H_{\mathbf{T}}(\mathbf{r}_o)$ for $\mathbf{D} = \mathbf{0}$. The last expression corresponds to a Fourier transformation with scaling and rotation (determined by $\mathbf{B}$).

If $\mathbf{C} = \mathbf{0}$, then

$$\mathcal{R}^{\mathbf{T}}\left[(f * h)(\mathbf{r}_i)\right](\mathbf{r}_o) = \int_{-\infty}^{\infty} F_{\mathbf{T}}(\mathbf{r}_o - \mathbf{A}\mathbf{v}) h(\mathbf{v}) d\mathbf{v}.$$

As we have seen from the shift and convolution theorems, the CTs with $\mathbf{A} = \mathbf{0}$ (and, consequently, $\mathbf{C}^t = -\mathbf{B}^{-1}$) describe Fourier transforming systems (besides the FT they produce also an appropriate scaling, rotation, and phase modulation); meanwhile, the CTs with $\mathbf{C} = \mathbf{0}$ (and thus $\mathbf{D}^t = \mathbf{A}^{-1}$) can be considered as generalized Fresnel transforms or convolution-type transforms.

1.3.4 Scaling theorem

Note that scaling itself belongs to the class of CTs. Therefore as it follows from the additivity property of the CTs, the scaling of the input function leads to a change of the parametrizing matrix. Indeed, the $\mathbf{T}$-parametrized CT of the scaled function $(\det \mathbf{W})^{1/2} f(\mathbf{W}\mathbf{r}_i)$, $\mathcal{R}^{\mathbf{T}}[(\det \mathbf{W})^{1/2} f(\mathbf{W}\mathbf{r}_i)](\mathbf{r}_o)$ with $\mathbf{W} = \mathbf{W}^t$, corresponds to the CT of $f(\mathbf{r}_i)$ itself, $\mathcal{R}^{\widehat{\mathbf{T}}}[f(\mathbf{r}_i)](\mathbf{r}_o)$, but parametrized by the matrix

$$\widehat{\mathbf{T}} = \mathbf{T} \begin{bmatrix} \mathbf{W}^{-1} & \mathbf{0} \\ \mathbf{0} & \mathbf{W} \end{bmatrix}. \quad (1.30)$$

In particular, the scaling theorems for two important uniparametric CTs—symmetric fractional FT and the gyrator transform—have been formulated in Refs. 2, 19, 20.

On the basis of the scaling theorem for the fractional FT, a method of the analysis of the fractal signals has been proposed.[24] Moreover the scaling theorem might be useful for the analysis of squeezed states evolution.[10]

1.3.5 Coordinates multiplication and derivation theorems

With $\nabla = (\partial/\partial x, \partial/\partial y)^t$ and $\det \mathbf{B} \neq 0$, the gradient of the CT can be written as

$$\begin{aligned}\nabla_o f_o(\mathbf{r}_o) &= \nabla_o \left\{ \mathcal{R}^{\mathbf{T}} \left[f_i(\mathbf{r}_i) \right] (\mathbf{r}_o) \right\} \\ &= i2\pi (\mathbf{B}^t)^{-1} \left\{ \mathbf{D}^t \mathbf{r}_o f_o(\mathbf{r}_o) - \mathcal{R}^{\mathbf{T}} \left[\mathbf{r}_i f_i(\mathbf{r}_i) \right] (\mathbf{r}_o) \right\}, \end{aligned} \quad (1.31)$$

where the symplecticity condition $\mathbf{B}^t \mathbf{D} = \mathbf{D}^t \mathbf{B}$ has been used. We thus obtain

$$\mathcal{R}^{\mathbf{T}} \left[\mathbf{r}_i f_i(\mathbf{r}_i) \right] (\mathbf{r}_o) = \left\{ \mathbf{D}^t \mathbf{r}_o - \frac{\mathbf{B}^t}{i2\pi} \nabla_o \right\} f_o(\mathbf{r}_o) \quad (1.32)$$

Analogously, the expression for $\mathcal{R}^{\mathbf{T}} \left[\nabla_i f_i(\mathbf{r}_i) \right] (\mathbf{r}_o)$ can be found[11]

$$\mathcal{R}^{\mathbf{T}} \left[\nabla_i f_i(\mathbf{r}_i) \right] (\mathbf{r}_o) = \left(-i2\pi \mathbf{C}^t \mathbf{r}_o + \mathbf{A}^t \nabla_o \right) f_o(\mathbf{r}_o). \quad (1.33)$$

We again stress the similarity of the CT to the Fresnel transform and the FT for cases $\mathbf{C} = \mathbf{0}$ and $\mathbf{A} = \mathbf{0}$, respectively.

1.4 Canonical Integral Transforms of Selected Functions

1.4.1 Plane wave, chirp, and Gaussian functions

Only for a limited number of functions, an analytical expression of their CTs can be found. Among them, we mention the function

$$f_i(\mathbf{r}) = \exp\left(i2\pi \mathbf{k}_i^t \mathbf{r} - \pi \mathbf{r}^t \mathbf{L}_i \mathbf{r} \right), \quad (1.34)$$

where $\mathbf{L}_i$ is a symmetric matrix with nonnegative definite real part, and $\mathbf{k}_i$ is a real vector. Following the calculations done in Ref. 25, it has been proved[11] that the CT of the function (1.34) takes the form

$$\begin{aligned} f_o(\mathbf{r}) &= \mathcal{R}^{\mathbf{T}}[f_i(\mathbf{r}_i)](\mathbf{r}) = \left[\det \left(\mathbf{A} + i\mathbf{B}\mathbf{L}_i \right) \right]^{-1/2} \\ &\times \exp\left[-i\pi \mathbf{k}_i^t \left(\mathbf{A} + i\mathbf{B}\mathbf{L}_i \right)^{-1} \mathbf{B} \mathbf{k}_i + i2\pi \mathbf{k}_o^t \mathbf{r} - \pi \mathbf{r}^t \mathbf{L}_o \mathbf{r} \right], \end{aligned} (1.35)$$

where $\mathbf{k}_o^t = \mathbf{k}_i^t \left(\mathbf{A} + i\mathbf{B}\mathbf{L}_i \right)^{-1}$ and $i\mathbf{L}_o = \left(\mathbf{C} + i\mathbf{D}\mathbf{L}_i \right) \left(\mathbf{A} + i\mathbf{B}\mathbf{L}_i \right)^{-1}$.

If $\mathbf{L}_i = -i\mathbf{H}_i$ is imaginary, then $\mathbf{L}_o = -i\mathbf{H}_o = -i(\mathbf{C} + \mathbf{D}\mathbf{H}_i)(\mathbf{A} + \mathbf{B}\mathbf{H}_i)^{-1}$ is imaginary, too, which implies that $f_i(\mathbf{r}_i)$ [Eq. (1.34)] and $f_o(\mathbf{r})$ [Eq. (1.35)] are the generalized chirp functions, which include as particular cases the plane, elliptic, hyperbolic, and parabolic waves. For example, the CT of a plane wave $f_i(\mathbf{r}) = \exp(i2\pi \mathbf{k}_i^t \mathbf{r})$ ($\mathbf{L}_i = \mathbf{0}$) is given by

$$f_o(\mathbf{r}) = (\det \mathbf{A})^{-1/2} \exp\left(-i\pi \mathbf{k}_i^t \mathbf{A}^{-1} \mathbf{B} \mathbf{k}_i + i2\pi \mathbf{k}_i^t \mathbf{A}^{-1} \mathbf{r} + i\pi \mathbf{r}^t \mathbf{C} \mathbf{A}^{-1} \mathbf{r} \right). \quad (1.36)$$

Then, a plane wave remains a plane wave under the CT only if $\mathbf{C} = \mathbf{0}$, which again stresses the interpretation of this case as a generalized Fresnel transformation.

For $\mathbf{k}_i = \mathbf{0}$, and thus $f_i(\mathbf{r}) = \exp(-\pi \mathbf{r}^t \mathbf{L}_i \mathbf{r})$, the CT takes the simple form

$$f_o(\mathbf{r}) = [\det(\mathbf{A} + i\mathbf{B}\mathbf{L}_i)]^{-1/2} \exp(-\pi \mathbf{r}^t \mathbf{L}_o \mathbf{r}). \quad (1.37)$$

A Gaussian beam $\exp[-\pi(l_{11}x^2 + 2l_{12}xy + l_{22}y^2)]$ appears when $\mathbf{L}_i$ is real and positive definite.

Equation (1.35) allows us to design systems that transform a certain type of wave into another one or maintain its structure during the transformation.

Moreover, the CTs convert a generalized chirp signal into the Dirac δ-function and vice versa. Indeed the CT of the Dirac δ-function, which corresponds to the point-spread function of the related first-order optical system, is a generalized chirp function

$$\mathcal{R}^{\mathbf{T}}[\delta(\mathbf{r}_i - \mathbf{v})](\mathbf{r}_o) = \frac{1}{\sqrt{\det i\mathbf{B}}} \exp\left[i\pi(\mathbf{v}^t \mathbf{B}^{-1}\mathbf{A}\mathbf{v} - 2\mathbf{v}^t \mathbf{B}^{-1}\mathbf{r}_o + \mathbf{r}_o^t \mathbf{D}\mathbf{B}^{-1}\mathbf{r}_o)\right]. \quad (1.38)$$

Then applying the CT parametrized by the matrix $\mathbf{T}^{-1}$ to this chirp function, we obtain $\delta(\mathbf{r}_i - \mathbf{v})$.

1.4.2 Periodic functions

Representing a periodic function $f_i(\mathbf{r})$ with periods p_x and p_y with respect to the x and y coordinates as a superposition of plane waves,

$$f_i(\mathbf{r}) = \sum_{m,n=-\infty}^{\infty} a_{mn} \exp(i 2\pi \mathbf{k}_{mn}^t \mathbf{r}), \quad (1.39)$$

with $\mathbf{k}_{mn}^t = (m/p_x, n/p_y)$ and using Eq. (1.36), we get its CT in the form

$$\begin{aligned} f_o(\mathbf{r}) &= (\det \mathbf{A})^{-1/2} \exp\left(i\pi \mathbf{r}^t \mathbf{C}\mathbf{A}^{-1}\mathbf{r}\right) \\ &\times \sum_{m,n=-\infty}^{\infty} a_{mn} \exp\left(-i\pi \mathbf{k}_{mn}^t \mathbf{A}^{-1}\mathbf{B}\mathbf{k}_{mn} + i 2\pi \mathbf{k}_{mn}^t \mathbf{A}^{-1}\mathbf{r}\right). \end{aligned}$$

If $\mathbf{k}_{mn}^t \mathbf{A}^{-1}\mathbf{B}\mathbf{k}_{mn} = \mathbf{k}_{mn}^t \mathbf{X}^{-1}\mathbf{Y}\mathbf{k}_{mn} = 2j$, where j is an integer, then the generalized Talbot imaging[11] is obtained

$$f_o(\mathbf{r}) = (\det \mathbf{A})^{-1/2} \exp\left(i\pi \mathbf{r}^t \mathbf{C}\mathbf{A}^{-1}\mathbf{r}\right) f_i(\mathbf{A}^{-1}\mathbf{r}). \quad (1.40)$$

It includes a rotation, scaling of the coordinates described by the matrix $\mathbf{A}^{-1}$, and phase modulation associated with the matrix product $\mathbf{C}\mathbf{A}^{-1}$. We stress again the similarity with the Fresnel transformation for $\mathbf{C} = \mathbf{0}$, in which case only an affine transformation of the input function remains.

1.4.3 Eigenfunctions for the canonical integral transforms

The search for the eigenfunctions of a certain CT is related to the eigenvalues analysis of the corresponding ray transformation matrix.[26] It has been shown that a 4×4 real symplectic ray transformation matrix $\mathbf{T}$ is similar to a real symplectic nucleus matrix $\mathbf{T}_n$, $\mathbf{T} = \mathbf{M}\mathbf{T}_n\mathbf{M}^{-1}$, where the matrix $\mathbf{M}$ is also real and symplectic. Two-dimensional CTs can be classified into ten (that corresponds to the degree of freedom of symplectic ray transformation matrix) classes, and for each class, a simple nucleus has been found. Once the eigenfunctions $\Phi(\mathbf{r})$ of a nucleus are known, the eigenfunctions of the related CT are given by their CTs parametrized by the matrix $\mathbf{M}$: $\mathcal{R}^\mathbf{M}[\Phi(\mathbf{r}_i)](\mathbf{r})$. Expressions for the eigenfunctions of a CT associated with nucleus matrix with real eigenvalues have been derived in Ref. 26.

More useful for information processing CTs relate to the separable fractional FT nucleus matrix.[27] The Hermite–Gaussian (HG) functions $\mathcal{H}_{m,n}(\mathbf{r}) = \mathcal{H}_m(x)\mathcal{H}_n(y)$, where $\mathcal{H}_n(x) = 2^{1/4}(2^n n!)^{-1/2} H_n(\sqrt{2\pi}x)\exp(-\pi x^2)$ and $H_n(\cdot)$ denotes the Hermite polynomials, are eigenfunctions for the separable fractional FT for any angles γ_x and γ_y with eigenvalues $\exp[-i(m+1/2)\gamma_x - i(n+1/2)\gamma_y]$ (see, for example, Ref. 2). Then the functions obtained from $\mathcal{H}_{m,n}(\mathbf{r}_i)$ by the CT parametrized by $\mathbf{M}$: $\mathcal{H}^\mathbf{M}_{m,n}(\mathbf{r}) = \mathcal{R}^\mathbf{M}[\mathcal{H}_{m,n}(\mathbf{r}_i)](\mathbf{r})$ are eigenfunctions for the CT described by the ray transformation matrix $\mathbf{T} = \mathbf{M}\mathbf{T}_f(\gamma_x, \gamma_y)\mathbf{M}^{-1}$ with eigenvalues $\exp[-i(m+1/2)\gamma_x - i(n+1/2)\gamma_y]$. The modes $\mathcal{H}^\mathbf{M}_{m,n}(\mathbf{r})$ for the same $\mathbf{M}$ and different indices $m, n \in [0, \infty)$ form a complete orthonormal set, and therefore, any function can be represented as their linear superposition.

Using again the matrix decomposition (1.4) and the additivity of the CTs $\mathcal{H}^\mathbf{M}_{m,n}(\mathbf{r}) = \mathcal{R}^{\mathbf{M}_L \times \mathbf{M}_S}[\mathcal{H}^{\mathbf{M}_O}_{m,n}(\mathbf{r}_i)](\mathbf{r})$ we observe that $\mathcal{H}^\mathbf{M}_{m,n}(\mathbf{r})$ can be derived from $\mathcal{H}^{\mathbf{M}_O}_{m,n}(\mathbf{r})$ by corresponding scaling and phase modulation.

The orthosymplectic modes $\mathcal{H}^{\mathbf{M}_O}_{m,n}(\mathbf{r})$, modes obtained from the HG ones $\mathcal{H}_{m,n}(\mathbf{r})$ by the CT associated with the orthogonal ray transformation matrix $\mathbf{M}_O$, have the following closed-form expression[27]

$$\mathcal{H}^{\mathbf{M}_O}_{m,n}(\mathbf{r}) = \frac{(-1)^{m+n}\exp[\pi(x^2+y^2)]}{2^{m+n-1/2}\left(\pi^{m+n}m!n!\det\mathbf{U}\right)^{1/2}}$$
$$\times \left(U^*_{11}\frac{\partial}{\partial x} + U^*_{21}\frac{\partial}{\partial y}\right)^m \left(U^*_{12}\frac{\partial}{\partial x} + U^*_{22}\frac{\partial}{\partial y}\right)^n \exp[-2\pi(x^2+y^2)],$$
(1.41)

where U_{jk} ($j, k = 1, 2$) are parameters of the unitary matrix $\mathbf{U}$ [see Eq. (1.7)] corresponding to $\mathbf{M}_O$.

For a more complete characterization of the orthosymplectic mode $\mathcal{H}^{\mathbf{M}_O}_{m,n}(\mathbf{r})$, we mention that the component of its orbital angular momentum (OAM) in the propagation direction z, measured per photon in units $\hbar$, is given by[28]

$$\mathrm{L}_z^{m,n} = 2\mathrm{Im}\left\{m U_{11} U_{21}^* - n U_{22} U_{12}^*\right\}. \qquad (1.42)$$

For $\mathbf{U} = \mathbf{U}_f(\gamma_x, \gamma_y)$ the HG modes $\mathcal{H}_{m,n}$ with $\mathrm{L}_z^{m,n} = 0$ are obtained from Eq. (1.41). Note that here we define modes up to a constant phase factor. For gyrator transform matrix $\mathbf{M}_O = \mathbf{T}_g(\vartheta)$ the modes with $\mathrm{L}_z^{m,n} = -(m-n)\sin 2\vartheta$ are generated. They reduce to the HG ones for $\vartheta = \pi l/2$ with integer l. If $\vartheta = \pm \pi/4 + \pi l$, $\mathcal{H}_{m,n}^{\mathbf{M}_O}(\mathbf{r})$ correspond to the Laguerre–Gaussian (LG) modes

$$\begin{aligned}\mathcal{H}_{m,n}^{\mathbf{M}_g(\mp \pi/4 + \pi l)}(\mathbf{r}) &= \mathcal{L}_{m,n}^{\pm}(\mathbf{r}) \\ &= 2^{1/2}\left[\frac{(\min\{m,n\})!}{(\max\{m,n\})!}\right]^{1/2}(\sqrt{2\pi}r)^{|m-n|} \\ &\quad \times \exp[\pm i(m-n)\psi]\, L_{\min\{m,n\}}^{(|m-n|)}(2\pi r^2)\exp(-\pi r^2),\end{aligned} \qquad (1.43)$$

where $L_n^{(\alpha)}(\cdot)$ denotes the generalized Laguerre polynomials and spatial coordinates are represented by the two-dimensional column vector $\mathbf{r} = (x,y)^t = (r\cos\psi, r\sin\psi)^t$. The LG mode $\mathcal{L}_{m,n}^{\pm}(\mathbf{r})$ is an eigenfunction for the rotator operation (1.15) and possesses the integer OAM projection $\mathrm{L}_z^{m,n} = \pm(m-n)$, also known as a topological charge.

The application of the orthosymplectic modes for the information processing will be discussed in Section 1.6.

1.5 Generalized Convolution for Analog Optical Information Processing

1.5.1 Analog optical information processing

There are two important operations that make the coherent optical processing possible: addition and multiplication. The addition operation is based on the superposition principle; meanwhile, the multiplication operation is related to diffraction or reflection phenomena.

To proceed with the information optically, we first have to incorporate it into the optical beam. We can do this in different manners, for example, by location of a transparency, hologram, or spatial light modulator (SLM) that carry the information on the way of an optical beam, often approximated by a plane wave. The reflection mode of operation is also possible. Then we have to modify the input signal by its propagation through an optical system in order to improve its quality or perform various operations on it, such as differentiation, integration, rotation, etc., extract signal characteristics or detect a specific pattern, and encrypt or decrypt the information. An optical signal, generally a two-dimensional complex function associated with a complex field amplitude, that enters a system, we call an input signal, and an outgoing one is an output signal. To detect the outgoing optical signal, the photographic film, holographic plate, or electro-optical devices, such as CCD and CMOS cameras, are used. We will focus our attention on the processing part.

It has been shown in Section 1.2 that the CTs produce affine transformation in phase space, such as rotation, scaling, skew, twisting, etc., and can be easily implemented in optics. We have already discussed the design of optical systems that perform the main orthosymplectic CTs (fractional FT, rotator, and gyrator), and therefore, we refer to these systems when mentioning the corresponding CTs of a signal. The combinations of the CTs produce new transforms. Thus, linear superpositions of the fractional FTs lead to fractional sine, cosine, Hartley transforms, and other types of the fractional transforms which are not canonical.[3] The cascades of the CTs with appropriate masks between them produce convolution, correlation operations, wavelet, Hilbert transforms, etc., which can be expressed in the framework of a generalized convolution operation.

As we have mentioned in Section 1.1, analog information processing is based on the convolution operation performed by optical systems. Indeed, signal filtering, pattern recognition tasks, etc., can be expressed in the form of signal convolution. There are many good books devoted to this subject. In this section, we discuss the advantages of application of the generalized canonical convolution for information processing.

1.5.2 Generalized convolution: Definition

By analogy with the alternative representation of the convolution operation via the Fourier domain[1],

$$\mathcal{C}_{f,h}(\mathbf{r}) = (f * h)(\mathbf{r}) = \mathcal{F}^{-1}\left\{\mathcal{F}[f(\cdot)](\mathbf{u})\,\mathcal{F}[h(\cdot)](\mathbf{u})\right\}(\mathbf{r}), \qquad (1.44)$$

we can introduce the generalized canonical convolution (GC) operation as[3,29]

$$\mathcal{GC}_{f,h}(\mathbf{T}_1, \mathbf{T}_2, \mathbf{T}_3, \mathbf{r}) = \mathcal{R}^{\mathbf{T}_3}\left\{\mathcal{R}^{\mathbf{T}_1}[f(\cdot)](\mathbf{u})\,\mathcal{R}^{\mathbf{T}_2}[h(\cdot)](\mathbf{u})\right\}(\mathbf{r}), \qquad (1.45)$$

which reduces to common convolution if the ray transformation matrices correspond to the direct/inverse FT ones $\mathbf{T}_1 = \mathbf{T}_2 = \mathbf{T}_3^{-1} = \mathbf{T}_f(\pi/2, \pi/2)$. Besides that the GC includes as particular cases: the correlation operation $\mathcal{GC}_{f,h^*}\left[\mathbf{T}_f(\pi/2, \pi/2), \mathbf{T}_f(-\pi/2, -\pi/2), \mathbf{T}_f(-\pi/2, -\pi/2), \mathbf{r}\right]$, used as a measure of similarity between two signals;[1] the fractional convolution $\mathcal{GC}_{f,h}\left[\mathbf{T}_f(\gamma_x, \gamma_y), \mathbf{T}_f(\beta_x, \beta_y), \mathbf{T}_f(\alpha_x, \alpha_y), \mathbf{r}\right]$, applied for shift-variant filtering and pattern recognition;[2,4] the Wigner distribution $2\mathcal{GC}_{f,f^*}$ $\left[\mathbf{T}_f(\gamma_x + \pi/2, \gamma_y + \pi/2), \mathbf{T}_f(-\gamma_x + \pi/2, -\gamma_y + \pi/2), \mathbf{T}_f(-\pi/2, -\pi/2), 2\rho\right]$ expressed in polar coordinates $\rho = (\rho_x, \rho_y)$ (more information about the importance of this distribution can be found in Chapter 2); the Radon–Wigner transform $\mathcal{GC}_{f,f^*}\left[\mathbf{T}_f(\gamma_x, \gamma_y), \mathbf{T}_f(-\gamma_x, -\gamma_y), \mathbf{I}, \mathbf{r}\right]$, which is a set of projections of the Wigner distribution; the CT power spectrum $\mathcal{GC}_{f,f^*}(\mathbf{T}, \widetilde{\mathbf{T}}, \mathbf{I}, \mathbf{r})$, corresponding to the squared modulus of the CT of the signal, etc.[3]

The generalized canonical convolution $\mathcal{GC}_{f,h}(\mathbf{T}_1, \mathbf{T}_2, \mathbf{T}_3, \mathbf{r})$ of two-dimensional signals f and h is a function of two variables ($\mathbf{r}$) and 30 parameters, defined by the ray transformation matrices $\mathbf{T}_1, \mathbf{T}_2$, and $\mathbf{T}_3$. Note that some of the parameters can also play a role of variables. Thus, the wavelet transform of f with mother wavelet h can be expressed in the form of the GC $\mathcal{GC}_{f,h^*}\left[\mathbf{T}_f(\pi/2, \pi/2), \mathbf{T}_S \mathbf{T}_f(\pi/2, \pi/2), \mathbf{T}_f(-\pi/2, -\pi/2), \mathbf{r}\right]$, where the parameters of the scaling matrix together with $\mathbf{r}$ are considered as variables. The choice of the parameters and the number of variables of the GC depends on the particular application as we will subsequently see. Thus, if we are interested in the improvement of image quality or in its manipulation for some feature extraction (for example edge enhancement or image deblurring), then we have to choose $\mathbf{T}_3 = \mathbf{T}_1^{-1}$, in order to represent the result of filtering in the position domain.

A typical optical scheme for GC is a straightforward generalization of the Van der Lugt processor and consists of a cascade of two first-order systems described by the ray transformation matrices $\mathbf{T}_1$ and $\mathbf{T}_3$ with diffraction/reflection screen between them corresponding to multiplication of the passing/reflecting beam by $\mathcal{R}^{\mathbf{T}_2}[h(\cdot)]$. Then with $f(\cdot)$ in the input of this system we have its GC with $h(\cdot)$ at the output plane. The common convolution operation $\mathcal{C}_{f,h}(\mathbf{r})$ (1.44) arrises when $\mathbf{T}_1$ and $\mathbf{T}_2$ correspond to the direct FT matrices and $\mathbf{T}_3$ to the inverse one. In optical realization, $\mathbf{T}_3$ is usually the direct FT, then we have $\mathcal{C}_{f,h}(-\mathbf{r})$ at the output plane.

Since two-dimensional canonical transforms include such operations as rotation, scaling, shearing, fractional FT, etc., the generalized canonical convolution can be helpful for resolving the problem of scale-, rotation-, and shift-invariant or partially variant filtering and pattern recognition. In this section, we discuss several applications of the GCs.

1.5.3 Filtering in fractional Fourier domains

A signal can be represented as a superposition of harmonics or saying in optical language—plane waves. The directions of these waves are related to the spatial frequencies. Different frequencies correspond to the different signal features. Thus, if we speak about an image, the high frequencies are responsible for the fine details; meanwhile, the low ones are for the large-scale structure. It can be shown by the elimination of the low- or high-frequency part of the Fourier spectrum and applying the inverse Fourier transform or, in other words, performing filtering, a convolution operation $\mathcal{C}(\mathbf{r})$, with high (H) pass or low (L) pass frequency filter masks respectively. Replacement the FT to the symmetric fractional FT at angle φ leads to the fractional convolution $\mathcal{R}^{\mathbf{T}_f(-\varphi,-\varphi)}\left\{\mathcal{R}^{\mathbf{T}_f(\varphi,\varphi)}[f(\mathbf{r}_i)](\mathbf{u})\, h(\mathbf{u})\right\}(\mathbf{r})$.

The amplitudes of the cameraman image filtered in the fractional domains for $\varphi = 0, 3\pi/8, \pi/2$ are displayed in Fig. 1.1, where the upper (lower) row corresponds to the H (L) filtering, respectively. For demonstration purposes we have chosen two circular binary amplitude filters $h(\mathbf{u})$: H-filter (opaque circle of radius

Figure 1.1 Results of the cameraman image filtering in fractional FT domains for $\varphi = 0, 3\pi/8, \pi/2$ (from left to right). The upper (lower) row corresponds to the filtering with H (L) mask, respectively.

R on the transparent background) and L-filter (transparent circle of radius R on the opaque background), which is clearly indicated at the images of the first column where the filtering in the position domain ($\varphi = 0$) is displayed.

The application of the H-filter in the fractional domain leads to the local high-frequency pass filtering, where the area of its action depends on the position of the filter, its size R, and parameter φ. In particular, such types of filters can be used for selective edge enhancement. For FT domain ($\varphi = \pi/2$), the edges of the entire image are underlined.

The L-filter produces image cutting together with smoothing. The area of image reproduction and the degree of smoothing are enlarged with an increase of φ.

Finally, we conclude that filtering is not shift-invariant if angle φ differs from $\pi/2$, which follows from the shift theorem [Eq. (1.27)]. Similar results have been obtained for the filtering in the different gyrator domains.[22]

The filtering in fractional Fourier domains demonstrated here has an illustrative character. Certainly, more sophisticated filters must be developed for every particular task such as noise reduction, edge enhancement, etc.

1.5.4 Pattern recognition

The correlation operation (the convolution between a signal f and the reversed complex conjugate of another one h^*) $\text{Cor}_{f,h^*}(\mathbf{r}) = \mathcal{GC}_{f,h^*}[\mathbf{T}_f(\pi/2, \pi/2), \mathbf{T}_f(-\pi/2, -\pi/2), \mathbf{T}_f(-\pi/2, -\pi/2), \mathbf{r}]$ is a measure of the similarity between two signals f and h. The mathematical verification of this statement is related to inequality of Schwarz, which permits one to discriminate two signals of equal energy, because in this case the autocorrelation peak $|\text{Cor}_{f,f^*}(\mathbf{0})|$ is larger than the cross-correlation one $|\text{Cor}_{f,h^*}(\mathbf{0})|$. Note that $|\text{Cor}_{f,f^*}(\mathbf{0})|$ has a maximum in the origin of the coordinates $\mathbf{r} = \mathbf{0}$. Then, applying the appropriate threshold to the correlation map $|\text{Cor}_{f,h^*}(\mathbf{r})|$, the pattern associated with h can be found on the investigated scene f. Moreover, because the correlation is shift-invariant, $\text{Cor}_{f(\mathbf{r}_i-\mathbf{v}),h^*(\mathbf{r}_i)}(\mathbf{r}) = \text{Cor}_{f(\mathbf{r}_i),h^*(\mathbf{r}_i)}(\mathbf{r}-\mathbf{v})$, the positions of all patterns h, if there are several, can be localized. This operation is also performed by the Van der Lugt processor using $\mathcal{F}^{-1}[h^*(\cdot)](\mathbf{u})$ as a filter mask.

Let us consider as an example a piece of text presented in Fig. 1.2(a). The amplitude of the cross-correlation between this image and the reference one, where only **O** is written in the center, is given in the Fig. 1.2(b). The largest peaks are observed at the positions where **O** is written, which permits its localization. Note that the value of the cross-correlation peaks depends on the similarity between **O** and other letters. Thus, a relatively large peak is also observed in the end of the bottom line where **G** is written. More sophisticated filters are usually used for better object discrimination.

If the pattern has to be detected only in a certain region of scene, then we must apply the fractional FT convolution with $\mathbf{T}_1 = \mathbf{T}_2^{-1} = \mathbf{T}_f(\gamma_x, \gamma_y)$, and $\mathbf{T}_3 = \mathbf{T}_f(-\pi/2, -\pi/2)$, which provides the shift-variant pattern recognition. The shift tolerance condition is usually written in the form[2,4] $\pi v_{x,y} \sigma_{x,y} \cot \gamma_{x,y} \ll 1$, where $v_{x,y}$ is the allowed shift of the pattern on the scene with respect to the reference one used for filter design and $\sigma_{x,y}$ is the pattern width in the x and y directions, correspondingly.

Thus, if we choose different fractional angles for two orthogonal coordinates, such that $\gamma_x = \pi/2$ and $\gamma_y = \pi/4$ and the same filter as we have used before, then **O** will be recognized only on the middle line of the text, as it is shown in Fig. 1.2(c), where the amplitude of the fractional correlation is represented. Therefore, the fractional correlation is a useful tool for shift-variant pattern recognition. This operation can be performed by a fractional Van der Lugt correlator, which is a modification of the common one, where the first part is replaced by the fractional FT system.

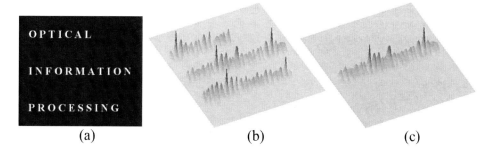

Figure 1.2 The input image (a) and the amplitude of its correlation with O, written in the center of the reference image, in Fourier domain (b) and in the fractional Fourier domain for $\gamma_x = \pi/2$ and $\gamma_y = \pi/4$ (c).

In order to maximize the Horner efficiency of the correlation operation, phase-only filters are often used. It was shown in Ref. 30 that, in general, the phase of the fractional FT for $\gamma \neq n\pi$ contains more information about the signal/image than the amplitude. Therefore, the phase-only filters can also be applied in the fractional FT domains.

If the pattern on the scene is rotated with respect to reference one, then we can apply the GC where $\mathbf{T}_1 = \mathbf{T}_f(\pi/2, \pi/2)\mathbf{T}_r(\alpha)$ and $\mathbf{T}_2 = \mathbf{T}_3 = \mathbf{T}_f(-\pi/2, -\pi/2)$, or $\mathbf{T}_1 = \mathbf{T}_3^{-1} = \mathbf{T}_f(\pi/2, \pi/2)$ and $\mathbf{T}_2 = \mathbf{T}_f(-\pi/2, -\pi/2)\mathbf{T}_r(\alpha)$. The identification of the largest correlation peak as a function of α indicates the right orientation of the pattern. This analysis for the two mentioned cases, respectively, can be done adding the flexible rotator system[15] just before the common correlator with an invariable filter mask or using the Van der Lugt correlator with a variable filter mask, which can be obtained by application of the SLM.

For rotation-invariant pattern recognition, the reference image is decomposed into a linear sum of the LG modes (1.43) and a superposition of them with the same topological charge (usually equals 1 or −1), known as a radial harmonic, is used as its substitute.

For recognition of the pattern of unknown scale, the GC defined by the following ray transformation matrices $\mathbf{T}_1 = \mathbf{T}_f(\pi/2, \pi/2)\mathbf{T}_S$ and $\mathbf{T}_2 = \mathbf{T}_3 = \mathbf{T}_f(-\pi/2, -\pi/2)$, or $\mathbf{T}_1 = \mathbf{T}_3^{-1} = \mathbf{T}_f(\pi/2, \pi/2)$ and $\mathbf{T}_2 = \mathbf{T}_f(-\pi/2, -\pi/2)\mathbf{T}_S$ can be used for the analysis of the correlation peak as a function of the scaling parameters.

Other affine transformations of the pattern, such as skew, which often can be treated as a projection of the studied object, can be incorporated into pattern recognition tasks via generalized convolution formalism.

1.5.5 Localization of the generalized chirp signals

Chirp detection, localization, estimation, and, if necessary, elimination are important tasks in processing of medical and industrial signals. The chirp signal, given, for example, by the expression in the right-hand side of Eq. (1.38), can be easily localized applying the CT parametrized by $\mathbf{T}^{-1}$ because the output signal becomes a δ-function. In particular, the application of the FT, the fractional FT, and the GT allows one to localize a plane, elliptic, and hyperbolic waves, respectively. Thus, the GC corresponding to the CT spectra $\left|\mathcal{R}^{\mathbf{T}_1}[f(\mathbf{r}_i)](\mathbf{r})\right|^2$ with modifying parameters of $\mathbf{T}_1$, associated to the intensity distributions of the output signal, is suitable for the detection of chirps presented in the signal $f(\mathbf{r}_i)$. Here, $\mathbf{r}$ and parameters of $\mathbf{T}_1$ are variables of the GC.

For example, if $\mathbf{T}_1 = \mathbf{T}_f(\gamma_x, \gamma_y)$, then elliptic-type chirps can be detected as a local maxima of function $\left|\mathcal{R}^{\mathbf{T}_f(\gamma_x, \gamma_y)}[f(\mathbf{r}_i)](\mathbf{r})\right|^2$ for $\gamma_x, \gamma_y \in [0, \pi]$, known as a Radon–Wigner transform map. The appropriate filtering in the fractional FT domains has been used for elimination of elliptic chirplike noise and, therefore, image quality improvement.² Analogously, the hyperbolic chirps can be localized by analyzing the gyrator power spectra $\left|\mathcal{R}^{\mathbf{T}_g(\vartheta)}[f(\mathbf{r}_i)](\mathbf{r})\right|^2$, Ref. 22.

1.5.6 Security applications

The algorithm for optical image encryption by random phase filtering in the position and FT domains[31] has been recently generalized to the case of random phase filtering in different fractional Fourier[5] and gyrator[22] domains. In these cases, not only the random phase masks but also the orders of the phase space domains (fractional or gyrator angles) where they are located play the role of encryption keys. It was demonstrated that it is impossible to reconstruct the image using the correct masks but the wrong phase space domains.

In general form the encryption procedure of signal f consists of a cascade of N operations: the CT transform parametrized by matrix $\mathbf{T}_n$ with further result multiplication at a random phase mask $\exp(i\phi_n)$ for $n = 1, 2, \ldots, N$, which can be summarized as

$$F = \exp(i\phi_N) R^{\mathbf{T}_N} \left[\ldots \left[\exp(i\phi_2) R^{\mathbf{T}_2} \left[\exp(i\phi_1) R^{\mathbf{T}_1} [f] \right] \right] \right]. \quad (1.46)$$

The decryption procedure is written correspondingly as

$$f = R^{\mathbf{T}_1^{-1}} \left[\exp(-i\phi_1) \ldots \left[R^{\mathbf{T}_{N-1}^{-1}} \left[\exp(-i\phi_{N-1}) R^{\mathbf{T}_N^{-1}} \left[\exp(-i\phi_N) F \right] \right] \right] \right]. \quad (1.47)$$

The randomness of the phase masks together with a large number of encryption parameters $\mathbf{T}_n$ provide a high security of the encryption procedure.

1.6 Other Optical Computing Approaches via Orthosymplectic Transforms

Besides the traditional analog optical information processing considered in Section 1.5, the application of the first-order optical systems for optical computing, where information is encoded into the orbital angular momentum carried by beam or into the geometric phase accumulated by beam under a cyclic transformation, is now widely discussed.[8,9,32] We briefly consider how the first-order optical systems associated with the CTs parametrized by the orthosymplectic matrix $\mathbf{T}_O$ can be used for this purpose.

1.6.1 Mode presentation on orbital Poincaré sphere

An orthosymplectic Gaussian mode $\mathcal{H}_{m,n}^{\mathbf{M}_O}(\mathbf{r})$, Eq. (1.41), can be obtained from the HG one $\mathcal{H}_{m,n}(\mathbf{r})$ by application of the CT associated with the orthogonal ray transformation matrix $\mathbf{M}_O$ or alternatively from the LG mode as $\mathcal{H}_{m,n}^{\mathbf{M}_O}(\mathbf{r}) = R^{\mathbf{T}_O}[\mathcal{L}_{m,n}^\pm(\mathbf{r}_i)](\mathbf{r})$, where $\mathbf{T}_O = \mathbf{M}_O \times \mathbf{M}_g(\pm\pi/4)$ [see Eq. (1.43)]. The modes here are defined up to the constant phase factor. As we have mentioned in Section 1.2, an orthosymplectic matrix $\mathbf{T}_O$ is described by four free parameters and can be represented as a separable fractional FT matrix embedded into two rotator ones Eq. (1.20). Because of the additivity of the separable fractional FT, it can be decomposed into symmetric and antisymmetric parts $\mathbf{T}_f(\gamma_x, \gamma_y) = \mathbf{T}_f(\varphi, \varphi)\mathbf{T}_f(\gamma, -\gamma)$, where $\gamma_x = \varphi + \gamma$ and $\gamma_y = \varphi - \gamma$. The symmetric fractional FT commutes with any CT described by the orthosymplectic matrix. It means[28] that the orthosymplectic Gaussian modes $\mathcal{H}_{m,n}^{M_O}(\mathbf{r})$ are eigenfunctions of the symmetric fractional FT $R^{\mathbf{T}_f(\varphi,\varphi)}[\mathcal{H}_{m,n}^{M_O}(\mathbf{r}_i)](\mathbf{r}) = \exp[-i(m+n+1)\varphi]\mathcal{H}_{m,n}^{M_O}(\mathbf{r})$ and also that $\mathbf{T}_O$ can be written as $\mathbf{T}_O = \mathbf{T}_r(\alpha)\mathbf{T}_f(\gamma,-\gamma)\mathbf{T}_r(\beta)\mathbf{T}_f(\varphi,\varphi)$. Because the LG modes are eigenfunctions for the symmetrical fractional FT and the rotator transform, we conclude that all different orthosymplectic Gaussian modes $\mathcal{H}_{m,n}^{M_O}(\mathbf{r})$ can be generated from the LGs by two parametric CTs described by matrix $\mathbf{T}_r(\alpha)\mathbf{T}_f(\gamma,-\gamma)$.

Therefore, any orthonormal set $\{\mathcal{H}_{m,n}^{M_O}(\mathbf{r})\}$ with integer $m,n \in [0,\infty)$ is characterized by only two parameters and can be associated with a certain direction in three-dimensional parametric space. Correspondingly, the modes with the same indices m,n belonging to the different sets can be represented on the sphere, called the orbital (m,n)-Poincaré sphere,[33,34] (see also Chapter 4) which is similar to the one used for presentation of polarized light. For example, starting from the LG mode $\mathcal{L}_{m,n}^+(\mathbf{r}) = \mathcal{L}_{m,n}^{(0,\cdot)}(\mathbf{r})$, living on the north pole of the (m,n)-Poincaré sphere, and applying the CT associated with two parametric matrix $\mathbf{T}_O(\theta,\psi) = \mathbf{T}_r(-\pi/4 + \psi/2)\,\mathbf{T}_f(\theta/2, -\theta/2)\,\mathbf{T}_r(\pi/4 - \psi/2)$ to this mode, the entire sphere can be populated by the different orthosymplectic modes $\mathcal{L}_{m,n}^{(\theta,\psi)}(\mathbf{r}) = R^{\mathbf{T}_O(\theta,\psi)}[\mathcal{L}_{m,n}^+(\mathbf{r}_i)](\mathbf{r})$, where the parameters $\theta \in [0,\pi]$ and $\psi \in [-\pi,\pi]$ indicate the colatitude of a parallel and the longitude of a meridian on the sphere, respectively. The HG modes $\mathcal{H}_{m,n}(\mathbf{r})$ and $\mathcal{H}_{n,m}(\mathbf{r})$ are located at the intersection of the main meridian and equator at points $(\theta,\psi) = (\pi/2, 0)$ and $(\pi/2, \pi)$, respectively. Moreover, it has been shown[28] that the transformation along the main meridian $\psi = 0$ corresponds to the gyrator transform (1.18), along the meridian with $\psi = \pi/2-$ to the antisymmetric fractional FT, and along the equator–to the rotator transform (1.15). Thus the HG mode $\mathcal{H}_{m,n}(\mathbf{r})$ rotated at angle $\psi/2$ lives on the equator at the longitude ψ.

1.6.2 Orbital angular momentum manipulation

In order to derive a closed-form expression for the mode $\mathcal{L}_{m,n}^{(\theta,\psi)}(\mathbf{r})$ in the form (1.41) and to define the z component of its OAM from Eq. (1.42), we again use the additivity property of the CTs and the fact that $\mathcal{H}_{m,n}^{\mathbf{T}_g(\mp\pi/4)}(\mathbf{r}) = \mathcal{L}_{m,n}^{\pm}(\mathbf{r})$. By doing this,[28] we find that the modes living on the same colatitude of the parallel θ have the same z component of the OAM, $\mathrm{L}_z^{m,n} = (m-n)\cos\theta$. In particular, for the HG modes $(\theta = \pi/2)\,\mathrm{L}_z^{m,n} = 0$. We observe that the OAM manipulation is related to the mode transformation, which can be performed by the orthosymplectic first-order optical systems. For example, applying the gyrator transform with scanning angle to the HG or LG (m,n) modes we can generate the modes $\mathcal{L}_{m,n}^{(\theta,0)}(\mathbf{r})$ with z component of the OAM from the interval $[-|m-n|,|m-n|]$. As we have mentioned before, the flexible optical setup has been designed for this purpose.[21]

1.6.3 Geometric phase accumulation

A movement on the orbital Poincaré sphere corresponds to a series of the orthosymplectic transformations of the mode associated with a starting point. If the path is closed we have a cyclic transformation. It has been shown[33,34] that the cyclic transformation of the (m,n) mode on the Poincaré sphere leads to the accumulation of the geometric phase $-(m-n)\Omega/2$, where Ω is a solid angle enclosed by the path. In other words, it means that the mode is an eigenfunction of the orthosymplectic transform associated with this path and represented as a cascade of the rotator and the antisymmetric fractional Fourier transforms. The phase of the related eigenvalue corresponds to the acquired geometric phase. We can easily define the optical systems suitable for the geometric phase accumulation for the modes from the main meridian of the orbital Poincaré sphere.[28] Thus the $\mathcal{L}_{m,n}^{(\theta,0)}(\mathbf{r})$ is an eigenfunction with eigenvalue $\exp[-i(m-n)\alpha]$ for the system described by the ray transformation matrix $\mathbf{T}_{\text{eigen}}(\theta,0) = \mathbf{T}_g(-\pi/4+\theta/2)\,\mathbf{T}_f(\alpha,-\alpha)\,\mathbf{T}_g(\pi/4-\theta/2)$. In particular, for the LG modes the related transform reduces to the rotator; meanwhile, for the HG ones it reduces to the antisymmetric fractional FT.

We conclude that the basic orthosymplectic operations: antisymmetric fractional FT, gyrator and rotator, whose flexible optical schemes have been recently proposed,[15,21] as well as their combinations allow one to easily control the information-encoding parameters: OAM projections and geometric phase accumulation.

Acknowledgments

I am pleased to express my gratitude to M.J. Bastiaans for stimulating discussions and critical comments during the manuscript preparation. I also thank him, M.L. Calvo, and J.A. Rodrigo for fruitful collaboration in many of the discussed topics.

References

1. Van der Lugt, A., (Ed.), *Optical Signal Processing*, Wiley, New York (1992).

2. Ozaktas, H.M., Zalevsky, Z., and Kutay, M.A., *The Fractional Fourier Transform with Applications in Optics and Signal Processing*, Wiley, New York (2001).

3. Alieva, T., Bastiaans, M.J., and Calvo, M.L., "Fractional transforms in optical information processing," *EURASIP J. Appl. Signal Process.*, **2005**, 1498–1519, 2005.

4. García, J., Mendlovic, D., Zalevsky, Z., and Lohmann, A.W., "Space-variant simultaneous detection of several objects by the use of multiple anamorphic fractional-Fourier transform filters," *Appl. Opt.*, **35**, 3945–3952, 1996.

5. Unnikrishnan, G., Joseph, J., and Singh, K., "Optical encryption by double-random phase encoding in the fractional Fourier domain," *Opt. Lett.*, **25**, 887–889, 2000.

6. Beijersbergen, M.W., Allen, L., Van der Veen, H.E.L.O., and Woerdman, J.P., "Astigmatic laser mode converters and transfer of orbital angular momentum," *Opt. Commun.*, **96**, 123–132, 1993.

7. Abramochkin, E.G., and Volostnikov, V.G., "Generalized Gaussian beams," *J. Opt. A.: Pure Appl. Opt.*, **6**, S157–S161, 2004.

8. Vaziri, A., Weihs, G., and Zeilinger, A., "Experimental two-photon, three-dimensional entanglement for quantum communication," *Phys. Rev. Lett.*, **89**, 240401, 2002.

9. Ekert, A., Ericsson, M., Hayden, P., Inamori, H., Jones, J.A., Oi, D.K.L., and Vedral, V., "Geometric quantum computation," *J. Mod. Opt.*, **47**, 2501–2513, 2000.

10. Schleich, W., *Quantum Optics in Phase Space*, Wiley-VCH, New York (2001).

11. Alieva, T., and Bastiaans, M.J., "Properties of the linear canonical integral transformation," *J. Opt. Soc. Am. A*, **24**, 3658–3665, 2007.

12. Simon, R., and Wolf, K.B., "Structure of the set of paraxial optical systems," *J. Opt. Soc. Am. A*, **17**, 342–355, 2000.

13. Wolf, K.B., *Geometric Optics in Phase Space*, Springer, New York (2004).

14. Collins, Jr., S.A., "Lens-system diffraction integral written in terms of matrix optics," *J. Opt. Soc. Am.*, **60**, 1168–1177, 1970.

15. Rodrigo, J.A., Alieva, T., and Calvo, M.L., "Optical system design for ortho-symplectic transformations in phase space," *J. Opt. Soc. Am. A*, **23**, 2494–2500, 2006.

16. Macukow, B., and Arsenault, H.H., "Matrix decompositions for nonsymmetrical optical systems," *J. Opt. Soc. Am.*, **73**, 1360–1366, 1983.

17. Lohmann, A., "Image rotation, Wigner rotation, and the fractional order Fourier transform," *J. Opt. Soc. Am. A*, **10**, 2181–2186, 1993.

18. Mendlovic, D., and Ozaktas, H.M., "Fractional Fourier transforms and their optical implementation: I," *J. Opt. Soc. Am. A*, **10**, 1875–1881, 1993.

19. Alieva, T., Lopez, V., Agullo-Lopez, F., and Almeida, L.B., "The fractional Fourier transform in optical propagation problems," *J. Mod. Opt.*, **41**, 1037–1044, 1994.

20. Rodrigo, J.A., Alieva, T., and Calvo, M.L., "Gyrator transform: properties and applications," *Opt. Express*, **15**, 2190–2203, 2007.

21. Rodrigo, J.A., Alieva, T., and Calvo, M.L., "Experimental implementation of the gyrator transform," *J. Opt. Soc. Am. A*, **24**, 3135–3139, 2007.

22. Rodrigo, J.A., Alieva, T., and Calvo, M.L., "Applications of gyrator transform for image processing," *Opt. Commun.*, **278**, 279–284, 2007.

23. Alieva, T., and Bastiaans, M.J., "Alternative representation of the linear canonical integral transform," *Opt. Lett.*, **30**, 3302–3304, 2005.

24. Alieva, T., "Fractional Fourier transform as a tool for investigation of fractal objects," *J. Opt. Soc. Am. A*, **13**, 1189–1192, 1996.

25. Bastiaans, M.J., and Alieva, T., "Propagation law for the generating function of Hermite-Gaussian-type modes in first-order optical systems," *Optics Express*, **13**, 1107–1112, 2005.

26. Bastiaans, M.J., and Alieva, T., "Classification of lossless first-order optical systems and the linear canonical transformation," *J. Opt. Soc. Am. A*, **24**, 1053–1062, 2007.

27. Alieva, T., and Bastiaans, M.J., "Mode mapping in paraxial lossless optics," *Opt. Lett.*, **30**, 1461–1463, 2005.

28. Alieva, T., and Bastiaans, M.J., "Orthonormal mode sets for the two-dimensional fractional Fourier transformation," *Opt. Lett.*, **32**, 1226–1228, 2007.

29. Akay, O., and Boudreaux-Bartels, G.F., "Fractional convolution and correlation via operator methods and an application to detection of linear FM signals," *IEEE Trans. Signal Process.*, **49**, 979–993, 2001.

30. Alieva, T., and Calvo, M.L., "Importance of the phase and amplitude in the fractional Fourier domain," *J. Opt. Soc. Am. A*, **20**, 533–541, 2003.

31. Towghi, N., Javidi, B., and Luo, Z., "Fully phase encrypted image processor," *J. Opt. Soc. Am. A*, **16**, 1915–1927, 1999.

32. Galvez, E.J., Crawford, P.R., Sztul, H.I., Pysher, M.J., Haglin, P.J., and Williams, R.E., "Geometric phase associated with mode transformations of optical beams bearing orbital angular momentum," *Phys. Rev. Lett.*, **90**, 203901, 2003.

33. Padgett, M.J., and Courtial, J., "Poincaré-sphere equivalent for light beams containing orbital angular momentum," *Opt. Lett.*, **24**, 430–432, 1999.

34. Calvo, G.F., "Wigner representation and geometric transformations of optical orbital angular momentum spatial modes," *Opt. Lett.*, **30**, 1207–1209, 2005.

**Tatiana Alieva** received her MSc in physics from M. V. Lomonosov Moscow State University, Russia, in 1983 and her PhD in physics from the Autonomous University of Madrid, Spain, in 1996. From 1983 until 1999, she was with the Radiotechnical Institute, Academy of Science, Moscow. She spent several years as a postdoctoral fellow at the Catholic University of Leuven, Belgium, and the Eindhoven University of Technology, the Netherlands. Currently, she is an associate professor, faculty of physics, at Complutense University of Madrid, Spain. Her current research interests are in optical information processing, characterization of optical fields, theory of integral transforms, phase retrieval, and fractal signal analysis. She is the author and coauthor of more than 120 scientific publications in books, international journals, and conference proceedings.

Chapter 2
Applications of the Wigner Distribution to Partially Coherent Light Beams

Martin J. Bastiaans
Technische Universiteit Eindhoven, The Netherlands

2.1 Introduction
2.2 Description of Partially Coherent Light
2.3 Wigner Distribution
    2.3.1 Definition of the Wigner distribution
    2.3.2 Some properties of the Wigner distribution
    2.3.3 Some examples of Wigner distributions
2.4 Modal Expansions
    2.4.1 Modal expansion of the power spectrum
    2.4.2 Modal expansion of the Wigner distribution
    2.4.3 Some inequalities for the Wigner distribution
2.5 Propagation of the Wigner Distribution
    2.5.1 Ray-spread function of an optical system
    2.5.2 Luneburg's first-order optical systems
    2.5.3 Transport equations for the Wigner distribution
    2.5.4 Geometric-optical systems
2.6 Second- and Higher-Order Moments
    2.6.1 Second-order moments of the Wigner distribution
    2.6.2 Second-order moments of Gaussian signals with twist
    2.6.3 Invariants for the second-order moments
    2.6.4 Higher-order moments and systems to measure them
References

2.1 Introduction

In 1932, Wigner introduced a distribution function in mechanics[1] that permitted a description of mechanical phenomena in a phase space. Such a Wigner distribution was introduced in optics by Walther[2] in 1968, to relate partial coherence to radiometry. A few years later, the Wigner distribution was introduced in optics again[3] (especially in the area of Fourier optics), and since then, a great number of applications of the Wigner distribution have been reported. It is the aim of this chapter to review the Wigner distribution and some of its applications to optical problems, especially with respect to partial coherence and first-order optical systems. The chapter is roughly an extension to two dimensions of a previous review paper[4] on the application of the Wigner distribution to partially coherent light, with additional material taken from some more recent papers on the twist of partially coherent Gaussian light beams[5,6] and on second- and higher-order moments of the Wigner distribution.[7-12] Some parts of this chapter have already been presented before[13-15] and have also been used as the basis for a lecture on "Representation of signals in a combined domain: Bilinear signal dependence" at the Winter College on Quantum and Classical Aspects of Information Optics, The Abdus Salam International Centre for Theoretical Physics, Trieste, Italy, January 2006. In order to avoid repeating a long list of ancient references, we will often simply refer to the references in Ref. 4; these references and the names of the first authors are put between square brackets like [Wigner,Walther,1,2].

We conclude this Introduction with some remarks about the signals with which we are dealing, and with some remarks about notation conventions. We consider scalar optical signals, which can be described by, say, $\tilde{f}(x,y,z,t)$, where x, y, z denote space variables and t represents the time variable. Very often we consider signals in a plane $z = $ constant, in which case we can omit the longitudinal space variable z from the formulas. Furthermore, the transverse space variables x and y are combined into a two-dimensional column vector $\mathbf{r}$. The signals with which we are dealing are thus described by a function $\tilde{f}(\mathbf{r},t)$.

We will throughout denote column vectors by boldface, lowercase symbols, while matrices will be denoted by boldface, uppercase symbols; transposition of vectors and matrices is denoted by the superscript t. Hence, for instance, the two-dimensional column vectors $\mathbf{r}$ and $\mathbf{q}$ represent the space and spatial-frequency variables $[x,y]^t$ and $[u,v]^t$, respectively, and $\mathbf{q}^t\mathbf{r}$ represents the inner product $ux + vy$. Moreover, in integral expressions, $d\mathbf{r}$ and $d\mathbf{q}$ are shorthand notations for $dx\,dy$ and $du\,dv$, respectively.

2.2 Description of Partially Coherent Light

Let (temporally stationary) partially coherent light be described by its mutual coherence function [Wolf, Papoulis, Bastiaans, 4–7] $\tilde{\Gamma}(\mathbf{r}_1,\mathbf{r}_2,t_1 - t_2) = \langle \tilde{f}(\mathbf{r}_1,t_1)\tilde{f}^*(\mathbf{r}_2,t_2)\rangle$, where the angular brackets denote time (or ensemble) averaging of the temporally stationary and ergodic stochastic process $\tilde{f}(\mathbf{r},t)$.

The mutual power spectrum [Papoulis, Bastiaans, 6, 7] or cross-spectral density function [Mandel, 8] $\Gamma(\mathbf{r}_1, \mathbf{r}_2, \omega)$ is defined as the temporal Fourier transform of the coherence function

$$\Gamma(\mathbf{r}_1, \mathbf{r}_2, \omega) = \int \tilde{\Gamma}(\mathbf{r}_1, \mathbf{r}_2, \tau) \exp(i\omega\tau) \, d\tau. \tag{2.1}$$

(Unless otherwise stated, all integrations in this chapter extend from $-\infty$ to $+\infty$.) The basic property [Bastiaans, Mandel, 7, 8] of the power spectrum is that it is a nonnegative definite Hermitian function of $\mathbf{r}_1$ and $\mathbf{r}_2$, i.e., $\Gamma(\mathbf{r}_1, \mathbf{r}_2, \omega) = \Gamma^*(\mathbf{r}_2, \mathbf{r}_1, \omega)$, where the asterisk denotes complex conjugation, and $\iint g(\mathbf{r}_1, \omega) \Gamma(\mathbf{r}_1, \mathbf{r}_2, \omega) g^*(\mathbf{r}_2, \omega) \, d\mathbf{r}_1 \, d\mathbf{r}_2 \geq 0$ for any function $g(\mathbf{r}, \omega)$.

Instead of describing partially coherent light in a space domain by means of its power spectrum $\Gamma(\mathbf{r}_1, \mathbf{r}_2, \omega)$, we can represent it equally well in a spatial-frequency domain by means of the spatial Fourier transform $\bar{\Gamma}(\mathbf{q}_1, \mathbf{q}_2, \omega)$ of the power spectrum

$$\bar{\Gamma}(\mathbf{q}_1, \mathbf{q}_2, \omega) = \iint \Gamma(\mathbf{r}_1, \mathbf{r}_2, \omega) \exp[-i\, 2\pi\, (\mathbf{q}_1^t \mathbf{r}_1 - \mathbf{q}_2^t \mathbf{r}_2)] \, d\mathbf{r}_1 \, d\mathbf{r}_2. \tag{2.2}$$

(Throughout, we represent the spatial Fourier transform of a function by the same symbol as the function itself, but marked with an overbar.) Unlike the power spectrum $\Gamma(\mathbf{r}_1, \mathbf{r}_2, \omega)$, which expresses the coherence of the light at two different positions $\mathbf{r}_1$ and $\mathbf{r}_2$, its spatial Fourier transform $\bar{\Gamma}(\mathbf{q}_1, \mathbf{q}_2, \omega)$ expresses the coherence of the light in two different directions $\mathbf{q}_1$ and $\mathbf{q}_2$. Therefore, we prefer to call $\Gamma(\mathbf{r}_1, \mathbf{r}_2, \omega)$ the positional power spectrum and $\bar{\Gamma}(\mathbf{q}_1, \mathbf{q}_2, \omega)$ the *directional* power spectrum of the light [Bastiaans, 9]. It is evident that the directional power spectrum $\bar{\Gamma}(\mathbf{q}_1, \mathbf{q}_2, \omega)$ is a nonnegative definite Hermitian function of the two spatial-frequency (or direction) variables $\mathbf{q}_1$ and $\mathbf{q}_2$.

Apart from the pure space or the pure spatial-frequency representation of a stochastic process by means of its positional or its directional power spectrum, respectively, we can describe a stochastic process in space and spatial frequency, simultaneously. In this chapter, we therefore use the Wigner distribution. Since in the present discussion the explicit temporal-frequency dependence is of no importance, we shall, for the sake of convenience, omit the temporal-frequency variable ω from the formulas in the remainder of this chapter.

2.3 Wigner Distribution

It is sometimes convenient to describe an optical signal not in a space domain by means of its positional power spectrum $\Gamma(\mathbf{r}_1, \mathbf{r}_2)$, but in a spatial-frequency domain by means of its directional power spectrum $\bar{\Gamma}(\mathbf{q}_1, \mathbf{q}_2)$. The directional auto power spectrum, also called directional intensity, $\bar{\Gamma}(\mathbf{q}, \mathbf{q})$ shows globally how the energy of the signal is distributed as a function of direction (i.e., spatial frequency) $\mathbf{q}$. However, instead of in this global distribution of the energy, one is often more interested in the local distribution of the energy as a function of spatial frequency. A similar local distribution occurs in music, for instance, in which a signal is usually described not by a time function nor by the Fourier transform of that function, but by its musical score.

The score is indeed a picture of the local distribution of the energy of the musical signal as a function of frequency. The horizontal axis of the score clearly represents a time axis, and the vertical one a frequency axis. When a composer writes a score, he prescribes the frequencies of the tones that should be present at a certain time. We see that the musical score is something that might be called the local frequency spectrum of the musical signal.

The need for a description of the signal by means of a local frequency spectrum arises in other disciplines, too. Geometrical optics, for instance, is usually treated in terms of rays, and the signal is described by giving the directions of the rays that should be present at a certain position. It is not difficult to translate the concept of the musical score to geometrical optics: we simply have to consider the horizontal (time) axis as a position axis and the vertical (frequency) axis as a direction axis. A musical note then represents an optical light ray passing through a point at a certain position and having a certain direction.

Another discipline in which the idea of a local frequency spectrum can be applied, is in mechanics: the position and the momentum of a particle are given in a phase space. It was in mechanics that Wigner introduced in 1932 a distribution function[1] that provided a description of mechanical phenomena in this phase space.

In this section, we define the Wigner distribution in optics, give some of its properties, and elucidate its concept by some simple examples.

2.3.1 Definition of the Wigner distribution

The (real-valued) Wigner distribution $W(\mathbf{r}, \mathbf{q})$ of a stochastic process can be defined in terms of the (Hermitian) positional power spectrum or, equivalently, in terms of the (Hermitian) directional power spectrum,

$$W(\mathbf{r}, \mathbf{q}) = \int \Gamma(\mathbf{r} + \tfrac{1}{2}\mathbf{r}', \mathbf{r} - \tfrac{1}{2}\mathbf{r}') \exp(-i\,2\pi\,\mathbf{q}^t \mathbf{r}')\, d\mathbf{r}', \qquad (2.3)$$

$$W(\mathbf{r}, \mathbf{q}) = \int \bar{\Gamma}(\mathbf{q} + \tfrac{1}{2}\mathbf{q}', \mathbf{q} - \tfrac{1}{2}\mathbf{q}') \exp(i\,2\pi\,\mathbf{r}^t \mathbf{q}')\, d\mathbf{q}'. \qquad (2.4)$$

A distribution according to definitions (2.3) and (2.4) was first introduced in optics by Walther,[2] who called it the generalized radiance.

The Wigner distribution $W(\mathbf{r}, \mathbf{q})$ represents a stochastic signal in space and (spatial) frequency simultaneously and is thus a member of a wide class of phase-space distributions[16,17] [Cohen, 10]. It forms an intermediate signal description between the pure space representation $\Gamma(\mathbf{r}_1, \mathbf{r}_2)$ and the pure frequency representation $\bar{\Gamma}(\mathbf{q}_1, \mathbf{q}_2)$. Furthermore, this simultaneous space-frequency description closely resembles the ray concept in geometrical optics, in which the position and direction of a ray are also given simultaneously. In a way, $W(\mathbf{r}, \mathbf{q})$ is the amplitude of a ray passing through the point $\mathbf{r}$ and having a frequency (i.e., direction) $\mathbf{q}$.

2.3.2 Some properties of the Wigner distribution

We consider some properties of the Wigner distribution that are specific for partially coherent light. Additional properties of the Wigner distribution, especially of the Wigner distribution in the completely coherent case, to be defined later by Eq. (2.11), can be found elsewhere; see, for instance, Refs. 18–22 and the many references cited therein.

2.3.2.1 Fourier transformation

The definition (2.3) of the Wigner distribution $W(\mathbf{r}, \mathbf{q})$ has the form of a Fourier transformation of the positional power spectrum $\Gamma(\mathbf{r}+\frac{1}{2}\mathbf{r}', \mathbf{r}-\frac{1}{2}\mathbf{r}')$ with $\mathbf{r}'$ and $\mathbf{q}$ as conjugate variables and with $\mathbf{r}$ as a parameter. The positional power spectrum can thus be reconstructed from the Wigner distribution simply by applying an inverse Fourier transformation; a similar rule holds for the directional power spectrum. The latter property follows from the general remark that space and frequency, or position and direction, play equivalent roles in the Wigner distribution: if we interchange the roles of $\mathbf{r}$ and $\mathbf{q}$ in any expression containing a Wigner distribution, then we get an expression that is the dual of the original one. Thus, when the original expression describes a property in the space domain, the dual expression describes a similar property in the frequency domain, and vice versa.

2.3.2.2 Radiometric quantities

Although the Wigner distribution is real, it is not necessarily nonnegative; this prohibits a direct interpretation of the Wigner distribution as an energy density function (or radiance function). Fibers has shown [Fibers, 24] that it is not possible to define a radiance function that satisfies all the physical requirements from radiometry; in particular, as we mentioned, the Wigner distribution has the physically unattractive property that it may take negative values.

Nevertheless, several integrals of the Wigner distribution have clear physical meanings and can be interpreted as radiometric quantities. The integral over the frequency variable, for instance, $\int W(\mathbf{r}, \mathbf{q}) \, d\mathbf{q} = \Gamma(\mathbf{r}, \mathbf{r})$, represents the positional intensity of the signal, whereas the integral over the space variable, $\int W(\mathbf{r}, \mathbf{q}) \, d\mathbf{r} = \bar{\Gamma}(\mathbf{q}, \mathbf{q})$, yields its directional intensity, which is, apart from the usual factor $\cos^2 \theta$ (where θ is the angle of observation with respect to the z-axis), proportional to the radiant intensity [Carter, Wolf, 12, 13]. The total energy E of the signal follows from the integral over the entire space-frequency domain: $E = \iint W(\mathbf{r}, \mathbf{q}) \, d\mathbf{r} \, d\mathbf{q}$.

The real symmetric 4×4 matrix $\mathbf{M}$ of normalized second-order moments, defined by

is a well-known example of quasi-homogeneous light. For such a source, we have $p(\mathbf{r}) = A_r(\mathbf{r})$ and $\bar{s}(\mathbf{q}) = A_q(\mathbf{q})/\cos\theta$, where $\cos\theta = \sqrt{k^2 - (2\pi)^2\,\mathbf{q}^t\mathbf{q}}/k$, with θ the angle of observation with respect to the z-axis, again.

2.3.3.4 Coherent light

Completely coherent light is our next example. Its positional power spectrum $\Gamma(\mathbf{r}_1, \mathbf{r}_2) = f(\mathbf{r}_1)\,f^*(\mathbf{r}_2)$ has the form of a product of a function with its complex-conjugated version [Bastiaans, 7], and the same holds, of course, for the directional power spectrum. The Wigner distribution of coherent light thus takes the form

$$w(\mathbf{r}, \mathbf{q}) = \int f(\mathbf{r} + \tfrac{1}{2}\mathbf{r}')\,f^*(\mathbf{r} - \tfrac{1}{2}\mathbf{r}')\,\exp(-i\,2\pi\,\mathbf{q}^t\mathbf{r}')\,d\mathbf{r}'. \qquad (2.11)$$

We denote the Wigner distribution of coherent light throughout by the lowercase character w.

Two examples of coherent light are obvious. The Wigner distribution of a point source $f(\mathbf{r}) = \delta(\mathbf{r} - \mathbf{r}_\circ)$ reads $w(\mathbf{r}, \mathbf{q}) = \delta(\mathbf{r} - \mathbf{r}_\circ)$, and we observe that all the light originates from one point $\mathbf{r} = \mathbf{r}_\circ$ and propagates uniformly in all directions $\mathbf{q}$. Its dual, a plane wave $f(\mathbf{r}) = \exp(i\,2\pi\,\mathbf{q}_\circ^t\mathbf{r})$, also expressible in the frequency domain as $\bar{f}(\mathbf{q}) = \delta(\mathbf{q} - \mathbf{q}_\circ)$, has as its Wigner distribution $w(\mathbf{r}, \mathbf{q}) = \delta(\mathbf{q} - \mathbf{q}_\circ)$, and we observe that for all positions $\mathbf{r}$ the light propagates in only one direction $\mathbf{q}_\circ$.

An important example of coherent light is the quadratic-phase signal $f(\mathbf{r}) = \exp(i\,\pi\,\mathbf{r}^t\mathbf{H}\mathbf{r})$, which represents, at least for small $\mathbf{r}$, i.e., in the paraxial approximation, a spherical wave whose curvature is described by the real symmetric 2×2 matrix $\mathbf{H} = \mathbf{H}^t$. The Wigner distribution of such a signal takes the simple form $w(\mathbf{r}, \mathbf{q}) = \delta(\mathbf{q} - \mathbf{H}\mathbf{r})$, and we conclude that at any point $\mathbf{r}$ only one frequency $\mathbf{q} = \mathbf{H}\mathbf{r}$ manifests itself. This corresponds exactly to the ray picture of a spherical wave.

As a one-dimensional example of coherent signals, important for the description of laser modes, we consider the Hermite functions $\mathcal{H}_m(\xi)$ ($m = 0, 1, \ldots$) that can be expressed in the form

$$\mathcal{H}_m(\xi) = 2^{1/4}\,(2^m\,m!)^{-1/2}\,H_m(\sqrt{2\pi}\,\xi)\,\exp(-\pi\xi^2) \quad (m = 0, 1, \ldots), \qquad (2.12)$$

where H_m are the Hermite polynomials. We remark that the Hermite functions satisfy the orthonormality relation $\int \mathcal{H}_m(\xi)\,\mathcal{H}_n^*(\xi)\,d\xi = \delta_{m-n}$, where $\delta_{m-n} = 1$ for $m = n$, and $\delta_{m-n} = 0$ for $m \neq n$. It can be shown [Janssen, 32] that the Wigner distribution of the signal $f(x) = \mathcal{H}_m(x/\rho)$ takes the form

$$w(x, u) = 2\,(-1)^m\,\exp\left[-2\pi\left(\frac{x^2}{\rho^2} + \rho^2 u^2\right)\right] L_m\left[4\pi\left(\frac{x^2}{\rho^2} + \rho^2 u^2\right)\right], \qquad (2.13)$$

where L_m ($m = 0, 1, \ldots$) are the Laguerre polynomials.

2.3.3.5 Gaussian light

Gaussian light is our final example. The positional power spectrum of the most general partially coherent Gaussian light can be written in the form

$$\Gamma(\mathbf{r}_1, \mathbf{r}_2) = 2\sqrt{\det \mathbf{G}_1}$$
$$\times \exp\left(-\frac{\pi}{2}\begin{bmatrix}\mathbf{r}_1+\mathbf{r}_2\\ \mathbf{r}_1-\mathbf{r}_2\end{bmatrix}^t \begin{bmatrix}\mathbf{G}_1 & -i\mathbf{H}\\ -i\mathbf{H}^t & \mathbf{G}_2\end{bmatrix} \begin{bmatrix}\mathbf{r}_1+\mathbf{r}_2\\ \mathbf{r}_1-\mathbf{r}_2\end{bmatrix}\right), \quad (2.14)$$

where we have chosen a representation that enables us to determine the Wigner distribution of such light in an easy way. The exponent shows a quadratic form in which a four-dimensional column vector $[(\mathbf{r}_1+\mathbf{r}_2)^t, (\mathbf{r}_1-\mathbf{r}_2)^t]^t$ arises, together with a symmetric 4×4 matrix. This matrix consists of four real 2×2 submatrices $\mathbf{G}_1, \mathbf{G}_2, \mathbf{H}$, and $\mathbf{H}^t$, where, moreover, the matrices $\mathbf{G}_1$ and $\mathbf{G}_2$ are positive definite symmetric. The special form of the matrix is a direct consequence of the fact that the power spectrum is a nonnegative definite Hermitian function. The Wigner distribution of such Gaussian light takes the form[9,24]

$$W(\mathbf{r}, \mathbf{q}) = 4\sqrt{\frac{\det \mathbf{G}_1}{\det \mathbf{G}_2}} \exp\left(-2\pi \begin{bmatrix}\mathbf{r}\\ \mathbf{q}\end{bmatrix}^t \begin{bmatrix}\mathbf{G}_1 + \mathbf{H}\mathbf{G}_2^{-1}\mathbf{H}^t & -\mathbf{H}\mathbf{G}_2^{-1}\\ -\mathbf{G}_2^{-1}\mathbf{H}^t & \mathbf{G}_2^{-1}\end{bmatrix} \begin{bmatrix}\mathbf{r}\\ \mathbf{q}\end{bmatrix}\right). \quad (2.15)$$

In a more common way, the positional power spectrum of general Gaussian light (with ten degrees of freedom) can be expressed in the form

$$\Gamma(\mathbf{r}_1, \mathbf{r}_2) = 2\sqrt{\det \mathbf{G}_1} \exp\left\{-\tfrac{1}{2}\pi(\mathbf{r}_1-\mathbf{r}_2)^t \mathbf{G}_0 (\mathbf{r}_1-\mathbf{r}_2)\right\}$$
$$\times \exp\left\{-\pi \mathbf{r}_1^t \left[\mathbf{G}_1 - i\tfrac{1}{2}(\mathbf{H}+\mathbf{H}^t)\right]\mathbf{r}_1\right\}$$
$$\times \exp\left\{-\pi \mathbf{r}_2^t \left[\mathbf{G}_1 + i\tfrac{1}{2}(\mathbf{H}+\mathbf{H}^t)\right]\mathbf{r}_2\right\}$$
$$\times \exp\left\{-i\pi \mathbf{r}_1^t (\mathbf{H}-\mathbf{H}^t)\mathbf{r}_2\right\}, \quad (2.16)$$

where we have introduced the real, positive definite symmetric 2×2 matrix $\mathbf{G}_0 = \mathbf{G}_2 - \mathbf{G}_1$. Note that the asymmetry of the matrix $\mathbf{H}$ is a measure for the twist[25-29] of Gaussian light and that general Gaussian light reduces to zero-twist Gaussian Schell-model light[30] [Gori, 15], if the matrix $\mathbf{H}$ is symmetric, $\mathbf{H} - \mathbf{H}^t = \mathbf{0}$. In that case, the light can be considered as spatially stationary light with a Gaussian power spectrum $2\sqrt{\det \mathbf{G}_1} \exp\{-\tfrac{1}{2}\pi(\mathbf{r}_1-\mathbf{r}_2)^t \mathbf{G}_0 (\mathbf{r}_1-\mathbf{r}_2)\}$, modulated by a Gaussian modulator with modulation function $\exp\{-\pi \mathbf{r}^t (\mathbf{G}_1 - i\mathbf{H})\mathbf{r}\}$. We remark that such Gaussian Schell-model light (with nine degrees of freedom) forms a large subclass of Gaussian light; it applies, for instance, in the following:

- The completely coherent case, i.e., $\mathbf{H} = \mathbf{H}^t$, $\mathbf{G}_0 = \mathbf{0}$, $\mathbf{G}_1 = \mathbf{G}_2$.
- The (partially coherent) one-dimensional case, i.e., $g_0 = g_2 - g_1 \geq 0$.
- The (partially coherent) rotationally symmetric case, i.e., $\mathbf{H} = h\mathbf{I}$, $\mathbf{G}_1 = g_1\mathbf{I}$, $\mathbf{G}_2 = g_2\mathbf{I}$, $\mathbf{G}_0 = (g_2 - g_1)\mathbf{I}$, with $\mathbf{I}$ the 2×2 identity matrix.

Gaussian Schell-model light reduces to so-called symplectic Gaussian light,[9] if the matrices $\mathbf{G}_0$, $\mathbf{G}_1$, and $\mathbf{G}_2$ are proportional to each other, $\mathbf{G}_1 = \sigma \mathbf{G}$, $\mathbf{G}_2 = \sigma^{-1}\mathbf{G}$, and thus $\mathbf{G}_0 = (\sigma^{-1} - \sigma)\mathbf{G}$, with $\mathbf{G}$ a real, positive definite symmetric 2×2 matrix and $0 < \sigma \leq 1$. The Wigner distribution then takes the form

$$W(\mathbf{r}, \mathbf{q}) = 4\sigma^2 \exp\left(-2\pi\sigma \begin{bmatrix} \mathbf{r} \\ \mathbf{q} \end{bmatrix}^t \begin{bmatrix} \mathbf{G} + \mathbf{H}\mathbf{G}^{-1}\mathbf{H} & -\mathbf{H}\mathbf{G}^{-1} \\ -\mathbf{G}^{-1}\mathbf{H} & \mathbf{G}^{-1} \end{bmatrix} \begin{bmatrix} \mathbf{r} \\ \mathbf{q} \end{bmatrix}\right). \quad (2.17)$$

The name symplectic Gaussian light (with six degrees of freedom) originates from the fact that the 4×4 matrix that arises in the exponent of the Wigner distribution (2.17) is symplectic. We will return to symplecticity later on in this chapter. We remark that symplectic Gaussian light forms a large subclass of Gaussian Schell-model light; it applies again, for instance, in the completely coherent case, in the (partially coherent) one-dimensional case, and in the (partially coherent) rotationally symmetric case. And again, symplectic Gaussian light can be considered as spatially stationary light with a Gaussian power spectrum, modulated by a Gaussian modulator, cf. Eq. (2.16), but now with the real parts of the quadratic forms in the two exponents described—up to a positive constant—by the same real, positive definite symmetric matrix $\mathbf{G}$.

2.4 Modal Expansions

To derive more properties of the Wigner distribution, we introduce modal expansions for the power spectrum and the Wigner distribution.

2.4.1 Modal expansion of the power spectrum

We represent the positional power spectrum $\Gamma(\mathbf{r}_1, \mathbf{r}_2)$ by its modal expansion [Wolf, 26], see also, for instance, [Gori, Gamo, 15, 27], in which a modal expansion of the (nonnegative definite Hermitian) mutual intensity $\tilde{\Gamma}(\mathbf{r}_1, \mathbf{r}_2, 0)$ is given,

$$\Gamma(\mathbf{r}_1, \mathbf{r}_2) = \frac{1}{\rho} \sum_{m=0}^{\infty} \mu_m f_m\left(\frac{\mathbf{r}_1}{\rho}\right) f_m^*\left(\frac{\mathbf{r}_2}{\rho}\right); \quad (2.18)$$

a similar expansion holds for the directional power spectrum. In the modal expansion (2.18), the functions f_m are the eigenfunctions, and the numbers μ_m are the eigenvalues of the integral equation $\int \Gamma(\mathbf{r}_1, \mathbf{r}_2) f_m(\mathbf{r}_2/\rho) d\mathbf{r}_2 = \mu_m f_m(\mathbf{r}_1/\rho)$ ($m = 0, 1, \ldots$); the positive factor ρ is a mere scaling factor. Because the kernel $\Gamma(\mathbf{r}_1, \mathbf{r}_2)$ is Hermitian and under the assumption of discrete eigenvalues, the eigenfunctions f_m can be made orthonormal: $\int f_m(\boldsymbol{\xi}) f_n^*(\boldsymbol{\xi}) d\boldsymbol{\xi} = \delta_{m-n}$ ($m, n = 0, 1, \ldots$). Moreover, because the kernel $\Gamma(\mathbf{r}_1, \mathbf{r}_2)$ is nonnegative definite Hermitian, the eigenvalues μ_m are nonnegative. Note that the light is completely coherent if there is only *one* nonvanishing eigenvalue. As a matter of fact, the modal expansion (2.18) expresses the partially coherent light as a superposition of coherent modes.

2.4.2 Modal expansion of the Wigner distribution

When we substitute the modal expansion (2.18) into the definition (2.3), the Wigner distribution can be expressed as

$$W(\mathbf{r},\mathbf{q}) = \sum_{m=0}^{\infty} \mu_m \, w_m\left(\frac{\mathbf{r}}{\rho}, \rho\mathbf{q}\right), \tag{2.19}$$

where $w_m(\xi, \eta)$ ($m = 0, 1, \ldots$) are the Wigner distributions of the eigenfunctions $f_m(\xi)$, as in the completely coherent case [see definition (2.11)]. By applying relation (2.9) and using the orthonormality property for $f_m(\xi)$, it can easily be seen that the Wigner distributions $w_m(\xi, \eta)$ satisfy the orthonormality relation

$$\iint w_m(\xi,\eta)\, w_n(\xi,\eta)\, d\xi\, d\eta = \left| \int f_m(\xi)\, f_n^*(\xi)\, d\xi \right|^2 = \delta_{m-n} \quad (m, n = 0, 1, \ldots). \tag{2.20}$$

2.4.3 Some inequalities for the Wigner distribution

The modal expansion (2.19) allows us to formulate some interesting inequalities for the Wigner distribution.

2.4.3.1 De Bruijn's inequality

Using the expansion (2.19), it is easy to see that De Bruijn's inequality [De Bruijn, 34]

$$\iint \left[g(\mathbf{e}^t\mathbf{r})^2 + g^{-1}(\mathbf{e}^t\mathbf{q})^2 \right]^m W(\mathbf{r},\mathbf{q})\, d\mathbf{r}\, d\mathbf{q} \geq \frac{m!}{(2\pi)^m} \iint W(\mathbf{r},\mathbf{q})\, d\mathbf{r}\, d\mathbf{q}, \tag{2.21}$$

with $\mathbf{e}$ a two-dimensional unit column vector, g a positive scalar, and m a positive integer, holds not only in the completely coherent case, but also for the Wigner distribution of partially coherent light. In the special case $m = 1$, and choosing $\mathbf{e} = [\cos\phi, \sin\phi]^t$, relation (2.21) reduces to $g\,(d_x^2 \cos^2\phi + d_y^2 \sin^2\phi) + g^{-1}(d_u^2 \cos^2\phi + d_v^2 \sin^2\phi) \geq 1/2\pi$, which leads to the uncertainty relation $(d_x^2 \cos^2\phi + d_y^2 \sin^2\phi)^{1/2}(d_u^2 \cos^2\phi + d_v^2 \sin^2\phi)^{1/2} \geq 1/4\pi$ by choosing $g = (d_u^2 \cos^2\phi + d_v^2 \sin^2\phi)^{1/2}(d_x^2 \cos^2\phi + d_y^2 \sin^2\phi)^{-1/2}$. The special cases [Papoulis, 6] $d_x d_u \geq 1/4\pi$, $d_y d_v \geq 1/4\pi$, and $(d_x^2 + d_y^2)(d_u^2 + d_v^2) \geq 1/4\pi^2$ arise for $\phi = 0$, $\phi = \pm\pi/2$, and $\phi = \pm\pi/4$, respectively. The equality signs in these uncertainty relations occur for completely coherent Gaussian light; for all other signals, the products of the effective widths in the space and the frequency direction are larger.

2.4.3.2 Positive average

Using the relationship (2.9) and expanding the power spectra $\Gamma_1(\mathbf{r}_1, \mathbf{r}_2)$ and $\Gamma_2(\mathbf{r}_1, \mathbf{r}_2)$ in the form (2.18), it can readily be shown that

$$\iint W_1(\mathbf{r},\mathbf{q})\, W_2(\mathbf{r},\mathbf{q})\, d\mathbf{r}\, d\mathbf{q} \geq 0. \tag{2.22}$$

Thus, as we remarked before, averaging one Wigner distribution with another one always yields a nonnegative result. In particular, the averaging with the Wigner distribution of completely coherent Gaussian light is of some practical importance [De Bruijn, Mark, Janssen, 34–36], because this Gaussian Wigner distribution occupies the smallest possible area in the space-frequency domain, as we concluded before.

2.4.3.3 Schwarz' inequality

An upper bound for the expression that arises in relation (2.22) can be found by applying Schwarz' inequality [Papoulis, 6]:

$$\iint W_1(\mathbf{r},\mathbf{q}) W_2(\mathbf{r},\mathbf{q}) \, d\mathbf{r} \, d\mathbf{q} \leq \sqrt{\iint W_1^2(\mathbf{r},\mathbf{q}) \, d\mathbf{r} \, d\mathbf{q}} \sqrt{\iint W_2^2(\mathbf{r},\mathbf{q}) \, d\mathbf{r} \, d\mathbf{q}}$$

$$\leq \left(\iint W_1(\mathbf{r},\mathbf{q}) \, d\mathbf{r} \, d\mathbf{q} \right) \left(\iint W_2(\mathbf{r},\mathbf{q}) \, d\mathbf{r} \, d\mathbf{q} \right), \quad (2.23)$$

where the latter expression is simply the product of the total energies of the two signals and where we have used the important inequality

$$\iint W^2(\mathbf{r},\mathbf{q}) \, d\mathbf{r} \, d\mathbf{q} \leq \left(\iint W(\mathbf{r},\mathbf{q}) \, d\mathbf{r} \, d\mathbf{q} \right)^2. \quad (2.24)$$

To prove this important inequality, we first remark that, by using the modal expansion (2.19), the identity $\iint W(\mathbf{r},\mathbf{q}) \, d\mathbf{r} \, d\mathbf{q} = \sum_{m=0}^{\infty} \mu_m$ holds. Second, we observe the identity $\iint W^2(\mathbf{r},\mathbf{q}) \, d\mathbf{r} \, d\mathbf{q} = \sum_{m=0}^{\infty} \mu_m^2$, which can be easily proved by applying the modal expansion (2.19) and by using the orthonormality property (2.20). Finally, we remark that, because all eigenvalues μ_m are nonnegative, the inequality

$$\sum_{m=0}^{\infty} \mu_m^2 \leq \left(\sum_{m=0}^{\infty} \mu_m \right)^2 \quad (2.25)$$

holds, which completes the proof of relation (2.24). Note that the equality sign in relation (2.25), and hence in relation (2.24), holds if there is only one nonvanishing eigenvalue, i.e., in the case of complete coherence. The ratio of the two expressions that arise in relation (2.24) or (2.25) can therefore serve as a measure of the overall degree of coherence of the light [Bastiaans, 31, 54]. Other measures for the overall degree of coherence can be defined; we mention, in particular, the expression [Bastiaans, 53, 55] $(\sum_{m=0}^{\infty} \nu_m^p)^{1/(p-1)}$ with $\nu_m = \mu_m / \sum_{n=0}^{\infty} \mu_n$, which is a nondecreasing function of the parameter p ($p > 1$), whose maximum value (for $p \to \infty$) is equal to the largest (normalized) eigenvalue $\nu_{\max}$ and whose minimum value (for $p \to 1$) is equal to the informational entropy [Bastiaans, 53] $\exp\left(\sum_{m=0}^{\infty} \nu_m \ln \nu_m \right)$. The case $p = 2$ leads to the ratio of the two expressions that arise in relation (2.25), as mentioned before.

2.5 Propagation of the Wigner Distribution

In this section, we study how the Wigner distribution propagates through linear optical systems. In Section 2.5.1, we therefore consider an optical system as a black box, with an input plane and an output plane, whereas in Section 2.5.3, we consider the system as a continuous medium, in which the signal must satisfy a certain differential equation.

2.5.1 Ray-spread function of an optical system

We consider the propagation of the Wigner distribution through linear systems. A linear system can be represented in four different ways, depending on whether we describe the input and output signals in the space or in the frequency domain. We thus have four equivalent input-output relationships, which for completely coherent light read as

$$f_o(\mathbf{r}_o) = \int h_{rr}(\mathbf{r}_o, \mathbf{r}_i) \, f_i(\mathbf{r}_i) \, d\mathbf{r}_i, \qquad (2.26)$$

$$\bar{f}_o(\mathbf{q}_o) = \int h_{qr}(\mathbf{q}_o, \mathbf{r}_i) \, f_i(\mathbf{r}_i) \, d\mathbf{r}_i, \qquad (2.27)$$

$$f_o(\mathbf{r}_o) = \int h_{rq}(\mathbf{r}_o, \mathbf{q}_i) \, \bar{f}_i(\mathbf{q}_i) \, d\mathbf{q}_i, \qquad (2.28)$$

$$\bar{f}_o(\mathbf{q}_o) = \int h_{qq}(\mathbf{q}_o, \mathbf{q}_i) \, \bar{f}_i(\mathbf{q}_i) \, d\mathbf{q}_i. \qquad (2.29)$$

The first relation (2.26) is the usual system representation in the space domain by means of the coherent point-spread function h_{rr}; we remark that the function h_{rr} is the response of the system in the space domain when the input signal is a point source. The last relation (2.29) is a similar system representation in the frequency domain, where the function h_{qq} is the response of the system in the frequency domain when the input signal is a plane wave; therefore, we can call h_{qq} the wave-spread function of the system. The remaining two relations (2.27) and (2.28) are hybrid system representations, since the input and output signals are described in different domains; therefore, we can call the functions h_{qr} and h_{rq} hybrid spread functions. The system description for partially coherent light is similar; in terms of the point-spread function, it takes the form

$$\Gamma_o(\mathbf{r}_1, \mathbf{r}_2) = \iint h_{rr}(\mathbf{r}_1, \boldsymbol{\rho}_1) \, \Gamma_i(\boldsymbol{\rho}_1, \boldsymbol{\rho}_2) \, h_{rr}^*(\mathbf{r}_2, \boldsymbol{\rho}_2) \, d\boldsymbol{\rho}_1 \, d\boldsymbol{\rho}_2, \qquad (2.30)$$

and there are similar expressions for the other system descriptions.

Unlike the four system representations (2.26)–(2.29) described above, there is only one system representation when we describe the input and output signals by their Wigner distributions. Indeed, combining the system representations (2.26)–(2.29) with the definitions (2.3) and (2.4) of the Wigner distribution, results in the relation

$$W_o(\mathbf{r}_o, \mathbf{q}_o) = \iint K(\mathbf{r}_o, \mathbf{q}_o, \mathbf{r}_i, \mathbf{q}_i) W_i(\mathbf{r}_i, \mathbf{q}_i) \, d\mathbf{r}_i \, d\mathbf{q}_i, \qquad (2.31)$$

in which the Wigner distributions of the input and the output signal are related through a superposition integral. The function K is completely determined by the system and can be expressed in terms of the four system functions h_{rr}, h_{qr}, h_{rq}, and h_{qq}. We have

$$K(\mathbf{r}_o, \mathbf{q}_o, \mathbf{r}_i, \mathbf{q}_i) = \iint h_{rr}(\mathbf{r}_o + \tfrac{1}{2}\mathbf{r}'_o, \mathbf{r}_i + \tfrac{1}{2}\mathbf{r}'_i) \, h_{rr}^*(\mathbf{r}_o - \tfrac{1}{2}\mathbf{r}'_o, \mathbf{r}_i - \tfrac{1}{2}\mathbf{r}'_i)$$
$$\times \exp[-i\, 2\pi \,(\mathbf{q}_o^t \mathbf{r}'_o - \mathbf{q}_i^t \mathbf{r}'_i)] \, d\mathbf{r}'_o \, d\mathbf{r}'_i \quad (2.32)$$

and similar expressions for the other system functions [Bastiaans, 37]. Relation (2.32) can be considered as the definition of a double Wigner distribution; hence, the function K has all the properties of a Wigner distribution, for instance, the property of realness.

Let us think about the physical meaning of the function K. In a formal way, the function K is the response of the system in the space-frequency domain when the input signal is described by a product of two Dirac functions $W_i(\mathbf{r}, \mathbf{q}) = \delta(\mathbf{r} - \mathbf{r}_i)\,\delta(\mathbf{q} - \mathbf{q}_i)$; only in a formal way, because an actual input signal yielding such a Wigner distribution does not exist. Nevertheless, such an input signal could be considered as a single ray entering the system at the position $\mathbf{r}_i$ with direction $\mathbf{q}_i$. Hence, the function K might be called the ray-spread function of the system.

2.5.2 Luneburg's first-order optical systems

We will now concentrate on Luneburg's first-order optical system [Luneburg, 40], for which the input-output relation (2.26) takes the form of a linear canonical integral transformation, $f_i(\mathbf{r}) \to f_o(\mathbf{r})$, parametrized by the ray transformation matrix $\mathbf{T}$, to be defined below. A first-order optical system can, of course, be characterized by its system functions h_{rr}, h_{qr}, h_{rq}, and h_{qq}; they are all quadratic-phase functions [Bastiaans, 37]. (Note that a Dirac function can be considered as a limiting case of such a quadratically varying function.) A system representation in terms of a Wigner distribution, however, is far more elegant. The ray-spread function of a first-order optical system takes the form of a product of two Dirac functions,

$$K(\mathbf{r}_o, \mathbf{q}_o, \mathbf{r}_i, \mathbf{q}_i) = \delta(\mathbf{r}_o - \mathbf{A}\mathbf{r}_i - \mathbf{B}\mathbf{q}_i)\,\delta(\mathbf{q}_o - \mathbf{C}\mathbf{r}_i - \mathbf{D}\mathbf{q}_i), \qquad (2.33)$$

where $\mathbf{A}$, $\mathbf{B}$, $\mathbf{C}$, and $\mathbf{D}$ denote real 2×2 matrices, and the input-output relationship reads very simply as

$$W_o(\mathbf{A}\mathbf{r} + \mathbf{B}\mathbf{q}, \mathbf{C}\mathbf{r} + \mathbf{D}\mathbf{q}) = W_i(\mathbf{r}, \mathbf{q}). \qquad (2.34)$$

From the ray-spread function (2.33) we conclude that a single input ray, entering the system at the position $\mathbf{r}_i$ with direction $\mathbf{q}_i$, will yield a single output ray, leaving the system at the position $\mathbf{r}_o$ with the direction $\mathbf{q}_o$. The input and output positions and directions are related by the matrix relationship

$$\begin{bmatrix} \mathbf{r}_o \\ \mathbf{q}_o \end{bmatrix} = \begin{bmatrix} \mathbf{A} & \mathbf{B} \\ \mathbf{C} & \mathbf{D} \end{bmatrix} \begin{bmatrix} \mathbf{r}_i \\ \mathbf{q}_i \end{bmatrix} = \mathbf{T} \begin{bmatrix} \mathbf{r}_i \\ \mathbf{q}_i \end{bmatrix}. \qquad (2.35)$$

Relation (2.35) is a well-known geometric-optical matrix description of a first-order optical system [Luneburg, 40]; the 4×4 **ABCD**-matrix in this relationship is known as the ray transformation matrix [Deschamps, 41]. We again observe a perfect resemblance to the geometric-optical behavior of a first-order optical system (see also, for instance, [Walther, 43]).

We recall that a common way to represent the linear canonical integral transformation is by means of Collins integral,[31] which holds for a nonsingular submatrix **B**, in which case the integral kernel, i.e., the point spread function $h_{rr}(\mathbf{r}_o, \mathbf{r}_i)$, takes the form $(\det i\mathbf{B})^{-1/2} \exp\left[i\pi\left(\mathbf{r}_i^t \mathbf{B}^{-1} \mathbf{A} \mathbf{r}_i - 2\mathbf{r}_i^t \mathbf{B}^{-1} \mathbf{r}_o + \mathbf{r}_o^t \mathbf{D} \mathbf{B}^{-1} \mathbf{r}_o\right)\right]$; for the case $\mathbf{B} = \mathbf{0}$, we get $h_{rr}(\mathbf{r}_o, \mathbf{r}_i) = |\det \mathbf{A}|^{-1/2} \exp\left(i\pi \mathbf{r}_o^t \mathbf{C} \mathbf{A}^{-1} \mathbf{r}_o\right) \delta(\mathbf{r}_o - \mathbf{A}^{-1} \mathbf{r}_i)$. We stress again that the good thing about the Wigner distribution is that we do not need any integral, but that the simple coordinate transformation (2.34) suffices.

We remark that the ray transformation matrix is symplectic [Luneburg, Deschamps, Bastiaans, 40–42]. To express symplecticity in an easy way, we introduce the 4×4 matrix $\mathbf{J}$ according to

$$\mathbf{J} = i \begin{bmatrix} \mathbf{0} & -\mathbf{I} \\ \mathbf{I} & \mathbf{0} \end{bmatrix}. \qquad (2.36)$$

The matrix $\mathbf{J}$ has the properties $\mathbf{J} = \mathbf{J}^{-1} = \mathbf{J}^\dagger = -\mathbf{J}^t$, where $\mathbf{J}^{-1}$, $\mathbf{J}^\dagger = (\mathbf{J}^*)^t$, and $\mathbf{J}^t$ are the inverse, the adjoint, and the transpose of $\mathbf{J}$, respectively. Note, in particular, that $\det \mathbf{J} = 1$. Symplecticity of the ray transformation matrix can then be expressed by the relationship

$$\begin{bmatrix} \mathbf{A} & \mathbf{B} \\ \mathbf{C} & \mathbf{D} \end{bmatrix}^{-1} = \mathbf{T}^{-1} = \mathbf{J}\mathbf{T}^t\mathbf{J} = \begin{bmatrix} \mathbf{D}^t & -\mathbf{B}^t \\ -\mathbf{C}^t & \mathbf{A}^t \end{bmatrix}. \qquad (2.37)$$

The symplecticity condition, which can also be written in the form $\mathbf{T}^t \mathbf{J} \mathbf{T} = \mathbf{J}$, leads directly to the well-known Smith–Helmholtz–Lagrange invariant [Luneburg, Deschamps, Born, 40, 41, 52] $\mathbf{r}_1^t \mathbf{q}_2 - \mathbf{r}_2^t \mathbf{q}_1$ in first-order optical systems.

We remark that spherical waves are intimately related to first-order optical systems, because both the Wigner distribution of a spherical way and the ray-spread function of a first-order optical system consist of Dirac functions. If a spherical wave with curvature matrix $\mathbf{H}_i$, thus having a Wigner distribution of the form $w_i(\mathbf{r}, \mathbf{q}) = \delta(\mathbf{q} - \mathbf{H}_i \mathbf{r})$, forms the input signal of a first-order **ABCD**-system, then the Wigner distribution of the output signal reads

$$w_o(\mathbf{r}, \mathbf{q}) = \frac{1}{\det(\mathbf{A}^t + \mathbf{H}_i \mathbf{B}^t)} \delta\left(\mathbf{q} - [\mathbf{A}^t + \mathbf{H}_i \mathbf{B}^t]^{-1}[\mathbf{C}^t + \mathbf{H}_i \mathbf{D}^t]\mathbf{r}\right). \qquad (2.38)$$

We conclude that the output signal is again a spherical wave with curvature matrix

$$\mathbf{H}_o = (\mathbf{C} + \mathbf{D}\mathbf{H}_i)(\mathbf{A} + \mathbf{B}\mathbf{H}_i)^{-1}. \tag{2.39}$$

The latter bilinear relationship is known as the **ABCD**-law. In Section 2.6, we will show that the **ABCD**-law can also be applied to describe the propagation of symplectic Gaussian light through first-order optical systems.

Special cases of first-order optical systems are as follows:

1. A (possibly anamorphic) lens, with $\mathbf{A} = \mathbf{D} = \mathbf{I}$, $\mathbf{B} = \mathbf{0}$, and $\mathbf{C} = \mathbf{C}^t$ a real symmetric matrix.
2. The dual of such a lens: a section of "anamorphic" free space in the Fresnel approximation, with $\mathbf{A} = \mathbf{D} = \mathbf{I}$, $\mathbf{C} = \mathbf{0}$, and $\mathbf{B} = \mathbf{B}^t$ a real symmetric matrix.
3. A (possibly anamorphic) Fourier transformer, with $\mathbf{A} = \mathbf{D} = \mathbf{0}$ and $\mathbf{B}\mathbf{C}^t = \mathbf{C}^t\mathbf{B} = -\mathbf{I}$.
4. A (possibly anamorphic) magnifier, with $\mathbf{B} = \mathbf{C} = \mathbf{0}$ and $\mathbf{A}\mathbf{D}^t = \mathbf{A}^t\mathbf{D} = \mathbf{I}$; for the two special choices

$$\mathbf{A} = \mathbf{D} = \begin{bmatrix} \cos\alpha & \sin\alpha \\ -\sin\alpha & \cos\alpha \end{bmatrix} \quad \text{and} \quad \mathbf{A} = \begin{bmatrix} 1 & s \\ 0 & 1 \end{bmatrix},$$

we get the two special cases of a rotator with a rotation through the angle α and a shearer with a shearing ratio s, respectively.

5. A separable fractional Fourier transformer, with

$$\mathbf{A} = \mathbf{D} = \begin{bmatrix} \cos\gamma_x & 0 \\ 0 & \cos\gamma_y \end{bmatrix} \quad \text{and} \quad \mathbf{B} = -\mathbf{C} = \begin{bmatrix} \sin\gamma_x & 0 \\ 0 & \sin\gamma_y \end{bmatrix}. \tag{2.40}$$

6. A gyrator, with

$$\mathbf{A} = \mathbf{D} = \begin{bmatrix} \cos\theta & 0 \\ 0 & \cos\theta \end{bmatrix} \quad \text{and} \quad \mathbf{B} = -\mathbf{C} = \begin{bmatrix} 0 & \sin\theta \\ \sin\theta & 0 \end{bmatrix}. \tag{2.41}$$

We recall that the separable fractional Fourier transformer, which corresponds to rotations in the xu plane through an angle γ_x and in the yv plane through an angle γ_y,[32] has been used as a basis for measuring the Wigner distribution by phase-space tomographic methods.[33-35]

For a detailed treatment of the linear canonical transformation and first-order optical systems, we refer to Chapter 1, by Tatiana Alieva.[36]

2.5.3 Transport equations for the Wigner distribution

With the tools of the previous section, we could study the propagation of the Wigner distribution through free space by considering a section of free space as an optical system with an input plane and an output plane. It is possible, however, to find the propagation of the Wigner distribution through free space directly from the differential equation that the signal must satisfy. We therefore let the longitudinal variable z enter into the formulas and remark that the propagation of coherent light in free space (at least in the Fresnel approximation) is governed by the differential equation (see [Papoulis, 6, p. 358])

$$-i\frac{\partial f}{\partial z} = \left(k + \frac{1}{2k}\frac{\partial^2}{\partial \mathbf{r}^2}\right) f, \qquad (2.42)$$

with $\partial^2/\partial \mathbf{r}^2$ representing the scalar operator $\partial^2/\partial x^2 + \partial^2/\partial y^2$; partially coherent light must satisfy the differential equation

$$-i\frac{\partial \Gamma}{\partial z} = \left[\left(k + \frac{1}{2k}\frac{\partial^2}{\partial \mathbf{r}_1^2}\right) - \left(k + \frac{1}{2k}\frac{\partial^2}{\partial \mathbf{r}_2^2}\right)\right] \Gamma. \qquad (2.43)$$

The propagation of the Wigner distribution is now described by a transport equation [Bremmer, McCoy, Besieris, Bastiaans, 44–49], which in this case takes the form

$$\frac{2\pi \mathbf{q}^t}{k}\frac{\partial W}{\partial \mathbf{r}} + \frac{\partial W}{\partial z} = 0, \qquad (2.44)$$

with $\partial/\partial \mathbf{r}$ representing the vectorial operator $[\partial/\partial x, \partial/\partial y]^t$. The transport equation (2.44) has the solution

$$W(\mathbf{r}, \mathbf{q}; z) = W\left(\mathbf{r} - \frac{2\pi \mathbf{q}}{k}z, \mathbf{q}; 0\right) = W(\mathbf{r} - \lambda_\circ z\, \mathbf{q}, \mathbf{q}; 0), \qquad (2.45)$$

which corresponds to case 2 in Section 2.5.2, with the special choice $\mathbf{B} = \lambda_\circ z\, \mathbf{I}$.

The differential equation (2.43) is a special case of the more general equation

$$-i\frac{\partial \Gamma}{\partial z} = \left[L\left(\mathbf{r}_1, -i\frac{\partial}{\partial \mathbf{r}_1}; z\right) - L^*\left(\mathbf{r}_2, -i\frac{\partial}{\partial \mathbf{r}_2}; z\right)\right] \Gamma, \qquad (2.46)$$

where L is some explicit function of the space variables $\mathbf{r}$ and z and of the partial derivatives of Γ contained in the operator $\partial/\partial \mathbf{r}$. The transport equation that corresponds to this differential equation reads as[4] [Besieris, Bastiaans, 46, 48, 49]

$$-\frac{\partial W}{\partial z} = 2\,\mathrm{Im}\left[L\left(\mathbf{r} + \frac{i}{2}\frac{\partial}{2\pi \partial \mathbf{q}}, 2\pi \mathbf{q} - \frac{i}{2}\frac{\partial}{\partial \mathbf{r}}; z\right)\right] W, \qquad (2.47)$$

in which Im denotes the imaginary part. In the Liouville approximation (or geometric-optical approximation), the transport equation (2.47) reduces to

$$-\frac{\partial W}{\partial z} = 2\,(\mathrm{Im}\,L)\,W + \left(\frac{\partial \mathrm{Re}\,L}{2\pi \partial \mathbf{r}}\right)^t \frac{\partial W}{\partial \mathbf{q}} - \left(\frac{\partial \mathrm{Re}\,L}{2\pi \partial \mathbf{q}}\right)^t \frac{\partial W}{\partial \mathbf{r}}, \qquad (2.48)$$

in which Re denotes the real part. Relation (2.48) is a first-order partial differential equation, which can be solved by the method of characteristics [Courant, 50]: along a path described in a parameter notation by $\mathbf{r} = \mathbf{r}(s)$, $z = z(s)$, and $\mathbf{q} = \mathbf{q}(s)$, and defined by the differential equations

$$\frac{d\mathbf{r}}{ds} = -\frac{\partial \operatorname{Re} L}{2\pi \, \partial \mathbf{q}}, \quad \frac{dz}{ds} = 1, \quad \frac{d\mathbf{q}}{ds} = \frac{\partial \operatorname{Re} L}{2\pi \, \partial \mathbf{r}}, \qquad (2.49)$$

the partial differential equation (2.48) reduces to the ordinary differential equation

$$-\frac{dW}{ds} = 2 \left(\operatorname{Im} L \right) W. \qquad (2.50)$$

In the special case that $L(\mathbf{r}, 2\pi\mathbf{q}; z)$ is a real function of $\mathbf{r}, \mathbf{q}$, and z, Eq. (2.50) implies that along the path defined by relations (2.49), the Wigner distribution has a constant value (see also, for instance, [Fibers, 51]).

Let us now concentrate on a weakly inhomogeneous medium. In such a medium, a coherent signal must satisfy the Helmholtz equation, whereas the propagation of partially coherent light is governed by the differential equation

$$-i \frac{\partial \Gamma}{\partial z} = \left[\sqrt{k^2(\mathbf{r}_1, z) + \frac{\partial^2}{\partial \mathbf{r}_1^2}} - \sqrt{k^2(\mathbf{r}_2, z) + \frac{\partial^2}{\partial \mathbf{r}_2^2}} \right] \Gamma; \qquad (2.51)$$

note that $k = k(\mathbf{r}, z)$. In this case, the function L reads as $L(\mathbf{r}, 2\pi \mathbf{q}; z) = \sqrt{k^2(\mathbf{r}, z) - (2\pi)^2 \mathbf{q}^t \mathbf{q}}$. We can again derive a transport equation for the Wigner distribution. The exact transport equation is rather complicated, but in the Liouville approximation it takes the simple form

$$\frac{2\pi \mathbf{q}^t}{k} \frac{\partial W}{\partial \mathbf{r}} + \frac{\sqrt{k^2 - (2\pi)^2 \mathbf{q}^t \mathbf{q}}}{k} \frac{\partial W}{\partial z} + \left(\frac{\partial k}{2\pi \, \partial \mathbf{r}} \right)^t \frac{\partial W}{\partial \mathbf{q}} = 0, \qquad (2.52)$$

which, in general, cannot be solved explicitly. With the method of characteristics we conclude that along a path defined by

$$\frac{d\mathbf{r}}{ds} = \frac{2\pi \mathbf{q}}{k}, \quad \frac{dz}{ds} = \frac{\sqrt{k^2 - (2\pi)^2 \mathbf{q}^t \mathbf{q}}}{k}, \quad \frac{d\mathbf{q}}{ds} = \frac{\partial k}{2\pi \, \partial \mathbf{r}}, \qquad (2.53)$$

the Wigner distribution has a constant value. When we eliminate the frequency variable $\mathbf{q}$ from Eqs. (2.53), we are immediately led to

$$\frac{d}{ds} \left(k \frac{d\mathbf{r}}{ds} \right) = \frac{\partial k}{\partial \mathbf{r}}, \quad \frac{d}{ds} \left(k \frac{dz}{ds} \right) = \frac{\partial k}{\partial z}, \qquad (2.54)$$

which are the equations for an optical ray in geometrical optics [Born, 52]. We are thus led to the general conclusion that in the Liouville approximation the Wigner distribution has a constant value along the geometric-optical ray paths.

When we apply this procedure to a rotationally symmetric medium that extends along the z-axis with $k = k(\sqrt{\mathbf{r}^t \mathbf{r}})$, a graded-index fiber, for instance, and introduce the ray invariants $h = vx - uy$ and $w = \sqrt{(k/2\pi)^2 - u^2 - v^2}$, we will find that these ray invariants do not enter the transport equation. This means that h and w are indeed invariant along a ray: $dh/dz = dw/dz = 0$.

Note that in a homogeneous medium, i.e., $\partial k/\partial \mathbf{r} = \mathbf{0}$ and $\partial k/\partial z = 0$, the ray paths become straight lines. The transport equation can then again be solved explicitly, and the solution reads as

$$W(\mathbf{r}, \mathbf{q}; z) = W\left(\mathbf{r} - \frac{2\pi \mathbf{q}}{\sqrt{k^2 - (2\pi)^2 \mathbf{q}^t \mathbf{q}}} z, \mathbf{q}; 0\right). \quad (2.55)$$

The difference from the previous solution (2.45), in which we considered the Fresnel approximation, is that the sine $2\pi \mathbf{q}/k$ has been replaced by the tangent $2\pi \mathbf{q}/\sqrt{k^2 - (2\pi)^2 \mathbf{q}^t \mathbf{q}}$. When we now integrate the transport equation over the frequency variable $\mathbf{q}$ and use the definitions (2.7) and (2.8), we get the relation $(\partial/\partial \mathbf{r})^t \mathbf{j}_r + \partial j_z/\partial z = 0$, which shows that the geometrical vector flux $\mathbf{j}$ has zero divergence [Winston, 25].

2.5.4 Geometric-optical systems

Let us start by studying a modulator described, in the case of partially coherent light, by the input-output relationship $\Gamma_o(\mathbf{r}_1, \mathbf{r}_2) = m(\mathbf{r}_1) \Gamma_i(\mathbf{r}_1, \mathbf{r}_2) m^*(\mathbf{r}_2)$. The input and output Wigner distributions are related by the relationship

$$W_o(\mathbf{r}, \mathbf{q}_o) = \int W_i(\mathbf{r}, \mathbf{q}_i) \, d\mathbf{q}_i$$
$$\times \int m(\mathbf{r} + \tfrac{1}{2}\mathbf{r}') m^*(\mathbf{r} - \tfrac{1}{2}\mathbf{r}') \exp[-i 2\pi (\mathbf{q}_o - \mathbf{q}_i)^t \mathbf{r}'] \, d\mathbf{r}'. \quad (2.56)$$

This input-output relationship can be written in two distinct forms. On the one hand, we can represent it in a differential format reading as

$$W_o(\mathbf{r}, \mathbf{q}) = m\left(\mathbf{r} + \frac{i}{2} \frac{\partial}{2\pi \partial \mathbf{q}}\right) m^*\left(\mathbf{r} - \frac{i}{2} \frac{\partial}{2\pi \partial \mathbf{q}}\right) W_i(\mathbf{r}, \mathbf{q}). \quad (2.57)$$

On the other hand, we can represent it in an integral format that reads as

$$W_o(\mathbf{r}, \mathbf{q}) = \int w_m(\mathbf{r}, \mathbf{q} - \mathbf{q}_i) W_i(\mathbf{r}, \mathbf{q}_i) \, d\mathbf{q}_i, \quad (2.58)$$

where $w_m(\mathbf{r}, \mathbf{q})$ is the Wigner distribution of the modulation function $m(\mathbf{r})$. Which of these two forms is superior, depends on the problem.

We now confine ourselves to the case of a pure phase modulation function $m(\mathbf{r}) = \exp[i 2\pi \varphi(\mathbf{r})]$. We then get

$$m\left(\mathbf{r} + \tfrac{1}{2}\mathbf{r}'\right) m^*(\mathbf{r} - \tfrac{1}{2}\mathbf{r}') = \exp\left\{i2\pi \left[\varphi\left(\mathbf{r} + \tfrac{1}{2}\mathbf{r}'\right) - \varphi\left(\mathbf{r} - \tfrac{1}{2}\mathbf{r}'\right)\right]\right\}$$
$$= \exp\left\{i2\pi \left[(\partial \varphi/\partial \mathbf{r})^t \mathbf{r}' + \text{higher-order terms}\right]\right\}. \quad (2.59)$$

If we consider only the first-order derivative in relation (2.59), we arrive at the following expressions:

$$m\left(\mathbf{r} + \frac{i}{2}\frac{\partial}{2\pi\partial\mathbf{q}}\right) m^*\left(\mathbf{r} - \frac{i}{2}\frac{\partial}{2\pi\partial\mathbf{q}}\right) \simeq \exp\left[-\left(\frac{\partial\varphi}{\partial\mathbf{r}}\right)^t \left(\frac{\partial}{\partial\mathbf{q}}\right)\right], \quad (2.60)$$

$$w_m(\mathbf{r},\mathbf{q}) \simeq \delta\left(\mathbf{q} - \frac{\partial\varphi}{\partial\mathbf{r}}\right), \quad (2.61)$$

and the input-output relationship of the pure phase modulator becomes $W_o(\mathbf{r},\mathbf{q}) \simeq W_i(\mathbf{r},\mathbf{q} - \partial\varphi/\partial\mathbf{r})$, which is a mere coordinate transformation. We conclude that a single input ray yields a single output ray.

The ideas described above have been applied to the design of optical coordinate transformers[37] [Jiao, 56] and to the theory of aberrations [Lohmann, 57]. Now, if the first-order approximation is not sufficiently accurate, i.e., if we have to take into account higher-order derivatives in relation (2.59), the Wigner distribution allows us to overcome this problem. Indeed, we still have the exact input-output relationships (2.57) and (2.58), and we can take into account as many derivatives in relation (2.59) as necessary. We thus end up with a more general differential form [Frankenthal, 58] than expression (2.60) or a more general integral form [Janssen, 59] than expression (2.61). The latter case, for instance, will yield an Airy function instead of a Dirac function, when we take not only the first but also the third derivative into account.

We concluded that a single input ray yields a single output ray. This may also happen in more general—not just modulation-type—systems; we call such systems geometric-optical systems. These systems have the simple input-output relationship $W_o(\mathbf{r},\mathbf{q}) \simeq W_i(\mathbf{g}_r(\mathbf{r},\mathbf{q}), \mathbf{g}_q(\mathbf{r},\mathbf{q}))$, where the approximately equal sign ($\simeq$) becomes an equal sign ($=$) in the case of linear functions $\mathbf{g}_r$ and $\mathbf{g}_q$, i.e., in the case of Luneburg's first-order optical systems, which we have considered in Section 2.5.2. There appears to be a close relationship to the description of such geometric-optical systems by means of the Hamilton characteristics [Bastiaans, 37].

Instead of the black-box approach of a geometric-optical system, which leads to the input-output relationship mentioned before, we can also consider the system as a continuous medium and formulate transport equations, as we did in Section 2.5.3. For geometric-optical systems, this transport equation takes the form of a first-order partial differential equation [Marcuvitz, 60], which can be solved by the method of characteristics. In Section 2.5.3, we reached the general conclusion that these characteristics represent the geometric-optical ray paths and that along these ray paths the Wigner distribution has a constant value.

The use of the transport equation is not restricted to deterministic media; Bremmer [Bremmer, 47] has applied it to stochastic media. The transport equation is not restricted to the scalar treatment of wave fields either; Bugnolo and Bremmer [Bugnolo, 61] have applied it to study the propagation of vectorial wave fields. In the vectorial case, the concept of the Wigner distribution leads to a Hermitian matrix rather than to a scalar function and permits the description of nonisotropic media as well.

2.6 Second- and Higher-Order Moments

In this final section, we study some miscellaneous topics related to second-order moments of the Wigner distribution and to their propagation through first-order optical systems.[7-9]

2.6.1 Second-order moments of the Wigner distribution

The propagation of the matrix $\mathbf{M}$ of second-order moments of the Wigner distribution through a first-order optical system with ray transformation matrix $\mathbf{T}$, can be described by the input-output relationship[38] [Bastiaans, 42] $\mathbf{M}_o = \mathbf{T}\mathbf{M}_i\mathbf{T}^t$. This relationship can readily be derived by combining the input-output relationship (2.35) of the first-order optical system with the definition (2.5) of the moment matrices of the input and the output signal. Because the ray transformation matrix $\mathbf{T}$ is symplectic, we immediately conclude that symplecticity of the moment matrix is preserved in a first-order optical system: if $\mathbf{M}_i$ is proportional to a symplectic matrix, then $\mathbf{M}_o$ is proportional to a symplectic matrix as well, with the same proportionality factor.

If we multiply the moment relation $\mathbf{M}_o = \mathbf{T}\mathbf{M}_i\mathbf{T}^t$ from the right by $\mathbf{J}$ and use the symplecticity property (2.37) and the properties of $\mathbf{J}$, then the input-output relationship can be written as[8] $\mathbf{M}_o\mathbf{J} = \mathbf{T}(\mathbf{M}_i\mathbf{J})\mathbf{T}^{-1}$. From the latter relationship, we conclude that the matrices $\mathbf{M}_i\mathbf{J}$ and $\mathbf{M}_o\mathbf{J}$ are related to each other by a similarity transformation. As a consequence of this similarity transformation, and writing the matrix $\mathbf{MJ}$ in terms of its eigenvalues and eigenvectors according to $\mathbf{MJ} = \mathbf{S}\mathbf{\Lambda}\mathbf{S}^{-1}$, we can formulate the relationships $\mathbf{\Lambda}_o = \mathbf{\Lambda}_i$ and $\mathbf{S}_o = \mathbf{T}\mathbf{S}_i$. We are thus led to the important property[8] that the eigenvalues of the matrix $\mathbf{MJ}$ (and any combination of these eigenvalues) remain invariant under propagation through a first-order optical system, while the matrix of eigenvectors $\mathbf{S}$ transforms in the same way as the ray vector $[\mathbf{r}^t, \mathbf{q}^t]^t$ does. A similar property holds for the matrix $\mathbf{JM}$, but in this chapter we will concentrate on the matrix $\mathbf{MJ}$. In Section 2.6.3, we will use the invariance property to derive invariant expressions in terms of the second-order moments of the Wigner distribution.

It can be shown[8] that the eigenvalues of $\mathbf{MJ}$ are real. Moreover, if λ is an eigenvalue of $\mathbf{MJ}$, then $-\lambda$ is an eigenvalue, too; this implies that the characteristic polynomial $\det(\mathbf{MJ} - \lambda\mathbf{I})$, with the help of which we determine the eigenvalues, is a polynomial of λ^2. Furthermore, if $\mathbf{M}$ is proportional to a symplectic matrix, then it can be expressed in the form

$$\mathbf{M} = m \begin{bmatrix} \mathbf{G}^{-1} & \mathbf{G}^{-1}\mathbf{H} \\ \mathbf{H}\mathbf{G}^{-1} & \mathbf{G} + \mathbf{H}\mathbf{G}^{-1}\mathbf{H} \end{bmatrix}, \qquad (2.62)$$

with m a positive scalar, $\mathbf{G}$ and $\mathbf{H}$ real symmetric 2×2 matrices, and $\mathbf{G}$ positive-definite; the two positive eigenvalues of $\mathbf{MJ}$ are now equal to $+m$ and the two negative eigenvalues are equal to $-m$.

Let us consider the case of a symplectic moment matrix (2.62) in more detail. From the equality of the positive and negative eigenvalues of $\mathbf{MJ}$, we conclude that these eigenvalues read as $\lambda = \pm m$, where each value of λ is twofold. The two eigenvectors that correspond to the positive eigenvalue $+m$ can be combined into the 4×2 matrix

$$\begin{bmatrix} \mathbf{R}^+ \\ \mathbf{Q}^+ \end{bmatrix}; \qquad (2.63)$$

a similar matrix (with superscripts $-$) can be associated with the two negative eigenvalues $-m$. The eigenvector equation for the matrix $\mathbf{MJ}$, $\mathbf{MJS} = \mathbf{S\Lambda}$, now takes the form

$$i \begin{bmatrix} \mathbf{G}^{-1} & \mathbf{G}^{-1}\mathbf{H} \\ \mathbf{HG}^{-1} & \mathbf{G} + \mathbf{HG}^{-1}\mathbf{H} \end{bmatrix} \begin{bmatrix} \mathbf{0} & -\mathbf{I} \\ \mathbf{I} & \mathbf{0} \end{bmatrix} \begin{bmatrix} \mathbf{R}^\pm \\ \mathbf{Q}^\pm \end{bmatrix} = \pm \begin{bmatrix} \mathbf{R}^\pm \\ \mathbf{Q}^\pm \end{bmatrix}, \qquad (2.64)$$

and the first block row of this relationship reads $i(-\mathbf{G}^{-1}\mathbf{Q}^\pm + \mathbf{G}^{-1}\mathbf{HR}^\pm) = \pm \mathbf{R}^\pm$, which yields the relation $\mathbf{Q}^\pm = (\mathbf{H} \pm i\mathbf{G})\mathbf{R}^\pm$. We remark that, because the eigenvectors satisfy the input-output relationship (2.35), and thus

$$\begin{bmatrix} \mathbf{R}_o^\pm \\ \mathbf{Q}_o^\pm \end{bmatrix} = \begin{bmatrix} \mathbf{A} & \mathbf{B} \\ \mathbf{C} & \mathbf{D} \end{bmatrix} \begin{bmatrix} \mathbf{R}_i^\pm \\ \mathbf{Q}_i^\pm \end{bmatrix}, \qquad (2.65)$$

the bilinear relationship

$$\mathbf{H}_o \pm i\mathbf{G}_o = [\mathbf{C} + \mathbf{D}(\mathbf{H}_i \pm i\mathbf{G}_i)][\mathbf{A} + \mathbf{B}(\mathbf{H}_i \pm i\mathbf{G}_i)]^{-1} \qquad (2.66)$$

holds. This bilinear relationship, together with the invariance of $\det \mathbf{M}$, completely describes the propagation of a symplectic matrix $\mathbf{M}$ through a first-order optical system. Note that the bilinear relationship (2.66) is identical to the $\mathbf{ABCD}$-law (2.39) for spherical waves: for spherical waves we have $\mathbf{H}_o = [\mathbf{C} + \mathbf{DH}_i][\mathbf{A} + \mathbf{BH}_i]^{-1}$, and we have only replaced the (real) curvature matrix $\mathbf{H}$ by the (generally complex) matrix $\mathbf{H} \pm i\mathbf{G}$. We are thus led to the important result that if the matrix $\mathbf{M}$ of second-order moments is symplectic (up to a positive constant) as described in Eq. (2.62), then its propagation through a first-order optical system is completely described by the invariance of this positive constant and the $\mathbf{ABCD}$-law (2.66).

2.6.2 Second-order moments of Gaussian signals with twist

Let us consider Gaussian light again, described by the Wigner distribution (2.15). For the matrix of second-order moments $\mathbf{M}$ for such light, we find

$$2\pi \mathbf{M} = \begin{bmatrix} \mathbf{R} & \mathbf{P} \\ \mathbf{P}^t & \mathbf{Q} \end{bmatrix} = \frac{1}{2} \begin{bmatrix} \mathbf{G}_1^{-1} & \mathbf{G}_1^{-1}\mathbf{H} \\ \mathbf{H}^t\mathbf{G}_1^{-1} & \mathbf{G}_2 + \mathbf{H}^t\mathbf{G}_1^{-1}\mathbf{H} \end{bmatrix}; \qquad (2.67)$$

hence, the matrices $\mathbf{G}_1$, $\mathbf{G}_2$, and $\mathbf{H}$ follow directly from the submatrices $\mathbf{P}$, $\mathbf{Q}$, and $\mathbf{R}$ of the matrix $2\pi \mathbf{M}$: $\mathbf{G}_1 = \mathbf{G}_1^t = \frac{1}{2}\mathbf{R}^{-1}$, $\mathbf{G}_2 = \mathbf{G}_2^t = 2\left(\mathbf{Q} - \mathbf{P}^t \mathbf{R}^{-1}\mathbf{P}\right)$, and $\mathbf{H} = \mathbf{R}^{-1}\mathbf{P}$. For symplectic Gaussian light, with $\mathbf{H} = \mathbf{H}^t$, $\mathbf{G}_1 = \sigma \mathbf{G}$, and $\mathbf{G}_2 = \sigma^{-1}\mathbf{G}$, we have

$$2\pi\mathbf{M} = \frac{1}{2\sigma}\begin{bmatrix} \mathbf{G}^{-1} & \mathbf{G}^{-1}\mathbf{H} \\ \mathbf{H}\mathbf{G}^{-1} & \mathbf{G} + \mathbf{H}\mathbf{G}^{-1}\mathbf{H} \end{bmatrix}, \qquad (2.68)$$

and the matrices $\mathbf{G}$ and $\mathbf{H}$ follow the $\mathbf{ABCD}$-law (2.66). The matrix decomposition described in Eq. (2.67) is, in fact, a general decomposition that holds for any symmetric matrix $\mathbf{M}$. From the positive definiteness of the matrix $\mathbf{M}$, we conclude that both $\mathbf{G}_1$ and $\mathbf{G}_2$ are positive definite.[5]

We remark that the twistedness T of Gaussian light—related to the skewness of the matrix $\mathbf{H}$—can directly be expressed in terms of the moment matrices $\mathbf{R}$ and $\mathbf{P}$; we have $\mathbf{PR} - \mathbf{RP}^t = \mathbf{R}(\mathbf{H} - \mathbf{H}^t)\mathbf{R} = (1/4)(\mathbf{H} - \mathbf{H}^t)\det \mathbf{G}_1^{-1}$ and might thus define T as the skewness of the matrix $\mathbf{PR}$: $T = m_{xu}m_{xy} + m_{xv}m_{yy} - (m_{xx}m_{yu} + m_{xy}m_{yv})$. The well-known propagation of the moments through first-order optical systems then leads to some useful information about the propagation of the twist.[5]

2.6.3 Invariants for the second-order moments

From the invariance of the eigenvalues of $\mathbf{MJ}$, we can derive invariants for the second-order moments of the Wigner distribution. The eigenvalues λ of $\mathbf{MJ}$ follow from the characteristic equation, which reads as $\det(\mathbf{MJ} - \lambda\mathbf{I}) = 0$ and thus $\lambda^4 - [(m_{xx}m_{uu} - m_{xu}^2) + (m_{yy}m_{vv} - m_{yv}^2) + 2(m_{xy}m_{uv} - m_{xv}m_{yu})]\lambda^2 + \det \mathbf{M} = 0$. We remark that the coefficient that arises with the term λ^2 is the sum of four of the 2×2 minors of $\mathbf{M}$,

$$\begin{vmatrix} m_{xx} & m_{xu} \\ m_{xu} & m_{uu} \end{vmatrix} + \begin{vmatrix} m_{yy} & m_{yv} \\ m_{yv} & m_{vv} \end{vmatrix} + \begin{vmatrix} m_{xy} & m_{xv} \\ m_{yu} & m_{uv} \end{vmatrix} + \begin{vmatrix} m_{xy} & m_{yu} \\ m_{xv} & m_{uv} \end{vmatrix}. \qquad (2.69)$$

Because the eigenvalues are invariant under propagation through a first-order optical system, the same holds for any combination of the eigenvalues, in particular, for the coefficients that arise in the characteristic polynomial. Therefore, both the determinant of $\mathbf{M}$ and the sum (2.69) of the four minors of $\mathbf{M}$, mentioned above, are invariant under propagation. See also Ref. 39.

2.6.4 Higher-order moments and systems to measure them

Because the Wigner distribution of a two-dimensional signal is a function of four variables, it is difficult to analyze. Therefore, the signal is often represented not by the Wigner distribution itself, but by its global moments. Beam characterization based on the second-order moments of the Wigner distribution thus became the basis of an International Organization for Standardization standard.[40]

Some of the Wigner distribution moments can directly be determined from measurements of the intensity distributions in the image plane or the Fourier plane, but most of the moments cannot be determined in such an easy way. In order to calculate such moments, additional information is required. Because first-order optical systems produce affine transformations of the Wigner distribution in phase space, the intensity distributions measured at the output of such systems can provide such additional information.[33–35,38,39,41–44]

In Ref. 12, it was shown how all global Wigner distribution moments can be determined from measurements of only intensity distributions in an appropriate number of (generally anamorphic) separable first-order optical systems. To do so, normalized moments of the Wigner distribution are considered, where the normalization is with respect to the total energy E of the signal. The normalized moments μ_{pqrs} of the Wigner distribution are thus defined by

$$\mu_{pqrs} E = \iiiint W(x,u;y,v)\, x^p\, u^q\, y^r\, v^s\, dx\, du\, dy\, dv \quad (p,q,r,s \geq 0); \quad (2.70)$$

note that for $q = s = 0$ we have intensity moments, which can easily be measured,

$$\mu_{p0r0} E = \iiiint W(x,u;y,v)\, x^p\, y^r\, dx\, du\, dy\, dv$$
$$= \iint x^p\, y^r\, \Gamma(x,x;y,y)\, dx\, dy \quad (p,r \geq 0). \quad (2.71)$$

The Wigner distribution moments μ_{pqrs} provide valuable tools for the characterization of optical beams (see, for instance, Ref. 20). First-order moments yield the position of the beam (μ_{1000} and μ_{0010}) and its direction (μ_{0100} and μ_{0001}). Second-order moments give information about the spatial width of the beam (the shape μ_{2000} and μ_{0020} of the spatial ellipse and its orientation μ_{1010}) and the angular width in which the beam is radiating (the shape μ_{0200} and μ_{0002} of the spatial-frequency ellipse and its orientation μ_{0101}). Moreover, as we already mentioned, they provide information about its curvature (μ_{1100} and μ_{0011}) and its *twist* (μ_{1001} and μ_{0110}), with a possible definition of T as[5]

$$T = \mu_{0020}\mu_{1001} - \mu_{2000}\mu_{0110} + \mu_{1010}(\mu_{1100} - \mu_{0011}) \quad (2.72)$$

(see Sec 2.6.2). Many important beam characterizers, such as the overall beam quality[39]

$$(\mu_{2000}\mu_{0200} - \mu_{1100}^2) + (\mu_{0020}\mu_{0002} - \mu_{0011}^2) + 2(\mu_{1010}\mu_{0101} - \mu_{1001}\mu_{0110}) \quad (2.73)$$

(see Section 2.6.3), are based on second-order moments. Also the orbital angular momentum $\Lambda = \Lambda_a + \Lambda_v \propto (\mu_{1001} - \mu_{0110})$ [see Eq. (3) in Ref. 45] and its antisymmetrical part Λ_a and vortex part Λ_v,

$$\Lambda_a \propto \frac{(\mu_{2000} - \mu_{0020})(\mu_{1001} + \mu_{0110}) - 2\mu_{1010}(\mu_{1100} - \mu_{0011})}{\mu_{2000} + \mu_{0020}}, \quad (2.74)$$

$$\Lambda_v \propto \frac{\mu_{0020}\mu_{1001} - \mu_{2000}\mu_{0110} + \mu_{1010}(\mu_{1100} - \mu_{0011})}{\mu_{2000} + \mu_{0020}} = \frac{T}{\mu_{2000} + \mu_{0020}} \quad (2.75)$$

[see Eqs. (22) and (21) in Ref. 45] are based on these moments.[6] Higher-order moments are used, for instance, to characterize the beam's symmetry and its sharpness.[20]

In the case of a separable first-order optical system, i.e., with diagonal submatrices **A**, **B**, **C**, and **D** [with diagonal entries (a_x, a_y) and so on], the normalized moments μ_{pqrs}^{out} of the output Wigner distribution are related to the normalized moments $\mu_{pqrs}^{in} = \mu_{pqrs}$ of the input Wigner distribution as[12]

$$\mu_{pqrs}^{out} E = E \sum_{k=0}^{p} \sum_{l=0}^{q} \sum_{m=0}^{r} \sum_{n=0}^{s} \binom{p}{k} \binom{q}{l} \binom{r}{m} \binom{s}{n} a_x^{p-k} b_x^k c_x^l d_x^{q-l}$$
$$\times a_y^{r-m} b_y^m c_y^n d_y^{s-n} \mu_{p-k+l,q-l+k,r-m+n,s-n+m}, \quad (2.76)$$

and for the intensity moments in particular (i.e., $q = s = 0$), we have

$$\mu_{p0r0}^{out} = \sum_{k=0}^{p} \sum_{m=0}^{r} \binom{p}{k} \binom{r}{m} a_x^{p-k} b_x^k a_y^{r-m} b_y^m \mu_{p-k,k,r-m,m}. \quad (2.77)$$

Note that only the parameters a and b enter the latter equation; the parameters c and d can be chosen freely, as long as the symplecticity condition $a_x d_x - b_x c_x = a_y d_y - b_y c_y = 1$ is satisfied.

It can be derived[12] that all ten second-order moments can be determined from the knowledge of the output intensities of four first-order optical systems, where one of them has to be anamorphic. In the case of fractional Fourier transformation systems, we could choose, for instance,[10,11] $a_x = a_y = 1$ and $b_x = b_y = 0$, $a_x = a_y = b_x = b_y = 1/\sqrt{2}$, $a_x = a_y = 0$ and $b_x = b_y = 1$, and the anamorphic combination $a_x = b_y = 0$ and $a_y = b_x = 1$. If we decide to determine the moments using free space propagation, we should be aware of the fact that an anamorphic free space system cannot be realized by mere free space, but can only be simulated by using a proper arrangement of cylindrical lenses.

Of course, optical schemes to determine all ten second-order moments have been described before (see, for instance Refs. 39 and 41–44), but the way to determine these moments as presented in Ref. 12 is based on a general scheme that can also be used for the determination of arbitrary higher-order moments. For the determination of the 20 third-order moments, we thus find the need of using a total of six first-order optical systems: four isotropic systems and two anamorphic systems. The 35 fourth-order moments can be determined from the knowledge of the output intensities of nine first-order optical systems spectra, where four of them have to be anamorphic. For the details of how to construct appropriate measuring schemes, we refer to Ref. 12.

To find the number of nth-order moments N, and the total number of first-order optical systems N_t (with N_a the number of anamorphic ones) that we need to determine these N moments, use can be made of the following triangle, which can easily be extended to higher order:

n	number of nth-order moments	N	N_t	N_a
0	1	1	1	0
1	2 + 2	4	2	0
2	3 + 4 + 3	10	4	1
3	4 + 6 + 6 + 4	20	6	2
4	5 + 8 + 9 + 8 + 5	35	9	4
⋮	⋮	⋮	⋮	⋮

Note that N (the number of nth-order moments) is equal to the sum of the values in the nth row of the triangle, $N = (1/6)(n+1)(n+2)(n+3)$; that N_t (the total number of first-order optical systems) is equal to the highest value that appears in the nth row of the triangle, $N_t = (1/4)(n+2)^2$ for n = even, and $N_t = (1/4)(n+3)(n+1)$ for n = odd; that the number of isotropic systems is $n+1$; and that N_a (the number of anamorphic systems) follows from $N_a = N_t - (n+1)$.

References

1. Wigner, E., "On the quantum correction for thermodynamic equilibrium," *Phys. Rev.*, **40**, 749–759, 1932.

2. Walther, A., "Radiometry and coherence," *J. Opt. Soc. Am.*, **58**, 1256–1259, 1968.

3. Bastiaans, M.J., "The Wigner distribution function applied to optical signals and systems," *Opt. Commun.*, **25**, 26–30, 1978.

4. Bastiaans, M.J., "Application of the Wigner distribution function to partially coherent light," *J. Opt. Soc. Am. A*, **3**, 1227–1238, 1986.

5. Bastiaans, M.J., "Wigner distribution function applied to twisted Gaussian light propagating in first-order optical systems," *J. Opt. Soc. Am. A*, **17**, 2475–2480, 2000.

6. Alieva, T., and Bastiaans, M.J., "Evolution of the vortex and the asymmetrical parts of orbital angular momentum in separable first-order optical systems," *Opt. Lett.*, **29**, 1587–1589, 2004.

7. Bastiaans, M.J., "Propagation laws for the second-order moments of the Wigner distribution function in first-order optical systems," *Optik*, **82**, 173–181, 1989.

8. Bastiaans, M.J., "Second-order moments of the Wigner distribution function in first-order optical systems," *Optik*, **88**, 163–168, 1991.

9. Bastiaans, M.J., "ABCD law for partially coherent Gaussian light, propagating through first-order optical systems," *Opt. Quant. Electron.*, **24**, 1011–1019, 1992.

10. Bastiaans, M.J., and Alieva, T., "Wigner distribution moments in fractional Fourier transform systems," *J. Opt. Soc. Am. A*, **19**, 1763–1773, 2002.

11. Bastiaans, M.J., and Alieva, T., "Moments of the Wigner distribution of rotationally symmetric partially coherent light," *Opt. Lett.*, **28**, 2443–2445, 2003.

12. Bastiaans, M.J., and Alieva, T., "Wigner distribution moments measured as intensity moments in separable first-order optical systems," *EURASIP J. Appl. Signal Process.*, **2005**, 1535–1540, 2005.

13. Bastiaans, M.J., "Wigner distribution function applied to partially coherent light," *Proceedings of the Workshop on Laser Beam Characterization*, P.M. Mejías, H. Weber, R. Martínez-Herrero, and A. González-Ureña, Eds., pp. 65–87, SEDO, Madrid (1993).

14. Bastiaans, M.J., "Application of the Wigner distribution function in optics," *The Wigner Distribution—Theory and Applications in Signal Processing*, W. Mecklenbräuker and F. Hlawatsch, Eds., pp. 375–426, Elsevier Science, Amsterdam (1997).

15. Bastiaans, M.J., "Application of the Wigner distribution function to partially coherent light," *Optics and Optoelectronics, Theory, Devices and Applications, Proc. ICOL'98, the International Conference on Optics and Optoelectronics, Dehradun, India, 9–12 Dec. 1998*, O.P. Nijhawan, A.K. Gupta, A.K. Musla, and K. Singh, Eds., pp. 101–115, Narosa Publishing House, New Delhi (1998).

16. Cohen, L., "Time-frequency distributions—A review," *Proc. IEEE*, **77**, 941–981, 1989.

17. Cohen, L., *Time-Frequency Analysis*, Prentice Hall, Englewood Cliffs, NJ (1995).

18. Mecklenbräuker, W., and Hlawatsch, F., Eds., *The Wigner Distribution—Theory and Applications in Signal Processing*, Elsevier Science, Amsterdam (1997).

19. Dragoman, D., "The Wigner distribution function in optics and optoelectronics," *Progress in Optics*, E. Wolf, Ed., Vol. 37, pp. 1–56, North-Holland, Amsterdam (1997).

20. Dragoman, D., "Applications of the Wigner distribution function in signal processing," *EURASIP J. Appl. Signal Process.*, **2005**, 1520–1534, 2005.

21. Torre, A., *Linear Ray and Wave Optics in Phase Space*, Elsevier, Amsterdam (2005).

22. Testorf, M.E., Ojeda-Castañeda, J., and Lohmann, A.W., Eds., *Selected Papers on Phase-Space Optics, SPIE Milestone Series*, Vol. **MS 181**, SPIE, Bellingham, WA (2006).

23. Moyal, J.E., "Quantum mechanics as a statistical theory," *Proc. Cambridge Philos. Soc.*, **45**, 99–132, 1949.

24. Simon, R., Sudarshan, E.C.G., and Mukunda, N., "Anisotropic Gaussian Schell-model beams: Passage through optical systems and associated invariants," *Phys. Rev. A*, **31**, 2419–2434, 1985.

25. Simon, R., and Mukunda, N., "Twisted Gaussian Schell-model beams," *J. Opt. Soc. Am. A*, **10**, 95–109, 1993.

26. Simon, R., Sundar, K., and Mukunda, N., "Twisted Gaussian Schell-model beams. I. Symmetry structure and normal-mode spectrum," *J. Opt. Soc. Am. A*, **10**, 2008–2016, 1993.

27. Sundar, K., Simon, R., and Mukunda, N., "Twisted Gaussian Schell-model beams. II. Spectrum analysis and propagation characteristics," *J. Opt. Soc. Am. A*, **10**, 2017–2023, 1993.

28. Friberg, A.T., Tervonen, E., and Turunen, J., "Interpretation and experimental demonstration of twisted Gaussian Schell-model beams," *J. Opt. Soc. Am. A*, **11**, 1818–1826, 1994.

29. Ambrosini, D., Bagini, V., Gori, F., and Santarsiero, M., "Twisted Gaussian Schell-model beams: a superposition model," *J. Mod. Opt.*, **41**, 1391–1399, 1994.

30. Schell, A.C., "A technique for the determination of the radiation pattern of a partially coherent aperture," *IEEE Trans. Antennas Propag.*, **AP-15**, 187–188, 1967.

31. Collins Jr., S.A., "Lens-system diffraction integral written in terms of matrix optics," *J. Opt. Soc. Am.*, **60**, 1168–1177, 1970.

32. Lohmann, A.W., "Image rotation, Wigner rotation, and the fractional Fourier transform," *J. Opt. Soc. Am. A*, **10**, 2181–2186, 1993.

33. Raymer, M.G., Beck, M., and McAlister, D.F., "Complex wave-field reconstruction using phase-space tomography," *Phys. Rev. Lett.*, **72**, 1137–1140, 1994.

34. McAlister, D.F., Beck, M., Clarke, L., Mayer, A., and Raymer, M.G., "Optical phase retrieval by phase-space tomography and fractional-order Fourier transforms," *Opt. Lett.*, **20**, 1181–1183, 1995.

35. Gitin, A.V., "Optical systems for measuring the Wigner function of a laser beam by the method of phase-spatial tomography," *Quantum Electron.*, **37**, 85–91, 2007.

36. Alieva, T., "First-order optical systems for information processing," *Advances in Information Optics and Photonics*, A. Friberg and R. Dändliker, Eds., pp. 1–26, SPIE, Bellingham, WA (2008).

37. Bryngdahl, O., "Geometrical transformations in optics," *J. Opt. Soc. Am.*, **64**, 1092–1099, 1974.

38. Simon, R., Mukunda, N., and Sudarshan, E.C.G., "Partially coherent beams and a generalized ABCD-law," *Opt. Commun.*, **65**, 322–328, 1988.

39. Serna, J., Martínez-Herrero, R., and Mejías, P.M., "Parametric characterization of general partially coherent beams propagating through ABCD optical systems," *J. Opt. Soc. Am. A*, **8**, 1094–1098, 1991.

40. International Organization for Standardization, TC172/SC9, "Lasers and laser-related equipment—test methods for laser beam parameters—beam widths, divergence angle and beam propagation factor," ISO 11146, International Organization for Standardization, Geneva, Switzerland (1999).

41. Nemes, G., and Siegman, A.E., "Measurement of all ten second-order moments of an astigmatic beam by the use of rotating simple astigmatic (anamorphic) optics," *J. Opt. Soc. Am. A*, **11**, 2257–2264, 1994.

42. Eppich, B., Gao, C., and Weber, H., "Determination of the ten second order intensity moments," *Opt. Laser Technol.*, **30**, 337–340, 1998.

43. Martínez, C., Encinas-Sanz, F., Serna, J., Mejías, P.M., and Martínez-Herrero, R., "On the parametric characterization of the transversal spatial structure of laser pulses," *Opt. Commun.*, **139**, 299–305, 1997.

44. Serna, J., Encinas-Sanz, F., and Nemes, G., "Complete spatial characterization of a pulsed doughnut-type beam by use of spherical optics and a cylindrical lens," *J. Opt. Soc. Am. A*, **18**, 1726–1733, 2001.

45. Bekshaev, A.Ya., Soskin, M.S., and Vasnetsov, M.V., "Optical vortex symmetry breakdown and decomposition of the orbital angular momentum of the light beams," *J. Opt. Soc. Am. A*, **20**, 1635–1643, 2003.

**Martin J. Bastiaans** received his MSc in electrical engineering and a PhD in technical sciences from the Technische Universiteit Eindhoven, The Netherlands, in 1969 and 1983, respectively. In 1969, he became an assistant professor and, since 1985, he has been an associate professor with the Department of Electrical Engineering, Technische Universiteit Eindhoven, currently in the Signal Processing Systems Group, where he teaches electrical circuit theory, signal theory, digital signal processing, and Fourier optics and holography. His research covers different aspects in the general field of signal and system theory and includes a signal-theoretical approach of all kinds of problems that arise in Fourier optics, such as partial coherence, computer holography, optical signal processing, and optical computing. His main current research interest is in describing signals by means of a local frequency spectrum and related issues. Dr. Bastiaans is a Fellow of the Optical Society of America and a senior member of the Institute of Electrical and Electronics Engineers. He is the author and coauthor of more than 160 papers in international scientific journals, books, and proceedings of scientific conferences.

Chapter 3
Characterization of Elliptic Dark Hollow Beams

Julio C. Gutiérrez-Vega
Optics Center Tecnológico de Monterrey, México

3.1 Introduction
3.2 Astigmatic Elliptic Dark Hollow Beams
    3.2.1 Hollow elliptical Gaussian beam
    3.2.2 Elliptic Hermite-Gaussian beam
    3.2.3 Controllable elliptic dark hollow beam
    3.2.4 Elliptic Laguerre-Gaussian beam
    3.2.5 Elliptic Bessel beam
3.3 Elliptic Dark Hollow Beams in Elliptic Coordinates
    3.3.1 Nondiffracting Mathieu beams
    3.3.2 Mathieu-Gauss beams
    3.3.3 Ince-Gaussian beams
References

3.1 Introduction

A dark hollow beam (DHB) is defined, in general, as a ring-shaped light beam with a null intensity center on the beam axis. Some known examples of DHBs are the Laguerre-Gaussian beams,[1,2] high-order Bessel and Bessel-Gauss beams,[3–6] hollow Gaussian beams,[7] helical Mathieu and Mathieu-Gauss beams,[8–10] high-order linearly polarized output fiber modes,[11] among others.[12] DHBs have interesting physical properties, such as a helical wavefront, a center vortex singularity, doughnut-shaped transverse intensity distribution. They may carry and transfer orbital and spin angular momentum,[13–16] and may also exhibit a nondiffracting behavior upon propagation.[17] All these characteristics make DHBs particularly useful in optical tweezers,[18,19] micromanipulation,[20] optical trapping and cooling,[21] Bose-Einstein condensation, optical metrology, computer-generated holography, and so on. A variety of methods have been proposed to generate DHBs, including passive optical methods,[22–26] mode selection in laser resonators,[27,28] passive mode conversion,[29–32] holographic methods,[33–36] and liquid crystal displays.[37,38]

Most of the known theoretical models to describe DHBs consider axially symmetric transverse intensity distributions.[12] Recently, several works have been devoted to explore the properties and applications of DHBs for which the circular symmetry no longer exists, with particular emphasis in the rectangular and elliptical geometry. For example, Cai and Zhang [39] and Cai and Ge [40] based on a suitable superposition of a finite series of fundamental Gaussian beams, proposed a model of a rectangular DHB that seems to be more suitable than a circular DHB to guide atom laser beams with a particular mode.[41] If the beam width in one direction of a rectangular DHB is much larger than that in the other direction, the rectangular DHB can be visualized as a one-dimensional beam, which may be applied to achieve one-dimensional Bose-Einstein condensates.[42]

DHBs with elliptic symmetry can be regarded as transition beams between circular and rectangular DHBs. For example, the high-order modes emitted from resonators with neither completely rectangular nor completely circular symmetry, but in between them, cannot be described by the known Hermite-Gaussian or Laguerre-Gaussian beams. In recent years, there has been an increasing interest in developing models to describe DHBs with elliptic symmetry. An important characteristic of the waves exhibiting elliptic geometries is the possibility of independently choosing the propagation parameters along the two orthogonal transverse directions. For example, in the case of the waveguides with an elliptic cross section, the fundamental propagating mode splits into two independent orthogonally polarized modes.[43] Such geometrical birefringence makes the elliptical fibers particularly useful in applications where the polarization plays a major role. This splitting feature is not restricted to propagation in waveguides, but is also present in free-space propagation of waves with elliptic symmetry.

The elliptic DHBs can be roughly classified into two types according to their mathematical construction:

1. Astigmatic elliptic DHBs: Beams resulting from astigmatic deformations of known circular DHBs. For example, the hollow elliptical Gaussian beam, the elliptic Laguerre-Gaussian beam, and the elliptic Bessel beam, among others.

2. Eigenmodes of the wave equations: Beams resulting from the exact separation of the wave equations in elliptic coordinates. Examples of this category are the nondiffracting Mathieu beams and the Ince-Gaussian beams.

In this chapter, we review the current state of research on elliptic DHBs. Section 3.2 summarizes the main results that have been reported thus far for the astigmatic elliptical DHBs. In Section 3.3, we discuss the physical and mathematical properties of the eigenmodes of the Helmholtz wave equation and the paraxial wave equation, namely, the Mathieu and Ince-Gaussian beams, respectively. Throughout the chapter, we consider that the optical fields are scalar and monochromatic with time dependence $\exp(-i\omega t)$, where ω is the angular frequency of the field.

3.2 Astigmatic Elliptic Dark Hollow Beams

The astigmatic elliptic DHBs are constructed using skew deformations of conventional circular DHBs. This section is devoted to describe the definitions and basic properties of this kind of elliptic beams.

3.2.1 Hollow elliptical Gaussian beam

The hollow elliptical Gaussian beam is perhaps the simplest elliptic DHBs.[44] Its complex field amplitude is defined at the initial plane $z = 0$ as follows:

$$U_m(\mathbf{r}_t) = \exp\left(-\frac{ik}{2}\mathbf{r}_t \mathbf{Q}^{-1} \mathbf{r}_t^T\right)\left(\frac{ik}{2}\mathbf{r}_t \mathbf{Q}^{-1} \mathbf{r}_t^T\right)^m, \qquad (3.1)$$

where $m = 0, 1, 2, \ldots$ denotes the order, k is the wave number, $\mathbf{r}_t = (x, y)$ is a row vector denoting the transverse coordinates, the superindex T means transpose operation. The matrix $\mathbf{Q}^{-1}$ is an astigmatic generalization of the complex beam parameter $1/q$ given by

$$\mathbf{Q}^{-1} = \begin{pmatrix} q_{xx}^{-1} & q_{xy}^{-1} \\ q_{xy}^{-1} & q_{yy}^{-1} \end{pmatrix}, \qquad (3.2)$$

where the matrix elements are related to the initial beam waist sizes in the x direction, y direction, and coupled direction, by $q_{xx}^{-1} = -i\lambda/\pi w_{0x}^2$, $q_{yy}^{-1} = -i\lambda/\pi w_{0y}^2$, and $q_{xy}^{-1} = -i\lambda/\pi w_{0xy}^2$.

For $m = 0$, Eq. (3.1) reduces to a simple elliptic Gaussian beam with a maximum at center. As shown in Fig. 3.1(a), for $m > 0$, the transverse intensity distributions of the hollow elliptical Gaussian beams at the initial plane is composed by a single and tilted elliptic ring whose dark area increases with increasing m. By use of a coordinate transformation, it can be demonstrated that the angle between the axis of the beam spot and the coordinate is given by[44]

$$\alpha = \frac{1}{2}\arctan\left(\frac{2w_{0xy}}{w_{0x} - w_{0y}}\right). \qquad (3.3)$$

Upon propagation, the initial elliptic ring diffracts loosing its elliptic nature. In the far field, the dark region disappears and the on-axis intensity becomes a maximum. This behavior can be explained by the fact that the hollow elliptical Gaussian beam is not a pure mode, but results from a superposition of elliptical Hermite-Gaussian beams with different longitudinal phase shifts.[44]

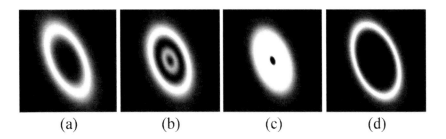

(a) (b) (c) (d)

Figure 3.1 Transverse intensity distributions at $z = 0$ for (a) a hollow elliptical Gaussian beam with $m = 2$, (b) an elliptical Hermite-Gaussian beam with $m = 3$, and (c) (d) a controllable elliptical dark hollow beam with $\varepsilon = 0.1$ and $\varepsilon = 0.9$, respectively. For all cases, $\lambda = 632.8$ nm, $\mathbf{Q}^{-1} = -i0.001 * [20; 5; 5; 9]$ m^{-1}, and the physical dimensions of the images are 8 mm × 8 mm.

3.2.2 Elliptic Hermite-Gaussian beam

The hollow elliptical Gaussian beam given by Eq. (3.1) exhibits only a single elliptic ring whose transverse size is adjusted by the parameter m. In order to have an elliptic DHB with several elliptic rings across the transverse initial plane, Cai and Lin[45,46] introduced a generalization of the hollow elliptical Gaussian beams, the so-called elliptic Hermite-Gaussian beam, whose complex field amplitude at $z = 0$ is defined by

$$U_m(\mathbf{r}_t) = \exp\left(-\frac{ik}{2}\mathbf{r}_t \mathbf{Q}_c^{-1} \mathbf{r}_t^T\right) H_m\left(\sqrt{ik\mathbf{r}_t \mathbf{Q}_h^{-1} \mathbf{r}_t^T}\right), \quad (3.4)$$

where H_m is the mth order Hermite polynomial, and $\mathbf{Q}_c^{-1}$ and $\mathbf{Q}_h^{-1}$ are two complex matrices of dimension 2×2 whose general form is given by Eq. (3.2).

A typical transverse intensity distribution of the elliptic Hermite-Gaussian beam is illustrated in Fig. 3.1(b). When the order index m is odd, the Hermite polynomial vanishes at the origin with the consequence that the beam center is dark. Therefore beams with odd m can be used to describe elliptic DHBs with $(m+1)/2$ elliptic rings. It can be verified that when $\mathbf{Q}_c^{-1} = \mathbf{Q}_h^{-1}$, beams of different orders are orthogonal.[45,46] If the beam parameters are chosen properly, the elliptical Hermite-Gaussian beams can be used to describe other complicated beams, such as the twisted Hermite-Gaussian beams.[47] Because the elliptic Hermite-Gaussian beams are not eigenmodes of the paraxial wave equation, then they are not shape invariant under propagation. In the far field, the dark region disappears and the on-axis intensity becomes a maximum. The propagation of decentered elliptical Hermite-Gaussian beams through nonaxisymmetric paraxial ABCD systems was also studied in detail in Ref. 46.

3.2.3 Controllable elliptic dark hollow beam

Mei and Zhao[48,49] proposed other generalization of the hollow elliptical Gaussian beams [Eq. (3.1)] in terms of the finite superposition

$$U(\mathbf{r}_t) = \sum_{n=1}^{N} \frac{(-1)^{n-1}}{N} \binom{N}{n}$$
$$\times \left[\exp\left(-\frac{ik}{2} n \, \mathbf{r}_t \mathbf{Q}^{-1} \mathbf{r}_t^T \right) - \exp\left(-\frac{ik}{2} \frac{n}{\varepsilon^2} \mathbf{r}_t \mathbf{Q}^{-1} \mathbf{r}_t^T \right) \right], \quad (3.5)$$

where $N = 1, 2, 3, \ldots$. The aim of this generalization is to introduce a new parameter ε that allows one to control not only the overall size of the ellipse but also the width of the elliptic ring. Similarly to the hollow elliptical Gaussian beams, the initial shape of the controllable elliptical DHBs is composed by a single elliptical ring whose ellipticity and ring width is adjusted simultaneously by the parameters N and ε.

Figures 3.1(c) and (d) show the transverse intensity distribution of a controllable elliptical DHB for different values of the parameter ε. Similarly to the hollow elliptic Gaussian beams, in the far field the dark region of the controllable elliptical DHBs disappears and the on-axis intensity becomes a maximum. The analytical propagation and transformation expressions of centered and decentered controllable elliptical DHBs passing through aligned and misaligned paraxial nonsymmetrical optical systems are reported in Ref. 49.

3.2.4 Elliptic Laguerre-Gaussian beam

The Laguerre-Gaussian beams are exact solutions of the paraxial wave equation in circular cylindrical coordinates that are shape invariant upon propagation with exception of a scaling factor.[1] The normalized Laguerre-Gaussian beam with radial number n and azimuthal number l is written as

$$\mathrm{LG}_{n,l}(r, \theta, z) = \sqrt{\frac{2n!}{(1+\delta_{0,l})\pi(n+l)!}} \frac{\exp(il\theta)}{w(z)} \left[\frac{\sqrt{2}r}{w(z)} \right]^l L_n^l\left(\frac{2r^2}{w(z)^2} \right)$$
$$\times \exp\left[\frac{-r^2}{w^2(z)} \right] \exp i \left[kz + \frac{kr^2}{2R(z)} - (2n+l+1)\psi_{\mathrm{GS}}(z) \right], \quad (3.6)$$

where $L_n^l(\cdot)$ is the generalized Laguerre polynomial, $w^2(z) = w_0^2\left(1 + z^2/z_R^2\right)$ describes the beam width, $R(z) = z + z_R^2/z$ is the radius of curvature of the phase front, $\psi_{\mathrm{GS}}(z) = \arctan(z/z_R)$ is the Gouy shift, $z_R = kw_0^2/2$ is the Rayleigh range, w_0 is the beam width at $z = 0$, and $\delta_{m,n}$ is the Kronecker δ. Laguerre-Gaussian beams are rotationally symmetric and exhibit an azimuthal angular dependence of the complex form $\exp(\pm il\theta)$. A linearly polarized Laguerre-Gauss beam possesses an orbital angular momentum (OAM) of $l\hbar$ per photon.

An elliptic Laguerre-Gaussian beam was proposed by Kotlyar et al.[50] via an oblique incidence of a plane wave onto an optical diffractive element with transmittance proportional to the function of a higher-order Laguerre-Gaussian mode. At the initial plane, the field is given by Eq. (3.6) evaluated at $z = 0$ and with the x variable replaced by the scaled coordinate x/μ, where the μ is the scaling factor. The initial pattern is composed by a set of concentric ellipses sharing the same scaling ratio between its major and minor axes.

The propagated elliptic beam can no longer be considered a mode exhibiting the shape-invariance property typical of the circular Laguerre-Gauss beams. Actually, the elliptic Laguerre-Gaussian beam does not preserve its structure and loses its ellipticity in the Fresnel diffraction zone. For $z \to \infty$ (far-field diffraction), the diffraction pattern is composed of a set of concentric rings rotated by 90 deg with respect to the initial diffraction pattern at $z = 0$. At the same time, the field of the elliptic Laguerre-Gaussian beam retains its vortex character. The phase singularity at the center of the order l appears broken down into s singularities numbered (ordered) l/s, with s points of zero intensity emerging in the beam-center neighborhood. The number s depends on the degree of ellipticity.

3.2.5 Elliptic Bessel beam

A mth-order Bessel beam is a nondiffracting DHB (except for the zero'th-order beam) whose ideal three-dimensional field distribution is given by[3,4]

$$\mathrm{BB}_m(\mathbf{r}) = J_m(k_t r) \exp(im\theta) \exp(ik_z z), \qquad (3.7)$$

where $\mathbf{r} = (x, y, z) = (r, \theta, z)$ denotes the position vector, J_m is the mth-order Bessel function, and k_t and k_z are the transverse and longitudinal components of the wave vector ($k = 2\pi/\lambda = \sqrt{k_t^2 + k_z^2}$). The angular spectrum of a Bessel beam reduces to a single circular ring of radius k_t in the spatial frequency space modulated by the angular spectrum $\exp(im\phi)$. The physical properties and applications of the Bessel beams are summarized in several review papers.[17,52]

The generation of an elliptic Bessel beam was reported recently by Chakraborty and Ghosh[51] by scaling the horizontal coordinate of a higher-order Bessel beam of the form in Eq. (3.7), namely, at $z = 0$ the elliptic Bessel beam becomes

$$e\mathrm{BB}_m(\mathbf{r}) = J_m\left(k_t\sqrt{\frac{x^2}{\mu^2} + y^2}\right) \exp\left[im \arctan\left(\frac{\mu y}{x}\right)\right], \qquad (3.8)$$

where μ is the stretching factor. From Fourier optics, it is well known that stretching coordinates in the space domain results in a contraction of the coordinates in the frequency domain, plus a change in the overall amplitude of the spectrum. Therefore, the original circular spectrum of the Bessel beam turns into elliptical for the elliptic Bessel beam, with the consequence that the elliptic Bessel beam is not strictly a nondiffracting beam.

3.3 Elliptic Dark Hollow Beams in Elliptic Coordinates

Let us turn our attention to the elliptic DHBs, which result from the separation of the wave equations in elliptic coordinates. Two known optical wave equations are identified as follows: (a) the scalar Helmholtz equation

$$\left(\frac{\partial^2}{\partial x^2} + \frac{\partial^2}{\partial y^2} + \frac{\partial^2}{\partial z^2} + k^2\right) E(\mathbf{r}) = 0, \tag{3.9}$$

for the monochromatic and linearly polarized electric field $E(\mathbf{r})$, and (b) the paraxial wave equation (PWE)

$$\left(\frac{\partial^2}{\partial x^2} + \frac{\partial^2}{\partial y^2} + 2ik\frac{\partial}{\partial z}\right) U(\mathbf{r}) = 0, \tag{3.10}$$

for the complex amplitude envelope $U(\mathbf{r})$ of the electric field $E(\mathbf{r}) = U(\mathbf{r}) \times \exp(ikz)$ traveling paraxially along the positive z direction. As we will see, the solution of Eqs. (3.9) and (3.10) leads to the complete and orthogonal families of optical beams expressed in terms of Mathieu functions and Ince polynomials, respectively.

Before proceeding to review the elliptic DHB solutions of Eq. (3.9) and (3.10), it is convenient to define the elliptic cylindrical coordinates (ξ, η). These coordinates are related to the Cartesian coordinates (x, y) by the transformation

$$x = f \cosh \xi \cos \eta, \quad y = f \sinh \xi \sin \eta, \quad z = z, \tag{3.11}$$

where $0 \leq \xi < \infty$ is the radial elliptic coordinate, $0 \leq \eta < 2\pi$ is the angular elliptic coordinate, $-\infty < z < \infty$ is the conventional longitudinal coordinate, and f is the semifocal separation of the system (see Fig. 3.2). The curves $\xi = $ const are confocal ellipses, and the curves $\eta = $ constant are orthogonal hyperbolas. The two families of conics intersect orthogonally and each intersection corresponds to a point on the (x, y) plane as shown in Fig. 3.2.

3.3.1 Nondiffracting Mathieu beams

Mathieu beams constitute a complete and orthogonal family of nondiffracting beams that are solutions of the wave equation [Eq. (3.9)] in elliptic coordinates.[8,9,53] These beams are fundamental in the sense that any optical field can be expanded in terms of Mathieu beams with appropriate weighting factors and spatial frequencies.[54] This result is perhaps the most distinctive property of the Mathieu beams when are compared to the astigmatic beams reviewed in Section 3.2. Mathieu beams have been applied for example to photonic lattices,[55] transfer of angular momentum using optical tweezers,[56–58] and localized X-waves.[59,60]

The application of the method of separation of variables to the three-dimensional Helmholtz equation [Eq. (3.9)] in elliptic coordinates leads to the general expression of the nondiffracting m'th order Mathieu DHBs[8]

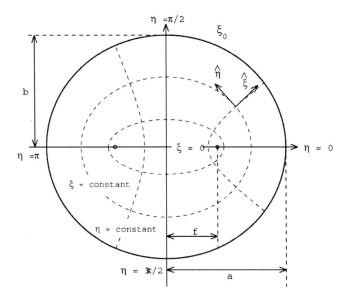

Figure 3.2 Elliptic cylindrical coordinate system.

$$\mathrm{MB}_m^\pm(\mathbf{r}) = [\mathcal{C}_m \mathrm{Je}_m(\xi, q)\, \mathrm{ce}_m(\eta, q) \pm i\, \mathcal{S}_m \mathrm{Jo}_m(\xi, q)\, \mathrm{se}_m(\eta, q)] \exp(ik_z z), \tag{3.12}$$

where $\mathrm{Je}_m(\cdot)$ and $\mathrm{Jo}_m(\cdot)$ are the mth order even and odd radial Mathieu functions, $\mathrm{ce}_m(\cdot)$ and $\mathrm{se}_m(\cdot)$ are the mth order even and odd angular Mathieu functions.[61]

The Mathieu beams are characterized by the ellipticity parameter $q = k_t^2 f^2 / 4$, which carries information about the transverse wave number k_t and the ellipticity of the coordinate system through f. When $q \to 0$, the foci of the elliptic coordinates collapse at the origin; therefore, the Mathieu beams reduce to the Bessel beams given in Eq. (3.7). In Eq. (3.12), $\mathcal{C}_m(q)$ and $\mathcal{S}_m(q)$ are normalization factors to ensure that the even and odd beam components carry the same power. It can be shown that for $m \geq 2$ and $q \lesssim m^2/2 - 1$, the normalization factors are numerically very similar, i.e., $\mathcal{C}_m(q) \approx \mathcal{S}_m(q)$, and hence, they can be factorized out.[62] The transverse and longitudinal wave numbers fulfill $k^2 = k_t^2 + k_z^2$, as expected.

Figure 3.3 shows the transverse intensity and phase distributions of Mathieu DHBs for $m = \{1, 4, 7\}$. The field intensity has a clear elliptic ringed structure, and its phase rotates following an elliptic trajectory on propagation.[62] For the total topological charge $m = 1$, the phase is formed by a single elliptic vortex at the axial line. For $m \geq 2$, the phase is formed by m in-line vortices, each with unitary topological charge such that the total charge (along a closed trajectory enclosing all the vortices) is m. The branch cuts lie on confocal hyperbolas, implying that, on propagation, a point in the phase front travels along an elliptic helix of constant ξ.

An intuitive understanding of these beams can be gained by thinking of them as a superposition of plane waves whose wave vectors lie on a cone around the z-axis, and whose amplitudes are modulated angularly by the angular spectrum

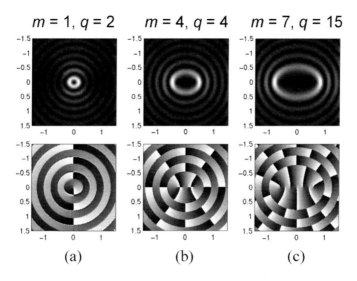

Figure 3.3 Intensity and phase transverse distributions of several Mathieu DHBs.

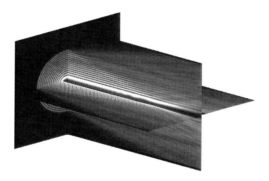

Figure 3.4 Propagation of an apertured sixth-order Mathieu DHB. Note the typical conical region of shape invariance.

$\mathrm{ce}_m(\phi; q) \pm i\, \mathrm{se}_m(\phi; q)$. All these plane waves have the same component of the wave vector along the z-axis. Accordingly, they all suffer the same phase change for any given pathlength along the z-axis.

Ideal Mathieu DHBs are propagation invariant across the whole space. But apertured Mathieu DHBs are diffraction-free only within the so-called conical propagation invariant region, as shown in Fig. 3.4. This is the well-known effect due to the finite size of the nondiffracting beams.[17] Fundamental Mathieu beams of zeroth order were produced by Gutiérrez-Vega et al.[9] employing the original Durnin setup based on the Fourier transform of a circular thin slit.[3,4] Higher-order Mathieu DHBs were generated by Chávez-Cerda et al.[56] with computer-generated holograms, obtaining propagation distances up to 50 cm.

Mathieu DHBs carry orbital angular momentum that can be transferred to trapped particles. The first experimental observation of dielectric particle dynamics

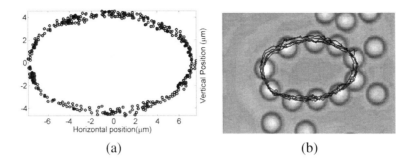

Figure 3.5 Elliptical movement of a trapped particle within the field of a Mathieu HDB. (Reprinted from Ref. 57 with permission of the authors and publisher.)

in the field of a nondiffracting Mathieu DHB was reported in 2005.[57,58] The experiment uses an optical tweezers setup with a Mathieu beam ($m = 7$, $q = 5$) generated with an off-axis blazed phase computer-generated hologram. The hologram was backlit by a collimated beam from a linearly polarized 1064 nm Ytterbium fiber laser. The sample chamber was filled with a diluted solution of monodisperse spherical polystyrene particles 3.0 μm in size, suspended in a mixture of anionic, nonanionic and amphoteric surfactants (1% in volume) in D_2O. Figure 3.5(a) shows the statistics of the position of one single particle over several revolutions in a 100 s time period. Note that the particle drifts away from their elliptic orbital position significantly more as their radial position decreases, which is a consequence of the overcoming of the gradient force by the Brownian motion. The graphic representation of the particles trajectory on the sample plane is depicted in Fig. 3.5(b).

Because of the mathematical structure of the Mathieu beams, it is not possible to determine a closed-form expression for their orbital angular momentum. However, within the paraxial regime, the z component of the orbital angular momentum per photon in unit length about the origin of a transverse slice of a beam $U(\mathbf{r})$ can be calculated numerically with[13]

$$J_z = \hbar \frac{\iint_{-\infty}^{\infty} \mathbf{r}_t \times \mathrm{Im}\left(U^* \nabla U\right) \, dx dy}{\iint_{-\infty}^{\infty} |U|^2 \, dx dy} \qquad (3.13)$$

where $\mathbf{r}_t$ is the transverse radius vector. Plots of J_z for several Mathieu beams of different orders are included in Ref. 56.

3.3.2 Mathieu-Gauss beams

Because ideal Mathieu DHBs are not square integrable, they cannot be normalized such that the integral of the intensities is unitary. In view of this, it was reported in 2005 a closed expression for the propagation of the so-called Mathieu-Gauss DHBs,[10] i.e., Mathieu DHBs apodized by a Gaussian transmittance, which carry a finite power and can be generated experimentally to a very good approximation.[63] Mathieu-Gauss DHBs constitute one of the simplest physical realizations of the exact nondiffracting Mathieu DHBs, and their propagation properties have been studied in free space[10,63] and through ABCD optical systems in the scalar[64] and the vector [65] formalisms.

The closed-form expression for the propagation of the mth-order even Mathieu–Gauss beams is found to be

$$\mathrm{MG}_m(\mathbf{r}) = \exp\left(-i\frac{k_t^2}{2k}\frac{z}{\mu}\right) \mathrm{GB}(\mathbf{r}) \tag{3.14}$$

$$\times \left[\mathcal{C}_m \mathrm{Je}_m(\overline{\xi}, q)\, \mathrm{ce}_m(\overline{\eta}, q) \pm i\, \mathcal{S}_m \mathrm{Jo}_m(\overline{\xi}, q)\, \mathrm{se}_m(\overline{\eta}, q)\right], \tag{3.15}$$

where $\mathrm{GB}(\mathbf{r})$ is the fundamental Gaussian beam

$$\mathrm{GB}(\mathbf{r}) = \frac{\exp(ikz)}{\mu} \exp\left(-\frac{r^2}{\mu w_0^2}\right), \tag{3.16}$$

and the parameter $\mu = \mu(z) = 1 + iz/z_R$ and $z_R = kw_0^2/2$ is the usual Rayleigh range of a Gaussian beam.[2] In a transverse z plane, the complex elliptic variables $(\overline{\xi}, \overline{\eta})$ are determined by the following relations:

$$x = f_0\left(1 + i\frac{z}{z_R}\right)\cosh\overline{\xi}\cos\overline{\eta}, \tag{3.17}$$

$$y = f_0\left(1 + i\frac{z}{z_R}\right)\sinh\overline{\xi}\sin\overline{\eta}, \tag{3.18}$$

where f_0 is the semifocal separation at the waist plane $z = 0$. Note that, although the elliptic variables $(\overline{\xi}, \overline{\eta})$ at the plane $z = 0$ are real, outside this plane they become complex in order to satisfy the requirement that the Cartesian coordinates (x, y) remain real in the entire space. Although the arguments of the Mathieu function at the plane $z = 0$ are real, outside this plane they become complex with the result that the initial shape defined by $\mathrm{MG}_m^e(\mathbf{r})$ changes its form on propagation. The Mathieu-Gauss beam will behave very similar to an ideal Mathieu beam within the range $z \in [0, z_{\max} = w_0 k/k_t]$.

Mathieu-Gauss DHBs ($m = 7$, $q = 16$) were produced using computer-generated holography.[63] The experimental transverse shapes recorded at planes $z = 0, 0.8 z_{\max}$ and $1.6 z_{\max}$ are shown in Figs. 3.6(a)–(c). The patterns consist of well-defined elliptic confocal rings with a dark elliptic spot on axis. The Mathieu-Gauss DHB is characterized by an elliptic-helical phase that rotates about this line as the beam propagates. Intensity evolution along the (x, z) and (y, z) planes is qualitatively similar to that of Bessel-Gauss beams of different orders, as shown in Fig. 3.6(d). In the limit $q \to 0$, they become identical and reduce to the case of the Bessel-Gauss DHBs. Note that due to the elliptical-helical rotation of the phase, the OAM density of a Mathieu beam will vary azimuthally across the elliptic nodes of the beam, in contrast to the case of Bessel beams, where the azimuthal symmetry yields constant OAM density for a fixed radial coordinate. The modulated circular ring in Fig. 3.6(e) is the intensity distribution at the Fourier plane. A plot of the beam intensity along the x-axis at the waist is shown in Fig. 3.6(f).

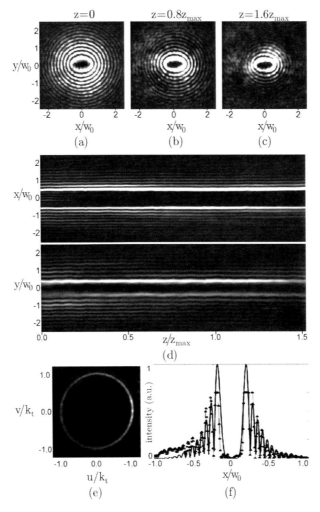

Figure 3.6 (a)–(c) Experimental transverse intensity distribution of a seventh-order Mathieu-Gauss DHB at different z planes, (d) propagation of the intensity along the (x, z) and the (y, z) planes, (e) intensity distribution of the power spectrum in the Fourier plane, and (f) transverse cut at $y = 0$ of the intensity profile of the beam at the waist. The dashed line is the theoretical intensity distribution. (Reprinted from Ref. 63 with permission of the authors and publisher.)

3.3.3 Ince-Gaussian beams

In addition to the well-known Hermite-Gaussian beams and Laguerre-Gaussian beams,[2] in recent papers the existence of the Ince-Gaussian beams, which constitute the third complete family of transverse eigenmodes of stable resonators, was theoretically[66,67] and experimentally[68] demonstrated. These new modes are exact and orthogonal solutions of the PWE in elliptic coordinates [Eq. (3.10)] and may be considered continuous transition modes between Hermite-Gaussian and Laguerre-Gaussian beams.

The Ince-Gaussian DHBs of order p, degree m, and ellipticity ε are defined by

$$\mathrm{HIG}_{p,m}^{\pm} = \mathrm{IG}_{p,m}^{e}(\mathbf{r}, \varepsilon) \pm i\, \mathrm{IG}_{p,m}^{o}(\mathbf{r}, \varepsilon), \qquad (3.19)$$

where

$$\left.\begin{array}{l} \mathrm{IG}_{p,m}^{e}(\mathbf{r}, \varepsilon) \\ \mathrm{IG}_{p,m}^{o}(\mathbf{r}, \varepsilon) \end{array}\right\} = \frac{w_0}{w(z)} \exp\left[\frac{-r^2}{w^2(z)}\right] \exp i \left[kz + \frac{kr^2}{2R(z)} - (p+1)\psi_{GS}(z)\right]$$

$$\times \left\{\begin{array}{l} \mathcal{C}\, \mathrm{C}_p^m(i\xi, \varepsilon)\, \mathrm{C}_p^m(\eta, \varepsilon) \\ \mathcal{S}\, \mathrm{S}_p^m(i\xi, \varepsilon)\, \mathrm{S}_p^m(\eta, \varepsilon) \end{array}\right., \qquad (3.20)$$

are the even and odd Ince-Gaussian beams. The functions $\mathrm{C}_p^m(\eta, \varepsilon)$ and $\mathrm{S}_p^m(\eta, \varepsilon)$ are the even and odd Ince polynomials of order p and degree m, respectively, where $0 \leq m \leq p$ for even functions, $1 \leq m \leq p$ for odd functions, and the indices (p, m) have the same parity, i.e., $(-1)^{p-m} = 1$. The constants $\mathcal{C}$ and $\mathcal{S}$ are normalization factors. In a transverse z-plane, the semifocal separation f of the elliptic coordinate system diverges in the same way as the width of the Gaussian beam envelope, that is $x = [f_0 w(z)/w_0] \cosh \xi \cos \eta$, $y = [f_0 w(z)/w_0] \sinh \xi \sin \eta$.

In order to fully describe the transverse distribution of an Ince-Gaussian DHB at the waist plane, it is needed to give the parity, the indices p, m, and two of the three next parameters: ε, w_0, f_0, where $\varepsilon = 2f_0^2/w_0^2$. In Fig. 3.7 we show the transverse magnitudes and phases of the even and odd stationary modes $\mathrm{IG}_{10,6}^{e,o}(\xi, \eta, \varepsilon = 1)$ and the corresponding Ince-Gaussian DHB at the waist plane. For this combination of indices, the pattern consists of three well-defined elliptic confocal rings with a dark elliptic spot on the axis. For the general case, the number of rings is given by the relation $1 + (p - m)/2$; thus, a single elliptic donut can be generated with modes for which $p = m$. On propagation, the phase of the Ince-Gaussian DHBs rotates elliptically around a line defined by $(|x| \leq f, 0, z)$ with a rotating direction defined by the sign in Eq. (3.19).

The first observation of even Ince-Gaussian modes directly generated in a stable laser resonator was reported by Schwarz et al.[68] In this experiment, several Ince-

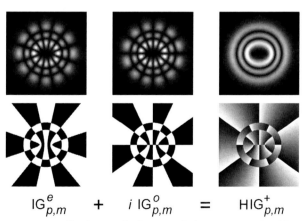

Figure 3.7 Transverse amplitudes and phases of the even and odd stationary modes $\mathrm{IG}_{10,6}^{\sigma}(\xi, \eta, \varepsilon = 1)$ and the corresponding Ince-Gaussian DHB at the waist plane.

Gaussian beams of even parity were excited by slightly breaking the symmetry of the cavity of a diode-pumped Nd:YVO4 laser. These observations were corroborated by Ohtomo et al.[69] using two laser-diode-pumped microchip solid-state lasers, including LiNdP$_4$O$_{12}$ (LNP) and Nd:GdVO$_4$, by adjusting the azimuthal symmetry of the short laser resonator.

The first generation of Ince-Gaussian DHBs carrying OAM was reported Bentley et al.[70] and Davis et al.[71], employing a novel interference technique for generating the object and reference beams with a single liquid crystal display. The experimental results are illustrated in Fig. 3.8. Unlike the Laguerre-Gaussian and Bessel beams, the Ince-Gaussian DHBs are dark elliptic hollow beams that exhibit an azimuthally asymmetric angular momentum density and intensity profile. This feature might be useful to control the rotational motion of microparticles along the elliptic rings of the Ince-Gaussian DHBs with the advantage that the ellipticity can be adjusted dynamically.

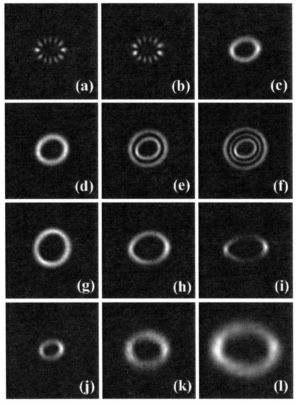

Figure 3.8 Experimental intensity results showing an (a) IG$_{6,6}^e$(r, 3) beam, (b) IG$_{6,6}^o$(r, 3) beam, (c) IG$_{6,6}^+$(r, 3) beam, (d) one intensity ring for the IG$_{8,8}^+$(r, 2) beam, (e) two intensity rings for the IG$_{10,8}^+$(r, 2) beam, (f) three intensity rings for the IG$_{12,8}^+$(r, 2) beam, (g) zero-ellipticity IG$_{12,12}^+$(r, 0) beam, (h) higher ellipticity IG$_{12,12}^+$(r, 6) beam, (i) high ellipticity IG$_{12,12}^+$(r, 12) beam. Intensity for the IG$_{4,4}^+$(r, 2) beam at (j) the waist plane, (k) 0.8 m, and (l) 1.0 m. Photographs show 600 × 600 camera pixels with sizes of 6.7 mm square. (Reprinted from Ref. 71 with permission of the authors and the publisher.)

Acknowledgments

The author acknowledges M. A. Bandres, C. López-Mariscal, R. I. Hernández-Aranda, M. Guizar-Sicairos, S. Chávez-Cerda, U. T. Schwarz, and J. Davies for fruitful collaborations. This work was partially supported by Consejo Nacional de Ciencia y Tecnología (Grant No. 42808), and by the Tecnológico de Monterrey (Grant No. CAT-007).

References

1. Kogelnik, H., and Li, T., "Laser beams and resonators," *Appl. Opt.*, **5**, 1550–1567, 1966.

2. Siegman, A.E., *Lasers*, University Science, San Diego (1986).

3. Durnin, J., "Exact solutions for nondiffracting beams. I The scalar theory," *J. Opt. Soc. Am. A*, **4**, 651–654, 1987.

4. Durnin, J., Micely, J.J., and Eberly, J.H., "Diffraction-Free Beams," *Phys. Rev. Lett.*, **58**, 1499–1502, 1987.

5. Gori, F., Guattari, G., and Padovani, C., "Bessel-Gauss beams," *Opt. Comm.*, **64**, 491–495, 1987.

6. Li, Y., Lee H., and Wolf, E., "New generalized Bessel-Gaussian beams," *J. Opt. Soc. Am. A*, **21**, 640–646, 2004.

7. Cai, Y., Lu, X., and Lin, Q., "Hollow Gaussian beam and its propagation," *Opt. Lett.*, **28**, 1084–1086, 2003.

8. Gutiérrez-Vega, J.C., Iturbe-Castillo, M.D., and Chávez-Cerda, S., "Alternative formulation for invariant optical fields: Mathieu beams," *Opt. Lett.*, **25**, 1493–1495, 2000.

9. Gutiérrez-Vega, J.C., Iturbe-Castillo, M.D., Ramírez, G.A., Tepichín, E., Rodríguez-Dagnino, R.M., Chávez-Cerda, S., and New, G.H.C., "Experimental demonstration of optical Mathieu beams," *Opt. Comm.*, **195**, 35–40, 2001.

10. Gutiérrez-Vega, J.C., and Bandres, M.A., "Helmholtz–Gauss waves," *J. Opt. Soc. Am. A*, **22**, 289–298, 2005.

11. Ghatak, A., and Thyagarajan, K., *Introduction to Fiber Optics*, Cambridge University Press, New York (1998).

12. Yin, J., Gao, W., and Zhu, Y., "Generation of dark hollow beams and their applications," *Progress in Opt.*, **45**, 119–204, 2003.

13. Allen, L., Beijersbergen, M.W., Spreeuw, R.J.C., and Woerdman, J.P., "Orbital angular momentum of light and the transformation of Laguerre-Gaussian laser modes," *Phys. Rev. A*, **96**, 8185–8189, 1992.

14. Volke-Sepúlveda, K.P., Garcés-Chávez, V., Chávez-Cerda, S., Arlt, J., and Dholakia, K., "Orbital angular momentum of a high-order Bessel light beam," *J. Opt. B: Quantum Semiclass. Opt.*, **4**, S82–S89, 2002.

15. Volke-Sepúlveda, K.P., Chávez-Cerda, S., Garcés-Chávez, V., and Dholakia, K., "Three-dimensional optical forces and transfer of orbital angular momentum from multiringed light beams to spherical microparticles," *J. Opt. Soc. Am. B*, **21**, 1749–1757, 2004.

16. Garcés-Chávez, V., McGloin, D., Padgett, M.J., Dultz, W., Schmitzer, H., and Dholakia, K., "Observation of the transfer of the local angular momentum density of a multiringed light beam to an optically trapped particle," *Phys. Rev. Lett.*, **91**, 093602-1, 2003.

17. Bouchal, Z., "Nondiffracting optical beams: physical properties, experiments, and applications," *Czech. J. Phys.*, **53**, 537–578, 2003.

18. Padgett, M.J., and Allen, L., "Optical tweezers and spanners," *Phys. World*, **10**, 35–40, 1997.

19. Molloy, J.E., Dholakia, K., and Padgett, M.J., "Optical tweezers in a new light," *J. Mod. Opt.*, **50**, 1501–1507, 2003.

20. Arlt, J., Garces-Chavez, V., Sibbett, W., and Dholakia, K., "Optical micromanipulation using a Bessel light beam," *Opt. Commun.*, **197**, 239–245, 2001.

21. Clifford, M.A., Arlt, J., Courtial, J., and Dholakia, K., "High-order Laguerre-Gaussian laser modes for studies of cold atoms," *Opt. Commun.*, **156**, 300–306, 1998.

22. Belanger, P.A., and Rioux, M., "Ring pattern of a lens-axicon doublet illuminated by a Gaussian beam (TE)," *Appl. Opt.*, **17**, 1080–1086, 1978.

23. Ito, H., Sakaki, K., Jhe, W., and Ohtsu, M., "Atomic funnel with evanescent light," *Phys. Rev. A*, **56**, 712–718, 1997.

24. Arlt, J., and Dholakia, K., "Generation of high-order Bessel beams by use of an axicon," *Opt. Commun.*, **177**, 297–301, 2000.

25. Arlt, J., Kuhn, R., and Dholakia, K., "Spatial transformation of Laguerre-Gaussian laser modes," *J. Mod. Opt.*, **48**, 783–787, 2001.

26. Liu, Z., Zhao, H., Liu, J., Lin, J., Ahmad, M.A., and Liu, S., "Generation of hollow Gaussian beams by spatial filtering," *Opt. Lett.*, **32**, 2076–2078, 2007.

27. Wang, X., and Littman, M.G., "Laser cavity for generation of variable-radius rings of light," *Opt. Lett.*, **18**, 767–769, 1993.

28. Zhang, L., Lu, X.-H., Chen, X.-M., and He, S.-L., "Generation of a dark hollow beam inside a cavity," *Chin. Phys. Lett.*, **21**, 298–301, 2004.

29. Beijersbergen, M.W., Allen, L., van der Veen, H.E.L.O., and Woerdman, J.P., "Astigmatic laser mode converters and transfer of orbital angular momentum," *Opt. Commun.*, **96**, 123–132, 1993.

30. Allen, L., Barnett, S.M., and Padgett, M.J., *Optical Angular Momentum*, Institute of Physics, Bristol (2003).

31. Courtial, J., and Padgett, M.J., "Performance of a cylindrical lens mode converter for producing Laguerre-Gaussian laser modes," *Opt. Commun.*, **159**, 13–18, 1999.

32. O'Neil, A.T., and Courtial, J., "Mode transformations in terms of the constituent Hermite-Gaussian or Laguerre-Gaussian modes and the variable-phase mode converter," *Opt. Commun.*, **181**, 35–45, 2000.

33. Vasara, A., Turunen, J., and Friberg, A.T., "Realization of general nondiffracting beams with computer-generated holograms," *J. Opt. Soc. Am. A*, **6**, 1748–1754, 1989.

34. Lee, H.S., Stewart, B.W., Choi, K., and Fenichel, H., "Holographic nondiverging hollow beam," *Phys. Rev. A*, **49**, 4922–4927, 1994.

35. Paterson, C., and Smith, R., "Higher-order Bessel waves produced by axicon-type computer-generated holograms," *Opt. Commun.*, **124**, 121–130, 1996.

36. Arlt, J., Dholakia, K., Allen, L., and Padgett, M. J., "The production of multi-ringed Laguerre-Gaussian modes by computer generated holograms," *J. Mod. Opt.*, **45**, 1231–1237, 1998.

37. Arrizón, V., Méndez, G., and Sánchez-de-La-Llave, D., "Accurate encoding of arbitrary complex fields with amplitude-only liquid crystal spatial light modulators," *Opt. Express*, **13**, 7913–7927, 2005.

38. Ohtake, Y., Ando, T., Fukuchi, N., Matsumoto, N., Ito, H., and Hara, T., "Universal generation of higher-order multiringed Laguerre-Gaussian beams by using a spatial light modulator," *Opt. Lett.*, **32**, 1411–1413, 2007.

39. Cai, Y., and Zhang, L., "Coherent and partially coherent dark hollow beams with rectangular symmetry and paraxial propagation properties," *J. Opt. Soc. Am. B*, **23**, 1398–1407, 2006.

40. Cai, Y., and Ge, D., "Propagation of various dark hollow beams through an apertured paraxial ABCD optical system," *Phys. Lett. A*, **347**, 72–80, 2006.

41. Ottl, A., Ritter, S., Kohl, M., and Esslinger, T., "Correlations and counting statistics of an atom laser," *Phys. Rev. Lett.*, **95**, 090404, 2005.

42. Parker, N.G., Proukakis, N.P., Leadbeater, M., and Adams, C.S., "Soliton-sound interactions in quasi-one-dimensional Bose-Einstein condensates," *Phys. Rev. Lett.*, **90**, 220401, 2003.

43. Dyott, R.B., *Elliptical Fiber Waveguides*, Artech House, Norwood, Mass., (1995).

44. Cai, Y., and Lin, Q., "Hollow elliptical Gaussian beam and its propagation through aligned and misaligned paraxial optical systems," *J. Opt. Soc. Am. A*, **21**, 1058–1065, 2004.

45. Cai, Y., and Lin, Q., "The elliptical Hermite-Gaussian beam and its propagation through paraxial systems," *Opt. Commun.*, **207**, 139–147, 2002.

46. Cai, Y., and Lin, Q., "Decentered elliptical Hermite-Gaussian beam," *J. Opt. Soc. Am. A*, **20**, 1111–1119, 2003.

47. Gao, C., Weber, H., and Laabs, H., "Twisting of three-dimensional Hermite-Gaussian beams," *J. Mod. Opt.*, **46**, 709–719, 1999.

48. Mei, Z., and Zhao, D., "Controllable elliptical dark-hollow beams," *J. Opt. Soc. Am. A*, **23**, 919–925, 2006.

49. Mei, Z., and Zhao, D., "Decentered controllable elliptical dark-hollow beams," *Opt. Commun.*, **259**, 415–423, 2006.

50. Kotlyar, V.V., Khonina, S.N., Almazov, A.A., Soifer, V.A., Jefimovs, K., and Turunen, J., "Elliptic Laguerre-Gaussian beams," *J. Opt. Soc. Am. A*, **23**, 43–56, 2006.

51. Chakraborty, R., and Ghosh, A., "Generation of an elliptic Bessel beam," *Opt. Lett.*, **31**, 38–40, 2006.

52. McGloin, D., and Dholakia, K., "Bessel beams: diffraction in a new light," *Contemporary Phys.*, **46**, 15–28, 2005.

53. Dartora, C.A., Zamboni-Racheda, M., Nobrega, K.Z., Recami, E., and Hernández-Figueroa, H.E., "General formulation for the analysis of scalar diffraction-free beams using angular modulation: Mathieu and Bessel beams," *Opt. Commun.*, **222**, 75–80, 2003.

54. Chafiq, A., Hricha, Z., and Belafhal, A., "Paraxial propagation of Mathieu beams through an apertured ABCD optical system," *Opt. Commun.*, **253**, 223–230, 2005.

55. Kartashov, Y.V., Egorov, A.A., Vysloukh, V.A., and Torner, L., "Shaping soliton properties in Mathieu lattices," *Opt. Lett.*, **31**, 238–240, 2006.

56. Chávez-Cerda, S., Padgett, M.J., Allison, I., New, G.H.C., Gutiérrez-Vega, J.C., O'Neil, A.T., MacVicar, I., and Courtial, J., "Holographic generation and orbital angular momentum of high-order Mathieu beams," *J. Opt. B: Quantum Semiclass. Opt.*, **4**, S52–S57, 2002.

57. López-Mariscal, C., Gutiérrez-Vega, J.C., Garcés-Chávez, V., and Dholakia, K., "Observation of the angular momentum transfer in the Mie regime using Mathieu beams," *Proc. SPIE*, **5930**, 59301U-1, 2005. [doi: 10.1117/12.616449].

58. López-Mariscal, C., Gutiérrez-Vega, J.C., Milne, G., and Dholakia, K., "Orbital angular momentum transfer in helical Mathieu beams," *Opt. Express*, **14**, 4182–4187, 2006.

59. Dartora, C.A., and Hernández-Figueroa, H.E., "Properties of a localized Mathieu pulse," *J. Opt. Soc. Am. A*, **21**, 662–667, 2004.

60. Davila-Rodriguez, J., and Gutiérrez-Vega, J.C., "Helical Mathieu and parabolic localized pulses," *J. Opt. Soc. Am. A*, **24**, 3449–3455, 2007.

61. Gutiérrez-Vega, J.C., Rodríguez-Dagnino, R.M., Meneses-Nava, M.A., and Chávez-Cerda, S., "Mathieu functions, a visual approach," *Am. J. Phys.*, **71**, 233–242, 2003.

62. Chávez-Cerda, S., Gutiérrez-Vega, J.C., and New, G.H.C., "Elliptic vortices of electromagnetic wave fields," *Opt. Lett.*, **26**, 1803–1805, 2001.

63. López-Mariscal, C., Bandres, M.A., and Gutiérrez-Vega, J.C., "Observation of the experimental propagation properties of Helmholtz-Gauss beams," *Opt. Eng.*, **45**, 068001, 2006.

64. Guizar-Sicairos, M., and Gutiérrez-Vega, J.C., "Generalized Helmholtz-Gauss beams and its transformation by paraxial optical systems," *Opt. Lett.*, **31**, 2912–2914, 2006.

65. Hernandez-Aranda, R.I., Gutiérrez-Vega, J.C., Guizar-Sicairos, M., and Bandres, M.A., "Propagation of generalized vector Helmholtz-Gauss beams through paraxial optical systems," *Opt. Express*, **14**, 8974–8988, 2006.

66. Bandres, M.A., and Gutiérrez-Vega, J.C., "Ince-Gaussian beams," *Opt. Lett.*, **29**, 144–146, 2004.

67. Bandres, M.A., and Gutiérrez-Vega, J.C., "Ince-Gaussian modes of the paraxial wave equation and stable resonators," *J. Opt. Soc. Am. A*, **21**, 873–880, 2004.

68. Schwarz, U.T., Bandres, M.A., and Gutiérrez-Vega, J.C., "Observation of Ince-Gaussian modes in stable resonators," *Opt. Lett.*, **29**, 1870–1872, 2004.

69. Ohtomo, T., Kamikariya, K., Otsuka, K., and Chu, S.-C., "Single-frequency Ince-Gaussian mode operations of laser-diode-pumped microchip solid-state lasers," *Opt. Express*, **15**, 10705–10717, 2007.

70. Bentley, J.B., Davis, J.A., Bandres, M.A., and Gutiérrez-Vega, J.C., "Generation of helical Ince-Gaussian beams with a liquid-crystal display," *Opt. Lett.*, **31**, 649–651, 2006.

71. Davis, J.A., Bentley, J.B., Bandres, M.A., and Gutiérrez-Vega, J.C., "Generation of helical Ince-Gaussian beams: beam shaping with a liquid crystal display," *Proc. SPIE*, **6290**, 62900R, 2006. [doi: 10.1117/12.679533].

**Julio C. Gutiérrez-Vega** is associate professor in the Physics Department and heads the Optics Center and the Photonics and Mathematical Optics Group at the Tecnológico de Monterrey, Monterrey, México. Julio C. Gutiérrez-Vega received a BS in physics (1991) and a MS in electric engineering (1995) from the Tecnológico de Monterrey. In 2000, he received his PhD in optics from the National Institute for Astrophysics, Optics, and Electronics in Puebla, México. He is the author and coauthor of more than 115 scientific publications in international journals, conference proceedings, and books. His research activities are focused on the nondiffracting propagation of wavefields, special solutions of the Helmholtz and paraxial wave equation: Mathieu, parabolic, and Ince-Gaussian beams, and laser resonators. Dr. Gutiérrez-Vega is a member of SPIE, OSA, and APS.

Chapter 4
Transfer of Information Using Helically Phased Modes

Miles Padgett, Graham Gibson, and Johannes Courtial
University of Glasgow, Scotland

4.1 Introduction
4.2 Generation of Helically Phased Beams
4.3 Transformations Between Modes
4.4 Measurement of Orthogonal States
4.5 A Communication System Based on Helical Beams
References

4.1 Introduction

In 1992, Les Allen and coworkers within Han Woerdman's group realized that light beams with an azimuthal phase dependence described by an additional $\exp(i\ell\theta)$ term, where ℓ can take any integer value and θ is the angular position within the beam, carry an orbital angular momentum.[1,2] This orbital angular momentum is additional to the spin angular momentum associated with circular polarization. These azimuthally phased beams comprise ℓ intertwined helical phase fronts, skewed with respect to the beam axis, meaning that at all positions within the beam cross section, the wavefront normal and associated momentum density has a well-defined azimuthal component. The insight of Allen and coworkers was that for all beams described in this way, when integrated over the beam cross section, the orbital angular momentum (OAM) is $\ell\hbar$ per photon. Laguerre-Gaussian and high-order Bessel beams are both examples of mode families having these properties. A further feature of these beams is that the phase singularity on the beam axis precludes a nonzero on-axis intensity and therefore all such beams with $\ell \neq 0$ have an annular intensity cross section (see Fig. 4.1). The circulation of the optical momentum and energy around the singularity leads them also to be called optical vortices.

The further recognition that beams carrying defined amounts of OAM could be created within the laboratory means that over the last 15 years they have

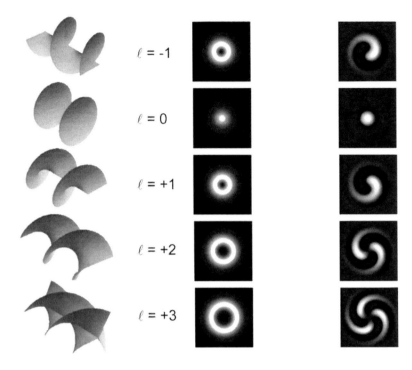

Figure 4.1 The phase structure of helically phased Laguerre-Gaussian beams (left), their corresponding intensity profile (center), and their interference pattern with a plane wave (right).

formed the basis of many investigations, ranging from the transfer of OAM to microscopic particles (in optical spanners or optical wrenches) to studies of the fundamental properties of light in both the classical and quantum regimes.[3] However, although OAM is an immensely useful concept, one must emphasize that all its properties are simply those predicted by standard electromagnetic theory.

The purpose of this chapter is to consider how sets of helical laser modes carrying OAM can be used for information transfer.

4.2 Generation of Helically Phased Beams

Phase singularity on the beam axis is an example of an optical vortex, present throughout natural light fields and studied extensively from the 1970s onward by Nye and Berry.[4] Helically phased laser beams had in fact been generated prior to the recognition that they carried OAM. In the early work of Tamm and Weiss,[5] helical beams were created from a laser in which the TEM Hermite Gaussian HG_{10} and HG_{01} were degenerate and interfered to give a Laguerre-Gaussian mode with $\ell = 1$. Subsequently, Harris et al. produced similar laser beams and showed that when they interfered with their own mirror image, or a plane wave, this gave characteristic spiral fringes.[6] The exact shape of these fringes, and how it changes with propagation, may be readily calculated.[7,8]

For the explicit purpose of creating OAM, it was realized by Beijersbergen et al. that any single Hermite-Gaussian mode with no angular momentum could be transformed into a Laguerre-Gaussian mode using a carefully selected pair of cylindrical lenses.[9] Beyond being a clever method for generating helical beams, this work established the elegant algebraic relationship between the Laguerre-Gaussian and Hermite-Gaussian modes, both forming complete orthonormal basis sets, any one mode of either set being expressible as an appropriate superposition of the other.

Although conversion with cylindrical lenses is highly optically efficient, each Laguerre-Gaussian mode requires a specific Hermite-Gaussian mode as the input, thus somewhat limiting the range of Laguerre-Gaussian modes that can be produced, and certainly restricting the ease at which the mode can be switched from one to another.

As an alternative to a cylindrical lens mode converter, a computer-generated hologram (i.e., diffractive optic) can be used to transform the normal plane-wave laser beam into one having any exotic phase structure—including the helical modes pertinent to this chapter.

The advantage of this holographic approach is the recent availability of high-quality spatial light modulators (SLMs. These pixilated liquid crystal devices take the place of the holographic film; the required hologram kinoform is calculated using a desktop PC and then displayed on the device. A further advantage is that the kinoform can be changed many times per second, controlling the form of the diffracted beam to meet the experimental requirements. It was 1990 when Soskin's group used the now classic "forked" pattern to produce a nearly pure Laguerre-Gaussian mode (see Fig. 4.2)[10] (although at this stage its angular momentum properties were not realized). The kinoforms can be designed to create elaborate superpositions of modes, for example, the combination of Laguerre-Gaussian modes required to form optical vortex links and knots.[11] Holograms can also be used to transform white-light beams, using a prism to compensate for the resulting chromatic dispersion, for example, the creation of achromatic Laguerre[12] or Bessel[13] beams.

4.3 Transformations Between Modes

Transformations of the polarization state on passage of light through an optical component fall into two distinct groups. The first includes those arising from the transmission of a birefringent wave plate, which introduces a phase shift between linear polarization states, having orientations defined with respect to the rotational position of the plate.[14] The second includes those arising from transmission through an optical active material that introduces a phase shift between circular polarization states. The result of the later is a rotation in transmitted polarization state, independent of the rotational orientation of the material.

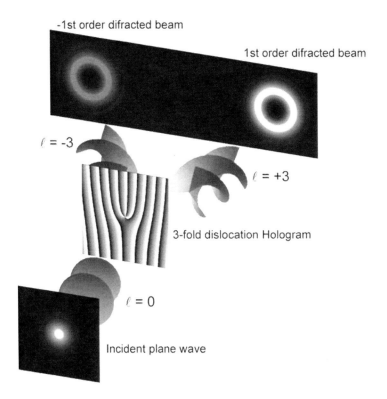

Figure 4.2 The "classic" forked hologram for the production of helically phased beams. Note that the blazing of the grating favors diffraction into one order, thereby maximizing the conversion efficiency.

For polarization, these transformations, often described by Jones matrices, may also be represented on the surface of a Poincaré sphere, on which all possible polarization states are represented. By convention, the north and south poles of the sphere correspond to left and right circular polarization and any other polarization state can be obtained by their appropriate superposition. Of particular significance is the equator, a superposition of equal magnitude of right and left circular polarization, the relative phase of the superposition determining the orientation of the resulting linear polarization. "Latitudes" between the equator and the poles correspond to elliptical polarization states. Starting with any polarization state, at a specific position on the sphere, a transmission through a birefringent component changes the latitude of the state, whereas transmission through an optically active material changes the longitude.

The Poincaré sphere is simply an example of the Block sphere, giving a geometric representation as to all possible complex superpositions of two orthogonal states. It is therefore no surprise that an equivalent geometrical construction can be made for Hermite-Gaussian and Laguerre-Gaussian modes (see Fig. 4.3).[15]

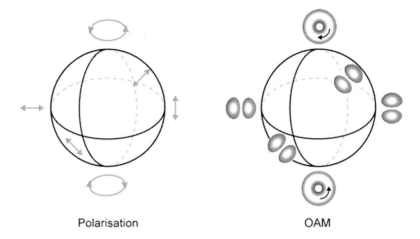

Figure 4.3 The Poincaré sphere (left) and its equivalent geometric construction for mode of *N* = 1 (right).

The principles of transformation extend beyond these $\ell = \pm 1$ modes to modes of any order, but rather than using the Poincaré sphere the higher dimensionality makes using the equivalent of Jones matrices more convenient.[16]

As stated in Section 4.2, Beijersbergen et al. recognized that the equivalent transformation to that of a wave plate may be performed on the OAM states by a cylindrical lens mode converter. A special case is for the equivalent transformation corresponding to that induced on polarization states by a half wave plate, which is equivalent to a mirror inversion. For helical beams this mirror inversion is readily accomplished using a Dove prism or similar inverter (see Fig. 4.4).

Less easy to understand than the OAM equivalent to birefringence is what constitutes the equivalent transformation to that resulting from optical activity. For polarization, a rotation corresponds to a phase shift between the circular polarization states. An equivalent phase shift between helical modes is simply a rotation of their interference pattern around the beam axis. This extends to modal superpositions of any order, and since any image can be formed by an appropriate superposition of Laguerre-Gaussian modes, one sees that optical activity is, for helically phased beams, analogous to image rotation.[17]

In general, a beam rotation can either be applied locally to each point in the beam, or globally about a specific axis. In the case of polarization, either mapping results in the same transformation of the beam. For helical modes, the rotation must be a global rotation around the beam axis. Similarly, whereas optical activity rotates the polarization at every point within the beam, an image rotation is a global mapping around the beam axis. Consistent with this understanding, it has been both predicted[18] and observed[19] that optically active materials do not alter the OAM state of the transmitted light.

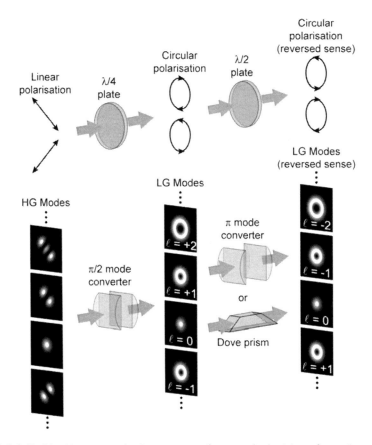

Figure 4.4 Cylindrical lenses and prisms can perform equivalent transformations on laser modes as wave plates do on polarization states.

4.4 Measurement of Orthogonal States

Any two orthogonal polarization states (diametrically opposite positions on the Poincaré sphere) can be separated with a combination of wave plate and polarizers. Most easily, a polarizing beam splitter can, with near 100% efficiency, resolve a single incident photon into one of two states, corresponding to orthogonal linear polarization states. By contrast, helically phased beams and their associated OAM have an arbitrary large number of orthogonal states and consequently are appealing applications in information transfer and optical processing. However, separating an incident beam into its orthogonal OAM states is not straightforward.

Previously, many different groups have shown that the OAM of a laser beam containing many photons in the same mode can be measured by interfering the mode with a plane wave to create spiral fringes, as shown in Fig. 4.1. Although this technique can discriminate between an arbitrarily large number of states, many photons are required to form the full interference pattern. The amount of information per photon is therefore small.

For individual photons, the same computer-generated holograms used to create helical beams, shown in Fig. 4.2, can also be used to measure the state. If the ℓ value of the incident beam/photon is equal in magnitude, but opposite in sign to that of the hologram, then the first-order diffracted beam has a planar phase front, without any phase singularity. The beam can now be focused into a single-mode fiber and detected.[20] If the ℓ value does not match that of the hologram, then the phase singularity is not removed and coupling efficiency into the fiber is low. Progressively incrementing the ℓ value of the hologram allows an arbitrarily high number of the different OAM states to be tested. However, each test requires at least one photon; therefore, the quantum efficiency of this approach can never exceed the reciprocal channel number to be tested.

A simplification of this holographic technique can be realized by designing the hologram kinoform to generate/detect several different OAM states simultaneously (see Fig. 4.5). Although this removes the need to switch between kinoforms, it does not improve the overall efficiency since the diffracted light is split equally among the various diffracted beams—only if there is a nonzero intensity is the ℓ value confirmed.

Beyond the use of holograms to sort OAM states, it is alternatively possible to apply an interferometric technique that can distinguish between an arbitrary large number of OAM states with no loss in efficiency.[21] As discussed in Section 4.3, a Dove prism mirror inverts the mode profile, inverting its handedness. A second Dove prism returns the mode to it original handedness, but the mode is

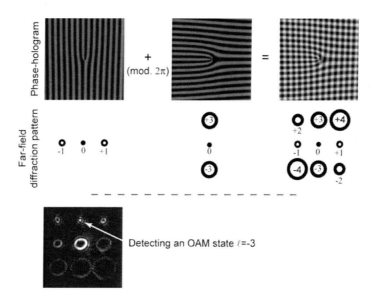

Figure 4.5 A single hologram can be designed to generate, or test for, a large number of OAM states but having a limiting quantum efficiency of the reciprocal channel number (top). Far-field diffraction pattern when hologram is used to detect an OAM state of $\ell = -3$ (bottom).

rotated by twice the angle between the prisms. For a helical mode this rotation corresponds to a phase shift. By placing Dove prisms in both arms of a Mach-Zehnder interferometer, the relative phase shift of any OAM state now depends on its ℓ value (see Fig. 4.6). Cascading interferometers allow any number of states to be separated, albeit at increasing complexity. The technique can also be extended by the inclusion of wave plates within the interferometer to sort modes on the basis of their total angular momentum.[22] However, although this technique offers in principle 100% efficiency, the associated technical challenges of maintaining an interferometric precision means that the technique has not been widely applied.

Other techniques for potentially measuring multiple OAM states are based on the rotational frequency shift, where the rotation of the beam around its own axis shifts its frequency proportionally to its total angular momentum.[23] Placing a beam rotator in the optical path would then allow the OAM state of a single photon to be determined based on its frequency shift.[24] However, as with interferometric sorting, the technical requirements for the precision of the optical system are difficult to meet.

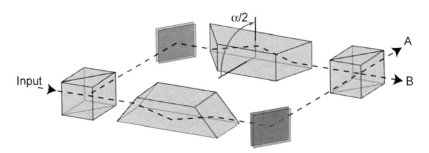

Figure 4.6 A Mach-Zehnder Interferometer within which there is a beam rotation, α, between the two arms, introducing a phase delay of $\Delta\phi = \alpha \times \ell$. OAM states can be separated depending on their rotational symmetry.

4.5 A Communication System Based on Helical Beams

We have previously reported the demonstration of a free-space optical communication system using OAM states onto which to encode the information.[25] It was based on the use of a computer-controlled spatial light modulator (SLM) (Boulder Nonlinear Systems, 512 × 512 pixels) for both the creation and measurement of the OAM states (see Fig. 4.7).

The encoding SLM was illuminated by a 6 mm diameter collimated beam from a HeNe laser. An afocal telescope expanded the encoded beam to approximately 40 mm diameter, ensuring eye-safe intensity levels. As shown in Fig. 4.7, the hologram kinoform was designed to operate on axis such that the specified first-order beam comprising the desired OAM state was superimposed on the residual zero-order beam, i.e., a fundamental Gaussian. This fundamental Gaussian beam provided a reference mode around which to perform the decoding. We chose an "alphabet" consisting of the values $\ell = -16, -12, -8, -4, 4, 8, 12,$ and 16; this separation minimized the crosstalk between channels while keeping the size of the beams within the aperture of the optical system.

The receiver unit comprised a similar telescope, SLM, and CCD array detector. The beam was collected by the telescope and reduced to 6 mm diameter before being reflected from the SLM, programmed with the analyzing hologram pattern of the type shown in Fig. 4.7. This analyzing hologram was designed to diffract the light into nine beams, each with a different helicity, arranged in a 3 × 3 grid. All nine beams were imaged onto the CCD array, from which we determined the on-axis intensity of each beam. One of the beams had high on-axis intensity, thereby establishing the OAM state of the transmitted beam. To compensate for small perturbations in alignment associated with atmospheric turbulence, etc., we defined the measurement axes relative to the observed position of the central $\ell = 0$ channel.

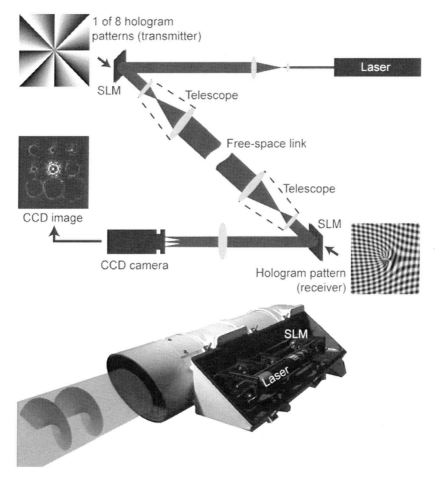

Figure 4.7 A free-space communication system based on encoding the information on helically phased beams; schematic (top), photograph of transmitter (bottom).

Although demonstrating that OAM can be used to transmit information, we must emphasize that even with perfect, lossless optics, this approach is limited in its quantum efficiency to the reciprocal channel number. This could be improved by using the interferometric technique discussed in Section 4.4, however, it is doubtful whether that technique is robust enough for implementation in the field. An ongoing challenge would be to devise a non-interferometric technique based on appropriately designed hologram kinoforms that can measure the OAM of a single photon—whether such a design is possible is unknown but none has yet been reported.

Even when used in this low-efficiency configuration, the purity of the OAM state can be significantly degraded by atmospheric aberrations, potentially introducing channel crosstalk.[26] However, much of this degradation would probably be eliminated by including an adaptive optical component for real-time aberration correction that has been widely applied to such free-space systems.[27]

At a single-photon level, Mair et al. have demonstrated that OAM is still a relevant quantum number.[20] This creates obvious applications within quantum information processing, but although many exciting experiments are possible using the holographic measurement technique, true quantum opportunities will probably only ever be fully enabled if a simple, yet quantum efficient, technique can be developed for the unambiguous measurement of single photons within any of a large number of OAM states.

References

1. Allen, L., Beijersbergen, M.W., Spreeuw, R.J.C., and Woerdman, J.P., "Orbital angular momentum of light and the transformation of Laguerre-Gaussian laser modes," *Phys. Rev. A*, **45,** 8185–8189, 1992.
2. Allen, L., Padgett, M.J., and Babiker, M., "The orbital angular momentum of light," *Progress in Optics,* 39, pp. 291–372, Elsevier Science, Amsterdam (1999).
3. Allen, L., Barnett, S.M., and Padgett, M.J., *Optical Angular Momentum*, Institute of Physics Publishing, Bristol (2003).
4. Nye, J.F., and Berry, M.V., "Dislocations in wave trains," *Proc. R. Soc. Lond. Ser. A*, **336,** 165–190, 1974.
5. Tamm, C., and Weiss, C.O., "Bistability and optical switching of spatial patterns in a laser," *J. Opt. Soc. Am. B*, **7,** 1034–1038, 1990.
6. Harris, M., Hill, C.A., and Vaughan, J.M., "Optical helices and spiral interference fringes" *Opt. Commun.* **106,** 161–166, 1994.
7. Padgett, M., Arlt, J., Simpson, N., and Allen, L., "An experiment to observe the intensity and phase structure of Laguerre-Gaussian laser modes," *Am. J. Phys.*, **64,** 77–82, 1996.
8. Soskin, M.S., Gorshkov, V.N., Vasnetsov, M.V., Malos, J.T., and Heckenberg, N.R., "Topological charge and angular momentum of light beams carrying optical vortices," *Phys. Rev. A*, **56,** 4064–4075, 1997.

9. Beijersbergen, M.W., Allen, L., van der Veen, H.E.L.O., and Woerdman, J.P., "Astigmatic laser mode converters and transfer of orbital angular momentum," *Opt. Commun.*, **96**, 123–132, 1993.

10. Bazhenov, V.Y., Vasnetsov, M.V., and Soskin, M.S., "Laser beams with screw dislocations in their wave-fronts," *JETP Lett.*, **52**, 429–431, 1990.

11. Leach, J., Dennis, M., Courtial, J., and Padgett, M., "Laser beams—Knotted threads of darkness," *Nature* **432**, 165, 2004.

12. Leach, J., and Padgett, M.J., "Observation of chromatic effects near a white-light vortex," *New J. Phys.*, **5**, 154.1–154.7, 2003.

13. Leach, J., Gibson, G.M., Padgett, M.J., Esposito, E., McConnell, G., Wright, A.J., and Girkin, J.M., "Generation of achromatic Bessel beams using a compensated spatial light modulator," *Opt. Expr.*, **14**, 5581–5587, 2006.

14. Hecht, E., and Zajac, A., *Optics,* Addison-Wesley, Reading, MA (1974).

15. Padgett, M.J., and Courtial, J., "Poincaré-sphere equivalent for light beams containing orbital angular momentum," *Opt. Lett.*, **24**, 430–432, 1999.

16. Allen, L., Courtial, J., and Padgett, M.J., "Matrix formulation for the propagation of light beams with orbital and spin angular momenta," *Phys. Rev. E*, **60**, 7497–7503, 1999.

17. Allen, L., and Padgett, M., "Equivalent geometric transformations for spin and orbital angular momentum of light," *J. Mod. Opt.,* **54**, 487–491, 2007.

18. Andrews, D.L., Dávila Romero, L.C., and Babiker M., "On optical vortex interactions with chiral matter" *Opt. Commun.*, **237**, 133–139, 2004.

19. Araoka, F., Verbiest, T., Clays, K., and Persoon, A., "Interactions of twisted light with chiral molecules: An experimental investigation," *Phys. Rev. A*, **71**, 055401, 2005.

20. Mair, A., Vaziri, A., Weihs, G., and Zeilinger, A., "Entanglement of the orbital angular momentum states of photons," *Nature*, **412**, 313–316, 2001.

21. Leach, J., Padgett, M.J., Barnett, S.M., Franke-Arnold, S., and Courtial, J., "Measuring the orbital angular momentum of a single photon," *Phys. Rev. Lett.*, **88**, 257901, 2002.

22. Leach, J., Courtial, J., Skeldon, K., Barnett, S.M., Franke-Arnold, S., and Padgett, M.J., "Interferometric methods to measure orbital and spin, or the total angular momentum of a single photon," *Phys. Rev. Lett.*, **92**, 013601, 2004.

23. Courtial, J., Robertson, D.A., Dholakia, K., Allen, L., and Padgett, M.J., "Rotational frequency shift of a light beam," *Phys. Rev. Lett.*, **81**, 4828–4830, 1998.

24. Molina-Terriza, G., Torres, J.P., and Torner, L., "Management of the angular momentum of light: preparation of photons in multidimensional vector states of angular momentum," *Phys. Rev. Lett.,* **88**, 013601, 2002.

25. Gibson, G., Courtial, J., Padgett, M.J., Vasnetsov, M., Pas'ko, V., Barnett, S.M., and Franke-Arnold, S., "Free-space information transfer using light beams carrying orbital angular momentum," *Opt. Expr.*, **12**, 5448–5456, 2004.
26. Paterson, C., "Atmospheric turbulence and orbital angular momentum of single photons for optical communication," *Phys. Rev. Lett.* **94**, 153901, 2005.
27. Tyson, R.K., "Bit-error rate for free-space adaptive optics laser communications," *J. Opt. Soc. Am. A*, **19**, 753–758, 2002.

Miles Padgett is a professor of optics in the Department of Physics and Astronomy at the University of Glasgow. He heads a 15-strong research team covering the full spectrum of blue-sky research to applied commercial development, funded by a combination of government, charity, and industry. In 2001, he was elected to the Fellowship of the Royal Society of Edinburgh. He is presently (2007–2008) supported by a Royal Society Leverhulme Trust Senior Research Fellowship. In 2008, he was awarded the Optics and Photonics Division Prize from the Institute of Physics.

Graham Gibson is a postdoctoral researcher working in the same group. Following his construction of the communication system discussed in this chapter, he has gone on to apply his expertise in spatial light modulators to the development of holographic optical tweezers for applications in bio and nano science.

Johannes Courtial is a Royal Society University Research Fellow also working in the same department. Beyond angular momentum, his research interests encompass the formation of optical fractals within certain kinds of laser resonators, and, most recently, how geometrical optics can be used to replicate some of the properties of negative refractive index materials.

Chapter 5
Microoptical Components for Information Optics and Photonics

Christof Debaes, Heidi Ottevaere, and Hugo Thienpont
Vrije Universiteit Brussel, Belgium

5.1 Introduction
5.2 Fabrication Techniques for Microoptical and Micromechanical Systems
    5.2.1 Microstructuring in a plane
    5.2.2 Fabrication methods for optical board-level interconnects
5.3 Deep Proton Writing as a Microoptical Prototyping Technology
    5.3.1 Ion interactions
    5.3.2 Irradiation process
    5.3.3 Etching process
    5.3.4 Swelling process
5.4 Microoptical Components for Board-Level Optical Interconnects
    5.4.1 Out-of-plane couplers for optical waveguides in standard FR4 printed circuit boards
    5.4.2 Two-dimensional single-mode fiber array couplers
    5.4.3 Intra-MCM interconnect module
5.5 Conclusions
References

5.1 Introduction

Creating performant communication channels between different parts of digital processing units is more then ever the limiting factor to further digital processing development. The confluence of increased performance of computing chips, huge off-chip interconnectivity requirements, and increasingly high channel speeds together with a looming issue of thermal and power management make that current galvanic links are without any doubt under high strain. It is indeed not unusual for the multigigabyte galvanic board links to require compensation on the high-frequency absorption for as much as 30 to 45 dB of attenuation,[1] while crosstalk, dispersion, and timing issues become more severe with each newly introduced complementary metal oxide semiconductor (CMOS) technology node.

Optical interconnects have therefore been often cited as a possible route to alleviate the current technology conundrum. Indeed, the use of photons for communication has some clear advantages from a physical point of view compared to its galvanic contenders[2] and is already now the default interconnect choice for link lengths above a few hundred meters. At a shorter distance, optics have also demonstrated their ability to enable high data transmission at a very high bitrate per channel and massive parallelism with single dependent electromagnetic interference (EMI).[3] This has lead to an enormous and successful development in optoelectronic devices and hybridization technologies. Yet, optical interconnect at such length scales have thus far failed to become a mainstream commercial reality mostly because of cost, uncertainties in reliability, and unsatisfying packaging solutions.

The incumbent technology, the printed circuit board (PCBs), has moreover relentlessly continued to become a cheaper and more mature technology, aided by the development in signal processing chips and new packaging technologies. Nevertheless, it is unclear with this technology where the future performance increases can be found without trading off too much in cost and complexity.

Its optical contender, the use of optical waveguides within PCB's, is still considered a serious challenger provided that the technology can be proved to be reliable and scalable for mass deployment. As a result of extensive development in materials and waveguide fabrication techniques,[4] it is now possible to fabricate large waveguide integrated printed circuit boards with waveguide losses below $0.1\,\mathrm{dB/cm}$. Recently, increasing emphasis in PCB-integrated waveguide technology is placed on the use of multilayer optical structures where each optical layer contains arrays of multimode waveguides and other passive optical elements. This is driven by the availability of two-dimensional (2D) arrays of optoelectronic elements such as VCSEL- and photodetector arrays. The use of multilayer structures additionally allows one to considerably increase the integration density and to simplify the routing schemes.

One of the greatest remaining challenges, however, is to efficiently couple light in and out of the waveguides to various packaged optoelectronic devices or fiber connectors and to be able to do that in a standard way. This requires a new breed of micromechanical and microoptical systems to align and guide the light from the optoelectronics to the waveguides on the board. The requirements are illustrated in Fig. 5.1. In our labs at the Vrije Universiteit Brussel, we are therefore focusing on the continuous development of a rapid prototyping technology for microoptical interconnect modules, which we call deep proton writing (DPW). The special feature of this prototyping technology is that it is compatible with commercial low-cost

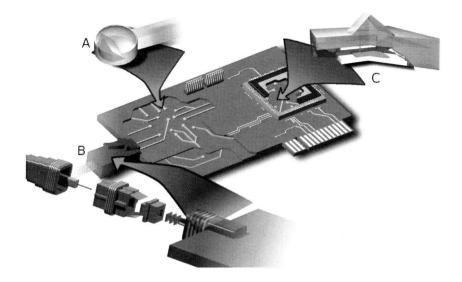

Figure 5.1 Examples of microoptical components for board-level photonic interconnects.

mass replication techniques, such as microinjection molding and hot embossing. With this DPW technology, we aim at overcoming most of the remaining microoptical hurdles to massively introduce photonic interconnects in digital systems.

As illustrated in Fig. 5.1 we currently focus on the prototyping of the following components: (A) out-of-plane couplers for optical waveguides embedded in PCB, (B) peripheral fiber ribbons and two dimensional singlemode and multimode fiber connectors for high-speed parallel optical connections, (C) intra-MCM level optical interconnections via free-space optical modules.

In what follows, we briefly review different microoptical fabrication techniques. We then introduce the DPW technology and highlight it main characteristics. Finally, we explain how we have applied this technology to prototype different microoptical components for board-level interconnects.

5.2 Fabrication Techniques for Microoptical and Micromechanical Systems

5.2.1 Microstructuring in a plane

By far, the most established technology for the fabrication of small structures is photolithography. In this process, a layer of photosensitive material is generally spin coated onto a substrate, which is subsequently patterned by exposing selected regions to light through a fine-structured mask by either projection lenses or direct contact methods. Since its invention, it has been a keydriver of the radical development in the microelectronics industry allowing to structure millions of transistors in parallel. To attain a higher resolution is in this field in high demand. An often cited route is the use of shorter wavelength exposure beams, leading to diminishing diffraction artefacts. This has led to the development of deep ultraviolet lithography ($\lambda_p = 183$ nm), which is now the standard technology to create the latest generation of integrated circuits and the widely investigated extreme UV lithography ($\lambda_p = 13.4$ nm).

Moreover, increasing attention is nowadays paid to lithographic techniques that rely on completely different physical processes. One approach is to use an exposure with electrically charged particles rather than photons. These methods rely on the extremely low associated wavelength of the charged particles; hence, they are in practical terms not limited by diffraction. The best-known lithographic technique of this type is electron beam lithography.[5] It uses electron beams with energies of 1–100 keV and can attain resolutions routinely below 100 nm. Other approaches include focused ion lithography[6] (usually with Gallium atoms), which exhibits, due to the higher particle mass, a smaller proximity effect than e-beam lithography, and electron projection lithography, which, in principle can overcome the very slow scanning speed of current e-beam and focused ion lithographic techniques.

5.2.2 Fabrication methods for optical board-level interconnects

All the above techniques are focused, however, on patterning a flat surface with very fine features in a plane. However, to create the necessary microoptics for the wire-replacing optics in board-level interconnects, one needs to fabricate deeper structures and thus rely on a more extensive set of techniques and even, in most cases, combine different technologies. Because solving the issue of waveguide coupling is crucial for optical board-level interconnect to become a standard, the manufacturing technologies have to be developed that can create or combine microlenses, titled (45 degree) micromirrors, micropillars, or flex waveguides.

Below we describe some of the most important candidate technologies to fabricate the required microoptics for board-level interconnects:

Photolithography. Because of its maturity thanks to the electronics industry, it is a highly preferred technology because it also allows structuring of all waveguides in parallel. There exist a large choice of photopatternable materials[7,8] that can either act as positive photoresist (exposed area will be removed) or negative photoresist (exposed areas will remain). After exposure and additional baking steps, the resist is then developed to create a waveguide structure. Often, the structure is then embedded into a new polymer to form the right index contrast between core and cladding for appropriate light guiding. Although the technique has shown to be able to create low-loss waveguides, it is not a straightforward method to create more complex structures, such as 45 degree tilted micromirrors and microlenses although one can use gray-mask photolithography, or used titled[9] or adapted[10] reactive ion etching (RIE) after photolithographic exposure.

Laser Direct Writing. In this process, an exposure with a focused laser spot is used to structure a photopatternable material. The technique is often applied to pattern the core of a waveguide by moving the substrate on a predetermined path under a fixed laser focus.[11,12] The scanning speed is usually a limiting factor for high-volume production. Out-of-plane coupling mirrors can be fabricated by using an inclined laser beam exposure. This can be combined with metalization steps, such as in Ref. 13, to create coupling structures.

Laser Photoablation. This is a method whereby a sample is exposed to such intense light pulses that some of the material at the surface is being spontaneously evaporated.[14,15] This manufacturing technique is very flexible because it can even be applied in the last steps of the board manufacturing process. In fact, laser ablation is already used in conventional board manufacturing for drilling microvias and trimming of resistors. It is also possible to create angled 45 degree micromirrors for out-of-plane coupling by tilting the incoming beam.[16]

X-Ray Lithography or LIGA. In this technique, a polymer substrate is exposed by a collimated beam of high-energy x-rays that penetrate deep into the substrate with negligible diffraction.[17] This highly involved technology allows millimeter-deep structures with high optical quality and can be adopted to make complex microoptical structures. To create deep structures, this technique requires, in fact, three steps: x-ray lithography, electroforming and molding. Hence, it is usually referred to as *LIGA* after the German acronym of lithografie, galvanoformung und abformung. The required equipment and resources to fabricate components with this method is high; therefore, it can only be applied to make master components which should in a later stage be replicated.

Mechanical Machining. Even mechanical machining can be used to create useful optical surfaces. For example, V-shaped 90 degree diamond blading has been used to create 45 degree micromirrors with excellent surface quality. However, the cut is generally much larger than the waveguide dimensions and, thus, the out-of-plane elements cannot placed on arbitrary locations.[18]

Drop-on-demand Printing. In this technology, small droplets of a photocurable polymer are deposited on substrates to create microlenses. The amount of polymer determines the focal number of the microlenses.[19] This technology has been successfully applied to create microlenses on top of boards and packages at various stages of the optical interconnect assembly. A good example is the OptoBump connection between surface mounted ball grid array (BGA) packages and the waveguides on the board, as described in Ref. 20.

Stereolithography. This prototyping technology is a novel approach to solidify selected regions from liquid photopolymers, layer by layer, using a scanning laser.[21] As a result, one can create prototypes of large and complex three-dimensional structures. Over the last few years, the writing speed of the technique has furthermore dramatically increased, bringing the technology into the realm of *rapid manufacturing*. However, stereolithography does not yet allow to create very high-quality optical surfaces in a direct way.

Some of the above-mentioned prototyping technologies (such as photolithography, x-ray lithography and mechanical machining) are readily compatible with replication methods for high volume-production. In these methods, one first has to create a suitable mold that can either be directly fabricated or can be fabricated by first creating a master component, which is then electroplated to create a mold. For the fabrication of this mold or the master, one can generally rely on the above manufacturing techniques.

Embossing. A first widely used replication method is hot embossing. Here, a mold is used to stamp a soft polymer that was raised just above its glass temperature. This method has been successfully applied to create optical waveguides in polymer layers.[22] Alternatively, on can also use ultraviolet (UV) embossing, where UV light exposure is used to polymerize a liquid polymer. The latter procedure requires thus one side of the mold structure to be UV-transparent.

Injection Molding. Another method that allows for mass-production of optical components is injection molding.[23,24] In this replication technique, a thermoplastic polymer is injected at high pressure into a mold. Although usually applied to manufacturing larger components, the method can also be applied to microsystems that need optical surface quality.

Furthermore, one can find a collection of inexpensive soft lithography replication techniques.[25] Here, the "soft" relates to the use of elastomers or organic materials for stamps or molds. It includes, among others, transferring lithographically created patterns in poly(dimethylsiloxane) (PDMS),[25] low-cost elastomeric vacuum casting with silicone molds[26] and rubber stamping, which can structure even highly curved surfaces.[27]

5.3 Deep Proton Writing as a Microoptical Prototyping Technology

At our labs we are optimizing yet another dedicated technology, which we call deep proton writing (DPW). Its concept finds its origin in the LIGA technology,[17] but differs on two important aspects. First, it is based on the use of protons rather than electromagnetic x-ray irradiation to shape polymer samples. Second, the DPW technology is using a direct write methodology as opposed to the projection lithography used in the LIGA process and thus eliminates the expensive mask creation process for each new design. Furthermore, in order to get the required thick masks, the LIGA has to be generally repeated with gradually higher energies resulting at each round in a thicker mask.

Both differences indicate that the DPW process requires less infrastructural demands and has the potential of being a more flexible technology for rapid prototyping while still attaining optical quality surfaces.

The basic concept of the deep proton writing process is based on the fact that irradiating swift protons onto a poly methyl methacrylate (PMMA) sample featuring long linear polymer chains (i.e., opposite of crosslinked) of high molecular mass will rupture the long chains. As a consequence, the molar mass of the irradiated material will be reduced and free radicals will be created in the polymer, resulting in material properties that are very different from those of unexposed material.

Two different chemical steps were developed that can be applied to the proton bombarded areas. The first consists of etching the exposed area with a specific developer to produce microholes, micromirrors and micromechanical structures. The second process involves the in-diffusion of a MMA monomer to locally swell the irradiated zones. This will result in microspherical and microcylindrical lens surfaces (see Fig. 5.2). Both processes can be applied to the same sample after a single irradiation session because the dose required for the etching or swelling is very different.

It is obvious that with a total cycle time of about a day per component and the required acceleration facilities, the DPW cannot be regarded as a mass fabrication technology as such. However, one of its assets is that it can be made compatible

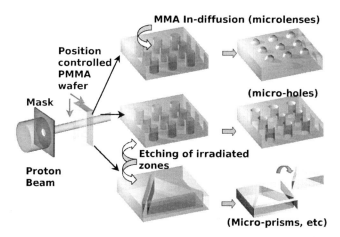

Figure 5.2 Basic fabrication processes of the deep lithography with protons. After a patterned irradiation we can either apply a binary chemical etching fluid to remove the irradiated regions or we can in-diffuse a monomer vapor to create microlenses through a swelling process.

with replication techniques. Indeed, once the master component has been prototyped with DPW, a metal mold can be generated from the master by applying electroplating. After removal of the master, this metal mold can be used as a shim in a final microinjection molding or hot-embossing step.

5.3.1 Ion interactions

The kernel process of DPW is the exposure of selected regions of the sample to accelerated protons. In comparison to photolithographic techniques where the exposure is governed by electromagnetic radiation, the energy transfer of protons with their target material is fundamentally different. For the UV-light lithography, the absorption of the beam is given by the well-known Lambert-Beer law, which states that the fractional absorption is constant along the penetrating axis. This will result in an exponential decay of absorbed light quanta along their path. With x-ray lithography, the released energy at the absorption of a photon is higher than the binding energy. Hence, such an event can create, apart from inner shell transition, secondary free electrons through the photoelectric effect. These electrons with high kinetic energy are capable of locally breaking the long polymer chains.

In contrast, the charged particles in an ion irradiation, or more specifically a proton irradiation, will only travel to an energy-dependent depth. Indeed, as the ions penetrate the substrate they gradually transfer energy to the host material mostly by interaction with the bonded electrons of the target material. This electric stopping power and associated range of ions in solids has been a domain of vivid research since the discovery of energetic particle emission from radioactive materials. In 1913, Bohr established a model, based on classical mechanics, to describe ion stopping in matter.[28,29] This was later refined to a quantum-mechanical approach in 1930 by Bethe.[30] Since then, various scientists have contributed to the subject. Extended reviews can be found in Refs. 29 and 31, 32.

It turns out that the stopping power (the energy transfer per unit of penetration depth, dE/dx) is small at the early part of the ion trajectory. This energy transfer will gradually slow the swift ions down while their interaction density with the PMMA molecules is increasing. This will result in a maximum energy transfer and absorbed dose when the impinging ions have velocities equal to those of the electrons in the amorphous host material. Below this energy level, the ions energy transfer will decrease immediately, and after a further penetration of some micrometers the ions will come to a full stop. The resulting absorbed dose profile for a proton irradiation can be found in Fig. 5.3. We can see that if we are using protons with an entrance energy of $8.3\,\text{MeV}$, then the range is $\sim750\,\mu\text{m}$. This allows us to cut through the standard substrate thickness of $500\,\mu\text{m}$. During the penetration in PMMA, the protons will cause electron excitation and ionization of the molecular chains and thereby induce stresses inside the molecular chains.[33] These stresses result in a degradation of the irradiated PMMA samples and make it possible to perform a selective chemical step. The reduced molar mass or molecular weight

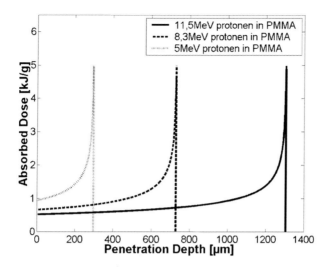

Figure 5.3 1D absorbed dose profile in PMMA after a proton irradiation of 1.2×10^6 particles/μm^2 for different entrance energies.

M_{irr} after the absorption of a dose D in joules per kilogram can be expressed in function of the initial molecular weight M_0 (=10^6 g/mol)

$$\frac{1}{M_{irr}} = \frac{1}{M_0} + \frac{GD}{100eN_a}, \qquad (5.1)$$

where e is the elementary charge unit and N_a is Avogadro's number. The factor G is the yield for main chain scissions per absorbed energy of 100 eV. Detailed studies of the molecular mass before and after irradiation via gel permeation chromatography (GPC) and a microthermal analyzer confirmed that the chain scission yield is equal to one.[34] The deposited dose D is related to the incoming proton fluence F, i.e., the number of impinged protons per unit surface, and the stopping power of the swift protons in PMMA by:.

$$D = \frac{F}{\rho_{PMMA}} \frac{dE}{dx}, \qquad (5.2)$$

where ρ_{PMMA} is the mass density of PMMA sample (1.19 g/cm^3).

Besides the gradual energy transfer of the protons to its amorphous host material, some spatial straggling of the impinging protons will occur while they are penetrating into the substrate. In contrast to the energy transfer, the straggling effects are primarily governed by multiple ion-ion (nuclear) interactions.[34] The straggling will result in a dose deposition slightly outside the targeted volume of the PMMA layer and thus will decrease the steepness of the optical surfaces of the fabricated microstructures after the etching process. We have developed an algorithm that can predict the 3D dose profile after a proton irradiation that includes both stopping power and the ion-ion scattering. In Fig. 5.4, we plot the resulting two-dimensional

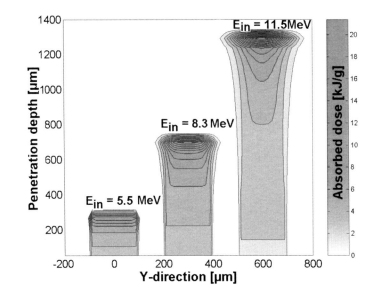

Figure 5.4 The 2D absorbed dose profiles are plotted for a proton irradiation in PMMA with fluences of 3.2×10^6 particles/μm^2 for different entrance energies.

(2D) absorbed dose profiles for a proton irradiation in PMMA through a 100 μm pinhole collimator with entrance energies of 5.5, 8.3 and 11.5 MeV. We can see that within the first 500 μm, the dose widening due to straggling is a few microns for protons with 8.3 MeV entrance energy. Recently, we have started to use proton beams of 16.5 MeV to reduce the effect even further.

Conventionally we would like to keep this straggling effect as small as possible to create deep optical surfaces with high aspect-ratio. However, in some cases we can use the straggling effect to our benefit. This is the case when fabricating conical-shaped fiber insertion holes to ease the fiber insertion. This will be explained in more detail in Section 5.4.2.

5.3.2 Irradiation process

The proton beam used for developing microoptical components is generated at the cyclotron facility of the VUB. The cyclotron (a CGR-MeV Model 560 Cyclotron) is capable of producing quasi monoenergetic ($\Delta E/E = 1\%$) proton beams in the energy range between 3 and 45 MeV. The accelerated protons are transferred via a set of focusing quadruple magnetic lenses and switching magnets to the DPW setup which is depicted in Fig. 5.5. In the figure, the protons enter the irradiation vacuum chamber from the right-hand side.

In order to avoid beam scattering and energy loss of the protons along their trajectory, we perform all the irradiations under vacuum (pressure below 10^{-4} mbar). Depending on the settings of the focusing magnet coils, the proton beam will enter the setup with a divergence of a few milliradians. From there, a set of collima-

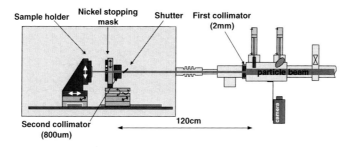

Figure 5.5 Schematic overview of the DPW irradiation setup.

tors will reduce the beam to a pencil-like uniform beam with selectable diameters. First, a fixed water-cooled aluminum collimator reduces the diameter of the entering beam from a few centimeters to 2 mm. The second collimator is a 10 cm long aluminum block with a $1 \times 1\,\text{mm}^2$ aperture hole. This design will restrict the divergence of the remaining proton beam and improves the beam pointing stability during irradiation. Part of this block is machined to include a mechanical shutter driven by a small electromotor which can block the beam within a 1 ms time span.

The final mask element is either 300 μm thick nickel stopping mask (fabricated with the LIGA technology) or a 500 μm thick tantalum mask (fabricated via electron discharge machining). The nickel mask contains different apertures with diameters ranging from 20 μm to 1 mm. By changing the position of the stopping mask, we can select different final proton beam sizes on the fly during one irradiation. The 300 μm mask is capable of stopping 8.3 MeV proton beams. With the 500 μm tantalum mask, we are able to stop 16.5 MeV proton beams. The mask of this thickness only provides circular holes with a 170 μm diameter but allows us to fabricate components with steeper walls.

We position the PMMA sample in a metal holder that is mounted on a biaxial translation stage with a 50 nm accuracy over a total travel range of 25.4 mm. Because the initial proton energy is high enough to pass through 500 μm thick PMMA samples, they induce a charge in the measurement probe located directly behind the target. It is therefore possible to monitor the proton current and the total amount of particles hitting the sample by integrating the proton current during the irradiation. The proton measurement is based on a precision-switched integrator trans-impedance amplifier and its concept is averse to any fluctuations in the proton current caused by instabilities of the cyclotron.[35] It provides us with a measurement resolution better than 250 fC, which is two orders of magnitude smaller than the minimum proton charges required for our purposes. For a point irradiation, the relation between the collected charge Q and the proton fluence F can be expressed as

$$Q = eF\frac{\pi d^2}{4}, \qquad (5.3)$$

where d is the aperture diameter.

Because PMMA is a positive resist, the exposed area can be developed in a subsequent chemical step. This means that we have to irradiate the entire contour of the designed component. To create the contour, the PMMA sample is quasi-continuously translated perpendicularly to the beam with steps Δx of 500 nm. At each step, the collected proton charge is measured at the measuring probe. If this value reaches the required proton charge, the microcontroller system will shift the sample to its new position one step away. The dose profile after a line diagram will not be uniform because it will be an overlap of different circular point irradiations. The peak proton fluence F_{max} in this case will be:

$$F_{max} = \frac{4Q_{step}}{e\pi d \Delta x}. \tag{5.4}$$

5.3.3 Etching process

As a next step, a selective etching solvent can be applied for the development of the irradiated regions. This allows for the fabrication of (2D arrays of) microholes, optically flat micromirrors, and microprisms with high optical quality, as well as alignment features and mechanical support structures.

For the etching process, we make use of a GG developer (diethylene glycol monobutyl ether 60%, morpholine 20%, 2-aminoethanol 5%, and DI water 15%) as the etching solvent. For standard components, etching lasts 1 h at an elevated temperature of 38°C. During the whole process the etching mixture is stirred by an ultrasonic stirrer. The etching is stopped by dipping the component in a stopping bath consisting of 20% water and 80% diethylene glycol monobutyl ether.

Following a study by Papanu et al.[36] the dissolution or etching rate can be expressed as

$$v_{etch} = \frac{c_0}{(M_{irr})^n} \exp\left(\frac{-E_a}{kT}\right), \tag{5.5}$$

where c_0 and n and the activation energy E_a are system-dependent parameters. Combining the above equation with Eq. (5.1) and (5.2), we get the following relation between the proton fluence and the etching rate:

$$v_{etch} = c_0 \left(\frac{1}{M_0} + \frac{GF/\rho_0 dE/dx}{100eN_a}\right) \exp\left(\frac{-E_a}{kT}\right). \tag{5.6}$$

Fitting the above relation to experimentally obtained etching rates of irradiated zones, results in the following values: $c_0 = 2.78 \times 10^{26}$, $E_a = 1.05 \, \text{eV}$ and $n = 2.9$.

To get an insight into the limits of the flatness of the created surfaces, we need to make a distinction between the direction along and perpendicular to the proton trajectory. Along the proton trajectory the most important parameters that are affecting the surface flatness are the divergence of the incoming beam and the straggling of protons along their path by multiple ion interactions as described in

Section 5.3.1. The flatness of surfaces in the direction perpendicular to the protons is limited by the precision on the movements of the translation stages (which have a closed-loop accuracy of 50 nm). The roughness of the obtained surfaces is mainly determined by the accuracy of the proton fluence measurement and the beam-pointing stability.

Recently, we have optimized the surface quality of the etched surface after irradiation with our smallest nickel aperture of 20 μm. With this proton beam size, an error in the position of the beam will be more pronounced and the signal-to-noise ratio of the measured proton current will be significantly smaller than with our standard 140 μm proton beam (as the proton current is 49 times smaller to obtain an equal proton fluence). Nevertheless, we succeeded in creating very high-quality surface profiles even with this aperture.[37] In Fig. 5.6, the resulting local surface rms roughness (R_q) and peak-to-valley flatness (R_t) are given. The graph shows the surface R_q and R_t as a function of the deposited particle charge per step. R_t was measured over a length of 500 μm along the proton trajectory and R_q was calculated by averaging several measurements over an area of 46 μm × 60 μm with a noncontact optical profilometer (WYKO NT2000). From this graph, we can conclude that the best results are obtained when we irradiate the sample with a collected proton charge of 8 pC per step of 50 nm corresponding to a peak proton fluence of 6.4×10^6 particles/μm^2 Fig. 5.7 shows the profiles of the created surfaces. On the bottom graph the profile is shown in the direction of the proton trajectory with a surface flatness R_t of 3.17 μm. The top graph is a zoomed in version of a profile perpendicular to the proton trajectory. The rms roughness R_q interval of 27.5 nm is indicated on the graph as well. These results are on par with the surface roughness and the flatness results we are obtaining with larger apertures at the same entrance energies.

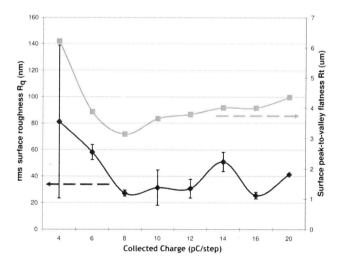

Figure 5.6 Surface roughness R_q and flatness R_t as a function of the deposited charge in each 0.5 μm step when using a 20 μm aperture.

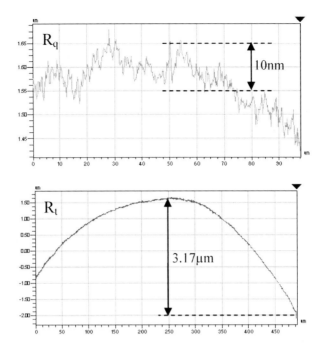

Figure 5.7 Detailed plots for R_q and R_t at the charge collection of $8\,\text{pC}$ per step of $500\,\text{nm}$.

5.3.4 Swelling process

The swelling process step will create spherical surfaces of the remaining irradiated zones that received a dose that was programmed to be too low for the etching. By exposing the sample to a controlled organic MMA vapor environment at an elevated temperature, the irradiated regions with a sufficiently low molecular weight will be receptive to an in-diffusion process of an organic monomer upon which their volume will expand. The combination of such a volume increase and surface tension, will transform the irradiated zones with a circular footprint, into microlenses with a nearly hemispherical shape.[38,39]

To swell the microlenses, we are currently using a diffusion reactor that is brought to an elevated temperature of 70°C. After the stabilization of the temperature (within 0.2°C), MMA monomer is injected into the chamber such that it will create a saturated MMA vapor. The monomer vapor will now diffuse into the irradiated zones to create hemispherically shaped microlenses. After 40 min, the sample is removed from the reactor and the in-diffused areas are stabilized by UV illumination during 1 hr while decreasing the temperature toward room temperature. The detailed physics behind the technological processing steps have been published before.[39–41] We demonstrated that DPW is a flexible technology to fabricate 2D matrices of spherical microlenses with different diameters between 120 and 200 μm and focal numbers ranging from 1 to 7 on the same PMMA substrate.

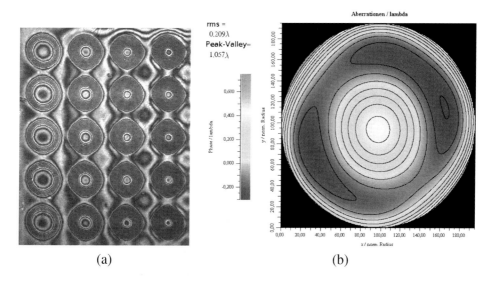

Figure 5.8 (a) Mach-Zehnder transmission interferogram of spherical microlenses with various lens sags (b) Contour plot of the wave aberrations (rms values 0.209λ, Peak-to-valley of 1.079λ, distance between lines 0.1λ) for a 200 μm diameter lens with lens sag 9.77 μm.

In Fig. 5.8(a) a planewave Mach-Zehnder interferogram is given of an array of spherical microlenses with a different lens sag (increasing from left to right).

We use an optical noncontact profilometer (WYKO NT2000) to measure the geometrical characteristics of the microlenses, such as lens sag and lens diameter. For the optical characteristics, we use a Mach-Zehnder interferometer constructed at the Erlangen Nürnberg University to measure the wave aberrations, such as the point spread function (PSF), the strehl ratio, and the modulation transfer function (MTF). A detailed comparative study of the obtained DPW microlens and some other fabrication technologies was recently published.[42] In Fig. 5.8(b) you can find the measured wave-aberration of a typical lens with a high focal number ($f/\# = 7$). The measured lens aberration of this microlens has a rms value of $\lambda/5$ which is above the Maréchal criterion to obtain diffraction limited lenses ($\phi_{rms} \leq \lambda/14$). We can conclude that although fabrication techniques exist that yield higher quality microlenses, the advantage of the DPW approach is that the lenses can be relatively fast prototyped and can be monolithically integrated in complex microsystems.

Within large arrays of DPW microlenses, we typically obtain a uniformity of the lens sag of 0.3%. However, when the microlenses are brought in close proximity we need to account for a small and deterministic change in the lens sag. This can be easily compensated by giving the peripheral lenses of the array a slightly lower dose. For example, when microlenses with a diameter of 140 μm are created on an array with a 250 μm pitch, the peripheral lenses need a 3% lower dose to obtain the same lens sags over the whole array.

Our sample-to-sample repeatability is however much lower. Therefore, we have started a new setup that will contain an in situ monitoring of the lens swelling behavior during the swelling process.[43] This new interferometric system will increase the repeatability as well as decrease the time and cost to calibrate the different microlens technologies. We calibrated our interferometric instrument and performed a benchmarking to compare our monitoring system with off-line characterization tools. The accuracy of the system is better than 5% and its repeatability better than 1.5%.

5.4 Microoptical Components for Board-Level Optical Interconnects

In what follows, we present the design and manufacturing of three case studies of prototyped components by the DPW technology. All components are related to solving the microoptical fabrication issues for board-level interconnects. We have also demonstrated the DPW microoptical prototyping technology for in other domains, such as biophotonics modules.[44]

5.4.1 Out-of-plane couplers for optical waveguides in standard FR4 printed circuit boards

As described in Section 5.1, one of the most critical elements in board-level optical interconnects is the manufacturing of the necessary microoptics for efficient light coupling. In order to connect the optical signals to surface-mounted components, a 90 degree turn is required, which can be accomplished either electrically or optically. The electrical turn is conventionally introduced by an extra-small flexible printed circuit board. A more common approach is to realize the turn optically by introducing 45° tilted micromirrors at the waveguide facets. If possible, these mirrors can be additionally be combined with microlenses to optimize the collimation and focus of the beams.

Various techniques exist today to fabricate such angled facets on the embedded waveguides (see also Section 5.2). Micro-machining, titled laser ablation,[15] or reactive ion etching of the waveguides[9] are among the techniques that are often pursued.

Our concept differs from most approaches because we propose a pluggable out-of-planing coupling component[37] instead of writing the micromirrors directly in the waveguides. We believe such components can be easily massreplicated from a master prototype. The replicated components can then readily be inserted into laser-ablated cavities on the printed circuit board to couple the light from the waveguide (see the top of Fig. 5.9). The micromirror can either use total internal reflection or a metal-coated mirror to bend the light.

For the fabrication of the structure with DPW, we have chosen to cut 500 μm thick PMMA plates by using a 8.3 MeV proton beam with a collimation aperture of 125 μm. This proton beam spot causes some rounding in the corners of the

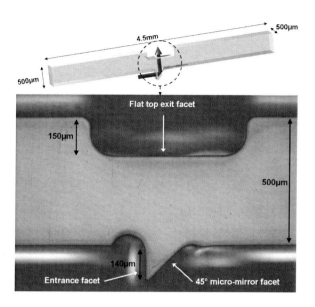

Figure 5.9 Designed (top) and fabricated (bottom) DPW out-of-plane coupler with integrated 45 degree micromirror of $140\,\mu m$ height.

component, as shown on the bottom of Fig. 5.9, but this does not affect the optical functionality in any way. The design is such that once the component is plugged in, the micromirrors will reach $140\,\mu m$ under the PCB surface. Nonsequential optical ray tracing predicts that when the component is interfaced with $50\,\mu m \times 50\,\mu m$ waveguides and with a numerical aperture (NA) of 0.3, the coupling efficiency will be 73% ($-1.37\,dB$). By using a metal coating over the micromirror, this efficiency can be further increased to 92% ($-0.36\,dB$). This loss is mostly related to the divergence of the beam and the necessary distance between waveguides and detector.

We have first measured the optical transmission of the out-of-plane couplers by using fibers to emulate the waveguide and detector. Here, the input fiber is a silica multimode fiber (MMF) with a core diameter of $50\,\mu m$ and an NA of 0.2. The output fiber is a silica MMF with a $100\,\mu m$ core and a 0.29 NA. With this setup we measure a coupling loss of 83% ($-0.77\,dB$). The reference measurement consisted of in-line butt coupling of both fibers.

Next, we have plugged the component into a PCB with integrated multimode optical waveguides. The Truemode Backplane™ polymer waveguides have a cross-section of $50\,\mu m \times 50\,\mu m$, an NA of 0.3 and a pitch of $250\,\mu m$. A detailed description of the waveguide fabrication and characterization can be found in Ref. 45. The microcavity that accommodated the DPW out-of-plane coupler was created using laser ablation. For the fabrication of the microcavity, the excimer laser for photoablation is tilted by 6 degrees such that a vertical endfacet of the waveguides can be created. The DPW out-of-plane coupler is inserted manually into the cavity by means of a tweezer. Figure 5.10 shows a top view of the DPW coupler inserted in the cavity and the light spot deflected by micromirror. Experimental characterization of the transmission resulted in an efficiency of 27%. This result now includes the loss of the buttcoupling into the PCB waveguide, a 5 cm transmission, coupling toward the out-of-plane coupler, and from the coupler to the detector. From Ref. 45, we know that the PCB-integrated waveguides exhibit a 0.13 dB/cm loss and an average coupling loss of 1.77 dB. Hence, the mirror loss can be estimated to be 3.27 dB (corresponding to a transmission efficiency of 47%).

One advantage of the discrete components presented in this work, is that the concept is easily extensible to boards with multilayer waveguide structures. A preliminary prototype of a two-layer out-of-plane coupler with monolithically integrated cylindrical microlenses at the input facet is shown in Fig. 5.11. The micromirror of this component is 300 µm high and the pitch between the upper and lower channel is 175 µm. Due to the increased optical path length in comparison to the single-layer out-of-plane coupler and to avoid crosstalk between the upper and

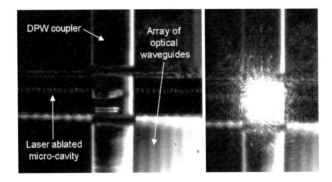

Figure 5.10 Top view of the DPW out-of-plane coupler inserted into a microcavity on a PCB with optical waveguides. The image on the right shows the light spot coupled out-of-plane when the laser is turned on.

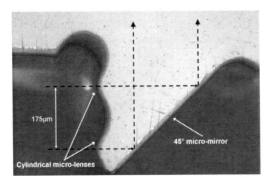

Figure 5.11 First prototype of a DPW multilayer out-of-plane coupler with monolithically integrated cylindrical microlenses at the input facet. The light paths for each layer are indicated by the dashed arrows. The total height of the micromirror is 300 µm.

the lower channel, we decided to monolithically integrate cylindrical microlenses to ensure collimation (in one direction) of the beam emitted by each waveguide layer. Although the design is not yet optimized, first measurements have shown coupling losses of 1.63 and 1.24 dB for, respectively, the upper and lower channel when using it in a similar fiber-to-fiber coupling scheme as the one described above. The crosstalk was smaller than −30 dB.

5.4.2 Two-dimensional single-mode fiber array couplers

High-precision two-dimensional fiber alignment modules can offer large benefits for high-density photonic interconnects at the board-to-board and chip-to-chip level, where parallel light signals have to be transferred between integrated dense 2D emitter and detector arrays. Even for the telecommunication infrastructure, the availability of highly accurate, low-cost, field-installable two-dimensional fiber couplers would boost the further integration of fiber-optics in future fiber-to-the-home networks.[46]

Although there are plenty of 1D connector components on the market, the existing connector fabrication techniques such as photoablation,[47] silicon micro-machining[48] and deep lithography with x-rays[49] have not yet been shown to be able to create 2D connectors for single mode fibres that can be massfabricated with the required high accuracies at low cost.

Using the DPW technology, we were able to fabricate such a connector, which features conical-shaped microholes to ease the insertion of single-mode fibers from the backside of the fiber holder into sub-micron precision holes. The conical shape of the holes has been obtained by taking advantage of the ion-ion scattering effect of protons. This effect becomes pronounced when the proton fluence is chosen in excess of 10^7 particles/μm^2. The optimized microholes for fiberinsertion feature a front-side diameter of 134 μm, an inner diameter to fit a fiber with cladding specification of 125 ± 0.7 μm, and a back side diameter of 165 μm. A schematic view of the obtained microholes can be viewed in Fig. 5.12.

The connector plate features a 4×8 array of holes with a pitch of 250 μm. With the DPW technology we are capable of integrating alignment holes to the design

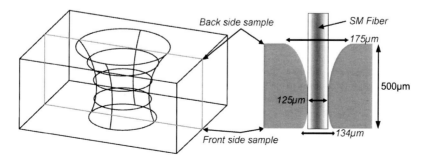

Figure 5.12 Schematic figure of the conical profile of the microholes for easy fiber insertion.

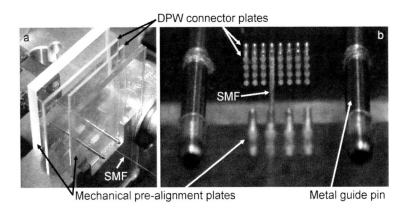

Figure 5.13 The assembled 2D SMF connector component in the setup for insertion loss measurements: Two prealignment plates for mechanical stability and two DPW connector plates with microholes are aligned using metal MT guide pins.

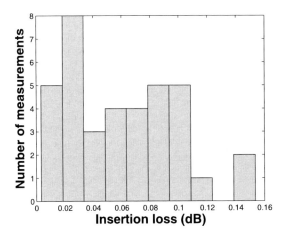

Figure 5.14 A histogram of the resulting coupling efficiencies of the 2D fiber connector.

via one single irradiation step. The fabricated alignment holes are three larger holes with a diameter of 700 µm, compatible with standard mechanical transferable (MT) ferrule pins. By measuring the rim of the microholes with an optical profilometer, we find a standard deviation on the hole positions below 0.8 µm (limited by the measurement apparatus). Although the conical holes allow for a tight control of the lateral position of the fibers, they still leave some angular freedom to the fibers. Therefore, we have opted to include in the DPW connector an additional mechanical prealignment plate, separated by ∼5 mm from the DPW plate, as illustrated in Fig. 5.13.

With this component, we measured an average in-line coupling loss of 0.06 dB in the telecom C and L bands. The maximum coupling loss over 37 link experiments was 0.15 dB.[50] The histogram of the connector losses over the whole array is shown in Fig. 5.14.

5.4.3 Intra-MCM interconnect module

Although the out-of-plane-coupler can be fabricated with only a single irradiation and etching step, other more complex interconnection components require a more intricate combination of high-quality optical surfaces, spherical microlenses, and micromechanical alignment features. An example of such a component is a prism-based intra-MCM interconnect module.

Density arguments can require free-space microoptical interconnections even at the intra-MCM interconnect level. Our approach here is based on a microprism reflector that transports and routes data-carrying light beams from a microemitter to a microdetector array, hence bridging intrachip interconnection throws ranging between a few tens of millimeters to only a few millimeters. On its way from source to detector, each of the multiple beams are collimated, bended at the 45 degree angled facets, and refocused by microlenses. We have shown that this type of interconnect module has the potential to provide the highly desirable massive parallel high-speed interconnects needed for future generation intra-MCM level interconnections.[41] We are currently working toward prototyping the necessary microoptics for a massively parallel intrachip interconnect with a channel density of > 4000 channels/cm^2. Furthermore, it is our intention to make this prototype replicatable such that it can serve as a low-cost, chip-compatible, plug-and-play optical interconnect solution.[51] Figure 5.15 shows the assembled interconnect module mounted above a dense optoelectronic chip.

In this work, our concern is not only to fabricate the module, but we are especially interested in solving how a component can be reliably attached above a dense optoelectronic chip.[52] We have therefore developed a solution consisting of a spacer plate accurately surrounding the optoelectronic chip. The spacer plate and optical interconnection module are attached to each other via precise microspheres.

Figure 5.15 The free-space intra-MCM interconnection module mounted on top of a dense optoelectronic chip.

Furthermore, We have developed a tolerance-aware design flow based on an extensive sensitivity analysis and Monte Carlo set of fabrication and misalignment tolerances. In this study, the complete packaging stack was taken into account by setting up a polygon-driven 3D model of the optomechanical system in Matlab^{TM}. These algorithms were then interfaced with the Zemax^{TM} optical simulation tool for an all-optical ray-trace analysis. The details of the design and a Monte Carlo analysis on the feasibility of the component can be found in Ref. 52.

5.5 Conclusions

One solution for the current pressure on galvanic interconnect is to use photonic links. However, for such a technique to become widely adopted, it is of paramount importance, which a low-cost, mass-producible, standard photonic interconnect solution can be fabricated. An approach for the optical interconnects that recently received much of attention, is the use of optical waveguides within PCBs because it can become a logic extension to the standard printing wiring boards which are ubiquitous.

Developments of polymer materials and waveguide manufacturing techniques now promise the integration of low-loss waveguides within and on top of these standard FR4 printing wiring boards with losses that are now regularly $< 0.1\,\mathrm{dB/cm}$. One important remaining issue is, however, the coupling to and from those waveguides. Although many manufacturing techniques exist, none of them yield satisfactory results in terms of manufacturability and cost.

In this work we have described a microoptical prototyping technology, called deep proton writing (DPW), that allows for prototyping of micromechanical and microoptical systems, which can be then replicated into many copies. We can conclude that the DPW is a viable technology for constructing prototypes of deep microoptical structures with high aspect ratio. The basic process of DPW consists of an irradiation of selected areas with swift protons. Two chemical steps can then be combined to either develop the high proton fluence areas by a selective developer or to swell circular footprints of the lower fluence into hemispherical microlenses. Through a series of optimizations for the irradiation setup, etching and swelling procedure, we are capable of making high-grade prototype microoptical elements. Although a typical cycle time of about one day does not allow for mass production, we have shown that the DPW technology can be used as a rapid prototyping technology, capable of fabricating master components. These can be electroplated to form molds for mass replication techniques, such as hot embossing and injection molding.

This was illustrated by showing a selection of DPW components that can be used to overcome the limitation of galvanic interconnects at the board-level. A first component is an insertable out-of-plane component for coupling between surface-mounted optoelectronics and the waveguides. The design includes a 45 degree micromirror. The second component consisted of a two-dimensional single-mode fiber array connector, with coupling losses below 0.15 dB. A third component features a free-space intra-MCM interconnection.

Acknowledgments

This work is financed by the EC 6th FP Network of Excellence on Micro-Optics NEMO, the FWO, IWT-GBOU, VUB-GOA, the DWTC IAP Photon network and the OZR of the Vrije Universiteit Brussel. C. Debaes and H. Ottevaere are supported by a postdoctoral fellowship from the FWO Vlaanderen.

References

1. Adamiecki, A., Duelk, M., and Sinsky, J., "25 Gb/s electrical duobinary transmission over fr-4 backplanes," *Electr. Let.*, **41**, 826–827, 2005.

2. Miller, D.A.B., "Rationale and challenges for optical interconnects to electronic chips," *Proc. of the IEEE*, **88**, 728–749, 2000.

3. Haney, M.W., Thienpont, H., and Yoshimura, T., "Introduction to the issue on optical interconnects," *J. of Sel. Top. of Quant. Elect.*, **9**, 347–349, 2003, and other papers in the volume.

4. Uhlig, S., and Robertsson, M., "Limitations to and solutions for optical loss in optical backplanes," *J. of Lightw. Tech.*, **24**, 1710–1724, 2006.

5. Vieu, C., Carcenac, A.P.F., Chen, Y., Mejias, M., Lebib, A., Manin-Ferlazzo, L., Couraud, L., and Launois, H., "Electron beam lithography: resolution limits and applications," *Applied Surface Science*, **164**, 111–117, 2000.

6. Melngailis, J., "Focused ion-beam lithography," *Nuclear Instruments and Methods in Physics Research*, **80**, 1271–1280, 1993.

7. Ma, H., Jen, A.K.-Y., and Dalton, L.R., "Polymer-based optical waveguides: Materials, processing, and devices," *Advanced Materials*, **14**, 1339–1365, 2002.

8. Uhlig, S., *ORMOCER Materials Characterization, LAP- Micro-Processing Applied to Optical Interconnects and High-Frequency Packaging*. PhD thesis (2006).

9. Liu, Y., Lin, L., Choi, C., Bihari, B., and Chen, R.T., "Optoelectronic integration of polymer waveguide array and metal-semiconductor-metal photodetector through micromirror couplers," *Phot. Techn. Let.*, **13**, 326–328, 2001.

10. Kagami, M., Kawasaki, A., and Ito, H., "A polymer optical waveguide with out-of-plane branching mirrors for surface-normal optical interconnections," *J. of Lightw. Tech*, **9**, 1949–1955, 2001.

11. Lunitz, B., Guttmann, J., Huber, H.-P., Moisel, J., and Rode, M., "Experimental demonstration of 2.5 gbit/s transmission with 1 m polymer optical backplane," *Electr. Let.*, **37**, 1079, 2001.

12. Tooley, F., Suyal, N., Bresson, F., Fritze, A., Gourlay, J., Walker, A., and Emmery, M., "Optically written polymers used as optical interconnects and for hybridisation," *Optical Materials*, **17**, 235–241, 2001.

13. McCarthy, A., Suyal, H., and Walker, A., "Fabrication and characterisation of direct laser-written multimode polymer waveguides with out-of-plane turning mirrors," in *CLEO/Europe*, 477–478 (2005).

14. Mihailov, S., and Lazare, S., "Fabrication of refractive microlens arrays by excimer laser ablation of amorphous teflon," *Applied Optics*, **32**(31), 6211–6218, 1993.

15. Van Steenberge, G., Geerinck, P., Van Put, S., Van Koetsem, J., Ottevaere, H., Morlion, D., Thienpont, H., and Van Daele, P., "Mt-compatible laser-ablated interconnections for optical printed circuit boards," *J. of Lighw. Techn.*, **22**, 2083–2090, 2004.

16. Hendrickx, N., Erps, J.V., and Thienpont, P.V.D.H., "Laser ablated micromirrors for printed circuit board integrated optical interconnections," *Phot. Tech. Lett.*, **19**, 822–824, 2007.

17. Becker, E.W., Erfield, W., Hagmann, P., Maner, A., and Munchmeyer, D., "Fabrication of microstructures with high aspect ratios and great structural heights by synchrotron radiation lithography, galvanoforming, and plastic moulding (liga process)," *J. Micromech. Microeng.*, **4**, 35–56, 1986.

18. Yoshimura, R., Hikita, M., Usui, M., Tomaru, S., and Imamura, S., "Polymeric optical waveguide with 45° mirrors formed with a 90° V-shaped diamond blade," *Electr. Let.*, **33**, 1311–1312, 1997.

19. Nallani, A.K., Ting, C., Hayes, D., Woo-Seong, C., and Jeong-Bong, L., "A method for improved vcsel packaging using mems and ink-jet technologies," *J. of Lightw. Techn.*, **24**, 1504–1512, 2006.

20. Ishii, Y., Tanaka, N., Sakamoto, T., and Takahara, H., "Fully smt-compatible optical-i/o package with microlens array interface," *J. of Lightw. Techn.*, **21**, 275–280, 2003.

21. Zhang, X., Jiang, X., and Sun, C., "Micro-stereolithography of polymeric and ceramic microstructures," *Sensors and Actuators*, 149–156, 1999.

22. Krabe, D., and Scheel, W., "Optical interconnects by hot embossing for module and pcb technology: the eocb approach," in *Electronic Components and Technology Conference (ECTC99)*, 1164–1166 (1999).

23. Bleich, W., "Injection molded polymer optics in the 21st-century," *Proc. of SPIE*, **5685**, 56850J, 2005.

24. Heckele, M., and Schomburg, W., "Review on micro molding of thermoplastic polymers," *J. Micromech. Microeng.*, **14**, R1–R14, 2004.

25. Love, J.C., Anderson, J.R., and Whitesides, G.M., "Fabrication of three-dimensional microfluidic systems by soft lithography," *MRS Bulletin*, 523–528, 2001.

26. Desmet, L., Overmeire, S.V., Erps, J.V., Ottervaere, H., Debaes, C., and Thienpont, H., "Fabrication of an array of concave refractive microlenses using elastomeric inverse moulding and vacuum casting," *J. Micromech. Microeng.*, **17**, 81–88, 2007.

27. Rogers, J.A., "Rubber stamping for plastic electronics and fiber optics," *MRS Bulletin*, 530–534, 2001.

28. Bohr, N., *Philosophical Magazine*, **25**(10), 1913.

29. Kumakhov, M., and Komarov, F., *Energy Loss and Ion Ranges in Solids*, Gordon and Breach Science Publishers, (1981).

30. Bethe, H.A., *Ann. Physik*, **5**(325), 1930.

31. Fano, U., *Annual Review of Nuclear Science*, **13**(1), 1963.

32. Ziegler, J.F., *Handbook of Stopping Cross Sections for Energetic Ions in All Elements*, vol. 5, Pergamon Press (1980).

33. Lee, E.H., *Radiation Physics and Chemistry*, **55**, 293–305, 1999.

34. Volckaerts, B., *Deep Lithography with Ions*. PhD thesis, Vrije Universiteit Brussel (2004).

35. Vynck, P., Volckaerts, B., Vervaeke, M., Ottevaere, H., Tuteleers, P., Cosentino, L., Finocchiaro, P., Pappalardo, A., Hermanne, A., and Thienpont, H., "Beam monitoring enhances deep proton lithography: towards high-quality micro-optical components," in *Proc. of the IEEE/LEOS Benelux Chapter*, 298–301, 2002.

36. Papanu, J., Soane, D., Bell, A., and Hess, D., "Transport models for swelling and dissolution of thin polymer films," *J. Applied Polymer Science*, **38**(5), 859–885, 1989.

37. Erps, J.V., Bogaert, L., Volckaerts, B., Debaes, C., and Thienpont, H., "Prototyping micro-optical components with integrated out-of-plane coupling structures using deep lithography with protons," *Proc. of SPIE*, **6185**, 33–46, 2006.

38. Kufner, M., Kufner, S., Chavel, P., and Frank, M., "Monolithic integration of microlens arrays and fiber holder arrays in poly(methylmethacrylate) with fiber self-centering," *Opt. Lett.*, **20**(3), 276–278, 1995.

39. Ottevaere, H., Volckaerts, B., Vervaeke, M., Vynck, P., and Thienpont, H., "Plastic microlens arrays by deep lithography with protons: Fabrication and characterization," *Proc. of the 9th Micro Optics Conference (MOC'03)*, 110–113, 2003.

40. Volckaerts, B., Ottevaere, H., Vynck, P., Debaes, C., Tuteleers, P., Hermanne, A., Veretennicoff, I., and Thienpont, H., "Deep lithography with protons: a generic fabrication technology for refractive micro-optical components and modules," *Asian Journal of Physics*, **10**(2), 195–214, 2001.

41. Debaes, C., Vervaeke, M., Baukens, V., Ottevaere, H., Vynck, P., Tuteleers, P., Volckaerts, B., Meeus, W., Brunfaut, M., Van Campenhout, J., Hermanne, A., and Thienpont, H., "Low-cost micro-optical modules for mcm level optical interconnections," *IEEE Journal of Selected Topics in Quantum Electronics, Special Issue On Optical Interconnects*, **9**, 518–530, 2003.

42. Ottevaere, H., Cox, R., Herzig, H.P., Miyashita, T., Naessens, K., Taghizadeh, M., Völkel, R., Woo, H.J., and Thienpont, H., "Comparing glass and plastic refractive microlenses fabricated with different technologies," *J. Opt. A: Pure Appl. Opt.*, **8**(7), S407–S429, 2006.

43. Gomez, V., Ottevaere, H., Volckaerts, B., and Thienpont, H., "Real-time in situ sag characterization of microlenses fabricated with deep lithography with protons," *Proc. of SPIE*, **5858**, 2005.

44. Overmeire, S.V., Ottevaere, H., Desmet, G., and Thienpont, H., "Miniaturized detection system for fluorescence and absorbance measurements in chromatographic applications," *J. of Sel. Top. on Quant. Elec.*

45. Steenberge, G.V., Hendrickx, N., Bosman, E., Erps, J.V., Thienpont, H., and Daele, P.V., "Laser ablation of parallel optical interconnect waveguides," *Phot. Techn. Let.*, **18**, 1106–1108, 2006.

46. Kim, J., Nuzman, C., Kumar, C., Lieuwen, D., Kraus, J., Weiss, A., Lichtenwalner, C.P., Papazian, A., Frahm, R., Basavanhally, N., Ramsey, D., Aksyuk, V., Pardo, F., Simon, M., Lifton, V., Chan, H., Haueis, M., Gasparyan, A., Shea, H., Arney, S., Bolle, C., Kolodner, P., Ryf, R., Neilson, D., and Gates, J., "1100×1100 port mems-based optical crossconnect with 4-db maximum loss," *Phot. Techn. Let.*, **15**, 1537–1539, 2003.

47. Proudley, G., Stace, C., and White, H., "Fabrication of two-dimensional fiber optic arrays for an optical crossbar switch," *Optical Engineering*, **33**(2), 627–635, 1994.

48. Suematsu, K., *et al.*, "Super low-loss, super high-density multi-fiber optical connectors," *Furukawa Review*, **23**, 53–58, 2003.

49. Dunkel, K., *et al.*, "Injection-moulded fiber ribbon connectors for parallel optical links fabricated by the liga technique," *J. Micromech. Microeng.*, **8**, 301–306, 1998.

50. Erps, J.V., Volckaerts, B., van Amerongen, H., Vynck, P., Krajewski, R., Debaes, C., Watté, J., Hermanne, A., and Thienpont, H., "High-precision 2d single mode fiber connectors fabricated through deep proton writing," *Phot. Techn. Let.*, **18**(10), 1164–1166, 2006.

51. Vervaeke, M., Desmet, L., Hermanne, A., and Thienpont, H., "Alignment features for micro-optical interconnect modules," in *Proceeding of the 2003 Symposium of the IEEE- LEOS Benelux Chapter*, 293–296, 2003.

52. Vervaeke, M., Debaes, C., Volckaerts, B., and Thienpont, H., "Opto-mechanical Monte Carlo tolerancing study of a packaged free-space intra-mcm optical interconnect system," *J. of Sel. Topics in Quant. Elec.*, **12**, 988–996, 2006.

Chapter 6
Intracavity Coherent Addition of Lasers

Vardit Eckhouse, Amiel A. Ishaaya, Liran Shimshi, Nir Davidson, and Asher A. Friesem
Weizmann Institute of Science, Israel

6.1 Introduction
6.2 Coherent Addition with Plane Parallel Intracavity Combiners
6.3 Coherent Addition of Single High-Order and Multimode Distributions
6.4 Improving the Beam Quality of Multimode Laser Resonators
6.5 Concluding Remarks
References

6.1 Introduction

The ability to obtain high output power with good beam quality from lasers has been, and still is, a major challenge, drawing considerable scientific attention. The two key parameters in a given size laser system, power and beam quality, are inversely related, whereby high output power results in low beam quality. Excellent output beam quality (i.e., low divergence) is obtained by operating a laser with the fundamental transverse TEM_{00} mode (transverse mode of a laser, with a Gaussian energy distribution). This is generally achieved by inserting an aperture of small diameter inside the resonator. Unfortunately, with such an aperture, the output power is reduced since because only a small volume of the gain medium is exploited. Increasing the diameter of the aperture, results in transverse multimode operation with inferior output beam quality but considerably higher output power. Heat dissipation problems and thermal effects, which are usually dominant in high-power laser systems, often degrade the beam quality even further. Thus, in an ordinary laser configuration of given size, there exists an inherent trade-off between the output power and the output beam quality, limiting the total optical brightness of the output beam.

The trade-off between output power and beam quality is not unique only to free-space conventional lasers. With semiconductor diode lasers, where the dimensions are typically small, the single-mode power from a single laser is rather low. Higher output powers are usually obtained by resorting to arrays of single lasers (diode bars and stacks). However, because the lasers are not coherent with each other the output beam quality is low, resulting in a combined output beam of low brightness. With the recent development of double-clad fiber lasers, relatively high CW (continuous wave) output powers with good beam quality are now readily achieved. The possibility to distribute the gain over very long single-mode waveguide fibers, with very good optical efficiency and low thermal distortions, already allow for single-mode output powers in the Kilowatt range. However, also with this type of lasers, single-mode output power is eventually limited by the small size of the core, whereby increasing the core size results in multimode operation with degraded output beam quality.

Various approaches have been investigated over the years to obtain lasers with high output powers and good beam quality. Among these are specialized laser resonator configurations, such as Talbot or Fourier transform resonators,[1,2] Vernier-Michelson resonators,[3-5] evanescent waves coupling,[6,7] the introduction of amplitude,[8,9] and phase[10] diffractive component elements into the resonator,[11,12] or insertion of interferometric combiners inside a laser resonator.[13,14] One of the most promising approaches for achieving good beam quality and high output power is to coherently add several low power and good beam quality laser distributions. In this approach several identical lasers, each with low power and excellent beam quality, are operated in a phase locked manner that enables their coherent addition. The output beam is a coherent superposition of all the individual beams; thus its power is the sum of all the powers of the individual beams, while the beam quality is as good as that of the individual beam. Because it is rather straightforward to obtain good beam quality with a low-power laser, it seems a very natural approach to try and coherently combine many such lasers. However, careful study of past investigations shows that phase locking of a large array of lasers is difficult in practice, and has been a constant challenge in the field of laser design.

In this chapter, we review and present our approach for achieving efficient intracavity coherent addition of several laser distributions with novel interferometric combiners. Section 6.2 describes our basic configuration for intra-cavity coherent addition, demonstrating coherent addition of 2, 4, 16 and 25 Gaussian laser distributions. It also includes an explanation on longitudinal mode considerations and upscaling limitations. Section 6.3 presents experimental results revealing how our approach can be exploited for coherent addition of several single high-order distributions and even multimode distributions. Section 6.4 describes our techniques for improving the output beam quality of multimode laser resonators, demonstrating increased output brightness. Finally, some concluding remarks are presented in Section 6.5.

6.2 Coherent Addition with Plane Parallel Intracavity Combiners

We begin with the basic configuration for coherent addition of two Gaussian laser distributions with an intracavity planar interferometric combiner, schematically presented in Fig. 6.1. The resonator is composed of a flat rear mirror, an output coupler which could be either flat or concave for stable laser operation, a common gain medium, two apertures with diameters suitable for independent fundamental TEM_{00} operation in each of the two channels, and a planar interferometric combiner. The use of interferometric combiners and common end mirrors is of great advantage. Together they alleviate the complexity of alignment and significantly improve the stability, thereby allowing for practical implementation in laser systems.

The interferometric combiner is comprised of a high-precision plane parallel plate, with specially designed coatings. For channels with equal gain, half of the front surface is coated with an anti-reflection (AR) layer, and the other half with a 50% beam splitter layer, while half of the rear surface is coated with a highly reflecting layer and the other half is coated with antireflection layer. In case of different gain in each channel, an appropriate beam splitter transmittivity should be chosen. The beam of the lower channel is directly incident on the beamsplitter coated-region, while the beam of the upper channel is transmitted through the AR coated region, reflected back from the rear surface, and then is incident on the beam splitter coated region; thus as to be collinear with the other beam. The thickness d of the interferometric combiner and its angle relative to the beams, are designed to match the distance between the two beams, so they optimally overlap and propagate collinearly after exiting the combiner. For an incident angle α, d is determined by the simple relation

$$d = x_0/2 \cos\alpha \tan[\arcsin(\sin\alpha/n)], \qquad (6.1)$$

where x_0 is the distance between the two beams and n is the refractive index of the combiner material.

In a simplified manner, the operation of our combined resonator can be explained as follows. If the two Gaussian beam distributions are incoherent (random

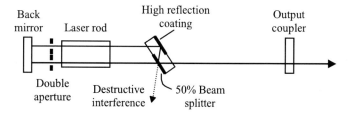

Figure 6.1 A configuration for intracavity coherent addition of two Gaussian beam distributions using a single interferometric combiner.

relative phase between the beams or different frequencies), then each beam will suffer a 50% loss passing through the interferometric combiner, thus, typically, no lasing will occur. On the other hand, if the beams add coherently, then the losses introduced by the interferometric combiner may be completely suppressed. Specifically, to the right of the combiner constructive interference occurs with little, if any, losses, while destructive interference occurs in the lower path from the combiner, as shown in Fig. 6.1. Indeed, the combined laser will tend to operate so that the losses are minimum, whereby the phases of the individual beams will be automatically matched (self-phase locking) such that coherent addition takes place. This of course can be achieved only for those longitudinal modes (frequencies) that are common in the two laser channels. Thus, care must be taken to imbalance the optical length of the two resonator channels in such a manner so as to obtain one or more common longitudinal modes.

Experiments performed with a pulsed Nd:YAG laser setup confirmed the ability to efficiently add coherently two Gaussian distributions.[13] The resonator was basically a 70 cm long planoconcave resonator, with a concave (R=3 m) output coupler of 40% reflectivity at 1064 nm and a high-reflective flat back mirror. The resonator included two apertures of 1.6 mm diameter each, positioned 2.4 mm apart (between centers), a high-quality thin-film polarizer, and a 3 mm thick interferometric combiner oriented at Brewster's angle. Experiments were performed both in free running and Q-switched operation. With this setup, two Gaussian channels, each with output power of 9.75 mJ per pulse when operated independently, were coherently added to obtain a combined output of 18 mJ, indicating 92% combining efficiency.[13]

Figure 6.2 shows the detected intensity distributions. Figures 6.2(a) and (b) show the near and far-field intensity distributions for one channel and Fig. 6.2(c) and (d) the near and far-field intensity distributions for the other channel, when operated independently. The calculated M^2 for these individual beams, was $M_x^2 = 1.15$ and $M_y^2 = 1.20$ for the first channel, and $M_x^2 = 1.14$ and $M_y^2 = 1.21$ for the second channel, indicating nearly pure Gaussian TEM_{00} distributions. The near and far field intensity distributions of the combined beam are shown in Fig. 6.2(e) and (f). The calculated M^2 values for the combined output beam were $M_x^2 = 1.12$ and $M_y^2 = 1.18$, indicating that the original, nearly Gaussian distribution was preserved. Slowly tilting the combiner at small angles, so as to slightly change the channel length difference, did not affect the output energy or its intensity distribution. This demonstrates the self-locking mechanism of the laser in our configuration, and its insensitivity to geometrical displacements of the combiner.

We obtained similar results when combining four Gaussian channels with a similar pulsed Nd:YAG configuration shown in Fig. 6.3. Here, two intracavity interferometric combiners, oriented along orthogonal axes, were used to coherently combine four Gaussian beam distributions. Figure 6.4 shows the output energy per pulse as a function of pump energy for coherent addition of four channels, and also

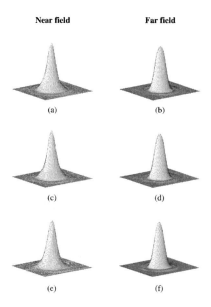

Figure 6.2 Experimental intensity distributions of the separate independent Gaussian beams, and the combined output beam obtained using the combiner: (a) and (b), near- and far-field intensity distributions of the first channel output beam; (c) and (d), near- and far-field intensity distributions of the second channel output beam; (e) and (f), near- and far-field intensity distributions of the combined channel output beam.

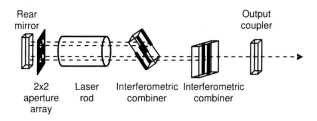

Figure 6.3 A laser resonator with two interferometric combiners resulting in phase locking and coherent addition of four laser distributions.

for a single channel. These results indicate that the combining efficiency slightly decreases with an increase in pump energy. The combining efficiency was more than 90%, and the combined output beam quality was as good as that of a single independent channel, for a pump energy of 7 J.[14]

We also considered the effect of thermal lensing on efficiency when coherently adding four laser distributions. The results are presented in Fig. 6.5, which shows the measured combining efficiency of four channels as a function of pulse repetition rate. As evident, as the pulse repetition increases i.e., increased thermal lensing, there is a degradation of the combining efficiency. We found how the thermal lensing effects can be suppressed by adding a proper compensating lens inside the laser cavity close to the laser rod.[15] Figure 6.5(b) shows the measured combining

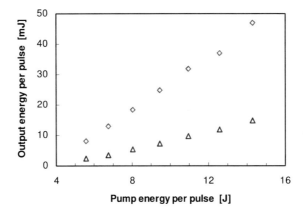

Figure 6.4 Measured output pulse energy as a function of the pump energy for a single Gaussian channel laser operation (triangles), and for intracavity coherent addition of four Gaussian channels (diamonds).

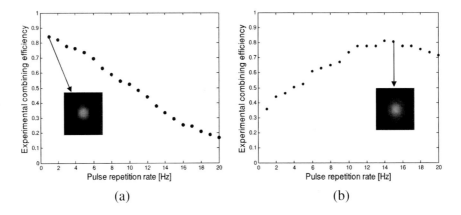

Figure 6.5 Measured combining efficiency as a function of the pulse repetition rate for coherent combining of four laser distributions. (a) without compensating lens and (b) with a compensating lens of focal length of 6 m. Insets: far-field distributions, indicating nearly Gaussian distributions for both cases.

efficiency for four channels as a function of pulse repetition rate with a compensating lens of focal length f = 6 meter. As evident, the combining efficiency is now, indeed, maximal at a much higher pulse repetition rate, where the deleterious effect of thermal lensing is now cancelled by the negative compensating lens.

As mentioned earlier, phase locking and coherent addition can occur only if all individual beam distributions have common frequencies (common longitudinal modes) within the gain bandwidth. To ensure that such common frequencies exist, the difference in the optical length between the individual resonator channels should be accurately controlled. When coherently adding two channels, the optical length difference ΔL need only be greater than a certain value, thus, it is easy to control for a Nd:YAG laser. But in general, with more resonator channels, each

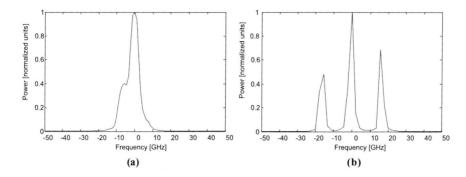

Figure 6.6 Experimental power spectrum measurements. (a) The power spectrum of the output beam for a single Gaussian channel operation (without interferometric combiners); (b) the power spectrum of the output beam for coherently added four-channel operation.

with a different optical length L, $L + \Delta L1$, $L + \Delta L2$, etc., the probability for having common frequencies within the gain bandwidth is drastically decreased. In our configurations, where the resonator channels end mirrors are common, the interferometric combiners introduce exact optical length differences, integer number of ΔL, (i.e ΔL, $2\Delta L$, $3\Delta L$, etc.), between the channels. For example, in our specific configuration for coherently adding four laser distributions, the resonator length of one channel is L, of two channels $L + \Delta L$, and of the remaining channel $L + 2\Delta L$. Thus, resorting to length differences of $n\Delta L$ ensures that common frequency bands equally exist in our coherent addition configuration. Representative output spectra for single channel operation and for combined four channel operation are shown in Fig. 6.6. The common longitudinal bands are clearly observed and their frequency separation indeed correspond to the actual path length difference between the channels in the experiments.

Scalability can be achieved by using additional pairs of interferometric beam combiners, where each additional pair increases the number of beams in the array by a factor of 4. The optical length differences between all channels can still be $n\Delta L$, thereby ensuring the existence of common longitudinal modes. For example, a 4×4 array of Gaussian beam distributions can be coherently added by using a total of four interferometric combiners. In this case, the thickness of the interferometric combiners in the added pair should be halved in order to obtain a displacement twice that of the first pair. In general, the number of channels in the array will scale as 2^N, where N is the number of interferometric combiners ($N = 2, 4, 6, \ldots$).

Two pairs of interferometric combiners were exploited for intra-cavity coherent addition of 16 Gaussian channels.[18] The basic configuration is presented schematically in Fig. 6.7. The configuration consists of a laser resonator with a large diameter gain medium, to enable sixteen separate channels, two common end mirrors and an aperture that selects only the Gaussian (TEM_{00} mode) distribution. Two pairs of interferometric combiners, similar to those previously described, were placed in the resonator. The thickness the combiners in the first pair was twice that of the combiners in the second pair. As shown in Fig. 6.7 the 16 laser distributions right after the gain medium are folded vertically after the first interferometric combiner to get eight distributions, which are then folded horizontally to form four distributions by the second interferometric combiner, and these four distributions are folded by the next pair to form a single output beam of Gaussian distribution. The special design of the interferometric combiners and the self organization property of the laser ensure that the sixteen separate Gaussian distributions are all phase locked, and coherently add into one high power output Gaussian beam distribution at the output.

The experimental results are shown in Fig. 6.8. Figure 6.8(a) shows the 16 separate laser distributions escaping from the back mirror. Figures 6.8(b) and (c) show the near field and far field intensity distributions of the output beam, respectively. As evident, the combined intensity distribution has the modal structure of the single TEM_{00} laser distribution. The output energy of a single channel operated independently was 2.6 mJ per pulse, while that of the sixteen coherently added

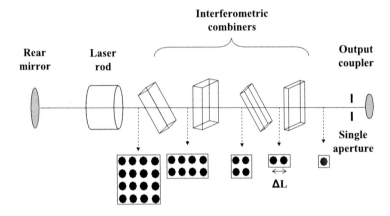

Figure 6.7 Basic configuration for coherent addition of sixteen laser distributions. The beam intensity distributions at five positions inside the resonator, composed of 1, 2, 4, 8, and 16 nearly Gaussian beams are schematically depicted.

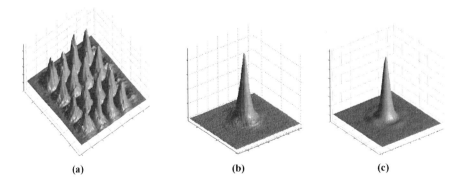

Figure 6.8 Experimental results for coherent addition of 16 laser distributions. (a) Detected intensity distribution of light escaping from the rear mirror; (b) detected near field intensity distribution of output, and (c) detected far field intensity distribution of output.

laser channels was 37 mJ per pulse, corresponding to 88% combining efficiency. The beam quality of the combined beam was calculated from the measured output to be $M^2 = 1.3$.

Scalability can also be achieved with less interferometric combiners providing they are somewhat more complex. Such an interferometric combiner for coherently adding five laser distributions is schematically shown in Fig. 6.9. The interferometric combiner is formed of a high-precision plane parallel plate, with specially designed coated regions. The first region in the front surface is coated with an antireflection layer, the second with a layer of 50% reflectivity (and 50% transmittivity), the third with a layer of 66.6% reflectivity, the fourth with a layer of 75% reflectivity, and the fifth region with a layer of 80% reflectivity. The rear surface is mainly coated with a highly reflecting layer, except for a small region that is coated with an anti-reflection layer. When placed in a laser resonator, this interferometric combiner will lead to sequential coherent addition of five beams. Moreover, with such a single interferometric combiner the path length differences between the channels will be an exact multiple of ΔL, where ΔL is the path length difference between two adjacent channels. This will ensure the existence of common longitudinal modes even when combining many channels sequentially. With two such interferometric combiners, oriented orthogonally with respect to each other, a two-dimensional array of 25 laser distributions can be coherently added in a compact fashion.[17]

Representative experimental results are shown in Fig. 6.10. Figure 6.10(a) shows the far-field Gaussian intensity distribution of the combined output. Figure 6.10(b) shows the near-field intensity distribution of the residual light escaping from the combined resonator back mirror, where 25 separate laser distributions are clearly detected.

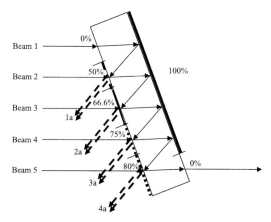

Figure 6.9 A five-channel interferometric combiner. Beams 1a, 2a, 3a, and 4a represent the loss channels, in which destructive interference occurs. Percentages indicates the amount of light reflected from each region.

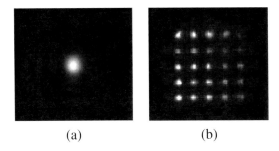

(a) (b)

Figure 6.10 Experimental coherent addition of 25 laser distributions: (a) far-field intensity distribution of the combined output and (b) near-field intensity distributions of the residual light leaking through the back mirror.

6.3 Coherent Addition of Single High-Order and Multimode Distributions

Coherent addition of laser beam distributions with intracavity interferometric combiners is not limited only to Gaussian beam distributions. Indeed, coherent addition of several single high-order modes and even multimode distributions were considered.[19–21] The coherent addition of several single high order mode distributions, whose amplitudes and phases are well defined, can potentially lead to greater combined laser output powers than addition of Gaussian distributions, while still retaining a reasonable output beam quality.

We obtained efficient intracavity coherent addition of two single high-order Laguerre-Gaussian modes (LG), using a configuration that is essentially similar to the basic configuration for coherent addition of two Gaussian beam distributions. Single high-order degenerate Laguerre-Gaussian mode operation in each channel was obtained by inserting inside the laser resonator cavity a binary phase element either along the combined channel and an aperture in each separate channel or along one of the separate channels, whereby the desired high-order mode operation will also be imposed on the other channel. The binary phase element consisted of π phase steps, corresponding to the uniform phase regions of the desired degenerate LG mode distributions.[19] The combination of the phase element and apertures ensures low losses for the desired modes and significant losses to all other modes.

With a TEM_{01} binary phase element placed in the combined channel of the resonator the measured combined output energy was 5.8 mJ per pulse, compared to 3 mJ for the independent single-channel operation with a TEM_{01} mode, and 2.1 mJ for a single channel Gaussian mode. These results indicate a 97% combining efficiency and an output energy almost three times that of the single Gaussian mode. The detected far field intensity distributions of the output beam, for the single-channel operation and for the combined two-channel operation, are shown in Fig. 6.11. Because the phase element is located near the output coupler, the TEM_{01} mode acquires a uniform phase when emerging from the resonator. Thus

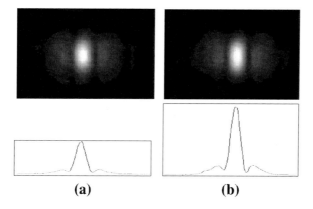

Figure 6.11 Experimental far-field intensity distributions and normalized cross section traces for operation with a binary TEM_{01} phase element placed along the combined channel path, near the output coupler. (a) Single-channel operation and (b) combined two-channel operation.

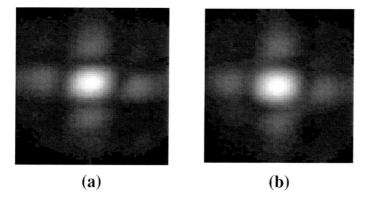

Figure 6.12 Experimental far-field intensity distributions with a binary TEM_{02} phase element placed along the combined channel path, near the output coupler; (a) Single channel operation and (b) combined two-channel operation.

the far field intensity distribution has a single central peak in the far field, which is almost identical to the distribution obtained with single-channel operation.

With a TEM_{02} binary phase element, the measured combined output energy was 7 mJ compared to 3.7 mJ for single-channel TEM_{02} operation. These results indicate a 95% combining efficiency, and an output energy in the combined TEM_{02} mode that is almost four times that of the single Gaussian TEM_{00} mode. The detected far field intensity distributions of the output beam, for a single TEM_{02} operation and for the combined two-channel operation, are shown in Fig. 6.12.

We also investigated how a channel with Gaussian distribution that is coherently added intra-cavity to one or more multimode distributions will impose its distribution on all the other channels. Experimentally, this was done, by inserting a small circular aperture in the path of one channel so as to obtain a nearly Gaussian mode distribution, and no aperture in the other channels so as to obtain multimode distribution. Superficially, we can argue that the phase of the Gaussian mode distribution in the channels with the multimode distributions will lock to the phase of the Gaussian distribution in the other channel, thereby suppressing losses of these Gaussian distributions at the interferometric combiner. Yet, the phases of the other modes in the channels with the multimode distributions will not lock to the Gaussian distribution in the other channel (because these modes are nearly orthogonal); thus these other modes will suffer a 50% loss at the combiner and consequently will be suppressed. Essentially, the Gaussian distribution of the one channel is thus imposed on the others.

Some representative results with a two channels configuration, where we coherently add a Gaussian distribution with a multimode distribution are shown in Fig. 6.13. Figures 6.13(a) and (b) show the near-field and far field intensity distributions of the channel with the aperture, for obtaining a Gaussian distribution. Figures 6.13(c) and (d) show the near-field and far-field intensity distributions of the channel with no aperture for obtaining the multimode distribution. The laser output energy for the channel with the aperture was 4.5 mJ per pulse, and the calculated M^2 was 1.13, indicating a nearly pure Gaussian TEM_{00} distribution. The calculated M^2 for the channel with no aperture was 4.9 indicating, as expected, that several high-order modes distributions are present in addition to the Gaussian distribution.

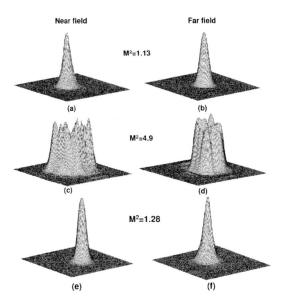

Figure 6.13 Experimental near- and far-field intensity distributions when coherently adding one Gaussian and one multimode distribution: (a) and (b), near- and far-field intensity distributions for channel with Gaussian distributions; (c) and (d), near- and far-field intensity distributions for channel with multimode distributions, and (e) and (f), near- and far-field intensity distributions of combined output.

Figures 6.13(e) and (f) show the near and far-field intensity distributions of the combined output beam. The calculated M^2 for the combined output was found to be 1.28. The slightly higher M^2 indicates that the combined output also contains some higher-order modes. The combined output energy was measured to be 8.5 mJ, indicating a combining efficiency of 94%.

We also coherently added four distributions, one Gaussian and three multimode by using a four channel configuration with two orthogonally-oriented identical interferometric combiners. The measured energy of the combined output was 17.2 mJ, indicating a combining efficiency of 95.5%. The corresponding calculated M^2 was 1.3, indicating that the output beam was a nearly Gaussian distribution, but again with some contribution from higher-order modes. This somewhat large M^2 indicates that the four-channel configuration has even a higher tendency to develop high-order modes than the two-channel configuration.

Thus far, we considered coherent addition of spatially coherent Gaussian TEM_{00} beam distributions as well as single high-order mode distributions, having well defined amplitudes and phase distributions. We now consider intracavity coherent addition of transverse multimode laser beam distributions. At first glance, the concept of efficient coherent addition of transverse multimode laser beam distributions does not seem plausible. Indeed, it is not feasible with independent multimode beams originating from independent lasers. However, with the multichannel laser cavity and a planar interferometric combiners, coherent addition of multimode field distributions becomes possible. This loss mechanism can cause simultaneous self-phase locking of all the corresponding transverse modes in the channels, enabling the coherent addition of the incoherent beam distributions. This approach, somewhat resembles passive longitudinal mode locking, where an intracavity nonlinear effect forces the various frequencies to phase lock such that short intense pulses are produced in the time domain.

The basic configuration for intracavity coherent addition of two transverse multimode field distributions is identical to that of combining two Gaussian beams (Fig. 6.1), except that the apertures are chosen such as to suit an identical multimode distribution in each of the two channels. In a simplified manner, the operation can be explained as follows. If the two multimode laser beam distributions are incoherent with respect to each other (random relative phase at each location in the beam or different frequencies), then each beam will suffer a 50% loss passing through the interferometric combiner; thus, typically, no lasing will occur. On the other hand, if the two multimode beam distributions have similar mode composition, and if each of the transverse modes in one distribution adds coherently with its counterpart in the other beam, then destructive interference occurs, thus the losses introduced by the combiner may be completely suppressed. The combined laser configuration tends to operate so that the losses are minimum, whereby the phases of the corresponding individual transverse modes automatically match, so that coherent addition takes place. The combined multimode beam is thus composed of many pairs of phase-locked modes, where the phase difference between the pairs is still completely undefined.

Coherent addition is achieved only for those longitudinal modes (frequencies) that are common in the two laser channels. Consequently, care must be taken to imbalance them in such a manner so as to obtain one or more mutual longitudinal modes.[3] If the aperture diameter in one channel is reduced then it is expected that this channel, with the lower transverse mode content, will imprint its modal content on the other channel, obtaining phase locking and coherent addition of the corresponding modes in the two distributions. This self-imprinting of the modal content occurs because the higher transverse modes in the channel with the larger aperture diameter do not have corresponding counterparts in the other channel, so they suffer considerable losses by the interferometric combiner. Figure 6.14 shows representative experimental near and far-field intensity distributions of each of the two channels (when operated independently) and of the combined laser output. The M^2 of each of the two laser channels, when operated separately, was 4.5, corresponding to roughly 13 transverse modes. We have measured 91% combining efficiency, where the combined output beam quality was not only preserved, but even slightly improved.

In principle, the basic resonator design can be scaled to coherent addition of more than two multimode beam distributions. This can be done by using several two-beam interferometric combiners for adding each pair of channels, or alternatively using a single interferometric combiner that includes several beamsplitter sections with appropriate reflectivity for sequentially adding multiple channels.

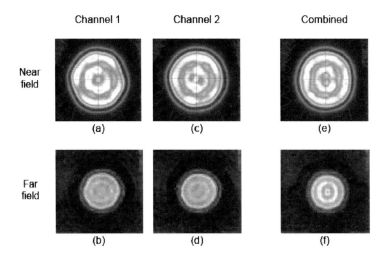

Figure 6.14 Experimental intensity distributions of the separate transverse multimode channels, and the combined laser output: (a) and (b), near- and far-field intensity distributions of the first channel; (c) and (d), near- and far-field intensity distributions of the second channel, and (e) and (f), near- and far-field intensity distributions of the combined laser output.

The ability to combine many multimode beam distributions can serve as another degree of freedom when designing lasers for applications where the required beam quality need not be strictly $M^2 = 1$. In these cases, combining multimode channels instead of Gaussian channels can reduce considerably the number of channels to be added, greatly simplifying the coherent addition configuration.

6.4 Improving the Beam Quality of Multimode Laser Resonators

Laser resonators operating with only the fundamental TEM_{00} mode provide excellent output beam quality but typically with relatively low output power. Increasing the output power can be readily achieved by resorting to multimode operation, where several transverse modes oscillate simultaneously and exploit a larger volume of the laser gain medium. Unfortunately, the multimode beam does not have well defined phase and amplitude distributions, thus the beam quality is relatively poor when compared to that of the TEM_{00} mode beam, and there is no increase of the optical brightness. Specifically, the output power P increases linearly with the aperture area, but in general, the one dimensional beam quality factor M_x^2, defined as the ratio between the space, bandwidth products of the beam to that of a Gaussian beam, also increases linearly with the aperture diameter. This means that although the output power increases with the diameter, the brightness of the output beam, which is proportional to $P/M_x^2 x M_y^2$, is not increased.

In order to increase the optical brightness of multimode laser resonators we resorted to an approach where coherent addition is involved. It is based on splitting the intracavity multimode beam into a tightly packed array of high-quality beam distributions, which are coherently added within the resonator to form a single high-power high-quality output beam. The coupling between several beam distributions, and their coherent addition, is achieved with intracavity planar interferometric beam combiners, similar to those described in the previous sections. As the individual distributions are more tightly "packed," i.e., with high fill factor, the brightness is found to increase considerably.

Experimentally we resorted to a configuration similar to that shown in Fig. 6.7, but in order to achieve a high fill factor we replaced the single small round aperture with a large square aperture, used a concave output coupler with appropriate radius, and changed the angular orientations of the interferometric combiners, so as to support a tight packed 4×4 array of Gaussian distributions. Representative experimental results are shown in Fig. 6.15. Figure 6.15(a) shows schematically the single large square aperture with the tightly packed 4×4 array. Operating the laser with the single square aperture and no interferometric combiners resulted in an output beam with a square multimode distribution shown in Fig. 6.15(b). Inserting the four interferometric beam combiners generated a 4×4 array of tightly packed Gaussian distributions, that were phase locked and coherently added to obtain a single Gaussian output beam shown in Fig. 6.15(c).

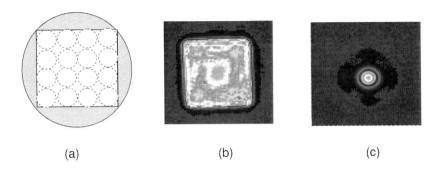

(a) (b) (c)

Figure 6.15 Experimental far-field intensity distributions of the multimode output beam and the Gaussian output beam after tight packing. (a) Single square aperture for generating a tightly packed 4×4 array; (b) square multimode beam output distributions without the intracavity interferometric beam combiners and (c) Gaussian beam output distribution after coherently adding a tightly packed 4×4 array of Gaussian distributions with four intracavity interferometric combiners.

The multimode output energy, measured with the large square aperture and no interferometric combiner was 147 mJ with a beam quality factor of $M^2 = 15$. The energy of the combined beam measured after insertion of the interferometric combiners was 32 mJ with beam quality factor of 1.3. Accordingly, the measured improvement in brightness was 30.

6.5 Concluding Remarks

We presented an approach in which several Gaussian, single high order modes and multimode laser distributions can be efficiently combined to obtain high output energies concomitantly with good output beam quality. It is based on intracavity coherent addition with novel interferometric combiners, which results in stable and robust overall laser operation. We also showed how the approach can be exploited for improving the brightness of multimode laser resonators, by splitting the multimode distribution into a tightly packed array of Gaussian distributions, which are phase locked and coherently added within the laser cavity.

We believe that upscaling the approach to coherently add many laser distributions is both feasible and practical. Moreover, the approach can be exploited in many types of lasers, especially the newly developed fiber lasers.

References

1. D'Amato, F.X., Siebert, E.T., and Roychoudhury, C., "Coherent operation of an array of diode lasers using a spatial filter in a Talbot cavity," *Phys. Lett.*, **55**, 816, 1989

2. Menard, S., Vampouille, M., Colombeau, B., and Froehly, C., "Highly efficient phase locking and extracavity coherent combination of two diode-pumped Nd:YAG laser beams," *Opt. Lett.*, **21**, 1996, 1996.

3. DiDomenico, Jr., M., "A single frequency TEM00 mode gas laser with high output power," *Appl. Phys. Lett.*, **8**, 29, 1966.

4. Sabourdy, D., Kermene, V., Desfarges-Berthelemot, A., Vampouille, M., and Barthelemy, A., "Coherent combining of Nd:YAG lasers in Vernier-Michelson type cavity," *Appl. Phys. B*, **75**, 503, 2002.

5. Xu, J., Li, S., Lee, K.K., and Chen, Y.C., "Phase locking in a two-element laser array: a test of the coupled-oscillator model," *Opt. Lett.*, **18**, 513, 1993.

6. Fabiny, L., Colet, P., Roy, R., and Lensta, D., "Coherence and phase dynamics of spatially coupled solid-state lasers," *Phys. Rev. A*, **47**, 4287, 1993.

7. Oka, M., Masuda, H., Kandeda, Y., and Kubot, S., "Laser diode pumped phase-locked Nd:YAG laser arrays," *IEEE J. Quantum Electron.*, **QE-28**, 1142, 1992.

8. Rutherford, T.S., and Byer, R.L., "Six beam phase-locked slab laser resonator," *CLEO/Europe-EQEC, the 15th International Conference on Lasers and Electrooptics, Munich Germany*, 2001.

9. Menard, S., Vampouille, M., Desfarges-Berthelemot, A., Kermene, V., Colombeau, B., and Froehly, C., "Highly efficient phase locking of four diode pumped Nd:YAG laser beams," *Opt. Comm.*, **160**, 344, 1999.

10. Leger, J.R., Swanson, G.J., and Veldkamp, W.B., "Coherent laser addition using binary phase grating," *Appl. Opt.*, **26**, 4391, 1987

11. Goodman, J.W., "Fan-in and fan-out with optical interconnections," *Optica Acta*, **32**, 1489, 1985.

12. Goodman, J.W., and Lam, J.C., "Fan-in loss for electrical and optical interconnections", *ICO Trends in Optics, Vol. 3*, ed. A. Consortini, Academic Press (1996).

13. Ishaaya, A.A., Davidson, N., Shimshi, L., and Friesem, A.A., "Intracavity coherent addition of Gaussian beam distributions using a planar Interferometric coupler," *Appl. Phys. Lett.*, **85**, 2187, 2004.

14. Ishaaya, A.A., Eckhouse, V., Shimshi, L., Davidson, N., and Friesem, A.A., "Improving the output beam quality of multimode laser resonators," *Opt. Exp.*, **13**, 2722, 2005.

15. Sedhgani, S., Eckhouse, V., Davidson, N., and Friesem, A.A., "Suppression of thermal lensing effects in intracavity coherent combining of lasers," *Opt. Comm.*, **276**, 139, 2007.

16. Shimshi, L., Ishaaya, A.A., Eckhouse, V., Davidson, N., and Friesem, A.A., "Passive intracavity coherent addition of nine laser distributions," *Appl. Opt. Lett.*, **88**, 041103, 2006.

17. Shimshi, L., Ishaaya, A.A., Eckhouse, V., Davidson, N., and Friesem, A.A., "Upscaling coherent addition of laser distributions," *Opt. Comm.*, **275**, 389, 2007.

18. Eckhouse, V., Ishaaya, A.A., Shimshi, L., Davidson, N., and Friesem, A.A., "Intracavity coherent addition of 16 laser distributions," *Opt. Lett.*, **31**, 350, 2006.

19. Ishaaya, A.A., Eckhouse, V., Shimshi, L., Davidson, N. and Friesem, A.A., "Intracavity coherent addition of single high-order modes," *Opt. Lett.*, **30**, 1770, 2005.

20. Ishaaya, A.A., Eckhouse, V., Shimshi, L., Davidson, N., and Friesem, A.A., "Coherent addition of spatially incoherent beams," *Opt. Exp.*, **12**, 1770, 2004.

21. Eckhouse, V., Ishaaya, A.A., Shimshi, L., Davidson, N. and Friesem, A.A., "Imposing a Gaussian distribution in multichannel laser resonators," *IEEE J. Quant. Elec.*, **41**, 686, 2005.

Vardit Eckhouse received her BSc in physics from the Technion in 1999, and her MSc in electrical engineering from Tel Aviv university, Israel. From 1999 to 2002 she was with the DOE laboratory as a researcher in the fields of diffractive optics, imaging, holography, and superresolution at the Tel Aviv University. From 2001 to 2003, she was with Civcom D&S Inc. as a senior optical engineer, experimenting in optical communication devices, simulating various network distortion effects, and working with various electro-optical materials and research involving the electro-optical effect. Since 2003, she has been studying for her PhD at the Weizmann Institute of Science, Rehovot, Israel, focusing on coherent addition of laser distributions in various laser configurations.

Amiel A. Ishaaya received his BSc and MSc in physics from Tel Aviv University, Israel, in 1987 and 1995 respectively, and a PhD in physics from the Weizmann Institute of Science, Israel, in 2005. From 1994 to 2001, he was with ELOP Ltd., Rehovot Israel, working on military laser R&D projects. From 2001 to 2005, during his PhD studies, he focused on high-order transverse mode selection and coherent beam combining in various laser configurations. From 2005 to 2007, he was a post-doctoral research associate at Cornell University, Ithaca NY, studying collapse dynamics of high-intensity optical beams and nonlinear interactions in photonic crystal fibers. In 2007, he joined the Dept. of Electrical and Computer Engineering at Ben Gurion University. His current research is mainly experimental and focuses on lasers and nonlinear optical devices based on photonic crystal fibers and silicon waveguides.

He has authored and coauthored over 50 journal and conference publications, 20 technical reports, and holds a US patent application, all in the field of optics and lasers. He is a member of the OSA and SPIE.

Liran Shimshi received his BSc in chemistry and in physics from Tel Aviv University, Tel Aviv, Israel, in 1997, and his MSc and PhD in physics from the Weizmann Institute of Science, Rehovot, Israel, in 2002 and 2007, respectively. From 1996 to 1999, he served in the Israeli Defense Forces, conducting research in space radiation and space environment effects. From 1999 to 2002, he was with the department of physics of complex systems, Weizmann Institute of Science, Rehovot, Israel, conducting research on high-order transverse mode selection and second harmonic generation. From 2002 to 2007, he was with the department of physics of complex systems, Weizmann Institute of Science, Rehovot, Israel, conducting research on phase locking and coherent beam combining in various laser configurations. Since 2007 he has been a senior researcher in the R&D Center of Samsung Electronics, Yakum, Israel, performing research in the areas of optics and electro-optics, new concepts, and applications. Dr. Shimshi is a member of the Optical Society of America, the American Physical Society, and the Israeli Physical Society.

Nir Davidson received his BSc in physics and mathematics from the Hebrew University, Jerusalem in 1983, his MSc in physics from the Technion in Haifa in 1988, and his PhD in physics from the Weizmann Institute of Science, Rehovot, in 1993. He was a postdoctoral fellow at Stanford University. He now holds the Peter and Carola Kleeman Professorial Chair of Optical Sciences in the Department of Physics of Complex Systems at the Weizmann Institute of Science in Rehovot, Israel. His research is in the areas of laser cooling and trapping of atoms, quantum chaos, quantum optics, Bose-Einstein condensation, laser physics, and physical optics. He has authored over 130 journal publications. Dr. Davidson has received the Allon award, the Yosefa and Leonid Alshwang Prize for Physics, and the Bessel award from the Alexander von Humboldt Foundation. He is the president of the Israeli Laser and Electro-Optics Society.

Asher A. Friesem received his BSc and PhD from the University of Michigan in 1958 and 1968, respectively. From 1958 to 1963 he was employed by Bell Aero Systems Company and Bendix Research Laboratories. From 1963 to 1969, at the University of Michigan's Institute of Science and Technology, he conducted investigations in coherent optics, mainly in the areas of optical data processing and holography. From 1969 to 1973 he was principal research engineer in the Electro-Optics

Center of Harris, Inc., performing research in the areas of optical memories and displays. In 1973 he joined the staff of the Weizmann Institute of Science, Israel and was appointed professor of optical sciences in 1977. He subsequently served as department head, chairman of the scientific council, and chairman of the professorial council. In recent years his research activities have concentrated on new holographic concepts and applications, optical image processing, electro-optic devices, and new laser resonator configurations. He has served on numerous program and advisory committees of national and international conferences. Among other posts, he served for many years as Vice President of the International Commission of Optics (ICO) and chairman of the Israel Laser and Electro-Optics Society. He is a Fellow of OSA, a life Fellow of IEEE, a member of SPIE, and a member of Sigma Xi. Over the years he has been a visiting professor in Germany, Switzerland, France, and the USA, has authored and coauthored more than 250 scientific papers, coedited four scientific volumes, and holds 30 international patents.

Chapter 7
Light Confinement in Photonic Crystal Microcavities

Philippe Lalanne and Christophe Sauvan
Université Paris-Sud, France

7.1 Introduction
7.2 Fabry-Perot Model and Photonic-Crystal Cavities
7.3 Recipes for High-Q Fabry-Perot Resonators
7.4 Beyond the Fabry-Perot Model
7.5 Conclusion
References

7.1 Introduction

Light is difficult to store in a small volume for a long time. Semiconductor cavities, which trap light efficiently, are an essential component of many important optical devices and effects, from optical processing to quantum light sources.[1] Cavities are characterized by two main quantities: the modal volume V and the quality factor Q. In many applications, high Qs and small Vs are highly desirable for the high finesse required for laser and filter applications, or for the large Purcell factor[2] required for controlling the spontaneous emission of atoms placed in resonance with the microcavity mode.[3] For a dipole linewidth much smaller than the cavity linewidth, a simple derivation shows that the Purcell factor is equal to $3/(4\pi^2)(\lambda/n)^3 Q/V$, where n is the refractive index of the medium and λ is the resonant wavelength matched with the emission wavelength. This formula holds for a perfect emitter placed at the antinode of the electromagnetic field and with its dipole parallel to the electric field. Thus the Q/V ratio is a figure of merit of the cavity alone, which describes the cavity capability to enhance light-matter interaction.

At optical frequencies, due to the lack of good metals, the last decade has seen intense research activity on a new generation of microresonator devices. Total internal reflection is solely exploited in spherical or disk-shaped resonators or in wire rings,[4–6] and a hybrid confinement that combines photonic bandgaps in one or two dimensions with index guiding is exploited in photonic-crystal (PhC) microcavities such as micropillars,[7] PhC cavities in semiconductor wires,[8–10] or in two-dimensional photonic-crystal membranes[11–12] (see Fig. 7.1).

With the progress of nanofabrication facilities, all types of microcavities are today facing tremendous increases of their performance. In general, the light confinement in cavities relying on pure refraction, like microdisks[13] or ring resonators,[14] is well understood as resulting from whispering gallery modes. Our understanding of the confinement in photonic crystal cavities, see Fig. 7.1, is much less mature, probably because the hybrid character of the confinement is conceptually difficult to apprehend in an intuitive way. This difficulty is reflected in the design approach that often relies on a complete resolution of the electromagnetic problem with fully vectorial numerical methods, followed by an optimization performed by repeatedly adjusting the cavity parameters, for instance, see Refs. 10, 12, and 15, for a-, b-, and c-type cavities (Fig. 7.1), respectively. In order to estimate the ultimate potential of these microcavities for future applications, it appears essential to understand the confinement mechanisms in these cavities. A good understanding is also important for new designs of microcavities in general.

In the following, we review the different mechanisms that can be used for designing high Qs and small Vs PhC microcavities. Although they represent the primary approach for understanding light confinement in classical cavities, Fabry-Perot models are rarely used to interpret the light confinement in PhC microcavities. In Section 7.2, we rehabilitate the classical Fabry-Perot description and we show that the model can be helpful to understand the confinement, even in single-hole-defect cavities in 2D PhC membranes. The model emphasizes two important quantities, namely, the group velocity of the mode bouncing between the mirrors and the mirror reflectance. Section 7.3 discusses general recipes that use these quantities for designing high-Q microcavities. Special attention is paid to the modal reflectance, for which we consider adding suitable interface layers to realize ultrahigh reflectors by tapering light from the waveguide defect into the mirror. Indeed, not all PhC cavities can be modeled as Fabry-Perot resonators. In Section 7.4, we briefly describe two physical mechanisms that significantly alter the Fabry-Perot predictions. The first mechanism is a recycling of radiation losses by leaky modes supported by the defect waveguide. It allows the realization of high-Q microcavities with poor-performance mirrors. The second one results from the existence of a higher-order propagative Bloch mode in the mirrors. It is shown to strongly alter the Q factor of pillar cavities for small diameters. As a whole, whether it provides quantitative predictions or not, the Fabry-Perot model is shown to be a useful tool for understanding the different mechanisms and recipes that can be used to design PhC microcavities.

7.2 Fabry-Perot model and Photonic-Crystal Cavities

Interpreting the light confinement in a resonator with a Fabry-Perot model consists of approximating the cavity mode as a stationary pattern formed by the

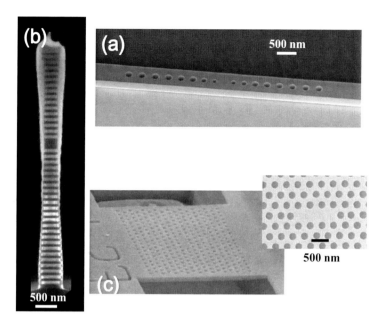

Figure 7.1 Different types of photonic-crystal microcavities. (a) Silicon ridge-waveguide cavity (courtesy David Peyrade, Laboratoire des Technologies de la Microélectronique) on a SiO_2 film on a silicon substrate. (b) A GaAs/GaAlAs micropillar on a GaAs substrate (courtesy Isabelle Robert-Philip, Laboratoire de Photonique et de Nanostructures). (c) Photonic-crystal-waveguide cavity in a suspended membrane in air (courtesy Sylvain Combrié, Thalès Research and Technology).

two counterpropagating defect-guided modes that are bouncing between two mirrors. In a- and b-type resonators, the defect mode is the fundamental guided mode of a z-invariant waveguide. In a c-type resonator, it is the fundamental Bloch mode of a z-periodic waveguide, a single-row-defect PhC waveguide for the example shown in Fig. 7.1(c). These modes all obey a dispersion relation $\omega(k_z)$, and are thus characterized by an effective index $n_{\text{eff}} = k_z/k_0$ and by a group index $n_g = c/v_g$, with v_g being the group velocity. For classical z-invariant waveguides such as in a- or b-type microcavities, n_{eff} and n_g only slightly differ, but for c-type microcavities, because of the periodicity, the waveguide potentially supports a slow Bloch mode and confinement regimes for which slow light ($n_g \gg n_{\text{eff}}$) rather than fast light is bouncing between two mirrors may be observed. Therefore the dispersion relation of the defect mode is an important quantity of the Fabry-Perot model.

The second important quantity is the modal reflectivity coefficient $r = |r| \exp(i\varphi)$ of the mirror. When the waveguide mode impinges onto the mirror, it is backreflected into the counterpropagating guided mode, which is again back-reflected onto the second mirror. Under the assumption (this assumption will be valid in the following) that the bouncing mode is truly guided, no radiation loss occurs when light propagates from one mirror to the other one. The cavity mode lifetime is simply limited by the imperfect mirror reflectivity, $R = |r|^2$, which is strictly smaller than 1. Either light can be transmitted into an output waveguide (often a desired effect for coupling into another channel), or light can be radiated out into the cladding (in general a detrimental effect) because of the termination experienced by the incident guided mode at the waveguide-mirror interface.[16]

Within the Fabry-Perot description, a resonance at a wavelength λ_0 ($k_0 = 2\pi/\lambda_0$) results from a phase-matching condition for the waveguide mode. The total phase delay $\Phi_T(\lambda_0)$ experienced by the guided mode along one-half cavity cycle has to be equal[17] to a multiple of π,

$$\Phi_T(\lambda_0) = k_0 n_{\text{eff}} L + \varphi(\lambda_0) = p\pi, \qquad (7.1)$$

where L is the physical cavity length (side-to-side separation distance between the two inner holes, for instance) and p is an integer. The mode lifetime is related to the cavity quality factor defined by $Q = \lambda_0/\Delta\lambda$, where $\Delta\lambda$ is the resonance width at half-maximum. For a Fabry-Perot resonator and under the legitimate assumption of a narrow resonance, $\Delta\lambda$ can be straightforwardly expressed as the derivative of $\Phi_T(\lambda)$ and one obtains

$$Q = \frac{\pi}{1-R} \left[\frac{2L n_g}{\lambda_0} - \frac{\lambda_0}{\pi} \left(\frac{\partial \varphi}{\partial \lambda} \right)_{\lambda_0} \right]. \qquad (7.2)$$

It is relevant to introduce the penetration length into the mirror, $L_p = -\lambda_0^2/(4\pi n_g)(\partial\varphi/\partial\lambda)_{\lambda_0}$, so that Eq. (7.2) is simply rewritten as

$$Q = \frac{k_0}{1-R} n_g L_{\text{eff}}, \qquad (7.3)$$

where $L_{\text{eff}} = L + 2L_p$ is the effective cavity length. From Eq. (7.3), one easily sees the important parameters that can be engineered to design high-Q and small-V microcavities. Considering that L_{eff} has to be kept as small as possible for small Vs, only two physical quantities, n_g and R, can be engineered to optimize Q.

PhC cavities exhibit so different geometries from one construct to another that, a priori, it is not obvious to predict whether a specific resonator behaves or not as a Fabry-Perot resonator, and it is not intuitive whether a classical Fabry-Perot model can be helpful or not for understanding the mechanisms of light confinement in the resonator. According to the classification table of the review article by Vahala,[18] only a- and b-type resonators would behave as Fabry-Perot style resonators, whereas the confinement method for c-type cavities in PhC membranes is understood as resulting from a different mechanism and deserves a specific column in the table. This classification that relies on geometrical considerations rather than on an in-depth analysis of the confinement properties is somewhat arbitrary in our opinion. As will be shown in Section 7.4, light confinement in micropillars generally does not follow a classical Fabry-Perot description. However, perhaps surprisingly, many important c-type constructs in PhC membranes may be considered as Fabry-Perot-style resonators.

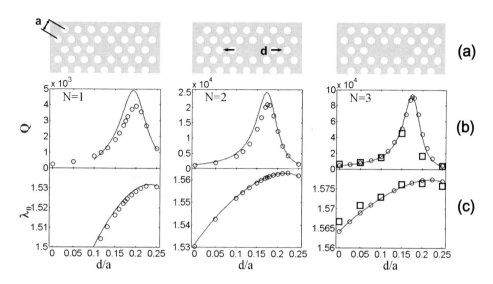

Figure 7.2 Validation of the Fabry-Perot model for two-dimensional PhC cavities. (a) Sketch of the cavities with N = 1, 2, and 3 missing holes. (b) and (c) Comparison between fully vectorial calculation data (blue circles) and Fabry-Perot model predictions (red solid curves) for the cavity Q (b) and for the resonance wavelength λ_0 (c) as a function of the normalized hole displacement d/a. The experimental data obtained in Ref. 12 are shown with squares for N = 3.

To illustrate our purpose, let us consider three PhC cavities obtained by removing N = 1, 2, and 3 holes in a 2D PhC slab; see Fig. 7.2(a) for a schematic view of the constructs along with a definition of the main parameters. Figures 7.2(b) and (c) show the Q factor and the resonance wavelength λ_0 as a function of the displacement d of the two inner cavity holes. The blue circles are obtained by computing the resonator complex pole $\tilde{\lambda}$, with a fully vectorial aperiodic-Fourier modal method[19] (a-FMM). From the complex pole, the resonant wavelength λ_0 = Re($\tilde{\lambda}$) and the cavity Q = Re($\tilde{\lambda}$)/[2Im($\tilde{\lambda}$)], are derived by assuming a Lorentzian resonance. For the sake of comparison, we also include experimental data[12] (squares) obtained for the cavity with N = 3. The calculations quantitatively agree with the experimental data, indicating that inevitable additional losses related to fabrication errors are kept at a rather small level in the experiment.

In Figs. 7.2(b) and (c), the solid red curves are Fabry-Perot model predictions obtained with Eqs. (7.1) and (7.2). The predictions rely on the sole knowledge of the dispersion curve of the single-row-defect waveguide and on the modal reflectivity $r(\lambda)$ shown in Fig. 7.3(a). Figure 7.3(b) shows the modal reflectance spectrum for two hole displacements, d = 0 and d = 0.18a, the latter value corresponding to the optimal Q value in Fig. 7.2(b). The net effect of a small hole shift by 0.18a is an increase of the mirror performance over the entire spectral domain of interest. The possibility of decreasing the out-of-plane radiation losses by hole tuning at the mirror-defect interface will be explained in the next section.

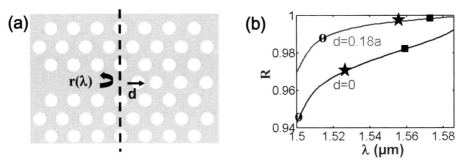

Figure 7.3 Modal reflectance for the cavities of Fig. 7.2. (a) Definition of $r(\lambda)$. (b) $R = |r(\lambda)|^2$ for two mirrors with $d = 0$ and $d = 0.18a$. The circle, star, and square marks indicate the reflectance associated to the cavities with $N = 1$, 2, and 3, respectively. R peaks up at 99.8% for the cavity with $N = 3$ and for $d = 0.18a$.

The quantitative agreement between the Fabry-Perot model predictions (solid red curves) and the experimental (squares) and numerical (circles) data suggests that light confinement in PhC slab microcavities can be largely understood as a Fabry-Perot mode formed by the bouncing of the fundamental single-row-missing PhC waveguide mode, and this even for ultrasmall cavities with a single hole missing ($N = 1$).

Usually, the confinement theories for light in 2D bandgap cavities are all nonpredictive since they all rely on an interpretation based on the knowledge of the cavity mode profile and thus they all require that the full 3D electromagnetic problem be solved before the analysis. In contrast, when it applies, the Fabry-Perot model provides a conceptually more helpful picture, since analyzing a mirror is much simpler than analyzing a whole cavity. It additionally provides new hints for improving cavity performance, such as engineering the mirror-waveguide interface to lower the radiation losses or slowing down the group velocity of the bouncing mode.

Indeed, not all PhC constructs can be modeled as Fabry-Perot resonators. But even when the Fabry-Perot model is not valid for a specific construct, it is important to be aware of that and to understand why. The gained understanding may again provide new routes or recipes for further improvements. Section 7.5 documents some examples of important constructs for which the Fabry-Perot model fails at predicting the confinement properties and describes the physical mechanisms responsible for the deviation from the Fabry-Perot-like resonator.

7.3 Recipes for high-*Q* Fabry-Perot resonators

In this section, we will assume that the Fabry-Perot description is valid. In general, the cavity volume linearly scales with L_{eff}, and although an increase of L_{eff} results in an increase of Q, the mode volume is altered and the Q/V ratio (the finesse) largely remains unchanged. As general design recipes, the model suggests that only two quantities, namely, R and n_g, can be used to design high-Q and small-V microcavities.

In general, the guided mode that is bouncing between the two mirrors possesses a group velocity that is comparable to the speed of light in the material. According to Eq. (7.3), the cavity mode lifetime can be increased by considering a slow mode bouncing between the mirrors. In this case, not only is the light trapped by the mirrors, but it also travels slowly in between them, still increasing the mode lifetime. The use of a slow wave to enhance a cavity Q has not been much studied in the literature.[20] Indeed, the use of any periodic waveguide operating in the slow light regime would potentially result in a Q enhancement. For instance, considering the geometry shown in Fig. 7.2(a), the single-row-defect waveguide supports a slow Bloch mode in the vicinity of the Brillouin zone edge. In fact, a detailed analysis of the case $N = 3$ shows that the 10-times increase of the Q is not only due to an increase of the modal reflectivity from 98.3 to 99.8, but also from a two-times slowing down of the cycling defect Bloch mode.[20] Although the slowdown effect is kept at a moderate level in the cavity construct with $N = 3$ ($n_g = 14$), further improvements appear possible. For instance, for the same geometry, the Fabry-Perot model predicts that by tuning the hole position in a cavity with $N = 4$, a peak Q factor in excess of 250,000 can be obtained for $n_g=25$. Other geometries are currently under study.[21,22]

The second quantity that affects the cavity Q is the modal reflectance. Indeed, according to Eq. (7.3), for $R = 1$ (perfect mirror), the light cannot escape from the cavity and is trapped in the defect for an infinitely long time. Before considering recipes for designing constructs offering a reflectivity close to one, let us first consider the physical reason leading to a nonperfect reflectivity ($R < 1$). Figure 7.4(a) shows an elementary scattering problem of light reflection from a semi-infinite z-invariant waveguide onto a semi-infinite PhC mirror. The total electromagnetic field in the z-invariant or periodic waveguides, $\Phi=|\mathbf{E}, \mathbf{H}\rangle$, can be expanded in terms of the complete sets of normal modes or normal Bloch modes. Let us denote by $\mathbf{M}^{(1)}$ the incident fundamental mode of the waveguide and by $\mathbf{M}^{(-1)}$ the associated backward propagating mode. For $z < z_0$, we have

$$\Phi = \mathbf{M}^{(1)} + r\mathbf{M}^{(-1)} + \sum_{p>1} r^{(1,p)} \mathbf{M}^{(-p)}, \qquad (7.4)$$

where $\mathbf{M}^{(-p)}$, $p > 1$, are the backward-propagating radiation modes and the $r^{(1,p)}$ are the modal amplitude coefficients of the backward-propagating modes. Similarly, in the mirror, only forward-propagating Bloch modes $\mathbf{B}^{(p)}$, $p > 0$, are excited, and the total field can be written for $z > z_0$,

$$\Phi = t\mathbf{B}^{(1)} + \sum_{p>1} t^{(1,p)} \mathbf{B}^{(p)}, \qquad (7.5)$$

where $t^{(1,p)}$ are the modal amplitude coefficients of the forward-propagating Bloch modes. In Eq. (7.5), a specific Bloch mode $\mathbf{B}^{(1)}$ has been isolated from the summation. In the perturbation regime with small holes, one may argue that every mode of the z-invariant waveguide becomes a Bloch mode of the perturbed

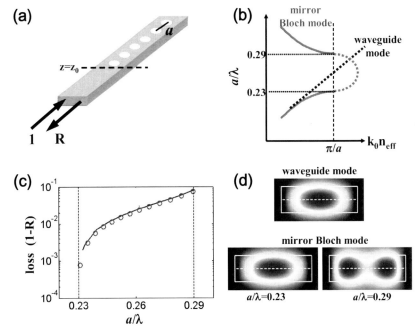

Figure 7.4 Modal reflectivity and transverse mode-profile mismatch. (a) Light reflection at the interface between a z-invariant waveguide and a periodic mirror. (b) Dispersion diagram of the two fundamental modes supported by the waveguide ($\mathbf{M}^{(1)}$ shown with a dotted black curve) and by the mirror ($\mathbf{B}^{(1)}$ shown with a solid blue curve). In the gap, the mirror Bloch mode $\mathbf{B}^{(1)}$ is guided and purely evanescent ($k_0 n_{eff} = \lambda/a + i\lambda$). (c) Loss spectrum $L(\lambda) = 1 - R(\lambda)$ calculated with electromagnetic theory (solid blue curve) and with the overlap integral model (red circles). The vertical dashed lines indicate the band edges. (d) Transverse mode profile mismatch. Top: dominant magnetic field of $\mathbf{M}^{(1)}$. Bottom: dominant magnetic field of $\mathbf{B}^{(1)}$ in a cross sectional plane located between two holes. The white solid lines indicate the semiconductor-air boundaries of the air bridge. The computations are performed for a 340 nm thick, 500 nm wide silicon air bridge, for a periodicity $a = 420$ nm and for a hole diameter of 230 nm.

periodic waveguide, so that to every $\mathbf{M}^{(p)}$ one may associate a Bloch mode $\mathbf{B}^{(p)}$. $\mathbf{B}^{(1)}$ is simply defined as the Bloch mode associated to $\mathbf{M}^{(1)}$. Even for infinitely small holes, a bandgap can be opened in the dispersion relation of the periodic waveguide. Thus the Bloch mode $\mathbf{B}^{(1)}$, which is guided at low frequencies, is either a propagative mode ($k_0 n_{eff} > 0$) outside the bandgap, or is a purely evanescent standing-wave mode ($k_0 n_{eff} = \pi/a + i\varepsilon$) inside the bandgap. Figure 7.4(b) shows the dispersion diagrams of $\mathbf{M}^{(1)}$ (dotted black curve) and $\mathbf{B}^{(1)}$ (solid blue curve). Note that the dotted blue line inside the bandgap represents the imaginary part ε of the propagation constant.

The mode $\mathbf{B}^{(1)}$ plays a central role in the backreflection process; its excitation is responsible for the backreflection of the incident light, and the nonperfection of the excitation is also responsible for the important radiation losses. The solid curve in Fig. 7.4(c) shows the calculated mirror modal reflectivity spectrum, $R = |r|^2$, over the entire bandgap from $a/\lambda = 0.23$ to $a/\lambda = 0.29$. The modal reflectivity does not reach unity, and decreases from the valence band edge to the conduction band edge. Since the mirror is semi-infinite, no light is transmitted and $\mathcal{L} = 1 - R$ simply represents radiation losses in the air-clad area. The losses are due to the transverse mode profile mismatch between the incident guided mode $\mathbf{M}^{(1)}$ and the evanescent mode $\mathbf{B}^{(1)}$. In the vicinity of the valence band, $a/\lambda = 0.23$, $\mathbf{B}^{(1)}$ and $\mathbf{M}^{(1)}$ are very similar and $\mathcal{L}$ is small, $\mathcal{L} = 10^{-3}$. As the frequency increases, the Bloch mode fields more and more penetrate into the low-index material (the hole) and $\mathcal{L}$ increases to culminate at the conduction band edge, where $\mathbf{B}^{(1)}$ and $\mathbf{M}^{(1)}$ are radically different [see Fig. 7.4(d)]. Indeed, it is possible to quantify the transverse mode profile mismatch by defining[16] an overlap integral η between $\mathbf{B}^{(1)}$ and $\mathbf{M}^{(1)}$. The circles in Fig. 7.4(c) simply represent $1 - \eta^2$. As shown, the approximate approach quantitatively predicts the radiation loss spectra calculated with the a-FMM, showing the crucial role played by the mode-profile mismatch problem at the mirror-waveguide interface.

From the previous analysis, one realizes that it is important to engineer the interface between the waveguide and the mirror in order to lower the mismatch. Figure 7.5 summarizes the general strategies that have been adopted so far. The first small-V cavities in semiconductor wafers have been fabricated mainly for laser applications in planar waveguides with PhC mirrors composed of slits and ridges,[23–25] and then in ridge waveguides with PhC mirrors composed of hole arrays.[8–10] Indeed, for abrupt interfaces, the mismatch is on the order of a few percent at the gap center frequency, and the Qs of cavities with abrupt interfaces (first row in Fig. 7.5) do not exceed several hundreds.

To lower the mismatch, a possible approach consists of finely tuning the geometry at the interface between the waveguide and the mirror to implement mirror tapers (second row in Fig. 7.5) to reduce the radiation losses at the interface. The single-hole-displacement approach of Fig. 7.3 simply represents a basic example. The tapers are in general composed of a progressive variation of several geometric parameters, which aims at implementing a gradual variation of the transverse mode profile between the modes $\mathbf{M}^{(1)}$ of ridge waveguide and $\mathbf{B}^{(1)}$ of the mirror.[16] The first attempt[26] to realize cavities with tapered mirrors was performed with ridges etched by slits, but the experimental evidence of the beneficial effect of the taper has been plugged by mode mixing in the ridge, as evidenced by further near-field measurements of the cavity mode. Today, typical Q values for cavities with tapered mirrors are in the range of 50,000–100,000 for single-defect resonators in semiconductor ridges on silica substrates[10,27,28] or in PhC waveguides in various semiconductor membranes[29,30] in air. Such high Q values correspond to an impressive modal reflectivity, $R \approx 0.9996$, for ridge cavities[10] and $R \approx 0.998$ for PhC cavities.[20]

A milestone improvement[31] has been recently achieved with matched interfaces in single-row-defect PhC waveguides (third row in Fig. 7.5). In the so-called heterostructure nanocavity, the waveguide and mirror modes are easier to match. The construct results in the realization of nanocavities with extremely high Q factors ($\approx 600,000$), more than one order of magnitude larger than in

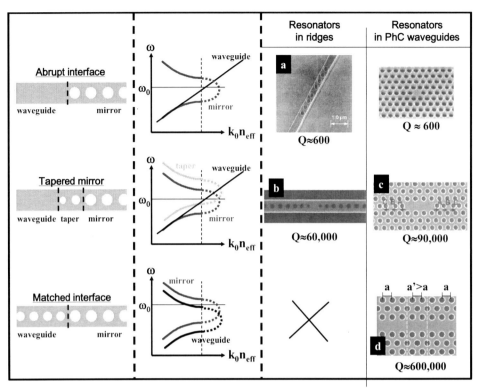

Figure 7.5 Strategies for designing a high-Q Fabry-Perot resonator. Upper row: the Q factors of cavities with abrupt interfaces is limited to a few hundreds. Second row: cavities with tapered mirrors. Third row: matched interface relying on a small mismatch between the waveguide mode and the mirror Bloch mode. (a) After Ref. 9; (b) After Ref. 28; (c) After Ref. 29; (d) After Ref. 31.

tapered cavities. Further engineering[31–33] of the interface by tapering and optimization leads to Q values slightly in excess of 10^6, with theoretical predictions greater than 10^7. The corresponding modal reflectance deduced from Eq. (7.3) is on the order of 0.9993, a value much larger than that achieved for PhC cavities with three missing holes, and comparable to that achieved for ridge cavities. For the sake of comparison, it is worth mentioning that state-of-the-art reflectance for dielectric Bragg mirrors manufactured for the Virgo project with up-to-date automated deposition techniques only provide an extra nine digit.[34] However, let us note that in thin-film coatings, one is currently concerned with controlling uniformity and roughness on tens of square centimeters scales, while the PhC high reflectors are dealing with guided modes with square micron cross sections.

7.4 Beyond the Fabry-Perot Model

Using a Fabry-Perot model is clearly the prime approach for analyzing the physics of light confinement in cavities. In general, microcavities are not expected to perfectly behave as Fabry-Perot resonators. As illustrated in Fig. 7.2, small deviations for the Q factor and for the resonance wavelength are obtained for the cases $N = 1$ and $N = 2$. The reason is that in the Fabry-Perot approach, the energy transport between the two mirrors is assumed to be solely ensured by the fundamental propagating modes of the defect waveguide, $\mathbf{M}^{(1)}$ and $\mathbf{M}^{(-1)}$, with all other energy-transport routes being neglected. The other routes are the modes $\mathbf{M}^{(p)}$, $p \neq 1$, and for monomode waveguides as in Fig. 7.2, these modes are all leaky in the spectral range of interest and their leakage guarantees that their impact on the cavity mode lifetime vanishes as N increases. This is clearly observed in Fig. 7.2 for $N = 3$.

Although kept at a negligible level for the PhC cavity shown in Fig. 7.2, the impact of the leaky modes on ultrasmall resonators can be remarkably important for specifically optimized constructs. In Ref. 35, fully vectorial calculations have shown that by finely tuning the position and the diameter of the two inner holes of a single-hole-defect PhC cavity in a ridge waveguide, the Q values can be boosted and can be two orders of magnitude greater than those predicted by a Fabry-Perot model. Although the mirror reflectance is kept at a modest level ($R < 0.99$) in the optimized cavity, the intrinsic Q factor is as high as 10^5. This effect has been modestly observed from transmission measurements performed for cavities formed in planar waveguides with mirrors composed of ridges and slits.[36] The physical mechanism at the origin of the effect is less intuitive than the tapering and matching processes shown in Fig. 7.5. It relies on a pure electromagnetism effect based on transient fields, which allows recycling a fraction of the mirror radiation losses $\mathcal{L}$. The recycling has been shown to be driven by a leaky mode and may be beneficial or detrimental for the resonator performance.[35] For specific subwavelength defect lengths, the recycling mechanism is phase matched with the cavity Fabry-Perot mode and drastically reduces the mirror radiation. Similar phenomena, which result in a reduction of the far-field radiation, are probably encountered for 2D PhC cavities usually designed by repetitively varying the positions and the diameters of the neighbor holes surrounding the defect to optimize the cavity Q. In general, the optimization is successful[37,38] and recycling and tapering processes may join together to achieve Q enhancements by at least a factor 10.

Strong deviations from the Fabry-Perot model predictions are also obtained for defect waveguides that support more than a single propagative guided mode. This situation commonly occurs for b-type pillar cavities because the cylindrical semiconductor spacer has a diameter in the range of the wavelength in free space for most applications. Pillar microcavities have played a major role in the development of several optoelectronic devices, including vertical cavity surface emitting lasers,[7] and in the realization of cavity quantum electrodynamic experiments in the solid state.[39-42]

For large diameters (planar cavity), the cavity mode is basically composed of two counter-propagating plane waves that are bouncing in the spacer region between the two distributed Bragg reflectors. In this limit, the finite mode lifetime is only due to the transmission through the top and bottom mirrors. Since the planar cavity is etched to fabricate pillars, one expects additional loss by radiation into the air clad and therefore a drop of the Q factor. This has been observed in many early experiments,[43] and it is only recently that theoretical predictions[44] have shown that this common point of view is not correct. In fact, perfectly cylindrical pillar microcavities present an intricate and unexpected electromagnetic behavior for diameters slightly smaller than their emission wavelength in vacuum. Due to the existence of a propagative higher-order Bloch mode $\mathbf{B}^{(2)}$ in the fundamental gap of the distributed Bragg mirror, the quality factors display a fast oscillatory variation as a function of the pillar diameter, and can reach values well in excess of the quality factor Q_p of the reference planar cavity. These surprising effects have been recently observed[45] by measuring the photoluminescence of InAs self-assembled quantum dots grown in the middle of the GaAs spacers in a series of micropillars fabricated for many different diameters varying from 550 nm to 1.3 μm.

7.5 Conclusion

The five last years have seen considerable progresses in microcavity performance using semiconductor materials. The Q factors of PhC microcavities with small volumes near the theoretical limit $(\lambda/2n)^3$ have been boosted by three orders of magnitude. Presently, cavities in two-dimensional PhC membranes suspended in air offer Qs that are ten times larger than those in ridge waveguides on a substrate. One may suspect that 2D PhC bandgaps appear more suitable than more traditional approaches based on light confinement in z-invariant ridges to achieve ultrahigh Qs. Although the former benefits from a natural small group velocity effect [Eq. (7.3)] to enhance the mode lifetime, there is no fundamental limitation that may preclude the realization of extremely high Qs in ridge waveguides. The 10-fold difference between state-of-the-art Qs for the two types of cavity constructs is likely to be due to the drastic attention that has been devoted to PhC cavities over the past few years. Additionally, one has to keep in mind that high Qs are much more difficult to achieve in constructs on a substrate than in membranes suspended in air. The realization of still higher Qs will be challenging because controlling the mirror reflectance to achieve very small radiation losses in the range of 10^{-4} requires an accurate control of the hole positions and diameters in the nanometer range. In our opinion, the group velocity effect that has been only weakly considered until now in resonator designs[20,46,47] may represent an interesting design strategy.

However, there might be no real need to achieve higher Qs in the future for real applications, and it is important to realize all the benefits of the present improvements for future electrically driven devices.[48] For instance, new design strategies that would allow the realization of cavities that are only weakly sensitive to various types of disorders (hole diameter, position, etc.) are highly desirable.[49] Such immunity would, for example, considerably enhance the device yield.

Acknowledgment

The authors acknowledge many interesting and stimulating discussions with their colleagues and collaborators Jean-Paul Hugonin, Guillaume Lecamp, Philippe Velha, Jean-Claude Rodier, David Peyrade, Isabelle Robert-Philip, Jean-Michel Gérard, Emmanuel Picard, and Emmanuel Hadji. This work is partly supported by the MIRAMAN Project No. PNANO06-0215 and by the NanoEPR project of the 2006 NanoSci-ERA program.

References

1. Shields, A.J., "Semiconductor quantum light sources," *Nature Photon.* **1**, 215–223, 2007.
2. Purcell, E.M., "Spontaneous emission probabilities at radio frequencies," *Phys. Rev.* **69**, 681, 1946.
3. Yokoyama, H., and Ujihara, K., *Spontaneous Emission and Laser Oscillation in Microcavities*, CRC Press, Boca Raton (1995).
4. Hagness, S.C., Rafizadeh, D., Ho, S.T., and Taflove, A., "FDTD microcavity simulations: design and experimental realization of waveguide-coupled single-mode ring and whispering-gallery-mode disk resonators," *IEEE J. Lightwave Technol.* **15**, 2154–2165, 1996.
5. Kippenberg, T.J., Spillane, S.M., and Vahala, K.J., "Demonstration of ultra-high-Q small mode volume toroid microcavities on a chip," *Appl. Phys. Lett.* **85**, 6113–6115, 2004.
6. Little, B.E., Haus, H.A., Foresi, J.S., Kimerling, L.C., Ippen, E.P., and Ripin, D.J., "Wavelength switching and routing using absorption and resonance," *IEEE Phot. Technol. Lett.* **10,** 816–818, 1998.
7. Iga, K., Koyama, F., and Kinoshita, S., "Surface emitting semiconductor-lasers," *IEEE J. Quantum Electron.* **24**, 1845–1855, 1988.
8. Zhang, J.P., Chu, D.Y., Wu, S.L., Bi, W.G., Tiberio, R.C., Joseph, R.M., Taflove, A., Tu, C.W., and Ho, S.T., "Nanofabrication of 1-D photonic bandgap structures along a photonic wire," *IEEE Photon. Technol. Lett.* **8**, 491–93, 1996.
9. Foresi, J.S., Villeneuve, P.R., Ferrera, J., Thoen, E.R., Steinmeyer, G., Fan, S., Joannopoulos, J.D., Kimerling, L.C., Smith, H.I., and Ippen, E.P., "Photonic-bandgap microcavities in optical waveguides," *Nature* **390**, 143–145, 1997.
10. Velha, P., Rodier, J.C., Lalanne, P., Hugonin, J.P., Peyrade, D., and Hadji, E., "Ultra-high-reflectivity photonic-bandgap mirrors in a ridge SOI waveguide," *New J. Phys.* **8**, 204, 2006.
11. Joannopoulos, J.D., Meade, R.D., and Winn, J.N., *Photonic Crystals*, Princeton University Press, Princeton (1995).

12. Akahane, Y., Asano, T., Song, B.S., and Noda, S., "High-Q photonic nanocavity in two-dimensional photonic crystal," *Nature* **425**, 944–947, 2003.
13. Peter, E., Senellart, P., Martrou, D., Lemaître, A., Hours, J., Gérard, J.M., and Bloch, J., "Exciton-photon strong-coupling regime for a single quantum dot embedded in a microcavity," *Phys. Rev. Lett.* **95**, 067401, 2005.
14. Xia, F., Sekaric, L., and Vlasov, Y., "Ultracompact optical buffers on a silicon chip," *Nature Photon* **1**, 65–71, 2007.
15. Vuckovic, J., Pelton, M., Scherer, A., and Yamamoto, Y., "Optimization of three-dimensional micropost microcavities for cavity quantum electrodynamics," *Phys. Rev. A* **66**, 023808, 2002.
16. Palamaru, M., and Lalanne, P., "Photonic crystal waveguides: out-of-plane losses and adiabatic modal conversion," *Appl. Phys. Lett.* **78**, 1466–1468, 2001.
17. Haus, H.A., *Waves and fields in optoelectronics*, Prentice-Hall International, London (1984).
18. Vahala, K.J., "Optical microcavities," *Nature* **424**, 839–846, 2003.
19. Silberstein, E., Lalanne, P., Hugonin, J.P., and Cao, Q., "On the use of grating theory in integrated optics," *J. Opt. Soc. Am. A.* **18**, 2865–2875, 2001.
20. Sauvan, C., Lalanne, P., and Hugonin, J.P., "Slow-wave effect and mode-profile matching in Photonic Crystal microcavities," *Phys. Rev. B* **71**, 165118, 2005.
21. Tsia, K.K., and Poon, A.W., "Dispersion-guided resonances in 2D photonic-crystal-embedded microcavities," *Opt. Expr.* **12**, 5711–5722, 2004.
22. Yang, X., and Wong, C.W., "Design of photonic band gap nanocavities for stimulated Raman amplification and lasing in monolithic silicon," *Opt. Expr.* **13**, 4723–4730, 2005.
23. Krauss, T.F., and De La Rue, R.M., "Optical characterization of waveguide based photonic microstructures," *Appl. Phys. Lett.* **68**, 1613–1665, 1996.
24. Baba, T., Hamasaki, M., Watanabe, N., Kaewplung, P., Matsutani, A., Mukaihara, T., Koyama, F., and Iga, K., "A novel short-cavity laser with deep-grating distributed Bragg reflectors," *Jpn. J. Appl. Phys.* **35**, 1390–1394, 1996.
25. Krauss, T.F., Painter, O., Scherer, A., Roberts, J.S., De La Rue, R.M., "Photonic microstructures as laser mirrors," *Opt. Eng.* **37**, 1143–1148, 1998.
26. Peyrade, D., Silberstein, E., Lalanne, P., Talneau, A., and Chen, Y., "Short Bragg mirrors with adiabatic modal conversion," *Appl. Phys. Lett.* **81**, 829–831, 2002.
27. Zain, A.R.M., Gnan, M., Chong, H.M.H., Sorel, M., and De La Rue, R.M., "Tapered Photonic Crystal Microcavities Embedded in Photonic Wire

Waveguides With Large Resonance Quality-Factor and High Transmission," *IEEE Photon. Technol. Lett.* **20**, 6–8, 2008.

28. Velha, P., Picard, E., Hadji, E., Rodier, J.C., Lalanne, P., and Peyrade, D., "Ultra-High Q/V Fabry-Perot microcavity on SOI substrate" *Opt Expr.* **15**, 16090–16096, 2007.

29. Akahane, Y., Asano, T., Song, B.S., and Noda, S., "Fined-tuned high-Q photonic-crystal," *Opt. Expr.* **13**, 1202–1214, 2004.

30. Herrmann, R., Sünner, T., Hein, T., Löffler, A., Kamp, M., and Forchel, A., "Ultrahigh-quality photonic crystal cavity in GaAs," *Opt. Lett.* **31**, 1229–1231, 2006.

31. Song, B.S., Noda, S., Asano, T., and Akahane, Y., "Ultra-high-Q photonic double-heterostructure nanocavity," *Nature Mater.* **4**, 207–210, 2005.

32. Kuramochi, E., Notomi, M., Mitsugi, S., Shinya, A., and Tanabe, T., "Ultrahigh-Q photonic crystal nanocavities realized by the local width modulation of a line defect," *Appl. Phys. Lett.* **88**, 041112, 2006.

33. Tanabe, T., Notomi, M., Kuramochi, E., Shinya, A., and Taniyama, H., "Trapping and delaying photons for one nanosecond in an ultrasmall high-Q photonic crystal nanocavity," *Nature Photon.* **1**, 49–52, 2007.

34. http://www.virgo.infn.it/

35. Lalanne, P., Mias, S., and Hugonin, J.P., "Two physical mechanisms for boosting the quality factor to cavity volume ratio of photonic crystal microcavities," *Opt Expr.* **12**, 458–467, 2004.

36. Riboli, F., Recati, A., Daldosso, N., Pavesi, L., Pucker, G., Lui, A., Cabrini, S., and Di Fabrizio, E., "Photon recycling in Fabry-Perot micro-cavities based on Si_3N_4 waveguides," *Photon. Nanostruct.: Fundamentals Appl.* **4**, 41–46, 2006.

37. Park, H.G., Hwang, J.K., Huh, J., Ryu, H.Y., Lee, Y.H., and Kim, J.S., "Nondegenerate monopole-mode two-dimensional photonic band gap laser," *Appl. Phys. Lett.* **79**, 3032–3034, 2001.

38. Vuckovic, J., Loncar, M., Mabuchi, H., and Scherer, A., "Design of photonic crystal microcavities for cavity QED," *Phys. Rev. E* **65**, 016608, 2002.

39. Gérard, J.M., Sermage, B., Gayral, B., Costard, E., and Thierry-Mieg, V., "Enhanced spontaneous emission by quantum boxes in a monolithic optical microcavity," *Phys. Rev. Lett.* **81**, 1110–1113, 1998.

40. Graham, L.A., Huffaker, D.L., and Deppe, D.G., "Spontaneous lifetime control in a native-oxide-apertured microcavity," *Appl. Phys. Lett.* **74**, 2408–2410, 1999.

41. Solomon, G., Pelton, M., and Yamamoto, Y., "Single-mode spontaneous emission from a single quantum dot in a three-dimensional microcavity," *Phys. Rev. Lett.* **86**, 3903–3906, 2001.

42. Reithmaier, J.P., Sek, G., Löffler, A., Hofmann, C., Kuhn, S., Reitzenstein, S., Keldysh, L.V., Kulakovskii, V.D., Reinecke, T.L., and Forchel, A., "Strong coupling in a single quantum dot–semiconductor microcavity system," *Nature* **432**, 197–200, 2004.

43. Gérard, J.M., "Solid-state cavity quantum electrodynamics with self-assembled quantum dots," *Single Quantum Dots: Physics and Applications*, Springer, Heidelberg (2003).

44. Lalanne, P., Hugonin, J.P., and Gérard, J.M., "Electromagnetic study of the Q of pillar microcavities in the small limit diameter," *Appl. Phys. Lett.* **84**, 4726–4728, 2004.

45. Lecamp, G., Hugonin, J.P., Lalanne, P., Braive, R., Varoutsis, S., Laurent, S., Lemaître, A., Sagnes, I., Patriarche, G., Robert-Philip, I., and Abram, I., "Submicron-diameter semiconductor pillar microcavities with very high quality factors," *Appl. Phys. Lett.* **90**, 091120, 2007.

46. Sauvan, C., Lalanne, P., and Hugonin, J.P., "Tuning holes in photonic-crystal nanocavities," *Nature* **429**(6988), 2004 (online only).

47. Bordas, F., Steel, M.J., Seassal, C., and Rahmani, A., "Confinement of band-edge modes in a photonic crystal slab," *Opt. Expr.* **15**, 10890–10902, 2007.

48. Park, H.G., Kim, S.H., Kwon, S.H., Ju, Y.G., Yang, J.K., Baek, J.H., Kim, S.B., and Lee, Y.H., "Electrically Driven Single-Cell Photonic Crystal Laser," *Science* **305**, 1444–1447, 2004.

49. Gerace, D., and Andreani, L.C., "Effect of disorder on propagation losses and cavity Q-factors in photonic crystal slabs," *Photon. Nanostruct.: Fundamentals Appl.* **3**, 120–128, 2005.

Chapter 8
Limits to Optical Components

David A. B. Miller
Stanford University

8.1 Introduction
8.2 Background
8.3 Mathematical Approach
    8.3.1 Communications modes
    8.3.2 New theorem for strong or multiple scattering
8.4 Limit to the Performance of Linear Optical Components
    8.4.1 Explicit limit for one-dimensional systems
    8.4.2 Slow light limit
    8.4.3 Limit to dispersion of pulses
8.5 Future Directions
8.6 Conclusions
References

8.1 Introduction

Is there a limit to the performance of linear optical components? Suppose we asked a specific question, such as how much glass would we need to make a device that would split 32 wavelengths in the telecommunications C-band near 1.5 µm wavelength. Intuitively, we would probably agree that 1 µm^3 of glass would not be enough. We could, however, certainly achieve this goal with a roomful of optics and, in fact, we know we could purchase a commercial arrayed waveguide grating device, with a scale of centimeters, to make such a splitter. So our intuitive experience suggests that there is some limit on performance of such optical components, though historically we have not had a general limit we could use.

    The need for a limit is not merely academic. With modern nanophotonic techniques, we can make a very broad variety of devices, some with little or no precedent. In part because of the high refractive index contrast available to us in photonic nanostructures, the design of such devices is often quite difficult, and we would at least like to know when to stop trying to improve the performance.

Classes of devices of interest to us could include dispersive structures, slow light elements, holograms, or any kind of device that separates different kinds of input beams or pulses to different positions in space or time.

Recently, we have been able to devise quite a general approach to limits for the performance of linear optical components.[1] This approach gives upper limits to performance that are quite independent of the details of the design, being dependent instead only on the overall geometry of the device and, for example, the largest dielectric constant variation anywhere in the structure at any wavelength. This overall limit has already been applied to calculate limits to dispersive devices[1] and to slow light.[2] Here we will introduce this limit, summarizing its derivation and the applications thus far.

This limit is based on the idea of counting possible orthogonal wave functions that can be generated when an optical component acts to "scatter" an incident wave into a receiving volume. This idea in turn is based on some earlier work[3] that is a generalization of diffraction theory to volumes, in which we can count the orthogonal "communications modes"—the best choices of sources in one volume and the resulting waves in another for communicating between the two.

In this chapter, after summarizing the background to the need for a new limit, especially in nanophotonics, we will then introduce the underlying mathematical methods, including a discussion of communications modes and their applications. We will give the proof of a new general theorem for strong and/or multiple scattering, a theorem that underlies our limit to optical components. Then we will summarize two applications of the new limit, namely, one to slow light devices and the other to dispersion of pulses, before indicating future directions and drawing conclusions.

8.2 Background

There has been extensive prior work on limits or at least design techniques for optical components that are intended to separate beams or pulses (see references in Ref. 1). That prior work, however, largely deduces limits for devices designed in a specific way, such as a simple resonator, coupled resonators, a periodic structure, or a grating. For example, we could deduce limits to group delay for a Fabry-Perot resonator, based on an explicit model of such a resonator. Such a limit would show a trade-off between the magnitude of the group delay and the bandwidth over which such a delay would be available. In such structures, we can often have relatively simple and sometimes intuitive models of how adjusting some part of the design will lead to a specific consequence. But there is a class of structures that can be made, and that can have relatively very good performance, for which there are no such models. Such structures can result from purely numerical optimization in design.

For example, we recently set out to design and test superprism wavelength splitters made from one-dimensional dielectric stacks.[4–6] In a superprism, the effective angle of propagation of a beam in a structure can vary very strongly with the wavelength of the incident light. Such superprism effects have been known for some time, at least for periodic structures,[7] and can be understood in that case in terms of the band structure of the periodic system. Better performance, both in the linearity of the shift with wavelength and in the magnitude of the shift, can, however, be obtained from nonperiodic structures in which the thickness of each layer is potentially adjusted during the design process.[4,5] It is easy to understand that such a structure might be able to give better performance—there are simply more engineering degrees of freedom available if we do not merely restrict ourselves to designing the unit cell of a periodic structure. We also found in this work that, for over 600 designs with different starting design concepts, different materials, and different optimizations, the performance of all of these lay near or below a specific line,[5] suggesting some underlying limit.

In work on designs of two-dimensional structures, in this case for mode splitting,[8,9] we were able to devise quite effective and very compact designs by numerical optimization based on removing or adding dielectric "rods" in a region of the structure, but in this case in particular we simply do not know "how" it works. We cannot say that a particular rod or group of rods does a specific function in a specific way. All the rods interact with the optical field in performing the final function. Because we do not know how it works, we also cannot say how well it could work, or what the limit to it should be based on some analysis of this specific device type. The "exhaustive search" approach of trying all possible designs to establish the best one is generally computationally very expensive, so we would very much like the guidance of some limit that was independent of device details.

The challenge is to devise a limit to the performance of optical components, completely independent of the design approach, so that we can bound the performance not only of the kinds of devices we have used up to now, but for any future device based on any kind of optical structure, including the many possibilities enabled by nanophotonics.

8.3 Mathematical Approach

Our approach to this limit[1] is based on counting the number of distinct available channels or "modes" for communicating with an incident wave through a "scattering" volume (the volume that contains our optical component structure of interest) to a receiving volume. As we will see below, with only simple information about the scattering structure, such as its size and shape and the largest dielectric constant variation within it, we can deduce upper limits to this available number of modes. When we are also able to state the number of channels needed for a given optical function, then we can deduce whether that function could possibly be performed by such a structure regardless of how we design it.

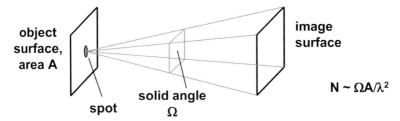

Figure 8.1 Illustration of diffraction from an "object" surface to an "image" surface, with the number of distinct spatial channels being given approximately by the number N of resolvable spots as deduced from a simple diffraction model.[3]

8.3.1 Communications modes

Before discussing the full problem of scattering waves from one volume to another, we can look at the simpler problem of communicating from sources in a transmitting volume to generate waves in a receiving volume.[3] If those volumes were simply thin parallel surfaces (see Fig. 8.1), and we were considering waves of a specific frequency only, in optics we would fall back on our understanding of diffraction to tell us how many distinct channels there are.[10] We would expect that there would be essentially one distinct channel possible for each resolvable spot, where the size of the resolvable spot is deduced from diffraction theory. Although such an approach is somewhat informal, it is essentially correct for such a problem.

There is a more formal and rigorous approach that was initially understood at least for simple (e.g., square or circular) parallel surfaces,[11,12] and that can be extended also to the case of volumes.[3] We sketch this approach here. Consider transmitting and receiving volumes,[3] as shown in Fig. 8.2, which are generalizations of the object surface and image surface, respectively. We presume that we have some source function in the transmitting volume V_T, and we write this function as $|\psi_T\rangle$. This source function generates a wave, leading specifically to a wave $|\phi_R\rangle = G_{TR} |\psi_T\rangle$ in the receiving volume V_R, where G_{TR} is the coupling operator (basically the Green's function of the wave equation).

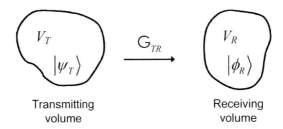

Figure 8.2 Illustration of a source function $|\psi_T\rangle$ in a transmitting volume V_T, giving rise to a wave $|\phi_R\rangle$ in a receiving volume V_R, formally through a coupling operator G_{TR} (the Green's function of the wave equation).

Note, incidentally, that here and below we are using Dirac's "bra-ket" notation as a convenient notation for linear algebra; the reader can think of a "ket" such as $|\psi_T\rangle$ as a column vector whose elements are the values of the function at the various values of the argument of the function, for example. The "bra" vector $\langle\psi_T|$ is then the row vector that is the Hermitian adjoint of $|\psi_T\rangle$. The Hermitian adjoint, which can also be indicated by a superscript dagger "†", is the transpose of a vector or matrix in which we also take the complex conjugate of all the elements (so, e.g., $\langle\psi_T| = |\psi_T\rangle^\dagger$). We use "san serif" letters, such as G, to represent operators, which we can think of as matrices.

Suppose we then wish to find the best possible set of distinct (i.e., mathematically orthogonal) source functions in the transmitting volume that would generate the largest possible amplitudes of wave functions in the receiving volume. The general solution to such a problem is known. It is well understood also in the context of imaging between surfaces (see Ref. 13, and references therein). We find the solution by formally performing the singular value decomposition of the coupling operator G_{TR} between these volumes. The best choices of source functions are the (orthogonal) eigenfunctions $|\psi_{Ti}\rangle$ of the operator $\mathsf{G}^\dagger_{TR}\mathsf{G}_{TR}$, with eigenvalues $|s_i|^2$. The corresponding wave functions in the receiving volume are the (orthogonal) eigenfunctions $|\phi_{Ri}\rangle$ of the operator $\mathsf{G}_{TR}\mathsf{G}^\dagger_{TR}$, also with eigenvalues $|s_i|^2$. Once we have solved these two eigenproblems, we can write G_{TR} in its singular value decomposition form as

$$\mathsf{G}_{TR} = \sum_i s_i |\phi_{Ri}\rangle\langle\psi_{Ti}|, \qquad (8.1)$$

where the s_i are called the singular values.

Specifically, the source function $|\psi_{Tm}\rangle$ leads to a corresponding wave $s_m|\phi_{Rm}\rangle$ in the receiving volume; such a pair of one of these source functions $|\psi_{Tm}\rangle$ in the transmitting volume and its corresponding wave function $|\phi_{Rm}\rangle$ in the receiving volume can be called a "communications mode."[3] These communications modes represent the best possible set of source and wave function pairs for establishing orthogonal communications channels from one volume to another. If we also choose the sets $|\psi_{Ti}\rangle$ and $|\phi_{Ri}\rangle$ to be normalized, then the singular values represent the coupling strengths of these modes.

When we take such an approach to the coupling between thin, plane-parallel square or circular transmitting (i.e., "object") and receiving (i.e., "image") volumes relatively far apart (so we can use a paraxial approximation in considering wave propagation), then there are specific so-called "prolate spheroidal" functions that are the eigenfunctions of these problems for each surface.[11,12] These functions are not simple small spots on one surface or the other—each of the prolate spheroidal functions covers the whole surface—but these functions are truly orthogonal on a given surface, in contrast to a set of spots that are only approximately orthogonal insofar as their "tails" do not overlap very much. These prolate spheroidal functions then form the communications modes of truly orthogonal channels. With such functions, it is known that the singular values are approximately all the same up to some critical number, after which they fall off drastically. That critical number corresponds in this case to the number N we would deduce from the idea of counting the number of resolvable spots.

With the general mathematical formalism above for singular value decomposition, we can also solve for communications modes between volumes[3] rather than merely surfaces, and we need not restrict ourselves to communications modes that can be calculated analytically; the numerical prescription for finding the communications modes for arbitrary volumes is quite clear in principle, merely requiring finding eigenfunctions of some specific operators. In this more general approach, we are also not restricted to paraxial approximations, and we can even find the communications modes in near-field problems.[3,14]

For more arbitrary volumes, it is not in general true that the singular values are all of similar size up to some critical number; that similarity in size of singular values arguably is a characteristic of volumes of uniform thickness.[3] There is, however, another relation that does bound the singular values. It is quite generally true for a linear operator that the sum of the squared moduli of its matrix elements is independent of the (orthonormal) basis sets used to represent it. One way of representing the coupling operator is in terms of its Green's function [which is technically the expansion of the operator on continuous (delta-function) basis sets, one set for each of the transmitting and receiving volumes]. Another representation is in terms of the singular value decomposition sets as in Eq. (8.1) above. Let us take as a concrete simple example the (retarded) Green's function for a free-space monochromatic scalar wave, which is, for a point source at position $\mathbf{r}_T$ in the transmitting volume and a resulting wave at position $\mathbf{r}_R$ in the receiving volume,

$$G_{TR}(\mathbf{r}_T, \mathbf{r}_R) = \frac{\exp(-ik|\mathbf{r}_T - \mathbf{r}_R|)}{4\pi|\mathbf{r}_T - \mathbf{r}_R|}. \tag{8.2}$$

We can then equate the sum of the moduli squared of the singular values, which is the sum of the moduli squared of the matrix elements of G_{TR} in its singular value decomposition form, to the integral of the moduli squared of $G_{TR}(\mathbf{r}_T, \mathbf{r}_R)$ over the volumes, i.e.,

$$\sum_i |s_i|^2 = \iint_{V_T, V_R} |G_{TR}(\mathbf{r}_T, \mathbf{r}_R)|^2 d^3\mathbf{r}_R d^3\mathbf{r}_T = \frac{1}{(4\pi)^2} \iint_{V_T, V_R} \frac{1}{|\mathbf{r}_T - \mathbf{r}_R|^2} d^3\mathbf{r}_R d^3\mathbf{r}_T. \tag{8.3}$$

Hence, performing a simple volume integral over the two volumes gives us an absolute upper bound to the sum of the squares of the coupling strengths between the volumes.[3] Even if the coupling strengths of the communications modes do not have the simple form of being approximately constant up to some value, there is still a limit on the sum of their squares. We can view this statement as being a generalization of the concept of the "diffraction limit," now for arbitrary three-dimensional volumes. This limit does agree with the specific results for planar surfaces above. Note then that if we try to defeat the diffraction limit, we will necessarily have to use communications modes that are very weakly coupled—in other words, we will need relatively very large source amplitudes.

This concept of communications modes has recently seen a number of applications in optics.[13-17] The theory has also been generalized to vector electromagnetic waves,[18] and it is helpful in understanding the limits to the synthesis of light fields in three dimensions.[19] It has also proved useful in analyzing wireless communications,[20-22] where the transmitting and receiving antennas and volumes are not plane surfaces.

The communications mode concept also illustrates some of the power of considering waves between volumes in terms of orthogonal source and wave functions. The theorem below expands on this approach for a quite different application.

8.3.2 New theorem for strong or multiple scattering

As mentioned above, we will think of our optical component in general as a "scatterer"—that is, there is some incident wave that is "scattered" by the optical component, generating a wave in the receiving volume. Our scattering problem here is particularly severe; we want to consider arbitrarily strong and/or multiple scattering by the object, including arbitrarily large and/or abrupt changes in dielectric constants. As a result, we cannot proceed simply by adding up all of the successive scatterings in some series—such a series will typically not converge. Below we will derive the core theorem that we use to deal with the problem. This derivation closely follows that of Ref. 1.

We consider (Fig. 8.3) two volumes, a scattering volume V_S that contains a scatterer that is our optical device (such as a dielectric with a dielectric constant that can vary strongly within the volume), and a receiving volume V_R in which we want to generate waves.

An incident wave (the input pulse or input beam in Fig. 8.3) will lead to some net wave within the scattering volume. That net wave will interact with the scatterer to generate some resulting effective source in the scatterer. A specific net source $|\psi_{Sm}\rangle$ in the scattering space will generate a wave $|\phi_{SCm}\rangle$ within the scattering space through the Green's function operator G_S within the scattering space, i.e.,

$$|\phi_{SCm}\rangle = G_S |\psi_{Sm}\rangle. \qquad (8.4)$$

Here $|\psi_{Sm}\rangle$ represents *all* sources in the volume. Hence, G_S is the "free-space" Green's function. We presume here that the only sources are the induced polarizations or currents generated as a result of the interaction between the net wave and the scattering material; we have no other sources in the scattering

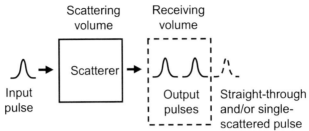
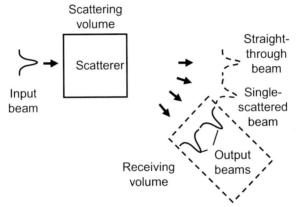

Figure 8.3 Illustration of scattered (a) pulses and (b) beams for temporal and spatial dispersers, respectively. The "straight-through" and "single-scattered" pulses or beams may not actually be present physically, but the theory first considers outputs that are orthogonal to what both of these would be mathematically. Here we show the case where the straight-through and single-scattered pulses or beams miss the receiving volume, though in general they may not.[1]

volume. We presume some incident wave $|\phi_{Im}\rangle$ caused all of these sources through its scattering. The net wave $|\phi_{Sm}\rangle$ in the space must be the sum of the incident and scattered waves, i.e.,

$$|\phi_{Sm}\rangle = |\phi_{Im}\rangle + |\phi_{SCm}\rangle. \tag{8.5}$$

Any wave $|\phi\rangle$, interacting with the scatterer, will in turn give rise to sources $|\psi\rangle$ through some other linear operator C, an operator that we can think of simply as representing the dielectric constant of the material, for example. Therefore,

$$|\psi\rangle = C|\phi\rangle. \tag{8.6}$$

We must have self-consistency, and so we require that the net source in the scattering volume, $|\psi_{Sm}\rangle$, is the one that would be generated by the net wave $|\phi_{Sm}\rangle$ interacting with the scatterer. Hence,

Limits to Optical Components 161

$$|\psi_{Sm}\rangle = C|\phi_{Sm}\rangle = C|\phi_{Im}\rangle + C|\phi_{SCm}\rangle,$$
$$= C|\phi_{Im}\rangle + CG_S|\psi_{Sm}\rangle = C|\phi_{Im}\rangle + A_S|\psi_{Sm}\rangle, \quad (8.7)$$

where

$$A_S = CG_S. \quad (8.8)$$

If we presume now that we have some specific source function $|\psi_{Sm}\rangle$ in the scattering space, then there also must be some linear operator G_{SR} (again a "free-space" Green's function, similar to the G_{TR} considered above for the communications mode problem) that we could use to deduce the resulting wave $|\phi_{Rm}\rangle$ in the receiving space, i.e.,

$$|\phi_{Rm}\rangle = G_{SR}|\psi_{Sm}\rangle. \quad (8.9)$$

Given that we want to separate light beams or pulses into some receiving space, we ask that the various waves we generate in the receiving space are mathematically orthogonal, just as we were considering above for the communications modes between two volumes. We will try to deduce some limit on the number of such orthogonal functions $|\phi_{Rm}\rangle$ that can be generated in the receiving space. In particular, here we will use the sets of functions $|\psi_{Sm}\rangle$ and $|\phi_{Rm}\rangle$ that are the communications modes between the scattering and receiving spaces, i.e., the singular value decomposition of G_{SR}, and specifically we will restrict these sets to those elements corresponding to nonzero singular values, i.e., we only want to consider source functions that give rise to nonzero wave amplitudes in the receiving space.

We will separate the counting of orthogonal waves into two parts. Specifically, we will come back later and consider the waves that correspond to "straight-through" or "single-scattered" waves. Straight-through waves are the waves that would exist in the receiving space in the absence of any scatterer; they correspond to propagation of the incident wave straight through the scattering volume. Single-scattered waves are the waves that would hypothetically arise from the scattering of the incident wave if it were imagined to be completely unchanged by its interaction with the scatterer, i.e., formally a wave $G_{SR}C|\phi_{Im}\rangle$. We will be interested for the moment only in waves in the receiving space that are formally orthogonal to both the straight-through and single-scattered hypothetical waves. These concepts are sketched in Fig. 8.3. This neglect of straight-through and single-scattered means we are only counting those orthogonal waves in the receiving volume that are the result of strong and/or multiple scattering in the scattering volume.

Our neglect of straight-through waves means that Eq. (8.9) gives the total wave in the receiving space, and thus that wave becomes

$$|\phi_{Rm}\rangle = G_{SR}|\psi_{Sm}\rangle = G_{SR}C|\phi_{Im}\rangle + G_{SR}A_S|\psi_{Sm}\rangle. \quad (8.10)$$

Because we presume we are only interested in scattered waves $|\phi_{Rm}\rangle$ that are orthogonal to the single-scattered wave $|\phi_{RIm}\rangle = G_{SR}C|\phi_{Im}\rangle$, by definition we have

$$\langle\phi_{Rm}|G_{SR}C|\phi_{Im}\rangle = 0. \tag{8.11}$$

Hence, from Eqs. (8.10) and (8.11),

$$\begin{aligned}\langle\phi_{Rm}|\phi_{Rm}\rangle &= \langle\psi_{Sm}|G_{SR}^{\dagger}G_{SR}|\psi_{Sm}\rangle, \\ &= 0 + \langle\psi_{Sm}|G_{SR}^{\dagger}G_{SR}A_S|\psi_{Sm}\rangle.\end{aligned} \tag{8.12}$$

Now, since the $|\psi_{Sm}\rangle$ are by definition complete and orthonormal for the source space of interest, we can introduce the identity operator for that space, which we can write as $I_{HS} = \Sigma_j |\psi_{Sj}\rangle\langle\psi_{Sj}|$, to obtain from Eq. (8.12)

$$\begin{aligned}\langle\psi_{Sm}|G_{SR}^{\dagger}G_{SR}|\psi_{Sm}\rangle &= \sum_j \langle\psi_{Sm}|G_{SR}^{\dagger}G_{SR}|\psi_{Sj}\rangle\langle\psi_{Sj}|A_S|\psi_{Sm}\rangle, \\ &= \langle\psi_{Sm}|G_{SR}^{\dagger}G_{SR}|\psi_{Sm}\rangle\langle\psi_{Sm}|A_S|\psi_{Sm}\rangle,\end{aligned} \tag{8.13}$$

where in the last step we have used the fact that

$$\langle\psi_{Sm}|G_{SR}^{\dagger}G_{SR}|\psi_{Sj}\rangle = \langle\phi_{Rm}|\phi_{Rj}\rangle = 0 \text{ unless } m = j, \tag{8.14}$$

because of the orthogonality we are enforcing for the waves $|\phi_{Rm}\rangle$ we want to generate in the receiving volume. Hence, we come to the surprising conclusion from Eq. (8.13) that for each m for which $|\psi_{Sm}\rangle$ gives rise to a nonzero wave in the receiving space, i.e., for which

$$\langle\psi_{Sm}|G_{SR}^{\dagger}G_{SR}|\psi_{Sm}\rangle \neq 0, \tag{8.15}$$

then

$$\langle\psi_{Sm}|A_S|\psi_{Sm}\rangle = 1, \tag{8.16}$$

and hence, trivially,

$$|\langle\psi_{Sm}|A_S|\psi_{Sm}\rangle|^2 = 1. \tag{8.17}$$

If we could separately establish a result of the form

$$\sum_i |\langle\psi_{Si}|A_S|\psi_{Si}\rangle|^2 \leq S_A \tag{8.18}$$

for some finite number S_A, then we would conclude that the maximum number M of possible orthogonal waves that could be generated in the receiving space by our strong scattering is

$$M \leq S_A. \tag{8.19}$$

An important additional result,[1] which we will not prove here, is that we can quite generally evaluate such a limit, and we can split it into two parts. Specifically, we can write

$$M \leq \sqrt{N_C N_{GS}}. \tag{8.20}$$

Here,

$$N_C \equiv Tr(\mathbf{C}^\dagger \mathbf{C}) \tag{8.21}$$

is essentially the integral of the modulus squared of the dielectric constant variation in the structure, and

$$N_{GS} \equiv Tr(\mathbf{G}_S^\dagger \mathbf{G}_S) \tag{8.22}$$

is essentially the integral of the modulus squared of the wave equation's Green's function within the scattering volume.

Hence, we can see the core of a remarkable bound on the possible performance of optical components. First, there is a number M that we can evaluate, based only on average properties of the dielectric medium and the shape of the scattering volume. Second, that number tells us an upper bound on the number of orthogonal functions that can be generated as a result of multiple and/or strong scattering into the receiving volume (technically, we are only counting those functions that are orthogonal to the single-scattered or straight-through waves, though for many strong scattering problems we will be able to deal with these separately or discount them altogether). For each such orthogonal function we want to be able to generate, we use up one unit of "strength" of the scatterer. Since the scatterer only overall has a finite "strength" M, there is a bound to the number of such orthogonal functions we can generate by this scattering into the receiving volume.

Note a key difference between this limit and the limit calculated above for the communications modes in Eq. (8.3). The communications modes limit was a bound on the sum of the strengths of the couplings between the transmitting and receiving volumes. It did not tell us directly the number of the communications modes that would be strongly coupled. We needed some other criterion to determine the actual coupling strengths of such strongly coupled modes. The new scattering limit we have proved here, Eq. (8.20), gives us an upper bound on the number of possible orthogonal functions that can be generated as a result of scattering from the scattering volume into the receiving volume, independent of the strengths of those scatterings (as long as they are nonzero). Once we can say how many functions we need to be able to control in the receiving volume for our optical component (scatterer) to have performed our desired optical operation, then we can use our new limit to tell us if that operation is impossible with an optical component of a given volume and given range of dielectric constants.

8.4 Limit to the Performance of Linear Optical Components

The use of this limit involves two steps. First, for some class of optical structures and problems of interest, we need to evaluate $\sqrt{N_C N_{GS}}$. Second, we need to think of how to express the optical operation we wish to perform in terms of the number of orthogonal functions we will need to be able to control in the receiving space. Thus far in our work, we have considered both of these steps for the simple case of one-dimensional systems, i.e., systems such as dielectric stack structures, a beam within a medium (such as an atomic vapor), or (with an appropriate renormalization to allow for mode overlap) single-mode guided wave structures, and we will summarize some results below.

8.4.1 Explicit limit for one-dimensional systems

As shown in Ref. 1, we can obtain simple explicit results for one-dimensional systems, i.e., any systems that can be described by a wave equation for a wave of frequency f_o that can be written as

$$\frac{d^2\phi}{dz^2} + k_o^2 \phi = -k_o^2 \eta(z, f_o) \phi. \tag{8.23}$$

Here, $k_o = 2\pi/\lambda_o = 2\pi f_o / v_o$, where v_o is the wave velocity and λ_o is the wavelength, both in the background medium. This is an appropriate equation for electromagnetic waves in one-dimensional problems in isotropic, nonmagnetic materials with no free charge or free currents. Then η is the fractional variation in the relative dielectric constant in the structure, i.e.,

$$\eta(z, f_o) \equiv \frac{\Delta\varepsilon(z, f_o)}{\varepsilon_{ro}}, \tag{8.24}$$

where ε_{ro} is the background relative dielectric constant, the wave velocity in the background medium is $v_o = c/\sqrt{\varepsilon_{ro}}$, where c is the velocity of light, and for a relative dielectric constant $\varepsilon(z, f_o)$, we define $\Delta\varepsilon = \varepsilon(z, f_o) - \varepsilon_{ro}$. (Note that ε may be complex.) With appropriate rescaling of the dielectric constant variation to include mode overlap, such an approach can also be taken for any single-mode system, such as a single-mode waveguide.

We will restrict ourselves here to situations where the frequency bandwidth δf of interest is much less than the center frequency f_c, and where the thickness L of the scattering medium is much larger than the wavelength $\lambda_c = v_o / f_c$ in the background material at the center frequency. Here we presume the scattering structure outputs pulses by transmission into a receiving space that is "behind" the slow light structure, as in Figs. 8.3 and 8.4. (The case of reflection rather than transmission is also easily handled, and gives similar results.[1,2]) We also allow the receiving space thickness, Δz_R, to be arbitrarily long so that it will capture any possible orthogonal function that results from scattering. With these simplifying restrictions, we can evaluate the quantities $Tr(C^\dagger C) \le n_{tot} \eta_{max}^2$,

Limits to Optical Components 165

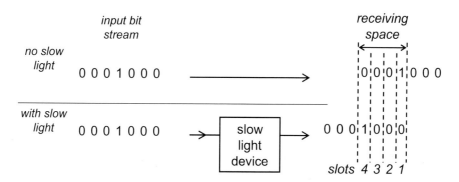

Figure 8.4 Illustration of a bit pattern delayed by three bit periods by scattering in a slow light device. In the design of the device, the scattering into a total of four bit slots has to be controlled so that the "1" appears in slot 4, and "0"s appear in each of slots 3, 2, and 1, hence requiring the control of four orthogonal functions in the receiving space.[2]

where η_{max} is the maximum value of $|\eta|$ at any frequency within the band of interest at any position within the scatterer, and $Tr(G_S^\dagger G_S) = n_{tot}(\pi^2/3)(L/\lambda_c)^2$. Here, $n_{tot} = 2\delta f \Delta z_R / v_o$ is the number of degrees of freedom required to define a function of bandwidth δf over a time $\tau = \Delta z_R / v_o$. The resulting M from Eq. (8.20) becomes

$$M \le n_{tot} \frac{\pi}{\sqrt{3}} \frac{L}{\lambda_c} \eta_{max}. \qquad (8.25)$$

If the scatterer has a similar range of variation of η over the entire scattering volume and if there are no dielectric constant resonances that are sharp compared to the frequency band of interest, then we can use the root-mean-square variation, η_{rms}, of the magnitude of η, averaged over position and frequency[1] instead of η_{max} in Eq. (8.25) and the expressions that follow below.

8.4.2 Slow light limit

A particularly clear and simple example of the application of this limit[2] is to the problem of slow light. Our approach here follows that of Ref. 2. We would like to understand for some linear optical system just what are the limits to the amount of delay we can get of some bit stream, in particular the number of bits of delay. We obtain quite a simple and general upper bound answer, an answer that does not even depend on the bandwidth of the slow light system. This approach does not require that we assume any particular pulse shape, and, unlike other approaches to slow light limits,[23–25] it does not rely on the concept of group velocity, a concept that has limited meaning within anything other than a uniform or periodic structure.

We note that we need N linearly independent functions to represent an arbitrary N-bit binary number in a receiving space, and so the basis set of physical wave functions used to represent the number in the receiving space must have at least N orthogonal elements. Suppose we have an incident bit stream with a logical "1" surrounded by logical "0"s (see Fig. 8.4). Without a slow light device, the bit stream propagates through to the receiving space. With the slow light device, however, we want the bit stream to be shifted, so that the "1" appears in a later bit period. To obtain a delay of S bit periods in the scattering in the slow light device, we need to be able to control the amplitudes of at least $S + 1$ orthogonal physical functions in the receiving space, so that we can center a function representing a "1" in the $(S + 1)$th bit period, and functions representing "0"s centered in the other S bit periods. We presumably control the amplitudes of these various functions by the design of the slow light device (the "scatterer").

In optics we typically consider pulses on a carrier frequency f_c. Then we note that there will be two different but almost identical pulses that have essentially the same amplitude envelope, but that are formally orthogonal only because they have a carrier phase that differs by 90 deg. Since typically we look only at the amplitude envelope, we then need to double the number of amplitudes we control in the bit periods containing logic "0"s, so that both of these pulses are "low" (i.e., logic "0"s). We likely do not care about the carrier phase of the pulse in the desired slot, so we need not add in another degree of freedom to control that. In this case, therefore, we need to control $2S + 1$ orthogonal functions in the receiving space to delay a pulse or bit stream by S bit periods.

One subtlety that we have to deal with in applying this limit is in the counting of available orthogonal functions as given by Eq. (8.25). Because we chose the receiving space to be arbitrarily long, we have included in M as separate possibilities the scattering not only of the pulse of interest, but also of every distinct delayed version of it. There are n_{tot} such different delayed versions of the same scattering that fit in the receiving space. Since we need consider only one of these, we can remove the factor n_{tot} below.

We also previously noted that this number M is the number of orthogonal waves possible in the receiving volume that are also orthogonal to the straight-through and single-scattered waves. At best, for a transmission device, considering these other two waves could at most add in two more available controllable degrees of freedom. Hence, we finally obtain for the upper limit to the number of accessible orthogonal functions in the receiving space for the scattering of a single pulse

$$M_{tot} \leq 2 + \frac{\pi}{\sqrt{3}} \frac{L}{\lambda_c} \eta_{max} . \qquad (8.26)$$

Given that we need to control at least $2S + 1$ amplitudes to delay by S bits, we must therefore have $M_{tot} \geq 2S + 1$. Hence, the maximum delay S_{max} in bit periods that we can have is

$$S_{max} \leq \frac{1}{2} + \frac{\pi}{2\sqrt{3}} \frac{L}{\lambda_c} \eta_{max} . \qquad (8.27)$$

Limits to Optical Components

Suppose, for example, we want to delay by 32 bits (i.e., $S = 32$). We choose to work with a layered structure of glass ($\varepsilon_r = 2.25$) and air ($\varepsilon_r = 1$)—thus, $\eta_{max} = 1.25$—and we choose a center wavelength of 1.55 µm. Then,

$$L \geq \left(S - \frac{1}{2}\right) \frac{2\sqrt{3}}{\pi \eta_{max}} \lambda_c = 43 \text{ µm} . \tag{8.28}$$

Note that this limit cannot be exceeded for a linear one-dimensional fixed glass/air optical structure, no matter how we design it; we would always need at least 43 µm of thickness.

8.4.3 Limit to dispersion of pulses

We can also calculate in a similar way an upper bound to a structure that we wish to have separate out pulses of different center wavelengths to different delays in time. The counting of the required number of orthogonal functions here is somewhat more complicated, so we omit the details in this summary. In the end, we achieve a rather similar result to that above for the slow light result. In this case, the bound for the number N_B of different wavelengths of pulses that can be separated in time is given by

$$N_B \leq \frac{3}{2} + \frac{\pi}{2\sqrt{3}} \frac{L}{\lambda_c} \eta_{max} \tag{8.29}$$

for a transmissive device. Hence, with a similar calculation to that above for the slow light structure, we would have, for a device to separate pulses of 32 different center wavelengths using a structure made of air and glass, that we must have the length $L \geq 41.7$ µm. Again, this limit is independent of how we design this glass/air layered structure.

8.5 Future Directions

The underlying limit to the performance of optical components, Eq. (8.20), is expressed in very general terms. It applies to any kind of linear wave interacting with a fixed medium, including scalar waves such as sound waves, vector electromagnetic waves, and even quantum mechanical waves. So far, we have only evaluated specific limits for one-dimensional structures here, but again the general form here should also give limits for two-dimensional and three-dimensional structures once the appropriate evaluation is performed for the quantity $N_{GS} \equiv Tr(G_S^\dagger G_S)$ of Eq. (8.22) for the two- and three-dimensional Green's functions for the relevant wave equations.

We can therefore expect additional specific limits for two- and three-dimensional optical structures. Since we have only considered one-dimensional structures so far, we have also only considered problems for optical pulses, because there is no different "position" other than a temporal one to which we can scatter waves when there is only one beam possible. With results for two- and three-dimensional situations, we could also examine limits to static scattering problems, such as limits to high-contrast holograms. Three-dimensional results for the vector electromagnetic case could also be particularly interesting for analyzing wireless communications in strongly scattering environments. In general, there is a broad range of additional wave problems to which we could apply this kind of approach to limits.

8.6 Conclusions

The body of work summarized here shows that there are many applications in optics for the idea of considering optics in terms of the number of orthogonal waves or communications modes that can be supported or generated in an optical system. Not only does this approach lead to a precise notion of the idea of diffraction limits, even beyond the simple plane-to-plane case of optical imaging or focusing to the case of volumes and near-field problems, but it also leads to well-defined and very general limits to the performance of optical systems and components. The new limit recently proposed[1] to the performance of optical devices, even those made with arbitrary structures of high index contrast, gives us for the first time a bounding limit to the performance of linear optical devices, completely independent of how we design them. We look forward to many novel applications and implications of this approach and the resulting bounds in optics.

References

1. Miller, D.A.B., "Fundamental limit for optical components," *J. Opt. Soc. Am. B*, **24**, A1–A18, 2007..
2. Miller, D.A.B., "Fundamental Limit to Linear One-Dimensional Slow Light Structures," *Phys. Rev. Lett.*, **99**, 203903, 2007 [doi:10.1103/PhysRevLett.99.203903]
3. Miller, D.A.B., "Communicating with waves between volumes—evaluating orthogonal spatial channels and limits on coupling strengths," *Appl. Opt.*, **39**, 1681–1699, 2000.
4. Gerken, M., and Miller, D.A.B., "Multilayer thin-film structures with high spatial dispersion," *Appl. Opt.*, **42**, 1330–1345, 2003.
5. Gerken, M., and Miller, D.A.B., "Limits to the performance of dispersive thin-film stacks," *Appl. Opt.*, **44**, 3349–3357, 2005.
6. Gerken, M., and Miller, D.A.B., "The Relationship between the superprism effect in one-dimensional photonic crystals and spatial dispersion in non-periodic thin-film stacks," *Opt. Lett.* **30**, 2475–2477, 2005.
7. Zengerle, R., "Light propagation in singly and doubly periodic planar waveguides," *J. Mod. Opt.* **34**, 1589–1617, 1987.
8. Jiao, Y., Fan, S.-H., and Miller, D.A.B., "Demonstration of systematic photonic crystal device design and optimization by low rank adjustments: an extremely compact mode separator," *Opt. Lett.* **30**, Issue 2, 141–143, 2005.

9. Jiao, Y., Fan, S.-H., and Miller, D.A.B., "Systematic photonic crystal device design: global and local optimization and sensitivity analysis," *IEEE J. Quant. Electron.* **42**, 266–279, 2006.

10. Gabor, D., "Light and information," *Progress in Optics I*, E. Wolf, Ed., pp. 109–153, North-Holland, Amsterdam (1961).

11. Toraldo di Francia, G.. "Degrees of freedom of an image," *J. Opt. Soc. Am.* **59**, 799–804 1969

12. Frieden, B.R., "Evaluation, design and extrapolation methods for optical signals, based on the use of the prolate functions," *Progress in Optics IX* E. Wolf, Ed., pp. 311–407, North-Holland, Amsterdam (1971).

13. Burvall, A., Barrett, H.H., Dainty, C., and Myers, K.J., "Singular-value decomposition for through-focus imaging systems," *J. Opt. Soc. Am. A*, **23**, 2440–2448, 2006.

14. Martinsson, P., Lajunen, H., and Friberg, A.T., "Scanning optical near-field resolution analyzed in terms of communication modes," *Opt. Expr.*, **14**, 11392–11401, 2006.

15. Thaning, A., Martinsson, P., Karelin, M., and Friberg, A.T., "Limits of diffractive optics by communications modes," *J. Opt. A*, **5**, 153–158, 2003.

16. Burvall, A., Martinsson, P., and Friberg, A., "Communication modes applied to axicons," *Opt. Expr.*, **12**, 377–383, 2004.

17. Burvall, A., Martinsson, P., and Friberg, A.T., "Communication modes in large-aperture approximation," *Opt. Lett.*, **32**, 611–613, 2007.

18. Piestun, R., and Miller, D.A.B., "Electromagnetic degrees of freedom of an optical system," *J. Opt. Soc. Am. A*, **17**, 892–902, 2000.

19. Piestun, R., and Shamir, J., "Synthesis of three-dimensional light fields and applications," *Proc. IEEE*, **90**, 222–224, 2002.

20. Godavarti, M., and Hero, III, A.O., "Diversity and degrees of freedom in wireless communications," *Proc. IEEE Int. Conf. on Acoustics, Speech, and Signal Processing (ICASSP 02)*, pp. III-2861–III-2864, IEEE, Piscataway, NJ (2002).

21. Xu, J., and Janaswamy, R., "Electromagnetic degrees of freedom in 2-D scattering environments," *IEEE Trans. Antennas Propagat.*, **54**, 3882–3894, 2006.

22. Hanlen, L., and Fu, M., "Wireless communication systems with spatial diversity: a volumetric model," *IEEE Trans. Wireless Commun.* **5**, 133–142 2006

23. Tucker, R.S., Ku, P.-C., and Chang-Hasnain, C.J., "Slow-light optical buffers: capabilities and fundamental limitations," *J. Lightwave Technol.*, **23**, 4046–4066, 2005.

24. Boyd, R.W., Gauthier, D.J., Gaeta, A.L., and Willner, A.E., "Maximum time delay achievable on propagation through a slow-light medium," *Phys. Rev. A*, **71**, 023801, 2005.

25. Khurgin, J.B., "Performance limits of delay lines based on optical amplifiers," *Opt. Lett.*, **31**, 948–950, 2006.

David A. B. Miller received his BS from St. Andrews University, and, in 1979, his PhD from Heriot-Watt University, both in physics. He was with Bell Laboratories from 1981 to 1996, as a department head from 1987, latterly of the Advanced Photonics Research Department. He is currently the W. M. Keck Professor of Electrical Engineering, the director of the Solid State and Photonics Laboratory, and a codirector of the Stanford Photonics Research Center at Stanford University. His research interests include physics and devices in nanophotonics, nanometallics, and quantum-well optoelectronics, and fundamentals and applications of optics in information sensing, switching, and processing. He has published more than 200 scientific papers, holds 62 patents, is a Fellow of OSA, IEEE, APS, and the Royal Societies of Edinburgh and London, holds honorary degrees from the Vrije Universiteit Brussel and Heriot-Watt University, and has received numerous awards.

Chapter 9
An Overview of Coherence and Polarization Properties for Multicomponent Electromagnetic Waves

Alfredo Luis
Universidad Complutense, Spain

9.1 Introduction
9.2 Coherence, Visibility, and Polarization for Two-Dimensional Waves
    9.2.1 Degree of polarization
    9.2.2 Visibility of interference
    9.2.3 Degree of local coherence
    9.2.4 Degree of global coherence
9.3 Degree of Polarization for Three-Dimensional Waves
9.4 Degree of Coherence for Pairs of Two-Dimensional Waves
9.5 Coherence for n-Dimensional Waves
9.6 Discussion
References

9.1 Introduction

Coherence is a key issue involved in many topics of quantum and classical physics. At a fundamental level, coherence appears in two basic optical items, namely, interference and polarization. More specifically, the polarization state is the result of coherence for the superposition of waves with orthogonal directions of vibration, while interference is the result of coherence for the superposition of waves with nonorthogonal (usually parallel) directions of vibration.

    The objective of this contribution is to develop a coherence-based relationship between interference and polarization focusing on the proper assessment of the amount of coherence.

We devote the following section to elaboration of the simpler two-dimensional case, i.e., the superposition of two scalar waves or two field components, as a suitable ground to sustain more complex situations. We examine several alternative formulations for the degrees of coherence, visibility, and polarization that are equivalent in the two-dimensional case but lead to diverging results for larger dimensions. Then we extend these ideas to more involved situations with a larger number of field components, such as the degree of polarization for three-dimensional waves and the degree of coherence for pairs of two-dimensional waves, where several approaches to the corresponding degrees of polarization and coherence currently coexist.

The multiplicity of degrees of coherence for dimensions larger than two need not be disturbing, and the coexistence can be peaceful and harmonious. Increasing the number of field components increases the degrees of freedom and the complexity of the problem so that a single measure of coherence may not be enough to account for all phenomena displayed by a richer situation.

9.2 Coherence, Visibility, and Polarization for Two-Dimensional Waves

For definiteness we focus on the spatial-frequency domain (without specifying, for simplicity, the spatial and frequency variables) where all second-order optical phenomena are described by the Hermitian and positive semidefinite covariance matrix M, i.e., the cross-spectral density tensor,[1-3]

$$M = \begin{pmatrix} \langle |E_1|^2 \rangle & \langle E_1^* E_2 \rangle \\ \langle E_2^* E_1 \rangle & \langle |E_2|^2 \rangle \end{pmatrix}, \qquad (9.1)$$

where the angle brackets denote ensemble averages. The field components E_j can represent the following two different physical situations:

1. Two-beam interference in which E_j represent the complex amplitudes of two interfering waves $\vec{E}_j = E_j \vec{\varepsilon}$ with the same state of polarization $\vec{\varepsilon}$.
2. Two-dimensional polarization in which E_j represent the complex amplitudes of the two orthogonal components $\vec{E}_j = E_j \vec{\varepsilon}_j$ of a single transversal wave $\vec{E} = \vec{E}_1 + \vec{E}_2$ with $\vec{\varepsilon}_1^* \cdot \vec{\varepsilon}_2 = 0$.

In both cases, we will find it useful to expand M in the matrix basis consisting of the three Pauli matrices $\vec{\sigma}$ and the identity σ_0,

$$M = \frac{1}{2}(s_0 \sigma_0 + \vec{s} \cdot \vec{\sigma}), \quad s_j = \text{tr}(M \sigma_j), \qquad (9.2)$$

where $s_0, \vec{s}$ are four real parameters.

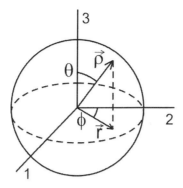

Figure 9.1 Unit sphere and the vectors $\vec{\rho}$ and $\vec{r}$.

The positive character of M implies that $|\vec{s}| \leq s_0$ so that for $s_0 \neq 0$ we can define the three-dimensional real vector $\vec{\rho} = \vec{s}/s_0$ with $|\vec{\rho}| \leq 1$ that lies in the interior of a unit-radius sphere as illustrated in Fig. 9.1, where we have also represented the two-dimensional real vector $\vec{r}$ obtained by projection of $\vec{\rho}$ on the 1-2 plane $\vec{r} = (\rho_1, \rho_2)$.

9.2.1 Degree of polarization

The degree of polarization represents many different concepts, especially in the case of an arbitrary number of field components, as we shall discuss in more detail in Section 9.5. Focusing on the idea of two-dimensional polarization, $s_0, \vec{s}$ are the Stokes parameters and the sphere enclosing $\vec{\rho}$ is the Poincaré sphere.[1,3] There are at least three different approaches to assess the amount of second-order polarization of two-dimensional waves, all leading to the same degree of polarization P.

Stokes parameters. The degree of polarization can be defined as the length of the intensity-normalized Stokes vector $\vec{\rho}$,

$$P = |\vec{\rho}| = \frac{|\vec{s}|}{s_0}, \qquad (9.3)$$

which can be understood as the distance of the point $\vec{\rho}$ to the origin of the sphere $\vec{\rho} = \vec{0}$ associated with second-order fully unpolarized light.

Hilbert-Schmidt distance. The same degree of polarization P can be understood as the Hilbert-Schmidt distance between the normalized correlation matrix

$M_N = M/\mathrm{tr}M$ and the normalized 2×2 identity matrix $I_N = I/2$ representing second-order fully unpolarized light,

$$P^2 = 2\mathrm{tr}\left[\left(M_N - I_N\right)^2\right] = 2\left[\frac{\mathrm{tr}(M^2)}{(\mathrm{tr}M)^2} - \frac{1}{2}\right]. \tag{9.4}$$

Seminorm. Since M is a 2×2 positive semidefinite and Hermitian matrix it has two nonnegative eigenvalues $\lambda_{1,2}$, with $\lambda_1 \geq \lambda_2 \geq 0$, and we have

$$P = \frac{\lambda_1 - \lambda_2}{\lambda_1 + \lambda_2} = \|M\|_s, \tag{9.5}$$

where $\|M\|_s$ is a matrix seminorm. This is not a norm since for $\lambda_1 = \lambda_2 \neq 0$ we have $\|M\|_s = 0$ with $M \neq 0$.

U(2) invariance. These approaches are invariant under U(2) transformations, i.e., $P(\vec{E}') = P(\vec{E})$ for $\vec{E}' = U\vec{E}$ with $\vec{E} = (E_1, E_2)$, where U are arbitrary 2×2 unitary matrices representing the action of linear, energy-conserving devices such as transparent phase plates and lossless beam splitters, that transform the correlation matrix as $M \to U^+ M U$. These transformations produce rotations on the Poincaré sphere that change the state of polarization without altering the degree of polarization.

9.2.2 Visibility of interference

Standard definition. Let us consider the interference fringes produced by the superposition of two waves $E \propto e^{i\phi_1} E_1 + e^{i\phi_2} E_2$ leading to an intensity distribution $I = \langle |E|^2 \rangle \propto I_1 + I_2 + 2|\langle E_1 E_2^* \rangle|\cos(\phi + \delta)$ where $\phi = \phi_2 - \phi_1$ is the phase difference, $I_j = \langle |E_j|^2 \rangle$ are the individual intensities, and $\delta = \arg\langle E_1^* E_2 \rangle$. The standard definition of visibility is

$$V = \frac{I_{\max} - I_{\min}}{I_{\max} + I_{\min}} = \frac{2|\langle E_1 E_2^* \rangle|}{I_1 + I_2} = \frac{\sqrt{s_1^2 + s_2^2}}{s_0} = |\vec{r}|, \tag{9.6}$$

so that V is the norm of the projection $\vec{r}$ of the vector $\vec{\rho}$ on the equatorial plane of the Poincaré sphere. Because $|\vec{r}| \leq |\vec{\rho}|$, we have always $V \leq P$ and the degree of polarization is an upper bound for the visibility. The projection takes place in the equatorial plane because the transformations $E_j \to e^{i\phi_j} E_j$ produce a rotation of the vector $\vec{\rho}$ of angle $\phi = \phi_2 - \phi_1$ around the third axis in Fig. 9.1.

Hilbert-Schmidt distance. Alternatively, the visibility can be expressed as the Hilbert-Schmidt distance between $M_N = M/\text{tr}M$ and the normalized diagonal matrix D_N obtained from M_N by removing the nondiagonal terms,[4]

$$V^2 = 2\text{tr}\left[(M_N - D_N)^2\right], \quad D_N = \frac{1}{I_1 + I_2}\begin{pmatrix} I_1 & 0 \\ 0 & I_2 \end{pmatrix}. \tag{9.7}$$

Distinguishability. The visibility can be regarded as a measure of the distance $|\vec{\rho}_\phi - \vec{\rho}|$ between the original $\vec{\rho}$ and rotated $\vec{\rho}_\phi$ vectors after a rotation of the Poincaré sphere produced by the phase shifts $E_j \to e^{i\phi_j} E_j$

$$V = \left.\frac{d|\vec{\rho}_\phi - \vec{\rho}|}{d\phi}\right|_{\phi=0}. \tag{9.8}$$

This expresses the usefulness of the light beam for metrological interferometric applications aimed to detect small phase shifts ϕ.

Lack of U(2) invariance. V is not invariant under U(2) transformations $V(\vec{E}') \ne V(\vec{E})$ for $\vec{E}' = U\vec{E}$, where U are arbitrary 2×2 complex matrices, since the rotation generated by U need not conserve the projection of $\vec{\rho}$ on the equatorial plane.

9.2.3 Degree of local coherence

Standard definition. The standard definition of the (local) degree of coherence is

$$\mu = \frac{|\langle E_1 E_2^* \rangle|}{\sqrt{I_1 I_2}} = \sqrt{\frac{s_1^2 + s_2^2}{s_0^2 - s_3^2}} = \frac{|\vec{r}|}{\sqrt{1-\rho_3^2}} = \frac{|\vec{r}|}{|\vec{r}|_{\max}}, \tag{9.9}$$

where $|\vec{r}|_{\max}$ is the maximum of $|\vec{r}|$ for fixed ρ_3 (see Fig. 9.2). We can appreciate that we have always $P \ge \mu \ge V$.[1]

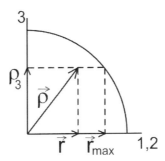

Figure 9.2 Geometrical picture of the degree of coherence.

Hilbert-Schmidt distance. μ can be formulated as the Hilbert-Schmidt distance between the 2×2 identity matrix I and a normalized version m of the correlation matrix M obtained by replacing in Eq. (9.1) E_j by $\varepsilon_j = E_j / \sqrt{\langle |E_j|^2 \rangle}$,

$$\mu^2 = \frac{1}{2}\mathrm{tr}\left[(m-I)^2\right]. \qquad (9.10)$$

Seminorm. A seminorm expression analogous to Eq. (9.5) is available in the form

$$\mu = \frac{\ell_1 - \ell_2}{\ell_1 + \ell_2} = \|m\|_s, \qquad (9.11)$$

where $\ell_1 \geq \ell_2$ are the eigenvalues of m.

Lack of U(2) invariance. There is also lack of invariance under U(2) transformations $\mu(\vec{E}') \neq \mu(\vec{E})$ for $\vec{E}' = U\vec{E}$ where U are arbitrary 2×2 complex matrices. In order to illustrate this point let us consider, for example, two totally incoherent scalar waves $\langle E_1 E_2^* \rangle = 0$. By a simple unitary transformation we can get two new waves $E_1' \propto E_1 + E_2$, $E_2' \propto E_1 - E_2$ with $\mu' = |\langle |E_1|^2 \rangle - \langle |E_2|^2 \rangle| / (\langle |E_1|^2 \rangle + \langle |E_2|^2 \rangle)$, so that the degree of coherence equals the degree of polarization in Eq. (9.4), being also the intensity contrast between the original components $E_{1,2}$. This example also illustrates that the seminorm expression in Eq. (9.5) may be regarded in general as an intensity contrast between uncorrelated components rendering the matrix M diagonal.

9.2.4 Degree of global coherence

Let us consider that $E_{1,2} \equiv E(x_{1,2})$ are scalar waves at two specific points $x_{1,2}$ so that $\mu(x_1, x_2)$ in Eq. (9.9) is the corresponding local degree of coherence. It is possible to provide a global degree of coherence μ_g as an average of $\mu(x_1, x_2)$ weighted by the intensities at points $x_{1,2}$, as proposed in Refs. 5 and 6,

$$\mu_g^2 = \frac{\int dx_1' dx_2' I(x_1') I(x_2') \mu^2(x_1', x_2')}{\left[\int dx' I(x')\right]^2}. \tag{9.12}$$

If we apply this global definition to the field after a Young interferometer with small enough apertures centered at points $x_{1,2}$ we naturally recover a local measure of coherence. However, this is not $\mu(x_1, x_2)$ but the degree of polarization P for $\vec{E} = (E_1, E_2)$.[7] To see this, let us describe the two apertures by identical field-amplitude transmission coefficients $t(x)$ centered at $x_{1,2}$. Since the apertures do not overlap, $t(x - x_1) t(x - x_2) = 0$. For small enough apertures, the field is approximately constant within each aperture, so that the field distribution immediately after the apertures is $E(x) \approx t(x - x_1) E_1 + t(x - x_2) E_2$. Then, the application of Eq. (9.12) to $E(x)$ leads to

$$\mu_g^2 = \frac{\text{tr}(M^2)}{(\text{tr} M)^2}, \tag{9.13}$$

where M is the correlation matrix in Eq. (9.1). It can be seen that $1 \geq \mu_g^2 \geq 1/2$. The maximum $\mu_{g,\max} = 1$ occurs when M has only one nonvanishing eigenvalue, while the minimum $\mu_{g,\min} = 1/\sqrt{2}$ occurs when M is proportional to the identity. A proper normalization leads to

$$\mu_{g,N}^2 = \frac{\mu_g^2 - \mu_{g,\min}^2}{\mu_{g,\max}^2 - \mu_{g,\min}^2} = 2\left[\frac{\text{tr}(M^2)}{(\text{tr} M)^2} - \frac{1}{2}\right] = P^2, \tag{9.14}$$

where P is the degree of polarization for $\vec{E} = (E_1, E_2)$.

9.3 Degree of Polarization for Three-Dimensional Waves

There are reasons justifying a three-dimensional analysis of polarization. For nonparaxial fields without a well-defined propagation direction, the plane of the polarization ellipse may vary from point to point.[8] Also, in most practical situations plane-wave behavior and the vanishing of the longitudinal component are approximations.[9] Moreover, quantum fluctuations affect the three field components even when they are in the vacuum state, so that the quantum electric field varies unavoidably in a three-dimensional region.[10]

Let us examine four different proposals for the degree of polarization of three-dimensional waves that follow essentially the possibilities examined in the two-dimensional case.

Stokes parameters. In this case the correlation matrix M is a 3×3 matrix and admits an expansion of the form in Eq. (9.2) in terms of a suitable matrix basis consisting of the identity Λ_0 and eight trace-orthogonal Gell-Mann matrices $\vec{\Lambda}$,

$$M = \frac{1}{3} s_0 \Lambda_0 + \frac{1}{2} \vec{s} \cdot \vec{\Lambda}, \quad s_j = \mathrm{tr}(M \Lambda_j), \qquad (9.15)$$

that defines nine Stokes parameters $s_0, \vec{s}$ and a degree of polarization analogous to the one in Eq. (9.3),[3,8,11–14]

$$P = \sqrt{\frac{3}{4}} \frac{|\vec{s}|}{s_0}. \qquad (9.16)$$

Hilbert-Schmidt distance. We can define the degree of polarization as the Hilbert-Schmidt distance between the normalized correlation matrix $M_N = M/\mathrm{tr}M$ and the normalized identity $I_N = I/3$ representing second-order unpolarized light[15] as

$$P^2 = \frac{3}{2} \mathrm{tr}\left[(M_N - I_N)^2\right] = \frac{3}{2}\left[\frac{\mathrm{tr}(M^2)}{(\mathrm{tr}M)^2} - \frac{1}{3}\right], \qquad (9.17)$$

that coincides with Eq. (9.16), and can be expressed also as

$$P = \frac{\sqrt{(\lambda_1 - \lambda_2)^2 + (\lambda_2 - \lambda_3)^2 + (\lambda_1 - \lambda_3)^2}}{\sqrt{2}(\lambda_1 + \lambda_2 + \lambda_3)}, \qquad (9.18)$$

where λ_j are the eigenvalues of M. This kind of average of seminorms is analogous to Eq. (9.5).

Seminorm. If we arrange the eigenvalues of M in the form $\lambda_1 \geq \lambda_2 \geq \lambda_3 \geq 0$, the degree of polarization can be defined also as a matrix seminorm similar to the one in Eq. (9.5),[16]

$$P' = \frac{\lambda_1 - \lambda_2}{\lambda_1 + \lambda_2 + \lambda_3}, \tag{9.19}$$

which is derived from the following decomposition of M as a superposition of a fully polarized wave, a two-dimensional unpolarized wave, and a three-dimensional unpolarized wave:[16]

$$M = (\lambda_1 - \lambda_2)\begin{pmatrix} 1 & 0 & 0 \\ 0 & 0 & 0 \\ 0 & 0 & 0 \end{pmatrix} + (\lambda_2 - \lambda_3)\begin{pmatrix} 1 & 0 & 0 \\ 0 & 1 & 0 \\ 0 & 0 & 0 \end{pmatrix} + \lambda_3 \begin{pmatrix} 1 & 0 & 0 \\ 0 & 1 & 0 \\ 0 & 0 & 1 \end{pmatrix}. \tag{9.20}$$

Two indexes. The prescription of a single parameter may not be enough to convey the complexity of coherence in the three-dimensional case. This idea is implemented by the two-index approach in Refs. 17 and 18,

$$P' = \frac{\lambda_1 - \lambda_2}{\lambda_1 + \lambda_2 + \lambda_3}, \quad P'' = \frac{\lambda_1 + \lambda_2 - 2\lambda_3}{\lambda_1 + \lambda_2 + \lambda_3}, \tag{9.21}$$

with $1 \geq P'' \geq P' \geq 0$. This conclusion has been reached also in Ref. 19 after analyzing the polarimetric contrast of a state of light with totally unpolarized light with the same intensity. It is shown that in general two different parameters are necessary to completely characterize the polarimetric contrast of three-dimensional waves. The actual choice of the parameters depends on the method of characterization.

Comparison. Unlike in the two-dimensional situation, in this case there is no complete coincidence between approaches. While the modulus of the Stokes vector and the Hilbert-Schmidt distance coincide exactly (which actually holds for arbitrary dimensions as shown in Section 9.5), the seminorm leads in general to different results $P \neq P'$, $P \neq P''$ as exemplified by the case $\lambda_1 = \lambda_2 \neq \lambda_3$, while they coincide $P = P' = P''$ if and only if $\lambda_2 = \lambda_3$.

U(3) invariance. All these approaches are invariant under U(3) transformations, i.e., P, P', P'' take the same value for $\vec{E}$ and $\vec{E}' = U\vec{E}$, where U is any unitary 3×3 matrix, since all of them can be defined in terms of the spectrum of M.

Two-dimensional reduction. When one of the field components E_j vanishes, i.e., $\lambda_3 = 0$, the Stokes parameters and the Hilbert-Schmidt distance P_{3D} in Eqs. (9.16)–(9.18) do not reproduce the two-dimensional degree of polarization P_{2D} in Eqs. (9.3)–(9.5), since we have $P_{3D}^2 = (3/4) P_{2D}^2 + 1/4$. From a three-dimensional perspective, the extra factor ¼ incorporates the knowledge that a field component vanishes without fluctuations, as illustrated by the fact that two-dimensional unpolarized light, contained always in a given plane, is not equivalent to three-dimensional unpolarized light. On the other hand, for $\lambda_3 = 0$ we have that the seminorm and two-index approaches in Eqs. (9.19) and (9.21) reproduce the two-dimensional definition in Eq. (9.5) so that $P'_{3D} = P_{2D}$.[8,14,16]

9.4 Degree of Coherence for Pairs of Two-Dimensional Waves

The complete picture of interference requires the inclusion of the vectorial character of electromagnetic waves. This basic and nontrivial issue has not been addressed in depth until recent times, and different and controversial proposals for the degree of coherence of a pair of vectorial waves currently coexist.

The coherence between two transversal waves (orthogonal to the *z*-axis) at points $x_{1,2}$ involves four field components $E_j(x_\ell)$, $j = x, y$, and $\ell = 1, 2$. Thus M is a 4×4 matrix,

$$M = \begin{pmatrix} M_1 & \tilde{M} \\ \tilde{M}^+ & M_2 \end{pmatrix}, \tag{9.22}$$

where $M_j = \Gamma_{j,j}$, $j = 1, 2$, and $\tilde{M} = \Gamma_{1,2}$, are 2×2 matrices,

$$\Gamma_{i,j} = \begin{pmatrix} \langle E_x^*(x_i) E_x(x_j) \rangle & \langle E_x^*(x_i) E_y(x_j) \rangle \\ \langle E_y^*(x_i) E_x(x_j) \rangle & \langle E_y^*(x_i) E_y(x_j) \rangle \end{pmatrix}. \tag{9.23}$$

Next we overview some possible particular approaches to the assessment of the amount of coherence of pairs of two-dimensional waves.

Stokes parameters in a Young interferometer. Two main proposals focus on the Young interferometer illustrated in Fig. 9.3.

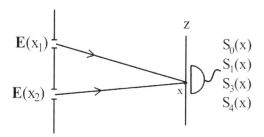

Figure 9.3 Young interferometer.

In this case the interference term depends exclusively on the correlations expressed by $\tilde{M}$, which may be expanded as in Eq. (9.2) as $\tilde{M} = (S_0 \sigma_0 + \vec{S} \cdot \vec{\sigma})/2$ with $S_j = \text{tr}(\tilde{M}\sigma_j)$ where S_j are four complex parameters.[20,21] Four independent measurements can be performed at each point on the interference plane (measuring the four Stokes parameters) so that we have four independent systems of fringes with modulations proportional to $|S_j|$. This has motivated the following two definitions of the degree of coherence:

$$\mu_1 = \frac{|\text{tr}\tilde{M}|}{\sqrt{\text{tr}M_1 \text{tr}M_2}} = \frac{|S_0|}{\sqrt{s_0^{(1)} s_0^{(2)}}}, \qquad (9.24)$$

and

$$\mu_2 = \sqrt{\frac{\text{tr}(\tilde{M}\tilde{M}^+)}{\text{tr}M_1 \text{tr}M_2}} = \sqrt{\frac{|S_0|^2 + |\vec{S}|^2}{2 s_0^{(1)} s_0^{(2)}}}, \qquad (9.25)$$

where $s_0^{(j)} = \text{tr}(M_j \sigma_0)$ is the intensity of each individual wave $\vec{E}(x_{1,2})$. In the first definition μ_1 corresponds to the measurement of the modulation of the total intensity exclusively,[22–24] while in the second one μ_2 is an average of the modulations of the four systems of fringes.[25–28]

U(2) × U(2) invariance. It is worth examining the potential invariance of $\mu_{1,2}$ under U(2) transformations, $\tilde{M} \to U_1^+ \tilde{M} U_2$, $M_j \to U_j^+ M_j U_j$, where U_j are 2×2 unitary complex matrices corresponding to the application at each point of energy-conserving polarization changing devices (i.e., transparent phase plates). The invariance is a desirable property since such transformations do not alter the statistical relations between field components. It can be checked that μ_2 is U(2) × U(2) invariant while μ_1 is not.[24,29,30]

In order to correct the lack of invariance of μ_1, two main solutions have been proposed. In the first one the idea is to consider the maximum μ_1 when arbitrary

U(2) transformations are applied to both apertures,[24] while the second one considers the application of arbitrary phase plates and polarizers,[29,30] leading to, respectively,

$$\mu_{1,\max} = \frac{\tilde{\lambda}_1 + \tilde{\lambda}_2}{\sqrt{s_0^{(1)} s_0^{(2)}}}, \quad \mu'_{1,\max} = \mathrm{msv}\left(M_1^{-1/2} \tilde{M} M_2^{-1/2}\right), \qquad (9.26)$$

where $\tilde{\lambda}_{1,2}$ are the singular values of $\tilde{M}$ and msv represents the maximum singular value. Incidentally, in this same language we have that

$$\mu_2 = \sqrt{\frac{\tilde{\lambda}_1^2 + \tilde{\lambda}_2^2}{2 s_0^{(1)} s_0^{(2)}}}. \qquad (9.27)$$

Degree of polarization. Since in the two-dimensional case the degree of coherence is bounded from above by the degree of polarization $\mu \leq P$, we may consider as a suitable measure of coherence the degree of polarization of the four-dimensional wave $\vec{E} = [E_x(x_1), E_y(x_1), E_x(x_2), E_y(x_2)]$, which can be expressed as the normalized Hilbert-Schmidt distance between the whole correlation matrix $M_N = M/\mathrm{tr}M$ in Eq. (9.22) and the identity $I_N = I/4$ representing fully second-order unpolarized light,[7]

$$P^2 = \frac{4}{3} \mathrm{tr}\left[(M_N - I_N)^2\right] = \frac{4}{3}\left[\frac{\mathrm{tr}(M^2)}{(\mathrm{tr}M)^2} - \frac{1}{4}\right], \qquad (9.28)$$

equivalent to the purity criterion of optical systems introduced in Refs. 31 and 32.

Other approaches based on the same idea of a degree of coherence determined by the full M instead of only $\tilde{M}$ can be found in Ref. 33 in terms of a normalized coherence matrix obtained from M by replacing E_j by $\varepsilon_j = E_j / \sqrt{\langle |E_j|^2 \rangle}$, analogous to matrix m in Eq. (9.10).

Stokes parameters. The same result in Eq. (9.28) can be derived in terms of the length of the Stokes vector defined by an expansion of M in a suitable trace-orthogonal matrix basis, as shown in more detail in Section 9.5.

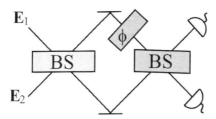

Figure 9.4 An arbitrary Mach-Zehnder interferometer.

Visibility of arbitrary interferometers. The approaches leading to Eqs. (9.24) and (9.25) are based on the Young interferometer, where the fringe visibility depends on the 2×2 correlation matrix $\tilde{M}$. However, arbitrary interferometers will mix the four input components so that the visibility will depend on the 4×4 correlation matrix M in Eq. (9.22). In particular, for two-beam interferometers such as the one illustrated in Fig. 9.4, it has been shown that the visibility V is bounded by[34]

$$V \leq \frac{\lambda_{max} - \lambda_{min}}{\lambda_{max} + \lambda_{min}}, \qquad (9.29)$$

where $\lambda_{max,min}$ are the maximum and minimum eigenvalues of M. This is very reminiscent of the seminorm in Eq. (9.5).

U(4) invariance. For four field components, the most natural group of transformations is U(4), i.e., unitary energy-conserving transformations mixing the four field components $E_j(x_\ell)$ on an equal footing that includes the U(2) $\times$ U(2) transformations as a particular case. The definitions $\mu_{1,2}$ in Eqs. (9.24) and (9.25) lack U(4) invariance, while Eqs. (9.28) and (9.29) fulfill it.

9.5 Coherence for *n*-Dimensional Waves

Most of the formulas in the preceding sections for two-, three-, and four-dimensional waves admit a proper generalization to an arbitrary number n of components.

This is the case of the degree of polarization defined as the Hilbert-Schmidt distance between the $n\times n$ correlation matrix $M_N = M/\mathrm{tr}M$ and the $n\times n$ identity $I_N = I/n$ representing second-order unpolarized light,[12,13,35]

$$P^2 = \frac{n}{n-1}\mathrm{tr}\left[(M_N - I_N)^2\right] = \frac{n}{n-1}\left[\frac{\mathrm{tr}(M^2)}{(\mathrm{tr}M)^2} - \frac{1}{n}\right]. \qquad (9.30)$$

The maximum degree of polarization $P = 1$ occurs when M has a single nonvanishing eigenvalue, while the minimum $P = 0$ occurs when M is proportional to the identity.

There is an equivalent relation in terms of Stokes parameters defined by an expansion of M in a suitable basis of trace orthogonal matrices consisting of the identity Λ_0 and $n^2 - 1$ traceless trace-orthogonal matrices Λ_j with $\text{tr}(\Lambda_i \Lambda_j) = 2\delta_{i,j}$, except for $i = j = 0$, with $\text{tr}(\Lambda_0^2) = n$,[12,13,35,36]

$$M = \frac{1}{n} s_0 \Lambda_0 + \frac{1}{2} \sum_{j=1}^{n^2-1} s_j \Lambda_j, \quad s_j = \text{tr}(M \Lambda_j). \tag{9.31}$$

Then we have that Eq. (9.30) is equivalent to

$$P^2 = \frac{n}{2(n-1)s_0^2} \sum_{j=1}^{n^2-1} s_j^2. \tag{9.32}$$

Seminorm formulas in Eqs. (9.5), (9.19), and (9.21) are valid in the general case with $\lambda_1 \geq \lambda_2 \geq \lambda_j$, $j = 3, \ldots n$.

It can be seen in Ref. 34 that Eq. (9.29) holds for the maximum visibility achievable in arbitrary two-beam interferometers illuminated with n field components. A multiple-parameter characterization of polarization by the scalar invariants of the coherence matrix can be found in Ref. 37. All these generalizations satisfy the desirable U(n) invariance.

The quantity P can be endowed with multiple meanings both in quantum and classical physics. In classical optics, when $n \to \infty$ we have that P coincides with the global degree of coherence.[6,36] In the quantum domain, P represents the total information,[38] or purity, conveyed by the quantum state with density matrix proportional to M. This enters into the equality $V^2 + C^2 = P^2$, where V is the n-dimensional generalization of the visibility in Eq. (9.7),[4]

$$V^2 = \frac{n}{n-1} \text{tr}\left[(M_N - D_N)^2\right], \quad C^2 = \frac{n}{n-1} \text{tr}\left[(D_N - I_N)^2\right], \tag{9.33}$$

and C represents the dispersion of the intensity distribution $I_j = \langle |E_j|^2 \rangle$,[39] where $D_{N,i,j} = I_j \delta_{i,j} / \text{tr} M$ is the normalized diagonal intensity matrix. The formula $V^2 + C^2 = P^2$ has been regarded as a kind of quantum complementarity relation between interference (wave behavior represented by V) and path (corpuscular behavior represented by C).[40] From a classical perspective, this expresses that optimum visibility requires both coherence and equality of intensities of the interfering beams. Moreover, we can appreciate that $V = 0$ does not imply $P = 0$. In such a case P is equal to the kind of intensity contrast represented by C between field components rendering M diagonal. This agrees with the two-dimensional example at the end of Section 9.2.3 and with Eq. (9.5).

Incidentally, the relation $V^2 + C^2 = P^2$ is the Pythagorean theorem since it holds that $M_N - D_N$ and $D_N - I_N$ are actually trace-orthogonal matrices $\text{tr}\left[(M_N - D_N)(D_N - I_N)\right] = 0$, as illustrated in Fig. 9.5.

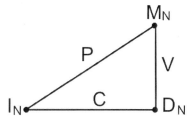

Figure 9.5 Orthogonal relation between M_N, D_N, I_N.

9.6 Discussion

We have elaborated on an approach to coherence for multicomponent waves by developing and generalizing simple approaches to coherence, visibility, and polarization for two-dimensional waves. They are equivalent for dimension two, but lead to different results for larger dimensions. This multiplicity of approaches is interesting since different definitions may grasp complementary aspects of coherence that otherwise cannot be embodied in a single parameter. For the interference of vectorial waves, multiple systems of independent fringes coexist whose visibilities may not be fully described by a single parameter.

Acknowledgment

This work has been supported by Project No. PR1-A/07-15378 of the Universidad Complutense.

References

1. Born, M., and Wolf, E., *Principles of Optics*, Cambridge University Press, Cambridge, UK (1999).
2. Mandel, L., and Wolf, E., *Optical Coherence and Quantum Optics*, Cambridge University Press, Cambridge, UK (1995).
3. Brosseau, Ch., *Fundamentals of Polarized Light*, Wiley, New York (1998).
4. Herbut, F., "A quantum measure of coherence and incompatibility," *J. Phys. A*, **38**, 2959–2974, 2005.
5. Tervo, J., Setälä, T., and Friberg, A.T., "Theory of partially coherent electromagnetic fields in the space-frequency domain," *J. Opt. Soc. Am. A*, **21**, 2205–2215, 2004.

6. Bastiaans, M.J., "Application of the Wigner distribution function to partially coherent light," *J. Opt. Soc. Am. A*, **3**, 1227–1238, 1986.

7. Luis, A., "Degree of coherence for vectorial electromagnetic fields as the distance between correlation matrices," *J. Opt. Soc. Am. A*, **24**, 1063–1068, 2007.

8. Setälä, T., Shevchenko, A., Kaivola, M., and Friberg, A.T., "Degree of polarization for optical near fields," *Phys. Rev. E*, **66**, 016615, 2002.

9. Simon, R., Sudarshan, E.C.G., and Mukunda, N., "Cross polarization in laser beams,"*Appl. Opt.*, **26**, 1589–1593, 1987.

10. Pollet, J., Méplan, O., and Gignoux, C., "Elliptic eigenstates for the quantum harmonic oscillator," *J. Phys. A*, **28**, 7287–7297, 1995.

11. Dennis, M.R., "Geometric interpretation of the three-dimensional coherence matrix for nonparaxial polarization," *J. Opt. A: Pure Appl. Opt.*, **6**, S26–S31, 2004.

12. Barakat, R., "Degree of polarization and the principal idempotents of the coherency matrix," *Opt. Commun.*, **23**, 147–150, 1977.

13. Samson, J.C., and Olson, J.V., "Generalized Stokes vectors and generalized power spectra for second-order stationary vector-processes, *SIAM J. Appl. Math.*, **40**, 137–149, 1981.

14. Luis, A., "Quantum polarization for three-dimensional fields via Stokes operators," *Phys. Rev. A*, **71**, 023810, 2005.

15. Luis, A., "Degree of polarization for three-dimensional fields as a distance between correlation matrices," *Opt. Commun.*, **253**, 10–14, 2005.

16. Ellis, J., Dogariu, A., Ponomarenko, S., and Wolf, E., "Degree of polarization of statistically stationary electromagnetic fields," *Opt. Commun.*, **248**, 333–337, 2005.

17. Gil, J.J., "Polarimetric characterization of light and media. Physical quantities involved in polarimetric phenomena," *Eur. Phys. J. Appl. Phys.*, **40**, 1–47, 2007.

18. Gil, J.J., Correas, J.M., Melero, P.A., and Ferreira, C., "Generalized polarization algebra," *Monografías del Seminario Matemático García de Galdeano,* **31**, 161–167, 2004; http://www.unizar.es/galdeano/actas_pau/-PDFVIII/pp161-167.pdf

19. Réfrégier, P., Roche, M., and Goudail, F., "Invariant polarimetric contrast parameters of light with Gaussian fluctuations in three dimensions," *J. Opt. Soc. Am. A*, **23**, 124–133, 2006.

20. Ellis, J., and Dogariu, A., "Complex degree of mutual polarization," *Opt. Lett.*, **29**, 536–538, 2004.

21. Korotkova, O., and Wolf, E., "Generalized Stokes parameters of random electromagnetic beams" *Opt. Lett.*, **30**, 198–200, 2005.

22. Karczewski, B., "Degree of coherence of the electromagnetic field," *Phys. Lett.*, **5**, 191–192, 1963.
23. Wolf, E., "Unified theory of coherence and polarization of random electromagnetic beams" *Phys. Lett. A*, **312**, 263–267, 2003.
24. Gori, F., Santarsiero, M., and Borghi, R., "Maximizing Young's fringe visibility through reversible optical transformations," *Opt. Lett.*, **32**, 588–590, 2007.
25. Tervo, J., Setälä, T., and Friberg, A.T., "Degree of coherence for electromagnetic fields," *Opt. Expr.*, **11**, 1137–1143, 2003.
26. Setälä, T., Tervo, J., and Friberg, A.T., "Complete electromagnetic coherence in the space-frequency domain," *Opt. Lett.*, **29**, 328–330, 2004.
27. Setälä, T., Tervo, J., and Friberg, A.T., "Stokes parameters and polarization contrasts in Young's interference experiment," *Opt. Lett.*, **31**, 2208–2210, 2006.
28. Setälä, T., Tervo, J., and Friberg, A.T., "Contrasts of Stokes parameters in Young's interference experiment and electromagnetic degree of coherence," *Opt. Lett.*, **31**, 2669–2671, 2006.
29. Réfrégier, P., and Goudail, F., "Invariant degrees of coherence of partially polarized light," *Opt. Expr.*, **13**, 6051–6060, 2005.
30. Réfrégier, P., and Roueff, A., "Coherence polarization filtering and relation with intrinsic degrees of coherence," *Opt. Lett.*, **31**, 1175–1177, 2006.
31. Gil, J.J., "Characteristic properties of Mueller matrices," *J. Opt. Soc. Am. A*, **17**, 328–334, 2000.
32. Gil, J.J., and Bernabeu, E., "Depolarization and polarization indices of an optical system," *Opt. Acta*, **33**, 185–189, 1986.
33. Ozaktas, H.M., Yüksel, S., and Kutay, M.A., "Linear algebraic theory of partial coherence: discrete fields and measures of partial coherence," *J. Opt. Soc. Am. A*, **19**, 1563–1571, 2002.
34. Luis, A., "Maximum visibility in interferometers illuminated by vectorial waves," *Opt. Lett.*, **32**, 2191–2193, 2007.
35. Vahimaa, P., and Tervo, J., "Unified measures for optical fields: Degree of polarization and effective degree of coherence," *J. Opt. A: Pure Appl. Opt.*, **6**, S41–S44, 2004.
36. Cloude, S.R., "Group theory and polarisation algebra," *Optik*, **75**, 26–36, 1986.
37. Barakat, R., "n-Fold polarization measures and associated thermodynamic entropy of N partially coherent pencils of radiation," *Opt. Acta*, **30**, 1171–1182, 1983.
38. Brukner, Č., and Zeilinger, A., "Conceptual inadequacy of the Shannon information in quantum measurements," *Phys. Rev. A*, **63**, 022113, 2001.

39. Luis, A., "Complementarity and duality relations for finite-dimensional systems," *Phys. Rev. A*, **67**, 032108, 2003.

40. Dürr, S., "Quantitative wave-particle duality in multibeam interferometers," *Phys. Rev. A*, **64**, 042113, 2001.

Alfredo Luis received his MS in physics from the University of Zaragoza, Spain, in 1986, and his PhD in physics from the University Complutense in 1992. Besides polarization and coherence for classical vectorial waves, his current research interests include quantum optical topics such as metrology, interferometry, nonclassical states of light, uncertainty relations, complementarity, and the Zeno effect.

Chapter 10
Intrinsic Degrees of Coherence for Electromagnetic Fields

Philippe Réfrégier and Antoine Roueff
Institut Fresnel, Aix-Marseille Université, France

10.1 Introduction
10.2 Background
    10.2.1 Second-order statistical properties
    10.2.2 Coherence and visibility of interference fringes
    10.2.3 Factorization condition at order one
    10.2.4 Coherence, invariance, and ability to interfere
10.3 Intrinsic Coherence and Visibility of Interference Fringes
    10.3.1 Intrinsic degrees of coherence
    10.3.2 Relation with interference experiments with double polarization optimization
    10.3.3 Relation with interference experiments with single polarization optimization
10.4 Physical Interpretations
    10.4.1 Mixing experiment
    10.4.2 General statistical interpretation
10.5 Conclusion
Appendix A Interference and Factorization Condition at Order One
Appendix B Intrinsic Degrees of Coherence of Mixing Uncorrelated Perfectly Polarized Lights
References

10.1 Introduction

These last few years, the analysis of partial coherence of partially polarized light has been the subject of intensive research.[1–16] In classical optics, the standard approach consists of considering the light as a realization of a random function and in analyzing its second-order statistical properties. In the case of partially polarized light, the physical situation is characterized by the knowledge of the polarization matrix of each field and of the mutual coherence matrix. Each polarization matrix is described by four real parameters while the mutual coherence matrix is described by eight real parameters. The second-order statistical properties of a couple of two electric fields of the light is thus characterized by 16 real parameters. In order to get a physically meaningful interpretation of the correlation properties, it can be interesting to define parameters that allow one to check if some properties are satisfied. In classical optics, the concept of interference plays a central role. It is thus interesting[1] to characterize the coherence properties in relation to the visibility of the interference fringes that can be obtained with the couple of electric fields of the light beams. It is worth noting that in this chapter only interference fringes relative to the intensity are considered. In particular, the modulation of the different Stokes parameters that appear as discussed in Refs. 11 and 12 in the interference plane are not analyzed. The intrinsic degrees of coherence[7,8] allow one to get two parameters that can be easily determined from the 16 parameters mentioned above and that can be useful in order to analyze some practical properties of the light. We propose to discuss in the following some theoretical motivations, some basic properties, and some physical interpretations of this approach.

After a presentation of some background concepts, the discussion first points out the invariance properties of the theory of intrinsic degrees of coherence. In particular, this notion is related to the invariance properties of the factorization condition at order one introduced in the quantum theory of coherence.[17] Then, it is shown that the intrinsic degrees of coherence are useful to predict the ability of light to interfere when the polarization states of both interfering fields are optimized[18,19] or when only the polarization state of one field is optimized.[20] Afterward, an experimental situation of mixing two uncorrelated perfectly polarized lights with different temporal coherence properties is presented. Finally, a general statistical interpretation will be discussed.

10.2 Background

10.2.1 Second-order statistical properties

The electric field of an electromagnetic radiation at point $\mathbf{r}$ and at time t is classically represented by a complex random vector[21,22] $\mathbf{E}(\mathbf{r}, t)$. One will consider that this vector is two-dimensional, although this is not a necessary assumption because the theory can easily be generalized in dimension three. This electrical field will be written

$$\mathbf{E}(\mathbf{r}, t) = \begin{bmatrix} E_X(\mathbf{r}, t) \\ E_Y(\mathbf{r}, t) \end{bmatrix}. \tag{10.1}$$

The second-order statistical properties of the polarization of the field at point $\mathbf{r}$ and at time t is represented by the polarization matrix[21,22]

$$\mathbf{\Gamma}(\mathbf{r},t) = \begin{bmatrix} \langle E_X(\mathbf{r},t)E_X^*(\mathbf{r},t)\rangle & \langle E_X(\mathbf{r},t)E_Y^*(\mathbf{r},t)\rangle \\ \langle E_Y(\mathbf{r},t)E_X^*(\mathbf{r},t)\rangle & \langle E_Y(\mathbf{r},t)E_Y^*(\mathbf{r},t)\rangle \end{bmatrix}, \qquad (10.2)$$

where $\langle . \rangle$ is the ensemble average and a^* is the complex conjugate of a. If one introduces the hermitian conjugate $\mathbf{A}^\dagger$ of $\mathbf{A}$, then one can write $\mathbf{\Gamma}(\mathbf{r},t) = \langle \mathbf{E}(\mathbf{r},t)\mathbf{E}^\dagger(\mathbf{r},t)\rangle$. Because this matrix is hermitian, it corresponds to four real scalar parameters. However, it can be shown that one can introduce interesting quantities that are invariant under unitary transformations of the electric field. The first one is the trace $I(\mathbf{r},t) = \text{tr}\,[\mathbf{\Gamma}(\mathbf{r},t)]$, which is equal to the intensity of the electric field. The second one is the degree of polarization, which is defined by[21,22]

$$\mathcal{P} = \sqrt{1 - 4\,\frac{\det[\mathbf{\Gamma}(\mathbf{r},t)]}{\text{tr}[\mathbf{\Gamma}(\mathbf{r},t)]^2}}, \qquad (10.3)$$

where $\det[\mathbf{\Gamma}(\mathbf{r},t)]$ is the determinant of $\mathbf{\Gamma}(\mathbf{r},t)$.

For the sake of simplicity, in the following, one will consider wide sense stationary fields.[21–23] In that case the polarization matrix is independent of time [i.e. $\mathbf{\Gamma}(\mathbf{r},t) = \mathbf{\Gamma}(\mathbf{r})$]. Although the intensity of a field is a familiar description, let us recall what is the physical meaning of the degree of polarization. If $\mathcal{P} = 1$, then the light is totally polarized. A simple example of totally polarized light is obtained when the electric field of the light can be written

$$\mathbf{E}(\mathbf{r},t) = E(\mathbf{r},t)\,\mathbf{u}(\mathbf{r}), \qquad (10.4)$$

where $E(\mathbf{r},t)$ is a scalar complex random field while $\mathbf{u}(\mathbf{r})$ is a deterministic vector field that describes the trajectory of the oscillation in the transverse plane.

If $\mathcal{P} = 0$, then the light is said totally unpolarized. In that case,[21–23] the polarization matrix has two equal eigenvalues and the polarization matrix is proportional to the identity matrix in dimension two. It is easy to check that the light is totally unpolarized if the field can be written

$$\mathbf{E}(\mathbf{r},t) = E_1(\mathbf{r},t)\,\mathbf{u_1}(\mathbf{r}) + E_2(\mathbf{r},t)\,\mathbf{u_2}(\mathbf{r}), \qquad (10.5)$$

where $E_1(\mathbf{r},t)$ and $E_2(\mathbf{r},t)$ are uncorrelated scalar complex random fields with the same intensity and $\mathbf{u_1}(\mathbf{r})$ and $\mathbf{u_2}(\mathbf{r})$ are orthonormal deterministic vector fields.

Finally, if $1 > \mathcal{P} > 0$, the light is said to be partially polarized. One thus sees that the degree of polarization allows one to get information on the polarization state of the light independently of linear unitary transformations that can be applied to the electric field. Let us recall that a unitary matrix can represent the action of a birefringent plate on the light.[24]

More generally, the second-order statistical coherence properties of a couple of electric fields $\mathbf{E}(\mathbf{r_1},t_1)$ and $\mathbf{E}(\mathbf{r_2},t_2)$ at points $\mathbf{r_1}$ and $\mathbf{r_2}$ and at times t_1 and t_2 is represented by the mutual coherence matrix[21,22]

$$\Omega(\mathbf{r_1},\mathbf{r_2},t_1,t_2) = \begin{bmatrix} \langle E_X(\mathbf{r_2},t_2)E_X^*(\mathbf{r_1},t_1)\rangle & \langle E_X(\mathbf{r_2},t_2)E_Y^*(\mathbf{r_1},t_1)\rangle \\ \langle E_Y(\mathbf{r_2},t_2)E_X^*(\mathbf{r_1},t_1)\rangle & \langle E_Y(\mathbf{r_2},t_2)E_Y^*(\mathbf{r_1},t_1)\rangle \end{bmatrix},$$
(10.6)

which can also be written $\Omega(\mathbf{r_1},\mathbf{r_2},t_1,t_2) = \langle \mathbf{E}(\mathbf{r_2},t_2)\mathbf{E}^\dagger(\mathbf{r_1},t_1)\rangle$. Because this matrix is a priori non-hermitian, it corresponds to eight real scalar parameters. The polarization matrix[21,22] corresponds to $\Gamma(\mathbf{r}_i,t_i) = \Omega(\mathbf{r}_i,\mathbf{r}_i,t_i,t_i)$. The second-order statistical properties of the electric fields $\mathbf{E}(\mathbf{r_1},t_1)$ and $\mathbf{E}(\mathbf{r_2},t_2)$ are thus described by 16 real parameters that are included in the matrices $\Omega(\mathbf{r_1},\mathbf{r_2},t_1,t_2)$ $\Gamma(\mathbf{r_1},t_1)$ and $\Gamma(\mathbf{r_2},t_2)$. In the case of wide sense stationary fields, the matrix $\Omega(\mathbf{r_1},\mathbf{r_2},t_1,t_2)$ can be written $\Omega(\mathbf{r_1},\mathbf{r_2},t_2-t_1)$.

10.2.2 Coherence and visibility of interference fringes

As discussed in Ref. 1, one can introduce in the Fourier domain a degree of coherence that is related to the visibility of interference fringes. For wide sense stationary and quasi-monochromatic light, one can also introduce a Wolf degree of coherence in the space-time domain

$$\mu_W(\mathbf{r_1},\mathbf{r_2},\tau) = \frac{\operatorname{tr}[\Omega(\mathbf{r_1},\mathbf{r_2},\tau)]}{\sqrt{\operatorname{tr}[\Gamma(\mathbf{r_1})]\operatorname{tr}[\Gamma(\mathbf{r_2})]}}.$$
(10.7)

To illustrate the relationship between this quantity and the visibility of interference fringes, let us consider the experiment schematically shown in Fig. 10.1. The total field is $\mathbf{E}_T(t,\tau) = \mathbf{E}(\mathbf{r_2},t+\tau) + \mathbf{E}(\mathbf{r_1},t)$ and the intensity is thus $I_T(\tau) = \langle \|\mathbf{E}(\mathbf{r_2},t+\tau) + \mathbf{E}(\mathbf{r_1},t)\|^2\rangle$ that can be written

$$I_T(\tau) = I_1 + I_2 + \operatorname{tr}[\Omega(\mathbf{r_1},\mathbf{r_2},\tau)] + \operatorname{tr}\left[\Omega^\dagger(\mathbf{r_1},\mathbf{r_2},\tau)\right],$$
(10.8)

where $I_i = \langle \|\mathbf{E}(\mathbf{r_i},t)\|^2\rangle = \operatorname{tr}[\Gamma(\mathbf{r_i})]$.

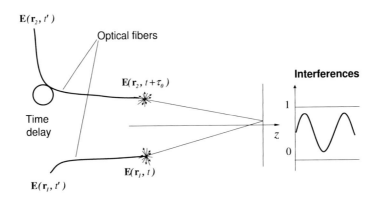

Figure 10.1 Schematic illustration of an interference experiment between partially polarized lights. Optical fibers are present in order to introduce time delays between the interfering fields.

In the case of quasi-monochromatic light, the visibility of the interference fringes is approximately constant in the area of interest in the interference plane. Thus, for small delays around a mean delay τ_0 (i.e., when $|\tau - \tau_0| \ll 1$), the local fringes can be described with the approximate mutual coherence matrix,

$$\Omega(\mathbf{r_1}, \mathbf{r_2}, \tau) \simeq \Omega(\mathbf{r_1}, \mathbf{r_2}, \tau_0) \, e^{i 2\pi \nu_0 (\tau - \tau_0)}, \tag{10.9}$$

where ν_0 is the mean frequency. Let us now introduce

$$\mu_W(\mathbf{r_1}, \mathbf{r_2}, \tau_0) = |\mu_W(\mathbf{r_1}, \mathbf{r_2}, \tau_0)| \, e^{i \phi_0(\mathbf{r_1}, \mathbf{r_2}, \tau_0)}; \tag{10.10}$$

one thus gets

$$I_T(\tau) \simeq I_1 + I_2 + \sqrt{I_1 I_2} \, |\mu_W(\mathbf{r_1}, \mathbf{r_2}, \tau_0)| \cos\left[2\pi\nu_0(\tau - \tau_0) + \phi_0(\mathbf{r_1}, \mathbf{r_2}, \tau_0)\right]. \tag{10.11}$$

The previous equation implies that when $I_1 = I_2$, the visibility of the interference fringes are equal to $|\mu_W(\mathbf{r_1}, \mathbf{r_2}, \tau_0)|$. Moreover, it is interesting to note that the Wolf degree of coherence of Eq. (10.7) can be written

$$\mu_W(\mathbf{r_1}, \mathbf{r_2}, \tau) = \frac{\langle \mathbf{E}^\dagger(\mathbf{r_1}, t) \, \mathbf{E}(\mathbf{r_2}, t + \tau) \rangle}{\sqrt{I_1 I_2}}. \tag{10.12}$$

This equation clearly shows that $\mu_W(\mathbf{r_1}, \mathbf{r_2}, \tau)$ is a normalized scalar product that results from the combination of two familiar scalar products. The first one is the geometrical scalar product $\mathbf{E}^\dagger(\mathbf{r_1}, t) \, \mathbf{E}(\mathbf{r_2}, t + \tau)$ between the electric fields. The second one is the statistical scalar product $\langle a^* b \rangle$ between random variables a and b, which results from the application of the average operator $\langle \rangle$ to the product $a^* b$. As a consequence, the Wolf degree of coherence can be equal to 0 in two different situations. The first one corresponds to totally uncorrelated fields. In that case, $\Omega(\mathbf{r_1}, \mathbf{r_2}, \tau) = \mathbf{0}$, where $\mathbf{0}$ is the matrix with zero elements. Whatever the linear transformation applied to each field $\mathbf{E}(\mathbf{r_1}, t_1)$ or $\mathbf{E}(\mathbf{r_2}, t_2)$, the Wolf degree of coherence will be equal to 0 in that case. The second situation corresponds to the case where the electric fields are geometrically orthogonal. Note that in the general case where the mutual coherence matrix is not equal to zero, there always exists a unitary transformation that can be applied to one of the electric fields $\mathbf{E}(\mathbf{r_1}, t_1)$ or $\mathbf{E}(\mathbf{r_2}, t_2)$ so that one gets $\mu_W(\mathbf{r_1}, \mathbf{r_2}, \tau) = 0$.

10.2.3 Factorization condition at order one

In the quantum theory of optical coherence introduced in 1963, Glauber proposed[17] to define a succession of correlation functions for the complex electromagnetic field from which different stated conditions have to be fulfilled in order to get complete coherence at different orders. When one concentrates on the second statistical order properties, one can simply analyze the definition of coherence at order one, which is related to the factorization condition for any $\mathbf{r_1}, \mathbf{r_2}, t_1, t_2$,

$$\Omega(\mathbf{r_1}, \mathbf{r_2}, t_1, t_2) = \mathbf{\Psi}(\mathbf{r_2}, t_2) \, \mathbf{\Psi}^\dagger(\mathbf{r_1}, t_1), \tag{10.13}$$

where $\mathbf{\Psi}(\mathbf{r}, t)$ is a deterministic vectorial function.

A direct consequence of this definition is that at order one the factorization condition is fulfilled only if the light is totally polarized. This is easily understandable in the context of the quantum theory of coherence proposed in Ref. 17 because the absence of randomness is required between each component of the fields.

Let us consider a field that satisfies Eq. (10.13) where $\mathbf{\Psi}(\mathbf{r_2}, t)$ is vertically polarized while $\mathbf{\Psi}(\mathbf{r_1}, t)$ is horizontally polarized for all t. Then, in that case, one has $|\mu_W(\mathbf{r_1}, \mathbf{r_2}, \tau)| = 0$. This result is the consequence of the invariance property satisfied by the factorization condition while Eq. (10.12) is sensitive to the rotation of one of the fields. More generally, one can note that if the electric field $\mathbf{E}(\mathbf{r}, t)$ satisfies the factorization condition at order one, this is also the case of the field

$$\mathbf{A}(\mathbf{r}, t) = \mathbf{J}(\mathbf{r}) \, \mathbf{E}(\mathbf{r}, t), \tag{10.14}$$

where $\mathbf{J}(\mathbf{r})$ is a non-singular deterministic Jones matrix that can depend on $\mathbf{r}$. Thus, if $\mathbf{E}(\mathbf{r}, t)$ is Glauber coherent at order one, then $\mathbf{A}(\mathbf{r}, t)$ is also Glauber coherent at order one. One will see below that this invariance property is also satisfied by the intrinsic degrees of coherence.

10.2.4 Coherence, invariance, and ability to interfere

On the one hand, the analysis of the Wolf degree of coherence [Eq. (10.7)] shows that one can observe interference fringes of unit visibility with partially polarized light. We have also seen that in order to observe such interference fringes of unit visibility the light has to satisfy two conditions that are respectively related to the geometrical and the statistical scalar product that appears in Eq. (10.12).

On the other hand, the factorization condition is invariant by deterministic linear transformations applied to each interfering electric field. It is a sufficient condition to be able to observe interference fringes with unit visibility after polarization optimization of the interfering fields (see Appendix A), but it is not necessary because a partially polarized light can lead to interference fringes of unit visibility. Actually, the factorization condition at order one has been introduced in order to go beyond the ability of two values of the field at distantly separated points or at separated times to provide interference fringes with unit visibility.

It can thus be interesting to determine sufficient and necessary conditions in order to characterize the ability of lights to interfere when the polarization states of the interfering fields are optimized. There exists several possibilities in order to optimize these polarization states.

One can restrict the optimization to local unitary transformations[2,13,15] (i.e., to unitary transformations that are applied separately to each interfering field). In particular, it has been proposed in Refs. 2 and 3 to look at a quantity that can be determined from the mutual coherence matrix and the polarization matrices of each field and that is invariant by unitary transformations. Another possibility[13,15] is to analyze the maximal value of the modulus of the Wolf degree of coherence when different unitary Jones matrices are applied to each interfering field.

It has also been proposed[14] to consider a measure invariant to unitary transformations that are applied to the four-dimensional vector

$$\mathbf{\Phi}(\mathbf{r_1}, t_1, \mathbf{r_2}, t_2) = \begin{bmatrix} E_X(\mathbf{r_1}, t_1) \\ E_Y(\mathbf{r_1}, t_1) \\ E_X(\mathbf{r_2}, t_2) \\ E_Y(\mathbf{r_2}, t_2) \end{bmatrix}. \qquad (10.15)$$

Its covariance matrix is $\mathbf{\Upsilon}(\mathbf{r_1}, t_1, \mathbf{r_2}, t_2) = \langle \mathbf{\Phi}(\mathbf{r_1}, t_1, \mathbf{r_2}, t_2) \mathbf{\Phi}^\dagger(\mathbf{r_1}, t_1, \mathbf{r_2}, t_2) \rangle$. It has been shown that this quantity plays a central role in the analysis of the Shannon entropy of partially polarized and coherent light.[9] In Ref. 14 a new measure of coherence has been introduced as a distance between $\mathbf{\Upsilon}(\mathbf{r_1}, \mathbf{r_2}, \tau)$ and the identity matrix in dimension four.* One can note that, local unitary transformations conserve the factorization condition but do not include nonunitary ones, which also conserve it.

In the following, one will consider optimization of the interference fringes visibility by local polarization transformations. From a practical point of view, the idea is that one analyzes the ability of light to interfere when one can modify each interfering field with arbitrary deterministic polarization devices. It will be shown that this approach, which leads to the intrinsic degrees of coherence, allows one to get practical interesting results for the characterization of the ability of the light to interfere. From a theoretical point of view, the basic concept is to consider parameters that do not vary when one applies linear transformations that conserve the global coherence characterized by the factorization condition at order one. It can be shown[25] that one needs two parameters to describe the fundamental symmetry that can appear. One will nevertheless concentrate here on practical results that can be obtained with the intrinsic degrees of coherence.

10.3 Intrinsic Coherence and Visibility of Interference Fringes

10.3.1 Intrinsic degrees of coherence

It has been shown in Ref. 7 that the two intrinsic degrees of coherence $\mu_S(\mathbf{r_1}, \mathbf{r_2}, t_2 - t_1)$ and $\mu_I(\mathbf{r_1}, \mathbf{r_2}, t_2 - t_1)$ between $\mathbf{E}(\mathbf{r_1}, t_1)$ and $\mathbf{E}(\mathbf{r_2}, t_2)$ are the same for any couple of fields $[\mathbf{J}_1 \mathbf{E}(\mathbf{r_1}, t_1), \mathbf{J}_2 \mathbf{E}(\mathbf{r_2}, t_2)]$, where $\mathbf{J}_1$ and $\mathbf{J}_2$ are non-singular

---
*In Ref. 14 the theory has been developed in the space frequency domain, but it can also be generalized in the space time domain.

deterministic Jones matrices. These intrinsic degrees of coherence are the singular values of the normalized mutual coherence matrix[7,8]

$$\mathbf{M}(\mathbf{r_1}, \mathbf{r_2}, \tau) = \mathbf{\Gamma}^{-1/2}(\mathbf{r_2})\, \mathbf{\Omega}(\mathbf{r_1}, \mathbf{r_2}, \tau)\, \mathbf{\Gamma}^{-1/2}(\mathbf{r_1}), \qquad (10.16)$$

and the singular value decomposition will be noted

$$\mathbf{M}(\mathbf{r_1}, \mathbf{r_2}, \tau) = \mathbf{U_2}(\mathbf{r_1}, \mathbf{r_2}, \tau)\, \mathbf{D}(\mathbf{r_1}, \mathbf{r_2}, \tau)\, \mathbf{U_1}^\dagger(\mathbf{r_1}, \mathbf{r_2}, \tau), \qquad (10.17)$$

where

$$\mathbf{D}(\mathbf{r_1}, \mathbf{r_2}, \tau) = \begin{bmatrix} \mu_S(\mathbf{r_1}, \mathbf{r_2}, \tau) & 0 \\ 0 & \mu_I(\mathbf{r_1}, \mathbf{r_2}, \tau) \end{bmatrix} \qquad (10.18)$$

and where $\mathbf{U_1}(\mathbf{r_2}, \mathbf{r_2}, \tau)$ and $\mathbf{U_2}(\mathbf{r_2}, \mathbf{r_2}, \tau)$ are unitary matrices. It can be shown[19] that

$$1 \geq \mu_S(\mathbf{r_1}, \mathbf{r_2}, \tau) \geq \mu_I(\mathbf{r_1}, \mathbf{r_2}, \tau) \geq 0. \qquad (10.19)$$

For the sake of clarity, and when no confusion is possible, the dependency on $(\mathbf{r_1}, \mathbf{r_2}, \tau)$ will not be mentioned in the following. Using

$$\mathbf{T} = \mathbf{M}\, \mathbf{M}^\dagger, \qquad (10.20)$$

one easily gets

$$\begin{aligned} \mu_S^2 &= \mathcal{T}\, [1 + \mathcal{Q}], \\ \mu_I^2 &= \mathcal{T}\, [1 - \mathcal{Q}], \end{aligned} \qquad (10.21)$$

where

$$\mathcal{T} = \frac{\operatorname{tr}[\mathbf{T}]}{2} \qquad (10.22)$$

and

$$\mathcal{Q} = \sqrt{1 - 4\frac{\det[\mathbf{T}]}{\operatorname{tr}[\mathbf{T}]^2}}. \qquad (10.23)$$

One can remark that the parameter $\mathcal{Q}$ has a mathematical expression that is analogous to the degree of polarization [see Eq. (10.3)]. One will see below that it is related to anisotropic coherence properties of the light. It can be shown that $\mathcal{Q} = (\mu_S^2 - \mu_I^2)/(\mu_S^2 + \mu_I^2)$ and that $\mathcal{T} = (\mu_S^2 + \mu_I^2)/2$. A direct consequence of this last expression is that

$$\begin{aligned} \mathcal{T} = 0 &\Rightarrow \mathbf{\Omega} = \mathbf{0}, \\ \mathcal{T} = 1 &\Rightarrow \mu_S = \mu_I = 1. \end{aligned} \qquad (10.24)$$

10.3.2 Relation with interference experiments with double polarization optimization

The intrinsic degrees of coherence may appear as only a mathematical development if they were not intimately related to practical experiments. In fact, the intrinsic coherence of partially polarized light has been developed to easily analyze interference properties that one can expect when the polarization states of the interfering fields are optimized. A similar approach has been used in the context of radar imagery.[26]

Let us first consider the simple interference experiment in which one analyzes the maximal value of the interference fringes visibility between the fields

$$\mathbf{A}(\mathbf{r}_1, t_1) = \mathbf{J_1}\, \mathbf{E}(\mathbf{r}_1, t_1) \tag{10.25}$$

and

$$\mathbf{A}(\mathbf{r}_2, t_2) = \mathbf{J_2}\, \mathbf{E}(\mathbf{r}_2, t_2) \tag{10.26}$$

when one optimizes the Jones matrices $\mathbf{J_1}$ and $\mathbf{J_2}$. This situation is schematically represented in Fig. 10.2. Maximizing the interference visibility necessitates one to maximize the modulus of the Wolf degree of coherence. It has been demonstrated in Refs. 7 and 18 that the largest intrinsic degrees of coherence μ_S is the maximal value of the modulus of the Wolf degree of coherence when the polarization state of the interfering fields can be modified arbitrarily by deterministic polarization modulators. Furthermore, it has been shown[18] that the polarization modifications that maximize the modulus of this Wolf degree of coherence correspond to a coherence polarization filtering.

It can be shown[19] that a not totally polarized light with a Wolf degree of coherence in the space-time domain of modulus equal to 1 has intrinsic degrees of coherence of unit values (i.e., $\mu_S = \mu_I = 1$). Furthermore,[19] the maximal value

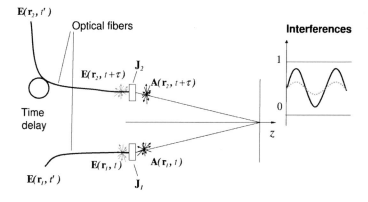

Figure 10.2 Schematic illustration of an interference experiment between partially polarized lights when the polarization states of both interfering fields are optimized with the Jones matrices $\mathbf{J_1}$ and $\mathbf{J_2}$.

of the modulus of the Wolf degree of coherence is obtained without polarizing the light only if $Q = 0$ (i.e., if $\mu_S = \mu_I$).† These results can be summarized as follows:

- If $Q = 0$, then the maximal value of $|\mu_W|$ can be obtained without having to polarize each interfering field.

- If $Q > 0$, then the maximal value of $|\mu_W|$ is obtained after each interfering field has been totally polarized.

A direct consequence of these results is that no interference fringes can be observed after polarization optimization if and only if $T = 0$.

One can also obtain a sufficient and necessary condition in order to be able to observe interference fringes with unitary visibility. Indeed, the previous result let us know that:

- Interference fringes with unit visibility can be observed after polarization optimization‡ if and only if $\mu_S = 1$.

- Interference fringes with unit visibility can be observed after polarization optimization with partially polarized light if and only if $T = 1$.

One can note that $T = 1$ implies that $Q = 0$ and is equivalent to $\mu_S = \mu_I = 1$.

One thus gets necessary and sufficient conditions in order to be able to observe interference fringes with unit visibility after polarization state optimization of each interfering electric field. One also has a necessary and sufficient condition to know if the light has to be perfectly polarized or not in order to get unitary visibility. Because the matrices Ω, Γ_1, and Γ_2 can be easily measured,[27] it is easy to determine the intrinsic degrees of coherence and thus to know if interference fringes with unit visibility can be observed without making an exhaustive experimental or numerical optimization of the visibility over the possible polarization states of the interfering beams.

10.3.3 Relation with interference experiments with single polarization optimization

It is also interesting to analyze the optimal visibility one can obtain in interference experiments with partially polarized light when one modifies the polarization state of only one of the two interfering beams. This is a practical situation that can appear when one does not want to modify or attenuate one of the beam, such as in

---

†In Ref. 19 the proof of property B concerns the case were both intrinsic degrees of coherence are equal to 1. This proof can nevertheless be easily generalized in order to get the general property that the maximal value of the modulus of the Wolf degree of coherence is obtained without polarizing the light only if $Q = 0$.

‡The polarization optimization one has to consider in order to get visibility optimization also includes intensity (i.e., it is a global optimization of the polarization matrix).

homodyne detection. The considered interference experiment is shown schematically in Fig. 10.3. In this interference experiment, one analyzes the maximal value of the interference fringes visibility between the fields

$$\mathbf{A}(\mathbf{r}_1, t_1) = \mathbf{J}_1 \mathbf{E}(\mathbf{r}_1, t_1) \quad (10.27)$$

and

$$\mathbf{A}(\mathbf{r}_2, t_2) = \mathbf{E}(\mathbf{r}_2, t_2) \quad (10.28)$$

when one optimizes only the Jones matrix $\mathbf{J}_1$. For simplicity reasons, $\mathbf{E}(\mathbf{r}_2, t_2)$ will be denominated the signal beam and $\mathbf{A}(\mathbf{r}_1, t_1)$ the reference beam. Here again, maximizing the interference visibility necessitates maximizing the modulus of the Wolf degree of coherence when the polarization state of the field $\mathbf{E}(\mathbf{r}_1, t_1)$ can be modified arbitrarily by a deterministic polarization modulator.

It can be shown[20] that in such an interference experiment the optimal value $|\mu_{W,0}(\mathbf{r}_1, \mathbf{r}_2, \tau)|^2$ of the square modulus of the Wolf degree of coherence is given by $|\mu_{W,0}(\mathbf{r}_1, \mathbf{r}_2, \tau)|^2 = \text{tr}[\mathbf{T}\,\mathbf{\Gamma}_2]/\text{tr}[\mathbf{\Gamma}_2]$, where the matrix $\mathbf{T}$ is defined with Eq. (10.20). The optimal Jones matrix (i.e., the Jones matrix that maximizes the modulus of the Wolf degree of coherence) is given[20] by $\mathbf{J}_1^{\text{opt}} = \mathbf{\Gamma}_2^{1/2} \mathbf{M}\, \mathbf{\Gamma}_1^{-1/2}$. The polarization matrix of the field $\mathbf{A}(\mathbf{r}_1, t_1)$ after the polarization modulator is $\mathbf{\Gamma}_1' = \mathbf{\Gamma}_2^{1/2}\,\mathbf{T}\,\mathbf{\Gamma}_2^{1/2}$ and it can be easily seen[20] that its degree of polarization can be written

$$\mathcal{P}_1' = \sqrt{1 - 4 \frac{\det[\mathbf{T}\,\mathbf{\Gamma}_2]}{\text{tr}[\mathbf{T}\,\mathbf{\Gamma}_2]^2}}. \quad (10.29)$$

One thus sees that, in general, the optimal configuration corresponds to a degree of polarization of the optimal reference beam, which is different from the one of the signal beam. It is however interesting to note that when $\mathcal{Q} = 0$ (i.e., when $\mu_S = \mu_I$) without restriction on the general form of $\mathbf{\Gamma}_2$, one has $\mathbf{T} = \mu_S^2(\mathbf{r}_1, \mathbf{r}_2, \tau)\,\text{Id}$,

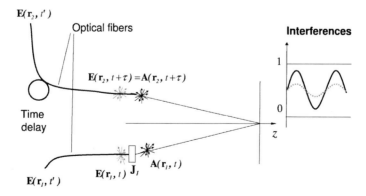

Figure 10.3 Schematic illustration of an interference experiment between partially polarized lights when the polarization state of only the field $\mathbf{E}_1(\mathbf{r}_1, t)$ is optimized with the Jones matrix $\mathbf{J}_1$.

where **Id** is the identity matrix in dimension 2. The maximal value of the modulus of the Wolf degree of coherence is then equal to $\mu_S(\mathbf{r_1}, \mathbf{r_2}, \tau)$. Furthermore, the optimal reference beam and the signal beam have, in that case, the same degree of polarization.

It is interesting to discuss the general situation in a simple and illustrative example. Let us assume that $\mathbf{E}(\mathbf{r}_2, t_2)$ is totally unpolarized. In that case, the polarization matrix of the field $\mathbf{A}(\mathbf{r}_1, t_1)$ after polarization optimization is

$$\mathbf{\Gamma}'_1 = \frac{I_2}{2}\,\mathbf{T}. \tag{10.30}$$

The degree of polarization $\mathcal{P}'_1$ of the optimized reference beam is in that case[20]

$$\mathcal{P}'_1 = \mathcal{Q}, \tag{10.31}$$

while the one of $\mathbf{E}(\mathbf{r_2}, t)$ is $\mathcal{P}_2 = 0$. This result exhibits a simple physical interpretation of the parameter $\mathcal{Q}$ because it is the degree of polarization of the reference beam that optimizes the fringes visibility when the signal beam is totally unpolarized.

In summary, we see that the visibility one can obtain in interference experiments with partially polarized light when one modifies the polarization state of only one of the two interfering beams can lead to the following different behaviors that can be analyzed efficiently using the intrinsic degrees of coherence:

- The polarization state that optimizes the modulus of the Wolf degree of coherence is, in general, different from the polarization state of the signal beam when $\mathcal{Q} \neq 0$. A simple interpretation can be obtained when the signal beam is totally unpolarized.

- When $\mathcal{Q} = 0$ the optimal reference beam has the same degree of polarization as the one of the signal beam.

One can thus conclude that $\mathcal{Q}$ characterizes the potential loss of symmetry between the optimal reference beam and the signal beam. It can be shown that this behavior is due to anisotropy of the coherence properties.[25]

10.4 Physical Interpretations

10.4.1 Mixing experiment

We have seen in the previous section that the intrinsic degrees of coherence allow one to predict the ability of two lights to interfere after polarization optimization. In this section, we propose to consider a simple experiment of mixing two perfectly polarized lights that are wide sense stationary and partially temporally coherent. For that purpose, let us consider the simple experiment described in Fig. 10.4. Because one considers only temporal coherence properties, the dependency on the

Intrinsic Degrees of Coherence for Electromagnetic Fields

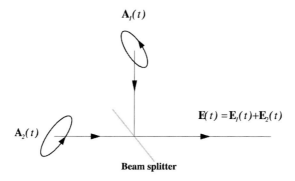

Figure 10.4 Schematic representation of the mixing experiment considered in Section 10.4.1. The fields $\mathbf{A}_j(t)$ and $\mathbf{E}_j(t)$ with $j = 1, 2$ are proportional.

spatial coordinates will not be mentioned. The total field can thus be written[§]

$$\mathbf{E}(t) = \mathbf{E_1}(t) + \mathbf{E_2}(t). \qquad (10.32)$$

The fields $\mathbf{E_1}(t)$ and $\mathbf{E_2}(t)$ are assumed statistically uncorrelated and perfectly polarized with arbitrary but non-parallel normalized polarization states. More precisely, one assumes that, for $i = 1, 2$, one can write

$$\mathbf{E_i}(t) = E_i(t)\,\mathbf{u_i}, \qquad (10.33)$$

where $E_i(t)$ are scalar complex random functions and where $\mathbf{u_1}$ and $\mathbf{u_2}$ are deterministic non-parallel vectors. The temporal coherence of each field is unambiguously described by the standard degree of coherence[21]

$$\mu_i(\tau) = \frac{\langle E_i^*(t)\,E_i(t+\tau)\rangle}{I_i}, \qquad (10.34)$$

where $I_i = \langle |E_i(t)|^2\rangle$ is the intensity of beam i. It can be shown (see Appendix B) that the intrinsic degrees of coherence $\mu_S(\tau)$ and $\mu_I(\tau)$ are simply related to the standard degrees of coherence $\mu_1(\tau)$ and $\mu_2(\tau)$. Indeed, one has for all τ

$$\begin{aligned}\mu_S(\tau) &= \max[|\mu_1(\tau)|, |\mu_2(\tau)|] \\ \mu_I(\tau) &= \min[|\mu_1(\tau)|, |\mu_2(\tau)|]\end{aligned}, \qquad (10.35)$$

where $\max[a, b]$ (respectively $\min[a, b]$) is the maximal (respectively, the minimal) value between a and b. For example, if one has $|\mu_1(\tau)| > |\mu_2(\tau)|$, then one gets $\mu_S(\tau) = |\mu_1(\tau)|$ and $\mu_I(\tau) = |\mu_2(\tau)|$. This result shows that the decomposition of the coherence properties with the intrinsic degrees of coherence allows one to recover the modulus of the standard degree coherence of each field when the light is a mixing of two uncorrelated and perfectly polarized lights. It thus clearly provides a simple physical interpretation of the intrinsic degrees of coherence.[28]

[§] Attenuation and phase factors between the fields $\mathbf{A}_j(t)$ and $\mathbf{E}_j(t)$ with $j = 1, 2$ are not mentioned because they do not influence the results of the proposed analysis.

10.4.2 General statistical interpretation

Let us consider two wide sense stationary electric fields $\mathbf{E}(\mathbf{r}_1, t)$ and $\mathbf{E}(\mathbf{r}_2, t)$ with polarization matrices respectively equal to $\mathbf{\Gamma}(\mathbf{r}_1)$ and $\mathbf{\Gamma}(\mathbf{r}_2)$. One can introduce the fields

$$\mathbf{A}(\mathbf{r_i}, t) = \mathbf{\Gamma}^{-1/2}(\mathbf{r_i})\, \mathbf{E}(\mathbf{r_i}, t) \tag{10.36}$$

with $i = 1, 2$. One has $\langle \mathbf{A}(\mathbf{r_i}, t) \mathbf{A}^\dagger(\mathbf{r_i}, t) \rangle = \mathrm{Id}$ and

$$\langle \mathbf{A}(\mathbf{r}_2, t + \tau) \mathbf{A}^\dagger(\mathbf{r}_1, t) \rangle = \mathbf{M}(\mathbf{r}_1, \mathbf{r}_2, \tau). \tag{10.37}$$

The singular value decomposition of $\mathbf{M}(\mathbf{r}_1, \mathbf{r}_2, \tau)$ can be written

$$\mathbf{M}(\mathbf{r}_1, \mathbf{r}_2, \tau) = \mathbf{U}_2(\mathbf{r}_1, \mathbf{r}_2, \tau)\, \mathbf{D}(\mathbf{r}_1, \mathbf{r}_2, \tau)\, \mathbf{U}_1^\dagger(\mathbf{r}_1, \mathbf{r}_2, \tau) \tag{10.38}$$

with

$$\mathbf{D}(\mathbf{r}_1, \mathbf{r}_2, \tau) = \begin{bmatrix} \mu_S(\mathbf{r}_1, \mathbf{r}_2, \tau) & 0 \\ 0 & \mu_I(\mathbf{r}_1, \mathbf{r}_2, \tau) \end{bmatrix}, \tag{10.39}$$

and where $\mathbf{U}_1(\mathbf{r}_2, \mathbf{r}_2, \tau)$ and $\mathbf{U}_2(\mathbf{r}_2, \mathbf{r}_2, \tau)$ are unitary matrices. Let us also introduce the fields

$$\boldsymbol{\mathcal{E}}^{(\mathbf{r}_1, \mathbf{r}_2, \tau)}(\mathbf{r_i}, t) = \mathbf{U}_i^\dagger(\mathbf{r}_1, \mathbf{r}_2, \tau)\, \mathbf{A}(\mathbf{r_i}, t) \tag{10.40}$$

with $i = 1, 2$. The complete notation is introduced in order to emphasize that, in general, each field $\boldsymbol{\mathcal{E}}^{(\mathbf{r}_1, \mathbf{r}_2, \tau)}(\mathbf{r}_1, t)$ or $\boldsymbol{\mathcal{E}}^{(\mathbf{r}_1, \mathbf{r}_2, \tau)}(\mathbf{r}_2, t)$ depends on both $\mathbf{r}_1, \mathbf{r}_2$, and on τ. Let us write

$$\boldsymbol{\mathcal{E}}^{(\mathbf{r}_1, \mathbf{r}_2, \tau)}(\mathbf{r_i}, t) = \begin{bmatrix} \mathcal{E}_X^{(\mathbf{r}_1, \mathbf{r}_2, \tau)}(\mathbf{r_i}, t) \\ \mathcal{E}_Y^{(\mathbf{r}_1, \mathbf{r}_2, \tau)}(\mathbf{r_i}, t) \end{bmatrix}, \tag{10.41}$$

one has

$$\left\langle \mathcal{E}_X^{(\mathbf{r}_1, \mathbf{r}_2, \tau)}(\mathbf{r}_2, t + \tau) \left[\mathcal{E}_X^{(\mathbf{r}_1, \mathbf{r}_2, \tau)}(\mathbf{r}_1, t)\right]^* \right\rangle = \mu_S(\mathbf{r}_1, \mathbf{r}_2, \tau), \tag{10.42}$$

$$\left\langle \mathcal{E}_Y^{(\mathbf{r}_1, \mathbf{r}_2, \tau)}(\mathbf{r}_2, t + \tau) \left[\mathcal{E}_X^{(\mathbf{r}_1, \mathbf{r}_2, \tau)}(\mathbf{r}_1, t)\right]^* \right\rangle = 0, \tag{10.43}$$

$$\left\langle \mathcal{E}_X^{(\mathbf{r}_1, \mathbf{r}_2, \tau)}(\mathbf{r}_2, t + \tau) \left[\mathcal{E}_Y^{(\mathbf{r}_1, \mathbf{r}_2, \tau)}(\mathbf{r}_1, t)\right]^* \right\rangle = 0, \tag{10.44}$$

$$\left\langle \mathcal{E}_Y^{(\mathbf{r}_1, \mathbf{r}_2, \tau)}(\mathbf{r}_2, t + \tau) \left[\mathcal{E}_Y^{(\mathbf{r}_1, \mathbf{r}_2, \tau)}(\mathbf{r}_1, t)\right]^* \right\rangle = \mu_I(\mathbf{r}_1, \mathbf{r}_2, \tau), \tag{10.45}$$

and

$$\left\langle \mathcal{E}_X^{(\mathbf{r}_1, \mathbf{r}_2, \tau)}(\mathbf{r_i}, t) \left[\mathcal{E}_X^{(\mathbf{r}_1, \mathbf{r}_2, \tau)}(\mathbf{r_i}, t)\right]^* \right\rangle = 1, \tag{10.46}$$

$$\left\langle \mathcal{E}_Y^{(\mathbf{r}_1, \mathbf{r}_2, \tau)}(\mathbf{r_i}, t) \left[\mathcal{E}_X^{(\mathbf{r}_1, \mathbf{r}_2, \tau)}(\mathbf{r_i}, t)\right]^* \right\rangle = 0, \tag{10.47}$$

$$\left\langle \mathcal{E}_X^{(\mathbf{r}_1, \mathbf{r}_2, \tau)}(\mathbf{r_i}, t) \left[\mathcal{E}_Y^{(\mathbf{r}_1, \mathbf{r}_2, \tau)}(\mathbf{r_i}, t)\right]^* \right\rangle = 0, \tag{10.48}$$

Intrinsic Degrees of Coherence for Electromagnetic Fields

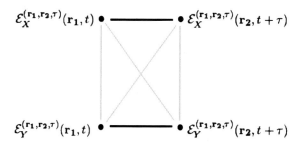

Figure 10.5 Schematic representation of the statistical correlation between the components of the vectors $\mathcal{E}^{(r_1,r_2,\tau)}(r_1,t)$ and $\mathcal{E}^{(r_1,r_2,\tau)}(r_2,t+\tau)$ considered in Section 10.4.2. Black lines represent the existence of correlation, and gray lines represent statistical uncorrelation.

$$\left\langle \mathcal{E}_Y^{(r_1,r_2,\tau)}(r_i,t) \left[\mathcal{E}_Y^{(r_1,r_2,\tau)}(r_i,t) \right]^* \right\rangle = 1, \qquad (10.49)$$

for $i = 1, 2$. These results are schematically represented in Fig. 10.5. Their physical interpretation is that one gets a decomposition of the fields so that $\mathcal{E}^{(r_1,r_2,\tau)}(r_1,t)$ and $\mathcal{E}^{(r_1,r_2,\tau)}(r_2,t+\tau)$ have their cross terms uncorrelated.

It is interesting to note that if $\mathcal{Q} = 0$, the previous relations are also valid for any fields $\mathcal{E}'^{(r_1,r_2,\tau)}(r_i,t) = U \, \mathcal{E}^{(r_1,r_2,\tau)}(r_i,t)$ where U is any unitary matrix. This is no more the case if $\mathcal{Q} \neq 0$. That property also shows that the parameter $\mathcal{Q}$ is a measure of anisotropy (see also Ref. 25 for a more complete analysis).

10.5 Conclusion

In the context of the second-order statistics, the coherence properties of two partially polarized fields are characterized by the 16 scalar real parameters that are included in the mutual coherence and the polarization matrices. As it has been done with the degree of polarization in order to analyze polarization properties, it can be interesting to determine fewer parameters that allow one to predict some physical properties related to the ability of the light to interfere independently to some geometrical transformations of the fields. The situation is nevertheless complex and probably a unique parameter is not sufficient to analyze it. We have discussed here some characteristics of the theory of intrinsic degrees of coherence.

We have first shown that the invariance properties of the intrinsic degrees of coherence correspond to transformations that conserve the factorization condition at order one introduced in the quantum theory of coherence. We have then illustrated the usefulness of the intrinsic degrees of coherence in the characterization of the ability of light to interfere when the polarization states of both fields are optimized and when the polarization state of only one of the fields is optimized. We have shown that in both cases the intrinsic degrees of coherence provide useful predictive information that allows one to avoid a numerical or an experimental exhaustive search over all the different possible polarization states of the interfering fields.

The physical interpretation of the intrinsic degrees of coherence has been also discussed in a simple experiment of mixing uncorrelated perfectly polarized light with different temporal coherence properties and a general statistical interpretation has also been discussed. There exist many perspectives using the intrinsic degrees of coherence. Some of the most interesting are related to the analysis of the evolution of these quantities in some practical experiments.

Appendix A Interference and Factorization Condition at Order One

Let us assume that the fields $\mathbf{E}(\mathbf{r_1}, t_1)$ and $\mathbf{E}(\mathbf{r_2}, t_2)$ satisfy

$$\Omega(\mathbf{r_1}, \mathbf{r_2}, t_1, t_2) = \mathbf{\Psi}(\mathbf{r_2}, t_2)\, \mathbf{\Psi}^\dagger(\mathbf{r_1}, t_1).$$

Let us apply the deterministic linear transformation $\mathbf{J}$ to $\mathbf{E}(\mathbf{r_1}, t_1)$ so that $\mathbf{A}(\mathbf{r_1}, t_1) = \mathbf{J}\,\mathbf{E}(\mathbf{r_1}, t_1)$ and $\mathbf{A}(\mathbf{r_2}, t_2) = \mathbf{E}(\mathbf{r_2}, t_2)$. The mutual coherence matrix thus becomes

$$\langle \mathbf{A}(\mathbf{r_2}, t_2)\mathbf{A}^\dagger(\mathbf{r_1}, t_1)\rangle = \mathbf{\Psi}(\mathbf{r_2}, t_2)\,\mathbf{\Psi}^\dagger(\mathbf{r_1}, t_1)\, \mathbf{J}^\dagger.$$

If one chooses $\mathbf{J}^\dagger = \mathbf{\Psi}(\mathbf{r_1}, t_1)\,\mathbf{\Psi}^\dagger(\mathbf{r_2}, t_2)$, then one sees that

$$\langle \mathbf{A}(\mathbf{r_2}, t_2)\mathbf{A}^\dagger(\mathbf{r_1}, t_1)\rangle = \rho(\mathbf{r_1}, t_1)\,\mathbf{\Psi}(\mathbf{r_2}, t_2)\,\mathbf{\Psi}^\dagger(\mathbf{r_2}, t_2),$$

with $\rho(\mathbf{r_1}, t_1) = \mathbf{\Psi}^\dagger(\mathbf{r_1}, t_1)\,\mathbf{\Psi}(\mathbf{r_1}, t_1)$. Furthermore, one has

$$\mathrm{tr}\left[\langle \mathbf{A}(\mathbf{r_2}, t_2)\mathbf{A}^\dagger(\mathbf{r_2}, t_2)\rangle\right] = \mathbf{\Psi}^\dagger(\mathbf{r_2}, t_2)\mathbf{\Psi}(\mathbf{r_2}, t_2),$$

$$\mathrm{tr}\left[\langle \mathbf{A}(\mathbf{r_1}, t_1)\mathbf{A}^\dagger(\mathbf{r_1}, t_1)\rangle\right] = \rho^2(\mathbf{r_1}, t_1)\,\mathbf{\Psi}^\dagger(\mathbf{r_2}, t_2)\mathbf{\Psi}(\mathbf{r_2}, t_2),$$

$$\mathrm{tr}\left[\langle \mathbf{A}(\mathbf{r_2}, t_2)\mathbf{A}^\dagger(\mathbf{r_1}, t_1)\rangle\right] = \rho(\mathbf{r_1}, t_1)\,\mathbf{\Psi}^\dagger(\mathbf{r_2}, t_2)\mathbf{\Psi}(\mathbf{r_2}, t_2).$$

Using these expressions, one gets

$$\left|\frac{\mathrm{tr}\left[\langle \mathbf{A}(\mathbf{r_2}, t_2)\mathbf{A}^\dagger(\mathbf{r_1}, t_1)\rangle\right]}{\sqrt{\mathrm{tr}\left[\langle \mathbf{A}(\mathbf{r_2}, t_2)\mathbf{A}^\dagger(\mathbf{r_2}, t_2)\rangle\right]\mathrm{tr}\left[\langle \mathbf{A}(\mathbf{r_1}, t_1)\mathbf{A}^\dagger(\mathbf{r_1}, t_1)\rangle\right]}}\right| = 1.$$

This result shows that the factorization condition implies that a simple polarization optimization leads to interference fringes with unit visibility.

Appendix B Intrinsic Degrees of Coherence of Mixing Uncorrelated Perfectly Polarized Lights

Let us introduce

$$\mathbf{u_i} = \begin{bmatrix} u_{i,X} \\ u_{i,Y} \end{bmatrix} \qquad (\text{B-1})$$

and
$$\mathbf{e}(t) = \begin{bmatrix} E_1(t) \\ E_2(t) \end{bmatrix}. \tag{B-2}$$

Because
$$\mathbf{E}(t) = E_1(t)\,\mathbf{u_1} + E_2(t)\,\mathbf{u_2}, \tag{B-3}$$

one can write
$$\mathbf{E}(t) = \mathbf{K}\,\mathbf{e}(t), \tag{B-4}$$

where
$$\mathbf{K} = \begin{bmatrix} u_{1,X} & u_{2,X} \\ u_{1,Y} & u_{2,Y} \end{bmatrix}. \tag{B-5}$$

Because of the invariance properties of the intrinsic degrees of coherence, the intrinsic degrees of coherence of $\mathbf{E}(t)$ and of $\mathbf{e}(t)$ are equal. In other words, the intrinsic degrees of coherence determined from

$$\boldsymbol{\Omega}_{\mathbf{EE}}(\tau) = \langle \mathbf{E}(t+\tau)\,\mathbf{E}^\dagger(t)\rangle = \mathbf{K}\,\langle \mathbf{e}(t+\tau)\,\mathbf{e}^\dagger(t)\rangle\,\mathbf{K}^\dagger \tag{B-6}$$

or from
$$\boldsymbol{\Omega}_{\mathbf{ee}}(\tau) = \langle \mathbf{e}(t+\tau)\,\mathbf{e}^\dagger(t)\rangle \tag{B-7}$$

are equal. One has

$$\boldsymbol{\Omega}_{\mathbf{ee}}(\tau) = \begin{bmatrix} \langle E_1(t+\tau)E_1^*(t)\rangle & \langle E_1(t+\tau)E_2^*(t)\rangle \\ \langle E_2(t+\tau)E_1^*(t)\rangle & \langle E_2(t+\tau)E_2^*(t)\rangle \end{bmatrix}. \tag{B-8}$$

Because $\langle E_1(t+\tau)E_2^*(t)\rangle = \langle E_2(t+\tau)E_1^*(t)\rangle = 0$ and $\langle E_i(t+\tau)E_i^*(t)\rangle = I_i\,\mu_i(\tau)$, one gets

$$\boldsymbol{\Omega}_{\mathbf{ee}}(\tau) = \begin{bmatrix} I_1\,\mu_1(\tau) & 0 \\ 0 & I_2\,\mu_2(\tau) \end{bmatrix} \tag{B-9}$$

and
$$\boldsymbol{\Gamma}_{\mathbf{ee}}(\tau) = \langle \mathbf{e}(t)\,\mathbf{e}^\dagger(t)\rangle = \begin{bmatrix} I_1 & 0 \\ 0 & I_2 \end{bmatrix}. \tag{B-10}$$

It is thus easy to see that the normalized mutual coherence matrix is

$$\mathbf{M}_{\mathbf{ee}}(\tau) = \begin{bmatrix} \mu_1(\tau) & 0 \\ 0 & \mu_2(\tau) \end{bmatrix}, \tag{B-11}$$

which demonstrates the result of Section 10.4.1.

References

1. Wolf, E., "Unified theory of coherence and polarization of random electromagnetic beams," *Phys. Lett. A*, **312**, 263–267, 2003.

2. Tervo, J., Setälä, T., and Friberg, A.T., "Degree of coherence of electromagnetic fields," *Opt. Expr.*, **11**, 1137–1142, 2003.

3. Tervo, J., Setälä, T., and Friberg, A.T., "Theory of partially coherent electromagnetic fields in the space-frequency domain," *J. Opt. Soc. Am. A*, **21**, 2205–2215, 2004.

4. Setälä, T., Tervo, J., and Friberg, A.T., "Complete electromagnetic coherence in the space-frequency domain," *Opt. Lett.*, **29**, 328–330, 2004.

5. Gori, F., Santarsiero, M., Simon, R., Piquero, G., Borghi, R., and Guattari, G., "Coherent-mode decomposition of partially polarized, partially coherent sources," *J. Opt. Soc. Am. A*, **20**, 78–84, 2003.

6. Agarwal, G.S., Dogariu, A., Visser, T., and Wolf, E., "Generation of complete coherence in young's interference experiment with random mutually uncorrelated electromagnetic beams," *Opt. Lett.*, **30**(2), 120–122, 2005.

7. Réfrégier, P., and Goudail, F., "Invariant degrees of coherence of partially polarized light," *Opt. Expr.*, **13**(16), 6051–6060, 2005.

8. Réfrégier, P., "Mutual information-based degrees of coherence of partially polarized light with gaussian fluctuations," *Opt. Lett.*, **30**(23), 3117–3119, 2005.

9. Réfrégier, P., and Morio, J., "Shannon entropy of partially polarized and partially coherent light with gaussian fluctuations," *J. Opt. Soc. Am. A*, **23**(12), 3036–3044, 2006.

10. Gori, F., Santarsiero, M., Borghi, R., and Wolf, E., "Effect of coherence on the degree of polarization in a young interference pattern," *Opt. Lett.*, **31**(6), 688–670, 2006.

11. Setälä, T., Tervo, J., and Friberg, A.T., "Stokes parameters and polarization contrasts in young's interference experiment," *Opt. Lett.*, **31**, 2208–2210, 2006.

12. Setälä, T., Tervo, J., and Friberg, A.T., "Contrasts of stokes parameters in young's interference experiment and electromagnetic degree of coherence," *Opt. Lett.*, **31**, 2669–2671, 2006.

13. Gori, F., Santarsiero, M., and Borghi, R., "Maximizing young's fringe visibility through reversible optical transformations," *Opt. Lett.*, **32**(6), 588–590, 2007.

14. Luis, A., "Degree of coherence for vectorial electromagnetic fields as the distance between correlation matrices," *JOSA A*, **24**(4), 1063–1068, 2007.

15. Martinez-Herrero, R., and Mejias, P., "Maximum visibility under unitary transformations in two-pinhole interference for electromagnetic fields," *Opt. Lett.*, **32**(11), 1471–1473, 2007.

16. Martinez-Herrero, R., and Mejias, P., "Relation between degrees of coherence for electromagnetic fields," *Opt. Lett.*, **32**(11), 1504–1506, 2007.

17. Glauber, R.J., "The quantum theory of optical coherence," *Phys. Rev.*, **130**(6), 2529–2539, 1963.

18. Réfrégier, P., and Roueff, A., "Coherence polarization filtering and relation with intrinsic degrees of coherence," *Opt. Lett.*, **31**(9), 1175–1177, 2006.

19. Réfrégier, P., and Roueff, A., "Linear relations of partially polarized and coherent electromagnetic fields," *Opt. Lett.*, **31**(19), 2827–2829, 2006.

20. Réfrégier, P., and Roueff, A., "Visibility interference fringes optimization on a single beam in the case of partially polarized and partially coherent light," *Opt. Lett.*, **32**(11), 1366–1368, 2007.

21. Mandel, L., and Wolf, E., *Optical Coherence and Quantum Optics*, ch. Second-order coherence theory of scalar wavefields, 160–170. Cambridge University Press, New York, 1995.

22. Goodman, J.W., *Statistical Optics*, ch. Some first-order properties of light waves, 116–156. John Wiley and Sons, Inc., New York, 1985.

23. Réfrégier, P., *Noise Theory and Application to Physics: From Fluctuations to Information*, ch. Fluctuations and covariance, 28–32. Springer, New-York, 2004.

24. Huard, S., *Polarization of light*, ch. Propagation of states of polarization in optical devices, 86–130. Wiley, Masson, Paris, 1997.

25. Réfrégier, P., "Symmetries in coherence theory of partially polarized light," *J. of Math. Physics*, **48**(3), 033303.1–033303.14, 2007.

26. Cloude, S.R., and Papathanassiou, K., "Polarimetric SAR interferometry," *IEEE Trans. Geoscience and Remote Sensing*, **36**(5), 1151–1565, 1998.

27. Roychowdhury, H., and Wolf, E., "Determination of the electric cross-spectral density matrix of a random electromagnetic beam," *Opt. Commun.*, **226**, 57–60, 2003.

28. Roueff, A., and Réfrégier, P., "Separation technique of a mixing of two uncorrelated and perfectly polarized lights with different coherence and polarization properties," *J. Opt. Soc. Am. A*, **25**(4), 838–845, 2008.

Philippe Réfrégier is presently full professor at the Ecole Centrale de Marseille and is a member of the Fresnel Institute. He obtained his PhD in 1987 in Paris in statistical physics, and he was with Thomson-CSF from 1987 to 1994. His fields of interest concern statistical optics, including polarization, low photon signal, and coherence from both theoretical and applied aspects. He also works in signal and image statistical processing and, in particular, in segmentation and restoration applications and image characterization.

Antoine Roueff graduated from Grenoble Polytechnic National Institute, France, in 2000 with a degree in electrical engineering. He earned his PhD from the Images and Signals Laboratory in 2003. From 2003 to 2005, he worked at the Commissariat à l'Energie Atomique in Paris. Since 2006, he has been associate professor at the Ecole Centrale de Marseille and is a member of the Fresnel Institute. He is especially involved in the team working on physics and image processing. His research interests include multicomponent signal analysis, statistics, geophysics, and optics.

Chapter 11
Digital Computational Imaging

Leonid Yaroslavsky
Tel Aviv University, Israel

11.1 Introduction
11.2 Opticsless Imaging Using "Smart" Sensors
11.3 Digital Video Processing for Image Denoising, Deblurring, and Superresolution
    11.3.1 Image and video perfecting: denoising and deblurring
    11.3.2 Perfecting and superresolution of turbulent vides
11.4 Computer-Generated Holograms and 3D Video Communication
    11.4.1 Computer-generated display holograms
    11.4.2 Feasible solutions for generating synthetic display holograms
References

11.1 Introduction: Present-Day Trends in Imaging

Imaging has always been the primary goal of informational optics. The whole history of optics is, without any exaggeration, a history of creating and perfecting imaging devices. Starting more than 2000 years ago from ancient magnifying glasses, optics has been evolving with ever increasing speed from Galileo's telescope and van Leeuwenhoek's microscope, through mastering new types of radiations and sensors, to the modern wide variety of imaging methods and devices of which most significant are holography, methods of computed tomography, adaptive optics, synthetic aperture and coded aperture imaging, and digital holography. The main characteristic feature of this latest stage of the evolution of optics is integration of physical optics with digital computers. With this, informational optics is reaching its maturity. It is becoming digital and imaging is becoming computational.

    The following qualities make digital computational imaging an ultimate solution for imaging:
- Processing versatility. Digital computers integrated into optical information processing and imaging systems enable them to perform not only element wise and integral signal transformations such as spatial Fourier analysis,

signal convolution, and correlation, which are characteristic for analog optics, but any operations needed. This eliminates the major limitation of optical information processing and makes optical information processing integrated with digital signal processing almost almighty.

- Flexibility and adaptability. No hardware modifications are necessary to reprogram digital computers for solving different tasks. With the same hardware, one can build an arbitrary problem solver by simply selecting or designing an appropriate code for the computer. This feature makes digital computers also an ideal vehicle for processing optical signals adaptively since, with the help of computers, they can easily be adapted to varying signals, tasks, and end-user requirements.
- Universal digital form of the data. Acquiring and processing quantitative information carried by optical signals and connecting optical systems to other informational systems and networks is most natural when data are handled in a digital form. In the same way that in economics money is a general equivalent, digital signals are the general equivalent in information handling. Thanks to its universal nature, the digital signal is an ideal means for integrating different informational systems.

Present-day main trends in digital computational imaging are as follows:

- Development and implementation of new digital image acquisition, image formation, and image display methods and devices
- Transition from digital image processing to real-time digital video processing
- Widening the front of research toward 3D imaging and 3D video communication

Currently there is a tremendous amount of literature on digital computational imaging, which is impossible to comprehensively review here, and the present-day trends can only be illustrated on some examples. In this chapter, we use an example of three developments, in which the present author was directly involved. In Section 11.2 we describe a new family of image sensors that are free of diffraction limitations of conventional lens-based image sensors and base their image formation capability solely on numerical processing of radiation intensity measurements made by a set of simple radiation sensors with natural cosine law spatial sensitivity. In Section 11.3 we describe real-time digital video processing for perfecting visual quality of video streams distorted by camera noise and atmospheric turbulence. For the latter case, the processing not only produces good quality video for visual analysis, but in addition, makes use of atmospheric turbulence-induced image instabilities to achieve image super-resolution beyond the limits defined by the camera sampling rate. In Section 11.4 we present a computer-generated display hologram based 3D video communication paradigm.

11.2 Opticsless Imaging Using "Smart" Sensors

One can treat images as data that indicate location in space and intensities of sources of radiation. In conventional optical imaging systems, the task of determining positions of sources of light is solved by lenses, and the task of measurement of light source intensities is solved by light-sensitive plane sensors such as photographic films or CCD/CMOS electronic sensor arrays. A lens directs light from each of the light sources to a corresponding place on the sensor plane, and the intensity sensor's output signal at this place provides an estimate of the light source intensity.

Lenses are wonderful processors of directional information carried by light rays. They work in parallel with all light sources in their field of view and with the speed of light. However, their high perfection has its price. Light propagation from the lens to the sensor's plane is governed by the diffraction laws. They limit the capability of the optical imaging system to distinguish light radiated from different light sources and to discriminate closely spaced sources. According to the theory of diffraction, this capability, called imaging system resolving power, is determined by the ratio of the light wavelength times the lens's focal distance and the dimensions of the lens. Therefore, good imaging lenses are large and heavy. Perfect lenses that produce low aberrations are also very costly. In addition, lens-based imaging systems have limited field of view and lenses are not available for many practically important kinds of radiation such as, for instance, x-rays and radioactive radiation. This motivates search for optics-less imaging devices.

Recently, H. J. Caulfield and the present author suggested a concept of a new family of opticsless radiation sensors[1,2] that exploits the idea of combining the natural cosine law directional sensitivity of radiation sensors with the computational power of modern computers and digital processors to secure the sensor's spatial selectivity required for imaging. These opticsless "smart" (OLS) sensors consist of an array of small elementary subsensors with the cosine law angular selectivity supplemented with a signal processing unit (the sensor's "brain") that processes the subsensors' output signals to produce maximum likelihood (ML) estimates of spatial locations of radiation sources and their intensities.

Two examples of possible designs of the OLS sensors are sketched in Fig. 11.1. Figure 11.1(a) shows an array of elementary subsensors arranged on a curved surface, such as spherical one. Such an array of subsensors, together with its signal processing unit, is capable of localizing sources of radiation and measuring their intensities at any distances and in a 4π steradian solid angle. Shown in Fig. 11.1(b) is an array of elementary subsensors on a flat surface together with its signal processing unit that is capable of measuring coordinates and intensities of radiation sources at close distances.

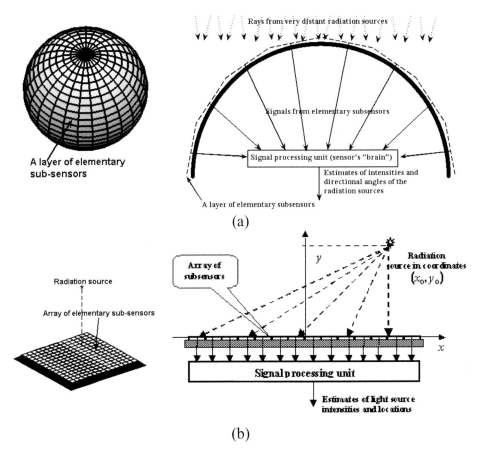

Figure 11.1 Two examples of possible designs of opticsless "smart" radiation sensors and their corresponding schematic diagrams.

The work principle of opticsless smart sensors can be explained using a simple special case of locating a certain number K of very distant radiation sources by an array of N elementary sensors placed on a curved surface. Consider a 1D model of sensor's geometry sketched in Fig. 11.2. For an nth elementary sensor with the cosine law spatial selectivity placed at angle φ_n with respect to the sensor's "optical" axis, its output signal $s_{n,k}$ to a ray of light emanating from the kth source under angle θ_k with respect to the sensor's "optical" axis is proportional to the radiation intensity I_k and cosine of angle $\theta_{n,k}$ between the vector of electrical field of the ray and the normal to the elementary sensor surface. Additionally, this signal contains a random component ν_n that describes the subsensor's immanent noise as

$$s_n = A_k \overline{\cos\theta}_{n,k} + \nu_n = A_k \overline{\sin}(\varphi_n + \theta_k) + \nu_n, \tag{11.1}$$

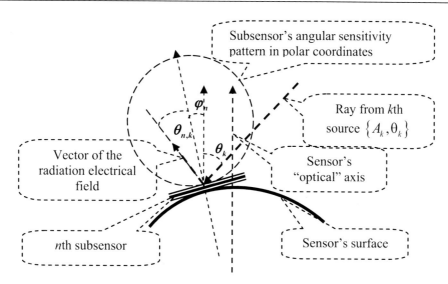

Figure 11.2 Geometry of an elementary subsensor of the sensor array.

where $\overline{\cos}(.)$ and $\overline{\sin}(.)$ are truncated cosine and sine functions equal to zero, when their corresponding cosine and sine functions are negative.

The sensor's "brain" collects output signals $\{s_n\}$ of all N elementary subsensors and estimates on this base intensities $\{A_k\}$ and directional angles $\{\theta_k\}$ of the given number K of distant radiation sources. In view of the statistical nature of the random sensor noise, statistically optimal estimates will be the maximum likelihood estimates. In the assumption that noise components of different subsensors are statistically independent and normally distributed with the same standard deviation, maximum likelihood estimates $\{\hat{A}_k, \hat{\theta}_k\}$ of intensities and directional angles of the sources can be found as solutions of the following equation:

$$\{\hat{A}_k, \hat{\theta}_k\} = \arg\min_{\{\hat{A}_k, \hat{\theta}_k\}} \left\{ \sum_{n=1}^{N} \left| s_{n,k} - \left[\sum_{k=1}^{K} \hat{A}_k \overline{\sin}(\varphi_n + \theta_k) \right] \right|^2 \right\}. \quad (11.2)$$

For a single source, an analytical solution of this equation is possible, which means that the computational complexity of estimating intensity and directional angles of a single light source is of the order of the number of sensors and that the computations can be implemented in quite simple hardware. For a larger number K of sources, solution of this equation requires optimization in K-dimensional space, and the computational complexity of the estimation of source parameters grows dramatically with the number of sources.

For locating multiple sources, "smart" sensors can be used in the following three modes:

- "General localization" mode for localization and intensity estimation of a known number of radiation sources
- "Constellation localization" mode for localization and intensity estimation of a given number of groups of sources with known mutual positions and intensity relations of individual sources
- "Imaging" mode for estimating intensities of the given number of radiation sources in the given locations such as, for instance, in nodes of a regular grid on certain known distances from the sensor

Some illustrative simulation results of localization of the given number of very distant and very proximal radiation sources by arrays of elementary subsensors are presented in Figs. 11.3 and 11.4, correspondingly. Figure 11.3(a) shows results of 100 statistical runs of a model of a spherical array of 30×30 subsensors with the signal-to-noise ratio (the ratio of the signal dynamic range to the noise standard deviation, or SNR) 20 used for localization and estimating intensities of three very distant point sources. Clouds of dots on the plot show the spread of the estimates. The simulation results show that for each source, standard deviations of the estimates of its angular position and intensity are, for reasonably high sensor SNRs, inversely proportional to SNR, to the source intensity, to the number of sources, and to the square root of the number of elementary subsensors.

Figure 11.3(b) illustrates the work of a spherical OLS sensor in the imaging mode and shows the distribution of estimates of intensities of an array of sources that form pattern "OLSS" (opticsless smart sensors). In this experiment, a model of the spherical sensor consisted of $15 \times 20 = 300$ subsensors arranged within spatial angles $\pm\pi$ longitude and $\pm\pi/2.05$ latitude, and the array of radiation sources consisted of 19×16 sources with known directional angles within spatial angles $\pm\pi/2$ longitude and $\pm\pi/3$ latitude. Each subsensor had a noise standard deviation of 0.01, and source intensities were 0 or 1.

Figure 11.4 illustrates the performance of a flat OLS sensor in the imaging mode for imaging of a pattern of 10×16 sources on different distances, from one to five inter-subsensor distances, from the sensor. The results are obtained on a 1D model of the OLS sensor consisting of 30 elementary subsensors. The source array was scanned by the sensor columnwise. Standard deviation of subsensors' noise was 0.01 (in units of the source intensity dynamic range [0 1]). Simulation results show that flat OLS sensors are capable of localization and imaging of radiation sources in their close proximity on distances of about half of the sensor's dimensions.

Digital Computational Imaging

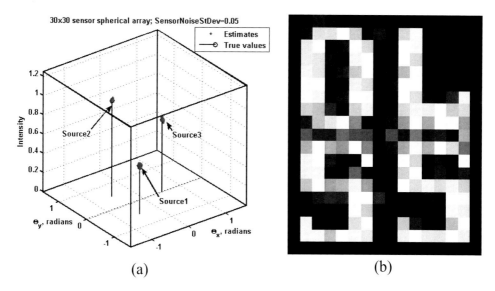

(a) (b)

Figure 11.3 Spherical "smart" sensor in the localization and imaging modes: (a) Simulation results of locating angular positions and intensities of three distant radiation sources using an array of 30 × 30 subsensors uniformly distributed over the surface of a hemisphere; (b) distribution of estimates of intensities of an array of sources that form pattern "OLSS."

As always, there is a trade-off between good and bad features of the OLS sensors. The advantages are as follows:

- No optics are needed, making this type of sensor applicable to virtually any type of radiation and wavelength.
- The angle of view (of spherical sensors) is unlimited.
- The resolving power is determined ultimately by the subsensor size and not by diffraction-related limits.

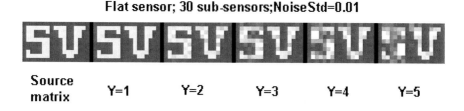

Figure 11.4 Simulation results of imaging of sources that pattern "SV" by a flat OLS sensor placed at different distances, from one to five intersubsensor distances, from the sources.

The cost for these advantages is the high computational complexity, especially when good imaging properties for multiple sources are required. However, the inexorable march of Moore's law makes such problems less hampering each year. Furthermore, the required computations can be parallelized, so the computational aspects are not expected to hinder usage.

In conclusion, it is interesting to note that one can find quite a number of examples of opticsless vision among live creatures. Obviously, plants, such as sunflowers, that feature heliotropism must have a sort of "skin" vision to determine direction to sunlight in order to be able to direct their flowers or leaves accordingly. There are also many animals that have extraocular photoreception. One can also find a number of reports on the phenomenon of "skin" vision in humans, though some of them that refer to paranormal phenomena may provoke skepticism. Properties of the described OLS sensors show that opticsless vision can, in principle, exist and also cast light on its possible mechanisms.

11.3 Digital Video Processing for Image Denoising, Deblurring, and Superresolution

11.3.1 Image and video perfecting: denoising and deblurring

Common sources of degradations of image quality in video sequences frequently are video camera noise and aperture distortions. These degradations may be considerably well compensated using spatial and temporal redundancy present in the video streams. One of the best video processing algorithms that allow reaching this goal is a 3D extension of the sliding window local adaptive discrete cosine transform (DCT) domain filtering, originally developed for image denoising and deblurring.[3,4] A flow diagram of such a filtering is shown in Fig. 11.5.

According to this flow diagram, filtering is carried out in a sliding, spacewise and timewise, window that, for each pixel in the currently processed video frame, contains the pixel's spatial neighborhood in the current input video frame and in a certain number of preceding and following frames. The size of the 3D space-time window is a user-defined parameter defined by spatial and time correlations in the video frames. At each position of the 3D window, the 3D DCT transform coefficients of the window data are recursively computed from those in the previous position of the window. The obtained signal spectral coefficients are then nonlinearly modified according to the principles of empirical Wiener filtering.[3] In its simplest implementation, this modification consists of the transform coefficient thresholding and zeroing the coefficients that, by their magnitude, do not exceed a certain threshold determined by the noise variance. For aperture correction, the remaining coefficients are multiplied by the inverse of the camera frequency response. The modified spectral coefficients are then

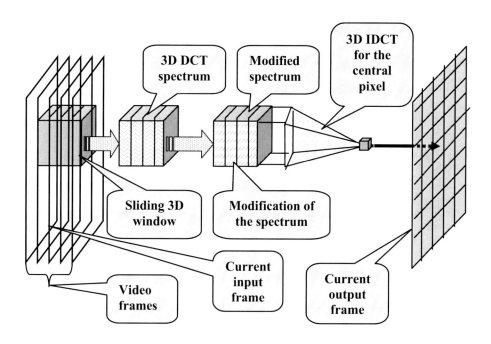

Figure 11.5 Sliding 3D window transform domain filtering.

subjected to inverse 3D DCT for computing the central pixel of the window. In this way, pixel by pixel, all pixels in the current frame are obtained; then the window starts sliding over the next frame, and so on.

Figures 11.6 and 11.7 illustrate denoising and deblurring of test and real-life video sequences using 5 × 5 × 5 sliding-window video processing. Corresponding demonstrative movies can be found on the Internet.[5]

11.3.2 Perfecting and superresolution of turbulent vides

In long-distance observation systems, images and video are frequently damaged by atmospheric turbulence, which causes spatially and temporally chaotic fluctuations in the index of refraction of the atmosphere[6] and results in chaotic, spatial, and temporal geometrical distortions of the neighborhoods of all pixels. This geometrical instability of image frames heavily worsens the quality of videos and hampers their visual analysis. To make visual analysis possible, it is first of all required to stabilize images of stable scenes while preserving real motion of moving objects that might be present in the scenes.

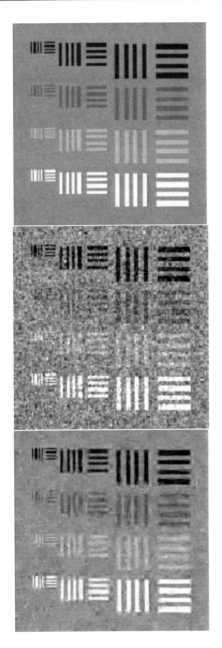

Figure 11.6 Local adaptive 3D sliding-window DCT domain filtering for denoising video sequences. Top first and second rows: examples of frames of initial and noisy test video sequence. The third row: examples of frames of restored video obtained using 3D sliding 5 × 5 × 5 spatial/temporal window filtering.

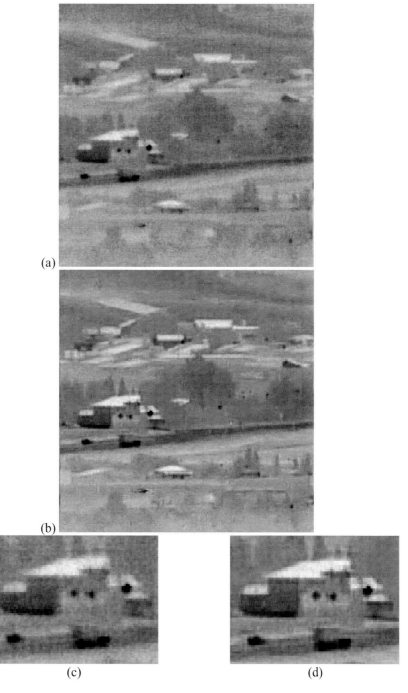

Figure 11.7 Examples of frames of (a) initial and (b) restored thermal real-life video sequence, and (c) and (d) corresponding magnified fragments of the images. Note enhanced sharpness of the restored image.

In Ref. 7, methods of generating stabilized videos from turbulent videos were reported. It was also found that along with image stabilization, image superresolution on stable scenes can be achieved.[8] The core of the stabilization and superresolution method is elastic pixelwise registration, with subpixel accuracy, of available video frames of stable scenes followed by resampling of the frames according to the registration results. For achieving the required elastic registration of frames, for each current video frame, a time window of several preceding and following frames of the video sequence is analyzed. For each pixel of the frame, its x-y displacements in all remaining frames of the window are found using the methods of block matching or optical flow methods. Then the displacement data are analyzed to derive their statistical parameters for distinguishing pixels that were displaced due to the atmospheric turbulence from those that belong to real moving objects. This distinction can be made on an assumption that turbulence-induced pixel displacements are relatively small and irregular in time, while displacements caused by real movement are relatively large and, what is more important, contain a regular, in time, component.

On the basis of these measurements, the stabilized and superresolved output frame is generated on a sampling grid built by subdivisions of the initial sampling grid. Nodes of the latter correspond, for each pixel, to mean values of found pixel displacements. Formation of the output frame consists of two steps.

In the first step, corresponding pixels from all time window frames are placed at the nodes of the subpixels' grid according to their found displacements minus displacement mean values. Because pixel displacements are chaotic, it may happen in this process that two or more corresponding pixels from different frames have to be placed in the same position in the output frame. In these cases, a robust to outliers estimate of average, such as median, can be taken as a replacement of those pixels. In the second step, subpixels that remain empty because of a possible shortage of data in the selected time window should be interpolated from available data. For the interpolation, different available methods for interpolation of sparse data can be used, of which discrete sinc interpolation proved to be the most suitable.[9]

Pixels retrieved from the set of the time window frames contain, in the form of aliasing components, high frequencies outside the image base band defined by the original sampling rate of the input frames. Placing them into proper subpixel positions results in dealiasing these frequencies and, therefore, in widening image bandwidth beyond the base band. The more pixels in different subsampling positions are available, the higher degree of superresolution is achieved.

The efficiency of the entire processing is illustrated in Fig. 11.8 by the results of computer simulations and in Fig. 11.9 for a real-life video.[8] As one can see from Fig. 11.8, even from as small a number of frames as 15, a substantial resolution enhancement is potentially possible.

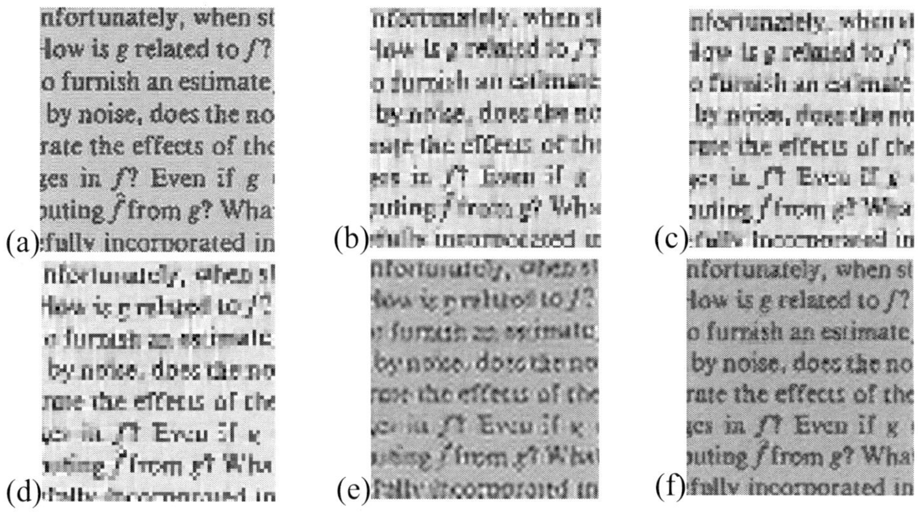

Figure 11.8 Illustrative simulation results of resolution enhancement of turbulent video frames: (a) Initial high-resolution frame; (b)–(d) examples of low-resolution frames distorted by simulated random local displacements with standard deviation 0.5 interpixel distance; (e) resolution-enhanced frame obtained by the described fusion process from 15 low-resolution frames; and (f) final output frame obtained by interpolation of samples that are missing in frame (e).

Figures 11.9(a) and (b) show a 256 × 256 pixel fragment of a stabilized frame of real-life video obtained as a temporal median over 117 frames, and the same fragment obtained after replacing its pixels, as described above, by pixels taken from those 117 frames according to their displacements. Since resolution improvement can be appreciated visually only on a high-quality display, Fig. 11.9(c) presents the difference between these two images that clearly shows edges enhanced in Fig. 11.9(b) compared to Fig. 11.9(a). Figure 11.9(d) presents the final result of the processing after reinterpolation and aperture correction are implemented in the assumption that the camera sensor has a fill factor, or the ratio of the size of individual sensing cells to the interpixel distance is close to 1.

In the evaluation of the results obtained for real-life video, one should take into consideration that substantial resolution enhancement in the described superresolution fusion process can be expected only if the video acquisition camera fill factor is small enough. The camera fill factor determines the degree of low-pass filtering introduced by the camera. Due to this low-pass filtering, image high frequencies in the base band and aliasing high-frequency components that come into the base band due to image sampling are attenuated. Those aliasing components can be recovered and returned back to their true frequencies outside the base band in the described superresolution process, but only if they have not been lost due to the camera low-pass filtering. The larger the fill factor, the harder it is to recover resolution losses. In the described simulation experiment, the camera fill factor is 0.05; whereas in reality fill factors of monochrome cameras are usually close to 1.

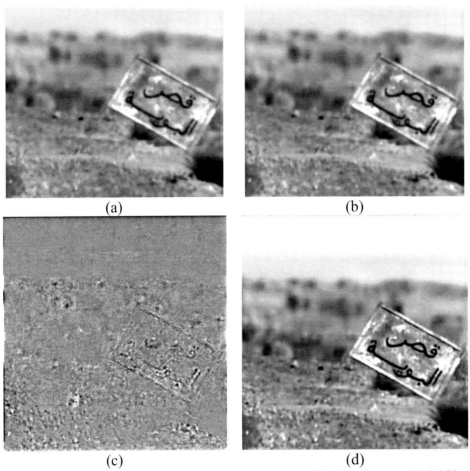

Figure 11.9 Illustration of resolution enhancement of real-life video frames. (a) A 256 × 256 pixels fragment of the stabilized frame; (b) the same fragment after resolution enhancement; (c) difference between images (a) and (b) that shows edges enhanced in image (b) as compared to image (a); and (d) aperture-corrected image (b).

11.4 Computer-Generated Holograms and 3D Video Communication

11.4.1 Computer-generated display holograms: an ultimate solution for 3D visualization and communication

There are no doubts that the ultimate solution for 3D visualization is holographic imaging. This is the only method capable of reproducing, in the most natural viewing conditions, 3D images that have all the visual properties of the original objects including full parallax, and are visually separated from the display device. Three-dimensional visual communication and display can be achieved through generation, at the viewer side, of holograms out of data that contain all relevant information on the scene to be viewed. Digital computers are ideal means for converting data on 3D scenes into optical holograms for visual perception.[10,11]

At the core of the 3D digital holographic visual communication paradigm is the understanding that, for 3D visual communication, one does not need to record, at the scene site, a physical hologram of the scene and that the hologram has to be generated at the viewer side. To this goal, one needs to collect, at the scene side, and to transmit to the viewer site a set of data that will be sufficient to generate, at the viewer site, a synthetic hologram of the scene for viewing. The major requirement for computer-generated display holograms is that they should provide natural viewing conditions for the human visual system and, in particular, separation of reconstructed images from the display device.

A crucial issue in transmitting data needed for the synthesis, at the viewer site, of display holograms is the volume of data to be collected and transmitted, and the computational complexity of the hologram synthesis. The upper bound of the amount of data needed to be collected at the scene side and transmitted to the viewer site is, in principle, the full volumetric description of the scene geometry and optical properties. However, a realistic estimation of the amount of data needed for generating a display hologram of the scene is by orders of magnitude lower than the upper bound due to the limitations of the human visual system. This also has a direct impact on the computational complexity of the hologram synthesis.

11.4.2 Feasible solutions for generating synthetic display holograms that fit human vision

Several computationally inexpensive and at the same time quite sufficient solutions for creating 3D visual sensation with synthetic display holograms can be considered.[10]

11.4.2.1 Multiple view compound macroholograms

In this method, the scene to be viewed is described by means of multiple view images taken from different directions in the required view angle; and for each image, a hologram is synthesized separately with an account of its position in the viewing angle (see Fig. 11.10). The size of each hologram has to be of about the size of the viewer's eye pupil. These elementary holograms will reconstruct different aspects of scenes from different directions, which are determined by their position in the viewing angle. The set of such holograms is then used to build a composite, or mosaic, macrohologram. It is essential that, for scenes given by their mathematical models, well-developed methods and software/hardware instrumentation tools of the modern computer graphics can be used for fast generating multiple view images needed for computing elementary holograms.

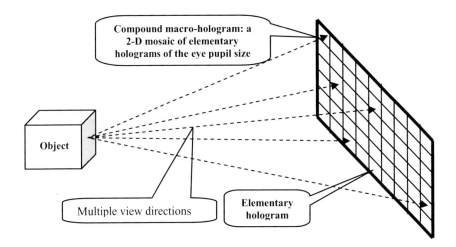

Figure 11.10 The principle of the synthesis of composite holograms.

In Ref. 12, an experiment on the synthesis of such a composite macrohologram composed of 900 elementary holograms of 256 × 256 pixels was reported. The hologram contained 30 × 30 views, in spatial angle –90 deg to 90 deg, of an object in the form of a cube. The synthesized holograms were encoded as purely phase holograms, or kinoforms,[10,11] and recorded with the pixel size 12.5 µm. The physical size of the elementary holograms was 3.2 × 3.2 mm. Each elementary hologram was repeated, in the process of recording, 7 × 7 times to the size 22.4 × 22.4 mm. The size of the entire hologram was 672 × 672 mm. Being properly illuminated, the hologram can be used for viewing the reconstructed scene from different angles such as, for instance, through a window (Fig. 11.11). Looking through the hologram with two eyes, viewers are able to see a 3D image of a cube (Fig. 11.11, bottom) floating in the air.

11.4.2.2 Composite stereoholograms

A special case of multiple view mosaic macroholograms is composite stereoholograms. Composite stereoholograms are synthetic Fourier holograms that reproduce only horizontal parallax.[10,13,15] When viewed with two eyes, they are capable of creating 3D sensation thanks to stereoscopic vision. With such holograms arranged in a circular composite hologram, a full 360 deg view of the scene can be achieved. Figure 11.12 shows such a 360 deg computer-generated hologram and examples of images from which it was synthesized.[14] The entire hologram was composed of 1152 fragmentary kinoform holograms of 1024 × 1024 pixels recorded with pixel size 12.5 µm. A viewer looking through the hologram from different positions sees a 3D image of an object in a form of a "molecule" of six "atoms" differently arranged in space. When the hologram is rotated, the viewer sees a 3D image of "atoms" floating in the air and continuously rotating in space, and easily recognize the rotation direction. A sample of the described 360 deg computer-generated hologram is stored at the MIT Museum.[14]

Figure 11.11 Viewing a compound computer-generated hologram using as an illumination source a miniature white-light lamp (left) and one of the views reconstructed from the hologram (right).

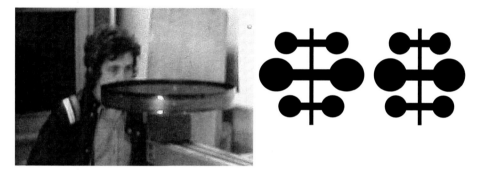

Figure 11.12 Synthetic computer-generated circular stereo-macrohologram (left) and two views of the scene (right).

11.4.2.3 "Programmed diffuser" holograms

The "programmed diffuser" method for synthesis of a Fourier display hologram was suggested for synthesis of computer-generated display holograms capable of reconstructing different views of 3D objects whose surfaces scatter light diffusely.[10,16] This method assumes that objects to be viewed are specified in the object coordinate system (x, y, z) by their "macro" shape $z(x, y)$, by the magnitude of the object reflectivity distribution $A(x, y)$ in the object plane (x, y) tangent to the object surface and by the directivity pattern of the diffuse component of its surface. The diffuse light scattering from the object surface is simulated by adding to the object wave front phase, as defined by the object macroshape, a pseudo-random phase component (a "programmable diffuser"), whose correlation function corresponds to the given directivity pattern of the object diffuse surface. This pseudo-random phase component is combined with the deterministic phase component defined by the object shape to form the distribution of the phase of the object wavefront.

Holograms synthesized with this method exhibit a spatial inhomogeneity that is directly determined by the geometrical shape and diffuse properties of the object surface. This allows imitation of viewing the object from different direction by means of reconstruction of different fragments of its "programmed diffuser" hologram as is illustrated in Fig. 11.13 in an example of an object in the form of a hemisphere.

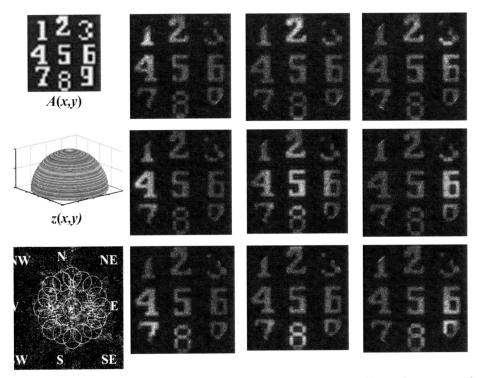

Figure 11.13 An object's image (upper left), its shape (center left), its "programmed diffuser" hologram (bottom left), and nine images reconstructed from northwest, north, northeast, west, center, east, southwest, south, and southeast fragments of the hologram (right). Circles on the image of the hologram encircle different fragments of the hologram.

References

1. Caulfield, H.J., Yaroslavsky, L.P., and Ludman, J., "Brainy light sensors with no diffraction limitations," http://arxiv.org/abs/physics/0703099.

2. Caulfield, H.J., and Yaroslavsky, L.P., "Flat accurate nonimaging point locator," Digital Holography and Three-Dimensional Imaging 2007 Conference, Vancouver, June 18–20, 2007.

3. Yaroslavsky, L., *Digital Holography and Digital Image Processing*, Kluwer, Boston (2004).

4. Yaroslavsky, L., "Space variant and adaptive transform domain image and video restoration methods," *Advances in Signal transforms: Theory and*

Applications, J. Astola and L. Yaroslavsky, Eds., Hindawi Publishing Corporation, http://www.hindawi.com (2007).

5. Shteinman, A., ttp://www.eng.tau.ac.il/~yaro/Shtainman/house_t18_b555.avi

6. Roggermann, M.C., and Welsh, B., *Imaging Through Turbulence*, CRC Press, Boca Raton (1996).

7. Fishbain, B., Yaroslavsky, L.P., and Ideses, I.A., "Real time stabilization of long range observation system turbulent video," *Journal of Real-Time Image Process.*, **2**, Number 1, October 2007, pp. 11-22.

8. Yaroslavsky, L., Fishbain, B., Shabat, G., and Ideses, I., "Super-resolution in turbulent videos: Making profit from damage," Opt. Lett., **32**(21), pp. 3038-3040, 2007.

9. Yaroslavsky, L., "Fast discrete sinc-interpolation: A gold standard for image resampling," *Advances in Signal Transforms: Theory and Applications*, J. Astola, and L. Yaroslavsky, Eds., EURASIP Book Series on Signal Processing and Communications, Hindawi Publishing Corporation, http://www.hindawi.com (2007).

10. Yaroslavkii, L., and Merzlyakov, N., *Methods of Digital Holography*, Consultants Bureau, New York (1980).

11. Lee, W.H., "Computer generated holograms: techniques and applications," *Progress in Optics*, E. Wolf, Ed., Vol. XVI, p. 121–231, North Holland, Amsterdam (1978).

12. Karnaukhov, V.N., Merzlyakov, N.S., Mozerov, M.G., Yaroslavsky, L.P., Dimitrov, L.I., and Wenger, E., "Digital Display Macro-holograms." Comput. Opt., 14-15, Pt 2, pp. 31-36, 1995.

13. Karnaukhov, V.N., Merzlyakov, N.S., and Yaroslavsky, L.P., "Three dimensional computer generated holographic movie," *Sov. Tech. Phys. Lett.*, **2**, 169–172, 1976.

14. Yaroslavskii, L.P., 360-degree computer generated display hologram, http://museumcollections.mit.edu/FMPro?-lay=Web&wok=107443&-format=detail.html&-db=ObjectsDB.fp5&-find.

15. Yatagai, T., "Stereoscopic approach to 3-D display using computer-generated holograms," *Appl. Opt.*, **15**, 2722–2729, 1976.

16. Merzlyakov, N.S., and Yaroslavsky, L.P., "Simulation of light spots on diffuse surfaces by means of a 'programmed diffuser'," *Sov. Phys. Tech. Phys.*, **47**, 1263–1269, 1977.

Chapter 12
Superresolution Processing of the Response in Scanning Differential Heterodyne Microscopy

Dmitry V. Baranov and Evgeny M. Zolotov
General Physics Institute of Russian Academy of Sciences,
Russia

12.1 Introduction
12.2 Image Formation in a Scanning Differential Heterodyne Microscope
12.3 Solution of the Inverse Problem
    12.3.1 Formulation of the inverse problem
    12.3.2 Evaluation of the microscope nondifferential response
    12.3.3 Diffraction resolution reconstruction of the optical and topographic profile
    12.3.4 Profile reconstruction with superresolution
12.4 Conclusion
References

12.1 Introduction

Heterodyne far-field optical microscopy, from the beginning of its development, has been an efficient tool for surface characterization with a high resolution.[1–3] The use of a nondestructive optical imaging to collect metrological information is of ever-increasing importance. The importance follows from its simplicity of use and potentially high accuracy.

    Optical heterodyne microscopes made possible the observing and measuring of submicron structures. This is due to a possibility of precise measurement of phase and amplitude of the reflected signal at an intermediate frequency.[4–6] In a conventional heterodyne microscope, the path of the reference and signal beams are separated, and the instrument is sensitive to vibrations and other environmental influences. Differential microscopes are relatively free of this shortcoming because the reference and signal components of the probe beam propagate along the same path (common path scheme).[7,8] But this system requires a complex algorithm for processing the signal reflected from an object because the signal is not linear with respect to the surface profile. The microscope scan represents a nonlinear function of the convolution of the point-spread function and the optical profile. The interpretation of a complex response demands special superresolving processing in the analysis of microscopic objects with submicron size based on image formation theory.

A resolution of the problem of image formation in modern microscopy as usual can be divided into two parts: first, the interaction of the illuminating beam with the object surface; second, the signal formation in the microscope optical system. The resolving process at the first stage depends on the object profile. In our notation, a profile can be partitioned approximately into three categories according to its depth: rather shallow, shallow, and deep. This classification depends also on the lateral scale of an object. The first category corresponds to depth $h \ll \lambda/4$. In this case, an object is visualized as a layer of negligible thickness resulting in a phase shift proportional to the height variation of the surface. If the object profile is shallow ($h \leq \lambda/4$), the interaction between the optical field and the object surface can be described by the local reflection coefficient approach. This model is based on local properties of the surface and neglects multiple-scattering effects. But it takes into account the finite thickness of an object, and both phase and amplitude components are needed for response interpretation in the scalar approach. If the profile is deep ($h > \lambda/4$), a rigorous electromagnetic theory is applied.

Rigorous studies of the diffraction for deep profile objects have been carried out with differential equation methods[9] and integral methods.[10] A review of these methods can be found in Ref. 11. Rigorous diffraction methods applying to differential heterodyne microscopy have been considered in Refs. 12 and 13. But the inverse optical problem in differential microscopy has not been solved in part because of the problem of response factorization for this kind of microscope.

The aim of this chapter is a consideration of methods for superresolution processing of a signal in a scanning differential heterodyne microscope (SDHM) in order to resolve the inverse problem for shallow structures in the presence of noise. For a shallow profile, as noted above, the scalar diffraction model is available. The resolution of the inverse problem is based on an image formation theory applied to the principal scheme of SDHM.[14]

The image formation theory of conventional optical microscope has been considered in terms of a transfer function approach by Hopkins.[15] This approach was later extended by Sheppard and Wilson[16] in order to analyze the image formation in scanning optical microscopes, including the confocal type. Later, Somekh developed the theory for the differential heterodyne profilometer and considered its depth discrimination properties to be used for 3D imaging.[17] The image formation theory for Linnik-type phase-shifting microscopes (version of heterodyne microscope) was used for resolving the direct optical problem including a superresolution in a number of papers.[18-20] But this type of microscope requires substantial protection against vibrations and fluctuations.

The scan response in a differential microscope is nonlinear relative to the optical profile. Therefore, the profile restoration (also with superresolution) requires the preprocessing of the microscope response. The differential response can be factorized at the appropriate adjustment of the microscope optical scheme with coherent registration in the Fourier plane. The factorized differential response is reduced to the nondifferential one by the finite difference method.[21] The superresolution processing of the nondifferential response is then applied for profile reconstruction. Because the factorization and differential response processing are key factors for resolving the SDHM inverse problem, they will be outlined here.

The inverse problem for a scanning differential phase-sensitive heterodyne microscope was first considered by Bozhevolnaya et al.[22] for recovering the refractive index profile of a waveguide structure. The algorithm was based on a resolution of the integral equation with corresponding regularization. The method of resolving the SDHM inverse problem was considered also in Ref. 23 using optimal filtering for a noisy signal.

The solution of the inverse problem in modern microscopy is directly related to the possibility of achieving a lateral superresolution, i.e., a resolution exceeding the classic Abbe-Rayleigh diffraction limit. The problem of a superresolution in relation to the signal-to-noise ratio was discovered in 1950–1960 and applied to modulation optics, spectroscopy, and astronomy.[24-26] The general aspects of a superresolution were analyzed in Refs. 16 and 27–29 and discussed also in Ref. 30. In microscopy, the phenomenon of a superresolution was observed by Tychinsky with coauthors.[31] They used an amplitude-phase heterodyne microscope based on the Linnik interferometer. The algorithm of the superresolving processing of the inverse problem was considered in Ref. 23. Its simulation was performed in Ref. 21 for a shallow groove.

The methods of a superresolution processing are based in part on the analytic continuation of the response spectrum beyond the aperture limits.[26] This follows from the invariance theorem for the information capacity of an imaging system. The theorem founded on the Shannon information theory[32] was formulated by Lukosz[33] and developed later by Cox and Sheppard.[34] According to this theorem, the spectrum of a finite object can be extrapolated beyond the system bandwidth, and the bandwidth extension depends on the signal-to-noise ratio.[35,36]

The algorithm of analytic continuation of the spectrum proposed for communication theory is based mainly on the expansion in terms of prolate spheroidal wave functions.[37] But these functions with double orthogonality have a shortcoming because of the fast decrement of eigenvalues.[28] Therefore, another set of functions was proposed for analytic continuation of image processing. In particular, the method of object restoration using Gabor function expansion was proposed for a nondifferential microscopy.[38] The possibility of extrapolating the response spectrum beyond the aperture limit with the use of polynomial expansion was demonstrated in Ref. 21. But this method is not optimal due to the divergence of the corresponding series with a finite number of expansion terms. An algorithm of solving the basic integral equation by expanding both the response function and the object function in terms of basis sampling functions gave a more correct approximation.[39]

For superresolution resolving the inverse problem for SDHM with an additional noise we have broken down the process into four consecutive steps: (1) formulation of the image formation theory for SDHM, (2) evaluation of the response of an absolute phase microscope[*] by using the Shannon sampling theorem, (3) resolution of the linear integral equation with the proper regularization, and (4) extrapolation of the response spectrum out of the aperture band. The second and third stages of solving can be performed by different methods. If the signal-to-noise ratio is high, the optimal Wiener filtering and the polynomial expansion of the response spectrum can be applied. For a low signal-to-noise ratio, the special processing is proposed based on the expansion of the object optical profile and the nondifferential response in terms of sampling functions. The quasi-optimal regularization also is applied, resulting in efficient smoothing of the response spectrum. This algorithm is available if the response spectrum overlaps the noise spectrum.

12.2 Image Formation in a Scanning Differential Heterodyne Microscope

The scanning differential heterodyne microscope (SDHM) to be used for modeling represents a Mach-Zender interferometer based on the common-path scheme (Fig. 12.1). The linear-polarized input laser beam at frequency ω_0 (λ = 0.63 µm) is passed through acousto-optic Bragg cell. The beam is diffracted at the cell into two first-order beams. Transmitted beams at different optical frequencies with a difference ω (heterodyne frequency) are focused on a sample to form two overlapped probe spots. The sample is placed on one-dimensional micropositioner. Reflected beams are directed by a beam splitter to a point photodetector (PD) at the center of the Fourier plane (FP). A response of the SDHM represents the scanning coordinate dependence of photocurrent at heterodyne frequency ω with the amplitude and the phase to be modulated by sample features.

Near the object surface, the reflected wave is generally related to the incident wave by an integral relationship between the reflected-wave amplitude at point (x, y) in the object plane and the amplitude of the incident wave in the vicinity of this point. In the spectral representation, the reflected wave is related by an integral expression to the incident wave via a scattering function. The relationship between the reflection coefficient and the scattering function is defined by scattering theory. However, under certain conditions of object illumination and response recording, it is possible to characterize an object by the

---

[*] We will call this response the "nondifferential response."

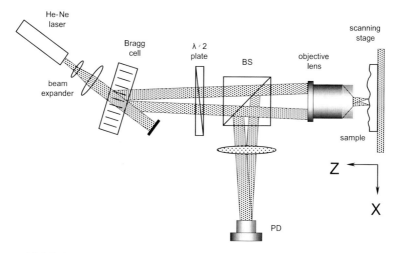

Figure 12.1 Schematic diagram of a scanning differential heterodyne microscope.

local reflection coefficient $r(x, y)$ if the detection system registers only one plane wave by a point detector. In this case $r(x, y) = r_0[1 + f(x, y)]$, where r_0 is the reflectivity of the substrate and $f(x, y)$ is the object optical profile, which is the Fourier transform of the scattering function.

In order to define the relationship between a microscope response and a surface profile to be investigated, it is necessary to consider the image formation theory for SDHM. If the object is illuminated with two beams separated by the distance ε and different in frequency by ω, then the incident field may be written as

$$E_0(x,y,t) = \sqrt{\frac{I_0}{4}} g\left(x+\frac{\varepsilon}{2},y\right)\exp\left[i\left(\omega_0+\frac{\omega}{2}\right)t\right]$$
$$+\sqrt{\frac{I_0}{4}} g\left(x-\frac{\varepsilon}{2},y\right)\exp\left[i\left(\omega_0-\frac{\omega}{2}\right)t\right]. \quad (12.1)$$

Here, I_0 is the illumination wave intensity and $g(x, y)$ is the field distribution of the probing beam. In our consideration, an object will be characterized by the reflectivity, which is a function of one direction only (e.g., x). According to the image formation theory of a scanning microscope with a differential coherent scheme, the one-dimensional scan response (i.e., the complex amplitude of the photodetector current at the frequency ω) is given by

$$F(x_s) = \int_{-\infty}^{\infty}\int_{-\infty}^{\infty} H(x,x',\varepsilon)r(x-x_s)r^*(x'-x_s)dxdx', \quad (12.2)$$

where $H(x, x', \varepsilon)$ is the characteristic function that is specified by parameters of the optical and registration system of the microscope, $r(x)$ is the local reflection coefficient, and x_s is the object position. The complexity of the function H complicates reconstruction processing, but if the signal is detected in the FP of the optical system, the function H can be simplified and represented by[14]

$$H(x,x',\varepsilon) = \int_{-\infty}^{\infty}\int_{-\infty}^{\infty} D(q,p)dqdp \int_{-\infty}^{\infty}\int_{-\infty}^{\infty} \exp[iq(x-x')]\exp[ip(y-y')] \\ \times g\left(x+\frac{\varepsilon}{2},y\right)g^*\left(x'-\frac{\varepsilon}{2},y'\right)dydy', \quad (12.3)$$

where $D(q, p)$ is the filtering function of the PD, $q = kx_d/f$ and $p = ky_d/f$ are the spatial frequencies in the FP, x_d and y_d are the coordinates in the FP, and f is the objective's focal length. If the reflected field is detected at the center of the FP with the point filtering ($q = p = 0$), the function H is simplified to

$$H(x,x',\varepsilon) = H_0 g\left(x+\frac{\varepsilon}{2}\right)g^*\left(x'-\frac{\varepsilon}{2}\right), \quad (12.4)$$

where $g(x)$ is the line spread function and H_0 is a constant. This simplification allows factorizing the differential response by separating the variables x and x'. In this case, the response can be expressed in terms of the convolution of the reflection coefficient and $g(x)$ as

$$F(x_s) = H_0 \int_{-\infty}^{\infty} g\left(x+\frac{\varepsilon}{2}-x\right)r(x)dx \int_{-\infty}^{\infty} g^*\left(x-\frac{\varepsilon}{2}-x'\right)r^*(x')dx'. \quad (12.5)$$

To exclude the reflectivity of a substrate, one normalizes the function $r(x)$ by introducing the optical profile $f(x)$ related to the reflection coefficient by the expression as

$$r(x) = r_0[1+f(x)]. \quad (12.6)$$

The function $f(x)$ defines the deviation of the reflection coefficient $r(x)$ from the reflectivity of the substrate r_0. Substituting Eq. (12.6) in Eq. (12.5) and introducing the nondifferential response as the convolution of the object profile with the line spread function

$$R(x_s) = \int_{-\infty}^{\infty} g(x_s - x)f(x)dx, \quad (12.7)$$

we have the following for coherent detection at the center of the FP:

$$F(x_s) = H_0 |r_0|^2 \left[1 + R\left(x_s + \frac{\varepsilon}{2}\right)\right] \cdot \left[1 + R^*\left(x_s - \frac{\varepsilon}{2}\right)\right], \quad (12.8)$$

i.e., the differential response F is expressed in terms of the nondifferential one.

In the back focal plane, the incident beam is spatially filtered by the input aperture of the objective and forms a diffraction-limited spot in the front focal plane. The function $g(x)$ is defined by Fourier transform of the incident optical field as

$$g(x) = \frac{1}{\lambda f} \int_{-f \cdot NA}^{f \cdot NA} \exp\left[i\frac{k}{f} x x_d\right] dx_d = \frac{1}{2\pi} \int_{-\pi/\delta}^{\pi/\delta} \exp[iqx] dq = \frac{1}{\pi x} \sin\frac{\pi x}{\delta}, \quad (12.9)$$

where $\delta = \lambda/2NA$ is the diffraction resolution of an objective with numerical aperture NA.

We illustrate the formation of the microscope response to the rectangular groove with the width w and the depth h_0 [Fig. (12.2(a)]. The reflectivity of the object is given by

$$r(x) = r_0 \exp[2ikh(x)].$$

According to definition (12.6), the optical profile is

$$f(x) = \exp[2ikh(x)] - 1, \quad (12.10)$$

and its real and imaginary parts have a rectangular form as well. The calculation was performed for a shallow groove with $w = 0.4$ μm and $h_0 = 0.1$ μm. The field distribution of the incident beams calculated with Eq. (12.9) is shown in Fig. 12.2(b) for the interval $\varepsilon = \delta/2 = 0.4$ μm (if NA = 0.4). Figure 12.2(c) shows the microscope normalized response calculated by Eq. (12.8). Since the groove depth $h_0 > \lambda/8$, the modulus component of the response is not small and the real part of $f(x)$ cannot be neglected. Therefore, both components (the phase and the modulus) of the complex response must be measured for the reconstruction of the shallow object profile.

12.3 Solution of the Inverse Problem

In the previous section we have obtained the solution of the direct problem for the SDHM in the form of the differential response. The simulated response will be used for reconstruction of the object profile while solving the inverse problem with superresolution.

12.3.1 Formulation of the inverse problem

We now formulate the inverse optical problem for the SDHM as the reconstruction of the object profile $h(x)$ from the complex function of differential response $F(x)$. For easy interpretation, the solution of the inverse problem is split into three main steps.

Step I. Evaluation of the nondifferential response $R(x)$ from the differential response $F(x)$ by solving Eq. (12.8).

Step II. Solving the linear integral Eq. (12.7) with the kernel $g(x)$ for reconstruction of the optical profile $f(x)$ without exceeding a diffraction-limited resolution. Extrapolation of the response spectrum for exceeding a diffraction-limited resolution.

Step III. Determination of the object profile $h(x)$ from the optical profile $f(x)$ by solving Eq. (12.10).

The main steps of the solution of the inverse problem are considered in the next sections.

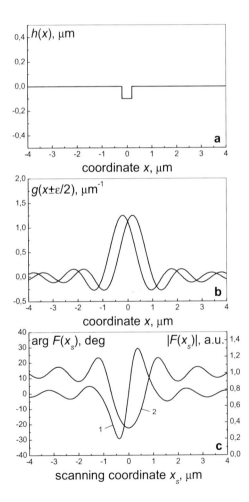

Figure 12.2 Successive stages of imaging in the SDHM. (a) Initial profile of the rectangular groove. (b) Probing beam profiles for $\delta = 0.8$ μm and $\varepsilon = 0.4$ μm. (c) The phase (1) and modulus (2) components of the differential response.

12.3.2 Evaluation of the microscope nondifferential response

To perform the first step of the inverse problem, let us solve Eq. (12.8) using the Shannon sampling theorem,[40] according to which a band-limited function can be defined from its values at a sequence of uniformly spaced discrete points (samples). Then the functions F and R are given by

$$F(x) = H_0 |r_0|^2 \sum_{n=-\infty}^{\infty} F_n S_n(x), \qquad (12.10)$$

$$R(x) = \sum_{n=-\infty}^{\infty} [R_n - 1] S_n\left(x - \frac{\varepsilon}{2}\right), \qquad (12.11)$$

where

$$R_n = R\left(x_n + \frac{\varepsilon}{2}\right) + 1, \quad F_n = \frac{F(x_n)}{H_0 |r_0|^2}, \quad S_n(x) = \frac{\sin \alpha(x - x_n)}{\alpha(x - x_n)}. \qquad (12.12)$$

Here S_n is the sampling function, $\alpha = \pi/\Delta x$ and $x_n = \Delta x n$.

The spacing Δx between the samples is defined by the spectrum bandwidth α of the function R or F, which is wider in the spatial-frequency domain. The spectrum of the nondifferential response is restricted by the objective pupil function: $P(x_d) = 1$ if $|x_d| \leq f\text{NA}$ and $P(x_d) = 0$ otherwise; therefore, the spatial-coordinate bandwidth in the FP is $2f\text{NA}$ and spatial-frequency bandwidth is $2\pi/\delta$. According to the convolution theorem in the frequency domain,[40] the bandwidth of the differential response is twofold, i.e., $4\pi/\delta$. By the Shannon sampling theorem, the spacing between the points must be $\Delta x \leq \pi/(2\pi/\delta) = \delta/2$. If we set $\Delta x = \varepsilon$, then the evaluation of the differential response is reduced to solving recurrent relations for response samples. Combining these conditions, we find the optimal condition, namely, $\Delta x = \varepsilon = \delta/2$.

The differential response (12.8) at the point x_n may be written as

$$\frac{F(x_n)}{H_0 |r_0|^2} = \left[1 + R\left(x_n + \frac{\varepsilon}{2}\right)\right]\left[1 + R^*\left(x_n - \frac{\varepsilon}{2}\right)\right]. \qquad (12.13)$$

Using the condition $\Delta x = \varepsilon = \delta/2$ and the designation (12.13), Eq. (12.14) can be rewritten in recurrent form,

$$F_n = R_n R_{n-1}^*. \qquad (12.14)$$

To solve this equation we represent the complex samples F_n and R_n in the exponential form

$$F_n = |F_n| \exp[i\Phi_n], \quad R_n = |R_n| \exp[i\varphi_n].$$

Substituting this in Eq. (12.15), we have

$$|F_n|\exp[i\Phi_n] = |R_n||R_{n-1}|\exp[i(\varphi_n - \varphi_{n-1})]. \tag{12.15}$$

Equating moduli and phases of the left- and right-hand sides of the last expression, Eq. (12.16) can be represented as

$$|R_n| = \frac{F_n}{|R_{n-1}|}, \qquad \varphi_n = \Phi_n + \varphi_{n-1}. \tag{12.16}$$

To solve these equations, one should specify initial conditions, i.e., the values $|R_{-\infty}|$ and $\varphi_{-\infty}$, since for a single object $R(-\infty) = 0$, we have: $|R_{-\infty}| = 1$ and $\varphi_{-\infty} = 0$. Then we can write the solution of Eq. (12.17) only for a finite number of samples $(2N + 1)$ with the initial conditions $|R_{-N}| = 1$, $\varphi_{-N} = 0$ as

$$|R_n| = \frac{|F_n||F_{n-2}||F_{n-4}|\cdots}{|F_{n-1}||F_{n-3}|\cdots}, \qquad \varphi_n = \sum_{m=-N}^{n} \Phi_m, \quad n = -N, -N+1, \ldots, N. \tag{12.17}$$

Practically, the number of samples of the function $F(x)$ being measured is restricted by the scanning interval $[-L/2, L/2]$, within which the response exceeds the background noise. The function $R(x)$ can be reconstructed further at any point of the scanning interval by Eq. (12.12) with the samples $|R_n|$ and φ_n.

To demonstrate the processing of the differential response numerically, we used the functions $\Phi(x_s)$ and $|F(x_s)|$ evaluated in the previous section (see Fig. 12.2(c)). In our case, the scanning interval was specified as $L = 40$ μm and the number of samples within this interval is $2N + 1 = 101$ (if $\delta = 0.8$ μm). The sampled data $\Phi_n = \Phi(x_n)$ and $|F_n| = |F(x_n)|$, where $x_n = n\delta/2$ ($n = -N, \ldots, N$), were derived as the initial samples. The samples of the non-differential response were evaluated by Eq. (12.18), and the complex samples are derived by $R_n = |R_n|\exp[i\varphi_n]$. The real and imaginary parts of the reconstructed function $R(x)$ are plotted in Fig. 12.3(a).

12.3.3 Diffraction resolution reconstruction of the optical and topographic profile

The next step of the inverse optical problem to be considered is the resolution of the linear integral equation (12.7) with the known kernel $g(x)$. Related inverse problems are ill-posed. To solve the problem numerically, one must introduce some additional information about the solution. This process is known as regularization. The Wiener filter as a regularization function will be used.

Performing the Fourier transform for both parts of Eq. (12.7) and using the convolution theorem, we obtain

$$\hat{R}(q) = \hat{g}(q)\hat{f}(q), \tag{12.18}$$

where $\hat{R}(q)$, $\hat{g}(q)$, and $\hat{f}(q)$ are the spectra of the functions $R(x)$, $g(x)$, and $f(x)$, correspondingly. The formal solution of this equation may be written as

$$\tilde{f}(x) = \frac{1}{2\pi} \int_{-\infty}^{\infty} \hat{f}(q) \exp(iqx) dq = \frac{1}{2\pi} \int_{-\infty}^{\infty} \frac{\hat{R}(q) \hat{g}^*(q)}{|\hat{g}(q)|^2} \exp(iqx) dq . \quad (12.19)$$

This solution is incorrect, considering that $\hat{g}(q) = 0$ outside the aperture interval ($|q| \geq \pi/\delta$). The correct approximate solution of Eq. (12.7) can be obtained by Wiener filtering,[41]

$$\hat{f}(q) = \frac{\hat{R}(q) \hat{g}^*(q)}{|\hat{g}(q)|^2 + N(q)/S(q)} \equiv \Omega(q) R(q) , \quad (12.20)$$

where $S(q)$ and $N(q)$ are the signal spectrum and the noise spectrum, correspondingly. It can be seen from Eq. (12.21) that the noise spectrum regularizes the incorrect solution (12.20) because the denominator in Eq. (12.21) is not equal to zero anywhere. Another important peculiarity of this solution is that the Wiener filter $\Omega(q)$ suppresses the signal in the frequency domain if the noise rises.

Under the assumption that the ratio $S(q)/N(q)$ is large within the aperture interval, we have the approximate filtering solution:

$$\tilde{f}(x) = \frac{1}{2\pi} \int_{-\pi/\delta}^{\pi/\delta} \frac{\hat{R}(q)}{\hat{g}(q)} \exp(iqx) dq , \quad (12.21)$$

where the spectrum $\hat{R}(q)$ is defined by

$$\hat{R}(q) = \int_{-\infty}^{\infty} R(x) \exp(-iqx) dx .$$

The spectral representation of the response is illustrated in Fig. 12.3(b). Curves 1 and 2 show the real and imaginary parts of the spectrum $\hat{R}(q)$, which in accordance with Eq. (12.19) represents the product of the rectangular pupil function $\hat{g}(q)$ and the spectrum of the optical profile $\hat{f}(q)$. As a result, only the central part of the spectrum $\hat{f}(q)$ is presented in the spectrum of the nondifferential response. The spectral components outside the aperture interval are filtered by the objective. Weak oscillations and spikes in the response spectrum are due to the Gibbs phenomenon.

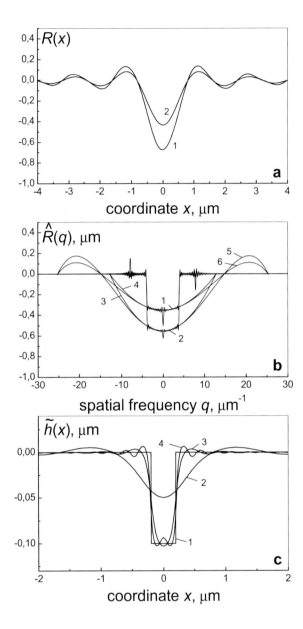

Figure 12.3 Successive stages of solving the inverse problem. (a) Nondifferential response: real (1) and imaginary (2). (b) Spectrum of the nondifferential response (1 real, 2 imaginary) and the approximating curves; second-order parabolas (3, 4), fourth-order parabolas (5, 6). (c) Reconstruction of the groove profile: initial profile (1), reconstruction with optimum filtering (2), reconstruction with the quadratic (3), and with the fourth-order parabolic approximation (4).

The geometrical profile may be reconstructed by

$$\tilde{h}(x) = \begin{cases} \dfrac{1}{2k}\arctan\dfrac{\tilde{f}''(x)}{\tilde{f}'(x)+1}, & \text{if } \tilde{f}'(x)+1 > 0, \\ \dfrac{1}{2k}\left[\arctan\dfrac{\tilde{f}''(x)}{\tilde{f}'(x)+1} - \pi\right], & \text{if } \tilde{f}'(x)+1 < 0, \end{cases} \qquad (12.22)$$

where $f'(x)$ and $f''(x)$ are the real and imaginary parts of the optical profile $f(x)$, respectively. The reconstructed profile is shown in Fig. 12.3(c) (curve 2). The rms deviation of this diffracted-limited function from the initial rectangular profile (curve 1) is $\sigma_h = 0.66$.

If the spatial-frequency components are restored outside the aperture limits, a sharp profile can be reconstructed with a higher accuracy. The methods of restoring high-frequency spectral components being lost by diffraction will be considered in the next section.

12.3.4 Profile reconstruction with superresolution

12.3.4.1 Spectrum approximation and extrapolation

In the previous section we solved the inverse problem and reconstructed the geometrical profile of an object with diffraction resolution ($\delta = 0.8$ μm, NA = 0.4). To enhance an optical resolution, different methods of response spectrum processing should be used.

With a priori information on optical profile (12.22), the limited number of object parameters can be evaluated with superresolution. To extrapolate the spectrum, we expand the scattering amplitude $\hat{f}(q)$ in terms of powers of q as

$$\hat{f}(q) \equiv \int_{-\infty}^{\infty} f(x)\exp(-iqx)dx = \sum_{n=0}^{\infty} \frac{(-iq)^n}{n!} m_n, \qquad (12.23)$$

where m_n are the moments of the optical profile, defined as

$$m_n = \int_{-\infty}^{\infty} x^n f(x)dx, \quad n = 0, 1, 2, \ldots \qquad (12.24)$$

The moments of the localized object define its integrated characteristics: averaged width or depth, shape asymmetry, tilt of the walls, etc. The restricted set of these parameters allows characterizing the object profile under study.

For objects with a symmetric profile (e.g., a rectangular groove), the odd moments vanish: $m_1 = m_3 = \cdots = 0$, and the scattering amplitude becomes the symmetric function

$$\hat{f}(q) = m_0 - \frac{1}{2}m_2 q^2 + \frac{1}{24}m_4 q^4 - \cdots. \tag{12.25}$$

Thus the problem of spectrum extrapolation is reduced to the determination of coefficients $m_0, m_2, m_4,\ldots$ by the approximation of $\hat{R}(q)$, which is equal to $\hat{f}(q)$ within the interval $[-\pi/\delta, \pi/\delta]$. If the spectrum shape within the aperture interval is distinguished from the straight line $\hat{f}(q) = m_0$, two coefficients (m_0 and m_2) can be determined. This truncated series cannot be extrapolated to an infinite interval because it diverges at infinity. Therefore, the extrapolation interval must be limited by a specified regularization condition. In our case, we used the conditions $\operatorname{Re}\hat{f}(\pm M\pi/\delta) = \hat{\sigma}_1$ and $\operatorname{Im}\hat{f}(\pm M\pi/\delta) = \hat{\sigma}_2$. Here, $\hat{\sigma}_{1,2}$ are the rms of real and imaginary parts of $\hat{R}(q)$, and M is the superresolution coefficient, which defines the extrapolation interval $[-M\pi/\delta, M\pi/\delta]$.

The extrapolation of the response spectrum is illustrated in Fig. 12.3(b). Leaving only the first two terms in Eq. (12.26), we find the zero and second moments according to the best fitting between the truncated power series $m_0 - m_2 q^2/2$ and the function $\hat{R}(q)$ within the interval $[-\pi/\delta, \pi/\delta]$. Curves 3 and 4 in Fig. 12.3(b) correspond to the real and imaginary parts of the approximation function for $m_0 = (0.55 + 0.35i)$ µm, $m_2 = (6.8 + 4.3i) \times 10^{-3}$ µm³. For the rms $\hat{\sigma}_1 = \hat{\sigma}_2 = 0.01$ µm, the extrapolation condition formulated above is completed at the superresolution coefficient $M_2 = 3.3$.

The width w and the depth h_0 of the geometrical profile $\tilde{h}(x)$ can be evaluated with the defined moments m_0 and m_2 by using a priori information on the object shape. Using Eqs. (12.10) and (12.25) for the rectangular profile, we have

$$m_0 = w\left[\exp(2ikh_0) - 1\right],$$

$$m_2 = \frac{1}{12} w^3 \left[\exp(2ikh_0) - 1\right].$$

Solving these equations, we obtain $h_0 = 0.103 \pm 0.007$ µm and $w = 0.38 \pm 0.05$ µm. The errors are due to the rms of $\hat{R}(q)$ within the aperture interval.

The partial sum of three terms in Eq. (12.26) allows us to evaluate the fourth moment m_4 and increase the extrapolation interval. However, the term with the factor m_4 is lost in spectrum noise within the aperture interval. Therefore, we have used another method based on the minimization of the reconstruction error σ_h. To find the geometrical profile with Eq. (12.23), the optical profile must be evaluated first. It is derived by the inverse Fourier transform of the spectrum $\hat{f}(q)$ as

$$\tilde{f}(x) = \frac{1}{2\pi} \int_{-M\pi/\delta}^{M\pi/\delta} \hat{f}(q) \exp(iqx) dq. \qquad (12.26)$$

Here $\hat{f}(q) = m_0 - m_2 q^2/2 + m_4 q^4/24$, and the domain of integration is defined by the extrapolation interval. The geometrical profile is evaluated by Eq. (12.23). Then the rms deviation of the function $\tilde{h}(x)$ from the initial rectangular profile is minimized by variation of the parameter m_4 with the coefficients m_0 and m_2 listed above. This gives $m_4 = -(9.4 + 6.0i) \times 10^{-5}$ µm^5 and the superresolution coefficient is $M_4 = 6.5$. The corresponding approximation with coefficients m_0, m_2, and m_4 is illustrated in Fig. 12.3(b) (curves 5 and 6). The reconstructed geometrical profiles are shown in Fig. 12.3(c). Curves 3 and 4 are the reconstructed profiles corresponding to the spectral approximations expanded by the factors of $M_2 = 3.3$ and $M_4 = 6.5$, respectively. The estimation of the reconstruction error gives $\sigma_h = 0.3$ for curve 3 and $\sigma_h = 0.2$ for curve 4.

Thus, the reconstruction of an object profile without spectrum extrapolation does not permit us to attain a superresolution. However, at high signal-to-noise ratio, the signal spectrum may be extrapolated outside the aperture interval, and the object profile can be reconstructed with superresolution.

12.3.4.2 Sampling expansion

In experimental measurements, the noise of the microscope response can be rather high and superresolution processing must take into account the noise component. Here we will consider the algorithm of the optical profile reconstruction of a subwavelength object with regard to the additive noise. The method described in the previous section is not optimal because of the divergence of the power series with a finite number of expansion terms. Therefore, we will use the algorithm of solving the linear integral equation by expanding both the nondifferential response and the optical profile in terms of the sampling functions. This algorithm also involves a quasi-optimal regularization of the solution.

The nondifferential response with the additive noise component $n(x)$ may be rewritten as

$$R(x) = \int_{-\infty}^{\infty} g(x-x') f(x') dx' + n(x). \qquad (12.27)$$

First, we will describe the algorithm for solving the problem, which employs sampling of the original integral equation and reduces it to a system of linear equations. Then we will construct a regularization function, which smoothes and averages a noisy spectrum.

We represent the solution as an expansion in terms of the sampling functions

$$f(x) = \sum_{m=-\infty}^{\infty} f_m S\left(x - \frac{m\delta}{M}, M\right), \qquad (12.28)$$

where $S(x, M) = \mathrm{sinc}(M\pi x/\delta)$ and $f_m = f(m\delta/M)$ are the samples of the optical profile, and M is the superresolution parameter. The expansion in series (12.29) corresponds to extrapolation of the scattering function $\hat{f}(q)$ over interval $[-M\pi/\delta, M\pi/\delta]$, which is M times wider than the aperture interval due to the spectral width of the function $S(x, M)$.

We use the expansion in terms of the sampling functions also for the nondifferential response

$$R(x) = \sum_{n=-\infty}^{\infty} R_n S(x - n\delta, 1), \qquad (12.29)$$

where $R_n = R(n\delta)$ are the samples including the noise component. Setting the limit N for the series in Eqs. (12.29) and (12.30), we derive the system of linear equations[39] $\mathbf{Gf} = \mathbf{R}$, where $\mathbf{R}$ and $\mathbf{f}$ are the column vectors of samples R_n and f_m, accordingly, and the matrix $\mathbf{G}$ has the elements

$$G_{mn} = \int_{-l}^{l} g(n\delta - x') S\left(x' - \frac{m\delta}{M}, M\right) dx'. \qquad (12.30)$$

Here the integration is performed over the interval of object localization $[-l, l]$ ($l = N\delta/M$). The system has the solution

$$f_m = \sum_{n=-N}^{N} R_n G_{nm}^{-1}, \quad m = -N, ..., N, \qquad (12.31)$$

where the elements G_{nm}^{-1} refer to the inverse matrix $\mathbf{G}^{-1}$.

In accordance with the information criterion of conservation of the number of degrees of freedom, the resolution of an optical system can be enhanced by reducing the localization interval of an object $[-l, l]$ with respect to response localization $[-L, L]$. Since the number of samples is retained, the decrease of the sampling interval reduces the localization interval. The number of samples $2N + 1$ of the nondifferential response is defined by the field of view $[-L, L]$, where the complex response exceeds the noise; therefore, the number of samples is equal to $2N = 2L/\delta$. The function to be restored has the same number of samples $2N + 1$. The decrease of the sampling interval between the samples f_m results in M-fold decreasing of the interval of object localization ($l = L/M$).

An important property of the solution (12.29) is that an infinitely high resolution cannot be achieved by letting M tend to infinity because of the limited accuracy of the initial data R_n. The increase of M reduces the sampling interval of $f(x)$ and enhances the resolution, but increases the reconstruction error. The object localization interval $[-l, l]$ should exceed the lateral size of the object. Thus, the superresolution coefficient M is defined by the number of initial samples and the noise magnitude.

The noise component of the nondifferential response samples R_n transforms to the samples f_m in Eq. (12.32) and the stability of the solution may not be achieved at $M > 1$. Tikhonov regularization[41] is not available in this case, since the optimal filtering algorithm can be used only if the noise and signal spectra are not overlapped. In our case, the spectra do overlap within the aperture interval at the Fourier plane. To efficiently suppress a noise component, we find the regularized solution using the regularization operator, which smoothes the spectra. The smoothing of the noisy samples by a linear filter operator as a stabilizing factor result in the regularized samples of the optical profile,

$$\tilde{f}_m = f_m K(b, x_m). \qquad (12.32)$$

Here, $x_m = m\delta/M$, $K(b, x)$ is the filtering function,[42] which averages the noisy spectrum of the function $f(x)$ over the interval $[-b, b]$, and b is the regularization parameter. By varying b, one can minimize the reconstruction error of the reconstructed profile. Its optimum value depends primarily on the spectrum extrapolation interval, signal-to-noise ratio, and the number of initial samples.

The reconstructed optical profile with smoothed-out samples is given by

$$\tilde{f}(x) = \sum_{m=-\infty}^{\infty} \tilde{f}_m S\left(x - \frac{m\delta}{M}, M\right). \qquad (12.33)$$

This quasi-optimal method gives less efficient noise suppression than Wiener filtering. However, in contrast to Wiener filtering, the method does not require the determination of the noise and signal spectral densities, and it more efficiently smoothes the noisy spectrum.

We now consider the model reconstruction of the rectangular groove with parameters given in Section 12.2. The nondifferential response is evaluated in accordance with Eq. (12.28), where the noise $n(x)$ is produced by a random number generator. The noise magnitude is uniformly distributed in the range $0 < n(x) < 0.1$. Figure 12.4 demonstrates the real part of the nondifferential response to the rectangular groove for three realizations of the noise. In our case, the field of view is $L = 10$ μm and the number of samples is $2N + 1 = 2L/\delta + 1 = 26$. Then, the samples f_m of the optical profile is evaluated by Eqs. (12.31) and (12.32) at given M and N. Without preliminary information on the function $K(b, x)$, the algorithm employs various smoothing functions with a band-limited or decaying at infinity spectrum. The function $K(b, x) = \text{sinc}(bx)$, proposed by Yegorov,[43] appears to be the most efficient and optimal in our case. The regularization procedure is reduced to varying the parameter b to minimize the reconstruction error of $\tilde{f}(x)$ in accordance with Eq. (12.34). The solution also should be stable for the optimal regularization parameter. Profile reconstruction begins with $M = 1$, for which the regularization parameter is $b = 2$ μm^{-1}. The reconstructed profiles are shown in Fig. 12.5. For $M = 2$, the minimum error is achieved at $b = $

10^7 µm^{-1}. The increase of M up to 3 leads to raising the rms, and the optimal solution for a specified signal-to-noise ratio is obtained at $M = 2$.

Thus, the stable solution of the inverse problem was obtained with a superresolution by the extrapolation of the spectrum of noisy response using quasi-optimal regularization.

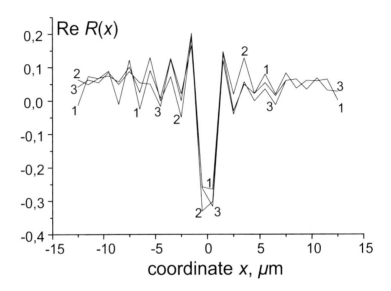

Figure 12.4 Real part of the nondifferential response to a submicron rectangular groove for three realizations of the additive noise.

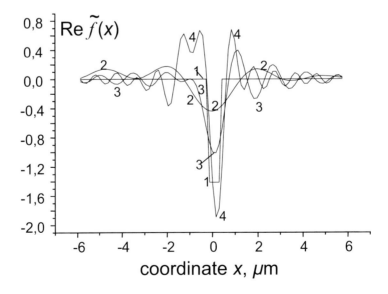

Figure 12.5 Real part of the reconstructed optical profile of a rectangular groove for $M = 1$ (curve 2), $M = 2$ (curve 3), and $M = 3$ (curve 4). Curve 1 represents the original rectangular profile.

12.4 Conclusion

An algorithm for superresolving reconstruction of a profile of a shallow submicron object from an SDHM complex response with additive noise is considered. The important feature of the processing is the evaluation of an absolute phase microscope response based on image formation theory for coherent detection in the Fourier plane. The quasi-optimal filtering was employed for a regularization of an incorrect solution of the linear integral equation. Superresolution processing was performed by using an expansion of the optical profile and a nondifferential response in terms of the sampling functions. The development of a more efficient algorithm for response processing including the statistics of experimental noise is required for increasing the accuracy of superresolving reconstruction. For deep objects, the rigorous diffraction theory must be used for correct modeling of the diffraction problem. In this case, the methods for a linearization of a highly nonlinear relationship between structure parameters and diffraction measurements must be developed for solving the inverse problem.

References

1. Jungerman, R.L., Hobbs, P.C.D., and Kino, G.S., "Phase sensitive scanning optical microscope," *Appl. Phys. Lett.*, **45**, 846–848, 1984.
2. Matsumoto, H., and Hirai, A., "3-D surface profile measurement by 2-D heterodyne low-coherence interferometer," *Proc. SPIE*, **4829**, 827–828, 2002.
3. Dandliker, R., "Heterodyne holographic interferometry," *Progress in Optics*, 17, E. Wolf, Ed., North-Holland, Amsterdam (1980).
4. Sommargren, G.E., "Optical heterodyne profilometry," *Appl. Opt.*, **20**, 610–618, 1981.
5. Massie, N.A., "Real-time digital heterodyne interferometry: a system," *Appl. Opt.*, **19**, 154–160, 1981.
6. Pantzer, D., Politch, J., and Ek, L., "Heterodyne profiling instrument for the angstrom region," *Appl. Opt.*, **25**, 4168–4172, 1986.
7. See, C.W., Vaez Iravani, M., and Wickramasinghe, H.K., "Scanning differential phase contrast optical microscope: application to surface studies," *Appl. Opt.*, **24**, 2373–2379, 1985.
8. Chung, H., Walpita, L.M., and Chang, W.S.C., "Simplified differential phase optical microscope," *Appl. Opt.*, **25**, 3014–3017, 1986.
9. Moharam, M.G., Grann, E.B., Pommet, D.A., and Gaylord, T.K., "Formulation for stable and efficient implementation of the rigorous coupled-wave analysis of binary gratings," *JOSA A* **12**, 1068–1076, 1995.
10. Aguilar, J.E., and Mendez, E.R., "Imaging optically thick objects in scanning microscopy: perfectly conducting surfaces," *JOSA A* **11**, 155–167, 1994.

11. Sheridan, J.T., and Sheppard, C.J.R., "Coherent imaging of periodic thick fine isolated structures," *JOSA A* **10**, 614–632, 1993.
12. Morgan, S.P., Choi, E., Somekh, M.G., and See, C.W., "Interferometric optical microscopy of subwavelength grooves," *Opt. Comm.* **187**, 29–38, 2001.
13. Akhmedzhanov, I.M., Baranov, D.V., and Zolotov, E.M., "Object characterization with a differential heterodyne microscope," *J. Opt. A: Pure Appl. Opt.* **5**, S200–S206, 2003.
14. Baranov, D.V., Egorov, A.A., Zolotov, E.M., and Svidzinskii, K.K., "Formation of the image of a microstep profile in a heterodyne differential-phase microscope," *Sov. J. Quantum Electron.* **26**, 360–364, 1996.
15. Hopkins, H.H., "On the diffraction theory of optical images," *Proc. Roy. Soc. London, Ser. A*, **217** (1130), 408–432, 1953.
16. Wilson, T., and Sheppard, C.J.R., *Theory and Practice of Scanning Optical Microscopy*, Academic Press, London (1984).
17. Somekh, M.G., "Defocus response of phase-sensitive heterodyne microscope," *Biomedical Image Processing and Three-Dimensional Microscopy*, R.S. Acharya, Ed., *Proc. SPIE* **1660**, 485–496, 1992.
18. Gale, D.M., Pether, M.I., and Dainty, J.C., "Linnik microscope imaging of integrated circuit structures," *Appl. Opt.* **35**, 131–148, 1996.
19. Tavrov, A., Totzeck, M., Kerwein, N., and Tiziani, H.J., "Rigorous coupled-wave analysis calculus of submicrometer interference pattern and resolving edge position versus signal-to-noise ratio," *Opt. Eng.* **41**, 1886–1892, 2002.
20. Huttunen, J., and Turunen, J., "Phase images of grooves in a perfectly conducting surface," *Opt. Comm.* **119**, 485–490, 1995.
21. Baranov, D.V., and Zolotov, E.M., "Direct and inverse problems of a differential heterodyne microscope," *Bull. Russian Acad. Sci.: Phys.*, **63**, 823–834, 1999.
22. Bozhevolnaya, E.A., Bozhevolnyi, S.I., Zolotov, E.M., Postnikov A.V., and Rad'ko P.S., "Index profile measurements of channel waveguides by using differential phase optical microscope," *Proc. SPIE*, **1932**, 41–50, 1993.
23. Baranov, D.V., Zolotov, E.M., and Svidzinsky, K.K., "Superresolution and response interpretation of a scanning differential microscope," *Laser Phys.*, **6**, 548–552, 1996.
24. Toraldo di Francia, G., "Resolving power and information," *JOSA*, **45**, 497–501, 1955.
25. Rautian, S.G., "Real spectral apparatus," *Sov. Phys. Usp.*, **1**, 245–273, 1958.
26. Harris, J.L., "Resolving power and decision theory," *JOSA* **54**, 606–611, 1964.
27. Goodman, J.W., *Introduction to Fourier Optics*, McGraw-Hill, New York (1968).

28. Marcuse, D., *Light Transmission Optics*, Van Nostrand Reinhold, New York (1972).

29. Schmidt-Weinmar, H.G., "Spatial resolution of subwavelength sources from optical far-zone data," *Inverse Source Problems in Optics*, H.P. Baltes, Ed., Topics in Current Physics **9**, Chap. 4, Springer-Verlag, Berlin (1978).

30. Terebizh, V.Y., "Image restoration with minimum a priori information," *Sov. Phys. Usp.*, **38**, 137–168, 1995.

31. Tychinsky, V.P., Masalov, I.N., Pankov, V.L., and Ublinsky D.V., "Computerized phase microscope for investigation of submicron structures," *Opt. Comm.*, **74**, 37–40, 1989.

32. Shannon, C., "A mathematical theory of communication," *Bell Syst. Tech. J.* **27**, 379–423, 623–656, 1948.

33. Lukosz, W., "Optical systems with resolving powers exceeding the classical limit," *JOSA*, **56**, 1463–1472, 1966.

34. Cox, I.J., and Sheppard, C.J.R., "Information capacity and resolution in an optical system," *JOSA A*, **3**, 1152–1158, 1986.

35. Kosarev, E.L., "Shannon's superresolution limit for signal recovery," *Inverse Prob.*, **6**, 55–76, 1990.

36. Bendinelli, M., Consortini, A., Ronchi, L., and Frieden, B.R., "Degrees of freedom, and eigenfunctions, for the noisy image," *JOSA*, **64**, 1498–1502, 1974.

37. Slepian, D., and Pollak, H.O., "Prolate spheroidal wave functions, Fourier analysis and uncertainty," *Bell Syst. Tech. J.*, **40**, 43–63, 1961.

38. Friedmann, M., and Shamir, J., "Resolution enhancement by extrapolation of the optically measured spectrum of surface profiles," *Appl. Opt.*, **36**, 1747–1751, 1997.

39. Baranov, D.V., Yegorov, A.A., and Zolotov, E.M., "Optical profile restoration with a superresolution using sampling expansions," *Laser Phys.*, **11**, 943–947, 2001.

40. Papoulis, A., *Systems and Transforms with Applications in Optics*, McGraw-Hill, New York (1968).

41. Tikhonov, A.N., and Arsenin, V.Y., *Methods of Solution of Ill-Posed Problems* (in Russian), Nauka, Moscow (1986).

42. Akhmanov, S.A., Dyakov, Y.E., and Chirkin, A.S., *Introduction to statistical Radiophysics and Optics* (in Russian), Nauka, Moscow (1981).

43. Baranov, D.V., Yegorov, A.A., and Zolotov, E.M., "Optical profile restoration from differential microscope response with additive noise," International Quantum Electronic Conference, IQEC-2002. Technical Digest, p. 353, June 22–27, 2002, Moscow.

Dmitry V. Baranov is a researcher at the General Physics Institute of Russian Academy of Sciences, Moscow. He received his MS in physics from Moscow State University in 1987. His current research interests include experimental as well as theoretical investigation of the image formation process in scanning heterodyne microscopy. He has published more than 30 journal papers on integrated optics and high-resolution microscopy.

Evgeny M. Zolotov is a head of the laboratory for optical investigation of surfaces at the General Physics Institute of Russian Academy of Sciences, Moscow. He received his MS in physics from Moscow State University, his PhD in laser spectroscopy from Lebedev Physics Institute, and his ScD in integrated optics from the General Physics Institute. He has written more than 180 publications on laser physics, integrated optics, optical metrology, and heterodyne microscopy.

Chapter 13
Fourier Holography Techniques for Artificial Intelligence

Alexander V. Pavlov
St. Petersburg State University for Information Technologies, Russia

13.1 Introduction
13.2 Holographic Implementation of the Linear Regression Model
    13.2.1 Neural network model of Fourier holography
    13.2.2 Holographic implementation of linear regression
    13.2.3 Selective attention and cognitive drift phenomenon realization
13.3 Chaotic Dynamics—Can Holography Help to Solve Creative Problems?
13.4 Experimental Illustration and Computer Simulation
13.5 Conclusion
References

13.1 Introduction

In the days immediately after the reinvention of holography by Leith and Upatnieks,[1,2] and independently by Denisyuk,[3] new applications of holography were being proposed. One of those was based on a number of deep analogies between the attributes of the human mind and brain, and also the properties of optical holography. Based on these analogies, the holographic paradigm was proposed and developed in cognitive science by Pribram.[4,5] Unfortunately, the term "holographic paradigm" was recently compromised regarding the so-called "holographic brain," "holographic universe," etc. in many near-scientific publications. Many of these papers are based on superficial knowledge of the outer effects of artistic holography and hold general conversations instead on the detailed analysis of particular physical phenomena and mathematical models. Thus, first of all it has to be underlined that there is no intention to maintain that the brain is a hologram—our aim is to analyze how holography can be utilized in the framework of artificial intelligence. Any anthropomorphisms and/or human analogs used in this chapter are for intuitive purposes; that is, to make the chapter more vivid and readable.

The first analogy between the brain and holography was the associative properties of both biological brain and optical hologram, which have been appreciated since the inception of holography.[6] Based on this analogy, a number of models for holographic associative memory were proposed and implemented.[7-10] The list of articles dedicated to holographic associative memories includes a few hundreds papers. The concept of associative memory has been developed as a paradigm for information processing by the brain, and incorporated in the paradigm of artificial neural networks. Associative memory is defined as fault-tolerant, content-addressable memory, which recalls a noise-free reference pattern when addressed by an erased and/or distorted input pattern. Associative memory plays a major role in thinking and forms a base of intelligence.

The concept of associative memory is closely linked with another attribute of the brain and mind, namely, the brain thinks not by operations on numbers, but by processing of the patterns of neural activity, also known as internal representation patterns. This attribute leads to the second analogy between brain and holography, i.e., both process information by patterns processing.

However, the concept of information associative processing is wider than associative memory only. For example, the main attribute of intelligence, tightly connected with the associativity of thinking, is the ability to predict the further development of events.[11,12] The importance of this attribute can hardly be overestimated since it is the ability to predict that determines both the ability of an individual to survive and to achieve life's goals. Prediction is the mechanism that allows a number of essential functions of the brain to be implemented, e.g., cognition, perception, attention, etc. Moreover, prediction is an essential mechanism for creative thinking to be realized.[13]

Indeed, as the brain solves problems by means of the patterns of internal representation processing, the procedure of problem solution can be represented as follows. The condition of a problem is a pattern. If the brain perceives the pattern as a completed one, then there is no problem. If the pattern is uncompleted, then there is a problem. To solve the problem is to complete the pattern.[13] By this approach, the problems can be classified as follows:

1. *Standard or routine problem.* This is the task of completing the pattern of a problem condition through relation to a reference one, i.e., to restore known information pieces by their part addressing. This sort of problem is solved by associative memory. No new information is created; the task is to remember known information only.

2. *Creative problem.* The problem cannot be solved by recalling any reference pattern from a memory since there is no such pattern in the memory. To solve the problem is to create new information. Two kinds of creative problem can be viewed, as follows:

    2.1. *"Simple" creative problem.* Existing knowledge is enough to solve the problem. In other words, to solve the problem is to create the new pattern, which is adequate for reality. This type of problem can be solved by using the mechanism of prediction; namely, the pattern condition is completed (extended) by using a model of extrapolation, i.e., prediction.

2.2. *"Complex" creative problem.* New knowledge is to be created because existing knowledge does not allow a pattern that is adequate for reality to be created. According to the results of psychological studies, the brain uses chaotic dynamics in order to solve this kind of problem.[13–20] It is the ability to be deep in chaos, and then to go back, that determines the ability of the brain to solve complex creative problems.[15,17]

This concept of problem solution includes associative memory as a way to solve the first sort of problem only. Thus, in order to develop applications of holography in artificial intelligence, it is vital to pay attention to developing holographic implementation of the models of information associative processing, which allow us to realize the abovementioned phenomena of the mind—first of all, the phenomenon of prediction, and the ability to realize chaotic dynamics.

In this chapter we restrict the consideration to Fourier holography only. In Section 13.2.1 we set the stage for the presentation of the developed model by discussing the neural network model of Fourier holography. In Section 13.2.2 we discuss the holographic implementation of the linear predictor, and in Section 13.2.2 we demonstrate how the phenomenon of cognitive drift can be implemented on the base of the predictor. In Section 13.3 we discuss how chaotic dynamics can be used in order to solve creative problems, determine some specific features that are needed for this task, and demonstrate how the idea can be implemented by the setup. In Section 13.4 we present some experimental and numerical results.

13.2 Holographic Implementation of Linear Regression Model

13.2.1 Neural network model of Fourier holography

In this section we analyze the neural network model of the Fourier holography setup with a nonlinear phase-conjugate mirror in the correlation domain, which is used to implement associative memory,[7–10] in order to find the possibility of associative processing that goes beyond the scope of associative memory. The setup is shown in Fig. 13.1.

The setup can be considered as a two-layered neural network, shown in Fig. 13.2. Pixels in both the image *Im* and correlation C planes are considered as the neurons, and Fourier hologram H implements the neurons' interconnection weights. Phase-conjugate mirrors PCM1 and PCM2 are placed into the image and correlation planes, respectively, lenses L_1 and L_2 are of the Fourier transform type. Beam splitter SM forms the output plane Out. Three stages of the neural network operation cycle can be viewed as follows:

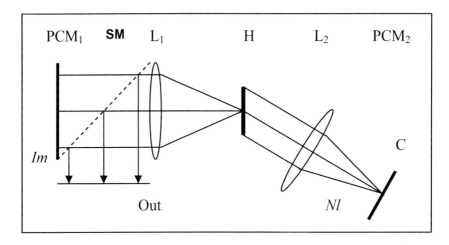

Figure 13.1 Holographic associative memory.

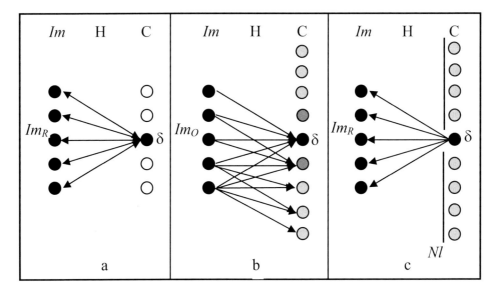

Figure 13.2 Neural network model of holographic associative memory.

1. *Learning.* Let that the Fourier hologram is recorded by the reference image Im_R, which is placed into the image plane, with an off-axial plane reference wave ; the last is a Fourier transform of the point reference source $R = F(\delta)$, where δ is the Kronecker delta function. For simplicity, the plane reference wave is not presented in the figure. The hologram recording can be considered from a "neural networks point of view" as learning of the neural network; the structure of the interconnections, formed by the learning procedure, is known as "Grossberg star", shown in Fig. 13.2(a)

2. *Forward propagation Im→C*. This is the well-known Van der Lught correlator.[10] If a thin hologram is restored by the object image Im_{obj} that is a distorted and/or erased version of the reference one, then activation of the neurons in the correlation layer C is described by the correlation function of the images $Im_{obj} \otimes Im_R$. Thus, if N_R and N_{obj} are the numbers of the neurons in layer *Im*, activated by Im_R and Im_{obj}, respectively, then $(N_R + N_{obj} - 1)$ neurons in the correlation layer are activated; one neuron is activated by the autocorrelation peak, other neurons are activated by the side maxima of the correlation function. In other words, due to the property of angular invariance, the hologram restores a "Grossberg star" from each neuron in the image layer, as is shown in Fig. 13.2(b). This is the point where the holographic neural network differs from electronic neural networks, because the latter are neural networks with fixed interconnections. It is this difference that gives the possibilities to make a step from associative memory to more complex models of associative processing, as will be discussed below.
3. *Back propagation C→Im*. By phase conjugating, the hologram is illuminated by a number of plane waves and restores the Grossberg star from each neuron in the correlation plane. Each Grossberg star restores the reference image in the output plane (image layer *Im*); the resulting image is a sum of these reference images, shifted one from another. In order to restore in the image layer *Im* the reference image Im_R, the hologram has to be illuminated by the plane wave, conjugated with the reference one; i.e., from the "optical point of view," the correlation function has to be transformed to the point source (delta function). From the "neural networks point of view," only the interconnections, which correspond to the initial Grossberg star [Fig. 13.2(a)], have to be used at this step; other interconnections have to be rejected. This transformation is implemented by using the nonlinear operator in the correlation domain, acting to implement the step $Nl(\text{Im}_{Obj} \otimes \text{Im}_R) \to \delta$. The nonlinear operator has to pass the autocorrelation peak only by rejecting the lateral correlation maxima, as is shown in Fig. 13.2(c). These side correlation maxima in the framework of associative memory are usually considered as noise. However, the lateral maxima of the correlation function are not noise, but information. This information is not used in the concept of associative memory, but this information can be used in order to implement more complex models of information associative processing, namely, the models based on the concept of regression

13.2.2 Holographic implementation of linear regression

Let $Im_R(x)$ and $Im_{obj}(x)$ be the realizations of a stationary stochastic process, and $Im_{obj}(x)$ is observed on the interval $[x_{min}, x_0]$, where x_0 is a point of observation or an instant. For simplicity of notations, but without loss of generality, we assume that the variables are separated in the functions that describe the images, and consider a 1D model. The best linear estimate (predictor) of the value of the random variable $Im_{pred}(x_k)$ with respect to $Im_R(x)$ at the point $x_k, k > 0$ is given by the expression[21]

$$Im_{pred}(x_k) = \int_{x_{min}}^{x_0} Im_{obj}(x_0 - x) a(x) dx, \qquad (13.1)$$

where the weight function $a(x)$ is found by solving the equation

$$\int_{x_{min}}^{x_0} a(x) C(|x - \xi|) dx = C(x_k + \xi), \qquad (13.2)$$

where C denotes the cross-correlation function of $Im_{obj}(x)$ and $Im_R(x)$.

To solve the task by the Fourier holography technique, we apply the Fourier transform to expression (13.1) twice, and, taking into account the inversion of the coordinates, we obtain

$$Im_{pred}(x_k) = F\{F[Im_{obj}(x_0 - x)] F[a(x)]\}. \qquad (13.3)$$

In the Fourier domain, Eq. (13.2) takes the form

$$F[a(x)] F[C(|x - \xi|)] = F[C(x_k + \xi)].$$

This implies that

$$F[a(x)] = \frac{F[C(x_k + \xi)]}{F[C(|x - \xi|)]}. \qquad (13.4)$$

Substitution of (13.4) into (13.3) yields

$$Im_{pred}(x_k) = F\left\{F[Im_{obj}(x_0 - x)] \frac{F[C(x_k + \xi)]}{F[C(|x - \xi|)]}\right\}.$$

Applying the Wiener-Khinchin theorem, we obtain the final expression for the best linear estimate of the function value at the point x_k, i.e., the linear predictor

$$Im_{pred}(x_k) = F\left\{\frac{F[C(x_k + \xi)]}{F^*[Im_R(x)]}\right\}, \qquad (13.5)$$

where the asterisk denotes the complex conjugation.

Expression (13.5) can be implemented by the Fourier holography technique by a procedure that consists of the following two stages:

1. The first one is to form the correlation function; this is implemented by a classical Van der Lught correlator[10].
2. The second stage is to obtain the prediction by propagation of the light from the part of the correlation function $C(x_k + \xi)$ in the reverse direction C→H→*Im* through the inverse hologram, i.e., the hologram with transfer function

$$H_{inv}(v) = \frac{1}{F[Im_R(x)]} .$$

This step is implemented by placing a phase-conjugate mirror in the correlation domain.

Thus, the model of the linear predictor, which is the best predictor for stationary stochastic processes, can be implemented by modification of the setup, presented in Fig. 13.1, namely:

1. Side maxima of the correlation function have to be used (instead of autocorrelation peak in associative memory) to restore the hologram by light back propagation.
2. The hologram for the second step has to be an inverse one.

From expression (13.5), it is obviously that only side maxima of the correlation function are used in the model, as is shown in Fig. 13.3 (only the down part of the scheme is shown in order to simplify the picture). Also, it is obvious that if the hologram has the property of angular invariance, which is needed to form the full correlation function at step 1, then, the requirement for selecting at the second step the part $C(x_k + \xi)$ from the correlation function automatically holds for all indexes $k \in [0, N_R]$, where N_R is the number of pixels in the reference image.

The implementation of expression (13.5) assumes that the condition $\xi \in [x_{min}, x_0]$ holds, or in the discrete formulation, $k \in [0, N_R]$. Figures 13.2(b) and 13.3 illustrate that the *k*th neuron in *Im* receives excitation from $(N_R - k)$ C neurons and it is connected through them with $(N_R - k)$ *Im* neurons activated by $Im_R(x)$. In other words, the condition $\xi_k \in [0, (N_R - k)]$ is implemented; i.e., the prediction for the depth k is calculated based on $(N_R - k)$.

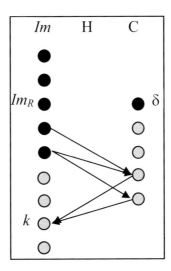

Figure 13.3 Neural network model of the holographic linear predictor.

13.2.3 Selective attention and cognitive drift phenomenon realization

Attention plays a central role in the phenomenon of perception.[22] It is known[13,15] that when a scene is analyzed, attention is to be paid to the new, unknown yet, or changed parts of the scene because these parts of the scene may bring either new information, or a danger. This phenomenon is known in cognitive psychology as "cognitive drift" and it is realized by using the mechanism of prediction[11] by comparing both the predicted and perceived scenes. In this section we demonstrate how the above-described model of the holographic predictor allows this phenomenon to be implemented.

Let that the analyzed scene $Im_{obj}(x)$ is described as follows:

$$Im_{obj}(x) = Im'_R(x) \cup Im_N(x),$$
$$Im'_R(x) \in Im_R(x),$$
$$Im_N(x) \notin Im_R(x).$$

The field, reconstructed by the setup in part of output plane, that corresponds to the "known or unchanged part of the scene" $Im'_R(x)$, can be represented as the sum of the interpolation $Im'_{Rinter}(x_k)$ and extrapolation (prediction) $Im'_{Npred}(x_k)$, i.e.,

$$Im'_R(x_k) = Im'_{Rinter}(x_k) + Im'_{Npred}(x_k) =$$
$$= F\left\{\frac{F[C_{R'R}(x_k+\xi)]}{F^*[Im_R(x)]}\right\} + F\left\{\frac{F[C_{NR}(x_k+\xi)]}{F^*[Im_R(x)]}\right\} \quad (13.6)$$
$$k \in [-N_{R'}, -N_N].$$

Accordingly, the field in the area of the output plane that corresponds to "changed, unknown" image $Im_N(x)$ can be represented as the sum of the extrapolation $Im'_{Rpred}(x_k)$ and interpolation $Im_{Ninter}(x_k)$, i.e.,

$$Im_N(x_k) = Im'_{Rpred}(x_k) + Im_{Ninter}(x_k) =$$
$$= F\left\{\frac{F[C_{R'R}(x_k+\xi)]}{F^*(Im_R(x))}\right\} + F\left\{\frac{F[C_{NR}(x_k+\xi)]}{F^*[Im_R(x)]}\right\}, \quad (13.7)$$
$$k \in [-N_N, 0].$$

By Eq. (13.1), the term $[C_{R'R}(x_k+\xi)]$ in expression (13.6) includes $[C_{R'R}(0)]$; i.e., it contains the autocorrelation peak. At the same time, on formation of the field described by (13.7), the estimate of $Im'_{Rpred}(x_k)$ with respect to $Im_R(x)$ does not contain the term $[C_{R'R}(0)]$. Since, according to the task conditions, both $Im'_R(x)$ and $Im_N(x)$ are realizations of the same stationary process, the means of (13.6) and (13.7) are the same. Hence, it follows that if the autocorrelation peak is rejected at the stage of the light back propagation C→H→Im, then the mean of (13.6) decreases, while the mean of (13.7) remains approximately the same. Therefore, the average amplitude of the field in the area of "new, unknown" $Im_N(x)$ is greater than in the area of the known part $Im'_R(x)$. In other words, the area of the new part of the scene is highlighted. However, the reconstructed field is neither $Im_N(x)$, nor $Im'_R(x)$.

In order to approximately estimate the ratio of mean intensities of the reconstructed fields in both areas, we consider the restoration of the inverse hologram by the correlation field as the computation of the cross-correlation of $C(x_k+\xi)$ and $F\{F[1/Im(x)]\}$. Taking into account the results, presented by Shubnikov,[21] to a first approximation, the average square of the cross-correlation is described by the expression

$$\mu^2 \approx 2kW_{obj}^2 \frac{S_C}{S_R},$$

where k is a coefficient depending on the correlation function, S_C and S_R are the areas of the reference and correlation, respectively, and W_{obj} is the energy of the object image. The signal-to-noise ratio in the correlation field is estimated by

$$V \approx \frac{1}{2k}\frac{S'_R}{S_C},$$

where S'_R is the area of the fragment $Im'_R(x)$. From this expression, omitting the simplest transformations, we can estimate the ratio of the average intensities of the reconstructed fields $Im'_R(x)$ and $Im_N(x)$ of the boundary of these areas by the following expression:

$$\left\langle \frac{(Im'_R)^2}{(Im_N)^2} \right\rangle \approx \left[\frac{S'_R \sqrt{2k(S_C/S_R)} + S_C}{S_R} \right]^2. \qquad (13.8)$$

It follows from (13.8) that the efficiency (with respect to energy) of the highlighting of the part $Im_N(x)$ in the reconstructed field depends both on its weight in the object image $Im_{obj}(x)$ and on the ratio S_C/S_R, which is the parameter inversed to the generalized spatial frequency; the latter is the main parameter describing the information capacity of the image.

It follows from (13.5) that if the global maximum of the autocorrelation function is rejected, a continuous estimate consisting of $Im_{Ninter}(x_k)$ and $Im_{Npred}(x_k)$ can be constructed; i.e., in the recovered image, the boundary (as an intensity drop) between the area of unknown part $Im_N(x)$ and the prediction $Im_{obj}(x_k)$ has less contrast than that between $Im'_R(x)$ and $Im_N(x)$ described by (13.8), and the mean intensity of the field decreases more gradually.

13.3 Chaotic Dynamics—Can Holography Help to Solve Creative Problems?

The above-considered model belongs to the class of dynamic systems with convergent dynamics, i.e., the system, once activated by an input pattern, converges to the stable state of two-pattern reverberation, namely, either reference pattern, prediction (13.5), or highlighted $Im_N(x)$ in the image plane, and correlation function in the correlation plane. The convergent dynamics allows either stale (associative memory) or "simple creative" problems (prediction) to be solved. However, convergent dynamics does not allow a "complex creative problem" to be solved because to solve the latter kind of problem is to create new knowledge. By the approach that knowledge is implemented by a hologram, both of the former types of solutions are reached by using existing knowledge, i.e., the hologram once being recorded is not rewritten. In this section we will try to take the next step—to find a way for application of the holographic technique to "complex creative problem" solving.

To demonstrate how the above-described holographic predictor can be applied to implement creative thinking, here we consider in brief the human way of "complex creative problem" solving.[15,17,24] Two ways are used to solve the problem, as follows:

1. *To view the problem from another point of view.* By the discussed technique, this means to change the pattern of the problem condition in order that the changed pattern be recognized by existing knowledge.

2. *To change the knowledge.* To try to change existing knowledge in order for the pattern condition to be recognized by the changed knowledge. And it has to be emphasized that the changing has to be a "virtual" one. This means that the hologram has not been rewritten until the reached solution is verified to be adequate by the real world. If the last condition is not met, then it is madness—this is the point where genius differs from madman.

It is useful to combine both methods in order to solve the problem more efficiently.[15,24] It is well known in cognitive psychology[13,16,17] that to solve a "complex creative problem" the brain generates a lot of patterns by using chaotic dynamics. These patterns have small differences from one another; the set of these patterns forms a "strange attractor."[25] The pattern solution is to be selected from the set under the criterion of being adequate to the real world.[24]

Thus, the method of problem solving is divided into two stages, namely, generating of the set of the patterns by chaotic dynamics, and selecting from the set of unique pattern solutions. We are concerned with holographic implementation of the first stage only.

One of the ways to change system dynamics from a convergent one to chaos and back is to use the Feigenbaum scenario.[25] This scenario is based on parametric mapping usage. Let $T(x,a)$ be a mapping $T: X \to X$, where a is a parameter. Thus, $x_{n+1} = T(x_n, a)$, where n is a number of iteration. By some nonlinear mappings, the changing of the parameter a leads to the dynamics of the system shift from a convergent one, to the cyclic dynamics, and then the chaotic one.

The scenario can be implemented by the previously discussed setup, presented in Fig. 13.1., by using a nonlinear phase-conjugate mirror in the correlation plane if the transfer function of the mirror meets the conditions on the mapping for the Feigenbaum scenario to be realized. Then, a pattern, generated by $n + 1$ iteration in the output plane, is described by the expression

$$Im_{n+1}(x_k) = F\left(\frac{F\{T[C_n(x_k + \xi), a]\}}{F^*[Im_R(x)]} \right), \qquad (13.9)$$

where parameter a is the writing-to-reading beams ratio.

This method meets the above-mentioned conditions; since the mapping is applied to the correlation function, both the pattern of the task condition and existing knowledge are changed, and the knowledge is changed while the hologram is not rewritten—the condition on the stability of existing knowledge, until the solution is verified, is met.

However, in order to be "intelligence," the system has to have the ability for self-tuning on the type of the dynamics, i.e., if the problem is a routine one, then the system has to realize either convergent or cyclic dynamics; if the problem is complex, then the system has to realize chaotic dynamics. The classical approach[23] is to change the value of parameter a in order to change the type of dynamics. However, in the discussed setup, the parameter does not depend on the kind of the task to be solved by the system; therefore, it is problematic to implement the idea of self-tuning on the kind of problem by the parameter changing. To implement the concept, we propose another approach.

Let us go back to the problem classification and analyze the predictor interconnections. By the discussed setup, if the problem is a routine one, then the signal-to-noise ratio in the correlation plane is high, and the system has to implement the model of autoassociative memory; in the correlation plane, only an autocorrelation peak is needed in order to implement the memory. If the problem belongs to the class of "simple creative task," then the ratio is smaller; this means the system has to realize the model of a linear predictor. In the correlation plane, side maxima of the correlation function are needed to reach this sort of solution. And, if the problem is "complex creative task," then the ratio is smallest and the system has to realize chaotic dynamics. In the terms of the Feigenbaum scenario, this means the type of needed dynamics depends on the signal-to-noise ratio in the correlation plane. Schematically, this is presented in the Fig. 13.4—if the process starts from the dynamic range, marked by "Conv," then it is convergent dynamics; if the process starts from the range marked by "Chaos," it is chaotic dynamics. Therefore, in order to self-tune on the needed type of dynamics, the kind of dynamics has to depend on the start point.

Let us define the conditions on the mapping $T : X \to X$ for the self-tuning on the needed type of dynamics to be implemented. Let for some values of a there is a range $X_a = X_{aC} \cup X_{aL} \cup X_{aR}$, and an orbit of period m $\{x_a^m\} \in X_a : x_{a,n+m} = x_{a,n}$. In other words, there is a cyclic dynamics for the values of a on the range X_a. For simplicity, but without loss of generality let $m=2$. This, the orbit $\{x_a^m\}$ consists of two points x_{aL}^2 and x_{aR}^2. Let $T^m(x_n, a)$ be a mapping of order m, that brings the orbit $\{x_a^m\}$. It means, that

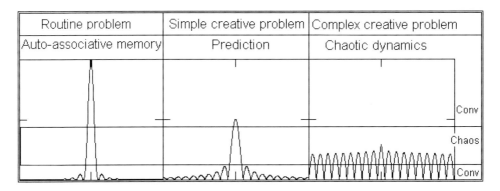

Figure 13.4 Schematic diagram of amplitude distribution in the correlation plane.

$$x_a^1 \in X_{aC}; \forall x \in X_{aC} : |T'(x_a)| > 1, T(x) \in X$$
$$x_{aL}^2 \in X_{aL}; \forall x \in X_{aL} : |T'(x_{aL})| < 1 \quad ,$$
$$x_{aR}^2 \in X_{aR}; \forall x \in X_{aR} : |T'(x_{aR})| < 1$$

Where $T'(x_a)$ - is a derivative in the point x_a. That means, the process, once starting from any point $x_1 \in X_a$, converges to the orbit x_a^2. Analogously, the conditions can be formulated for any m. Then, if there are such intervals X_{aLR} and X_{aRL} on the universe X, that

$$X_{aLR} \cap X_a = \emptyset$$
$$X_{aRL} \cap X_a = \emptyset \quad ,$$

and exists an orbit $\{x_a^k\} \in X_{aLR} \cup X_{aRL}$ with perion k, that Feigenbaum scenario is implemented by starting from the intervals X_{aLR} and X_{aRL}, then the properties of the mapping depend on the start point x_1. If the start point x_{a1} belongs to the interval X_a, then the dynamics will be either convergent, or cyclic one with periodт m. If the start point x_{a1} belongs to the range $X_{aLR} \cup X_{aRL}$, then, by the corresponding value of a, the dynamics of the system will be chaotic one.

Therefore, to implement the idea of self-tuning on the kind of problem, it is necessary to find the transfer function of the phase-conjugate mirror in the correlation plane, which meets the above-presented conditions. Fortunately, it was found that the transfer function of the spatial light modulator based on the photoconductor liquid-crystal structure, developed and investigated by Amosova et al.,[26] meets these conditions.

13.4 Experimental Illustration and Computer Simulation

An experimental setup was assembled using the scheme of Fig. 13.1. Since the drop in amplitudes in the correlation field substantially exceeds the dynamic range of the holographic recording medium, which is used to implement the phase-conjugate mirror in the correlation plane, then the phase-conjugate mirror was situated behind the correlation plane at an empirically chosen distance for which the amplitudes were distributed approximately in the same way over the beam section. The scheme adopted provided the ability for three models of the associative processing to be implemented by the same hologram and phase-conjugate mirror, by choosing a nonlinear operator in the correlation plane, as follows:

1. *Autoassociative memory.* The nonlinear filter passes only the autocorrelation peak, just as it is in classical works.[7-9]
2. *Linear predictor.* As follows from (13.5), only side maxima of the correlation function take part in the prediction constructing, At the same time, the brightness of the restored reference image exceeds the

brightness of the prediction, which fails the TV sensor. Therefore, in order to register the prediction, at the stage of the inverse hologram reconstruction by light back-propagation from the correlation plane, the autocorrelation peak was rejected.

3. *Selective attention and cognitive drift.* The nonlinear filter rejects the autocorrelation peak and passes side maxima of the correlation field only.

To provide the intensity of the reconstructed image that was sufficient for its registration by TV sensor, the Fourier hologram was recorded as a narrowband filter, instead of using both "direct" and inverse Fourier holograms. This led to narrowband filtering in the reconstructed image.

Aerial images of a homogenous forestland were used in the experiments, since it is this kind of natural object that is the closest to stationary stochastic fields in its statistic properties. The dimension of the reference image Im_R was 650×650 pixels. The object image consisted of both part of the reference one Im'_R that had dimension 650×450 pixels, and changeable part Im_N whose dimension was 650×200 pixels. The latter was another part of the same forestland.

The task of prediction was the task to restore the parts of the image that were not submitted to the setup, neither at the stage of the hologram recording by reference image nor at the stage of the hologram restoration by the object one, i.e., the parts were outside of the frame gate. Figure 13.5. presents the photometric profiles of the original image, which was outside the frame (white curve), and the predicted one, reconstructed by the setup (black curve). The x-axis presents pixels, while the y-axis specifies the brightness (in relative units). It is important to take into account the narrowband filtering of the reconstructed prediction by the setup, while the original image was not filtered.

Figure 13.6 (left-hand side) presents the image that was used as an object in order to implement the cognitive drift phenomenon; changeable part Im_N is at the top. The right-hand side presents the image of the field, reconstructed under a

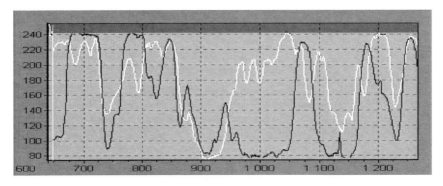

Figure 13.5 The photometric profiles of the real image (white curve) outside the frame and the prediction (black curve).

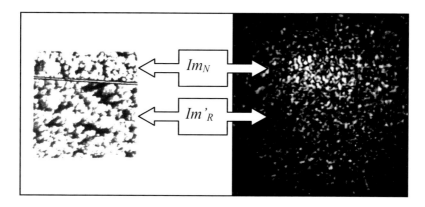

Figure 13.6 The image used in the experiment and an implementation of cognitive drift.

rejection of the autocorrelation peak. The part Im_N has a higher brightness against the background, including part Im'_R. And, as it was mentioned above, the image cannot be recognized. Also, as it was discussed, the effect of the new part highlighting (13.8) is visible mainly on the boundaries of the areas Im'_R and Im_N.

Analogous results were experimentally obtained for another kind of images, which cannot be treated as realizations of stationary stochastic fields.

In order to simulate chaotic dynamics, an experimentally obtained transfer function of a holographic reversible recording medium on the photoconductor liquid-crystal structure presented by Amosova et al.[26] was used. This experimentally obtained transfer function has the peculiarities discussed in Section 13.3 that are needed to implement the concept. Figure 13.7 presents the bifurcation diagram that was calculated for the transfer function; start point was $x_1 = 40$.

The same diagram for the same transfer function, but when the start point was 100, is presented in Fig. 13.8(a). Figure 13.8(b) shows the dependence of the fixed points for the first-order mapping (curve 1) and for the second-order

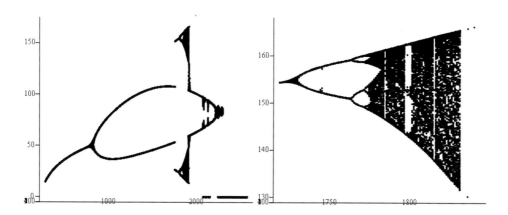

Figure 13.7 The bifurcation diagram and its part for start point $x_1 = 40$.

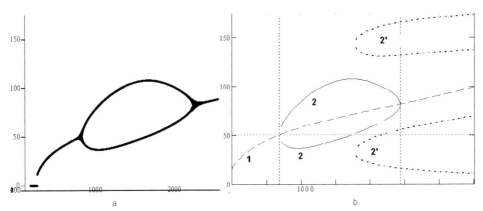

Figure 13.8 (a) The bifurcation diagram for start point $x_1 = 100$. (b) The fixed points' dependences on the parameter for second-order mapping.

mapping (curves 2 and 2') on the value of the mapping parameter. It is seen from the Fig. 13.8(b) that the dependence of the type of system dynamics on the start point is brought about by the additional four fixed points, presented by curves 2'. Although the "main" fixed points for the second-order mapping, presented by curves 2, are attractive, the new fixed points are attractors for some range of the parameter value, then some lose this property and become repellers, which leads to the chaotic dynamics of the Feigenbaum scenario. Thus, if the start point occurs within the main fixed points' (curves 2) neighborhood, then the dynamics will be cyclic with period 2. If a point occurs within the neighborhood of additional fixed points (curves 2'), then the dynamics may be chaotic.

In order to demonstrate the dependence of the dynamics type on the start point, Fig. 13.9 presents the convergence of the process for iterations 40 and 200 with dependence on the start point, which is scaled on the *x*-axis; the value of the mapping parameter was 1826. It shows, once starting from some points, that the process stabilizes in cycle with period two, and the dynamics is cyclic one. But once starting from other points, the process does not stabilize in any point, but fills all points in two limited areas, and the system demonstrates chaotic dynamics.

Thus, in order to allow the holographic system to self-tune on the type of solution, it has to provide for the hitting of the autocorrelation peak into the range, starting from which the system demonstrates the needed kind of dynamics. If cyclic dynamics, presented by the bifurcation diagram in Fig. 13.8, is needed, the peak has to hit into ranges either A or B in Fig. 13.9. If the peak is too weak, then it means the pattern condition is not recognized; therefore, lateral maxima of the correlation function have to hit into the range, starting from which the system demonstrates chaotic dynamics (C in Fig. 13.9). The last type of dynamics is presented by bifurcation diagram, shown in Fig. 13.7. Chaotic dynamics here fills in two subareas. According to neuropsychological studies[16], cyclic rather than convergent dynamics is demonstrated by biological systems.

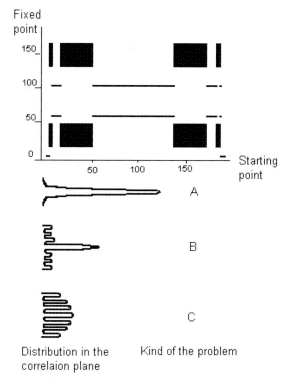

Figure 13.9 The convergence of the process.

13.5 Conclusion

Based on the consideration of Fourier holography with phase-conjugate mirrors in both image and correlation planes as a two-layer neural network, we have demonstrated theoretically and experimentally that the setup can implement a model of the linear predictor, which is the best predictor for stationary stochastic processes. This mode of the setup working can be considered as complementary to well-known holographic associative memory because only the autocorrelation peak is used in order to implement the memory, while only side maxima of the correlation function are used in order to implement the predictor. It has to be emphasized that a thin hologram is to be used in order to implement the model since angular invariance of the hologram is needed. At the same time, angular selectivity is the property that allows a high capacity of the memory to be reached. Here, there is some contradiction between the "intellectual" ability of the setup as an ability to predict, and ability to store a lot of information.

Based on the linear predictor implementation, we have demonstrated theoretically and experimentally an ability of the setup to implement the phenomenon of cognitive drift, which plays a major role in the phenomenon of perception. It is the ability to focus attention on the new information that determines the ability of an autonomous intellectual agent to adapt to the outer world.

Then, we have proposed an approach for the above-mentioned results to be applied to creative problems solving. We determined the type of transfer function needed in order to allow the system to implement the type of dynamics needed for the problem to be solved. Fortunately, we have found the experimentally obtained transfer functions, which are needed for this task, and used these functions for computer simulation.

Thus, we have made a few steps from holographic associative memory to a more complex model of information associative processing, which corresponds to the attributes of biological intelligence; namely, we implement abilities to predict, to focus attention on the new information, and to be deep in chaos but still come back to solve creative problems. The last was simulated computationally.

Acknowledgment

The author thanks Prof. O.P. Kuznetsov and Prof. I.B. Fominykh for useful discussions, and Dr. A.N. Chaika for experimental results used in computer simulation. Some parts of this research were supported by the Russian Foundation for Basic Research, grant 03-01-00825-a, and by the Russian Ministry for Science and Education.

References

1. Leith, E.N., and Upatnieks, J., "Reconstructed wavefronts and communication theory," *J. Opt. Soc. Am.*, **52**, 123–1130, 1962.
2. Leith, E.N., and Upatnieks, J., "Wavefront reconstruction with diffused illumination and three-dimensional objects," *J. Opt. Soc. Am.*, **4**, 1295–1301, 1964.
3. Denisyuk, Y.N., "Photographic reconstruction of the optical properties of an object in its own scattered field," *Sov. Phys. Dokl.*, **7**, 543–545, 1962.
4. Pribram, K.H., "Some dimensions of remembering; steps toward a neuropsychological model of memory," *Macromoiecules and Behavior*, J. Gaito, Ed., p. 165–187, Academic Press, New York (1966).
5. Pribram, K.H., Nuwer. M., and Baron, R., "The holographic hypothesis of memory structure in brain function and perception," *Contemporary Developments in Mathematical Psychology*, R.C. Atkmson, D.H. Krantz, R.C. Luce, and P. Suppes, Eds., p. 416–467, Freeman, San Francisco (1974).
6. Gabor, D., "Associative holographic memories," *IBM J. Res.Dev.* **13**, 156–159, 1969.
7. Farhat, N.H., Psaltis, D., Prata, A., and Paek, E., "Optical implementation of the Hopfield method," *Appl. Opt.*, **24**, 1469–1475, 1985.
8. Owechko, Y., Dunning, G.J., Marom, E., and Soffer, B.H., "Holographic associative memory with nonlinearities in the correlation domain," *Appl. Opt.*, **26**, 1900–1911, 1987.

9. Owechko, Y., "Nonlinear holographic associative memories," *IEEE J. Quant. Electron.*, **25**, 619–634, 1989.

10. Yu, F.T.S., and Suganda, J., *Optical Signal Processing, Computing, and Neural Networks*, Wiley, New York (1992).

11. Moeller, R., and Gross, H.-M., "Perception through anticipation," *Proc. of PerAc94*, p. 408–411, IEEE, Piscataway, NJ (1994).

12. Kay, L.M., Freeman, W.J., and Lancaster, L., "Limbic markers during olfactory perception," *Advances in Processing and Pattern Analysis of Biological Signals*, I. Gath and G. Inbar, Eds., Plenum, New York (1995).

13. Varela F.J., "Patterns of Life: Interwining Identity and Cognition," *Brain Cognition*, **34**, 72–87, 1997.

14. Tsuda, I., "Chaotic itinerancy as a dynamical basis of hermeneutics in brain and mind," *World Futures*, **32**, 167–184, 1991.

15. Golitsyn, G.A., and Petrov, V.M., *Information and Creation*. Birkhauser Verlag, Basel (1995).

16. Freeman, W.J., "The physiology of perception," *Sci. Am.*, **264**, 78–85, 1991.

17. Combs, A., "Psychology, chaos, and the process nature of consciousness," *Chaos Theory in Psychology*, F. Abraham and A. Gilden, Eds., Greenwood Publ., Westport, CT (1994).

18. Goertzel, B., *Chaotic Logic*, Plenum, New York (1994).

19. Freeman, W.J., Kozma, R., and Werbos, P.J., "Biocomplexity: adaptive behavior in complex stochastic dynamical systems," *Biosystems*, **59**, 109–123, 2001.

20. Perez Velazquez J. L., "Brain, behaviour and mathematics: Are we using the right approaches?" *Physica D*, **212**, 161–182, 2005.

21. Shubnikov, E.I., "On the signal-to-noise ratio for pattern recognition," *Opt. Spectrosk.*, **62**, 450–456, 1987.

22. Grimmet, G.R., and Sterzaker D.R., *Probability and Random Processes*, Oxford Sc. Publ., Claredon Press, Oxford (1992).

23. Nelson, C., *Attention and Memory. An Integrated Framework*, Oxford Psychology Series, 26, Oxford Sci. Publ., Oxford (1995).

24. Fominykh, I.B., "Creative problems and artificial intelligence systems," *Proceedings of 2002 IEEE International Conf. on Artificial Intelligence Systems*, Los Alamitos, CA, IEEE, Piscataway, NJ (2002).

25. Crownover, R.M., *Introduction to Fractals and Chaos*, Jones & Bartlett, London (1995).

26. Amosova, L.P., Pletneva, N.I., and Chaika, A.N., "Nonlinear regime of the reversible recording of holograms on photoconductor-liquid-crystal structures with high sensitivity to He-Ne laser radiation," *J. Opt. Technol.*, **72**, 469–473, 2005.

Chapter 14
Division of Recording Plane for Multiple Recording and Its Digital Reconstruction Based on Fourier Optics

Guoguang Mu and Hongchen Zhai
Nankai University, China

14.1 Introduction
14.2 RDM in Pulsed Holography Recording Ultrafast Events
    14.2.1 RDM implemented by ADM
    14.2.2 RDM implemented by WDM
14.3 RDM by Encoding Grating Recording Color Objects
    14.3.1 Principle and theoretical analysis
    14.3.2 Experiment and results
14.4. Conclusions
References

14.1 Introduction

Multiplexing techniques, such as angular division multiplexing,[1,2] wavelength division multiplexing,[3] etc.,[4] have been widely applied to different optical information processes for an efficient, optimized, or smart utilizing of the processing components. In this communication, we introduce a new concept, namely, recording-plane division multiplexing (RDM), if it is acceptable,[5] and its relevant experimental approaches, which can be implemented in different domains to carry out multiplex recordings of different parameters on a single recording plane.

As introduced in the Section 14.2, the multiple recordings will be implemented by pulsed holography, in which angular division multiplexing (ADM) or wavelength division multiplexing (WDM) can be applied for an ultrafast recording with a high resolution in the time domain. In Section 14.3, the implementation of multiple recordings of color patterns will be described, for which patterns with different element color will be recorded in a single monochromatic recording plane by the help of a specially designed color-encoding grating. Experiment and digital reconstructions results will be reported, respectively, in the above sections, in which Fourier transformation and filtering are used to obtain separated images, and display of a series of images with a high resolution in the time domain, or a fused image with multiple colors can be obtained at the end of the processing. A conclusion and discussion will be given in the last section.

14.2 RDM in Pulsed Holography Recording Ultrafast Events

14.2.1 RDM implemented by ADM

14.2.1.1 Principle and theoretical analysis

Figure 14.1 shows the principle of RDM recordings in pulsed holography on a single frame of a CCD. The reference beams composed of a series of subpulses with different incident spatial angles and object beams composed of a series of subpulses with the same incident spatial angle are employed to record a series of subholograms overlapped on a single frame of CCD, as shown in Fig. 14.2, if the time duration of all the recordings is shorter enough than the frame duration of the CCD.

Suppose that each of the object beams, S_n, on the recording plane at different time can be expressed as

$$S_n = |S_n|\exp(i\phi_n), \quad n = 1, 2, 3... \tag{14.1}$$

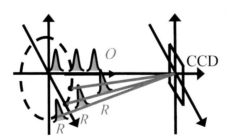

Figure 14.1 RDM implemented by ADM recording.

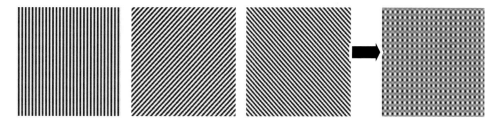

Figure 14.2 Subholograms of RDM recorded by ADM and their overlapped result.

and each of the reference beams, R_n, matching a corresponding subject beam at different time with different incident angles is

$$R_n = R_0 \exp[-j2\pi/\lambda(\sin\alpha_n x + \sin\beta_n y)] \ . \tag{14.2}$$

The intensity of the composite hologram after n times of holographic recordings will be

$$\begin{aligned}I &= \sum_n |S_n + R_n||S_n + R_n|^* = \sum_n \{|S_n|^2 + |R_n|^2 \\ &+ S_n \times R_0 \exp[j2\pi/\lambda(\sin\alpha_n x + \sin\beta_n y)] \\ &+ S_n^* \times R_0 \exp[-j2\pi/\lambda(\sin\alpha_n x + \sin\beta_n y)]\}\end{aligned} \tag{14.3}$$

In the reconstruction, the hologram will be Fourier transformed into its Fourier plane as separated spectra as shown in Fig. 14.3, which can be filtered and reconstructed by inverse Fourier transformation and be displayed as a series of independent images. The time resolution of the display depends on the pulse duration and the repeat frequency of the subpulses, which can be as high as femto seconds.

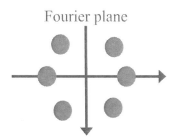

Figure 14.3 Fourier transformation of the RDM recording.

Ignoring the zero-order term, the Fourier transform of the composite hologram can be expressed as

$$\begin{aligned}F(I)=&\sum_n\left(R_0F\left\{S_n\exp\left[\frac{-j2\pi}{\lambda(\sin\alpha_nx+\sin\beta_ny)}\right]\right\}\right.\\&\left.+R_0F\left\{S_n^*\exp\left[\frac{j2\pi}{\lambda(\sin\alpha_n+\sin\beta_n)}\right]\right\}\right)\\=&\sum_n\left[R_0F(S_n)\otimes\delta\left(f_x-\frac{\sin\alpha_n}{\lambda},\ f_y-\frac{\sin\beta_n}{\lambda}\right)\right.\\&\left.+R_0F(S_n^*)\otimes\delta\left(f_x+\frac{\sin\alpha_n}{\lambda},\ f_y+\frac{\sin\beta_n}{\lambda}\right)\right]\\=&\sum_n\left[R_0\tilde{S}_n\left(f_x-\frac{\sin\alpha_n}{\lambda},\ f_y-\frac{\sin\beta_n}{\lambda}\right)\right.\\&\left.+R_0\tilde{S}_n^*\left(f_x+\frac{\sin\alpha_n}{\lambda},\ f_y+\frac{\sin\beta_n}{\lambda}\right)\right],\end{aligned}\qquad(14.4)$$

where $\tilde{S}_n$ is the Fourier transformation of s_n, f_i is the coordinate of the Fourier plane. By choosing a suitable window function, $\tilde{S}_n$ can be obtained after filtering, from which the individual subject wave s_n can be then constructed after an inverse Fourier transformation as

$$S_n=F^{-1}(\tilde{S}_n)=|S_n|\exp(i\phi_n),\qquad(14.5)$$

with both amplitude $|S_n|$ and phase ϕ_n.

14.2.1.2 Experiment and results

The pulsed holographic experimental system of RDM implemented by ADM^2 is shown in Fig. 14.4, where the laser pulse output from a Ti: sapphire laser amplifier system combined with a half-wave plate is divided by a polarizing beam splitter into two parts, the stimulating beam and the recording beam. The former can be focused by lens L to stimulate an ultrafast event adjusted by a time delayer, and the latter will be further divided by the beam splitter BS_1 into two parts, for SPG_1 to generate the subpulse train of the object beam and for SPG_2 to generate that of the reference beam, respectively. Beam splitter BS_2 is used to couple the object beam and the reference beam optically into the recording plane of the CCD. Each of the SPGs are composed of a set of mirrors and beam splitters, which can be adjusted with the same time delays or not, but the orientation of each is so adjusted that all the incident subpulses on the recording plane from SPG_1 will be at different spatial angles, while that on the event to be recorded as well as that on the recording plane from SPG_2, however, will be kept

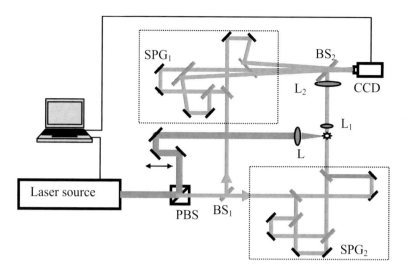

Figure 14.4 Pulsed holographic recording system.

the same, as shown in Fig. 14.1. This can ensure a successful RDM recording object at the same viewing angle. Furthermore, by tuning the mechanical stage over a range of microns to millimeters, the time delay between the subpulses in a pulse train can be adjusted from femtosecond to picosecond order, unless there is no overlap between them within their coherence time. The 4-f system composed of L_1 and L_2 is used to record amplified image holograms at the recording plane of a single frame of the CCD.

In the experiment, a Ti: sapphire laser amplifier system (Spitfire HP 50), with the maximum output energy of a single pulse of 2 mJ, a repeated frequency of 1 kHz, and an FWHM (full width at half maximum) of 50 fs at a wavelength of 800 nm, is used as an optical source for both inducing the ultrafast event of air ionization and its holographic recording. The focus lengths of lenses L1 and L2 are 1.5 cm and 15 cm, respectively, with which a 10 times magnified image hologram of the ionization region can be recorded by a CCD with pixel number of 576 × 768 and pixel size of 10.8 μm × 10 μm. The laser amplifier system and the CCD image capturer are synchronously controlled by a PC to have the output of a single laser pulse match the recording of a single frame of CCD in the time scale.

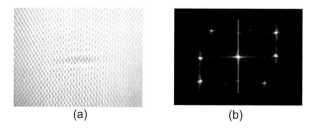

(a) (b)

Figure 14.5 Overlapped subholograms (a) and their Fourier frequency spectra (b).

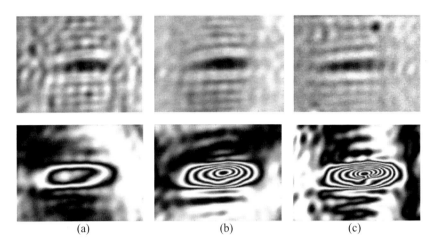

(a) (b) (c)

Figure 14.6 Intensity images (top) and their contour maps of phase difference (bottom) digitally reconstructed from the subhologram in Fig. 14.5(a) with an exposure time of 50 fs and frame interval $\geq t$ = 300 fs.

The overlapped subholograms and their Fourier frequency spectra are shown in Figs. 14.5(a) and 14.3(b), respectively. It is noticed that by employing the RDM with an ADM, three Fourier frequency spectra of the subholograms along different orientations are spatially well separated in the Fourier plane.

Their digitally reconstructed intensity images and their corresponding contour maps of phase difference are shown in Fig. 14.6, of which the exposure time is 50 fs and the frame intervals are 300 fs. This demonstrates clearly the gradient increase of electron density within the plasma on a femtosecond order.

14.2.2 RDM implemented by WDM

14.2.2.1 Principle and theoretical analysis

Figure 14.7 shows the principle of RDM recordings of WDM in pulsed holography on a single frame of a CCD, where a series of subpulses with different wavelengths and the same incident angles are employed to record a series of subholograms with different wavelengths overlapped on a single frame of a CCD, if the time duration of all the recordings is shorter enough than the frame duration of the CCD.

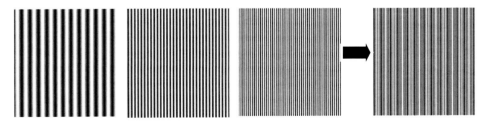

Figure 14.7 Subholograms of RDM recorded by WDM and their overlapped result.

Suppose each subject wave S_n recorded on the image plane of a CCD at different times can be expressed as

$$S_n = |S_n| \exp(i\phi_n), \qquad (14.6)$$

and the reference wave of each subject wave, R_n, at different times and the same incident angle is

$$R_n = R_0 \exp\left[-\frac{j2\pi \sin \alpha x}{\lambda_n}\right]. \qquad (14.7)$$

According to the theory of image holography, the intensity of the composite hologram after n times of holographic recording will be

$$I = \sum_n |S_n + R_n||S_n + R_n|^* = \sum_n \left\{ |S_n|^2 + |R_n|^2 \right.$$
$$+ S_n \times R_0 \exp\left[\frac{j2\pi \sin \alpha x}{\lambda_n}\right] \qquad (14.8)$$
$$\left. + S_n^* \times R_0 \exp\left[-\frac{j2\pi \sin \alpha x}{\lambda_n}\right] \right\}.$$

Ignoring the zero-order term, the Fourier transform of the composite hologram can be expressed as

$$F(I) = \sum_n \left\{ R_0 F\left[S_n \exp\left(-\frac{j2\pi \sin \alpha x}{\lambda_n}\right)\right] \right.$$
$$\left. + R_0 F\left[S_n^* \exp\left(\frac{j2\pi \sin \alpha}{\lambda_n}\right)\right] \right\}$$
$$= \sum_n \left[R_0 F(S_n) \otimes \delta\left(f_x - \frac{\sin a}{\lambda_n}, f_y\right) \right. \qquad (14.9)$$
$$\left. + R_0 F(S_n^*) \otimes \delta\left(f_x + \frac{\sin a}{\lambda_n}, f_y\right) \right]$$
$$= \sum_n \left[R_0 \tilde{S}_n\left(f_x - \frac{\sin a}{\lambda_n}, f_y\right) + R_0 \tilde{S}_n^*\left(f_x + \frac{\sin a}{\lambda_n}, f_y\right) \right],$$

where $\tilde{S}_n$ is the Fourier transformation of s_n. By choosing a suitable window function, $\tilde{S}_n$ can be obtained after filtering, from which the individual subject wave s_n can be then obtained after an inverse Fourier transformation,

$$S_n = F^{-1}(\tilde{S}_n) = |S_n|\exp(i\phi_n), \qquad (14.10)$$

with both amplitude $|S_n|$ and phase ϕ_n reconstructed.

14.2.2.2 Experiment and results

The experimental system of RDM implemented by WDM[3] is shown in Fig. 14.8, where the pulse output is divided by a polarizing beam splitter PBS into two parts, the inducing beam and the recording beam. The former can be focused by lens L_1 to induce the air ionization, and the latter is used to generate a harmonic wave, where wave plates P_1 and P_2 are used to adjust the polarization of the incident beam to obtain appropriate frequency-doubling efficiency from a piece of BBO crystal. The basic and doubled frequency waves will be separated by a dichromatic mirror DM_1 into two parts with their individual wavelengths and time delays, so that two subholograms based on WDM can be recorded at different times. From dichromatic mirror DM_2 on, the successive optical path based on Michelson's interferometer is equal for both of the two wavelengths, in which M_3 and M_4 are used to ensure the optical paths are exactly equal in both of the two arms, respectively, for object and reference. The 4-f system composed of L_2 and L_3 is used to obtain amplified image holograms of both amplitude and phase at the recording plane of the CCD.

In the experiment, the same pulsed laser source and micro-optical recording system as described above is used for ultrafast holographic recording, except that the recorded hologram at the image plane of the 4-f image system is composed of two separated subholograms in the same orientation and of different stripe spaces

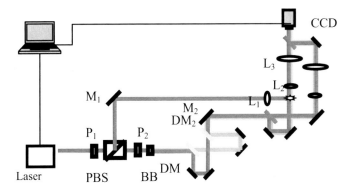

Figure 14.8 Pulsed digital holographic system of RDM implemented by WDM.

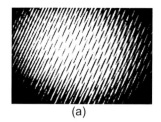 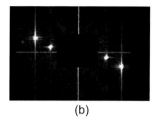

(a)　　　　　　　　　　　　　　(b)

Figure 14.9 Two overlapped subholograms of RDM by WDM and their Fourier frequency spectra. (a) Two overlapped subholograms. (b) Fourier frequency spectra of the subholograms shown in (a).

due to the different recording wavelengths. Therefore, their spatial resolution is only restricted by the pixel size of the CCD used in the experiment. That means that, in our case, the spatial resolution in each dimension is about 10 μm.

To show the RDM recording of the ultrafast process of air ionization by WDM, two overlapped subholograms are recorded by the system shown in Fig. 14.8. The overlapped subholograms and their Fourier frequency spectra are shown in Figs. 14.9 (a) and (b), respectively. It is noticed that by employing WDM, two Fourier frequency spectra of the subholograms are spatially well separated. After filtering the composed hologram in the Fourier frequency space, followed by an inverse Fourier transform, both the amplitude and phase are reconstructed, and their digitally reconstructed intensity and contour maps of phase difference are shown in Fig. 14.10, of which the exposure time is 50 fs and the frame interval is 400 fs, showing clearly the shadow area and distribution of the same density of electrons within the plasma induced by femtosecond laser pulse.

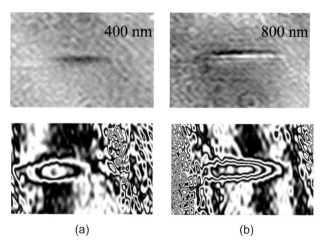

(a)　　　　　　　　　　　　　　(b)

Figure 14.10 Intensity images (top) and contour maps of phase difference (bottom). Exposure time: 50 fs ; frame interval: 400 fs.

14.3 RDM by Encoding Grating Recording Color Objects

14.3.1 Principle and theoretical analysis

A color-encoding grating[6] is a composition of three sub-Ronchi gratings with different element colors and oriented in different directions, as shown in Fig. 14.11, which can be recorded by holographic RDM with different color light sources on a piece of positive color film. This color-encoding grating can be employed to encode color images on a monochromatic recording plane for RDM recording by putting it at the surface of the monochromatic recording plane.[7]

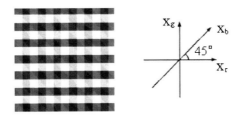

Figure 14.11 Color-encoding grating.

Suppose that the intensity transmittance of the color-encoding grating can be expressed in RGB space as

$$t(x,y;\lambda_r,\lambda_g,\lambda_b) = t_r(x,y;\lambda_r) + t_g(x,y;\lambda_g) + t_b(x,y;\lambda_b), \quad (14.11)$$

and the color object to be recorded can be expressed in RGB space as

$$O(x,y;\lambda_r,\lambda_g,\lambda_b) = O_r(x,y;\lambda_r) + O_g(x,y;\lambda_g) + O_b(x,y;\lambda_b). \quad (14.12)$$

The intensity of the encoded image, O_{en}, on the monochromatic recording plane by RDM can be then expressed as

$$O_{en}(x,y) = O_r(x,y)t_r(x,y) + O_g(x,y)t_g(x,y) + O_b(x,y)t_b(x,y), \quad (14.13)$$

where x and y are the coordinate parameters, and r, g, and b refer to one of the three element colors, respectively. The color object will be recorded by means of RDM on a monochromatic recording plane as overlapped subimages encoded by the sub-Ronchi grating of each element color, as shown in Fig. 14.12.

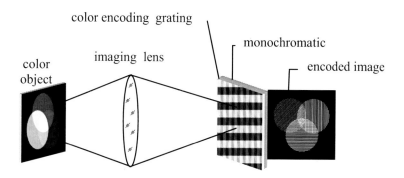

Figure 14.12 Recoding color object on a monochromatic recording plane by RDM.

During the decoding,[8] the encoded monochromatic image will be at first transformed as

$$F[O_{en}(x,y)] = F\left[O_r(x,y)t_r(x,y) + O_g(x,y)t_g(x,y) + O_b(x,y)t_b(x,y)\right]$$
$$= F(O_r) \otimes F(t_r) + F(O_g) \otimes F(t_g) + F(O_b) \otimes F(t_b)$$
$$= \tilde{O}_r \otimes \delta\left(f_x - \frac{1}{d}, f_y\right) + \tilde{O}_g \otimes \delta\left(f_x, f_y - \frac{1}{d}\right)$$
$$+ \tilde{O}_b \otimes \delta\left(f_x - \frac{1}{\sqrt{2}d}, f_y - \frac{1}{\sqrt{2}d}\right),$$

(14.14)

where $\tilde{O}_i$ is the Fourier transformation of the intensity of O_i, d is the grating period of the subgratings, and f_i is the coordinate in the Fourier plane. Obviously, the spectrum of each monochromatic subimage is located differently in the Fourier plane. It is possible, therefore, to put a filter with a corresponding element color, $C(\lambda_i)$, in the zero and the first order, respectively, and block all the other spectra of the higher frequencies in the Fourier transformation, so that the recorded monochromatic image can be decoded and fused into a color image by a followed inverse Fourier transformation, as expressed and shown in Eq. (14.15) and Fig. 14.13, respectively.

$$O(x,y;\lambda_r,\lambda_g,\lambda_b) = F^{-1}\left[(\tilde{O}_r)\right]C(\lambda_r)$$
$$+ F^{-1}\left[(\tilde{O}_g)\right]C(\lambda_g) + F^{-1}\left[(\tilde{O}_b)\right]C(\lambda_b)$$
$$= O_r(x,y;\lambda_r) + O_g(x,y;\lambda_g)$$
$$+ O_b(x,y;\lambda_b)$$

(14.15)

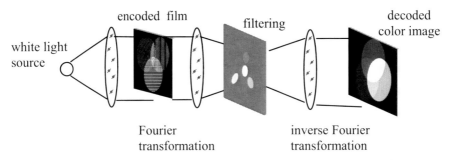

Figure 14.13 Decoding and fusion of a color image.

14.3.2 Experiment and results

In the experiment of the WDM for recording color objects,[9] an Airscape camera with an imaging lens of focus length 70 mm is used as the recording device, of which a specially designed color-encoding grating described in Fig. 14.11 with a size matching that of the recording film is placed on the plane of the black-and-white photo film of size 180 × 180 mm. The color object to be recorded is an overlook from a scout airplane. The recoded image by RDM of the color object encoded by the color-encoding grating in the black-and-white film is shown in Fig. 14.14(a), of which the encoding frequency of the subgratings on the photo film is measured at 30 lp/mm.

In the decoding of the image recorded by RDM, a digital reconstructing process based on an optical 4-f system with the same focal length, $f_1 = f_2 = 240$ mm, of the two lenses, L_1 and L_2, is carried out, of which P1 is the plane of the encoded image, P_2 is the Fourier plane for filtering, and the element wavelengths selected for the color filters are 700 nm, 550nm, and 430 nm. The decoded color image recovered from the RDM recording film[10] is shown in Fig. 14.14(b).

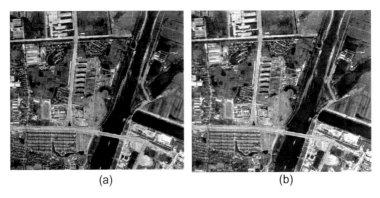

(a)   (b)

Figure 14.14 Encoded image on black-and-white film by RDM (a) and its decoded color object (b).

14.4 Conclusions

RDM can be applied to the implementation of multiple recordings on a single recording plane, for an ultrafast event, or for a recording of color objects on a piece of monochromatic recording material. ADM or WDM can be employed to record ultrafast events at subholograms with different orientations or that with the same orientation and different wavelength, on a single frame of the recording plane. In the recording of color objects, a specially designed color-encoding grating is employed to record multiple images on a single frame of a monochromatic recording material. In all of the above reconstruction processes of the RDM recording, Fourier transformation and frequency filtering in the Fourier plane have to be employed to separate the spatial spectra of the multiple recordings. In the case of the holographic recording of ultrafast events, each of the reconstructed images can be reconstructed by an inverse Fourier transformation and displayed as a series of images in an ultrafast process with a high resolution in the time domain on the order of femtoseconds; and in the case of color objects recording, the filters of element colors have to be inserted at the Fourier plane to decode and fuse the monochromatic subimages into a chromatic image by a followed inverse Fourier transformation.

In pulse holographic RDM recording, the time resolution of the exposure time is limited by the laser pulse, and that of the frame interval can be adjusted by any short value, on the condition that the adjacent subpulses arriving at the recording plane are not yet overlapped at all. In the RDM recording of color objects by a color-encoding grating, the spatial resolution of the decoded image is mainly limited by the spatial frequency and the fabricating quality of the color-encoding grating.

References

1. Liu, Z.W., Centurion, M., Panotopoulos, G., Hong, J., and Psaltis, D., "Holographic recording of fast events on a CCD camera," *Opt. Lett.* **27**, 22–24, 2002.
2. Wang, X., Zhai, H., and Mu, G., "Pulsed digital holography system recording ultrafast process of the femtosecond order," *Opt. Lett.*, **31**, 1636–1638, 2006.
3. Wang, X., and Zhai, H., "Pulsed digital micro-holography for femtosecond order ultra-fast process recording by using wavelength division multiplexing technology," *Opt. Comm.*, **275**, 42–45, 2007.
4. Keiser, G., *Optical Fiber Communcations*, Chap. 10, McGraw-Hill, New York (2000).
5. Hongchen Zhai, Xiaolei Wang, and Caojin Yuan, "Division of Recording Plane in Pulsed Digital Holography." *To be published in JOSA.*
6. Mu, G., Fang, Z., and Liu, F., "Color data image encoding method and apparent with image encoding method and apparatus with spectral zone-filter [p]," US Patent No. 545 2002 1995-2009 (1995).

7. Mu, G., Lin, L., and Wang, Z., "Multi-channel photography with digital Fourier optics," *Proc. ICO IV, International Trends in Optics and Photonics*, A. Toshimitsu, Ed., p. 86–94, Springer-Verlag, Berlin (1999).
8. Mu, G., Fang, Z., Liu, F., and Zhai, H., *A Physical Method for Colour Photography*, ICO Book, Vol. 3, Trends in Optics, A. Consoctini, Ed., p. 527–542, Academic Press, New York (1996).
9. Mu, G., Fang, Z., and Wang, X., "Color image display with blank-white film," *Electro-Optical Displays*, Ed. Mohammad A. Karim, Marcel Dekker Inc., New York, 187–210, 1992.
10. Mu, G., "White-light optical information processing and its color photography," *J. Optoelectron. Laser*, **12**, 285–292, 2001.

Chapter 15
Fundamentals and Advances in Holographic Materials for Optical Data Storage

María L. Calvo
Universidad Complutense de Madrid, Spain

Pavel Cheben
National Research Council Canada

15.1 Introduction
15.2 General Requirements for a Holographic Recording Medium
15.3 Photorefractive Inorganic Crystals
    15.3.1 Physical mechanisms
    15.3.2 Thermal fixing
    15.3.3 Electrical fixing
    15.3.4 Two-wavelength storage
15.4 Organic Photorefractive Materials
    15.4.1 Charge generation
    15.4.2 Charge transport
    15.4.3 Nonlinear optical properties
    15.4.5 Materials classification
15.5 Photopolymerizable Materials
    15.5.1 Physical mechanism
    15.5.2 DuPont photopolymers
    15.5.3 Polaroid photopolymers
    15.5.4 Lucent photopolymer
    15.5.5 Polymers functionalized with liquid crystals and optical chromophores
    15.5.6 Photopolymers developed in Russia

15. 6 Hybrid Organic-Inorganic Materials
   15.6.1 Vycor-type porous glasses impregnated with organic materials
   15.6.2 Sol-gel holographic materials
   15.6.3 Organic-inorganic nanocomposites incorporating high refractive index species
15.7 Conclusions
Acknowledgments
References

15.1 Introduction

The idea of using holograms for storage of information was first suggested by van Heerden[1] in 1963, who proposed to store data by recording the information carrying a light interference pattern in a holographic medium. He also predicted that the minimum volume necessary to record a bit of information is $\sim\lambda^3$, where λ is the wavelength used in the holographic recording. This involves an impressive density of data on the order of 10 Tbits/cm^3 for $\lambda \sim 400$ nm. In addition to high-density storage, holography permits short access times to the data since the direction of propagation of a light beam changes rapidly without inertia, unlike magnetic disk heads. Furthermore, a high data transfer speed is achieved since the complete sheet of information is recorded or read at the same time. Nevertheless, despite these advantages, after more than 40 years of research and development there are still not holographic drives in our personal computers. This is due principally to the lack of an adequate recording material.[2]

15.2 General Requirements for a Holographic Recording Medium

Until the present time, many holographic materials such as photopolymers, inorganic and organic photorefractive materials, dichromatic gelatin, silver halides, photoresists, sol-gel glasses, and thermoplastic, photochromic and photodichroic materials have been developed. Nevertheless, few of them potentially have the characteristics required for data storage applications. Among these, the most important are a sufficient thickness, high refractive index modulation, high sensitivity, excellent optical quality with low levels of scattering and absorption loss, dimensional stability during the recording of the hologram and its functioning in the memory, good thermal and chemical stability, and a moderate price.

Optical thickness of the material. A sufficient thickness of the material is needed, typically on the order of several hundreds of micrometers or greater, in order to ensure the diffraction in the Bragg regimen with high angular or spectral selectivity required for the multiplexing techniques. The volume holograms are very sensitive to deviations of the Bragg resonance, and even small changes in the refractive index or the thickness of material may destroy the latter. It is therefore important to ensure that the recording or development processes do not introduce significant changes in the average refractive index and sample thickness.

Refractive index modulation. In order to be able to multiplex the maximum number of holograms, materials capable of producing high refractive index modulation Δn, i.e., with a high dynamic range, are necessary. According to Kogelnik's coupled wave theory,[3] the diffraction efficiency of a phase grating is given by (in Bragg's condition)

$$\eta = \sin^2\left(\frac{\pi \Delta n d}{\lambda \cos\alpha}\right), \quad (15.1)$$

where d is the thickness of the material, α the angle of incidence of the reconstruction beam measured in the material, and λ the wavelength of the beam in a vacuum. For small refractive index modulations, $\eta^{1/2}$ grows nearly linearly with Δn. In holographic memories, instead of dedicating to a single grating the full dynamic range, the latter is divided between the N multiplexed gratings, assigning to each of them a small refractive index modulation, $\Delta n_i \ll \lambda\cos\alpha/2d$, obtaining a diffraction efficiency $\eta_i^{1/2} \sim d\Delta n_i$. It is observed that

$$\sum_{i=1}^{N} \eta_i^{1/2} \sim d \sum_{i=1}^{N} \Delta n_i = d\Delta n. \quad (15.2)$$

The quantity $\sum_{i=1}^{N} \eta_i^{1/2} = M\#$ is called the *M number*[4]. $M\#$ is a convenient parameter[i] in order to compare different materials since it is directly related to diffraction efficiencies of the multiplexed holograms and therefore with the energy of the reconstructed images. Unlike Δn, $M\#$ includes the contribution of the material thickness to the multiplexing capacity of the medium.

Sensitivity. The greater the sensitivity of the material, the fewer are the photons needed to record a hologram, which obviously reduces the recording time, thus increasing speeds or permitting low-power (thus less expensive) lasers to be used. Typically, the sensitivity is defined as the modulation of the refractive index induced by the exposure energy[ii] E per unit area

$$S_1 = \frac{\Delta n}{E} \; [\text{cm}^2/\text{J}]. \quad (15.3)$$

---

i. It is observed that $M\#$ is the number of holograms of efficiency $\eta = 1$ (100%) that can be recorded in a material of a given thickness.

ii. $E = It$, where I is the total intensity of the recording beams [in W/cm^2], and t the exposure time.

Optical quality. In commercial data storage systems, the probability that there is an error in the recovery of a bit, also called BER (bit error rate) is on the order of 10^{-15}, using algorithms of error correction (ECC, error correction codes). This corresponds to a BER $\sim 10^{-5}$ before using ECC. In order to reach this level of fidelity of data recovery in holographic memories, it is important to control various sources of noise.

The imperfections in surface or volume of the holographic medium can result in serious negative effects such as noise gratings and optical beam distortion. It is important to ensure a good homogeneity of the material both at the microscopic (minimized scattering) as well as at the macroscopic (minimizing the beam distortion) levels. Also, surfaces should have high optical quality and a minimum error of parallelism between both sides of the material, preferably less than a second of arc.[iii] The relevance of the different types of imperfections varies from one material to another or among the different families of materials. For example, in organic materials, the scattering is usually the principal factor limiting optical quality.

During the readout of a page of data selected from the various ones superimposed in the same volume of material by means of spectral or angular multiplexing, a small deviation of the Bragg condition may result in the partial readout (with lower efficiency), also called crosstalk, of the other images. Deviations of the Bragg condition can be caused by changes of the wavelength of the laser beam, angular errors of the tracking system, or changes of the refractive index and thickness of the holographic material; the latter being typical of photopolymers. The crosstalk also can deteriorate due to various types of noise gratings that are formed when the beams used during the recording of a hologram are diffracted by the holograms previously written and when the recording, diffracted, and reflected beams mutually interfere in the medium. Noise gratings can also result from the interference of the recording beams with the scattered field from defects in the surface of the medium, but also from the dopant molecules, such as sensitizers or the photoinitiator in a photopolymer. Similar effects are caused by fluctuations of the spatial field in a photorefractive crystal and the accompanying fluctuations of the refractive index since the ion centers that participate in the generation process and recombination of electrical charges are located randomly. Still other sources of noise are the harmonic gratings that originate due to the nonlinearities of holographic recording. The scattering by the imperfections of the surfaces or within the optical elements, the aberrations, or the misalignments of the optical system, as well as the diffraction caused by the optical apertures, also increase the level of holographic noise.

---

iii. For example, if the medium is in the form of a disk, the parallelism error will produce, upon turning the disk, an angular change of similar order (δ) in the direction of the propagation of the beam after passing through the material. These angular changes will produce transversal displacement of $f \times \delta$ in the image plane, f being the focal distance of the lens used in the holographic recording and readout.

15.3 Photorefractive Inorganic Crystals

In 1966, Ashkin and his collaborators at Bell Laboratories[5] observed changes optically induced in the refractive index of ferroelectric crystals of lithium niobate (LiNbO$_3$) and lithium tantalite (LiTaO$_3$). Two years after, Chen et al. suggested that this phenomenon, called the photorefractive effect, although originally considered as an optical damage, can actually be useful, for example, for holographic data storage.[6] From then, the photorefractive effect has been observed in different materials, such as BaTiO$_3$, KTa$_x$Nb$_{x-1}$O$_3$ (KTN), Ba$_2$NaNb$_5$O$_{15}$, Sr$_x$Ba$_{(1-x)}$Nb$_2$O$_6$ (SBN), sillenites Bi$_{12}${Si,Ge,Ti}O$_{20}$ (BSO, BGO, BTO, respectively), semiconductors InP and GaAs, PLZT ceramics, and other photoconductive materials that exhibit electro-optical effects. Among photorefractive materials, LiNbO$_3$ has been the most studied and frequently used in holographic memories. This is mainly due to its good holographic properties, such as the dynamic range and the sensitivity of the recording and erasing processes controllable via dopants (e.g., Fe, Cu, Mn, Zn), and that crystals with excellent optical quality can be grown. In the beginning of the 1970s, Amodei, Phillips, and Staebler at RCA Laboratories achieved a stable storage[iv] of 500 holograms angularly multiplexed with an angular selectivity of 0.1 deg, in a crystal of lithium niobate with 1 cm thickness doped with iron. During the 1990s, the number of holograms stored in LiNbO$_3$:Fe crystals has increased from 1000 at the beginning of the decade[7] up to 10,000 at the end of the same.[8]

15.3.1 Physical mechanisms

The basic physical mechanisms of the photorefractive effect involve optical excitation, migration, and trapping of charges. It requires the material be photosensitive, photoconductive, and electro-optical. Typically, the linear electro-optic effect (Pockel's effect) is involved, which requires a noncentrosymmetric media, a property that satisfies 21 of the 32 crystallographic groups.

Let us assume that an electro-optical photoconductor is exposed to a spatially nonuniform light distribution, for example, created by an interference pattern during holographic recording. Free charges (electrons or holes) are generated. The charges move in the crystal by means of diffusion current, drag, or by photovoltaic effect, until they eventually recombine with traps (impurities or defects in the crystal) in locations different from where the charges originated. In this way the charge is redistributed in the crystal,[9] yielding a nonhomogenous charge density. As a consequence, an electric field appears[v] in the material that possesses a spatial modulation similar to an interference pattern. The modulation of this electric field causes, through the linear electro-optical effect, a corresponding modulation in the dielectric permittivity (refractive index) tensor.

---

iv In order to stabilize the holograms thermal fixing was used; it will be discussed in Section 15.3.2.
v The electric field is related to the space charge distribution by the Poisson equation: $\nabla \cdot (\varepsilon E) = \rho/\varepsilon_0$, where ε is the dielectric tensor and ε_0 is the vacuum dielectric constant.

The change in the refractive index depends on the symmetry of the crystallographic group, the orientation of the axes (**a**,**b**,**c**) of the crystal, the polarization of the light, and the orientation of the spatial field. In crystals with symmetry group *4mm* (e.g., BaTiO$_3$), *3m* (e.g., LiNbO$_3$, LiTaO$_3$), and *mm2*, the change of the refractive index is

$$\Delta n = -\frac{1}{2} n^3 r E_{sp} , \qquad (15.4)$$

where $n = n_{a,b,c}$ and $r = r_{113,223,333}$ are the refractive indexes and electro-optical coefficients, respectively, for the light polarized in directions **a**, **b**, or **c**. Thus, the information pattern is recorded in the material in the form of a refractive index diffraction grating (phase-type hologram).

It should be noted, though, that every successive illumination of the recorded grating (for example, during its reading or during the recording/reading of the other gratings in the same volume) shall produce, by means of the mechanisms that we have just discussed, new spatial fields that will partially erase the field of the originally recorded grating. This causes serious problems in the applications of photorefractive crystals in holographic memories. In order to avoid the partial erasure of the previously recorded holograms, as well as increasing the storage times, various techniques are used that will be discussed below: thermal fixing, electrical fixing, and two-wavelength holography.

15.3.2 Thermal fixing

This technique was developed by Amodei and Staebler[10] in LiNbO$_3$:Fe and it involves the transformation of the spatial charge pattern into an ionic charge distribution, which is much more stable. Once the holographic recording is finalized, the LiNbO$_3$:Fe crystal is heated to ~160°C. At this temperature the ionic conductivity is greater than the electric conductivity in darkness and the ions present in the crystal move in the space charge field. The drag current originated by this spatial charge distribution redistributes the ions[vi] to screen the electronic charge field. This creates an ionic grating and a corresponding spatial ionic field. The two fields, the ionic and the electronic, tend to cancel each other, producing a significant reduction of the hologram diffraction efficiency. The decrease in crystal temperature freezes the ionic pattern, since at the room temperature the ionic conductivity is insignificant. The last step (also call the developing) of the process consists of illuminating homogeneously the hologram, which will partially erase the electronic grating thus breaking the compensation of the ionic field with the electronic field, developing the former. With this method, storage times can be increased to several months, which is typical for LiNbO$_3$:Fe congruent crystals (with nonstechiometric growth), to times beyond 10 years.[11] Using thermal fixing, multiplexing up to 10,000 holograms[8] in LiNbO$_3$:Fe has been achieved. Thermally fixed holograms have been tested in

---

vi Of the possible mechanisms of ionic conductivity, the most relevant in LiNbO$_3$ is the migration of the hydrogen ions H$^+$ [H. Vormann, G. Weber, S. Kapphan, and M. Wöhlecke, Solid State Commun. **57**, 543 (1981)]. The changes of the spatial distribution of H$^+$ can be monitored by means of infrared spectroscopy, for example, observing the line of the vibrational mode of OH$^-$ group in 2870 nm [R. G. Smith, D. B. Fraser, R. T. Denton, and T. C. Rich, *J. Appl. Phys.* **39**, 4600 (1968)].

optical storage systems based on hologram multiplexing with high angular selectivity. This method has also been demonstrated in other photorefractive crystals such as $LiTaO_3$, $BaTiO_3$, $KNbO_3$, $Ba_2NaNb_5O_{15}$, and $Bi_{12}SiO_{20}$ (BSO).

15.3.3 Electrical fixing

This method can be applied to photorefractive materials that exhibit the ferroelectric effect, particularly to those that have low coercitive fields E_c, such as $BaTiO_3$ ($E_c \sim 1.1$ kV/cm) and $Sr_{0.75}Ba_{0.25}Nb_2O_6$ [(SBN), $E_c \sim 970$ V/cm]. The photoelectrical effect, discovered by Vaselek in 1921 in Rochelle salt, is the electric analogy of the ferromagnetic effect. Below Curie's temperature, ferroelectric materials exhibit spontaneous alignment of dipolar moments, producing a spontaneous electrical polarization P_s. Applying an external electrical field E, the dependency P on E presents a typical hysteresis curve. In order to achieve $P = 0$, one needs to apply the external field E_c (coercitive field) in the opposite direction from that of spontaneous polarization (i.e., $E_c < 0$ in order to compensate $P_s > 0$, and vice versa). In electrical fixing, in order to eliminate the domains with spontaneous polarization and achieve alignment of the dipoles in the desired direction, an electric field is first applied (also called polarizing field) that is greater than the coercitive field. In SBN, a polarizing field of $E_p \sim 2E_c \sim 2$ kV/cm is usually sufficient. In the following step, that of the holographic recording, the spatial field E_{sp} is created. Then, applying an external field of a similar magnitude but with opposite orientation to the polarizing field, an inversion of polarization P and the corresponding change of the dielectric permittivity and the refractive index is achieved at the places where the sum of the polarizing field with the space charge field E_{sp} exceeds the coercitive field. For example, in SBN the inversion occurs in 1% of the domains.[12] This gives rise to a spatial pattern of oriented ferroelectric domains that follow the modulation of the pattern of the E_{sp} field, the two patterns partially compensating each other. The compensation results in a substantial diminution of the diffraction efficiency. Finally, as in the case of thermal fixing, the exposure of the hologram to light will redistribute the trapped electronic charges and erase the original E_{sp} field, revealing the pattern of ferroelectric domains. The method has been applied to crystals of $BaTiO_3$ (Ref. 13) and $Sr_xBa_{1-x}Nb_2O_6$ for $x = 0.75$ (Ref. 14), as well as for other values in the interval of $0 \leq x \leq 1$ (Ref. 15). Unfortunately, this interesting method has its limits in the spatial frequencies that can be recorded.

15.3.4 Two-wavelength storage

The underlying idea of this group of techniques is to use a reading beam with a wavelength to which the medium is not sensitive, thus eliminating the partial erasure caused during the hologram readout. The photorefractive crystals are typically sensitive to green and blue, though insensitive to red and infrared since for these wavelengths the energy of the photons is not sufficient to excite the charges from the traps. The simplest approach would seem to be to use one wavelength for writing (e.g., 514 nm) and another, longer one (e.g., 633 nm), for reconstruction of the hologram. However, this change would cause Bragg's condition to be breached in the volume grating, accompanied by a reduction in diffraction efficiency, deterioration in image quality, and an increase of crosstalk with other holograms multiplexed in the medium.

The problems mentioned may be eliminated by using a holographic technique based on the dual-wavelength method. The basis of this technique consists of using a light beam with wavelength λ_1 (for example, 488 nm from an Ar^+ laser, or GaN LED incoherent light) in order to excite the charge carriers, and another wavelength λ_2 (for example, 660 nm of a laser diode) to record and read out the hologram. The hologram is recorded with λ_2 in the medium excited with λ_1, and is read with λ_2 in the absence of λ_1. The reading does not produce the hologram erasure because the material is not sensitive to λ_2 in the absence of the exciting light. The method was first used by Linde et al.[16] in pure $LiNBO_3$ and also in $LiNBO_3:Cu^{2+}$ with an Nd-YAG mode-locked laser (~10 ps pulses) with $\lambda_1 = 1.06$ μm and the second harmonic $\lambda_2 = 0.53$ μm. It has also been demonstrated that the use of high-power pulsed lasers is not indispensable. Holograms have been recorded with continuous wave moderate power lasers (5 W/cm² or less),[17] thanks to optimization of the composition and the reduction state of lithium niobate crystals. This technique[18] uses the intrinsic defects of the crystal or the impurities such as Fe^{2+}/Fe^{3+} and is schematically illustrated in Fig. 15.1.

Level 1 is attributed to the bipolaron $Nb_{Li}Nb_{Nb}$ (ion Nb^{4+} instead of Li^+ and ion Nb^{4+} instead of Nb^{5+}) or to the impurities Fe^{2+}/Fe^{3+} and is responsible for the absorption of the gating light (λ_1, blue or green). The reduction in an oxygen-poor atmosphere at a temperature of ~950°C after the growth of the crystal induces more bipolarons since the Li and Nb ions diffuse to occupy the vacancies in the crystal. This is used to increase the sensitivity of the medium to the

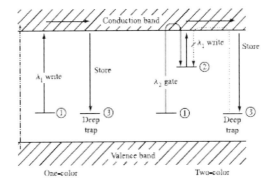

Figure 15.1 Schematic diagrams of one-wavelength and two-wavelength photorefractive effects. Level 1 is attributed to a bipolaron state $Nb_{Li}Nb_{Nb}$ or Fe^{2+}/Fe^{3+} state, level 2 to an Nb_{Li} state, and level 3 to an Fe^{3+} trap. [Reproduced with permission from H. Güenther, R. Macfarlane, Y. Furukawa, K. Kitamura, R. Neurgaonkar, *Appl. Opt.* **37**, 7611 (1998)].

exciting light. The illumination in blue-green (λ_1) results in the dissociation of bipolarons, freeing electrons to the conduction band to be finally trapped in the Nb sites forming Nb_{Li} (small polaron, Nb^{4+} ion instead of Li^+, level 2 in Fig. 15.1). The small polarons are responsible for the sensitivity of the medium to red and near infrared[19] (maximum sensitivity at $\lambda_2 = 852$ nm). The hologram is recorded by photoexciting the small polarons with λ_2, whereby the liberated electrons are redistributed in the conduction band and are eventually trapped in deep traps, typically Fe^{3+} (level 3 in Fig. 15.1). The sensitivity of the medium for λ_2 is proportional to the density of small polarons and, therefore, to their life span[vii]. The charge redistribution leads to the formation of a space charge field and, finally, to the modulation of the refractive index, as it has been explained in the previous section. Apart from the intrinsic defects, extrinsic levels may also be introduced through dopants, such as Mn^{2+} (level 1),[20] which may result in increased sensitivities and high diffraction efficiencies.

15.4 Organic Photorefractive Materials

In 1990, the photorefractive effect was observed for the first time in an organic material[21]—an organic nonlinear COANP crystal[viii] doped with TCNQ[ix]. However, given that organic crystals are difficult to grow since the majority of dopants are expelled during preparation of the crystal, research was focused on noncrystalline materials, particularly on polymers containing nonlinear chromophores. The first photorefractive polymeric material (developed at IBM Almaden) was a nonlinear epoxy polymer, the bisA-NPDA[x], doped with hole-transport agent DEH[xi] to make the material photoconductive.[22]

Organic photorefractive materials have a major advantage over inorganic materials: potentially, they have a high figure of merit, $Q = n^3 r/\varepsilon_r$, where n is the refractive index, r the electro-optical coefficient, and ε_r the static relative dielectric permittivity. It is observed that Q is proportional to the ratio between the optical nonlinearity and the screening of the spatial electric field by the polarization of the medium. In inorganic materials, optical nonlinearity originates primarily in the ionic polarizability of the medium, since high electro-optical coefficients are accompanied by high static dielectric constants and similar Q values are obtained for different materials. In organic materials, nonlinearity is, however, a molecular property resulting from the asymmetry of the electronic charge distributions in the fundamental and excited states of the nonlinear chromophore. In organic materials, high nonlinearities are not, therefore, accompanied by high dielectric constants and the Q values may be more than one order of magnitude greater than those of inorganic materials.

---

vii Congruent lithium niobate, which is deficient in Li (48.6% molar of Li_2O), and the resulting defects reduce the lifetime of the small polarons through nonradiative decay. In the quasi-stoichiometric crystals (~50% molar of Li_2O), the concentration of defects is significantly reduced and the life time of the small polarons can be on the order of tenths of a second, which is three orders of magnitude greater than for congruent crystals. This results in a high sensitivity to λ_2 in the presence of λ_1.
viii COANP: 2-cyclooctylamino-5-nitropyridine.
ix TCNQ: 7,7,8,8-tetracyanoquinodimethane.
x bisA-NPDA: bisphenol-A-diglycidylether 4-nitro-1,2-phenylenediamine.
xi DEH: diethylaminobenzaldehyde diphenylhydrazone.

In organic photorefractive materials, the properties necessary for photorefractivity, namely, photoconductivity, charge transport, and optical nonlinearity, are provided by several ingredients such as photosensitive charge generators, conducting species, and nonlinear optical chromophores.

15.4.1 Charge generation

Dyes are typically used as charge generators. These are reduced on light absorption and inject a hole in the material. It is desirable that these molecules absorb in the red or near infrared, which enables these wavelengths to be used in photorefractive experiments, since there the nonlinear chromophores possess minimum absorption. TNF (2,4,7-trinitro-9-fluorene) and C_{60} (fullerene) are frequently used, thanks to their high quantum efficiency of charge generation, wide spectral range, and good solubility. It should be noted that unlike in inorganic materials, the quantum efficiency of charge generation in organic materials may be increased by applying an external electric field (E_0). This is due to the following: absorption of a photon creates a correlated electron-hole pair (a Frenkel exciton), the separation of the hole competes with the recombination[xii], and the external field helps with charge separation[xiii].

15.4.2 Charge transport

The next step in the space charge field generation is the charge transport from lighter to darker zones, where the charges get trapped. Unlike crystalline photoconductor materials where charge transport can be described by means of the bands model, in amorphous materials the energy levels of the molecules are affected by their heterogeneous environment and this disorder separates the conduction bands into a distribution of localized electronic states. Charge transport occurs through leaps by the adjacent transport species and is usually described by means of Bässler's formalism.[23] Charge mobility[xiv] depends on separation between transport species, so that high concentrations of these are usually employed. Several photoconductor molecules may be used, such as hydrazone, carbazoles, arilamines, or many of those developed for electrophotography.[24]

15.4.3 Nonlinear optical properties

The nonlinear effect is a function of the induced polarization P of the nonlinear molecule (chromophore) in the presence of an optical field of intensity E, and it may be expressed in terms of a Taylor series as

---

[xii] In organic materials, the likelihood of charge recombination immediately after generation of the exciton is high. This is due to a low screening of the electrostatic interaction between electrons and holes since organic materials have a low dc electric permittivity.

[xiii] The dependence of the quantum efficiency of charge generation with the electric field is usually described by using Onsager's theory [L. Onsager, "Initial recombination of ions", *Phys. Rev.* **54**, 554 ,1938] for the dissociation of ion pairs in weak electrolytes.

[xiv] Unlike in inorganic photorefractive crystals, mobility µ in organic polymers significantly depends on the electric field, µ ~ $\ln(E_0)^{1/2}$.

$$P = \alpha E + \left(\frac{1}{2!}\right)\beta E^2 + \left(\frac{1}{3!}\right)\gamma E^3 + \cdots, \qquad (15.5)$$

where α is the linear polarizability, and β and γ are the first and second hyperpolarizabilities. The first hyperpolarizability β determines the second-order nonlinear effects, and it may be approximated using the two-state model,[25]

$$\beta \propto \frac{\mu_{fe}^2 (\mu_{ee} - \mu_{ff})}{E_{fe}^2}, \qquad (15.6)$$

where μ and E are the elements of the dipolar matrix and the transition energy, respectively, between the fundamental state f and the first permitted excited state e. The physical interpretation of the term $\mu_{ee} - \mu_{ff}$ is that during interaction with the light, the electrons move preferably according to a particular axis of the molecule. Typically, nonlinear chromophores (see Fig. 15.2) are conjugated aromatic systems asymmetrically terminated with electron donor or acceptor groups, in which case the electrons preferably move according to the long axis of the molecule. For nonlinear properties at the molecular level (β) to be observable macroscopically as second-order susceptibility $\chi^{(2)}$ and the resulting linear electro-optical effect, it is necessary to orient the nonlinear molecules in one preferential direction and so create a noncentrosymmetric material. This can be achieved through the use of polarizing external electric fields applied on a material[xv] contained between two transparent electrodes (e.g., of indium tin oxide), or by means of the corona effect. The photorefractive polymers with low glass transition temperatures ($T_g < 100°C$) require a permanently applied field to maintain chromophore orientation, since these would otherwise be randomly reoriented at $T \sim T_g$. In materials with a higher T_g (over 100°C), the polarizing field is applied to the material heated to $T \sim T_g$ to facilitate chromophore orientation. The orientation will be "frozen" and remain stable after reducing the temperature to $T < T_g$ and switching off the poling field.

Polymers with low T_g usually show the orientational photorefractive effect, as was first demonstrated by Moerner et al.[26] At temperatures close to T_g, the chromophore molecules can be oriented not only by an externally applied electric field but also by the space charge field. The result is that the sinusoidal spatial field causes a periodic modulation of the orientation of the chromophores, leading to a modulation of birefringence and of electro-optical coefficient, which are added to the modulation produced by the "conventional" electro-optical effect.

---

xv The poled polymers belong to the symmetry group ∞mm. Their second-order susceptibility tensor has, in general, three independent components or two independent components outside the chromophore resonances.

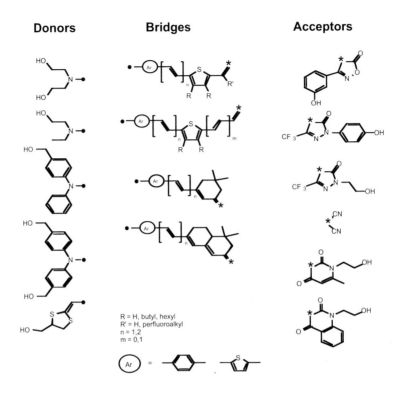

Figure 15.2 Electro-optical chromophores as bipolar charge transfer molecules comprising donor, bridge, and acceptor segments [Source: L. Dalton, A. Harper, A. Ren, F. Wang, G. Todorova, J. Chen, C. Zhang and M. Lee, "Polymeric electro-optic modulators: From chromophore design to integration with semiconductor very large scale integration electronics and silica fiber optics", *Ind. & Eng. Chem. Res.*, **38**, 8 (1999)].

15.4.5 Materials classification

The materials developed to date may be grouped in two categories: guest-host systems and fully functionalized systems. Guest-host systems contain at least one part of the components dispersed in the matrix of the host medium. The typical example is the matrices where the nonlinear chromophore forms a covalent link with the host, while the generation and charge transport molecules are dispersed in the latter. Moreover, the charge transport groups may form a covalent bond with the host medium or, as in the case of polyvinylcarbazole (PVK), the host medium itself is a photoconductor, while the charge generating molecules and the nonlinear chromophore are dispersed in the host. A typical example of guest-host materials is shown in Fig. 15.3. In this material,[27] DMNPAA:PVK:ECZ:TNF, diffraction efficiencies close to 100% and a high two-beam coupling coefficient ($\Gamma \sim 200$) were achieved for the first time.

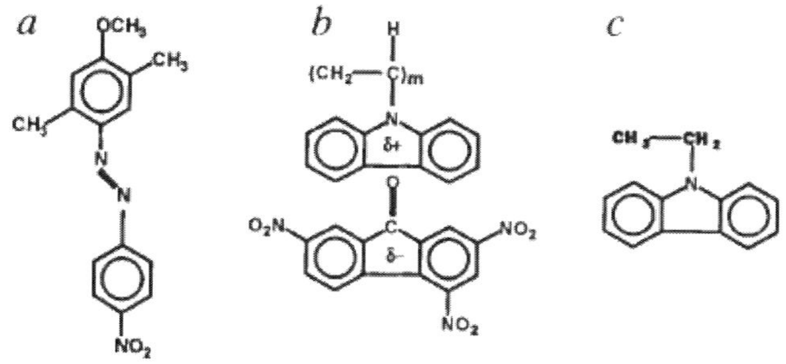

Figure 15.3 Chemical structure of the components used in the photorefractive composite DMNPAA:PVK:ECZ:TNF. (a) Nonlinear chromophore: 2,5-dimethyl-4-(p-nitrophenylazo)anisole (DMNPAA). (b) Charge transfer complex: poly(N-vinylcarbazole) (PVK) and 2,4,7-trinitro-9-fluorenone (TNF). (c) Plasticizer: N-ethylcarbazole (ECZ) [source: Ref 27].

The limited solubility of the guest material in the host medium and the metastability of the system is usually a problem in guest-host materials. For example, in order to achieve high optical nonlinearity, high chromophore concentrations are needed, which may cause crystallization of the latter in the host and thus degradation of the optical properties. These problems with stability have been resolved in completely functionalized systems formed by a single component. For holographic data storage, chromophores capable of forming glass, such as 2BNCM[xvi], seem promising. High refractive index modulation Δn ~ 0.01 and excellent optical quality was achieved in 2BNCM.[28] Because the different functionalities (charge transport and generation, optical nonlinearity) are included in the same molecule, the material is optimally used with a minimal inert volume. The main disadvantage of functionalized systems resides in the complexity of synthesis of such molecules and in a certain lack of flexibility, since a modification of the material requires a new chemical synthesis.

Another interesting class of emerging materials is mesogenic composites. In these materials, photorefractivity is based on reorientation of liquid-crystalline molecules in optical and electric fields, also called the "orientational photorefractive effect."[29]

Several interesting organic photorefractive materials have been developed recently[30-33] with high refractive index modulation, good optical quality, and a sufficiently fast response, including sensitivity extended to near infrared.[34-37]

15.5 Photopolymerizable materials

Photopolymerizable materials (photopolymers) are of great interest for the construction of read-only memory (ROM) and write-once read-many (WORM) holographic memories, since permanent phase holograms can be formed with high refractive index modulation Δn and can be recorded in photopolymers. These materials use to have high optical quality and, unlike silver halides or

---

xvi 2BNCM: N-2-butyl-2,6-dimethyl-4H-pyridone-4-ylidenecyanomethylacetate.

dichromate gelatins, do not require complicated developing processes. Variation of their properties through the design of different compositions is virtually unlimited and the cost is low.

The first photograph was produced in a photopolymerizable substance by Joseph N. Niépce in his experiments (1822 and 1827), where he used photoinduced changes in a natural polymer, Syrian asphalt, for photoengraving in stone, copper, and pewter. Cross-linking of the polymer in regions illuminated for several hours with sunlight converted the substance to being insoluble in solvents that normally dissolve the polymer. Later, in 1945 W.E.F. Gates formed shallow relief images in liquid monomers such as methyl methacrylate, by using a combination of luminous radiation and heat[38]. In the late 1950s, the first commercial photopolymer was introduced under the name Dycril for printing applications.[39] Close et al.,[40] at Hughes Aircraft, were the first to use (in 1969) a photopolymerizable system to record volume holograms. Since then, several groups of researchers have developed a large number of photopolymerizable systems (for early works, see Ref. 41), of which those of E. I. du Pont de Nemours and Polaroid were marketed.

15.5.1 Physical mechanism

Photopolymers typically comprise four basic components: a sensitizing dye, a polymerization initiator, and one or more monomers (usually liquid), all dispersed in a polymeric solid matrix called a binder. In order to enhance the optical or mechanical properties of the photopolymer, other components such as chain transfer agents and plasticizers are often included. To prevent thermal polymerization and thus increase polymer stability and shelf life prior to exposure, commercial systems usually include inhibitors of thermally induced polymerization.

Illumination of the photopolymer with light in the spectral range of the photosensitizer ("actinic radiation") causes photochemical reactions in the illuminated areas, such as cross-linking or solubilization of the polymer, or polymerization of the monomer. The latter is the mechanism most frequently used in holographic photopolymers and it involves several physical processes such as monomer diffusion and deplasticizing. The physical and chemical changes causes corresponding local refractive index changes that can be used for recording volume diffraction gratings.

The formation and propagation of the polymer chain may occur through polymerization by free radicals or by ionic polymerization. The latter may be performed by using several compounds (e.g., sulphonium and iodine salts) whose photoinduced decomposition generates strong protic acids that efficiently initiate polymerization of the monomers such as polyvinylcarbazole, vinyl ethers, and epoxides. Ionic polymerization is strongly inhibited by water. This is one of the reasons why the majority of photopolymerizable systems use polymerization by free radicals. An important exception is a photopolymer developed by Polaroid that is based on cationic polymerization (see Section 15.5.3).

Free radical photopolymerization is initiated with the absorption of the actinic radiation by the sensitizer S, giving rise to the excited molecule S^*,

$$S + h\nu \rightarrow S^*. \quad (15.7)$$

The sensitizer is indispensable when actinic radiation wavelength between 300 and 700 nm is used, since the majority of monomers do not absorb in this spectral range. In the next stage, the energy of the sensitizer is transferred to the initiator I which is excited to I^*,

$$S^* + I \rightarrow S + I^*. \quad (15.8)$$

The excited initiator may decompose in a pair of free radicals $R_1\bullet + R_2\bullet$ as

$$I^* \rightarrow R_1\bullet + R_2\bullet. \quad (15.9)$$

Some systems also include the so-called chain transfer agents RH (hydrogen donors, also called co-initiators), which react with the excited initiator and form secondary free radicals[xvii] as

$$I^* + RH \rightarrow IH\bullet + R_3\bullet. \quad (15.10)$$

The free radicals react with the monomer or the oligomer causing their polymerization through propagation and chain transfer until its eventual completion as

$$R\bullet + \text{monomer/oligomer} \rightarrow \text{polymer}. \quad (15.11)$$

The reactions involved are chain type, so that a single photon may cause the polymerization of up to $\sim 10^5$ monomer molecules. This chemical amplification of the image is the main cause of high photopolymer sensitivities compared to photorefractive materials.

Free radical polymerization has a major negative implication in holographic data storage applications. This is the reduction in volume of the material during polymerization. Whenever a new molecule of monomer is added to the growing polymer chain, then the total volume decreases since a new covalent link replaces the previous van der Waals contact. The shrinkage may be ~10% in acrylates. This distorts the holograms, ultimately reducing the data storage capacity.

Another negative feature (though less significant than the one above) is that oxygen is an efficient inhibitor of free radical polymerization. It reacts with the active radicals and converts them into non-active peroxyradicals, so that until the oxygen dissolved in the material has been consumed, polymerization will not

---

[xvii] The secondary radicals may be more efficient polymerization initiators than the original radicals $R_1\bullet + R_2\bullet$.

commence[xviii]. Although techniques exist for reducing this induction period during which exposure does not produce a hologram, inhibition by oxygen is a basic mechanism that ultimately limits the sensitivities achievable in free radical photopolymerizable systems.

During recording of a holographic grating, polymerization proceeds faster at regions of maximum illumination than in the less illuminated areas[xix]. The resulting monomer concentration gradient originates a flow of the monomer from less illuminated areas toward the more illuminated zones. Also, the monomer flow toward areas of maximum illumination is facilitated by the creation of free volume in the illuminated areas since there the monomer molecules, originally separated at van der Waals distances, become polymer with covalent bonds. Monomer diffusion induces microscopic redistribution of the material and eventually creates areas rich in monomer-derived polymer (in more illuminated areas) and areas rich in binder (in less illuminated areas). The difference between monomer/polymer and binder refractive indices produces the desired spatial modulation of the refractive index. The latter can be increased by using compositions with large differences between monomer and binder refractive indices. This fact was first verified in a photopolymer developed at E. I. du Pont de Nemours and Company.[42] The transport process ends when there is no more monomer available in the less illuminated areas, or if the increasing rigidity of the polymer (deplasticizing, vitrification) inhibits diffusion so that monomer molecules are unable to reach the reactive (radical) centers in the polymeric chain[xx].

As with monomer molecules, other small molecules present in the composition (sensitizers or initiators) respond to holographic recording by the formation of their own concentration gradients and the corresponding diffusion flows, forming undesired "dual" gratings. In order to prevent the formation of such gratings, moderate sensitizer or initiator concentrations should be used. Moreover, in thick media, a high concentration of the sensitizer and thus a high absorption coefficient α would lead during exposure to a light intensity gradient $I(z) \sim \exp(-\alpha z)$ in the propagation direction. This would result in different exposures and refractive index modulations at different depths of the material.[xxi] The sensitizer and initiator concentrations must, however, be sufficient to ensure the efficiency of the photoinduced polymerization and the desired sensitivity of the medium. At the end of the holographic exposure, the residual monomer is usually polymerized by a uniform exposure using an ultraviolet light lamp. This last step also bleaches the sensitizer, converting it into its non-photosensitive

---

xviii Typically, an initial exposure of ~10 mJ/cm² is needed to consume the O_2 present in the material.

xix Polymerization reaction kinetics is usually studied by means of techniques such as DSC (differential scanning photocalorimetry), in which the heat produced in this exothermic reaction is monitored, or by near-infrared spectroscopy where diminishing with polymerization of the absorption bands of the acrylic group is monitored.

xx Inhibition of monomer diffusion may be diminished with plasticizers (e.g., triethylene glycol diacetate, dibutyl phtalate).

xxi From the coupled wave equations it can be demonstrated that the dependence $\Delta n(z)$ causes the diffraction efficiency η to be nonzero in the first minima of the angular selectivity curve (in contrast to the case Δn = const). We recall that the basis for angular multiplexing resides in the diffraction efficiency of hologram k recorded in the angular position of these first minima to be $\eta_k = 0$, which permits to record the next hologram $k + 1$ under the same angle without crosstalk. The minima elevation ($\eta_k \neq 0$) will deteriorate crosstalk between the multiplexed holograms. Furthermore, the minima will be shifted $(1/4\pi)(\alpha d)^2$ from the position $m\pi$ corresponding to Δn = const. in the increasing angular deviation direction.

product, thus stabilizing ("fixing") the hologram. For the majority of photopolymers, further developing processes are not needed, although in some systems an increase of Δn has been observed by heating the hologram to 100–150°C over several hours.[43]

We shall outline below some examples of holographic photopolymers relevant for holographic data storage.

15.5.2 DuPont photopolymers

In the 1980s, DuPont (abbreviation of E. I. du Pont de Nemours and Co.) developed and commercialized a family of holographic photopolymers (HRF series) that became very popular, thanks to their versatility, good holographic properties, and ease of use. There are several optimized formulas for transmission holograms, reflection holograms, or both, sensitized at wavelengths from ultraviolet to near infrared (~700 nm). They are capable of producing holograms with refractive index modulation of up to 0.07 with typical exposures of 10–100 mJ/cm^2 (Ref. 44). These photopolymers have the following generic composition: polymeric binder 46–65%, acrylic monomer(s) 28–46%, plasticizer 0–15%, chain transfer agent 2–3%, initiator 1–3%, and sensitizer 0.1–0.2%[xxii]. The photopolymer film is deposited on a sheet of Mylar and covered with another similar one. The latter is removed before exposure so that the polymer (which is slightly sticky) can adhere "face down" to the substrate. After exposure, the hologram is stabilized by uniform UV illumination. Thermal treatment up to 100–150°C can be used to increase Δn.

The maximum refractive index modulation was achieved in either compositions that combined an aliphatic binder[xxiii] having a refractive index of n ~ 1.47 with aromatic monomers [e.g., 2-phenoxyethyacrylate, n = 1.514], or compositions combining an aromatic binder [75:25 poly(styrene-acrilonitril), n = 1.57; 70:30 poly(styrene-methyl methacrylate), n = 1.56] with aliphatic monomers [e.g., 2-(2-(etoxy)etoxyethylacrylate), n = 1.436; or diacrylate of triethylenglycol, n = 1.459].

Such combinations maximize the difference between the binder and monomer refractive indexes. As sensitizers, various ketones with blue, green, and red absorption are used.[45] A HABI[xxiv] with UV absorption is used as an initiator. The MBO[xxv] is used as a chain transfer agent.

Mok et al.[46] demonstrated the storage densities of 40 bit/µm^2 by multiplexing 80 holograms, each with 640 × 480 pixels, in the DuPont polymer with a thickness of 100 µm, laminated on a glass disk with a 12 cm diameter, to obtain a total storage capacity of 340 Gbits. The limiting factors in the use of this photopolymer family in holographic memories are the limited thickness of the medium (~100 µm) and its shrinkage (3–10%) during holographic recording.

---

xxii Percents expressed with respect to the total weight.
xxiii Typically CAB: cellulose acetate butyrate.
xxiv HABI: 2,2',4,4',5,5'-hexaarylbisimidazole.
xxv   MBO: 2-mercaptobenzoxazole.

15.5.3 Polaroid photopolymers

At the end of the 1980s, Polaroid introduced the DMP-128 photopolymer based on a polymeric binder, acrylic monomer, photoinitiator, and cross-linking agent. The DMP-128 is highly sensitive (4–8 mJ/cm^2 for transmission holograms and 15–30 mJ/cm^2 for reflection holograms) and has a high refractive index modulation. Unlike the DuPont polymers, the DMP-128 needs humid processing conditions. Its maximum thickness is limited to 20 µm.

Polaroid later developed a ULSH series photopolymer based on CROP-type cationic polymerization (cationic ring-opening polymerization).[47] CROP can be initiated with strong protic acids, which are usually generated, for example, from the photodecomposition of sulphonium or iodinium salts. For example, the derivatives of cycloaliphatic epoxy, whose ring opens during polymerization, can be used as monomers. The opening of the monomer ring produces an increase in molecular volume, which partially compensates the reduction in volume that typically accompanies polymerization, as mentioned in the previous section. In the case of the ULSH polymer, shrinkage during holographic recording can be very small (~ 0.1%).

The generic formulation of ULSH consists of a sensitizing colorant (0.2–0.02%), acid photogenerator (3–10%), monomer(s) (40–75%), and binder (40–70%). The monomers are composed of two (bifunctional monomer) or more groups of cyclohexane oxide connected through segments of a siloxane chain, the bifunctional monomer being DiEPOX[xxvi]. High modulation of the index is obtained, for example, from the combination of DiEPOX with the DOW 705 binder and the MPIB acid photogenerator[xxvii].

Characteristics such as low shrinkage, a high sensitivity of 10–50 mJ/cm^2 (thanks in part to the fact that cationic polymerization is not inhibited by oxygen), a good dynamic range ($\Delta n \sim 0.006$), and thicknesses of up to 200 µm make this polymer particularly suitable for holographic memories. In 1999, Waldman et al.[48] multiplexed 100 holograms, each with 262 Kbits of data and with diffraction efficiencies of ~10^{-4}, on an ULSH-500 photopolymer with a thickness of 200 µm. In the same year, Aprilis Inc. was founded to commercialize the Polaroid polymer for data storage applications.[49]

15.5.4 Lucent photopolymer

A photopolymer optimized specifically for holographic data storage applications was developed by Bell Laboratories Lucent Technologies.[50] In order to reduce shrinkage during holographic recording, the photopolymer uses two independent polymerization systems, one for the formation of the matrix (binder) and the other for the holographic recording. The first is an oligomer of di(urethane-acrylate) with a molecular weight of ~1700 and a refractive index of 1.49 (ALU-351, Echo Resins, Inc.), while the second is composed of various acrylic

---

xxvi DiEPOX: 1,3-bis[2-(3{7-oxabicyclo[4.1.0]heptyl})ethyl]-tetramethyl disiloxane.
xxvii MPIB: bis(4-methylphenyl)iodonium tetrakis(pentafluorophenyl)borate.

monomers, the ones that allow high modulation of the refractive index are, e.g., IBA (isobornil acrylate) and VNA (vinyl napthoate): $\Delta n \sim 0.002$ for the IBA(25%):VNA(10%):ALU-351(60%) formulation, as a percentage of total weight. Among the different photoinitiation systems examined, CGI-784 (Ciba-Geigy) was found to be particularly effective, with a typical concentration of ~1%. Prior to holographic exposure, the liquid resin, composed of oligomer, monomer(s), and initiator (contained between the two glass plates separated from each other with teflon spacers), is pre-exposed with incoherent light from the 546 nm Hg line, which results in the cross-linking of the oligomer, thus forming a solid matrix. During this pre-exposure, the oligomer and also part of the monomer polymerizes[xxviii], but most of the latter remains intact and available for the next stage, which is holographic recording. Holographic exposure (e.g., with green light at $\lambda = 532$ nm from a frequency-doubled Nd-YAG) causes polymerization and diffusion of the monomer as explained in Section 15.5.1, and a phase grating is produced.

Using shift multiplexing, Curtis et al. multiplexed 4000 holograms in a polymer with a similar composition, attaining a storage density of 45 bits/μm^2, corresponding to ~50 Gbytes on a 5¼ in. disk. Later, new formulations were developed with a higher dynamic range (4×), aimed at reaching 150 Gbytes on a disk with the same format.[51] In 2001, Lucent New Ventures Group founded InPhase Technologies, dedicated to the development and commercialization of holographic memories based on this family of polymers. At the National Association of Broadcasters (NAB) Convention in April 2007, InPhase announced the first commercial holographic storage systems for broadcasters. The product, Tapestry 300r, is a 300 GB archival WORM drive providing file-based data access with a 50-year media life.

15.5.5 Polymers functionalized with liquid crystals and optical chromophores

Polymers functionalized with liquid crystals[52] were originally developed at Moscow State University. Following the first holographic experiments with these types of materials,[53] various functionalized systems with liquid crystals and/or optical chromophores were developed (known in the literature as photoaddressable polymers, PAPs), one of the most relevant being the polymer developed by Bayer AG.[54] The Bayer PAP contains two different groups, the chromophoric and the mesogenic, which are anchored to the principal polymer chain.

Azobenzene chromophores exist in two isomeric configurations, i.e., the *trans* and the *cis* form (Fig. 15.4). In the presence of polarized light, the chromophoric groups are pumped from the *trans* configuration to the *cis* state, with a probability proportional to $\cos^2(\theta)$, thanks to the interaction of the µ

[xxviii] Partial polymerization of the monomer during pre-exposure obviously reduces the amount of monomer available for holographic recording and, as a consequence, it limits the dynamic range of the photopolymer. Therefore, it is important to use monomers with a significantly lower reactivity than that of the binder-forming oligomer.

dipolar moment of the molecule with the electric field of the polarized light[55] (θ is the angle between the light electric field vector and the long axis of the chromophore).

Because the *cis* state is thermodynamically less stable than the *trans* state, thermal relaxation converts the *cis* isomer (produced by photoexcitation of the *trans* isomer) back to the energetically favored *trans* form. Given the mentioned angular dependency, it is unlikely that those chromophores that remain oriented perpendicular to the polarization of the beam will be pumped again to the *cis* state. As a result, after several *trans-cis-trans* cycles, the number of chromophoric groups oriented in the $\theta = 90°$ direction will gradually increase, producing changes in the orientational distribution of the chromophore and in the associated optical birefringence. At the same time, the interaction between the chromophoric group and the mesogenic group (liquid crystal), typically through their respective dipolar moments, causes the mesogenic group to follow the orientational distribution of the chromophore, thus amplifying and stabilizing the changes in birefringence. Extremely high birefringence (as much as $\Delta n \sim 0.5$) and a high optical quality[56] have been obtained. However, owing to the lack of an efficient image amplification mechanism, high exposure energies (on the order of 40 J/cm^2) are needed, which limits the use of these types of materials in holographic memories. Surface relief gratings have also been demonstrated in azo-polymers.[57,58] Recently, such gratings have been holographically recorded in a waveguide composite structure[59,60] including silicon-on-insulator waveguides of submicrometer thickness and an azo-polymer waveguide cladding with applications such as reconfigurable spectral filtering and off-chip coupling.

15.5.6 Photopolymers developed in Russia

Various original polymeric materials[61] were developed at the S.I. Vavilov State Institute of Optics in Leningrad (now St. Petersburg). Here we will discuss Reoxan and PQ-PMMA.

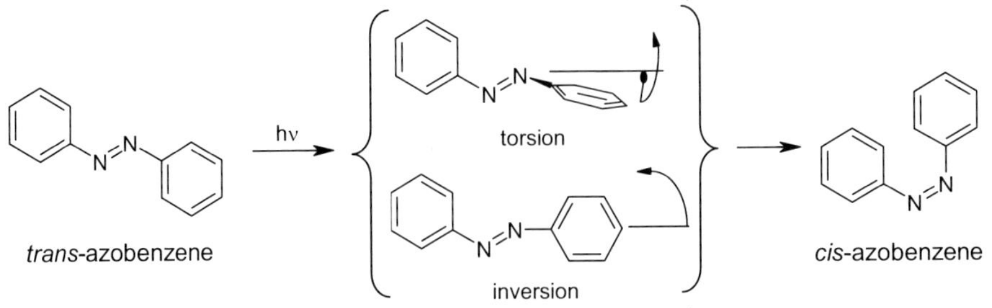

Figure 15.4 Azo-group photoisomerization [Source: A. Stolow, NRC Canada].

Reoxan. Reoxan (from recording oxidized medium with anthracene)[62] consists of a transparent polymeric matrix doped with anthracene and a colorant. Sensitization is achieved by impregnating the material with molecular oxygen at high pressure in an autoclave and it covers the entire visible spectrum up to 900 nm. During holographic recording, anthracene photo-oxide forms in the illuminated zones. This photo transformation is accompanied by changes in the refractive index of up to 0.02. After holographic recording, the oxygen in the matrix is left to diffuse, which desensitizes the material, allowing nondestructive reading. As well as its high dynamic range, Reoxan has high optical quality, with a resolution of > 5000 lines/mm. It can be prepared in thicknesses ranging from tens of micrometers to several millimeters. A problem with this material is thermal diffusion and consequent redistribution of the anthracene molecules and their photo-oxide. This produces grating degradation, accompanied by reduction of diffraction efficiency by 10% per year (room temperature).

PQ-PMMA. The problem of diffusion mentioned above is solved by PQ-PMMA (polymethylmetacrylate doped with phenathrenequinone).[63] Holographic illumination photoexcites the PQ molecules, creating a covalent link between the PQ photoproducts (semiquinone radicals) and the polymer matrix in illuminated areas. Then, changes in molar refraction and therefore in the refractive index arise in places where the reaction occurs. The resulting concentration gradient of free-moving nonphotoexcited PQ molecules causes diffusion of PQ from nonilluminated to illuminated zones[xxix]. Once the PQ molecules are redistributed by diffusion, homogeneous illumination is used to consume the nonexcited molecules and thus desensitize (fix) the material. The main disadvantage of the material is its low sensitivity, with saturation energy on the order of tens of J/cm^2. PQ-PMMA is discussed in detail in the chapter authored by S. H. Lin, M. Gruber, Y.-N. Hsiaso and K. Y. Hsu in this book.

15.6 Hybrid Organic-Inorganic Materials

15.6.1 Vycor porous glasses impregnated with organic materials

An interesting solution to the problem of shrinkage and limited polymer thickness is to use a matrix (binder) of porous glass[xxx] impregnated with photosensitive materials (e.g., photopolymers, silver halides, dichromated gelatine, etc.). Vycor-type porous glass originally developed by Corning Glass consists of a silica backbone with a network of interconnected nanoporosity. The large internal surface area (200–1000 m^2/g) permits absorption of high quantities of doping molecules. The available free volume is on the order of 30–40%. The size of the pores has to be small enough to minimize light scattering, but also to permit diffusion of doping molecules during impregnation of the material. Typically, matrices with pore diameters of several nanometers (1–5 nm) are used.

---

xxix To facilitate diffusion, it is necessary to heat the material to 50–60°C for 24 h after holographic exposure.
xxx   Porous glasses offer several additional advantages, such as high thermal and chemical stability, a low thermal expansion coefficient, excellent mechanical properties, and the possibility of high-quality optical polishing.

The first holographic material based on a porous matrix was developed by Bell Laboratories[64] in the late 1970s. The material consisted of a Vycor 7930 matrix (4 nm pore diameter, 28% free volume) impregnated with benzoin as a polymerization initiator. Holographic exposure with a 364 nm wavelength (from Ar$^+$ laser) causes destruction of the initiator in the illuminated regions, whereas in the dark areas of the interference pattern the photoinitiator remains intact. A spatial modulation in the concentration of the photoinitiator is then produced, forming a latent image. In the next stage (developing), the matrix is impregnated with a mixture of monomers and illuminated homogeneously. Illumination activates the remaining intact photoinitiator and causes polymerization of the monomer. Since the photoinitiator concentration has been modulated (through selective destruction) during holographic recording, the polymerization initiation rate is modulated alike, as well as modulation of the resulting polymer refractive index, thus "revealing" the latent image. The dynamic range of this material was modest ($\Delta n \sim 3 \times 10^{-4}$), but experiments carried out subsequently achieved modulation levels comparable to those obtained with photopolymers. For example, Sukhanov[65] demonstrated excellent holographic properties of porous glass impregnated with different compounds, such as fine-grained silver halide emulsions, photopolymers, and dichromated gelatines. The results reported included diffraction efficiencies near 100% for exposures of ~10 mJ/cm^2 in 1 mm thick media, resolution exceeding 6000 lines/mm, and large dynamic range. Schnoes et al.[66] multiplexed 25 holograms with diffraction efficiencies of 2% in Vycor glass with a ~1.5 mm thickness impregnated with the Lucent photopolymer.

15.6.2 Sol-gel holographic materials

Silica gels are common in nature in the form of opals and agates. The first synthetic silica[67] was prepared by the French chemist Jacques-Joseph Ebelmen in 1844, using silicon alkoxide, founding the basis of the sol-gel technique. The first industrial application of sol-gel was the manufacture of antireflective films densified at low temperatures and resistant to scratching[68] developed by Jena Glasswerk Schott in 1939. At the present time, the sol-gel process is used for synthesis of a large variety of gels, glasses, and ceramics.[69]

A simplified scheme of the sol-gel reaction can be represented by two main steps. First, the reaction with water (hydrolysis) of metal alkoxides $M(OR)_n$ (called precursors, M being a metal, e.g., Si, and R an alkyl group, e.g., CH_3, C_5H_5, etc.), and second, a polycondensation reaction:

$$M(OR)_n + nH_2O \rightarrow M(OH)_n + nR(OH) \qquad (15.12)$$
(hydrolysis)

$$M(OH)_n \rightarrow MO_{n/2} + {}^n/_2 H_2O \qquad (15.13)$$
(polycondensation)

Because metal alkoxides are insoluble in water but soluble in alcohols, a small quantity of alcohol (ROH, e.g., methanol or ethanol) is added in order to obtain a homogeneous initial solution. The product of the polycondensation reaction, the metal oxide[xxxi] (e.g., SiO_2), is contained in a colloidal solution (sol) of very small particles (~2 nm) in contact with each other. When the gelation point is reached, a solid gel is formed, typically containing a high concentration of pores. The reaction is speeded up using either acid catalysts, usually HF or HCl, or base catalysts, e.g., NH_4OH. For the preparation of optical gels, it is preferable to use acid catalysis, since this tends to produce interconnected polymeric networks[xxxii]. An example of a simple starting solution that produces transparent matrices with a high optical quality can be as follows: silicon tetraethoxide, ethanol, and water and hydrofluoric acid in molar concentrations of 1:4:4:0.05 (see Ref. 70 for details on synthesis).

Figure 15.5 shows, in a simple manner, a model for explaining the formation of silica gel from a solution. The fundamental motivation for using sol-gel materials is to replace the high-temperature glass and ceramic fabrication techniques by a process that can take place at lower temperatures, even at room temperature. Avoiding high temperatures allows incorporation of organic molecules with low thermal stability into inorganic matrices, resulting in hybrid organic-inorganic materials. Combining the properties of organic and inorganic components in a composite material opens up new opportunities for the development of innovative materials, including holographic recording media.

The first organically modified sol-gel material capable of recording volume holograms was demonstrated by Cheben et al.[71] in 1996. This material was developed in order to overcome problems with the limited maximum thickness of commercial holographic photopolymers and with material shrinkage during polymerization, typical for acrylic-based materials. The basic idea here was to disperse organic photopolymerizable species in an inorganic host matrix rather than in an organic binder typically used for this purpose. The inorganic host matrix significantly improves the physical properties of the holographic recording material, such as its rigidity, environmental stability, dimensional changes on holographic exposure, maximum achievable thickness, and the ability to accept an optical-grade polish. The support matrix of this organic-inorganic material, in contrast to the Vycor glass holographic materials, was formed by *in situ* polymerization (sol-gel reaction) of liquid silica precursors in the presence of dissolved photoinitiating and photopolymerizable species. The material was prepared in the form of monoliths a few millimeters thick and volume gratings with diffraction efficiencies > 90% were holographically recorded in it.

[xxxi] Hydrolysis and polycondensation reactions take place simultaneously and are usually incomplete; thus, the general formula of the final product can be expressed as $(MO)_x (OH)_y (OR)_z$.

[xxxii] Basic catalysis tends to produce colloidal particles.

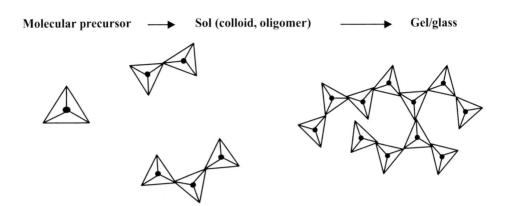

Figure 15.5 Formation of a silica gel from a solution. Polymerization of a silica precursor leads to a colloidal silica "sol," and then to a solid "gel" that can be further densified to a glass.

Following this strategy, a sol-gel glass was developed with a refractive index modulation of $\Delta n \sim 0.004$ and a diffraction efficiency of 98% for an exposure of 230 mJ/cm^2 at $\lambda = 514.5$ nm.[72] The material consists of a glassy host containing an ethylenic unsaturated monomer ethylene glycol phenyl ether acrylate and a free radical generating titanocene photoinitiator bis(μ^5-2,4-cyclopentadien-1-yl)-bis-[2,6-difluoro-3-1H-pyrrol-1-yl)phenyl]titanium (Irgacure-784). It was fabricated both as thick films and as monoliths. The results obtained with this material demonstrated that sol-gels are important candidates for holographic data storage, and various new sol-gel photopolymerizable compositions have subsequently emerged.[73-75]

In addition to photopolymerizable systems, various photorefractive ormosils (organically modified silicates) have been developed.[76-78] Photorefractive gratings with a refractive index modulation of 0.002 and a two-beam coupling gain of 444 cm^{-1} were demonstrated in an organically modified permanently poled sol-gel glass.[78] The azo-dye 2,5-dimethyl-4-(2-hydroxyethoxy)-4'-nitroazobenzene (DMHNAB) was used as a nonlinear optical chromophore. The chromophore molecules were covalently bonded to the silica glass backbone in order to achieve the high dye concentration needed for efficient nonlinear optical properties. This also avoids dye crystallization often observed in guest-host photorefractive polymers. 2,4,7-trinitro-9-fluorenone (TNF) was used as a photosensitizer and N-ethylcarbazole (ECZ) as the charge-transporting agent, both being present as guests in the glass, i.e., without being covalently attached to the matrix. Excellent resistance against chromophore crystallization was achieved by covalently bonding the chromophore. The high stability of the electric field–induced chromophore alignment is due to a gradual heat-induced densification of the gel with an initially low glass transition temperature (T_g) during electric field poling, eventually yielding a high-T_g hard glassy film. This densification process is essential for slowing down diffusive randomization of the chromophore alignment and for improving the glass mechanical, electrical, and thermal properties. Similar materials, with high glass transition temperatures, are desirable, for example, in long-term data storage and electro-optical modulators.

15.6.3 Organic-inorganic nanocomposites incorporating high refractive index species

Attempts have been made to improve the refractive index modulation of a photopolymerizable medium by incorporating diffusible high refractive index species (HRIS),[79-81] for example, titania nanoparticles, together with the regular photopolymerizable monomer. Volume holographic gratings with refractive index modulations of up to 0.015 have been reported.[81] However, the incorporation of nanoparticles tends to increase the scattering noise, due in part to nanoparticle agglomeration.

It has recently been demonstrated that the scattering can be substantially reduced by incorporating the HRIS at the molecular rather than the nanoparticle level. Sol-gel holographic material has been developed with a large Δn incorporating a high-refractive-index MA:Zr molecular complex, namely, zirconium isopropoxide chelated with (metha)acrylic acid (MA) in a sol-gel matrix.[82] The large Δn resides in the ability of the high-index MA:Zr complex to diffuse and thus contribute to grating formation on inhomogeneous illumination (Fig. 15.6). On photoinduced polymerization of the (metha)acrylic acid, diffusion driven by a concentration gradient of the MA:Zr complex takes place from the dark to the light regions of the interference pattern. By incorporating the high-index MA:Zr species in the host, the refractive index modulation of the material is increased to $\Delta n \sim 0.01$, compared to $\Delta n \sim 0.005$ in the sample without the high-index species [see Fig. 15.7(a)]. Furthermore, compared to the photopolymers with dispersed high-index (TiO$_2$) nanoparticles, scattering is markedly reduced as a consequence of the molecular nature of MA:Zr.

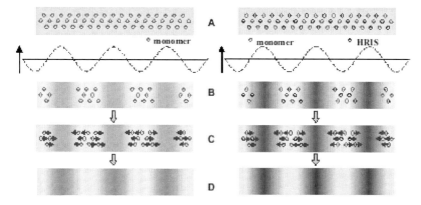

Figure 15.6 Formation of the refractive index grating in a photopolymerizable material based on a conventional (Colburn-Haines) monomer diffusion mechanism (left), and in a material modified with HRIS (right) (after Ref. 82).

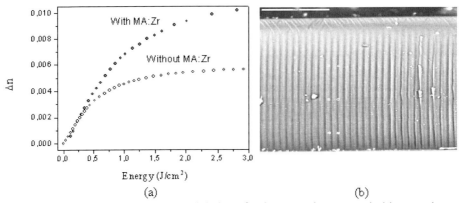

Figure 15.7 (a) Refractive index modulation of volume gratings recorded in samples with and without the HRIS (MA:Zr). Sample thickness 35 μm. (b) Scanning electron microscope image of a grating recorded in a sample containing HRIS. Sample thickness: 250 μm, grating spatial frequency: 100 lines/mm (after Ref. 82).

The recording of volume holographic gratings in this new photomaterial has permitted experimental detection of diffraction effects not previously reported in the optical domain. This is the case of the so-called Pendellösung fringes, first observed in 1968 by C.G. Shull for neutron diffraction by a thick perfect crystal of silicon. In hybrid organic-inorganic material incorporating HRIS, volume holographic gratings were recorded with high diffraction efficiency and large Δn, allowing for the first time experimental detection of the Pendellösung effect in the optical domain.[83]

15.7 Conclusions

We have reviewed fundamentals and recent advances in holographic recording materials with emphasis on applications in optical data storage. Significant progress has been achieved in this field and the first commercial holographic data storage systems are emerging. Today, several excellent holographic materials are available, and even if they are not fully meeting the demanding specifications in holographic data storage applications, these materials are finding interesting niche applications in a variety of fields in research and industry. Hybrid organic-inorganic materials appear particularly promising since they can benefit from synergies arising from a combination of different physical and chemical properties and specific optical functionalities in a single matrix.

Acknowledgments

We thank Oscar Martínez-Matos, Francisco del Monte, and José A. Rodrigo for invaluable help and many insightful discussions. We thank as well K. Y. Hsu for helpful comments and suggestions. The financial support of the Spanish Ministry of Education and Science (Project No. TEC2005-02180) is also acknowledged.

References

1. van Heerden, P.J., *Appl. Opt.*, **2**, 393 (1963).
2. Coufal, H., *Nature,* **393**, 628 (1998).
3. Kogelnik, H., *Bell. Syst. Tech. J.,* **48**, 2909 (1969).
4. Mok, F.H., G. W. Burr and D. Psaltis, *Opt. Lett.*, **21**, 896 (1996).
5. Ashkin A., G.D. Boyd, J.M. Dziedzic, R.G. Smith, A.A. Ballman, J.J. Levinste and K. Nassau, *Appl. Phys. Lett.*, **9**, 72 (1966).
6. Chen F.S., J.T. LaMacchia, and D. B. Fraser, *Appl. Phys. Lett.*, **13**, 223 (1968).
7. Mok F.H., *Opt. Lett.* **18**, 915 (1991).
8. An X., D. Psaltis and G. Burr, *Appl. Opt.* **38**, 386 (1999). See also Volk T., M. Wöhlecke, *Crit. Rev. Solid State Mater. Sci.*, **30**, 125 (2005).
9. Vinetskii V.L. and N.V. Kukhtarev, *Sov. Phys. Solid State,* **16**, 2414 (1975).
10. Amodei J.J. and D.L. Staebler, *Appl. Phys. Lett.* **18**, 540 (1971). See also Carrascosa, M., J.M. Cabrera, and F. Agulló-López, in: *Infrared Holography for Optical Communications: Techniques, Materials, and Devices*, Topics in Applied Physics, **86**, Chap. 5, Springer, Berlin (2003).
11. Yariv, A., S. Orlov and G. Rakuljic, *J. Opt. Soc. Am. B,* **13**, 2513 (1996).
12. Kewitsch, A.S., T.W. Towe, G.J. Salamo, A. Yariv, M. Zhang, M. Segev, E.J. Sharp and R. R. Neurgaonkar, *Appl. Phys. Lett.*, **66**, 1865 (1995).
13. Micheron, F. and G. Bismuth, *Appl. Phys. Lett.*, **20**, 79 (1972).
14. Micheron, F. and G. Bismuth, *Appl. Phys. Lett.*, **23**, 71 (1973).
15. Qiao, Y., S. S. Orlov, D. Psaltis and R.R. Neurgaonkar, *Opt. Lett.*, **18**, 1004 (1993).
16. von der Linde, D., Glass, A.M., and Rodgers, K.F., *Appl. Phys. Lett.*, **25**, 155 (1974).
17. Lande, D., S.S. Orlov, A. Akella, L. Hesselink and R.R. Neurgaonkar, *Opt. Lett.,* **22**, 1722 (1997).
18. Güenther, H., R. Macfarlane, Y. Furukawa, K. Kitamura, R.R. Neurgaonkar, *Appl. Opt.*, **37**, 7611 (1998).
19. Jermann, F., M. Simon, R. Bower, E. Kratzig, O.F. Schirmer, *Ferroelectrics,* **165**, 319 (1995).
20. Buse, K., A. Abidi, and D. Psaltis, *Nature,* **393**, 665-668 (1998).
21. Sutter, K., J. Hullinger and P. Guenter, *Solid State. Commun.*, **74**, 867 (1990).
22. Ducharme, S., J.C. Scott, R.J. Twieg and W.E. Moerner, *Phys. Rev. Lett.,* **66**, 1846 (1991).
23. Bässler, H., *Adv. Mater.* **5**, 662 (1993).

24. Borsenberger, P.M. and D.S. Weiss, *Organic Photoreceptors for Imaging Systems*, Marcel Dekker, New York (1993).
25. Oudar, J.L. and D.S. Chemla, *J. Chem. Phys.*, **66**, 2662 (1997).
26. Moerner ,W.E., M. Silence, F. Hache and G.C. Bjorklund, *J. Opt. Soc. Am. B,* **11**, 320 (1994).
27. Meerholz, K., B.L. Volodin, N.F.N. Sandalphon, B. Kippelen and N. Pyghambarian, *Nature,* **371**, 497 (1994).
28. Lundquist, P.M., R.Wortmann, C. Geletneky, R.J. Twieg, M. Jurich, V.Y. Lee, C.R. Moylan and D.M. Burland, *Science,* **274**, 1182 (1996).
29. Ono, H., A. Emoto and N. Kawatsuki, in *Photorefractive Materials and Their Applications*, P. Gunter, and J.-P. Huignard, Eds., Chap. 6, Springer, Berlin (2007).
30. Ostroverkhova, O., and W.M. We, *Chem. Rev.,* **104**, 3267–3314 (2004).
31. Eralp, M., J. Thomas, S. Tay, G.L.A. Schulzgen, R.A. Norwood, M. Yamamoto and N. Peyghambarian, *Appl. Phys.Lett.*, **89**, 114105 (2006).
32. Salvador, M., F. Gallego-Gomez, S. Köber and K. Meerholz, *Appl. Phys. Lett.*, **90**, 154102 (2007).
33. Gunter, P., and J.-P. Huignard, Eds., *Photorefractive Materials and Their Applications*, Springer, Berlin (2007).
34. Mecher E., F. Gallego-Gomez, H. Tillmann, H. H. Hörhold, J. C. Hummelen and K. Meerholz, *Nature,* **418,** 959 (2002).
35. Mecher E., F. Gallego-Gomez, K. Meerholz, H. Tillman, H. H. Hörhold and. J. C. Hummelen, *Chem.Phys. Chem.,* **5**, 277 (2004).
36. Tay S., J. Thomas, M. E. M, G. Li, et al., *Appl. Phys.Lett.,* **85**, 4561-4563 (2004).
37. Acebal P., S. Blaya, and L. Carretero, *Opt.Express*, **13**, 8296-8307 (2005).
38. Gates, W.E.F., British Patent No. 566,795 (1945).
39. Crawford, H.E., *Proceedings of the Technical Association of the Graphic Arts*, TAGA Abstracts, p. 193 (1960).
40. Close, D.H., A.D. Jacobson, J.D. Margerum, R.J. Brault and F.J. Mcclung, *Appl. Phys. Lett.*, **14**, 159 (1969).
41. Lessard, R.A., and B.J. Thompson, Eds., *Selected Papers on Photopolymers*, SPIE Milestone Series, Vol. MS 114, SPIE, Bellingham, WA (1995).
42. Monroe, B.M., and W.K. Smothers, Photopolymers for Holography and Waveguide Applications, in *Polymers for Lightwave and Integrated Optics*, L.A. Hornak, Ed., p.p. 145–169, Marcel Dekker, New York (1992).
43. Mothers, W.K., B.M. Monroe, A.M. Weber and D.E. Keys, Practical Holography IV, S.A. Benton, Ed., *Proc. SPIE* **1212**, p.p. 20-29 (1990).
44. Gambogi, W.J., A.M. Eber and T.J. Trout, Holographic Imaging and Materials, T.H. Jeong, Ed., *Proc. SPIE* **2043**, p.p. 2-13 (1994).

45. Monroe, B.M., and G.C. Weed, *Chem. Rev.*, **93**, 435 (1993).
46. Mok, F., G. Zhou and D. Psaltis, *Holographic Read-Only Memory*, in *Holographic Data Storage*, H.J. Coufal, D. Psaltis, and G.T. Sincerbox, Eds., p.p. 399–407, Springer, Berlin (2000).
47. Waldman, D.A., Ingwall, R.T., Dhal, P.K., M.G. Horner, E.S. Kolb, H.Y.S. Li, R.A. Minns and H.G. Schild, Diffractive and Holographic Optics Technology III, I. Cindrich and S.H. Lee, Eds., *Proc. SPIE* **2689**, p.p. 127-141 (1996).
48. Waldman, D.A., R.T. Ingwall, H.Y.S. Li and R.M. Shelby, Annual Meeting of International Society for Optical Engineering, SPIE, Denver (1999).
49. http://www.aprilisinc.com/datastorage.htm
50. Schilling, M.L., V.L. Colvin, L. Dhar, A.L. Harris, F.C. Schilling, H.E. Katz, T. Wysocki, A. Hale, L.L. Blyler and C. Boyd, *Chem. Mater.*, **11**, 247-254 (1999).
51. Curtis, K., W.L. Wilson, M.C. Tackitt., A.J. Hill and S. Campbell, *High-Density, High-Performance Data Storage via Volume Holography: The Lucent Technologies Hardware Platform*, in *Holographic Data Storage*, H.J. Coufal, D. Psaltis, and G.T. Sincerbox, Eds., p.p. 359–368, Springer, Berlin (2000).
52. Shibaev, V.P., I.Y. Yakovlev, S.G. Kostromin, S.A. Ivanova and T.I. Zverkova *Polym. Sci. USSR,* **32**, 1552 (1990).
53. Eich, M., J.H. Wendorff, B. Reck and H. Ringsdorf, *Makromol. Chem., Rapid Commun.*, **8**, 59 (1987).
54. Zilker, S.J., M.R. Huber, T. Bieringer and D. Haarer, *Appl. Phys. B,* **68**, 893 (1999).
55. Torodov, T., L. Nikolova and N. Tomova, *Appl. Opt.*, **23**, 4309 (1984).
56. Bieringer, T., *Photoaddressable Polymers*, in *Holographic Data Storage*, H.J. Coufal, D. Psaltis, and G.T. Sincerbox (Eds.), p.p. 209–228, Springer, Berlin (2000).
57. Natansohn, A., P.L. Rochon and E. Batalla, *Appl. Phys. Lett.*, **66**, 136 (1995).
58. Natansohn, A. and P.L. Rochon, *Chem. Rev.*, **102**, 4139 (2002).
59. Lausten, R., P.L. Rochon, M. Ivanov, P. Cheben, S. Janz, P. Desjardins, J. Ripmeester, T. Siebert and A. Stolow, *Appl. Opt.,* **44**, 7039 (2005).
60. Cheben, P., R. Lausten, M. Ivanov, S. Janz, B. Lamontagne, E. Post, J. Ripmeester, P.L. Rochon, A. Stolow and D.X. Xu, *Proc. ICO Topical Meeting on Optoinformatics/Information Photonics 2006*, St. Petersburg, Russia, 4-7 September 2006, ISBN 5-7921-0719-6, M.L. Calvo, A.V. Pavlov and J. Jahns, Eds., ITMO, St. Petersburg, Russia, 57–59 (2006).
61. Sukhanov, V.I., *J. Opt. Technol.*, **61**, 49–56 (1994).
62. Lashkov, G.I. and S. Cherkasov, *New Recording Media for Holography*, p.p. 89–101, Nauka, Leningrad (1983).

63. Veniaminov, A.V., V.F. Goncharov and A.P. Popov, *Opt. Spektrosk.*, **70**, 864 (1991).
64. Chandross, E.A., W.J. Tomlinson and G.D. Aumiller, *Appl. Opt.*, **17**, 566 (1978).
65. Sukhanov, V.I., Holographic Optics II. Principles and Applications, *Proc. SPIE* **1136**, p.p. 73-78 (1989).
66. Schnoes, M.G., L. Dhar, M.L. Schilling, S.S. Patel and P. Wiltzius, *Opt. Lett.*, **24**, 658 (1999).
67. Ebelmen, J.J., *C. R. Acad. Sci.*, **19**, 398 (1844).
68. Geffcken, W., and E. Berger, US Patent No. 2,366,5126 (1945).
69. Brinker, C.J., and G.W. Scherer, *Sol-Gel Science. The Physics and Chemistry of Sol-Gel Processing*, Academic Press, San Diego (1990).
70. Pope, E.J.A., and J.D. Mackenzie, *J. Non-Crystal. Solids,* **87**, 185 (1986).
71. Cheben, P., T. Belenguer, A. Núñez, F. Del Monte and D. Levy, *Opt. Lett.,* **21**, 1857 (1996).
72. Cheben, P., and M.L. Calvo, *Appl. Phys. Lett.* **78**, 1490 (2001). See also *Nature Sci. Update*: http://www.nature.com/nsu/010315/010315-7.html; *New York Times*: http://www.nytimes.com/2001/07/12/technology/circuits/12NEXT.html; *Nature*, **422**, 56–558 (2003); *Phys. Today*: http://www.physicstoday.org/pt/vol-54/iss-5/p9.html; *Chem. Eng. News:* http://pubs.acs.org/cen/topstory/7912/7912notw5.html; *Phys. News Update:* http://www.aip.org/physnews/update/529-3.html; *New Scientist:* http://www.newscientist.com/tech/holographic.jsp; *Optics. Org:* http://www.optics.org/article/news/07/5/13; *Laser Focus World,* **37**(5), May 2001; *Mater. Res. Soc. Bull.*, **26**(4), 277 (2001); *Chem. Phys. Chem.,,* **3**, 333 (2002).
73. Ramos, G., A. Alvarez-Herrero, T. Belenguer, F. Del Monte and D. Levy, *Appl. Opt.,* **43**, 4018 (2004).
74. Carretero L., A. Murciano, S. Blaya, M. Ulibarrena and A. Fimia, *Opt. Expr.,* **12**, 1780 (2004).
75. Park J. and E. Kim, *Key Engineering Materials,* **1039,** 277 (2005).
76. Zhang Y., R. Burzynski, S. Ghosal and M. Casstevens, *Adv. Mater.*, **8,** 111 (1996).
77. Darracq B., M. Canva, F. Chaput, et al., *Appl Phys. Lett.* **70,** 292 (1997).
78. Cheben, P., F. Del Monte, D.J. Worsfold, D.J. Carlsson, C.P. Grover and J.D. Mackenzie, *Nature.* **408**, 64 (2000).
79. Suzuki, N., and Y. Tomita, *Appl. Opt.*, **43**, 2125 (2004).
80. Suzuki, N., Y. Tomita and T. Kojima, *Appl. Phys. Lett.*, **81,** 4121 (2002).
81. Sanchez, C., M.J. Escuti, C. van Heesch, C.W.M. Bastiaansen, D.J. Broer, J. Loos and R. Nussbaumer, *Adv. Funct. Mater.*, **15**, 1623 (2005).

82. Del Monte, F., O. Martínez-Matos, J.A. Rodrigo, M.L. Calvo and P. Cheben, *Adv. Mater.*, **18**, 2014 (2006).

83. Calvo, M.L., P. Cheben, O. Martínez-Matos, J.A. Rodrigo and F. Del Monte, *Phys. Rev. Lett.*, **97**, 084801 (2006).

Maria L. Calvo is a professor of optics at the Universidad Complutense de Madrid (UCM), Spain. She received her MS in physics from the UCM in 1969, and her PhD in physics from the UCM in 1977. She also received a doctorate from the University of Paris VI, France, in 1971. She is the author of more than 100 journal papers, has written six book chapters, and has coordinated three textbooks in optics. Her current research interests include holographic materials and applications to optical computing, optical signal processing, light scattering, and optical waveguides. She is an elected Fellow of SPIE and OSA. She is currently Secretary General of the International Commission for Optics (ICO), term 2005–2008.

Pavel Cheben is a senior research officer at the Institute for Microstructural Sciences at the National Research Council Canada. His current research interests include planar waveguide circuits, holographic materials, and silicon photonics. He was a member of the team which started up Optenia, Inc. and developed the first commercial echelle grating wavelength demultiplexer for applications in WDM optical networks. He obtained his PhD in physics (optics) at the Complutense University of Madrid, Spain, and his MSc in microelectronics and optoelectronics at the Slovak Technical University in Bratislava, Slovakia. Cheben is a coauthor of 8 book chapters, 140 publications and 23 patent applications in the fields of integrated optics, photonics, and optical and photonic materials.

Chapter 16
Holographic Data Storage in Low-Shrinkage Doped Photopolymer

Shiuan Huei Lin
National Chiao Tung University, Taiwan

Matthias Gruber
Fern Universität Hagen, Germany

Yi-Nan Hsiao and Ken Y. Hsu
National Chiao Tung University, Taiwan

16.1 Introduction
16.2 Prospect and Challenges of Volume Holographic Data Storage
    16.2.1 Key issue: holographic storage materials
    16.2.2 Material requirements
    16.2.3 Shrinkage tolerance
16.3 Phenanthrenequinone-Doped Poly(methyl methacrylate) PQ:PMMA)
    16.3.1 Strategy to minimize material shrinkage
    16.3.2 Material fabrication of the PQ-doped PMMA photopolymer
    16.3.3 Optical transmission and surface flatness
16.4 Holographic Characterization of PQ:PMMA
    16.4.1 Material shrinkage coefficient
    16.4.2 Dynamic range and sensitivity
16.5 Holographic Data Storage in a PQ:PMMA Disk
    16.54.1 Optical setup
    16.5.2 Experimental demonstrations
16.6 Advanced Photopolymer Materials
    16.6.1 Irgacure 784-doped PMMA
16.7 Microintegration of the Optical Read/Write Head
    16.7.1 General design considerations
    16.7.2 Planar-optical system design

16.8 Conclusions
References

16.1 Introduction

With the rapid growth of the Internet, the amount of information available to users has been increasing explosively, generating a strong demand for information technology (IT) equipment that can keep pace with this development. In particular, storage technologies and devices with ultralarge capacity and ultrafast data access are required. Conventional mass storage technologies based on magnetic hard disk drives or on optical storage media such as CD, DVD, and successors are approaching fundamental physical limits. Also, the underlying serial, bit-oriented 2D storage concept constitutes an immanent readout bottleneck. Especially this latter disadvantage can be overcome with volume holographic storage techniques. Optical holography is inherently parallel and page oriented, and allows one to implement truly 3D storage concepts with huge storage capacity simply by using thick recording materials. In the search for future "ultra" storage technologies, volume holography is therefore considered as a potential candidate.[1] Currently, one can observe intensified research activities concerning various aspects of this field. One of these aspects, suitable volume holographic storage materials, is the topic of two chapters in this book. In Calvo and Cheben's chapter, the fundamental principles and advances in various holographic materials, both inorganic and organic, are presented. In this chapter, we report about the chemical and physical engineering of novel low-shrinkage photopolymers, evaluate their holographic storage characteristics, and discuss the optimal use in microintegrated mass storage devices.

16.2 Prospect and Challenges of Volume Holographic Data Storage

The basic principle of volume holographic data storage as depicted in Fig. 16.1 has been known for more than four decades.[2-4] A laser beam is encoded with a data page by means of a spatial light modulator (SLM). Denoted as the signal beam, this beam is relayed to the holographic storage medium, most commonly with a 2-f lens assembly that performs a free-space optical Fourier transformation, and superimposed with a coherent reference beam to form a volume hologram in the storage medium. The stored data page is retrieved by using the very same reference beam as the address beam. Diffraction from the hologram will then regenerate the signal beam, which can be relayed to and read out by a detector array.

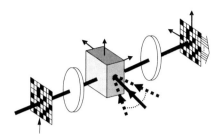

Figure 16.1 Basic principle of volume holographic data storage.

Of particular importance for holographic data storage applications is the Bragg selectivity property of volume holograms. Slight changes of the reference beam parameters during readout already lead to a complete detuning of the system, and no signal beam is regenerated. On the other hand, thanks to this property, multiplexing is possible, i.e., the recording of many different data pages in the same physical volume by use of different reference beams, which enables a selective retrieval of data pages with negligible crosstalk. A huge storage capacity can therefore be achieved; van Heerden[2] already estimated the (theoretical) upper limit at V/λ^3 bits, with V denoting the storage volume and λ the wavelength of the light.

Despite this huge potential, volume holography has not found its way into commercial storage devices in the past. This was largely due to a lack of crucial enabling technologies. Meanwhile, however, many of these technologies have matured and the situation is beginning to change. Several start-up companies[5] are now pursuing the commercialization of holographic memories and targeting the first mass production to appear within the next few years. Among the important technological milestones that have been passed are the development and theoretical analysis of powerful multiplexing schemes[6] that allow one to efficiently tap the huge storage capacity potential of volume holography. Equally important is the availability of suitable and affordable lasers, modulators, and detectors.[7-9] Examples are the second-harmonic diode-pumped solid-state (DPSS) lasers that emit coherent radiation with enough power and that are reasonably compact and efficient—small 2D SLM arrays in the megapixel range such as the ones used in digital cameras or mobile phones—and CMOS detector arrays.

16.2.1 Key issue: holographic storage materials

Probably the most important, however, is the technological progress in the field of holographic storage materials. Such materials need to show a photorefractive effect, i.e., a photoinduced (local) change of the refractive index that is used to memorize a holographic interference pattern in form of a phase grating. Depending on whether this photoinduced index change is reversible or permanent, the storage material is either of the rewritable or write-once-read-many (WORM) type. For a long time, (rewritable) inorganic crystals such as $LiNbO_3$ and $LiTaO_3$ were used almost exclusively. More recently, organic materials have moved into the focus of research because their fabrication is less complicated and costly, and because organic materials provide more possibilities for material engineering, i.e., the adjustment of desired (optical) properties by modifying the chemical composition of the material.

Photopolymers are part of this organic material group. The physical mechanism of holographic recording in these materials typically involves photoinduced polymerization of acrylic monomers in a polymeric film. Photopolymers are interesting write-once recording materials due to their high sensitivity and large modulation depth of the refractive index.[10] A number of materials for holographic applications have been developed since the first holographic photopolymer was reported in 1969.[11] However, only recently have companies such as Inphase and Aprilis announced offering such storage materials commercially.[5,12,13]

The main practical problems with photopolymer materials are limited thickness and high shrinkage during holographic exposure. Material shrinkage is the most severe problem for implementing high-performance holographic data storage via the volume, because it induces dimensional distortions of the recorded gratings and leads to a mismatch of the Bragg condition such that the recorded information cannot be retrieved completely. Thus, one of the fundamental issues for the success of holographic storage technology is the availability of thick recording materials with low shrinkage effect.

In our laboratory, we have developed a technique to fabricate phenanthrenequinone-doped poly(methyl methacrylate) polymers (shortly named PQ:PMMA) in centimeter-thick slabs and with satisfactory optical quality for holographic WORM memories.[14-17] Holographic experiments using these PQ:PMMA samples showed that the shrinkage coefficient of the material is less than 10^{-5}, which turns out to be a good figure with respect to holographic data storage applications. In order to develop and evaluate a suitable material that fulfills practical requirements of volume holographic data storage applications, it is important to know what specifications are realistically required and how an appropriate material can be developed and characterized. In other words, it is necessary to establish a methodology for defining and characterizing material parameters. This will be done in the following sections before we proceed to a detailed description of PQ:PMMA design and fabrication.

16.2.2 Material requirements

In general, a good holographic storage material should possess the following characteristics: high sensitivity with respect to optical exposure, large dynamic recording range, good homogeneity and surface quality to avoid scattering noise, and easy fabrication of large-area or large-volume samples. A detailed description of these parameters can be found in Calvo and Cheben's chapter in this book. From the systems point of view, the first two items, sensitivity and dynamic range, are most important.

High sensitivity of a holographic material allows one to use short exposure times and/or low-power lasers for the recording process, which is a prerequisite for obtaining a compact system and a practically acceptable recording speed. From a more physical point of view, the sensitivity S can be defined as the amount of photoinduced refractive index modulation Δn that is achieved with a certain amount of exposure energy E_i. However, in the context of optical storage it is more convenient to use the following alternative definition based on the diffraction efficiency η of a single recorded hologram:

$$S = \frac{\sqrt{\eta}}{E_i}. \qquad (16.1)$$

Both definitions are equivalent since to the first approximation of weak holograms the grating strength Δn is proportional to the square root of η.

The dynamic range of a holographic recording material indicates the maximum possible change in refractive index due to holographic exposures. This parameter obviously affects the diffraction efficiencies of holograms that are stored in the same volume element by means of a multiplexing scheme because if, as a consequence of previous hologram recording, this maximum index change has been reached at a certain physical location, the material can no longer respond to further hologram exposure at this particular location. Like sensitivity, the dynamic range of a material can therefore be defined via the diffraction efficiencies of recorded holograms. The commonly used system metric is called the $M\#$ (pronounced "M number")[18,19] and defined as

$$M\# = \sum_{i=1}^{N} \sqrt{\eta_i}, \qquad (16.2)$$

where N represents the total number of holograms that are recorded at one location. If all N individual diffraction efficiencies η_i are identical, this equation can be simplified to read $M\# = N\sqrt{\eta}$. Thus, the $M\#$ of a storage material should be large to be able to store a large number of (multiplexed) holograms with a given diffraction efficiency.

The question what $M\#$ is actually needed for a holographic storage application depends on the performance parameters of the peripheral devices, such as the sensitivity of the detector array and the pixel number of the input SLM. To elaborate this interrelation, we exemplarily consider a system with the following specifications:

Data recording/readout speed:	$R = 1$ Gbit/s
SLM pixels:	$D = 1$ Mbit per page
Laser power on recording spot:	$P = 20$ mW
Recording area per hologram spot:	$A = 5$ mm^2
CCD sensitivity:	$S_c = 10{,}000$ electrons/pixel
Data storage density:	$C_d = 320$ bits/μm^2 (corresponding to at least 400 GB per 5 in. disk)

Based on the above data-recording rate, the light energy density that impinges on the recording material per hologram recording is $E_i = (D/R) \cdot (P/A) = 0.4 \text{ mJ/cm}^2$. The above read-out data rate and CCD sensitivity and an assumed quantum efficiency of 1 lead to a required minimum diffraction efficiency of $\eta = S_c h \nu D (R/D)(1/P) = 1.94 \times 10^{-4}$ for each hologram. With these values, the material specifications can be estimated. With Eq. (16.1), the required sensitivity of the holographic material becomes $S = \sqrt{\eta}/E_i = 35 \text{ cm}^2/\text{J}$. An expression for the dynamic range of a recording material can be derived from the relation $C_d = (D/A)N$ and Eq. (16.2) as

$$M\# = \frac{C_d}{D} A \sqrt{\eta}. \qquad (16.3)$$

By inserting the above numerical values, the required dynamic range of recording material is found to be $M\# = 22.4$.

Actually, these requirements for the sensitivity and for the dynamic range of a recording medium are not so difficult to meet by current photopolymer materials.[20,21] The difficult part is the shrinkage problem. We will next describe how material shrinkage affects the quality of an image when it is retrieved from a holographic data storage system.

16.2.3 Shrinkage tolerance

Material shrinkage induced by photochemical reaction during holographic recording will result in a violation of the Bragg condition such that the recorded information cannot be read out completely. The thicker the material is, the more serious the shrinkage effect becomes.[22,23] Thus, one can apply an indirect approach for evaluating the material shrinkage effect by first recording an image into a volume hologram and then checking the quality of the retrieved image; the degree of image distortion provides an estimation of the shrinkage coefficient.

For a computer simulation of this method, we assumed a Fourier holographic setup with 90 deg geometry, as shown in Fig. 16.1. Reference and signal beams are incident on the recording medium, a photopolymer block with dimensions of $5 \times 5 \times 5$ mm, at adjoining sides. The input images are presented on a liquid crystal television (LCTV), and readout images shall be detected by a CCD camera. By using the first-order Born approximation theorem of a scattering medium, the 2D distribution of the reconstructed image can be expressed as[24]

$$g(x_1, y_1) \propto g\left[-\frac{(x_1 + \alpha_x f \sin\theta)}{(1+\alpha_x)}, -\frac{y_1}{(1+\alpha_y)}\right] \\ \times t \operatorname{sinc}\left(\frac{t}{2\pi}\left\{k\cos\theta\alpha_z + \frac{2\pi}{\lambda f^2}\left[\begin{array}{l} x_1^2 + y_1^2 - (x_1 + \alpha_x f \sin\theta)^2 \frac{(1+\alpha_z)}{(1+\alpha_x)^2} \\ -y_1^2 \frac{(1+\alpha_z)}{(1+\alpha_y)^2} + \alpha_z f^2 \end{array}\right]\right\}\right) \qquad (16.4)$$

Table 16.1 The reconstructed images under different material shrinkage coefficients (computer simulation).

Original image	Retrieved image			
	$\alpha=1\times10^{-5}$	$\alpha=5\times10^{-5}$	$\alpha=1\times10^{-4}$	$\alpha=5\times10^{-4}$
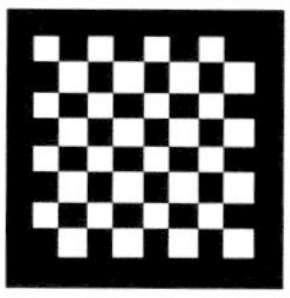 [checkerboard]	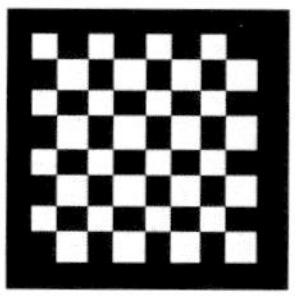 [checkerboard]	 [checkerboard]	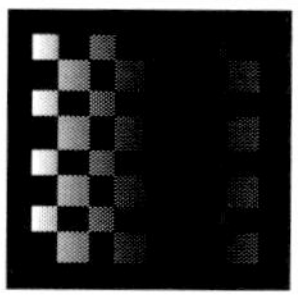 [distorted]	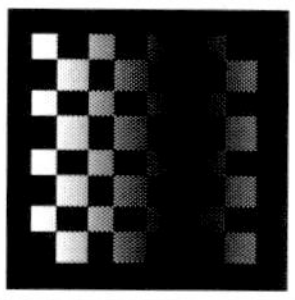 [distorted]

where α_x, α_y, and α_z are the material shrinkage coefficients in the x, y, and z directions in the material, respectively, λ is the wavelength, t is the thickness of the medium, f is the focal length of Fourier lenses L1 and L2, and θ is the angle between the z-axis and the wave vector of the reference beam. Assuming $t = 5$ mm, $f = 10$ cm, $\lambda = 514.5$ nm, the reconstructed image can be calculated if the shrinkage coefficients are given. Table 1 shows the simulation result. It can be seen that when the shrinkage coefficient is larger than 5×10^{-5}, the retrieved images become seriously distorted. Thus, in order to achieve high-performance holographic data storage, the material shrinkage coefficient should not be larger than 10^{-5} in a 5 mm thick medium. This number is already close to the linear thermal expansion coefficient of conventional polymer materials, which is typically around 10^{-6}–10^{-5}/K. Hence, it cannot be expected to be easy to fabricate a photopolymer material with such a small shrinkage. We will next describe our approach to solving this problem.

16.3 Phenanthrenequinone-Doped Poly(methyl methacrylate) (PQ:PMMA) Photopolymer

It has been pointed out earlier that doped photopolymers are promising volume holographic storage materials. Our investigations focus on acrylates, more specifically on phenanthrenequinone-doped poly(methyl methacrylate) (PQ:PMMA). In this material, PMMA serves as the host matrix that provides mechanical stability, and PQ is the photosensitive dopant. The chemical structures of both substances are shown in Fig. 16.2.

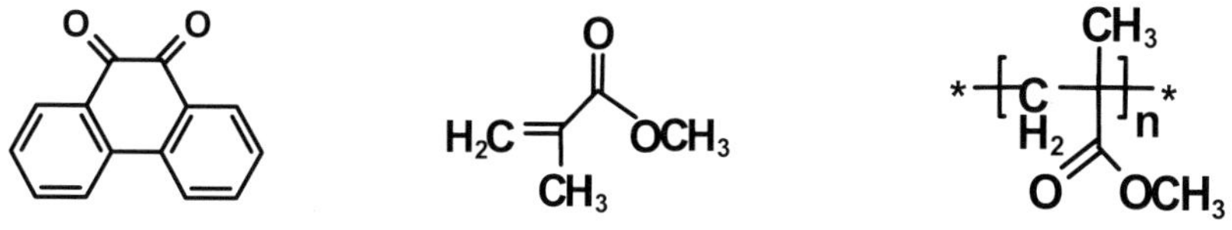

9,10-PhenanthreneQuinone   Methyl methacrylate   Poly(methyl methacrylate)

Figure 16.2 The chemical structures of PQ, MMA, and PMMA.

Through chemical and optical investigations of PQ:PMMA before and after light exposure, the physical mechanism of holographic recording in this material can be analyzed and summarized as follows:[25-28] in the bright regime of an exposing interference pattern, the quinone double bond on the carbonyl functional group of a PQ molecule is excited such that the PQ molecule becomes a radical. The radical reacts with the carbonic double bond on the vinyl group of a residual MMA molecule in the PMMA to form a new compound. This compound is less conjugated than the original molecular structure, and thus the refractive index of the material is changed locally. On a more global scale, the arising refractive index modulation follows the exposing modulated intensity pattern. In summary, the refractive index change for hologram recording is produced by the photochemical reaction of photosensitive elements (PQ) dispersed in the polymer matrix with residual monomers (MMA).

16.3.1 Strategy to minimize material shrinkage

The above recording mechanism gives us a strong hint of how to alleviate the shrinkage problem. The idea is to separate the photochemical reaction during holographic recording from the polymerization of the host monomer molecules during material preparation. The polymerization of monomers in our material forms a strong polymer matrix that can mechanically support the geometrical dimensions of the medium. This polymer structure will not be affected by holographic recording, and its geometrical dimensions remain stable during optical exposure. On the one hand, the material is photosensitive for hologram recording because of a small portion of PQ and the residual MMA that are dispersed within the PMMA structure. In other words, the material can be made photosensitive and the shrinkage problem can be reduced to a minimum at the same time.

To achieve this goal, it is essential to control the fabrication procedure such that most of the monomer molecules are polymerized to form the polymer matrix PMMA during the material preparation, while only a small portion of the MMA monomer molecules do not react and are left over for optical exposure usage during holographic recording.

16.3.2 Material fabrication of the PQ-doped PMMA photopolymer

In a series of experiments, a two-step fabrication procedure was developed that can produce PQ:PMMA samples of high quality. It requires MMA, PQ, and AIBN (2,2-Azo-bisisobutyrolnitrile) as main raw materials. In the first step, the thermoinitiator AIBN (~0.5%) and PQ molecules (up to 0.7%) are dissolved in solvent MMA. The solution is purified to remove undissolved particles so as to reduce light scattering centers. The purified solution is stirred (using a magneto stirrer) in an open glass bowl at room temperature for about 120 h until it becomes homogeneously viscoid. In the second step, the viscoid solution is poured into a glass mold, and then put in an oven that is kept at a temperature of 45°C to accelerate the polymerization process and the thermodecomposition rate of the AIBN. After 24 h, the sample has become a self-sustained solid block and can be removed from the mold. The final shape of the sample is determined by the geometry of this container. To form a disk, the viscoid solution is dispensed into a glass cell formed by two optically flat, disk-shaped glass plates with spacers in between.

16.3.3 Optical transmission and surface flatness

The finished PQ:PMMA samples have a yellowish appearance as one can see from the photograph in Fig. 16.3. Optical transmission measurements in the visible range show a strong absorption in the blue spectral range (< 450 nm), whereas the material becomes transparent for red and near-infrared wavelengths (> 540 nm); the absorption curve of Fig. 16.3 thereby stems from a sample with 2 mm thickness. Thus, an argon laser with wavelength 514.5 nm is a good choice as a light source for holographic experiments. At this wavelength the absorption coefficient is 2.7 cm^{-1}.

The above technique can produce large samples with high surface quality. Measurements of the surface flatness were carried out with a Mach-Zehnder interferometer. For a sample disk of 2 mm thickness, only about one fringe can typically be observed across the 5 in. diameter, which means that the samples are flat to within one wavelength over the diameter.

16.4 Holographic Characterization of PQ:PMMA

Next, we characterize the material properties that determine the suitability of PQ:PMMA for volume holographic data storage applications. According to Section 16.1, these are the shrinkage coefficient α, the holographic recording sensitivity S, and the dynamic range $M\#$.

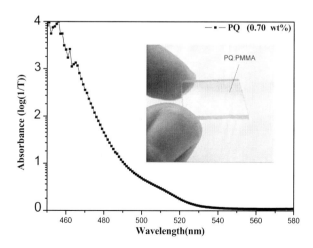

Figure 16.3 Typical absorption spectrum of 2 mm thick PQ/PMMA sample.

16.4.1 Material shrinkage coefficient

For the shrinkage evaluation, we have fabricated a 5 × 5 × 5 mm photopolymer cube and performed a Fourier holographic storage using an optical setup with 90 deg geometry that was essentially identical to the one on which the computer simulation of the shrinkage effect in Section 16.1.3 was based. The intensity of each beam was 2 mW/cm^2. An angle-multiplexing scheme was used in the experiment to record 250 Fourier holograms of a chessboard pattern in the same volume element of the cube. An LCTV with 320 × 240 pixels resolution served as the input device. To achieve equal diffraction efficiency, the exposure time of each hologram was set according to a recording schedule based on the use-and-exhaust behavior of photopolymer materials.[16]

The holograms were reconstructed and compared with the original chessboard pattern to assess the quality of the retrieved images. Figure 16.4 shows the original and 11 of the retrieved images. We also compared linear scans of the gray levels of the corresponding images. It was found that the retrieved images constructed the original chessboard pattern with high fidelity. Since images can be reconstructed completely from such a 5 mm thick block, the simulation results of Section 16.1.3 imply that shrinkage in this material is as small as 10^{-5}. Thus, our material is capable of recording multiplexed holograms in thick gratings. This result strongly supports our hypothesis that the two-step fabrication technique presented above is an effective way to minimize material shrinkage. It is interesting to note that the underlying design and fabrication strategy is not only applicable for PQ:PMMA; it is rather general and can be extended to other doped photopolymer systems. We anticipate that by suitably changing the doping composition, new PMMA-based photopolymers with further improved holographic recording characteristics will be developed.

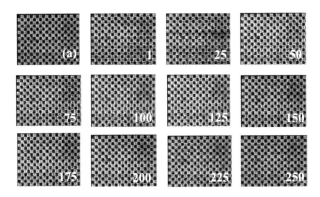

Figure 16.4 The original (a) and 11 of the retrieved images from a medium.

16.4.2 Dynamic range and sensitivity

Dynamic range and recording sensitivity are crucial parameters for the storage capacity and the read/write data rate that can be achieved with PQ:PMMA. To determine these parameters, one needs to perform multiple plane-wave hologram recordings at a single spot in the material by use of a multiplexing scheme. The series of holographic recordings is continued until the material is exhausted, showing no further photosensitive response. After each recording, the diffraction efficiency of each hologram is measured, and the summation of the square roots of the diffraction efficiencies forms a running curve of the cumulative grating strength as a function of cumulative exposure energy, i.e.,

$$C(E) = \sum_{i=1}^{n} \sqrt{\eta_i}, \qquad (16.5)$$

where E is the cumulative exposure energy and n is the total number of holograms that have been recorded by E. The curve indicates the dynamics of the buildup process of the multiple hologram recordings. When n approaches N, the material becomes exhausted and no more holograms can be recorded, thus $C(E)$ tends to be saturated, and its saturation value is equal to the $M\#$.

Figure 16.5 shows a typical experimental result for our photopolymer sample with 2 mm thickness. It can be estimated from the figure that the $M\#$ for this material is about 3.

The $M\#$, together with the exposure energy constant E_τ of a holographic storage material, can also be obtained from the running curve by fitting it with a exponential function of type $C(E) = M\#[1 - \exp(E/E_\tau)]$. Applying the above procedures to the data of Fig. 16.5, we obtain $M\# = 3.3$ and $E_\tau = 4.76$ J/cm^2.

By definition, the sensitivity can be calculated as the ratio of the change of the cumulative grating strength and the corresponding exposure energy of that hologram,

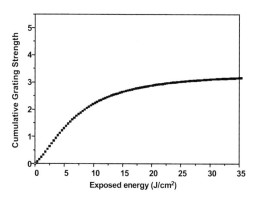

Figure 16.5 The running curve of the cumulative grating strength as a function of total exposure energy for a sample with thickness of 2 mm.

$$S = \frac{\Delta C(E)}{\Delta E}\bigg|_{\text{one hologram}} . \qquad (16.6)$$

Typically, the recording sensitivity is evaluated with a fresh material sample. During this stage, the cumulative exposure energy E is still small enough compared to E_τ, that the exponential function in the fitting expression for $C(E)$ can be linearized and the material sensitivity S is found to be

$$S = \frac{dC(E)}{dE}\bigg|_{E \to 0} \approx \frac{M\#}{E_\tau} . \qquad (16.7)$$

By inserting the above values into Eq. (16.7), the sensitivity is found to be 0.69 cm²/J. Using the same scheme, samples with different thicknesses have been characterized. In Fig. 16.6, the values for $M\#$ and the sensitivity are plotted as a function of thickness. It can be seen that both $M\#$ and S increase linearly with the thickness of the storage medium.

16.5 Holographic Data Storage in a PQ:PMMA Disk

In order to test PQ:PMMA under realistic conditions, a disk with 5 in. diameter and 2 mm thickness was fabricated with the objective of storing a digital data file on it using a read/write system that is controlled by a personal computer.[15]

16.5.1 Optical setup

A schematic drawing and a photograph of the optical setup used for this experiment are shown in Fig 16.7. The PQ:PMMA disk is mounted on a rotation stage and driven by a stepping motor. Light from an argon laser is delivered into the system through a single-mode fiber. A LCD SLM (with a resolution of 800 × 600 pixels and a pixel pitch of 33 × 33 µm) manufactured by Control Opto serves as data input device.

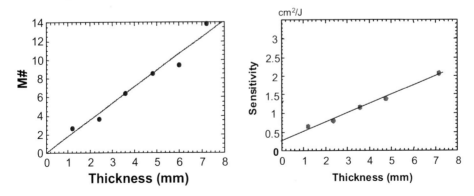

Figure 16.6 The $M\#$ and sensitivity of materials have been plotted as a function of sample thickness.

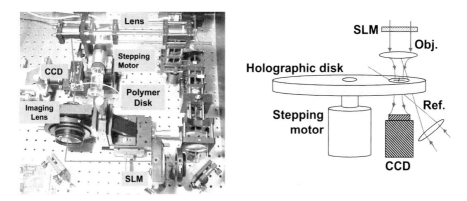

Figure 16.7 Photograph and schematic drawing of the holographic disk setup.

In the recording stage, the data file is presented on the LCD. Logic "1" bits are displayed as bright pixels, whereas logic "0" bits are displayed as dark pixels. In order to facilitate the optical disk servo, blocks of 3 × 3 LCD pixels are used as tracking marks presented in the four corners of each data page. The object beam is collimated before it illuminates the SLM and is projected to the CCD device by an imaging system. The disk is placed near the focal plane of the lens system. A divergent spherical beam with NA around 0.15 is used as the reference beam; it is incident on the disk from the back side to form a reflection-type hologram. Shift-multiplexing is used in this experiment, which requires the photopolymer disk to be rotated by a small angle to provide a small tangential shift between two hologram recordings. The whole rotational mechanism can be moved along the radial direction to different tracks by a motorized linear stage.

After the recording is completed, the holograms can be reconstructed individually by illuminating the disk with the same reference beam at the corresponding positions. The reconstructed images are captured by a digital CCD camera (with a resolution of 512 × 512 pixels and a pixel pitch of 7.4 × 7.4 µm), which is located at the image plane of the LCD device and connected to a frame grabber for postprocessing of the data. Since the pixel pitches of the input and output devices are different, it is necessary to have a suitable imaging system for demagnification from the LCD to the CCD. In order to simplify the imaging optics, we used only 432 × 432 pixels in the central part of the LCD. A commercially available photographic lens is sufficient for this purpose.

16.5.2 Experimental demonstration

The digital information to be stored is a color image encoded in BMP format. The 145 KB of data were split into 57 hologram pages and recorded on part of a track on the disk. After the recording, the holograms were optically reconstructed, and the raw data from the CCD camera were normalized, thresholded, and recombined to return a BMP file. Figure 16.8 shows the reconstructed and decoded BMP image, which has raw bit error rate (BER) of ~0.18%.

Figure 16.8 Four retrieved data pages (left) and reconstructed image from a PQ:PMMA photopolymer disk (right).

The achievable storage capacity of a holographic data storage system depends on two factors, namely, the Bragg selectivity of the optical system and the dynamic range of the material. The more stringent one of these two factors sets the actual limit. Bragg selectivity of a holographic disk along the tangential ($\Delta\delta_t$) and radial ($\Delta\delta_r$) directions can be expressed formulas as follows:[6]

$$\Delta\delta_t = \frac{\lambda z_o}{t \sin \theta_s}, \qquad \Delta\delta_r = \sqrt{\frac{\lambda}{t}} z_o, \qquad (16.8)$$

where t is the thickness of the disk, z_o is the distance from the focal point of the spherical reference beam to the disk, and θ_s is the intersection angle between the signal and the reference beam. In our system, $\Delta\delta_t$ and $\Delta\delta_r$ are calculated to be 2.96 µm and 2.7 mm, respectively. Because the recording area of each hologram has a diameter of 1.5 mm, it is safe to record tracks with a separation of 1.5 mm. On the other hand, in order to prevent crosstalk between neighboring holograms of the same track, we select the tangential shift between two holograms to be 10µm. With this recording density, a 5 in. disk can accommodate ~7.7 × 10^5 holograms. If the SLM delivers 1 Mbit per page, then the recording geometry sets the upper limit of the disk storage capacity at ~96 GB.

On the other hand, using the other perspective, the disk storage capacity can be estimated in terms of the *M#* of the recording material, which has a value of 3.3 for the 2 mm thick PQ/PMMA sample. According to the system parameters described in Section 16.1, this *M#* corresponds to ~60 GB on a 5 in. disk. Also, with the material sensitivity of 0.69 cm^2/J, this material is capable of supporting a data-recording rate of ~20 Mbit/s.

16.6 Advanced Photopolymer Materials

The above results show that the two-step thermopolymerization technique is capable of fabricating thick holographic recording materials with negligible shrinkage. In this section, we demonstrate exemplarily that suitable modifications of the basic two-step approach can lead to advanced material systems with custom-designed properties.

Our previous investigations have shown that the attachment of the PQ radical to the residual MMA molecule plays a key role for the holographic recording in PQ:PMMA. The basic idea is, therefore, to replace PQ with another photosensitive substance so that, for example, a higher material sensitivity, dynamic range, or different spectral sensitivity can be achieved.

16.6.1 Irgacure 784–doped PMMA

It has already been mentioned that the advancements in DPSS lasers at $\lambda = 532$ nm facilitate the development of compact holographic data storage systems. This has in turn evoked considerable interest in storage materials with maximum sensitivity at exactly this wavelength. Now we discuss how this design goal can be achieved with doped photopolymers.

Holographic recording in PQ:PMMA at $\lambda = 532$ nm reveals an $M\#$ of 1.7 and sensitivity of 0.009 cm^2/J, respectively, of which the $M\#$ is about half and the sensitivity is about two orders of magnitude lower of that at 514 nm (see Section 16.3.2). The reason for such low sensitivity and low dynamic range can be attributed to the fact that the resonance frequency of the PQ molecule is far away from 532 nm.

In order to fabricate a material with a better response at 532 nm, we start by changing the "ingredients" of the doped photopolymer system. The idea is to choose a new photosensitive molecule with a resonance frequency near 532 nm so that the recording sensitivity is properly enhanced. Irgacure molecules are a possible choice, since they are usually used for initiating the photopolymerization of acrylate monomer materials with green light. Figure 16.9(a) shows the molecular structure of one class of these photosensitive elements, named Irgacure 784. Figure 16.9(b) illustrates that under light exposure, the molecule is decomposed into two fragments. One of them is a radical with Ti ions that can react with an MMA monomer[29], thus enabling a recording process similar to that in PQ:PMMA. The absorption spectra of both PQ and Irgacure 784 molecules are shown in Fig. 16.10. It can be seen that Irgacure 784 has a side absorption band with a peak value close to 532 nm.

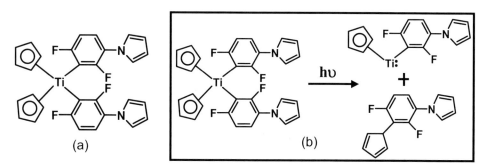

Figure 16.9 (a) The molecular structure of Irgacure 784 and (b) the photoreaction of this molecule.

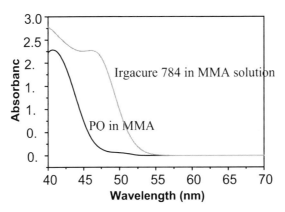

Figure 16.10 The absorption spectra of PQ and Irgacure 784 in MMA solutions.

For the fabrication of Irgacure-doped PMMA, we adopted a two-stage process. In stage one, ~0.7 wt% Irgacure 784 was added to a MMA monomer solution. The solution was stirred at room temperature until it became viscoid. Then, in the second stage, the sticky solution was poured into a glass mold with a 2 mm thick spacer, and the sample was baked at 40°C to form a solid polymer slab. Again, we fabricated a PQ:PMMA sample (with 0.7 wt% PQ) using an identical process for comparison. It was found that Irgacure 784–doped PMMA appears as a darker yellow color than PQ:PMMA, indicating that there is stronger absorption at the green wavelengths.

The holographic recording characteristics of both samples were measured according to the methodology described earlier. For each sample, 200 plane-wave holograms were recorded at one location of the material until the sample became saturated. In the experiment, exposure energies of 1.69 J/cm^2 and 4.46 J/cm^2 per hologram were used for Irgacure 784–doped and for PQ-doped

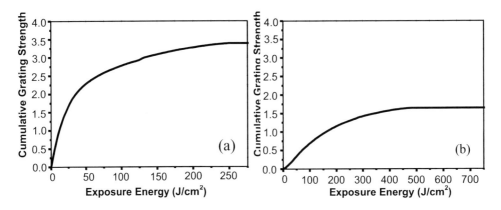

Figure 16.11 Typical running curves of the cumulative grating strength for a 2 mm thick sample by using a recording laser with a wavelength of 532 nm. The materials are (a) Irgacure 784– and (b) PQ -doped PMMA.

samples, respectively. The running curves for the cumulative grating strength of both materials are shown in Fig. 16.11. It can be seen that the Irgacure 784–doped PMMA sample possesses a much faster response.

As usual, the parameters $M\#$ and E_τ were extracted from an exponential function fitted to the running curves; then the sensitivities were calculated. It was found that whereas the $M\#$ of Irgacure 784–doped PMMA is only two times higher than that of PQ:PMMA (3.26 versus 1.7), there is a significant increase in sensitivity of about an order of magnitude (0.076 cm^2/J versus 9×10^{-3} cm^2/J, respectively). As a consequence, the material can support a much higher data recording rate at a wavelength of 532 nm. As for the $M\#$, the improvement is not (yet) significant. Further material engineering is needed to enhance the dynamic range of the material.

These results confirm again that the applied design and fabrication strategy is generally suitable to obtain low-shrinkage doped photopolymer systems for volume holographic data storage. Furthermore, by properly manipulating the components and synthesis conditions, it is possible to tailor the holographic characteristics of the recording material to fulfill specific requirements of a storage system.

16.7 Microintegration of the Optical Read/Write Head

The optical setups of Fig. 16.7 are well suited for fundamental investigations of the material aspects of PQ:PMMA; however, as typical experimental setups with numerous facilities for mechanical and optical adjustments, they are not optimized in terms of system complexity. In order to improve the commercial prospect of a holographic mass storage based on doped photopolymers, read/write setups with reduced system complexity and with the potential to be fabricated at low cost are necessary. Some general design aspects in this respect will be discussed below before we present the outline of a particularly promising system architecture that is based on the concept of planar-integrated free-space optics (PIFSO).

16.7.1 General design considerations

The first general design issue to be discussed here is whether to use a unidirectional or a bidirectional systems approach according to the definition of Fig. 16.12. In the first case, data are transferred into the memory on one side and retrieved from it on the opposite side, whereas memory access is from one side only in the second case. Since the signal paths for writing and reading do not overlap in a unidirectional setup, functional contentions of hardware components cannot occur. Furthermore, such setups are rather sparse in the sense that there is usually enough space for correctly positioning all hardware components, even if they are bulky and not custom designed. This is often a practical advantage, especially in laboratory experiments; on the other hand, it tends to be unfavorable for the construction of a compact setup.

Figure 16.12 (a) Unidirectional and (b) bidirectional system approaches.

Another disadvantage with regard to the construction of a commercial device is that in a unidirectional setup, especially in a holographic one, the memory needs to be accessed under precisely defined geometrical conditions from two sides, which significantly increases the mechanical complexity of the system. Therefore, we favor the bidirectional approach.

The question of how to implement a bidirectional volume holographic read/write head is not trivial since holography is basically a unidirectional method. As shown in Fig. 16.13(a), the signal beam S and the reconstructed beam R propagate in the same direction as long as the address beam Addr for data retrieval is identical to the reference beam Ref used during recording. However, holography can be made bidirectional by the use of a suitable reflection mechanism. Two possibilities are shown. In Fig. 16.13 (b), a phase-conjugate mirror (PCM) is used to generate an address beam A_{pc} with reversed propagation direction but identical wavefronts as the illuminating beam Addr. This Bragg-matched beam will then reconstruct the phase-conjugate version of the original signal beam. The PCM approach is elegant, but its implementation makes the system considerably more complex. A simpler alternative is based on a conventional mirror, as shown in Fig. 16.13(c). It is equivalent to the PCM approach if a suitable reference wave form (e.g., plane waves) is used and if the system is perfectly adjusted such that A_{mirr} becomes the counterpropagating version of the reference beam used during recording.

An important advantage of the above bidirectional systems approaches over unidirectional ones has to do with optical aberrations. Looking at the unidirectional setup of Fig. 16.7, we note that in the signal path the optical

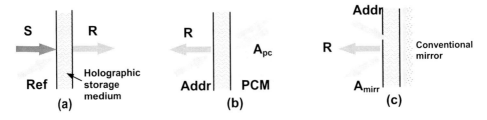

Figure 16.13 (a) Basic principle of holography with signal (S), reference (Ref), address (Addr), and reconstructed (R) beams. Bidirectional systems approaches based (b) on a phase-conjugate mirror (PCM) and (c) on a conventional mirror.

system contains a 4-f lens arrangement that realizes an imaging operation between the input array (LCTV, SLM) and the detector array (CCD chip). In the detector plane, one can therefore ideally observe a perfect copy of the intensity distribution just behind the LCTV. With the occurrence of optical aberrations, this image becomes degraded and in extreme cases (strongly aberrated imaging system, large field angle) these degradations may reach the point where the binary intensity level of a pixel can no longer be clearly identified. Since the holographic storage medium effectively reconstructs the wave field that was present during recording, the diffracted read-out beam will show the same optical aberrations as the original signal beam and it will produce the same degradations of the image in the detector plane as if the original input signal distribution was imaged by the optical system. This means that with a badly corrected 4-f lens system, a data page cannot be fully retrieved. To avoid a loss of information in a unidirectional system, aberrations need to be suppressed as much as possible. Usually this requires the use of well-corrected imaging objectives.

The situation is different for a bidirectional setup in which the reconstructed beam is effectively phase conjugate to the signal beam. Since both beams are perfect copies in terms of their waveforms, any aberrations that might build up in the optical system on the way from the input array to the storage medium are reversed and thus eliminated when the reconstructed beam takes the same way back.

The technical challenge with this system approach lies in the condition that reference and address beam need to be exactly phase conjugate. This condition implies that wave aberrations need to be kept at a level below the order of a wavelength. Although this is not a trivial task, it is alleviated by the fact that the waveform of the reference beam can, in principle, be chosen freely. If one chooses a simple type such as plane waves, then even with a single-lens system a perfect solution can be obtained because this single collimator lens only needs to image an object field that is reduced to a single point source (by contrast, the imaging system in the signal path needs to image the extended field of the input SLM).

The bidirectional approach of Fig. 16.13(c) with counterpropagating reference beam Ref and address beam A_{mirr} is a good starting point for the development of a commercial holographic read/write system because the requirements for the imaging/collimation optics are comparatively moderate. However, this system approach is not unrestrictedly compatible with standard holographic multiplexing schemes,[6] in particular with phase code multiplexing because this scheme requires a spatial modulation of the reference beams, and hence imaging objectives that are well corrected for an extended object field. Similar but less critical is the situation with angular multiplexing. The angular variation of the reference beams is equivalent to an extended object field, but since very small angular shifts are typically used it is possible to keep the total angular shift range limited to the paraxial region of the collimator lens. No problems arise with wavelength multiplexing, peristrophic multiplexing, and shift multiplexing. Especially the latter type is well suited if holographic disks are used as the storage medium because disk-based systems need to be equipped with a precision rotation stage anyway.

In addition to the above issues, considerable improvements of the optical read/write setup can be obtained through a system integration, which means to proceed from a component-based design and fabrication approach to a system-based one. As a consequence, fabrication complexity becomes largely independent of the number of components in the system, and most adjustment tasks are already settled during the design stage. Integrated systems thus have the potential to be fabricated with high quality at low cost. With regard to a commercial application, it is therefore almost mandatory to devise an integrated polymer-based holographic read/write setup. A highly suitable integration concept for this purpose is PIFSO.

16.7.2 Planar-optical system design

The idea of PIFSO[30–32] is to miniaturize and "fold" a free-space optical system with a certain desired functionality into a transparent substrate of a few millimeters thickness in such a way that all optical components fall onto the plane-parallel surfaces. Passive components such as lenses or beam deflectors can then be integrated into the surfaces, for example, through surface relief structuring, and the implementation as diffractive optical components offers an almost unlimited design freedom. Active components such as optoelectronic I/O devices can be bonded on top of the plane-parallel substrates. Reflective coatings ensure that optical signals propagate along zigzag paths inside the substrate. Since all passive components are arranged in a planar geometry, the optical system can be fabricated as a whole using mask-based techniques. Lithographic precision for the lateral positioning of components is thereby ensured.

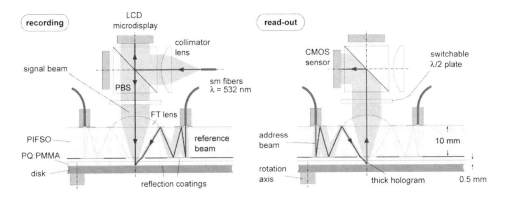

Figure 16.14 Schematic setup of the PIFSO-type reflection holographic read/write system depicting it in the recording and in the read-out mode. Reference and address beams are exactly counterpropagating along zigzag paths inside the PIFSO substrate. The FT lens performs an optical Fourier transformation from the LCD and the CMOS sensor to the PQ:PMMA layer on the storage disk.

Because of the monolithic integration into a rigid substrate, the optical system remains perfectly adjusted and long-term stable, and it is well protected against disturbing environmental influences. The application of replication techniques[33] and the use of plastic substrate materials[34] allow one to potentially keep the fabrication cost of PIFSO systems low.

We apply the PIFSO principle for the construction of a read/write head for holographic storage disks.[35] Figure 16.14 shows the proposed bidirectional Fourier optical system architecture in the recording and the readout mode. One can recognize an orthogonal signal beam, skew reference, and address beam paths that intersect at a target position on the reflective lower side of the photosensitive layer of the storage disk on which the hologram is recorded. All beams originate from the same laser source from which they are coupled into the PIFSO system by single-mode optical fibers. The relay of the signal beam from the fiber end to the disk is carried out by a 4-f system; in its Fourier plane, the expanded beam is 2D spatially modulated by a LCD microdisplay. To be able to record a complete signal page without loss, the diameter of the reference beams has to be matched to the width of the signal spectrum at the disk. Reference and address beams are furthermore perfectly collimated and counterpropagating so that they can be considered as mutually phase conjugate. Hence, if the reference beam is used for the recording of a hologram, then a readout with the address beam will generate the phase-conjugate version of the original signal beam; this reconstructed beam propagates through the 4-f system in the opposite direction and is projected onto a CMOS sensor. The reference/address beam relay is carried out by an assembly of diffractive lenses that are operated off axis to achieve a beam inclination of 30 deg. From Fig. 16.15, which depicts an unfolded version of this optical subsystem, one can recognize that the beam width is adjusted by a Galilei-type telescope formed by the two lenses next to the disk plane.

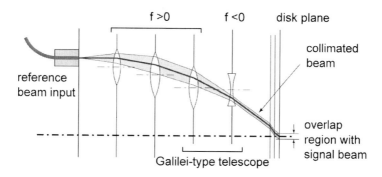

Figure 16.15 Unfolded version of the optical system that relays the reference beam to the holographic disk. The four lenses effectively implement a collimator and a Galilei-type telescope in series.

16.8 Conclusions

We have reviewed the progress and challenges of volume holographic data storage, and presented a methodology for the design and fabrication for a practical storage medium and a planar integration of a read/write head. A strategy using doped photopolymers to fulfill most of the material requirements has been proposed and demonstrated. The concept of using planar-integrated free-space optics to realize microintegration of the optical read/write head has been described. With these two innovations, we anticipate further accelerated advances in page-oriented holographic data storage techniques in the near future that may eventually lead to a commercial breakthrough.

References

1. Sarid, D., and Schechtman, B.H., "A roadmap for optical data storage applications," *Opt. Photon. News*, **18**, 34–37, 2007.
2. van Heerden, P.J., "Theory of optical information storage in solids," *Appl. Opt.*, **2**, 393–400, 1963.
3. Leith, E.N., Kozma, A., Upatnieks, J., Marks, J., and Massey, N., "Holographic data storage in three-dimensional media," *Appl. Opt.*, **5**, 1303–1311, 1966.
4. Mok, F.H., Tackitt, M.C., and Stoll, H.M., "Storage of 500 high-resolution holograms in a $LiNbO_3$ crystal," *Opt. Lett.*, **16**, 605–607, 1991.
5. See, for examples, InPhase technologies: http://www.inphase-technologies.com; Optware Corporation: http://www.optware.co.jp; Aprilis, Inc.: http://www.aprilisinc.com.
6. Barbastathis, G., and Psaltis, D., "Volume holographic multiplexing methods," *Holographic Data Storage*, H. Coufal, D. Psaltis, and G. Sincerbox (Eds.), p. 21–62, Springer-Verlag, New York (2000).
7. Psaltis, D., and Mok, F., "Holographic memories," *Sci. Am.*, **273**, 70–76, 1995.
8. Hesselink, L., Orlov, S.S., and Bashaw, M.C., "Holographic data storage systems," *Proc. IEEE*, **92**, 1231–1280, 2004.
9. Zhou, G., Mok, F., and Psaltis, D., "Beam deflectors and spatial light modulators for holographic storage applications," *Holographic Data Storage*, H. Coufal, D. Psaltis, G. Sincerbox (Eds.), p. 241–258, Springer-Verlag, New York (2000).
10. Lougnot, D.J., "Self-processing photopolymer materials for holographic recording," *C. R. Opt. Sci. Technol.*, **CR63**, 190–213, 1996.
11. Lessard, R.A., and Manivannan, G., Eds., *Selected Papers on Photopolymers*, Vol. MS 114, SPIE, Bellingham, WA (1995).
12. Schilling, M.L., Colvin, V.L., Dhar, L., Harris, A.L., Schilling, F.C., Katz, H.E., Wysocki, T., Hale, A., Blyer, L.L., and Boyd, C., "Acrylate oligomer-

based photopolymers for optical storage applications," *Chem. Mater.*, **11**, 247–254, 1999.

13. Waldman, D.A., Ingwall, R.T., Dal, P.K., Horner, M.G., Kolb, E.S., Li, H.-Y.S., Minns, R.A., and Schild, H.G., "Cationic ring-opening photopolymerization methods for holography," *Proc. SPIE*, **2689**, 127–141, 1996.

14. Lin, S.H., Hsu, K.Y., Chen, W.Z., and Whang, W.T., "Phenanthrenequinone-doped poly(methyl methacrylate) photopolymer bulk for volume holographic data storage," *Opt. Lett.*, **25**, 451–453, 2000.

15. Hsu, K.Y., and Lin, S.H., "Holographic data storage using photopolymer," *Proc. SPIE*, **5206**, 142–148, 2003.

16. Hsu, K.Y., Lin, S.H., Hsiao, Y.N., and Whang, W.T., "Experimental characterization of phenanthrenequinone-doped poly(methyl methacrylate) photopolymer for volume holographic storage," *Opt. Eng.*, **42**, 1390–1396, 2003.

17. Lin, S.H., Hsiao, Y.-N., Chen, P.-L., and Hsu, K.Y., "Doped poly(methyl methacrylate) photopolymers for holographic data storage" *J. Non. Opt. Phys. Mats.*, **15**, 239–247, 2006.

18. Mok, F., Burr, G., and Psaltis, D., "A system metric for holographic memory systems," *Opt. Lett.*, **21**, 886–888, 1996.

19. Pu, A., Curtis, K., and Psaltis, D., "Exposure schedule for multiplexing holograms in photopolymer films," *Opt. Eng.*, **35**, 2824–2829, 1996.

20. Dhar, L., Schnoes, M.G., Wysocki, T.L., Bair, H., Schilling, M., and Boyd, C., "Holographic storage of multiple high-capacity digital data pages in thick photopolymer," *Opt. Lett.*, **23**, 1710–1712, 1998.

21. Pu, A., and Psaltis, D., "High-density recording in photopolymer-based holographic three-dimensional disk," *Appl. Opt.*, **35**, 2389–2398, 1996.

22. Hsieh, M.L., and Hsu, K.Y., "Grating detuning effect on holographic memory in photopolymers," *Opt. Eng.*, **40**, 2125–2133, 2001.

23. Yi, X.M., Yeh, P., Gu, C., and Campbell, S., "Crosstalk in volume holographic memory," *Proc. IEEE*, **87**, 1912–1930, 1999.

24. Gu, C., Hong, J., McMichael, I., Saxena, R., and Mok, F., "Cross-talk-limited storage capacity of volume holographic memory," *J. Opt. Soc. Am. A*, **9**, 1978–1983, 1992.

25. Farid, S., Hess, D., Pfundt, G., Scholz, K.H., and Steffan, G., "Photoreactions of o-quinones with olefins: a new type of reaction leading to dioxole derivatives," *Chem. Commun.*, **434**, 638–639, 1968.

26. Franke, H., "Optical recording of refractive-index patterns in doped poly-(methyl methacrylate) films," *Appl. Opt.*, **23**, 2729–2733, 1984.

27. Hsiao, Y.-N., Whang, W.-T., and Lin, S.H., "Analyses on physical mechanism of holographic recording in phenanthrenequinone-doped

poly(methyl methacrylate) hybrid materials," *Opt. Eng.*, **43**, 1993–2002, 2004.

28. Mumbru, J., Solomatine, I., Psaltis, D., Lin, S.H., Hsu, K.Y., Chen, W.Z., and W.T. Whang, "Comparison of the recording dynamics of phenanthrenequinone-doped poly(methyl methacrylate) materials," *Opt. Commun.*, **194**, 103–108, 2001.

29. Degirmenci, M., Onen, A., Yagci, Y., and Pappas, S.P., "Photoinitiation of cationic polymerization by visible light activated titanocene in the presence of onium salts," *Polym. Bull.*, **46**, 443–449, 2001.

30. Jahns, J., and Huang, A., "Planar integration of free-space optical components," *Appl. Opt.*, **28**, 1602–1605, 1989.

31. Gruber, M., and Jahns, J., "Planar-integrated free-space optics – from components to systems," *Microoptics–From Technology to Applications*, Chap. 13, J. Jahns and K.-H. Brenner (Eds.), Springer, New York (2004).

32. Gruber, M., "Multichip module with planar-integrated free-space optical vector-matrix type interconnects," *Appl. Opt.*, **43**, 463–470, 2004.

33. Heming, R., Jahns, J., Gruber, M., Wittig, L.-C., and Kley, E.-B., "Combination of binary and analog lithography to fabricate efficient planar-integrated free-space optical interconnects," *Proc. DGaO* (2005), http://www.dgao-proceedings.de/download/106/106_p36.pdf

34. Wei, A.C., Gruber, M., Jarczynski, M., Jahns, J., and Shieh, H.-P., "Plastic Planar-Integrated Free-Space Optical Interconnector," *Jap. J. Appl. Phys.*, **46**, 5504–5507, 2007.

35. Gruber, M., Vieth, U., Hsu, K.-Y., .and Lin, H., "Design of a planar-integrated r/w-head for holographic data storage," *Proc. DGaO* (2007); http://www.dgao-proceedings.de/download/108/108_p47.pdf

Shiuan Huei Lin received his BSc in electrophysics in 1990, and a MSc and PhD in electro-optical engineering in 1992 and 1996, respectively, all from the National Chiao Tung University in Taiwan. He is currently an associate professor in the Department of Electrophysics at the National Chiao Tung University. His research interests are in the area of holographic storage, optical computing, optical devices, holographic materials, and holography for information processing. He is a member of SPIE.

Matthias Gruber received the Diplom in physics from the University of Erlangen, Germany, in 1991, and a PhD in electrical engineering from the University of Hagen, Germany, in 2003. He is currently an assistant professor of optical microsystems on the faculty of Mathematics and Computer Science of the University of Hagen. His research interests include optical information processing, micro-optics, and optical MEMS. He is a member of SPIE, OSA, and EOS.

Yi-Nan Hsiao received his BS in chemical engineering in 1998 from the Chung Yuan Christian University, and a MS and PhD in material science and engineering in 2000 and 2006, respectively, both from the National Chiao Tung University in Taiwan. He is currently a postdoctoral fellow at the Institute of Electro-Optical Engineering at the National Chiao Tung University. His research interests include photopolymer for holographic storage and organic material synthesis for optics.

Ken Y. Hsu received his BS in electrophysics in 1973 and his MS in electronic engineering in 1975, both from the National Chiao Tung University in Taiwan. He received his PhD in electrical engineering from the California Institute of Technology in 1989. He is currently a professor at the Display Institute and Department of Photonics at the National Chiao Tung University. His research interests are in the areas of optical computing, optical neural networks, optical devices, and holography for information processing. He is a Fellow of SPIE and OSA.

Chapter 17
Temporal Optical Processing Based on Talbot's Effects

Jürgen Jahns
FernUniversität Hagen, Germany

Adolf W. Lohmann
University of Erlangen, Germany

Hans Knuppertz
FernUniversität Hagen, Germany

17.1 Optical Signal Processing in Space and Time
17.2 William Henry Fox Talbot
17.3 Optical Tapped Delay Lines
17.4 Tapped Delay Line Based on Talbot Self-Imaging
17.5 Tapped Delay Line Using Talbot Band Experiments
17.6 Conclusion
References

17.1 Optical Signal Processing in Space and Time

For a long time, optical instruments were described in terms of spatial coordinates only while their temporal aspects were neglected. Typical examples are interferometers as well as systems for imaging and information processing. Those instruments are operating in a time-stationary fashion. In the time-stationary case, the *instantaneous intensity* $I(P,t)$ at a point P, with

$$I(P,t) = |u(P,t)|^2 \qquad (17.1)$$

and $u(P,t)$ being the optical field, is proportional to the time-averaged intensity

$$I(P) = \text{const} \langle I(P,t) \rangle, \qquad (17.2)$$

as described in Ref. 1. Here, t is the time coordinate and the angular brackets denote infinite time averaging. Time stationarity implies that the spatial coordinate along which propagation takes place, commonly denoted by the z coordinate, often only occurs as part of phase terms in the mathematical analysis. Hence, it is usually sufficient to describe time stationary optical systems in terms of the two spatial coordinates x and y.

However, the situation changes notably when we turn to nonstationary systems and the temporal dependence becomes an issue. In that case, Eq. (17.2) is no longer valid and z and t play an explicit role in the description of the optical system. Two recent developments have made it interesting to consider "optics in four dimensions":[2] laser light, with its narrow temporal frequency spectrum, has been available for a few decades and, more recently, practical laser systems, which generate short pulses with durations in the pico- and femtosecond range. This matches with the need for high-speed optical devices in communications and computing. Hence, it is worthwhile to consider in more detail the temporal aspects of optical instruments in order to investigate their capabilities for temporal processing.

Time does play an important role in optics. A few scientists paid attention to the temporal aspects of light already, long before it became a central issue in the context of optical communications and nonlinear optics. Already, in 1914, von Laue[3] wrote an article entitled "Die Freiheitsgrade von Strahlenbündeln" (The degrees of freedom of bundles of light rays). There he implicitly introduces what would now be called the "time-bandwidth product" of an optical signal. Even earlier, Talbot described a diffraction experiment in white light, where he discovered a band structure across the spectrum,[4] called after him "Talbot's bands." (Note: This phenomenon is not to be confused with the Talbot self-imaging effect that will also be discussed below.) In 1904, Schuster came up with an explanation of Talbot's bands based on the pulse theory of white light.[5] As a result, the occurrence of Talbot's bands has been considered to prove the pulse theory of white light. The pulse theory says that light from a broadband source consists of a sequence of uncorrelated short pulses. The Talbot band experiment is worth reviewing for historical reasons and because it offers an interesting approach to femtosecond-pulse processing.[6,7]

In order to categorize the different aspects of "time in optics," we suggest the scheme shown in Table 17.1, utilizing the analogy to spatial optics.[8] From this matrix, one can deduce four categories for the role of time in optics: first, topics

Table 17.1 An optical signal can be represented directly in the temporal and the spatial domain with the coordinates t and x. (For simplicity, a one-dimensional description is used here.) Equivalently, we may use a description for the respective frequency domains represented by the coordinates ν_t and ν_x.

Representation	Domain	
	Time	Space
Signal	$s(t)$	$u(x)$
Fourier spectrum	$\tilde{s}(\nu_t)$	$\tilde{u}(\nu_x)$

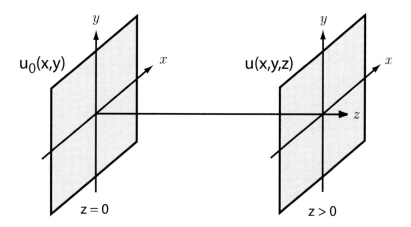

Figure 17.1 Spatial optics: near-field propagation. The field in the plane $z > 0$ is given by the diffraction integral in Eq. (17.4).

that can directly be described in the t domain; second, topics for which a frequency description in ν_t is appropriate; third, the space-time isomorphism with analogous phenomena in both domains; and fourth, the coupling of the spatial and temporal domain.

We will not describe the various aspects in detail here. The interested reader is referred to Ref. 8 for a detailed discussion. However, it is worthwhile comparing light propagation in the spatial and temporal dimension in order to clarify the various aspects. Propagation in space is usually described by Fresnel diffraction (Fig. 17.1). Diffraction represents spatial dispersion and is—in the paraxial case—mathematically described by a quadratic phase term $\nu_x^2 + \nu_y^2$ in the exponent of the diffraction integral,

$$u_0(x, y) = \int \int \tilde{u}_0(\nu_x, \nu_y) \exp\left[2\pi i \left(\nu_x x + \nu_y y\right)\right] d\nu_x d\nu_y, \quad (17.3)$$

$$u_0(x, y, z) = \int \int \tilde{u}_0(\nu_x, \nu_y) \\ \times \exp\left\{2\pi i \left[\nu_x x + \nu_y y - \left(\frac{\lambda z}{2}\right)\left(\nu_x^2 + \nu_y^2\right)\right]\right\} d\nu_x d\nu_y. \quad (17.4)$$

Propagation in the temporal domain (for example, of a light wave in an optical fiber, see Fig. 17.2) is described by a similar set of equations,[9]

$$s_0(t) = \int \int \tilde{s}_0(\nu_t) \exp\left(-2\pi i \nu_t t\right) d\nu_t, \quad (17.5)$$

$$s(t, L) = \int \int \tilde{s}_0(\nu_t) \exp\left[-2\pi i \left\{\nu_t \left(t + L/c\right) - D\nu_t^2\right\}\right] d\nu_t. \quad (17.6)$$

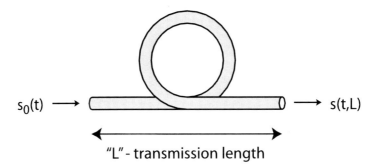

Figure 17.2 Propagation of a temporal signal: due to dispersion, the transmitted signal is spread out in time, and temporal dispersion is equivalent to spatial diffraction [see Eq. (17.6)].

The expression L/c is just a delay, the second term in the exponent, Dv_t^2, is the temporal dispersion. It is the analog to the quadratic expression of Eq. (17.4). Note that consider only quadratic and no higher phase terms for the sake of simplicity.

There are a number of interesting parallels between spatial and temporal optics. One is optical imaging. Usually, optical imaging is considered in the spatial domain. However, as is well known, one may consider one of the fundamental theorems of optics, Fermat's principle, to explain image formation in terms of the temporal properties of light: the optical path of a light ray is a minimum,[10]

$$\frac{1}{c}\int_{P_1}^{P_2} n\,dl = \int_{t(P_1)}^{t(P_2)} dt = \Delta t. \qquad (17.7)$$

Here, the integral on the left-hand side represents the optical path and the integral on the right-hand side, the duration for the light propagation. In this situation, the time coordinate comes into play in an implicit manner by the propagation delay of the light signal. However, it may also play an explicit role as in the case of a modulated light signal. Von Laue, mentioned earlier, discussed the temporal modulation of light and distinguished between "true" and "pseudo" modulation of light, depending on the ratio of the modulation frequency and the optical (carrier) frequency of the light signal.[11] The first case of true modulation of light was achieved by Connes et al. in 1962.[12]

Another interesting case where analogies occur between the temporal and the spatial domain is the self-imaging phenomenon. This description of this meanwhile well-known phenomenon is also due to Talbot, who reported it in 1836.[13] It occurs when a grating is illuminated by a plane wave (Fig. 17.3). In regular distances behind the grating images (so-called Talbot images) occur, i.e., the light distribution is the same as in the object plane. In the longitudinal separation of these Talbot images, the "Talbot distance" is determined by the period and the wavelength of the light. A second grating can be used to "probe" the periodic wavefield generated by the first in an interferometric setup.[14] The Talbot self-imaging effect is an example of Fresnel diffraction in the paraxial case, mathematically described by Eq. (17.4). Using the space-time analogy discussed above, a similar phenomenon must occur in the temporal domain. This case was, in fact, analyzed and demonstrated by using light propagation in a glass fiber by Jannson and Jannson.[15]

Thus far, we have talked about spatial phenomena and their temporal pendants. However, in the following, we are interested in the temporal properties of spatial structures. The reason is that we are interested in the implementation of ultrafast linear optical filters, the so-called tapped delay lines. Such ultrafast filters may become useful for all-optical signal processing in future communications. We will show that such tapped delay lines (TDLs) can be built based on the Talbot self-imaging effect and also based on the Talbot band experiment.

Because Talbot's work is so much in the focus of our activities, we will devote a short section to his life and work. Then we shall proceed with the Talbot self-imaging phenomenon and its use for implementing a TDL. After that, we will apply the same scheme to the Talbot band experiment. Here we will show that it can be extended by using microstructured elements to be useful for filtering purposes.

17.2 William Henry Fox Talbot

William Henry Fox Talbot was born on February 11, 1800. He is known to the optics community mostly for the self-imaging phenomenon of periodic wavefields.[13] However, to a wider audience he is known as one of the inventors of modern photography. Around 1834, he began to experiment with photosensitive chemistry. In January 1839, he announced his invention of the "photogenic" drawing, only two weeks after Daguerre's "daguerreotype" process debuted in France. Talbot's improved process, the "calotype" (derived from the Greek word καλοσ–beautiful) was introduced in 1840. This invention shortened exposure times and facilitated making multiple prints from a single negative. The latter feature, became the basis for modern photography. Talbot's work on photography, which is described here only briefly, took place simultaneously and in competition with Daguerre in France. The daguerreotype was based on a process in which the image is exposed directly onto a mirror-polished surface of silver bearing a coating of silver halide particles. The daguerreotype is a negative image, but the mirrored surface of the metal plate reflects the image and makes it appear positive in the proper light. Thus, daguerreotype is a direct photographic process without the capacity for duplication. The quality of the daguerreotype, however, exceeded the photographs taken by Talbot. Unlike Daguerre, who made his discovery available to the public at no fee, Talbot patented his invention in England and pursued infringers. This was one of the reasons why the daguerreotype was more popular, in addition to the sharpness of detail mentioned above. However, as we know now, the enormous success of photography is due to the possibility of creating a negative from which many prints can be generated.

Quite typical another Talbot story: Talbot owned a large piece of land located somewhere between London and Bristol. The Royal Railway system wanted its rails to cross Talbot's land. He agreed, with a modest demand: one penny for every railway passenger crossing his property. The Railway authorities smiled and agreed.

To the scientific optics community, Talbot is known for numerous contributions. For example, Talbot is the inventor of the half-toning technique, where gray levels are represented by periodically arranged binary (i.e., black and white) structures. Here, we want to consider two pieces of his scientific work that were published in short succession in 1836 and 1837. Well known is the self-imaging effect named after him the "Talbot effect".[13] The situation is described in Fig. 17.3. If a grating of period p is illuminated by a plane (quasi-)monochromatic wave of wavelength λ, then the wavefield repeats itself periodically behind the grating. The longitudinal period $z_T = 2p^2/\lambda$ is called the "self-imaging distance" or the "Talbot distance."

Interestingly, Talbot signed his papers often as "Henry Fox Talbot", which is the reason that one can often find references to his work under "W. H. F. Talbot" and "H. F. Talbot". Here, we show the beginning of the famous article from 1836[13] (Fig. 17.4).

Just one year later, in 1837, he published the already cited article, "An experiment on the interference of light" in which on just one page he reported the following observation in spectroscopic instruments:[4]

"Make a circular hole in a piece of card of the size of the pupil of the eye. Cover one half of this opening with an extremely thin film of glass (probably mica would answer the purpose as well, or better). Then view through this aperture a perfect spectrum formed by a prism of moderate dispersive power, and the spectrum will appear covered throughout its length with parallel obscure bands, resembling the absorptions produced by iodine vapour."

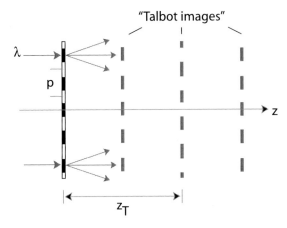

Figure 17.3 Talbot self-imaging: a monochromatic wavefield with lateral period p repeats itself in the longitudinal direction with a period z_T.

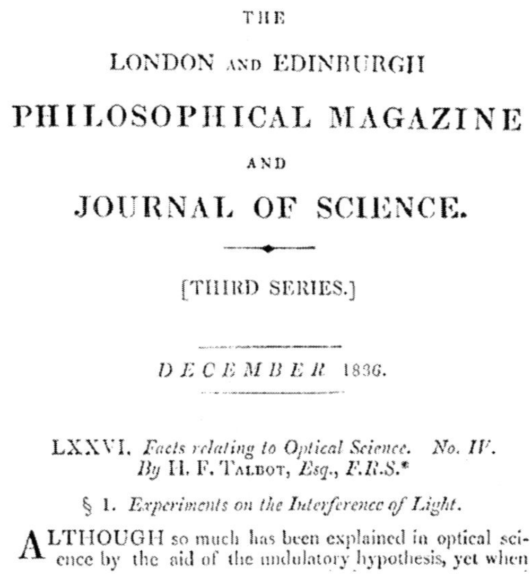

> THE
>
> LONDON AND EDINBURGH
>
> PHILOSOPHICAL MAGAZINE
>
> AND
>
> JOURNAL OF SCIENCE.
>
> [THIRD SERIES.]
>
> DECEMBER 1836.
>
> LXXVI. *Facts relating to Optical Science.* No. IV.
> By H. F. TALBOT, *Esq., F.R.S.*
>
> § 1. *Experiments on the Interference of Light.*
>
> ALTHOUGH so much has been explained in optical science by the aid of the undulatory hypothesis, yet when

Figure 17.4 Excerpt from Talbot's article on self-imaging. (Reprinted with permission from Taylor & Francis.)[13]

The situation is depicted in Fig. 17.5 with the difference that we show a grating as dispersive element. If the glass plate is inserted in the upper half of the illuminating beam as shown in the figure, then bands will appear in the upper, here called the "+1st order" but not in the "−1-st order." The occurrence of these bands (often called "Talbot's bands" or also "Talbot bands") is somewhat surprising, as is the asymmetry between the two diffraction orders. An explanation can be given in the temporal domain, which makes the Talbot band experiment an interesting object of investigation when talking about temporal optical filtering. We will come back to this issue in a later section.

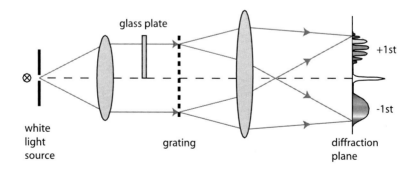

Figure 17.5 Talbot band experiment: a diffraction grating is illuminated by a collimated beam of white light from the source. A glass plate of suitable thickness is inserted halfway into the illuminating beam. In the output plane of the grating, spectrometer bands are observed only in the +1st order.

17.3 Optical Tapped Delay Lines

We are interested in implementing temporal filters for the processing of signals. Filtering may be of interest for various purposes: channel separation in optical communications systems, shaping and detection of femtosecond pulses, etc. This means that the frequency range at which the filtering operation is performed may range from the terahertz to the petahertz range. Temporal filtering may be done optically or electronically. Electronics runs up close to the terahertz limit; optical techniques range between several hundred terahertz down to a few terahertz. Both approaches are temporally compatible in the terahertz regime, which has recently gained in prominence.

Linear systems theory distinguishes between different filter types: finite-impulse resonse (FIR) and infinite-impulse response (IIR) filters. IIR filters are currently under investigation using a waveguide-optical implementation with ring resonator structures (e.g., Ref. 16). We want to build FIR filters. The general structure of an FIR filter is shown in Fig. 17.6. The input signal is split up into N branches. The nth branch is delayed by $n\tau$ and weighted with a factor a_n.

The output signal, $s_{\text{out}}(t)$, can be described mathematically by a discrete convolution

$$s_{\text{out}}(t) = \sum_{n=0}^{N-1} a_n s_{\text{in}}(t - n\tau) = [h(t)][s_{\text{in}}(t)] \quad \text{with } h(t) = \sum_{n=0}^{N-1} a_n \delta(t - n\tau) \tag{17.8}$$

A filter is represented mathematically in the time domain by its impulse response $h(t)$ or in the spectral domain by its transfer function, given as the Fourier transform of $h(t)$. For realizing a filter, it is necessary to implement well-defined weights a_n and time delays τ optically. For operation in the terahertz range, the delays need to be on the order of 10–1000 fs. In the following, we show two examples for optical tapped delay-line filters where the delays are implemented by using microstructured optics with structural depths of 1–100 μm.

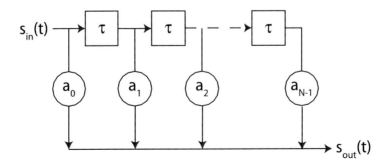

Figure 17.6 Graphical representation of a finite impulse response (FIR) filter (or tapped delay line).

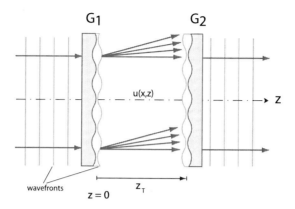

Figure 17.7 Principle of Talbot interferometer using two phase gratings: G_2 compensates the phase delays that occur in the Talbot image. Thus, a single plane wave emerges behind G_2. However, the temporal composition of the output wave is different from the "temporal structure" of the input wave (as indicated by the pulse symbols), because it includes contributions from different diffraction orders that have traveled different amounts of time between the two gratings.

17.4 Tapped Delay Line Based on Talbot Self-Imaging

Viénot and Froehly were probably the first to discuss the temporal processing of optical signals.[17] Later, Weiner et al.[18] and Sun et al.[19] carried the work further. It is based on the use of a grating interferometer and far-field diffraction. Here, we present an optical tapped delay-line filter that is using the Talbot effect, i.e., near-field diffraction. Specifically, the use of a Talbot interferometer[14] as a temporal filter was demonstrated[20] using the setup shown in Fig. 17.7. It uses a first beamsplitter grating to split an incoming signal into different diffraction orders. After propagation over a multiple of the Talbot distance, a second grating is used to combine the different diffraction orders into a single output beam. This works efficiently if both gratings are phase gratings and if G_2 is phase complementary to G_1.

For understanding the temporal behavior of the interferometer, it is sufficient to look at the propagation delays for the different diffraction orders. We denote the phase delay of the mth diffraction order with respect to the zeroth order as τ_m, it is given as

$$\tau_m = \frac{\Delta z}{c\sqrt{1-(m\lambda/p)^2}} - \Delta z \approx \frac{1}{2}\frac{\Delta z}{c}\left(\frac{m\lambda}{p}\right)^2. \tag{17.9}$$

The impulse response may be expressed by a discrete sum

$$h(t) = \sum_{m=0}^{N-1} |g_m|^2 \delta(t-\tau_m). \tag{17.10}$$

Here, g_m is the amplitude of the mth diffraction order. It is assumed that the gratings are designed in such a way that only N diffraction orders are significant. Notable is the quadratic behavior of the time delays, which was also confirmed experimentally[21] (Fig. 17.8). This makes the Talbot interferometer less suitable for general filtering purposes. We have shown, however, that the double-diffraction setup can be modified to generate linear delays. This is achieved if the more general case of self-imaging with Montgomery wavefields is used.[22] Whereas in the Talbot case, the spatial frequencies of the wavefield are given as $v_{x,n}^T = nv_1 = n/p$, in the case of Montgomery self-imaging, one has $v_{x,n}^M \approx \sqrt{n}v_1$. The latter expression holds in the case of the paraxial approximation. Unlike the Talbot effect, however, the Montgomery case of self-imaging is not limited to paraxial propagation. A Montgomery wavefield can be implemented by using a Fabry-Perot filter[23] or by using diffractive optics.[24,25] Going beyond the paraxial case does not represent any problem in theory, but possibly for the implementation of the diffractive optics. A detailed analysis of the use of double diffraction for temporal filtering was given in Ref. 26.

17.5 Tapped Delay Line Using Talbot Band Experiments

The classical Talbot band experiment[4] was shown earlier in Fig. 17.5. Here, we first want to explain the occurrence of the interference fringes and then extend it by using microstructured optics. A detailed mathematical description is found in Ref. 6.

A remarkable fact about the experiment is the asymmetry of the interference pattern in the +1st and the −1st diffraction order. It can be explained by using the pulse theory of white light, as mentioned above. Suppose, a short pulse hits the glass plate, then two pulses will emerge in the upper and lower halves of the aperture (Fig. 17.9). The delay (denoted here as τ_r) is, of course, given by the

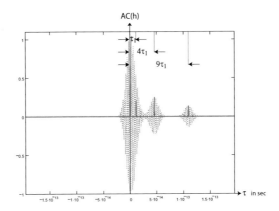

Figure 17.8 Temporal impulse response of the Talbot interferometer measured with femtosecond-pulse techniques.[21]

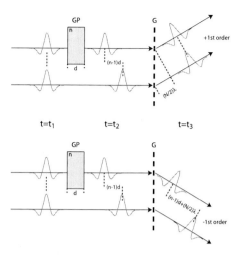

Figure 17.9 a) Three "snapshots" taken at different times for (a) the +1st order and (b) the −1st order. At $t = t_1$ both incoming pulses are at the same z position. At $t = t_2$, the lower pulse is delayed relative to the first by a distance $(n-1)d$, where d is the thickness of the glass plate. $t = t_3$: for (a) the detour due to grating diffraction is such that the lower pulse is delayed relative to the lower pulse so that both pulses are (approximately) in phase again. In the −1st order (b) the detour leads to a further delay of the upper pulse relative to the lower pulse, so that both do not overlap. (The dimensions are not at scale.)

optical path difference introduced by the glass plate, i.e., $\tau_r = \Delta n d / c$. Here, Δn is the index difference and d is the thickness of the plate. Here, the index r was chosen to indicate the refractive nature of the delay due to the glass plate. Behind grating G, the pulses experience a further time delay, now due to the optical detour caused by diffraction. This delay, denoted τ_d, varies with the diffraction order. In the +1st order, the pulse in the lower half is delayed with respect to the pulse in the upper half. If we assume a lateral offset between the two pulses of $(N/2)\,p$, the detour is $(N/2)\,\lambda$. Taking into account the diffraction order m, we can write for the delay due to diffraction

$$\tau_d = -m\left(\frac{N}{2}\right)\lambda. \qquad (17.11)$$

The minus sign reflects the fact that in the +1st order the time delays between lower and upper pulses have opposite signs. The total time delay between the lower and the upper pulses is

$$\tau = \tau_r + \tau_d. \qquad (17.12)$$

For $m = 1$, we obtain

$$\Delta n d = \left(\frac{N}{2}\right)\lambda. \qquad (17.13)$$

Hence, the delay due to the glass plate and the delay due to the grating diffraction compensate for the central wavelength if the thickness is

$$d = \frac{N\lambda}{2\Delta n}. \qquad (17.14)$$

For example, with $\lambda = 500$ nm, $N = 100$ and $\Delta n \approx 0.5$, one obtains $d \approx 50$ μm.

For the -1st order, the situation is different: here, the refractive delay is not compensated by the diffractive delay. Rather, it is increased. Hence, one cannot observe fringes. To be precise, however, one would have to say that there are fringes, but due to the large temporal separation between the two pulses the fringe spacing is extremely narrow.

Thus far, we have considered the case of the Talbot band experiment as originally described. It is possible, however, to modify and extend the setup. This can be achieved by using structured optical elements rather than a simple glass plate[7] and by using a reflective rather than refractive delay.[27] Here, we show the case of a reflective setup with a tilted retroreflector array used to implement multiple time delays (Fig. 17.10).

It can be shown that the situation is similar to the classical Talbot band experiment, and one can indeed observe a similar fringe pattern in the first diffraction order. The shape of the fringes is different, however, because now it is a case of multiple beam interferometry rather than two-beam interferometry. Thus, relatively sharp peaks occur rather than a sinusoidal fringe modulation.

This experiment demonstrates the usefulness of modified Talbot band experiments for implementing ultrabroadband tapped delay lines. The filter characteristics can be adjusted: a mask can be used for selecting the weights, the delays are controlled by the tilt. It is necessary, however, to fabricate the retroreflector array with suitable precision, fabrication tolerances should be small enough so that phase errors are smaller than $\lambda/4$, for example. This means that the shape of the array has

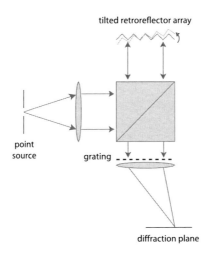

Figure 17.10 Reflective Talbot band experiment: the time delays are generated by a tilt of the retroreflector array.

to be precise to a few hundred nanometers, which is still a challenge. Ultraprecision cutting might be able to solve this challenge.

17.6 Conclusion

We have discussed temporal processing by using spatial microstructured devices. This allows one to implement filter devices that may be useful for processing terahertz signals. This is similar to the use of fiber optical tapped delay lines that are being used for the processing of radio frequency signals in the megahertz and gigahertz domain.

Temporal optics, however, has many other aspects that we did not discuss here. In this contribution, we were discussing *analog* filtering. In context with *digital* signal processing at ultrahigh speeds, additional aspects become relevant, such as latency, throughput, skew, etc. The fundamental trade offs and properties were discussed earlier in Ref. 29. Finally, we refer the reader to our Table 17.1, which includes all cases in which the time variable is a relevant parameter in an optical system.

References

1. Goodman, J.W., *Introduction to Fourier Optics*, 3rd ed., Roberts & Co., Englewood, CO (2005).

2. Lohmann, A.W., "Optics in four dimensions—Why?," *AIP Conf. Proc.*, **65**, 1–8, 1980.

3. v. Laue, M., "Die Freiheitsgrade von Strahlenbündeln," *Ann. Phys.*, **44**, 1197–1212, 1914.

4. Talbot, H.F., "An experiment on the interference of light," *Phil. Mag.*, **10**, 364–, 1837.

5. Schuster, A., "A simple explanation of Talbot's bands," *Phil. Mag. J. Sci.*, **6**, 1–8, 1904.

6. Lohmann, A.W., *Optical Information Processing*, Uttenreuth (1979) and Erlangen (2006).

7. Jahns, J., Lohmann A.W., and Bohling, M., "Talbot bands and temporal processing of optical signals," *J. Eur. Opt. Soc-RP*, **1**, 06001, 2006.

8. Jahns, J., and Lohmann, A.W., "Diffractive-optical processing of temporal signals, part I: basic principles," *Chin. Opt. Lett.*, **4**, 259–261, 2006.

9. Lohmann, A.W., and Mendlovic, D., "Temporal filtering with time lenses," *Appl. Opt.*, **31**, 6212–6219, 1992.

10. Born, M., and Wolf, E., *Principles of Optics, 7th (expanded) ed.*, Ch. 3.3, Cambridge University Press (1999).

11. Jaumann, J., "Die Formen der Lichtwelle, welche eine schwarze Temperaturstrahlung repräsentiert," *Z. Phys.*, **72**, 70, 1931.

12. Connes, P., Tuan, D.H., and Pinard, J., "Décomposition des raies spectrales par modulation en haute fréquence," *J. Phys. Radium*, **23**, 173–183, 1962.

13. Talbot, H.F., "Facts relating to optical sciences," *Phil. Mag.*, **9**, 401–407, 1836.

14. Lohmann, A.W., and Silva, D.E., "An interferometer based on the Talbot-effect," *Opt. Comm.*, **2**, 413–415, 1971.

15. Jannson, T., and Jannson, J., "Temporal self-imaging effect in single-mode fibers," *J. Opt. Soc. Am.*, **71**, 1373–1376, 1981.

16. Rasras, M.S., Madsen, C.K., Cappuzzo, M.A., Chen, E., Gomez, L.T., Laskowski, E.J., Griffin, A., Wong-Foy, A., Gasparyan, A., Kasper, A., Le Grange, J., and Patel, S.S., "Integrated resonance-enhanced variable optical delay lines," *IEEE Photon. Techn. Lett.*, **17**, 834–836, 2005.

17. Froehly, C., Colombeau, B., and Vampouille, M., "Shaping and analysis of picosecond light pulses," *Progress in Optics, vol. XX*, E. Wolf, Ed., 63–153, Elsevier Sciences, Amsterdam (1983).

18. Weiner, A.M., Heritage, J.P., and Kirschner, E.M., "Encoding and decoding of femtosecond pulses," *Opt. Lett.*, **13**, 300–302, 1988.

19. Sun, P.-C., Oba, K., Mazurenko, Y.T., and Fainman, Y., "Space-time processing with photorefractive volume holography," *Proc. IEEE*, **87**, 2086–2097, 1999.

20. Jahns, J., ElJoudi, E., Hagedorn, D., and Kinne, S., "Talbot interferometer as a time filter," *Optik*, **112**, 295–298, 2001.

21. Knuppertz, H., Jahns, J., and Grunwald, R., "Temporal impulse response of the Talbot interferometer," *Opt. Comm.*, **277**, 67–73, 2007.

22. Montgomery, W.D., "Self-imaging objects of infinite aperture," *J. Opt. Soc. Am.*, **57**, 772–778, 1967.

23. Indebetouw, G., "Polychromatic self-imaging," *J. Mod. Opt.*, **35**, 243–252, 1988.

24. Jahns, J., Knuppertz, H., and Lohmann, A.W., "Montgomery self-imaging effect using computer-generated diffractive optical elements," *Opt. Comm.*, **225**, 13–17, 2003.

25. Lohmann, A.W., Knuppertz, H., and Jahns, J., "Fractional Montgomery effect: a self-imaging phenomenon," *J. Opt. Soc. Am. A*, **22**, 1500–1508, 2005.

26. Jahns, J., and Lohmann, A.W., "Temporal filtering by double diffraction," *Appl. Opt.*, **43**, 4339–4344, 2004.

27. Sabatyan, A., and Jahns, J., "Retroreflector array as tapped delay-line filter for ultra-short optical pulses," *J. Eur. Opt. Soc. - RP*, **1**, 06022, 2006.

28. Lohmann, A.W., and Marathay, A.S., "Globality and speed of optical parallel processors," *Appl. Opt.*, **28**, 3838–3842, 1989.

Jürgen Jahns was born in Erlangen, Germany, in 1953. He received a diploma and doctorate in physics from the University of Erlangen-Nürnberg, Germany, in 1978 and 1982, respectively, where he was working in the group of A.W. Lohmann. From 1983 to 1986, he was with Siemens AG, Munich, Germany. From 1986 to 1994, he was a member of the technical staff in the Optical Computing Research Department, AT&T Bell Laboratories, Holmdel, NJ. Since 1994, he has been a full professor and chair of optical information technology at the Fernuniversität Hagen, Germany. With his group, he is working on microoptics and nanooptics for short pulses and interconnection. Jürgen Jahns has coauthored more than 80 journal articles and several textbooks on photonics and microoptics. He is a member of several optics societies and a fellow of the OSA since 1998 and of SPIE since 2007.

Adolf W. Lohmann studied physics in Hamburg, Germany, from 1946 until 1953. He held faculty positions in Germany, Sweden, USA, Mexico, Israel and finally again in Germany. He also worked in IBM California 1961–67 and NEC in Princeton 1992–93. Computer holography and optical information processing were his research areas. He is a member of several optics organizations, including the International Commission of Optics (ICO), where served as President in 1978–81. Adolf Lohmann is the recipient of the OSA 2008 Emmett N. Leith Medal and the SPIE 2008 A. E. Conrady Award.

Hans Knuppertz received his diploma in business administration in 1988. Until 2002 he acted as product and marketing manager for CAD and ERP software dedicated for electrotechnical companies. At the same time, he joined a distance learning study at the FernUniversität Hagen, Germany. There he received a diploma and doctorate in electrical engineering in 2001 and 2008, respectively. Since 2002, he has been a member of the scientific staff of the chair of optical information technology at the FernUniversität Hagen. His research interests lie in optical temporal filtering and microoptics for processing ultra short pulses. In these fields Hans Knuppertz has coauthored about 20 journal articles and conference contributions.

Chapter 18
Spectral Line-by-Line Shaping

Andrew M. Weiner, Chen-Bin Huang, Zhi Jiang, Daniel E. Leaird, and Jose Caraquitena
Purdue University

18.1 Introduction
18.2 Spectral Line-by-Line Shaping
18.3 Line-by-Line Shaping on a Mode-Locked Laser
    18.3.1 Optical and RF arbitrary waveform generation
    18.3.2 Pulse repetition rate multiplication
    18.3.3 Optical arbitrary pulse train generation using line-by-line shaping
18.4 Impact of Comb Frequency Instability on Shaped Waveforms
18.5 Shaping on Phase-Modulated Continuous-Wave Laser Comb
18.6 Summary
References

18.1 Introduction

Optical pulse shaping is a widely adopted technique in which intensity and phase manipulation of optical spectral components allow synthesis of user-specified ultrashort pulse fields according to a Fourier transform relationship.[1] Furthermore, mode-locked lasers producing combs of frequency-stabilized spectral lines have resulted in revolutionary advances in frequency metrology.[2] However, until recently pulse shapers addressed spectral lines in groups at low spectral resolution. Line-by-line pulse shaping,[3] in which spectral lines are resolved and manipulated individually, leads to a fundamentally new regime for optical arbitrary waveform generation (O-AWG)[4,5] in which the advantages of pulse shaping and of frequency combs are exploited simultaneously.

Bringing pulse shaping and frequency combs together allows O-AWG with both a controllable ultrafast time structure and long-term coherence. Intuitively, full control of individual frequency comb lines require (1) frequency-stabilized sources to generate stable spectral lines and (2) high-resolution pulse shapers to resolve and control individual spectral lines. O-AWG promises a broad impact both in optical science, allowing for example coherent control generalizations of comb-based time-frequency spectroscopies,[6] and in technology, enabling new truly coherent multiwavelength processing concepts for spread spectrum light wave communications and lidar.[7]

This chapter is organized as follows. The significance of spectral line-by-line shaping will be first compared to conventional pulse shaping. Shaper setup will be explained. Line-by-line shaping examples using an actively mode-locked fiber laser will be presented in the second section. Impact of frequency comb frequency instabilities over line-by-line shaped waveforms are explored in the third section. In the fourth section, shaping examples using alternative comb sources are given.

18.2 Spectral Line-by-Line Shaping

Previous pulse shapers have generally manipulated groups of spectral lines rather than individual lines, which results in waveform bursts that are separated in time with low duty factor and that are insensitive to the absolute frequency positions of the mode-locked comb. This is primarily due to the practical difficulty of building a pulse shaper capable of resolving each spectral line for typical mode-locked lasers with repetition rates below 1 GHz. When extending pulse shaping to independently manipulate the intensity and phase of individual spectral lines (line-by-line pulse shaping), the shaped pulses can overlap with each other, which leads to waveforms spanning the full time period between mode-locked pulses (100% duty factor). Waveform contributions arising from adjacent mode-locked pulses will overlap and interfere coherently in a manner sensitive to the offset of the frequency comb.[3] Such line-by-line control is an important step toward optical waveform generation (O-AWG) since the intensity and phase of each individual spectral line is independently controlled. Previous efforts toward spectral line-by-line control utilized a hyperfine filter but were limited within a narrow optical bandwidth—the free spectral range of this device.[4] Recently, we demonstrated spectral intensity/phase line-by-line pulse shaping[3] and thus O-AWG over a considerably broader band[5] based on high-resolution grating-based pulse shapers.

Group-of-lines pulse shaping is illustrated in Figure 18.1(a), where f_{rep} is the spacing between comb lines. Assuming that the pulse shaping occurs M lines at a time, the shaped pulses have maximum duration $1/Mf_{rep}$ and repeat with period $T = 1/f_{rep}$. Accordingly, the pulses are isolated in time. In contrast, for line-by-line pulse shaping, $M = 1$, as shown in Figure 18.1(b), the shaped pulses can overlap, which leads to interference between contributions from different input pulses in the overlapped region. Previously, a hyperfine wavelength-division multiplexing filter was used for spectral line-by-line phase manipulation with 5 GHz line spacing in an optical code division multiaccess system[5] and with 12.4 GHz spacing in photonic RF arbitrary waveform generation experiments[4] but without investigation of this pulse overlap issue. More importantly, the hyperfine wavelength-division device has a periodic spectral response, which means that

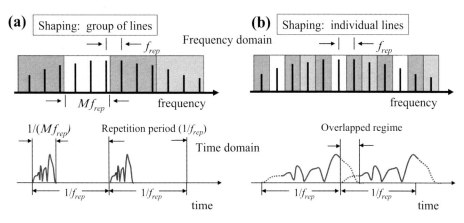

Figure 18.1 Illustration of pulse shaping with (a) group of lines and (b) line-by-line regime.

independent manipulation of the spectrum is possible only within one free spectral range, which was only 75–80 GHz in the experiments described in Ref. 4.

Our free-spaced grating-based shaper setup is illustrated in Fig. 18.2. A fiber-pigtailed collimator and subsequent telescope take the light out of the fiber and magnify the beam size to ~18 mm diameter on the 1200 grooves/mm grating in order to enhance the pulse shaper resolution. A fiberized polarization controller is used to optimize the grating diffraction efficiency. A polarization beam splitter (PBS) aligned for horizontal polarization is inserted between the collimator and the telescope. Discrete comb lines are diffracted by the grating and focused by a lens with 1000 mm focal length. The 3 dB spectral resolution of our shaper is 2.6 GHz, ensuring line-by-line shaping. A 2 × 128 pixel liquid crystal modulator (LCM) array is placed just before the lens focal plane to independently control both the amplitude and phase of individual spectral lines. A gold mirror is placed at the focal plane, leading to double-pass geometry, with all the spectral lines recombined into a single fiber and separated from the input via an optical circulator. Amplitude manipulation is realized by LCM control and the polarization extinction of the PBS in the recombination path back into the collimator. The total optical loss is around 10 dB. We note that in addition to LCMs, other spatial light modulators are also well known in pulse shaping.[1] Line-by-line pulse shaping experiments analogous to what is reported here should also be possible using other spatial light modulator technologies, provided that the key requirement of very high spectral resolution for clear separation of adjacent spectral lines is met.

An interesting point is that when individual comb line resolution is achieved, the field at the focal plane of the pulse shaper is stretched to a duration greater than the pulse repetition period. This places a limit on the speed at which the pulse shaper may be reprogrammed. This limit, which is fundamental in nature, restricts the ability to achieve simultaneous line-by-line resolution and pulse-by-pulse update. This effect has recently been investigated theoretically.[8] In the current experiments, in which the pulse shaping mask is reprogrammed only very slowly compared to the pulse period, such effects do not play a role.

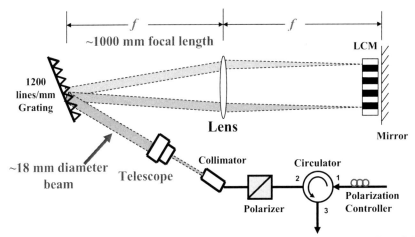

Figure 18.2 Schematic of a reflective line-by-line shaper. LCM: liquid crystal modulator.

18.3 Line-by-Line Shaping on a Mode-Locked Laser

18.3.1 Optical and RF arbitrary waveform generation

In this section we show shaping examples performed using a home-built harmonically mode-locked fiber laser producing 3 ps duration pulses at 10 GHz repetition rate with center wavelength of 1542.5 nm.[5] Each spectral line is spaced by 200 µm, corresponding to two LCM pixels. The frequency offset of the mode-locked comb is not actively stabilized; instead we exploit the passive frequency stability of the mode-locked comb, ~1 GHz over the time scale of our experiments. This suffices for these first proof-of-concept experiments on line-by-line pulse shaping control for O-AWG. To achieve a larger number of spectral lines, the 3 ps pulses are compressed to 400 fs by a commercially available dispersion-decreasing fiber soliton compressor. Either 3 ps or 400 fs pulses at 10 GHz repetition rate are used as input to the pulse shaper according to the required bandwidth for the specific demonstration.

Figure 18.3 shows power spectra (log scale) and resultant time-domain intensity waveforms when the pulse shaper is used to select just two spectral lines out of the mode-locked frequency comb. Data are shown for spectral line separations of 10 and 20 GHz [Figs. 18.3(a) and (b), 3 ps input pulses] and 400 and 500 GHz [Figs. 18.3(c) and (d), 400 fs input pulses]. The measured optical line widths are limited by the 0.01 nm resolution of the optical spectrum analyzer. Suppression ratios of the deselected lines are given for each case. Data shown in the figures reveal cosines with periods of 100, 50, 2.5, and 2 ps, respectively. These data are measured after an optical amplifier, using either a 50 GHz bandwidth photodiode and sampling oscilloscope (10 and 20 GHz) or standard short optical pulse intensity cross-correlation techniques (400 and 500 GHz). The pulse shaper itself is able to afford much higher frequency cosine waveforms, but

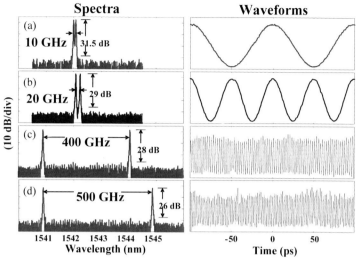

Figure 18.3. Selecting two spectral lines separated by (a) 10 GHz, (b) 20 GHz, (c) 400 GHz, and (d) 500 GHz and corresponding cosine waveforms (with periods of 100 ps, 50 ps, 2.5 ps, and 2 ps). Waveforms for (a) and (b) are measured using a sampling scope, while (c) and (d) are measured using cross-correlation.

here we are limited by the available optical bandwidth from the laser. This limitation can be relaxed or eliminated by utilizing short pulse compression techniques to achieve large bandwidth at high repetition rate.[5] The increased fluctuations and distortions in the cosine waveforms at higher frequencies (especially, 400 and 500 GHz) arise due to decreased optical power as the selected lines approach the edge of the input spectrum, which increases susceptibility to imperfect suppression of deselected lines and optical amplifier noise. Nevertheless, these data clearly demonstrate the potential to synthesize modulations over a very broad frequency range.

It is essential to note that strong suppression of deselected spectral lines is critical for accurate waveform generation. To validate this point, Fig. 18.4 shows the two-line selection waveforms viewed on a sampling scope with different suppression ratios of the deselected lines. In Fig. 18.4(a), the waveform evidently deviates from a cosine function with only 15 dB of suppression ratio. For suppression ratios of 25 and 38 dB, the scope traces reveal a better cosine waveform in Figs. 18.4(b) and (c), respectively.

Figure 18.5 shows another example of high fidelity O-AWG. Four spectral lines [five consecutive lines with the centerline blocked, Fig. 18.5(a)] are selected within a relatively narrow bandwidth to ease comparison with a theoretical calculation. By applying the same amplitude modulation, [1 1 0 1 1], but different phase modulation, [π 0 - 0 π] or [0 π - 0 π], two distinct waveforms are generated. The intensity cross-correlation measurements are in almost perfect agreement with and are essentially indistinguishable from the calculations (circles) [Figs. 18.5(b) and (c)]. It is clear that one can now synthesize high

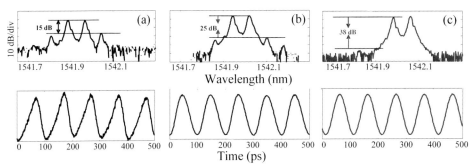

Figure 18.4 Waveforms with suppression ratio of the deselected lines: (a) 15 dB, (b) 25 dB, and (c) 38 dB.

fidelity, 100% duty factor optical waveforms with desired amplitude and phase by manipulating the individual spectral lines from a mode-locked frequency comb. As is well known, both intensity and phase gray level control can be readily achieved[1,3,9] using an LCM. Using this capability, the comb lines shown in Fig. 18.5(a) are intensity equalized in this demonstration.

One immediate application of line-by-line shaping is for radio-frequency arbitrary waveform generations (RF-AWGs). Figures 18.5(d) and (e) show sampling oscilloscope measurements of the electrical output generated when the

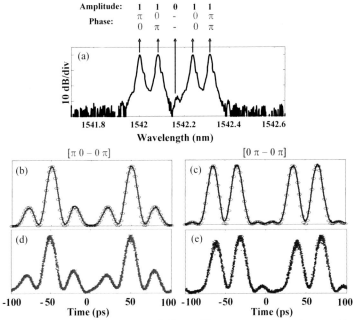

Figure 18.5 (a) Selecting four spectral lines (five consecutive lines with centerline blocked). (b) and (c) Waveforms measured by intensity cross-correlation with different applied spectral phases (red and blue curves). Calculations (black circles) are essentially indistinguishable from the data, showing the high fidelity of the generated waveforms. (d) and (e) Waveforms are detected by a 50 GHz photodiode and measured by sampling scope in persistent mode to demonstrate radio frequency arbitrary waveform generation (RF-AWG).

optical waveforms of Figs. 18.5(b) and (c) drive a 50 GHz photodiode. Such highly structured, high frequency, and broadband radio frequency waveforms are impossible to implement using current electrical AWG technology, which is typically limited to ~1 GHz.[10] RF-AWG has the potential to impact fields such as ultra-wideband (UWB) wireless,[11] which uses sub-nanosecond electrical bursts for communications and sensing, and impulse radar, where the use of highly structured transmit waveforms designed to optimize discrimination between different scattering targets has been proposed.[12]

18.3.2 Pulse repetition rate multiplication

Optical pulse trains with high repetition rates are very attractive for ultrahigh-speed optical communication systems and signal processing. Beyond those achievable by mode locking or direct modulation, one alternative is repetition-rate multiplication (RRM) of a lower rate source by applying amplitude[13–15] or phase[16–18] spectral filtering. A technique based on the temporal Talbot effect[19] is a simple and efficient method, since it simply requires the propagation of the pulse train in a first-order dispersive medium. This option, which is equivalent to a phase-only filtering process, has been traditionally demonstrated by using conventional optical fibers[16,17] or linearly chirped fiber Bragg gratings.[18] Here we demonstrate high-quality repetition-rate multiplication based on line-by-line shaping on a 9 GHz actively mode-locked fiber laser.[20]

Figure 18.6 illustrates the temporal Talbot effect for RRM.[19] This phenomenon occurs when periodic trains of optical pulses propagate through a first-order dispersive medium. An appropriate amount of dispersion leads to repetition-rate multiplication by an integer factor. For our purposes it is important to note that it is unnecessary to introduce continuous first-order dispersion.

In fact, only the spectral phases at frequencies equal to the discrete spectral lines, $\omega_n = \omega_0 + n\omega_{rep}$, are relevant, where ω_0 is the carrier frequency, ω_{rep} is the input repetition rate, and n is an integer. The Talbot condition provides the phase shifts to be applied to the spectral lines for RRM[19] as

$$\phi(\omega_n) = \frac{s}{r}\pi n^2, \quad (18.1)$$

where s and r are mutually prime integer numbers. The multiplication factor is given by the integer r. In practice, the actual phase shifts are applied by the pulse shaper modulo 2π, which yields a periodic phase filter,[21,22] as shown in Fig. 18.7.

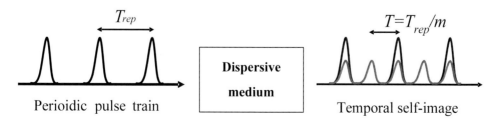

Figure 18.6 Schematic of temporal Talbot effect.

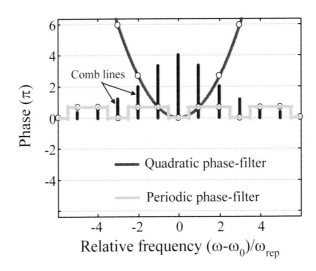

Figure 18.7 Talbot filter function. Periodic phase filter is applied in line-by-line shaping.

In Fig. 18.8 (top row), we show experimental oscilloscope traces both for the initial train and for the rate-multiplied trains at 18, 27, 36, and 45 GHz. We obtain ~1%, 1.5%, 1.5%, and 2% maximum peak-to-peak variations for the 18, 27, 36, and 45 GHz output trains, respectively. To optimize the intensity uniformity, we used an iterative correction algorithm that adaptively modifies the phases applied by the LCM.[20] Typically five or six iterations were required to minimize peak-to-peak pulse intensity variations in the output pulse train. In general, we are able to routinely generate multiplied trains with less than 3% peak-to-peak pulse intensity variations. For comparison, note that tunable pulse RRM experiments using dispersive fibers[23] and linearly chirped fiber Bragg gratings[24] yielded multiplied trains with significantly larger intensity variations in the ranges 8–17% and 7–17%, respectively. Note that RRM using phase-only filtering multiplies the repetition rate of the intensity but not of the field. This is because the phases of the multiplied pulses are not the same. Full RRM, in which the phases of all pulses are the same, is possible by performing spectral selection,[13,14] which may be achieved, for example, using our line-by-line shaper as an amplitude filter.

RF spectrum analyzer measurements of the multiplied pulse trains are plotted in Fig. 18.8 (bottom row) to illustrate the good intensity uniformities. For the original pulse train, RF tones at 9 GHz and its harmonics are observed, as expected. When the pulse shaper is programmed for 2× multiplication, Fig. 18.8(b), the now undesired tones at 9 and 27 GHz are suppressed almost into the noise floor, 46 dB below the 18 GHz tone. This large suppression factor is consistent with the degree of intensity variation from the scope traces. Figures 18.8(c)–(e) depict the RF spectra of the multiplied trains at 27, 36, and 45 GHz, respectively.

Spectral Line-by-Line Shaping

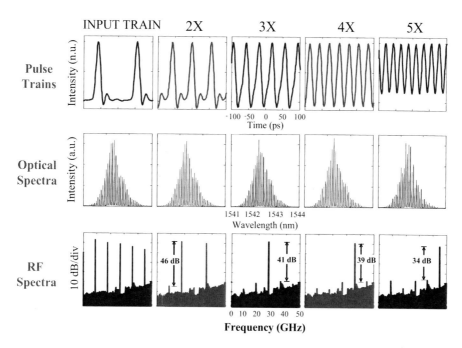

Figure 18.8 Experimental results of repetition-rate multiplication using phase filtering.

Our experimental demonstration of periodic spectral phase filtering via line-by-line pulse shaping produces excellent 2–5× RRM of a 9 GHz mode-locked laser. Higher repetition-rate multiplication factors can be readily achieved if trains with shorter pulses are used as the input. As a result, we have achieved unprecedented uniformity in pulse trains multiplied according to the temporal Talbot effect.

18.3.3 Optical arbitrary pulse train generation using line-by-line shaping

In addition to simple RRM demonstrated above, continuous pulse trains with controllable patterns and envelopes have also been proposed and demonstrated.[26–28] In this section we demonstrate optical arbitrary pulse train generation (OAPTG), in which a pulse train spanning the whole period can be generated and individual pulses can be independently manipulated to have different user-specified waveforms.[29] This is in contrast to previous work on pulse train generation using Fourier transform pulse shapers in the group-of-lines pulse shaping regime,[9,30,31] in which sequences of pulses with different shapes have been demonstrated, but only over a time aperture that is short compared to the repetition period.

For OAPTG, a generalized description of the spectral filter function $H(\omega)$ is needed in order to properly control the input comb lines through our line-by-line shaper. Mathematically, the filter function can be written as[29]

$$H(\omega) = \sum_k a_k \exp(-i\omega\tau_k) H_k(\omega), \quad (18.2)$$

where the complex amplitude a_k allows the control of individual amplitudes and phases, the delay of the kth pulse in the period is determined by τ_k, and the waveform of the kth pulse is determined by its individual filter function $H_k(\omega)$.

Figure 18.9(a) shows a pulse train with binary pattern [1 1 0 1] in each period, where pulses are located at 0, $T/4$, and $3T/4$. Pulse intensities for "1" are very close to equal and are negligible for "0" (25 dB lower). Both attributes still remain significant challenges in previous demonstrations.[28] An intensity ladder pattern [0 ¼ ½ 1] is presented in Fig. 18.9(b). Unlike simple RRM in Fig. 18.8 implemented by either a spectral intensity-only or phase-only filter function, the patterns in Fig. 18.9 generally require simultaneous spectral intensity and phase control. Figure 18.9(c) shows an example for arbitrary delay control. Three pulses in each period are located at 0 ps, –25 ps, and –75 ps, resulting in irregular pulse separations. As a result, the filter function is no longer periodic, as is clear from the spectrum in Fig. 18.9(c). These examples demonstrate the capability for arbitrary temporal intensity and delay control on individual pulses in each period.

Next we show that examples with pulses within a single period may also be controlled individually to have different user-specified waveforms. Figure 18.10(a) shows an example of three pulses in each period with one pulse broadened. The temporal broadening is realized by applying an additional spectral quadratic phase. Figure 18.10(b) shows an example of two pulses in each period with an additional spectral cubic phase applied on one of the pulses. This pulse shows an oscillatory tail as expected, which is a sign of a spectral cubic phase. Figure 18.10(c) shows an example with three pulses in each period, where pulse #1 has a linear spectral phase, pulse #2 has a quadratic spectral phase, and pulse #3 has a cubic spectral phase. These examples show the capability for OAPTG using line-by-line pulse shaping; in particular, the capability for achieving arbitrary waveform control on individual pulses in each period.

18.4 Impact of Comb Frequency Instability on Shaped Waveforms

The waveforms generated via spectral line-by-line shaping are sensitive to the optical comb frequency instability. The impact of frequency comb instability on optical waveform generation can be apparent even for the simple case of selecting only two spectral lines with applied phase control.[3] To provide a clear visualization, we compared the line-by-line phase control experiment and recorded both the optical spectra and sampling scope traces consecutively. With the harmonically mode-locked laser operating at 8.5 (relatively frequency stable) and at 10.5 (frequency unstable) GHz repetition rates, we can investigate the role of optical comb frequency fluctuations on line-by-line shaping.

Spectral Line-by-Line Shaping

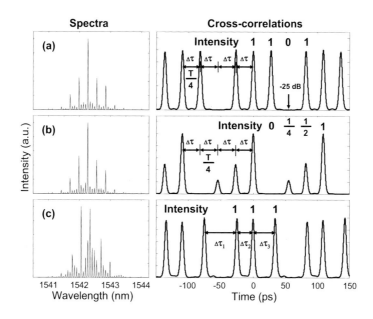

Figure 18.9 OAPTG examples. (a) Binary pattern [1 1 0 1]. (b) Ladder patten [0 ¼ ½ 1]. (c) Arbitrary delay ($\Delta\tau_1$ = 50 ps, $\Delta\tau_2$ = 25 ps, $\Delta\tau_3$ = 34.3 ps).

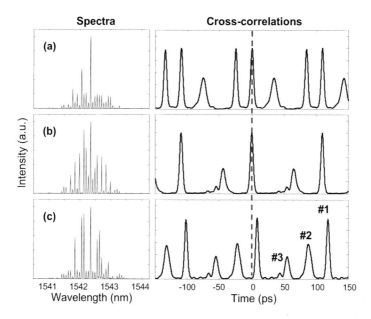

Figure 18.10 OAPTG examples. (a) One pulse width is broadened. (b) One pulse has a spectral cubic phase. (c) Pulse #1 has a spectral linear phase, pulse #2 has a spectral quadratic phase, pulse #3 has a spectral cubic phase.

Figure 18.11(a) shows an overlap of 100 scans for the two spectral lines and sampling scope traces for an 8.5 GHz pulse repetition rate, which show relatively stable features. If there is no pulse shaping ($\Phi = 0$), the sampling scope traces are clear, as shown in Fig. 18.11(b). If there is pulse shaping with $\Phi = \pi$ on one spectral line, the sampling scope traces become slightly noisy because of the small fluctuations of spectral lines. Nevertheless, the positions of the spectral lines are stable enough for line-by-line control, as demonstrated above.

When the laser repetition rate is tuned to 10.5 GHz, we observe empirically that the absolute frequency positions of the spectral lines become considerably less stable, with frequency fluctuations observable on a time scale of seconds, as shown in Fig. 18.11(c). We attribute the spectral line fluctuations in our harmonically mode-locked laser to comb-offset frequency fluctuations; we attribute the difference in optical frequency stability at different laser repetition rates to the frequency-dependent response of the microwave components used for feedback control of the cavity length. If there is no pulse shaping ($\Phi = 0$), under the conditions of our experiments the sampling scope traces are clear even if the spectral lines are relatively unstable, as shown in Fig. 18.11(d). However, if there is pulse shaping with a π phase shift on one spectral line, the sampling scope traces become extremely noisy because of the large fluctuations of the spectral line positions. This result can be understood from the time-domain overlap effect. For a π phase shift, the original laser pulses (corresponding to $\Phi = 0$) are reshaped to form waveforms with intensities in the temporal region where contributions from adjacent input pulses overlap. Since the adjacent original pulses have an unstable phase relationship (intimately related to unstable comb-offset frequency), their interference in the overlapped region as a result of pulse shaping leads to large intensity fluctuations. Much weaker fluctuations, if any, are observed at the time locations of the original input pulses, because there is little temporal overlap at those times. Clearly, this overlap effect leads to observation of a new time-dependent noise effect in pulse shaping that is directly linked to variations in the comb-offset frequency. Further studies of such noise phenomena, in which shaped waveforms are measured as the offset frequency of the comb is varied, are described in Ref. 32.

18.5 Shaping on Phase-Modulated Continuous-Wave Laser Comb

Optical frequency combs are usually generated by mode-locked lasers emitting periodic trains of ultrashort pulses. Highly stable, frequency-stabilized mode-locked lasers, such as self-referenced Ti:sapphire lasers,[33] are available at repetition rates (comb spacing) of ~1 GHz and below. Unfortunately current pulse shapers are unable to cleanly resolve such closely spaced spectral lines. Therefore, combs with larger line spacing are desired for line-by-line shaping; such combs also have the practical advantage of placing proportionally more power in individual lines. There are several alternative approaches for obtaining

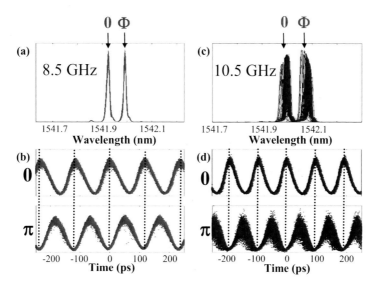

Figure 18.11 (a) Two relatively stable spectral lines at 8.5 GHz. (b) Sampling scope traces with phase modulation (0 and π) on one spectral line. The traces are scanned 100 times. (c) Two relatively unstable spectral lines at 10.5 GHz. (d) Sampling scope traces with phase modulation (0 and π) on one spectral line. The traces are scanned 100 times.

relatively frequency-stable combs with higher repetition rate. Harmonically mode-locked lasers are well known but often exhibit optical frequency instabilities, which lead to serious pulse shape noise when employed for O-AWG. Examples of frequency-stabilized harmonically mode-locked lasers have been demonstrated, but only with complicated control and/or compromised frequency tunability.[34,35] Approaches based on optical cavities are also relevant. For example, a Fabry-Perot cavity with free spectral range set to N times the comb spacing acts as a periodic transmission filter that passes only one of every N lines from an input comb source.[36] One may also generate a well-defined frequency comb by imposing a strong periodic phase modulation onto a continuous-wave (CW) laser, either without[37–39] or with the aid of a cavity.[40] This PMCW scheme has the significant advantage that the frequency offset of individual lines is controlled by the input CW laser and decoupled from the pulse generation process. In the following, we adopt such a modulation-of-CW source with additional spectral broadening via nonlinear fiber optics as the input for our O-AWG demonstrations.[41]

Figure 18.12(a) shows our experimental setup. A CW laser with specified 1 kHz line width centered at 1542 nm is modulated by two phase modulators, driven synchronously (with adjustable delay) by 20 GHz and 5 GHz cosine waveforms. The phase modulator driven at 20 GHz contributes to broad bandwidth, since generated bandwidth is proportional to modulation frequency. The modulator driven at 5 GHz determines the comb spacing and temporal periodicity. The resulting comb is manipulated by spectral line-by-line pulse shaper #1 to convert the broadband constant-intensity waveform to a pulse train.

Here we use a line-by-line shaper with one LC pixel to control one comb line, separated by 5 GHz. Our experiments are the first to resolve comb lines spaced by only 5 GHz in a pulse-shaping apparatus.

Figure 18.12(b) shows the measured spectrum after shaper #1; the discrete lines making up the spectrum are clear in the inset. Figure 18.12(c) shows the discrete spectral phases applied by shaper #1 onto the individual lines, in order to convert the phase-modulated but constant intensity field into bandwidth-limited 2.4 ps (FWHM) pulses [intensity autocorrelation shown in Fig. 18.12(d)]. Although not yet illustrating true O-AWG, such high-rate pulse generation starting from a CW source is already a powerful application of line-by-line pulse shaping.[42] The 2.4 ps pulses are then amplified by a fiber amplifier and directed into a dispersion-decreasing fiber (DDF) soliton compressor. The interplay of self-phase modulation and dispersion in the DDF yield pulse compression to durations as short as 270 fs [Fig. 18.12(e)]. Pulse compression is accompanied by strong spectral broadening. Figure 18.13 shows the spectrum after DDF compression obtained with slightly higher optical power coupled into the DDF. Over 1000 lines are generated between 1525 nm and 1565 nm.

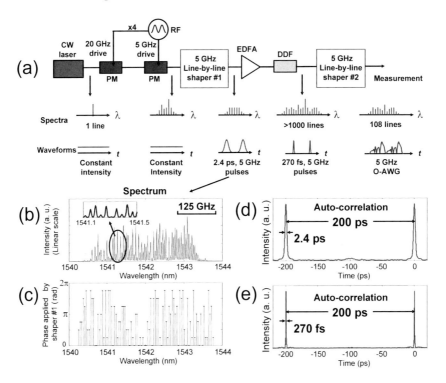

Figure 18.12 Experimental setup and high-rate ultrashort pulse generation. (a) Schematic diagram. (b) Spectrum (linear scale) after shaper #1. (c) Spectral phases applied by shaper #1. (d) and (e) Intensity autocorrelations after shaper #1 and DDF, respectively. PM: Phase modulator. RF: radio frequency. EDFA: Erbium-doped fiber amplifier. DDF: Dispersion decreasing fiber.

Spectral Line-by-Line Shaping 373

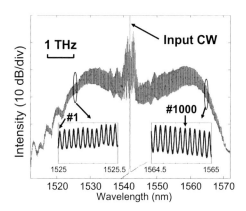

Figure 18.13 Generation of over 1000 stable spectral lines starting from one single line. Input CW is also shown for comparison.

Such a comb, created from a single CW laser, is highly useful for line-by-line pulse shaping studies. In our following O-AWG demonstrations we use a second line-by-line pulse shaper following the DDF [see Fig. 18.12(a)] to select and individually manipulate a set of 108 lines centered around 1537.5 nm and spanning a 540 GHz bandwidth. The spectrum is shown in Fig. 18.14(a). An intensity cross-correlation measurement of the resulting pulse train, demonstrating transform-limited 1.65 ps pulses at 5 GHz repetition rate, is also shown. In this and subsequent measurements, temporal intensity waveforms are obtained via cross-correlation with high-quality, pedestal-free, 1.02 ps reference pulses. Figure 18.14(b) demonstrates a simple example of line-by-line intensity control, where the LCM is programmed to block every other comb line. The resultant doubling of the comb spacing leads in the time domain to a doubling of the pulse repetition rate to 10 GHz.

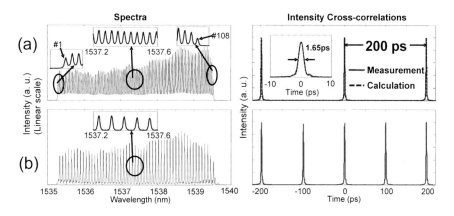

Figure 18.14 Line-by-line intensity control of 108 lines. (a) Spectrum and intensity cross-correlation. Calculated intensity cross-correlation is also shown for comparison. (b) Spectrum and intensity cross-correlation by blocking every other line.

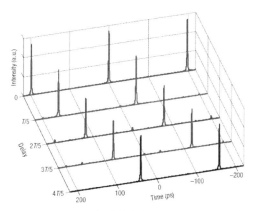

Figure 18.15 Line-by-line phase control of 108 lines: In a series of experiments, the pulse shaper applies various linear spectral phase ramps, resulting in delays proportional to the slope of the applied phase ramp. Intensity cross-correlation measurements are plotted for pulse trains delayed by 0, $T/5$, $2T/5$, $3T/5$, and $4T/5$.

We demonstrate the capability of line-by-line phase control for O-AWG by exploiting the relation $\tau(\omega) = -[\partial \Psi(\omega)/\partial \omega]$, where $\tau(\omega)$ and $\Psi(\omega)$ are the frequency-dependent delay and spectral phase, respectively. In the simplest case we apply linear spectral phase, which results in pure delay as shown in Fig. 18.15. Notice that the delay is scanned across the whole repetition period T (200 ps), which is only possible when individual spectral comb lines are independently manipulated. The waveforms remain clean, with only very small satellite pulses. This is further evidence that we are in the line-by-line regime. In the group-of-lines regime, a strongly stepped masking function gives rise to significant satellite pulses, with satellite intensity becoming equal to main pulse intensity at phase steps of π per pixel.[43]

Figure 18.16 shows examples of O-AWG with highly structured temporal features: each pulse is split into two pulses per period, one of which is delayed and the other of which has cubic spectral phase. In contrast to the data of Section 18.2.3, here we chose to program the shaper such that the delayed pulse and the cubic spectral phase pulse correspond to different halves of the spectrum. Figure 18.16(a) shows such an example with the two pulses still non-overlapping. The cubic spectral phase corresponding to quadratic frequency-dependent delay yields a strongly oscillatory tail in the time domain. Figure 18.16(b) shows an example with a larger cubic spectral phase. Here the waveform oscillations span the whole period with 100% duty cycle, which is one of the hallmarks of line-by-line pulse shaping. To confirm O-AWG fidelity, the calculated intensity cross-correlations are also shown for comparison. The agreement is excellent everywhere, even in the lowest intensity oscillations. Figures 18.16(c) and (d) show the unwrapped and wrapped discrete spectral phases applied to the 108 lines by shaper #2 in order to generate the waveform in Fig. 18.16(b). The linear and cubic spectral phases are clearly visible on the respective halves of the

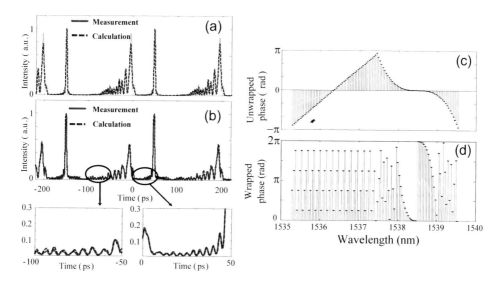

Figure 18.16 Line-by-line shaping of 108 lines; O-AWG with high temporal complexity. (a) The intensity cross-correlation; each pulse is split into two pulses per period, one of which is delayed and the other of which has a cubic spectral phase. Each pulse corresponds to one-half of the spectrum. The two pulses are still temporally separated. (b) A larger cubic phase on one pulse. Solid line: measured intensity cross-correlation. Dashed line: calculated intensity cross-correlation. (c) The unwrapped spectral phases applied to shaper #2 for waveform (b). (d) The wrapped spectral phases applied to shaper #2 for waveform (b).

spectrum. At some locations, the phase change per pixel is π or more—again a hallmark of operation in the line-by-line regime.

One important aspect of this setup is that it is scalable in both spectral line generation and spectral line manipulation. Our comb generator already provides > 1000 frequency-stable spectral lines available for future O-AWG investigations, and further nonlinear spectral broadening should be possible. By developing pulse shapers based on novel two-dimensional spectral dispersers,[44] a radically increased number of control elements should be possible, considering the large array technologies in use for two-dimensional display applications.

18.6 Summary

In summary, we demonstrated the capability of high-fidelity O-AWG with > 100% duty factor using spectral line-by-line pulse shaping. Line-by-line shaping on both a harmonically mode-locked fiber laser comb and a PMCW comb are shown. O-AWG examples of simple repetition-rate multiplication, optical arbitrary pulse train generation, radio-frequency waveform generation, and very complicated waveforms using more than 100 spectral lines are experimentally performed. Combining the detailed spectral control available from pulse shaping with the high coherence of short pulse frequency comb sources ushers in a new level of control over pulsed optical fields.

References

1. Weiner, A.M., "Femtosecond pulse shaping using spatial light modulators," *Rev. Sci. Inst.*, **71**, 1929–1960, 2000.
2. Udem, T., Holzwarth, R., and Hansch, T.W., "Optical frequency metrology," *Nature*, **416**, 233–237, 2002.
3. Jiang, Z., Seo, D.S., Leaird, D.E., and Weiner, A.M., "Spectral line-by-line pulse shaping," *Opt. Lett.*, **30**, 1557–1559, 2005.
4. Yilmaz, T., DePriest, C.M., Turpin, T., Abeles, J.H., and Delfyett, P.J., "Toward a photonic arbitrary waveform generator using a modelocked external cavity semiconductor laser," *IEEE Photon. Technol. Lett.*, **14**, 1608–1610, 2002.
5. Jiang, Z., Leaird, D.E., and Weiner, A.M., "Line-by-line pulse shaping control for optical arbitrary waveform generation," *Opt. Expr.*, **13**(25), 10431–10439, 2005.
6. Marian, A., Stowe, M.C., Lawall, J.R., Felint, D., and Ye, J., "United time-frequency spectroscopy for dynamics and global structure," *Science*, **306**, 2063–2068, 2004.
7. Swann, W.C., and Newbury, N.R., "Frequency-resolved coherent lidar using a femtosecond fiber laser," *Opt. Lett.*, **31**, 826–828, 2006.
8. Willits, J.T., Weiner, A.M., and Cundiff, S.T., "Theory of rapid-update line-by-line pulse shaping," *Opt. Expr.*, **16**(1), 315-327, 2008.
9. Wefers, M.M., and Nelson, K.A., "Generation of high-fidelity programmable ultrafast optical waveforms," *Opt. Lett.*, **20**, 1047–1049, 1995.
10. McKinney, J.D., Leaird, D.E., and Weiner, A.M., "Millimeter-wave arbitrary waveform generation with a direct space-to-time pulse shaper," *Opt. Lett.*, **27**(15), 1345–1347, 2002.
11. Win, M.Z., and Scholtz, R.A., "Ultra-wide bandwidth time-hopping spread-spectrum impulse radio for wireless multiple-access communications," *IEEE Trans. Commun.*, **48**, 679–691, 2000.
12. Rothwell, E., Nyquist, D.P., Chen, K.M., and Drachman, B., "Radar target discrimination using the extinctionpulse technique," *IEEE Trans. Antennas Propag.*, **AP-33**, 929–937, 1985.
13. Sizer II, T., "Increase in laser repetition rate by spectral selection," *IEEE J. Quant. Electron.*, **25**, pp. 97–103, 1989.
14. Petropoulos, P., Ibsen, M., Zervas, M.N., and Richardson, D.J., "Generation of a 40-GHz pulse stream by pulse multiplication with a sampled fiber Bragg grating," *Opt. Lett.*, **25**, 521–523, 2000.
15. Yiannopoulos, K., Vyrsokinos, K., Kehayas, E., Pleros, N., Vlachos, K., Avramopoulos, H., and Guekos, G., "Rate Multiplication by Double-Passing Fabry–Pérot Filtering," *IEEE Photon. Technol. Lett.*, **15**, 1294–1296, 2003.
16. Arahira, S., Kutsuzawa, S., Matsui, Y., Kunimatsu, D., and Ogawa, Y., "Repetition Frequency Multiplication of Mode-Locked Pulses Using Fiber Dispersion," *J. Lightwave Technol.*, **16**, 405–410, 1998.

17. IShake, ., Takara, H., Kawanishi, S., and Saruwatari, M., "High-repetition-rate optical pulse generation by using chirped optical pulses," *Electron. Lett.*, **34**, 792–793, 1998.
18. Longhi, S., Marano, M., Laporta, P., Svelto, O., Belmonte, M., Agogliati, B., Arcangeli, L., Pruneri, V., Zervas, M.N., and Ibsen, M., "40-GHz pulse-train generation at 1.5 mm with a chirped fiber grating as a frequency multiplier," *Opt. Lett.*, **25**(19), 1481–1483, 2000.
19. Azaña, J., and Muriel, M.A., "Temporal Self-Imaging Effects: Theory and Application for Multiplying Pulse Repetition Rates," *IEEE J. Sel. Top. Quant. Electron.*, **7**, 728–744, 2001.
20. Caraquitena, J., Jiang, Z., Leaird, D.E., and Weiner, A.M., "Tunable pulse repetition-rate multiplication using phase-only line-by-line pulse shaping," *Opt. Lett.*, **32**, 716–718, 2007.
21. Schroeder, M.R., *Number Theory in Science and Communication*, p. 180, Springer-Verlag, Berlin (1986).
22. Huang, C.-B., and Lai, Y.C., "Loss-Less Pulse Intensity Repetition-Rate Multiplication Using Optical All-Pass Filtering," *IEEE Photon. Technol. Lett.*, **12**, 167–169, 2000.
23. de Matos, C.J.S., and Taylor, J.R., "Tunable repetition-rate multiplication of a 10 GHz pulse train using linear and nonlinear fiber propagation," *Appl. Phys. Lett.*, **83**(26), 5356–5358, 2003.
24. Pudo, D., and Chen, L.R., "Tunable Passive All-Optical Pulse Repetition Rate Multiplier Using Fiber Bragg Gratings," *J. Lightwave Technol.*, **23**, 1729–1733, 2005.
25. Bolger, J.A., Hu, P.F., Mok, J.T., Blows, J.L., and Eggleton, B.J., "Talbot self-imaging and cross-phase modulation for generation of tunable high repetition rate pulse trains," *Opt. Commun.*, **249**, 431–439, 2005.
26. McKinney, J.D., Seo, D.S., Leaird, D.E., and Weiner, A.M., "Photonically assisted generation of arbitrary millimeterwave and microwave electromagnetic waveforms via direct space-to-time optical pulse shaping," *J. Lightwave Technol.*, **21**, 3020–3028, 2003.
27. Xia, B., and Chen, L.R., "Ring resonator arrays for pulse repetition rate multiplication and shaping," *IEEE Photonics Technol. Lett.*, **18**(19), 1999–2001, 2006.
28. Xia, B., Chen, L.R., Dumais, P., and Callender, C.L., "Ultrafast pulse train generation with binary code patterns using planar lightwave circuits," *Electron. Lett.*, **42**(19), 1119–1120, 2006.
29. Jiang, Z., Huang, C.-B., Leaird, D.E., and Weiner, A.M., "Spectral line-by-line pulse shaping for optical arbitrary pulse-train generation," *J. Opt. Soc. Am. B*, **24**, 2124–2128, 2007.
30. Weiner, A.M., and Leaird, D.E., "Generation of terahertz rate trains of femtosecond pulses by phase-only filtering," *Opt. Lett.*, **15**, 51–53, 1990.
31. Hillegas, C.W., Tull, J.X., Goswami, D., Strickland, D., and Warren, W.S., "Femtosecond laser pulse shaping by use of microsecond radio frequency pulses," *Opt. Lett.*, **19**, 737–739, 1994.

32. Huang, C.-B., Jiang, Z., Leaird, D.E., and Weiner, A.M., "The impact of optical comb stability on waveforms generated via spectral line-by-line pulse shaping," *Opt. Expr.*, **14**(26), 13164–13176, 2006.
33. Cundiff, S.T., "Phase stabilization of ultrashort optical pulses," *J. Phys. D*, **35**, R43–R59, 2002.
34. Gee, S., Quinlan, F., Ozharar, S., and Delfyett, P.J., "Simultaneous optical comb frequency stabilization and super-mode noise suppression of harmonically mode-locked semiconductor ring laser using an intracavity etalon," *IEEE Photon. Technol. Lett.*, **17**, 199–201,, 2005.
35. Yoshida, M., Yaguchi, T., Harada, S., and Nakazawa, M., "A 40 GHz regeneratively and harmonically mode-locked erbium-doped fiber laser and its longitudinal-mode characteristics," *IEICE Trans. Electron.*, **E87C**, 1166–1172, 2004.
36. Diddams, S.A., Hollberg, L., and Mbele, V., "Molecular fingerprinting with the resolved modes of a femtosecond laser frequency comb," *Nature,* **445**, 627–630, 2007.
37. Murata, H., Morimoto, A., Kobayashi, T., and Yamamoto, S., "Optical pulse generation by electrooptic-modulation method and its application to integrated ultrashort pulse generators," *IEEE J. Sel. Top. Quant. Electron.*, **6**, 1325–1331,, 2000.
38. Hisatake, S., Nakase, Y., Shibuya, K., and Kobayashi, T., "Generation of flat power-envelope terahertz-wide modulation sidebands from a continuous-wave laser based on an external electro-optic phase modulator," *Opt. Lett.*, **30**, 777–779,, 2005.
39. Jiang, Z., Leaird, D.E., and Weiner,A.M., "Optical processing based on spectral line-by-line pulse shaping on a phase modulated CW laser," *IEEE Quant. Electron.*, **42**, 657–665, 2006.
40. Imai, K., Kourogi, M., and Ohtsu, M., "30-THz span optical frequency comb generation by self-phase modulation in an optical fiber," *IEEE J. Quant. Electron.*, **34**, 54–60,, 1998.
41. Jiang, Z., Huang, C.-B., Leaird, D.E., Weiner, A.M., "Optical arbitrary waveform processing of more than 100 spectral comb lines," *Nature Photon.*, **1**, 463–467, 2007.
42. Huang, C.-B., Jiang, Z., Leaird, D.E., and Weiner, A.M., "High-rate femtosecond pulse generation via line-by-line processing of a phase-modulated CW laser frequency comb," *Electron. Lett.*, **42**(19), 1114–1115, 2006.
43. Weiner, A.M., Leaird, D.E., Patel,J.S., and Wullert, J.R., "Programmable shaping of femtosecond optical pulses by use of a 128-element liquid crystal phase modulator," *IEEE J. Quant. Electron.* **28,** 908–920, 1992.
44. Wang, S.X., Xiao, S., and Weiner, A.M., "Broadband, high spectral resolution 2-D wavelength-parallel polarimeter for dense WDM systems," *Opt. Expr.,* **13**(23), 9374–9380, 2005.

Andrew M. Weiner graduated from M.I.T. in 1984 with a ScD in electrical engineering. He joined Bellcore, first as a member of technical staff and later as manager of ultrafast optics and optical signal processing research. Prof. Weiner moved to Purdue University in 1992 and is currently the Scifres Distinguished Professor of Electrical and Computer Engineering. His research focuses on ultrafast optics signal processing and applications to high-speed optical communications and ultrawideband wireless. He is especially well known for pioneering work in the field of femtosecond pulse shaping. Prof. Weiner is a member of the National Academy of Engineering and a Fellow both of the Optical Society of America (OSA) and of the Institute of Electrical and Electronics Engineers (IEEE). He has won numerous awards for his research, including the Hertz Foundation Doctoral Thesis Prize (1984), the OSA Adolph Lomb Medal (1990), the Curtis McGraw Research Award of the American Society of Engineering Education (1997), the International Commission for Optics Prize (1997), the IEEE LEOS William Streifer Scientific Achievement Award (1999), the Alexander von Humboldt Foundation Research Award for Senior U.S. Scientists (2000), and the inaugural Research Excellence Award from the Schools of Engineering at Purdue (2003). Prof. Weiner has published six book chapters and over 200 journal articles and has been author or coauthor of over 350 conference papers. He holds 10 U.S. patents. Prof. Weiner has served as co-chair of the Conference on Lasers and Electro-optics and the International Conference on Ultrafast Phenomena and as associate editor of several journals. He has also served as Secretary/Treasurer of IEEE LEOS and as a Vice-President of the International Commission for Optics (ICO).

Chen-Bin Huang received a BS in electrical engineering from National Tsing Hua University, Hsinchu, Taiwan in 1997 and a MS in electro-optical engineering from National Chiao Tung University, Hsinchu, Taiwan in 1999. He is currently working towards a PhD at the School of Electrical and Computer Engineering, Purdue University. His current research interests include optical and RF arbitrary waveform generations and characterizations of optical frequency combs. From 1999 to 2003, he joined the Opto-Electronics & Systems Laboratories (OES), Industrial Technology Research Institute (ITRI), Taiwan, as a research engineer, developing passive fiber optical and photonic crystal devices. He was awarded with the Master's Thesis of the Year by the Optical Engineering Society of Republic of China in 1999, Personal Research Achievement Award by OES, and Personal Distinguished Research Achievement Award by ITRI, both in 2002. He received the Andrews and Mary I. Williams Fellowship at Purdue University, in 2004 and 2005. He was selected as a finalist for the LEOS 2007 Best Student Paper Award.

Zhi Jiang received his BS (highest honors) and MS in electronics engineering from Tsinghua University, Beijing, China, in 1999 and 2002, respectively, and a PhD in electrical and computer engineering from Purdue University in 2006. He is currently a Beckman Institute Postdoctoral Fellow, University of Illinois at

Urbana-Champaign (UIUC). His research interests include biomedical optics, ultrafast optics, and fiber optics. During his PhD study, Dr. Jiang received the Ross and Mary I. Williams Fellowship from Purdue University and the 2005 IEEE/LEOS Graduate Student Fellowship. He was selected as a finalist for the 2005 OSA New Focus/Bookham Student Award and a finalist for the Purdue-Chorafas Best Thesis Award for his PhD thesis work. He was a recipient of the 2005 Chinese Government Award for Outstanding Graduate Students Abroad. In 2007, he was selected as one of the four recipients of the Beckman Postdoctoral Fellowship.

Daniel E. Leaird received a BS in physics from Ball State University, Muncie, IN, in 1987, and a MS and PhD from the School of Electrical and Computer Engineering, Purdue University, in 1996 and 2000 respectively. He joined Bell Communications Research (Bellcore), Red Bank, NJ, as a senior staff technologist in 1987, and later advanced to member of technical staff. From 1987 to 1994, he worked in the Ultrafast Optics and Optical Signal Processing Research Group. Dr. Leaird is currently a senior research scientist and laboratory manager of the Ultrafast Optics and Optical Fiber Communications Laboratory in the School of Electrical and Computer Engineering, Purdue University, where he has been since 1994. He has coauthored approximately 80 journal articles, 100 conference proceedings, and has three issued U.S. patents. He is active in the optics community and professional organizations including the Optical Society of America and IEEE Lasers and Electro-Optics Society (LEOS) where he is the chair of the Ultrafast technical committee as well as serving as a consultant to venture capitalists by performing technical due diligence. He also serves as a frequent reviewer for *Optics Letters*, *Photonics Technology Letters*, *Applied Optics*, and *Journal of the Optical Society of America B* in addition to serving on National Science Foundation review panels in the SBIR program. Dr. Leaird has received several awards for his work in the ultrafast optics field including a Bellcore "Award of Excellence," a Magoon Award for outstanding teaching, and an Optical Society of America/New Focus Student Award.

José Caraquitena received his BS, MS, and PhD in physics from the Universidad de Valencia, Valencia, Spain, in 1999, 2001, and 2004, respectively. His thesis work focused on space-time dualities and propagation of ultrashort pulsed beams in free-space dispersion-compensated optical setups. From 2005 to 2007, he was a postdoctoral research fellow in the Ultrafast Optics and Optical Fiber Communications Laboratory at the School of Electrical and Computer Engineering, Purdue University, where he performed research in high-resolution line-by-line pulse shaping. Since July 2007, he is a postdoctoral researcher at the Nanophotonics Technology Center, Universidad Politécnica de Valencia, Valencia, Spain. His current research interests include optical signal processing and microwave photonics.

Chapter 19
Optical Processing with Longitudinally Decomposed Ultrashort Optical Pulses

Robert Saperstein and Yeshaiahu Fainman
University of California San Diego

19.1 Introduction
19.2 Longitudinal Spectral Decomposition
    19.2.1 Theoretical description
    19.2.2 Practical realization
19.3 Optical Pulse Shaping
    19.3.1 Operating principle
    19.3.2 Experimental demonstration
    19.3.3 Distortion from higher-order dispersion
    19.3.4 Waveform Detection
19.4 Microwave Spectrum Analysis
    19.4.1 Operating principle
    19.4.2 Experimental realization
    19.4.3 Waveform synthesis
    19.4.4 Distortion from higher-order dispersion
19.5 High-Speed Optical Reflectometry
    19.5.1 Introduction
    19.5.2 Experimental demonstration
19.6 Conclusion

19.1 Introduction

Fourier methods are widely used in optical information processing for time domain waveform synthesis and detection. In such approaches an input pulse is spectrally decomposed and modulated by an appropriately designed spectral filter. In general, various ultrafast applications may need either generation or characterization of waveforms depending whether their frequency response is or is not known a priori. Ultrafast pulses prove useful in driving or probing such systems because their broad, deterministic complex spectral amplitude "sees" a large continuous portion of the system transfer function. Traditionally, space domain spectral decomposition is used to perform optical Fourier processing, where pairs of gratings and lenses are used to decompose the spectral content of the pulsed, optical waveforms.[1] Liquid crystal and acousto-optic (AO) modulators provide a means for introduction of somewhat arbitrary system transfer functions.[2,3] More recently, compact waveguide-based approaches have used arrayed waveguide gratings (AWGs) to perform an analogous spectral decomposition to those achieved with traditional, free-space coupled gratings.[4]

Information processing arrangements using the space domain for transverse spectral decomposition benefit from the maturity of such techniques, with respect to basic understanding, component quality, and the excellent phase control that free-space manipulations provide. Component insensitivity to optical power permits the shaping and detecting of high peak power pulses used to drive nonlinear optical systems. Furthermore, nonlinear optical elements can be introduced directly into these processors to provide unique ultrafast waveform synthesis capabilities[5–12] with response times in the range of femtoseconds. Drawbacks to free-space approaches include scaling in volume to achieve large time bandwidth products (TBWPs) and limiting the temporal extent of waveforms coupled back into a single-mode fiber (SMF). The origin of this last restraint is the time/space interrelation inherent in traditional pulse shaping devices.[13] AWG approaches rely on complex waveguide elements to circumvent large processor volumes and time/space coupling issues, but they too are practically limited by fabrication requirements. The number of resolvable spots in the output waveform for both Fourier synthesis and direct space-to-time AWG approaches[4,14] is equal to the number of waveguide channels. An alternative approach to Fourier processing relying on a single transverse spatial mode can better integrate with fiber systems and photonic lightwave circuits and will scale in length only for the achievement of large TBWPs.

Fiber-based and fiber-integrated processors can manipulate optical waveforms in the time domain exploiting chromatic dispersion for longitudinal spectral decomposition and applying to it Fourier synthesis techniques. Drawing on the identical mathematical treatments of diffraction in space and dispersion in time,[15] an approximate Fourier transform (FT) of an incident optical signal is achieved via chromatic dispersion after reaching the so-called "far-field approximation." With such an amount of dispersion, the temporal waveform closely resembles the spectrum.[16] Such a waveform is called a longitudinal spectral decomposition wave (SDW). A time-variant element (e.g., modulator) filters this SDW. And it is possible to recompose the pulse waveform through propagation in a conjugate dispersion source matched (i.e., of opposite sign but equal magnitude) to the first dispersive element. The number of resolvable spots that such a processor offers relies on a suitable modulation scheme and the experimental ability to disperse a sub-picosecond pulse using second-order dispersion. Applications utilizing large accumulated chromatic dispersion include, among others, optical pulse shaping,[17-19] microwave spectrum analysis,[20] microwave signal generation,[21-24] temporal signal magnification,[25,26] coherent Raman spectroscopy,[27,28] and, for optical communications, coherent code division multiple access encoding/decoding[29] and delay buffering.[30,31] In this chapter an in-depth discussion is restricted to longitudinal SDW processing technologies and their use in optical pulse shaping and microwave spectrum analysis applications. The distorting effect of higher-order dispersion on longitudinal SDW processors is also introduced and is shown to be mitigated in a high-speed optical ranging system using the advanced dispersion technology of chirped fiber Bragg gratings.

19.2 Longitudinal Spectral Decomposition

19.2.1 Theoretical description

Longitudinal spectral decomposition provides a means to access and operate on the spectral components of broadband optical pulses in a single-mode, time-of-flight operation. Physically, pulse propagation in a chromatically dispersive medium is described through the accumulation of spectral phase. A sub-picosecond transform limited pulse at $z = 0$ with its complex spectral amplitude $U(0,\omega)$ propagates along the z-axis of an optical fiber with propagation constant $\beta(\omega)$, described by a Taylor expansion,[32]

$$\beta(\omega) = \beta_0 + \beta_1 \omega + \frac{1}{2}\beta_2 \omega^2 + \frac{1}{6}\beta_3 \omega^3 + \cdots, \quad (19.1)$$

yielding complex amplitude spectrum at distance z,

$$U(z,\omega) = U(0,\omega)\exp\left(-j\frac{1}{2}\beta_2 z \omega^2\right). \quad (19.2)$$

Here, ω represents a shifted spectrum around the carrier frequency, i.e., $\omega = \Omega - \omega_c$ if Ω is the absolute frequency. The quadratic spectral phase represents the phase term of interest for generating longitudinal SDWs. The linear phase term, with the coefficient β_1, causes a simple delay and is ignored in the moving reference frame of the carrier. Higher-order terms are revisited and further discussed below. The strength of the quadratic phase depends on the value of the coefficient β_2, and on the propagation length z. In the time domain, pulse propagation is represented by a Fresnel integral, which relaxes with the so-called "far-field approximation" when the condition $|\beta_2 z| \gg \Delta^2/2$ is satisfied; here Δ is the transform limited pulse width. In this regime the time domain signal represents the spectral amplitude closely and has an instantaneous carrier frequency varying linearly with time under the envelope,

$$U(z,t) = e^{j(t^2/2\beta_2 z)} \int_{-\infty}^{\infty} U(0,t')e^{-j(t/\beta_2 z)t'}dt' = e^{j(t^2/2\beta_2 z)}\text{F}\{U(0,\omega')\}_{\omega'=(t/\beta_2 z)}. \quad (19.3)$$

Note that the approximate Fourier transform maps the spectral content to time using the relationship $\omega' = t/\beta_2 z$. The far-field condition, $|\beta_2 z| \gg \Delta^2/2$, thus indicates the propagation requirements needed to achieve a longitudinal SDW in the time domain. As an example, with dispersion in SMF of $\beta_2 = -20$ ps^2/km and pulses of about 0.1 ps duration, the far field is achieved within only a few meters of fiber.

Higher-order dispersion is the principle aberration to longitudinal SDW processors. Most commonly available dispersion sources cannot effectively be described by Eq. (19.2) but must include additional spectral phase terms described by higher-order coefficients, e.g., β_3 and higher. For low-loss operation in the conventional telecommunications band and for initial pulse durations longer than 100 fs, β_3 proves to be the principle distortion source to longitudinal SDWs. Equation (19.2) is modified to reflect the cubic spectral phase as

$$U(z,\omega) = U(0,\omega)\exp\left(-j\frac{1}{2}\beta_2 z\omega^2\right)\exp\left(-j\frac{1}{6}\beta_3 z\omega^3\right). \tag{19.4}$$

Further higher-order terms can be included similarly, but are neglected here for simplicity. In the time domain description of the longitudinal SDW, the approximate FT implies that the higher-order spectral phase terms are mapped into temporal phase terms using $\omega' = t/\beta_2 z$. The general result of higher-order dispersion is the addition of spectral group delays with quadratic, cubic, and higher descriptions. The generated longitudinal SDWs no longer enjoy the simple linear mapping of frequency components to temporal position. The specific impact of the distortion is application dependent and will be treated through the highlighted technologies introduced in subsequent sections.

19.2.2 Practical realization

Single-mode fiber technologies allow for the practical implementation of longitudinal SDW processors. Table 19.1 details a list of useful components and their processing role. Chromatic dispersion is commonly accumulated for wideband sources through propagation in single-mode fibers or in chirped fiber Bragg gratings (CFBGs). SMF and dispersion compensating fiber (DCF) are the most easily obtained dispersion sources, and when they are matched to one another, they allow an operator to create conjugated spectral phase signals (i.e., signals with a flipped sign in the spectral phase). In conjunction, SMF and DCF can be used to create a longitudinal SDW and then recompress the signal back to the transform limit. However, these dispersive media contain higher-order dispersion, and a theoretical treatment of a system to which they contribute must follow from Eq. (19.4). In contrast, CFBGs are less commonly available but can be written to contain very low higher-order dispersion. Writing techniques for long CFBGs are improving, and commercially available gratings offer differential group delays well in excess of 100 ps (shortest path to longest path delay relates to grating length).[33,34] Gratings also offer the benefits of low loss, low latency, short interaction length, and small footprint. A number of works have demonstrated passive optical pulse shaping through the use of superstructures and spectral filters written directly into fiber Bragg gratings.[35–38] When using designed CFBGs, the resulting complications of β_3 can give way to the simplified propagator of Eq. (19.2). Experimental evidence of the improved dispersion performance of CFBGs with respect to SMF is given in Ref. 21. As an additional benefit to the processor, when access is available to both sides of the grating, one can easily create conjugated SDWs by reflecting from both the red and blue ends.

The deployment of CFBG technology for SDW generation is slowed by several technical challenges stemming from current fabrication methods. The achievable grating length limits total optical bandwidth and dispersion strength,

Table 19.1 Processing elements for longitudinal SDWs

Technology	Contribution	Advantage	Drawback
SMF *SDW generator*	Spectral phase filter	Broadband, easily trimmed in length	Higher-order dispersion terms present
DCF *SDW generator*	Conjugate spectral phase filter, matched to SMF (opposite sign)	Like SMF + no soliton support and lower loss per unit dispersion	Higher-order dispersion terms present
CFBG *SDW generator*	Quadratic spectral phase filter, choice of sign for phase	Can be written without higher-order dispersion, very low loss per unit dispersion	Amplitude and phase ripple are present, birefringence limits compatibility with modulators
Modulators	High-speed mixer, contributes temporal phases	Spectral phase and intensity manipulation, provides frequency shifts to SDW	Time-variant elements require synchronization with incoming signal
Temporal delay lines	Linear spectral phase filter	Due to the mapping of frequency to time, this filter works similarly to a frequency shifter	Generally performed in free space for ease of tuning

while phase and amplitude ripple may place a limit on achievable dynamic range in some longitudinal SDW processing methods. Most restrictive to wide utilization is the well-known problem of CFBG birefringence.[21,39] This birefringence is fundamental and is induced during the writing process. Because each region of the grating is responsible for reflecting a different spectral component, the output longitudinal SDW suffers from spectrally dependent polarization rotation or multiorder polarization mode dispersion (PMD).

Compensating for the polarization rotation within a generated longitudinal SDW requires either a very fast time domain polarization rotator or fine stress modifications to the grating. Both approaches prove experimentally very challenging. The net result of higher-order PMD is the poor efficacy of any subsequent polarization-dependent device or process. The most important limitation in the processors is the incompatibility of longitudinal SDWs from CFBGs with electro-optic modulators as discussed below. The polarization variation results in both spectrally dependent transmittance and spectrally dependent modulation depth. In order to freely use linear CFBGs to improve the performance of longitudinal SDW processors, a method for complete PMD mitigation through improved CFBG design is needed. Due to its lack of active modulation, the reflectometry technique discussed in Section 19.5 is the only application discussed in this work capable of taking advantage of CFBG technology.

A time-variant element is required to operate on the dispersed temporal spectral components of the longitudinal SDW. Fortunately, due to its basis in fiber technologies, longitudinal SDW processing is fundamentally compatible with advanced fiber-coupled electro-optic (EO) and acousto-optic (AO) modulators. With their broad industrial utilization, these modulators are reasonably priced and widely available. EO modulation technology currently pushes toward 100 GHz operation, implying that a large number of manipulations can be made to the optical spectrum of a longitudinal SDW of nanosecond to microsecond scale duration. Note that while intensity modulation of the optical spectrum is possible with such modulators, the frequency shifts associated with intensity and phase-only modulation are experimentally most relevant. Theoretically, each modulation frequency contributes a temporal linear phase to the time domain SDW. To complement this temporal phase term, additional phase manipulation is possible via signal delay. Controllable delays produce a tunable linear phase in the spectral domain for the SDW. However, the mapping of frequencies to time implies an ambiguity or coupling between the temporal and spectral linear phases. Except near the margins of the signal bandwidth, delays and frequency shifts to longitudinal SDWs prove indistinguishable.

19.3 Optical Pulse Shaping

19.3.1 Operating principle

Optical communication systems can potentially benefit from the incorporation of Fourier synthesis pulse shapers directly into fiber systems.[40] Longitudinal spectral decomposition is an ideal candidate to perform such processing because pulse shaping is achieved through the manipulation of signals in dispersion-compensated links (see the schematic in Fig. 19.1). Experimentally altering dispersion-induced SDWs to perform pulse shaping has been described[17] and shown in mixed fiber-free space systems[18] and demonstrated in an all-fiber realization.[19] Most recently, a practical variant of the synthesis method has been applied to communications for phase coding optical pulses in a coherent optical code-division multiple access scheme.[29]

To understand the pulse shaping technique of Fig. 19.1, consider first a sub-picosecond transform limited optical pulse propagating in an ideal dispersion medium according to Eq. (19.2). Modulating the temporal SDW with an ideal signal $v(t) = A + B\cos(\omega_0 t)$ using an electro-optic (EO) device and then propagating through a dispersion-compensating element providing matched standard dispersion leads to an output waveform in the time domain,[19]

$$U_{out}(z,t) = AU(0,t)\exp(j\omega_c t)$$
$$+ B\exp\left[-j\frac{1}{2}\beta_2 z\omega_0^2\right]U(0,t-\beta_2 z\omega_0)\exp\left[j(\omega_c+\omega_0)t\right] \quad (19.5)$$
$$+ B\exp\left[-j\frac{1}{2}\beta_2 z\omega_0^2\right]U(0,t+\beta_2 z\omega_0)\exp\left[j(\omega_c-\omega_0)t\right].$$

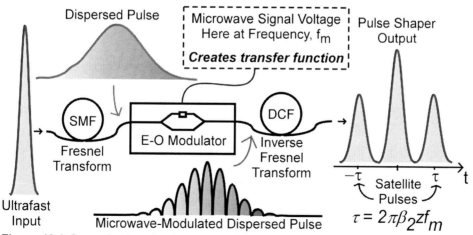

Figure 19.1 General pulse shaping approach using dispersive fiber and an electro-optic modulator for signal processing.

An assumption is made that the period of the modulating signal, $T_0 = 2\pi/\omega_0$, is less than the time aperture of the SDW. In Eq. (19.5), the formation of three pulses is recognized: one advanced and one delayed from a central pulse. Ignoring the constant phase terms, Eq. (19.5) shows that the temporal shifts of the satellite pulses are proportional to the strength of β_2, z, and ω_0. Larger chromatic dispersion or higher modulation frequency leads to greater shifts in time. Because the output of a longitudinal SDW FT pulse shaper is the input pulse convolved with the FT of the modulating spectral filter, intuitively, the pulses in Eq. (19.5) correspond to the three frequency components of the modulating signal, $v(t)$. This observation reaffirms the assertion made in Section 19.2 that in longitudinal SDW processing there exists a close similarity between temporal delays and frequency shifts. Most relevant for pulse shaping, by controlling the spectrum of the electronic signal applied to the modulator, an arbitrary temporal optical waveform can be created. A linear phase term modulating each satellite pulse is seen to upshift and downshift the temporal bandwidth of the delayed and advanced pulses, respectively, by the modulation frequency. With the use of an EO modulator for signal filtration, these "Doppler shifts" may approach tens of gigahertz. Although frequency shifts from the carrier are present in the satellite pulses created in traditional pulse shaping methods as well, such shifts are smaller because slower modulators are utilized. To reduce the frequency shifts of the satellite pulses in the longitudinal SDW technique, a larger dispersion-induced time aperture is needed. Though not critical to all pulse-shaping applications, this dependence of frequency shifts on temporal shifts requires consideration when performing signal processing.

The intensity waveform resulting from a simulation that follows the ideal model developed by Eq. (19.5) is shown in Fig. 19.2(a). The modulating signal is identical to $v(t)$ used in Eq. (19.5), where the coefficients are such that the DC component is stronger than the microwave signal ($A > B$). The launched 150 fs

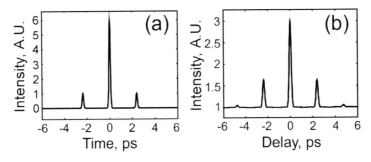

Figure 19.2 (a) Modeled intensity of a temporal waveform output from the system as described by Eq. (19.5). Time is designated relative to the central pulse corresponding to the constant (DC) component of the modulating signal $v(t)$. The satellite pulses relate to the microwave sidebands, ± 7 GHz. Here the microwave signal is weaker than the DC. (b) Intensity autocorrelation trace (with background) of the modeled optical waveform in (a). The small pulses occurring at ± 4.7 ps are cross-correlations of the pulses at ± 2.3 ps in (a). The cross-correlation of the DC and satellite pulses in (a) form the pulses at ± 2.3ps in (b).

pulse was stretched to and compressed from a nanosecond with purely second-order dispersion, and the modulating tone is set such that ω_0 is equal to $2\pi\cdot 7$ gigaradians per second. Satellite pulses should appear at temporal locations, $t = \pm\beta_2 z\omega_0$, which correspond to roughly ± 2.3 ps as $|\beta_2 z| = 52$ ps^2. Because of the difficulties in measuring optical waveforms with sub-picosecond features, Fig. 19.2(b) shows the modeled results of a practical solution to detection: an intensity autocorrelation (with background) of the signal in Fig. 19.2(a).

19.3.2 Experimental demonstration

The experimental approach to realize pulse shaping follows the schematic in Fig. 19.1. A Ti:sapphire pumped optical parametric oscillator delivers 150 fs transform limited pulses at 76 MHz and a nano-Joule of energy per pulse. The matched dispersive elements are spools of SMF and a DCF module. The presence of higher-order dispersion in these fibers will be treated subsequently. The operating wavelength is tuned to 1.55 µm to minimize attenuation and realize substantial dispersion. A grating stretcher precedes coupling into the SMF to avoid optical nonlinear effects associated with the highest peak intensities. Approximately 2.6 km of SMF are employed to disperse the pulse to a nanosecond FWHM. The DCF module provides −44 ps/nm of dispersion compensation. Experimentally, launching pulses on the order of 100 fs requires that all fibers/prechirping devices be matched with high precision. The length of the SMF must be tailored to within a few meters of the ideal length to measure a sub-picosecond pulse at the output. Spectral filtering via modulation is performed by a 10 GB/s LiNbO$_3$ Mach-Zehnder intensity modulator. The output waveform from the DCF is measured using a Michelson interferometer-based autocorrelator. To view the interferometric autocorrelation trace (ACT), an amplified silicon photodiode is operated in a nonlinear detection mode via two-photon absorption (TPA).[41] The TPA detection scheme provides useful phase sensitivity that gives experimental verification of full pulse recompression after propagation in the DCF. The nonlinear detection also facilitates detection of the time/frequency shifted satellite pulses since the cross-correlation of these pulses (within the autocorrelation of the entire waveform) will average to the background level with a slow linear detector. The intensity requirements of the TPA process, however, require that an erbium-doped fiber amplifier (EDFA) be used in the system.

Detected results from the experimental setup described above are given in Fig. 19.3. The two intensity autocorrelations are generated through digital postprocessing of measured interferometric ACTs. In both cases, the radio frequency (RF) signal generator supplied a 7 GHz sinusoid with 15 dBm electrical power. In the experiments represented in Fig. 19.3(a), the bias to the modulator is controlled such that a strong DC (unmodulated) component is allowed to pass. Relating to Eq. (19.5), the optical field can be described with coefficient A greater than coefficient B. The temporal shift, $t = \pm\beta_2 z\omega_0$, should

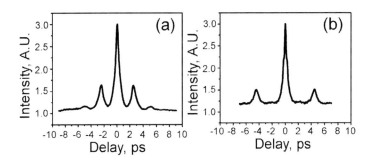

Figure 19.3 Experimental intensity autocorrelation traces of signals with 7 GHz modulation. RF power is maintained at 15 dBm while bias voltage is varied. (a) ACT in the presence of a strong DC component in the modulating signal $v(t)$. Experimental result shows agreement with the trace obtained from our linear model in Fig. 19.2(b). (b) ACT when the modulator was null biased. The autocorrelation process leads to the formation of three pulses as the inner two satellite pulses in the ACT signal dropout.

correspond to ± 2.3ps. Correspondingly, Fig. 19.3(a) shows close similarity to the modeling output of Fig. 19.2(b). Here again the small pulses at ± 4.7 ps in the ACT are formed by the cross-correlation of satellite pulses at ± 2.3 ps in the output optical waveform. The strong central pulse in the output field cross-correlating with a satellite pulse leads to the pulses at ± 2.3 ps in the ACT. In Fig. 193(b), the modulator is null biased to completely remove the unmodulated optical carrier component. The output field is described by setting A to zero in Eq. (19.5). Confirmation of the near full modulation depth is given by the complete removal of visible pulses at ± 2.3 ps in the detected intensity ACT. The elevation of the signal background in Fig. 19.3b from 1 to just less than 1.2 is evidence of an uncompressed background arising from self-phase modulation and dispersion. The background is estimated to contain approximately 75% as much power as the signal. Here the localized loss of the null-biased modulator led to a compensating increase of power above the nonlinear threshold at another location in the system. This background is undesirable and can be avoided, as seen in Fig 19.3(a), with stricter power budgeting and avoidance of nonlinear interactions. Of central importance, the trace in Fig. 19.3(b) contains a departure from the 2:1 ratio of the central pulse to side pulses. Two equally intense pulses should produce a 2:1 ratio when autocorrelated. This distortion is evidence of the higher-order dispersion in the fibers, which introduces signal-dependent broadening.

19.3.3 Distortion from higher-order dispersion

To better understand the effect of higher-order dispersion on the pulse shaper, it helps to consider again the operating principle. The processor creates delayed copies of the original pulse through frequency shifting the signal at the last precompensation point of a dispersion-compensated link. While the technique is introduced in this section for ultrafast optical waveform synthesis, the general approach is receiving attention as a method for "slow-light" optical buffering of longer, optical communication pulses.[30,31,42] This community likes to refer to frequency shifters (the EO modulators) as time prisms. Recall that the principle of operation is to introduce a variable group delay through upshifting or downshifting the carrier frequency before the compensating fiber, where β_2 gives the desired group velocity adjustment to the frequency shifted signal. Similarly, if β_3 is present, it results in a group velocity dispersion adjustment. Thus time delayed/advanced pulses will see residual uncompensated group velocity dispersion, which will cause pulse broadening. A detailed mathematical treatment of these processes is described in Ref. 19. The net result is that pulse time shifts greater than a certain duration will result in a convolution smearing of the satellite pulse energy to background levels. As a figure of merit, the pulse shift (delay or advance) that results in pulse width doubling is introduced as

$$T_s = \sqrt{3} T_o^2 \left| \frac{\beta_2}{\beta_3} \right|. \tag{19.6}$$

For standard SMF where $\beta_2 \approx -20$ ps^2/km and $\beta_3 \approx .1$ ps^3/km and $T_o = .1$ ps, then $T_s = 3.4$ ps. Thus, a pulse shifted by ~ 30 pulse widths will encounter a pulse width doubling. Note that longer initial pulse widths like those used in optical communications incur reduced β_3 spreading, but their time delay comes at the expense of larger $\beta_2 z$ products, and thus increased system latency and loss. Experimental evidence of the β_3 delay-induced spreading is shown in the four overlaid intensity autocorrelation traces of Fig. 19.4. These ACTs are similar to the result in Fig. 19.3(b) with modulation frequencies of 3, 4, 5, and 6 GHz. The EO modulator is null biased to produce two pulses in the output field and three pulses in the ACT. Incurred nonlinear penalties lead to an uncompressed background as in Fig. 19.3(b). In the ACTs of Fig. 19.4, larger frequency/temporal shifts introduce larger temporal spreads in the detected pulse pairs. The departure from the 2:1 central pulse to satellite pulse ratio in the ACTs gives evidence that the advanced pulse in the field was less broadened than the delayed pulse. This effect arose through interplay between the β_3 effect and residual uncompensated β_2. At the output, spreading from uncompensated β_2 is the same for all shifted pulses, but spreading due to β_3 is pulse position dependent. As the two effects superpose, a single pulse shift will see minimum spreading. Thus while β_3 is generally an aberration, it could be useful in creating asymmetric waveforms.

The figure of merit to describe the pulse shaper is its TBWP. Equivalent to the number of resolvable spots, the ultimate observable TBWP restriction comes from the presence of higher-order dispersion. This limit is given by the spreading time shift of Eq. 19.6. However, if the fiber-based pulse-shaping method were implemented without higher-order dispersion, the maximum TBWP would relate to the ratio of the time aperture of the SDW to the transform-limited pulse duration (bandwidth equivalent). For nanosecond-scale time apertures and pulses of about 0.1 ps, this number can well exceed 10^3. However, a maximum modulator speed of ~ 100 GHz limits the number of resolvable spots in the output

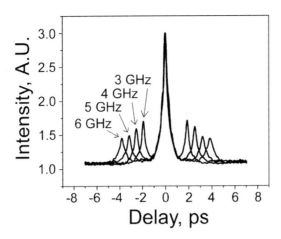

Figure 19.4 Overlaid intensity autocorrelations showing the effect of increased modulation frequency on the shaped, pulsed waveform. The temporal broadening of the 6 GHz up and down shifted pulses is apparent when comparing their FWHM to those pulses originating from 3 GHz modulation.

by reducing the maximum temporal shifts. In practice, the TBWP could then be defined as the ratio of maximum modulator speed to minimum achievable modulation speed (defined as the inverse of the time aperture of the SDW). For nanosecond-scale SDWs, the TBWP would exceed 100. To return to 1000 resolvable spots, > 10 ns time apertures are required. Limiting progress is the challenge of recompressing large time aperture SDWs with DCF.

19.3.4 Waveform Detection

The pulse shaper creates copies of the original pulse at temporal shifts corresponding to the spectral components of the modulating signal, indicating that the processor can function as a high-bandwidth microwave spectrum analyzer.[20] Here, the spectral resolution figure of merit is determined by two satellite pulses that can be temporally resolved. This requirement is satisfied to the first order if their modulation frequency difference is greater than the inverse of the time aperture of the SDW. For microwave spectrum analysis applications as described in Ref. 20, one can determine the microwave spectral content of an unknown signal $v(t)$ by examining the temporal waveform output from the system while scaling time to frequency by $\omega = t/(\beta_2 z)$.

19.4 Microwave Spectrum Analysis

19.4.1 Operating principle

Analysis of microwave signals in the gigahertz to terahertz range is of critical importance to a variety of fields including RADAR/LIDAR, imaging, and astronomical detection.[43,44] Commonly employed electronic techniques are limited in speed and rely on channelization of the signal bandwidth to provide high-resolution spectrum analysis. Microwave photonics methods for spectrum analysis offer the potential to achieve wide instantaneous bandwidth in compact designs.[20,45] Recognizing the utility of the pulse-shaping method in Section 19.3 for microwave spectrum analysis, the technique is modified to remove the technical challenge of recompressing ultrafast optical signals while continuing to exploit the bandwidth of longitudinal SDWs. Specifically, a straightforward method for microwave spectrum analysis is realized with ultimate potential for analyzing signals with bandwidths approaching 100 GHz and with resolution less than 10 MHz.[21]

A schematic diagram of the approach is shown in Fig. 19.5. The front end is similar to that of the fiber-based pulse shaper. After large initial dispersion, the SDW is separated into two copies, whereby one is modulated by the microwave signal of interest, and the other copy is given an adjustable temporal delay. On recombination and square-law detection, the delayed, dispersed pulse copies create an intermediate frequency (IF) in the generated photocurrent. To the first order, a unique temporal beat frequency exists between the two SDWs in relation to the time delay and strength of dispersion. For microwave signal processing, the IF serves as a delayed-tuned local oscillator (LO) sweeping over the modulation spectrum. Using a time-averaged detector as a low-pass filter (LPF), the system isolates that microwave signal component that is homodyned by the variable LO. A trace over all delays, equivalent to a cross-correlation of the two pulses, shows correlation peaks for all spectral components of the microwave signal. This arrangement is analogous to an RF super-homodyne receiver.

An analytic description of the microwave spectrum analyzer ignoring higher-order dispersion proceeds in the time domain. The longitudinal SDW of Eq.

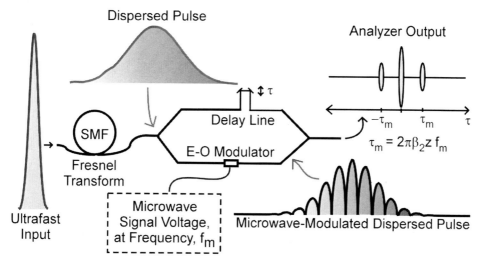

Figure 19.5 Schematic diagram of the microwave spectrum analyzer. The signal voltage delivered to the EO modulator is a single RF tone with a bias. The correlation output shows the two sidebands and the central DC spike.

(19.3) is split into two copies: The amplitude of one copy is modulated by a microwave signal, $s(t)$, while a delay, τ, is introduced onto the second one. Recombining the two pulses yields a signal,

$$U_r(z,t) = e^{j[(t-\tau)^2/2\beta_2 z]}\mathcal{F}\{U(z,\omega'')\}_{\omega''=[(t-\tau)/\beta_2 z]} + e^{j(t^2/2\beta_2 z)}s(t)\mathcal{F}\{U(z,\omega')\}_{\omega'=(t/\beta_2 z)}.\quad(19.7)$$

For small delays, the pulses overlap nearly completely, and the SDW envelope shape becomes of secondary importance. Detection of the received signal with a square-law photodiode produces a cross-term with a linear phase proportional to the delay. Time averaging the cross-term by the detector for time T leads to a signal,

$$\frac{1}{T}\int_0^T \left|\mathcal{F}\{U(0,\omega')\}_{\omega'=(t/\beta_2 z)}\right|^2 s(t)\cos\left[2\pi\left(\frac{\tau}{2\pi\beta_2 z}\right)t - \frac{\tau^2}{2\beta_2 z}\right]dt.\quad(19.8)$$

Thus as the time integration is lengthened, the detection will isolate the component of $s(t)$ oscillating at a microwave frequency, $\tau/(2\pi\beta_2 z)$. By varying the delays, all the spectral components of $s(t)$ can be determined from the correlation process. The additional constant phase related to τ^2 in the cosine term is negligibly small.

19.4.2 Experimental realization

Figure 19.6 shows experimental results from a proof-of-concept experiment. The system is implemented by first generating 150 fs pulses at 1.55 μm in a

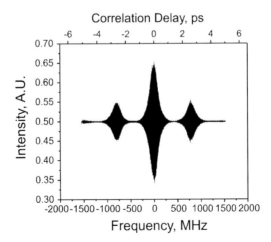

Figure 19.6 Correlation trace from the experimental proof of concept. The RF signal consists of DC and 760 MHz components. Results are filtered in postprocessing to remove background fluctuations. The upper axis gives temporal correlation delays as recorded. The lower axis shows a scaling of the delay axis to frequency ($f = \tau/2\pi\beta_2 z$).

Ti:sapphire/OPO laser system. The pulses are dispersed in a free-space grating stretcher to the picosecond scale to lower peak power before they are launched into 25.2 km of SMF fiber. An optical spectrum analyzer is used to monitor and confirm no change in the spectrum of the pulse at the output of the fiber to assure that the propagation is linear. The generated longitudinal SDW is over 6 ns in duration. The polarization of the SDW is set with a polarization controller and a fiber-coupled linear polarizer. This waveform enters a free-space Mach-Zehnder arrangement where it is split; one copy is controllably delayed while the other copy enters a 1 GHz LiNbO$_3$ modulator. The modulator is driven by an RF source, which produces a 760 MHz tone locked in phase to the 76 MHz repetition rate of the laser (see following discussion). A bias accompanies the RF tone. Detection of the cross-correlation is recorded in an InGaAs photodiode. The upper axis of Fig. 196 provides the correlation delays as recorded, while the lower axis makes use of the conversion of delay to frequency, $f = \tau/(2\pi\beta_2 z)$. The corresponding spectrum of $s(t)$, as read from the lower axis, confirms the DC term and oscillatory terms at ± 760 MHz. Viewing the upper axis, the trace has peaks at the central point and ± 2.4 ps ($|2\pi\beta_2 z f_m| = 2.4$ ps if $-\beta_2 = 20$ ps^2/km, $z = 25.2$ km, and $f_m = 7.6 \times 10^{-4}$ THz).

The requirement of time averaging introduces an experimental trade-off. In order that the LPF operation remains narrowband, long-time integration is necessary. This additional time not only introduces latency, but also implies that more pulses from the pulse train will fall on the detector in a given integration window. The spectrum of the ultrashort pulse subsequently becomes discretized where the comb function multiplying the optical spectrum has a finesse related to the integration time. With respect to the operation of the microwave spectrum analyzer, the LO is discretized because the beat between a pulse pair can only constructively add to that of the next pair in the train at frequencies corresponding to multiples of the laser repetition rate. Ultimately, spectrum analysis produces a sampled version (Fourier series) of the microwave spectrum of $s(t)$. To produce a measurable signal when multiple pulses are averaged, the demonstrator analyzes an RF tone derived from a 10× multiplication of the repetition rate of the laser. However, low repetition rate mode-locked laser pulses with large subsequent dispersion can generate SDWs that fill the entire integration window. If integration is matched to the pulse repetition period, the probability of intercept goes to unity while LO spectral discretization and the signal phasing requirements vanish.

The resolution of the method is locked to the time aperture created by the SDW and, hence, the length of propagation in the dispersive element. To achieve 10 MHz resolution, 100 ns SDWs are needed. While the LO can be tuned to frequencies approaching the bandwidth of the ultrashort pulse, at such delays the pulse pair is incapable of providing full LO resolution. The lack of pulse overlap reduces the effective time aperture and, by the uncertainty principle, degrades resolution. Current standard technologies limit EO modulation to the 100 GB/s and slower range. For such microwave signals, the amount of delay required to generate the LO is less than 10% of the pulse width, thus ensuring large pulse overlap. More significant than pulse overlap, higher-order dispersion can reduce the TBWP by orders of magnitude because the LO resolution degrades quickly at low frequencies. The effect is detailed further below. Note that were a CFBG induced SDW without polarization rotation used, the time bandwidth of such a device could exceed 10,000 resolvable spots. To avoid the need for scanning and to fully take advantage of the large instantaneous bandwidth that short pulses offer, the space domain must be exploited. In such a manner, variable delays are introduced in parallel and arrayed detection is employed to achieve correlation in space.[46]

19.4.3 Waveform synthesis

A simple system variation allows the processor to perform microwave signal generation as opposed to spectrum analysis. With a high-speed photodetection system present to covert the IF into oscillating photoelectric current, applications could exploit the interfering longitudinal SDWs to create delay-tunable microwave signal carriers.[21,47] The electro-optic modulator permits envelope control on top of the high-speed carrier. An example waveform captured using a 30 GHz sampling oscilloscope shows a 4.5 GHz carrier and 760 MHz envelope features [see Fig. 19.7(a)]. A high-pass filter with 3 GHz cutoff was used to remove the DC components, creating a bipolar waveform. The signal spectrum is generated through a fast Fourier transform (FFT) and presented in Fig. 19.7(b). Incorporation of more arms in the interferometric processor and/or a large dynamic-range EO modulator offers a path toward linear synthesis of ultra-wideband microwave signals with arbitrary waveform.

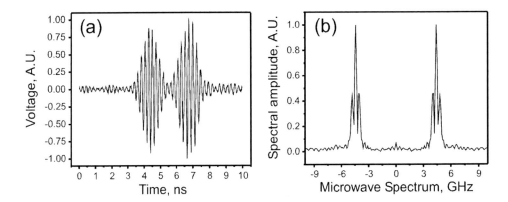

Figure 19.7 (a) Oscilloscope-recorded waveform from the modified processor designed for high-speed electrical waveform synthesis. A 3 GHz high-pass filter removes the slow DC signal producing a bipolar waveform. (b) FFT of recorded waveform shows the 4.5 GHz carrier and 760 MHz envelope feature.

19.4.4 Distortion from higher-order dispersion

The microwave spectrum analyzer and waveform synthesizer approaches operate by creating a single-frequency beat tone in the microwave range through the interference of delayed copies of the longitudinal SDW. The presence of β_3, taken as the dominant source of higher-order dispersion, leads to a cubic spectral phase, which produces a higher-order mapping of spectrum to time in the SDW. In this case the beat between delayed longitudinal SDWs does not form from a constant-frequency offset at all points in time under the SDW envelope. The resulting IF is time variant, and therefore the LO for the spectrum analyzer or the carrier for the synthesized microwave signal suffers an increased line width. Furthermore, when introducing larger delays to increase the LO or carrier frequency, the line width will increase as well. In the spectrum analyzer, the result is rapid resolution decrease and loss of signal strength. For the signal synthesizer, the result is carrier chirp. This issue is illustrated schematically in Fig. 19.8. For the two system demonstrations using SMF, the fractional bandwidth of the LO approaches 10% at all carrier frequencies. Thus, a 15 GHz carrier frequency has a line width in excess of 1 GHz.

19.5 High-Speed Optical Reflectometry

19.5.1 Introduction

For the restricted class of longitudinal SDW processors that do not require EO modulation, current linear CFBG technology is available to permit large TBWP realization. As an example, high-speed optical reflectometry (ranging) can be implemented using a processor geometry akin to the microwave-photonic systems of Section 19.4, but without a modulator. In this technique the range to target is deduced by analyzing the interference frequency between reference and target longitudinal SDWs under a delay corresponding to the round trip time of the target pulse to the target (see Fig. 19.9). A one-to-one correspondence between distance to the target and generated IF allows the target range to be deduced through a simple FFT operation on the digitized received waveform.

Such systems are well known in radar and lidar where actively swept single frequency sources are employed.[48] Here the passive dispersion of the longitudinal SDWs allows the operator to create and use extremely wideband, stable sweeps in relatively short, nanosecond- to microsecond-scale time apertures. These characteristics imply high depth resolution and fast depth read times. Achieving linear frequency sweeps is an important step toward the goal of pushing this reflectometry approach to micron-scale resolution over centimeter ranges. Recent demonstrations with spectrally broadened short pulses dispersed in fiber media with higher-order dispersion required the use of signal postprocessing to maintain fine depth resolution over extended ranges.[49,50] The procedure essentially narrows the IF line width in the digital domain. However, replacing these dispersion sources with a linear CFBG addresses the resolution problem in the physical layer, removing the fundamental need for signal postprocessing.

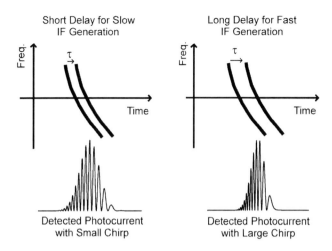

Figure 19.8 Time-frequency plot visualizations of longitudinal SDWs with β_3 distortion. The interference of two delayed copies (above) produces the beat tone under the SDW envelope (below). For larger delays, the detected photocurrent has a large microwave chirp. This chirp broadens the line width of the microwave signals used in the techniques of this section.

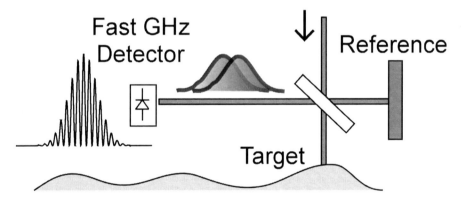

Figure 19.9 Conceptual diagram describing the high-speed reflectometry technique based on the interference of longitudinal SDWs. Target range is deduced by analyzing the interference frequency present in the detected photocurrent.

19.5.2 Experimental demonstration

A demonstration of the ranging system proceeds by launching 150 fs pulses at 1540 nm from the Ti:sapphire/OPO source into a free-space Michelson interferometer, where one arm contains focusing optics and a sample to be measured. The combined target and reference pulses are coupled to SMF. A spool of 10.5 km of SMF or a 1.1 m CFBG disperses the pulses to the nanosecond scale. The CFBG uniformly maps its spectrum to time as 2708 ps/THz. In SMF the center frequency sees ~1300 ps/THz, while a 400 ps/THz variation exists over the pulse bandwidth. The copropagating target and reference longitudinal SDWs interfere on a 30 GHz photodetector and are digitized in a sampling oscilloscope. The temporal delay to the target is acquired by analyzing the IF and scaling this frequency value to time using the deterministic chirp rate. A comparison of the ranges measured for the linear CFBG and SMF-induced SDWs is shown in Fig. 19.10. The profile of a staircase sample made from gold-coated microscope slide glasses of roughly 150 µm and 1 mm thicknesses is rendered accurately by the pulse dispersed with the CFBG [solid line in Fig. 19.10(a)]. The measurable height range and TBWP (effective number of resolvable spots) are limited only by the physical system parameters. The raw profile from the SMF SDW shows larger variations with growing degradation for taller levels (longer ranges) [dashed line in Fig. 19.10(a)]. The total usable range and system TBWP are clearly impacted. The system accuracy can be enhanced using digital postprocessing. By interpolating and resampling the data using a time-variant sampling rate deduced from the dispersion time/frequency map, the phase of the beat signal is made linear before the FFT operation, and the line width of the IF reaches its transform-limited performance. The approach is analogous to the procedure in spectral interferometry to resample spectral data with evenly spaced frequency samples when native data is obtained in wavelength samples.[51] Figure 19.10(b) shows a restoration of the system accuracy for the SMF data through appropriate signal interpolation. Although postprocessing employing signal-processing algorithms is generally an inexpensive approach, it should be noted that the use of interpolation removes the passive nature of the SDW ranging technique. The interpolation algorithm needs to not only match in accuracy the magnitude of the desired system TBWP, but also temporally synchronize with the incoming data[52]. Such a complication can be avoided through the use of the linear CFBG technology. Additional processing advantages of the linear CFBG for optical reflectometry, specifically with respect to simultaneous ranging and velocimetry, are highlighted in Saperstein et al.[52]

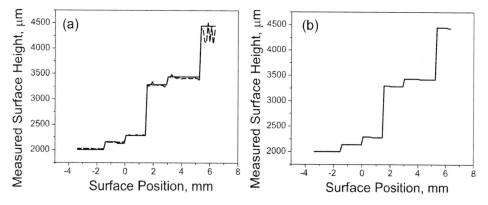

Figure 19.10 Measured surface profiles using high-speed detection of the interference between target and reference longitudinal SDWs. (a) Overlaid profiles taken by a linear CFBG-induced SDW (solid) and SMF-induced SDW (dashed). Higher-order dispersion in SMF limits the resolution of the system at larger ranges while the CFBG sees no such degradation. (b) Using digital postprocessing, it is possible to restore the number of resolvable spots captured by the SMF dispersed waveforms.

19.6 Conclusion

In summary, this chapter introduces and presents the state of the art in the optical signal processing methods based on longitudinal spectral decomposition of ultrashort pulses using chromatic dispersion. Key technologies, including optical modulators, optical fibers, and CFBGs, permit the generation and manipulation of longitudinal SDWs and are used in experimental demonstrations to implement ultrafast optical pulse shaping, microwave spectrum analysis, microwave electronic signal generation, and optical reflectometry. With the maturity of space-domain techniques for ultrafast optical signal processing, the motivation to pursue time-domain approaches stems from the ease of system scalability to large time apertures, the compatibility with fiber-optic equipment, and the achievement of high processing speeds. Ultimately the creation of nanosecond to microsecond, longitudinal SDWs implies huge TBWPs for these processors. Accessing the large time bandwidth product requires some technological progress. Processing advancements for longitudinal SDW systems will accompany improvements to stable wideband optical source generation, CFBG fabrication, and high-speed electronic A/D and D/A conversion.

Acknowledgments

This work was supported by the Defense Advanced Research Projects Agency, the UC Discovery Grants Program, the National Science Foundation, and the U.S. Air Force Office of Scientific Research.

References

1. Weiner, A.M., Heritage, J.P., and Kirschner, E.M., "High-resolution femtosecond pulse shaping," *J. Opt. Soc. Am. B*, **5**, 1563–1572, 1988.
2. Weiner, A.M., Leaird, D.E., Patel, J.S., and Wullert II, J.R., "Programmable shaping of femtosecond optical pulses by use of 128-element liquid crystal phase modulator," *J. Quant. Electron.*, **28**, 908–920, 1992.
3. Dugan, M.A., Tull, J.X., and Warren, W.S., "High-resolution acousto-optic shaping of unamplified and amplified femtosecond laser pulses," *J. Opt. Soc. Am. B*, **14**, 2348–2358, 1997.
4. Kurokawa, T., Tsuda, H., Okamoto, K., Naganuma, K., Takenouchi, H., Inoue, Y., and Ishii, M., "Time-space-conversion optical signal processing using arrayed-waveguide grating," *Electron. Lett.*, **33**, 1890–1891, 1997.
5. Sun, P.-C., Mazurenko, Y., Chang, W.S C., Yu, P.K.L., and Fainman, Y., "All optical parallel-to-serial conversion by holographic spatial-to-temporal frequency encoding," *Opt. Lett.*, **20**, 1728–1730, 1995.
6. Sun, P.C., Mazurenko, Y., and Fainman, Y., "Femtosecond pulse imaging: ultrafast optical oscilloscope," *JOSA A*, **14**, 1159–1169, 1997

7. Sun, P.C., Mazurenko, Y., Fainman,Y., "Real-time 1-D coherent imaging through single-mode fibers by space-time conversion processors," *Opt. Lett.*, **22**, 1861–1863, 1997

8. Marom, D.M., Panasenko, D., Sun, P.-C., and Fainman, Y., "Spatial-temporal wave mixing for space-to-time conversion," *Opt. Lett.*, **24**, 563–565, 1999.

9. Marom, D.M., Panasenko, D., Rokitski, R., Sun, P.-C., and Fainman, Y., "Time reversal of ultrafast waveforms by wave mixing of spectrally decomposed waves," *Opt. Lett.*, **25**, 132–134, 2000.

10. Marom, D., Panasenko, D., Sun, P.C., and Fainman, Y., "Femtosecond rate space-to-time conversion," *J. Opt. Soc. Am.: B*, **17**, 1759–73, 2000.

11. Marom, D., Panasenko, D., Sun, P.C., and Fainman, Y., "Linear and nonlinear operation of a time-to-space processor," *J. Opt. Soc. Am.: A*, **18**, 448–458, 2001.

12. Marom, D., Panasenko, D., Sun, P.C., Mazurenko, Y., and Fainman, Y., "Real-time spatial-temporal signal processing with optical nonlinearities," *IEEE J. Sel. Top. Quant. Electron.*, **7**, 683–693, 2002.

13. Wefers, M.M., and Nelson, K.A., "Space-time profiles of shaped ultrafast optical waveforms," *IEEE J. Quant. Electron.*, **32**, 161–172, 1996.

14. Leaird, D.E., Weiner, A.M., Kamei, S., Ishii, M., Sugita, A., and Okamoto, K., "Generation of flat-topped 500-GHz pulse bursts using loss engineered arrayed waveguide gratings," *IEEE Photon. Technol. Lett.*, **14**, 816–818, 2002.

15. Papoulis, A., "Pulse compression, fiber communications, and diffraction: a unified approach," *J. Opt. Soc. Am. A*, **11**, 3–13,, 1994.

16. Tong, Y.C., Chan, L.Y., and Tsang, H.K., "Fiber dispersion or pulse spectrum measurement using a sampling oscilloscope," *Electron. Lett.*, **33**, 983–985, 1997.

17. Weiner, A.M., and Heritage, J.P., "Optical systems and methods based upon temporal stretching, modulation, and recompression of ultrashort pulses," U.S. Patent No. 4,928,316,

18. Haner, M., and Warren, W.S., "Synthesis of crafted optical pulses by time domain modulation in a fiber-grating compressor," *Appl. Phys. Lett.*, **52**, 1548–1550, 1988.

19. Saperstein, R.E., Alic, N., Panasenko, D., Rokitski, R., and Fainman, Y., "Time-domain waveform processing using chromatic dispersion for temporal shaping of optical pulses," *J. Opt. Soc. Am. B*, **22**, 2427–2436, 2005.

20. Saperstein, R.E., Panasenko, D., and Fainman, Y., "Demonstration of a microwave spectrum analyzer based on time-domain optical processing in fiber." *Opt. Lett.*, **29**, 501–503, 2004.

21. Saperstein, R.E., and Fainman, Y., "Information processing with longitudinal spectral decomposition of ultrafast pulses," *Appl. Opt.*, **47**, A21–A31, 2008.

22. Chou, J., Han, Y., and Jalali, B., "Adaptive RF-photonic arbitrary waveform generator," IEEE Phot. *Tech. Lett.*, **15**, 581–583, 2003.

23. Lin, I., McKinney, J.D., and Weiner, A.M., "Photonic synthesis of broadband microwave arbitrary waveforms applicable to ultra-wideband communication," *IEEE Microw. Wireless Compon. Lett.*, **15**, 226–228, 2005.

24. McKinney, J.D., Lin, I.-S., and Weiner, A.M., "Shaping the power spectrum of ultra-wideband radio-frequency signals," *IEEE Trans. Micro. Theor. Tech.*, **54**, 4247–4255, 2006.

25. Azaña, J., Berger, N.K., Levit, B., Smulakovsky, V., and Fischer, B., "Frequency shifting of microwave signals by use of a general temporal self-imaging (Talbot) effect in optical fibers," *Opt. Lett.*, **29**, 2849–2851, 2004.

26. Kolner, B.H., "Space-time duality and the theory of temporal imaging," *IEEE J. Quant. Electron.*, **30**, 1951–1963, 1994.

27. Gershgoren, E., Bartels, R.A., Fourkas, J.T., Tobey, R., Murnane, M.M., and Kapteyn, H.C., "Simplified setup for high-resolution spectroscopy that uses ultrashort pulses," *Opt. Lett.*, **28**, 361–363, 2003.

28. Hellerer, T., Enejder, A.M.K., and Zumbusch, A., "Spectral focusing: High spectral resolution spectroscopy with broad-bandwidth laser pulses," *Appl. Phys. Lett.*, **85**, 25–27, 2004.

29. Wang, X., and Wada, N., "Spectral phase encoding of ultra-short optical pulse in time domain for OCDMA application," *Opt. Expr.*, **15**, 7319–7326, 2007.

30. Sharping, J., Okawachi, Y., van Howe, J., Xu, C., Wang, Y., Willner, A., and Gaeta, A., "All-optical, wavelength and bandwidth preserving, pulse delay based on parametric wavelength conversion and dispersion," *Opt. Expr.*, **13**, 7872–7877, 2005.

31. Ren, J., Alic, N., Myslivets, E., Saperstein, R.E., McKinstrie, C.J., Jopson, R.M., Gnauck, A.H., Andrekson, P.A., and Radic, S., "12.47ns Continuously-Tunable Two-Pump Parametric Delay," European Conference on Optical Communication 2006 Postdeadline **Th4.4.3.**, 2006.

32. Diels, J.-C., and Rudolph, W., *Ultrashort Laser Pulse Phenomenon*, Academic Press, San Diego (1996).

33. Advanced Optical Solutions, www.aos-fiber.com

34. Proximion Fiber Systems AB, http://www.proximion.com/

35. Petropoulos, P., Ibsen, M., Ellis, A.D., and Richardson, D.J., "Rectangular pulse generation based on pulse reshaping using a superstructured fiber Bragg grating," *J. Lightwave Technol.*, **19**, 746–752, 2001.

36. Longhi, S., Marano, M., Laporta, P., Svelto, O., and Belmonte, M., "Propagation, manipulation, and control of picosecond optical pulses at 1.5 m in fiber Bragg gratings," *J. Opt. Soc. Am. B*, **19**, 2742–2757, 2002.

37. Azaña, J., and Chen, L.R., "Synthesis of temporal optical waveforms by fiber Bragg gratings: a new approach based on space-to-frequency-to-time mapping," *J. Opt. Soc. Am. B*, **19**, 2758–2769, 2002.

38. Wang, X., Matsushima, K., Kitayama, K., Nishiki, A., Wada, N., and Kubota, F., "High-performance optical code generation and recognition by use of a 511-chip, 640-Gchip/s phase-shifted superstructured fiber Bragg grating," *Opt. Lett.*, **30**, 355–357, 2005.

39. Chou, P.C., Haus, H.A., and Brennan III, J.F., "Reconfigurable time-domain spectral shaping of an optical pulse stretched by a fiber Bragg grating," *Opt. Lett.*, **25**, 524–526, 2000.

40. Salehi, J.A., Weiner, A.M., and Heritage, J.P., "Coherent ultrashort light pulse code-division multiple-access communication systems," *J. Lightwave Technol.*, **8**, 478–491, 1990.

41. Takagi, Y., Kobayashi, T., Yoshihara, K., and Imamura, S., "Multiple- and single-shot autocorrelator based on two-photon conductivity in semiconductors," *Opt. Lett.*, **17**, 658–660, 1992.

42. van Howe, J., and Xu, C., "Ultrafast optical delay line by use of a time-prism pair," *Opt. Lett.*, **30**, 99–101, 2005.

43. Zmuda, H., and Toughlian, E.N., *Photonic Aspects of Modern Radar*, Artech House, Boston (1994).

44. Hu, B.B., and Nuss, M.C., "Imaging with terahertz waves," *Opt. Lett.*, **20**, 1716–1718, 1995.

45. Lavielle, V., Lorgeré, I., Le Gouët, J.-L., Tonda, S., and Dolfi, D., "Wideband versatile radio-frequency spectrum analyzer," *Opt. Lett.*, **28**, 384–386, 2003.

46. Panasenko, D., and Fainman, Y., "Interferometric correlation of infrared femtosecond pulses with two-photon conductivity in a silicon CCD," *Appl. Opt.*, **41**, 3748–3752, 2002.

47. Weling, A.S., Hli, B.B., Froberg, N.M., and Auston, D.H., "Generation of tunable narrow-band THz radiation from large aperture photoconducting antennas," *Appl. Phys. Lett.*, **64**, 137–139, 1994.

48. Rihaczek, A.W., *Principles of High-Resolution Radar*, McGraw-Hill, New York (1969).

49. Moon, S., and Kim, D.Y., "Ultra-high-speed optical coherence tomography with a stretched pulse supercontinuum source," *Opt. Expr.*, **14**, 11575–11584, 2006.

50. Park, Y., Ahn, T.-J., Kieffer, J.-C., and Azaña, J., "Optical frequency domain reflectometry based on real-time Fourier transformation," *Opt. Expr.*, **15**, 4598–4617, 2007.

51. Dorrer, C., Belabas, N., Likforman, J.-P., and Joffre, M., "Spectral resolution and sampling issues in Fourier transform spectral interferometry," *J. Opt. Soc. Am. B*, **17**, 1795–1802, 2000.

52. Saperstein, R.E., Alic, N., Zamek, S., Ikeda, K., Slutsky, B., and Fainman, Y., "Processing advantages of linear chirped fiber Bragg gratings in the time domain realization of optical frequency-domain reflectometry," *Opt. Expr.*, **15**, 15464–15479, 2007.

Robert E. Saperstein received his BSE in electrical engineering from Princeton University, Princeton, NJ, in 2001 and his MS and PhD in electrical engineering (photonics) from the University of California (UC), San Diego, La Jolla, in 2004 and 2007, respectively. He is currently working as a postdoctoral researcher at UC San Diego. He has been a member of Prof. Fainman's ultrafast and nanoscale optics group since 2001. His current research interests include signal processing based on temporal/longitudinal spectral decomposition of ultrafast pulses with application to imaging and metrology.

Yeshaiahu Fainman received his PhD from the Technion-Israel Institute of Technology, Haifa, Israel, in 1983. He is a professor of electrical and computer engineering at the Department of Electrical Engineering, University of California, San Diego, La Jolla. He has contributed more than 150 manuscripts in referred journals and more than 250 conference presentations and conference proceedings. His current research interests include optical network systems and devices, with special emphasis on ultrafast information processing with optical nonlinearities and the use of femtosecond laser pulses, near-field phenomena in optical nanostructures and nanophotonic devices, quantum communication, and multidimensional quantitative imaging. Dr. Fainman is a Fellow of the Optical Society of America (OSA), Institute of Electrical and Electronics Engineers (IEEE), and the Society of Photo-Optical Instrumentation Engineers (SPIE). He was a Chair of the IEEE/IEEE Lasers and Electro-Optics Society Subcommittee on Optical Interconnects and Signal Processing from 2004 to 2006 and a General Chair of the OSA topical meeting on nanophotonics for information systems in 2005. He also served on numerous conference program committees, organized symposia and workshops, and, between 1993 and 2001, served as a topical editor of the *Journal of the Optical Society of America A* on optical signal processing and imaging science. He was a recipient of the Miriam and Aharon Gutvirt Prize.

Chapter 20
Ultrafast Information Transmission by Quasi-Discrete Spectral Supercontinuum

Mikhail A. Bakhtin, Victor G. Bespalov, Vitali N. Krylov,
Yuri A. Shpolyanskiy, and Sergei A. Kozlov
State University of Information Technologies, Russia

20.1 Introduction
20.2 Theory
    20.2.1 Equation of field dynamics and its link with equations for pulse envelopes
    20.2.2 Interaction of femtosecond pulses with supercontinuum spectra in optical waveguides
    20.2.3 Binary data coding using the generated sequences of ultrashort pulses
    20.2.4 Interaction of femtosecond pulses with supercontinuum spectra in bulk optical media
20.3 Experiment
20.4 Conclusion
Acknowledgement
References

20.1 Introduction

Few-cycle femtosecond pulses steadily generated in many laboratories[1] attract not only theoretical, but also practical interest as a source of many prospective applications.[2,3] One of the unique features of few-cycle pulses is that optical media remain intact even at ultrahigh field intensities because of ultrashort pulse duration and, as a consequence, the limited pulse energy.[4] High intensities that cannot be achieved with longer pulses due to immediate breakdown of a material appear to be feasible with few-cycle fields and result in new characteristics of well-known effects. For example, self-phase modulation turns into the generation of spectral supercontinuum,[1–3] which becomes virtually immanent to intense few-cycle pulses and accompanies other phenomena such as pulse temporal broadening or compression, self-focusing, etc. These effects were extensively studied for single few-cycle pulses.[5–7] However, interactions of few-cycle pulses were beyond massive investigations. It was shown theoretically and experimentally in Refs. 5, 8, and 9 that the interaction of pulses with different spectral contents in nonlinear media can lead to significant enhancement of spectral ultrabroadening due to cross-phase modulation. In the present study, we demonstrate numerically that the pulse collision can become a source of quasi-discrete supercontinuum visible in the temporal domain as a regular pulse chain with an ultrahigh repetition rate. The opportunity for the generation of a quasi-discrete supercontinuum is confirmed experimentally.

20.2 Theory

20.2.1 Equation of field dynamics and its link with equations for pulse envelopes

The conventional theoretical approach based on the consideration of pulse envelopes, which was originally suggested for quasi-monochromatic fields,[10,11] becomes questionable for pulses with ultrabroad spectra propagating in nonlinear media.[7] Besides, the envelope approach is rather complicated for the interaction case since it requires solving not one, but a set of coupled nonlinear equations.[8] That is why we use here an alternative approach: just the electric field instead of the envelope is considered for the treatment of pulse collision.[5,6,12]

The equation for the dynamics of a transversely uniform and linearly polarized light field in homogeneous and isotropic transparent media with dispersion and nonresonant electronic nonlinearity can be written in the form[2]

$$\frac{\partial E}{\partial z} + \frac{N_0}{c}\frac{\partial E}{\partial t} - a\frac{\partial^3 E}{\partial t^3} + b\int_{-\infty}^{t} E d\tau + gE^2\frac{\partial E}{\partial t} = 0, \quad (20.1)$$

where E is the electric field, z is the propagation coordinate, t is time, and N_0, a, and b are constants of the medium dispersion relation

$$n(\omega) = N_0 + a\omega^2 - \frac{b}{\omega^2}, \quad (20.2)$$

where $n(\omega)$ is the linear refraction index as a function of radiation frequency, g is the parameter proportional to the nonlinear refractive index coefficient n_2 ($g = 2n_2/c$), and c is the light velocity in vacuum.

The initial field distribution of two copropagating pulses with different central frequencies at $z = 0$ will be defined by the formula

$$E(0,t) = E_1 \exp\left[-2\ln(2)\left(\frac{t}{\Delta t_1}\right)^2\right]\sin(\omega_1 t)$$

$$+ E_2 \exp\left[-2\ln(2)\left(\frac{t+\Delta\tau}{\Delta t_2}\right)^2\right]\sin[\omega_2(t+\Delta\tau)], \qquad (20.3)$$

where E_1, E_2 are the maximum field amplitudes of the two pulses at the medium input; ω_1, ω_2 are their central frequencies; Δt_1, Δt_2 are the initial pulse durations; and $\Delta\tau$ is the time interval between the field maximums.

Before numerically solving Eq. (20.1) with boundary conditions (20.3), let us show analytically that Eq. (20.1) generalizes the known envelope equations. For that we substitute the following representation into (20.1):

$$E(z,t) = \frac{1}{2}\left[\varepsilon_1(z,t)e^{i(\omega_1 t - k_1 z)} + \varepsilon_2(z,t)e^{i(\omega_2 t - k_2 z)}\right] + \text{c.c.}, \qquad (20.4)$$

where ω_1 and ω_2 are some constant frequencies; $k_1 = n(\omega_1)\omega_1/c$; $k_2 = n(\omega_2)\omega_2/c$; and $\varepsilon_1(z,t)$, $\varepsilon_2(z,t)$ are new variables. The substitution gives[13]

$$\left\{\frac{\partial \varepsilon_1}{\partial z} + \frac{1}{V_1}\frac{\partial \varepsilon_1}{\partial t} - \sum_{n=2}^{\infty}\beta_n^{(1)}\frac{i^{n+1}}{n!}\frac{\partial^n \varepsilon_1}{\partial t^n} + \right.$$

$$+ \frac{g}{4}\left[i\omega_1\left(|\varepsilon_1|^2 + 2|\varepsilon_2|^2\right)\varepsilon_1 + 2\left(|\varepsilon_1|^2 + |\varepsilon_2|^2\right)\frac{\partial \varepsilon_1}{\partial t} + \right.$$

$$\left.\left. + 2\left(\varepsilon_1\frac{\partial \varepsilon_1^*}{\partial t} + \varepsilon_2\frac{\partial \varepsilon_2^*}{\partial t} + \varepsilon_2^*\frac{\partial \varepsilon_2}{\partial t}\right)\varepsilon_1\right]\right\}\exp(i\alpha_1) +$$

$$+ \frac{g}{4}\left\{\left[i(\omega_1 + 2\omega_2)\varepsilon_1\varepsilon_2 + \varepsilon_2\frac{\partial \varepsilon_1}{\partial t} + 2\varepsilon_1\frac{\partial \varepsilon_2}{\partial t}\right]\varepsilon_2\exp[i(\alpha_1 + 2\alpha_2)] + \right. \qquad (20.5)$$

$$+ \left[i(2\omega_1 - \omega_2)\varepsilon_1\varepsilon_2^* + 2\varepsilon_2^*\frac{\partial \varepsilon_1}{\partial t} + \varepsilon_1\frac{\partial \varepsilon_2^*}{\partial t}\right]\varepsilon_1\exp[i(2\alpha_1 - \alpha_2)] +$$

$$\left. + \left(i\omega_1\varepsilon_1 + \frac{\partial \varepsilon_1}{\partial t}\right)\varepsilon_1^2\exp(i3\alpha_1)\right\} + S(1\leftrightarrow 2) + \text{c.c.} = 0$$

where

$$\alpha_1\omega_1 t - k_1 z;\ V_1 = \left(\frac{\partial k}{\partial \omega}\right)_{\omega_1}^{-1};\ \beta_n^{(1)} = \left(\frac{\partial^n k}{\partial \omega^n}\right)_{\omega_1};\ k(\omega) = \frac{n(\omega)}{c}\omega;$$

and $S(1\leftrightarrow 2)$ designates the sum of all preceding terms, but with all interreplaced indices 1 and 2. No limiting assumptions have been made by now in the derivation of Eq. (20.5): it is mathematically equivalent to Eq. (20.1) with substitution (20.4).

If we apply Eq. (20.5) to the treatment of the interaction of two pulses with different central frequencies, then it is natural to consider ω_1 and ω_2 as the carrier frequencies and associate $\varepsilon_1(z, t)$, $\varepsilon_2(z, t)$ with the pulse envelopes. Furthermore, (1) assuming that the envelopes are slowly varying $\omega_1 \varepsilon_1 \gg \partial \varepsilon_1 / \partial t$, $\omega_2 \varepsilon_2 \gg \partial \varepsilon_2 / \partial t$, (2) leaving only the β_2, β_3 terms in the dispersion relation, and (3) ignoring the generation of new frequencies, one obtains the following known set of equations[11] from (20.5):

$$\begin{cases} \dfrac{\partial \varepsilon_1}{\partial z} + \dfrac{1}{V_1}\dfrac{\partial \varepsilon_1}{\partial t} + i\dfrac{\beta_2^{(1)}}{2}\dfrac{\partial^2 \varepsilon_1}{\partial t^2} - \dfrac{\beta_3^{(1)}}{6}\dfrac{\partial^3 \varepsilon_1}{\partial t^3} + i\dfrac{g\omega_1}{4}\left(|\varepsilon_1|^2 + 2|\varepsilon_2|^2\right)\varepsilon_1 = 0, \\ \dfrac{\partial \varepsilon_2}{\partial z} + \dfrac{1}{V_1}\dfrac{\partial \varepsilon_2}{\partial t} + i\dfrac{\beta_2^{(2)}}{2}\dfrac{\partial^2 \varepsilon_2}{\partial t^2} - \dfrac{\beta_3^{(2)}}{6}\dfrac{\partial^3 \varepsilon_2}{\partial t^3} + i\dfrac{g\omega_2}{4}\left(|\varepsilon_2|^2 + 2|\varepsilon_1|^2\right)A_2 = 0. \end{cases} \quad (20.6)$$

It is clear that the constraints (1)–(3) are incompatible with ultrabroad spectra and few-cycle pulses. Therefore, our analysis of the interaction of ultrashort pulses is based on Eq. (20.1) rather than (20.6) because the former is more accurate and self-consistent.

20.2.2 Interaction of femtosecond pulses with supercontinuum spectra in optical waveguides

Let us simulate the collision of copropagating femtosecond pulses centered at fundamental and second-harmonic frequencies of a Ti:sapphire laser in fused silica fiber via numerical integration of Eq. (20.1), which is applicable to the description of field dynamics in optical waveguides.[5,13]

The pulse parameters written in the notation of Eq. (20.3) will be $\lambda_1 = 2\pi c/\omega_1$ = 780 nm; $\lambda_2 = 2\pi c/\omega_2 = 390$ nm; $\Delta t_1 = \Delta t_2 = 18$ fs; $\Delta \tau = 40$ fs. The dispersion and nonlinear parameters of fused silica fiber appearing in (20.1) were estimated in Ref. 14 as $N_0 = 1.4508$, $a = 2.7401 \times 10^{-44}$ s^3/cm, $b = 3.9437 \times 10^{17}$ 1/cm s, and $n_2 = 2.9 \times 10^{-16}$ cm^2/W. All computations are made for the retarded time $\tau = t - z/v$, where v is the group velocity at the average central frequency $\omega = (\omega_1 + \omega_2)/2$. This allows us to keep both pulses in the considered time window during their interaction.

Figure 20.1 presents the simulated evolution of the two pulses with very low intensities so that the nonlinearity of the fiber response can be ignored and only dispersion influences the field dynamics considered. This means that the pulses propagate independently without any interaction. As follows from Eq. (20.2), the dispersion is more pronounced for the high-frequency (second-harmonic) pulse and it broadens faster than the low-frequency (first-harmonic) one, but lags behind the latter due to different group velocities in the normal dispersion range.

As the intensity increases, the nonlinearity of fused silica becomes important: single pulses undergo self-action, while the pulse collision turns out to be inseparable from the nonlinear interaction. In Figs. 20.2–20.4 we give the dynamics of field and spectral profiles of the same pulses as in Fig. 20.1, but with the higher initial intensities $I_1 = I_2 = 5 \times 10^{12}$ W/cm^2 [I kW/cm^2 = $(3N_0/8\pi)E_0^2$ CGS electrostatic units] in Fig. 20.2 and $I_1 = I_2 = 2 \cdot \times 10^{13}$ W/cm^2 in Figs. 20.3 and 20.4. For visualization purposes, we present pulse envelopes computed artificially as a connection between field maximums, while just the electric field was obtained in the numerical solution of Eq. (20.1).

The figures illustrate that intense pulses are widening both temporally and spectrally due to self-phase modulation and then, while colliding, produce a complex temporal structure with an ultrabroad spectrum. The temporal duration and spectral width of the output field are more than an order of magnitude higher than those of the initial pulses. The structure includes the sequence of ultrashort subpulses with central frequencies slightly varying from one subpulse to another. Each peak in the quasi-discrete spectrum corresponds to a certain subpulse in the sequence.

Analysis of Figs. 20.2 and 20.3 shows that temporal and spectral patterns become more contrasted with the rise of intensities of interacting pulses. For the highest considered intensity, the subpulses have near-rectangular profiles (Fig. 20.3). This can be explained by dispersion-induced stretching of areas with sufficiently high, but near-linear, frequency modulation originating from strong self-phase modulation at high intensities.[11]

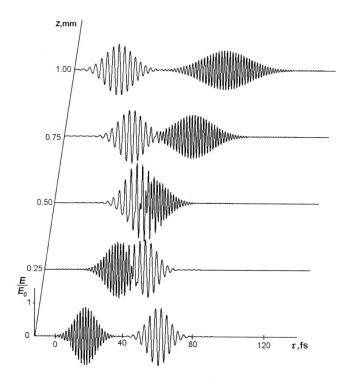

Figure 20.1 The dispersion-induced field dynamics of two 18 fs pulses spectrally centered on fundamental and second harmonics of a Ti:sapphire laser. The electric field $E(z,t)$ is normalized to the input amplitude E_0.

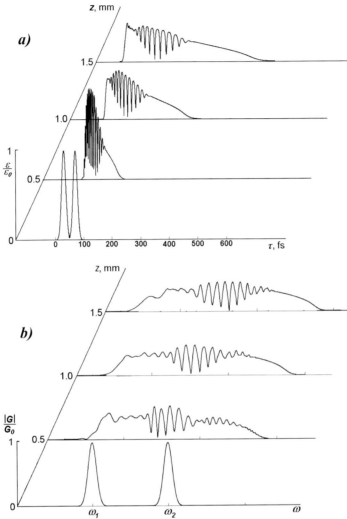

Figure 20.2 Temporal (a) and spectral (b) dynamics of interacting pulses with the initial intensities $I_1 = I_2 = 5 \times 10^{12}$ W/cm^2 in the fused silica fiber. The field envelope $\varepsilon(z,t)$ (a) and the spectral density $|G(z,t)|$ (b) are normalized to the maximum input values ε_0 and G_0, respectively.

In Fig. 20.4 we demonstrate that such a profile is the result of nonlinear interaction rather than linear interference of independently propagating pulses in the nonlinear medium. Figure 20.4 presents a comparison of the chain of quasi-rectangular subpulses generated at $z = 1.5$ mm (solid line), which corresponds to the last z layer of Fig. 20.3, and the chain of quasi-sinusoidal subpulses obtained as a result of the linear interference of two noninteracting pulses with the same parameters propagating independently in identical sections of fused silica fiber and interfering only after they leave the fibers (dashed line). A similar scheme was described in Ref. 15.

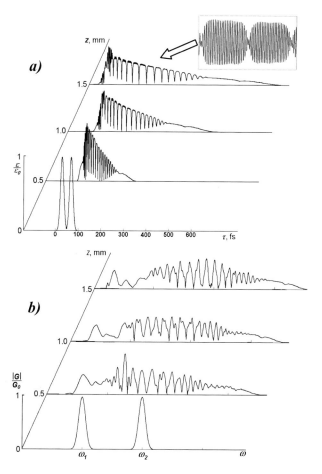

Figure 20.3 Temporal (a) and spectral (b) dynamics of interacting pulses with the initial intensities $I_1 = I_2 = 2 \times 10^{13}$ W/cm^2 in the fused silica fiber. The field envelope $\varepsilon(z,t)$ (a) and the spectral density $|G(z,t)|$ (b) are normalized to the maximum input values ε_0 and G_0, respectively. The inset gives the electric field in the center area.

20.2.3 Binary data coding using the generated sequences of ultrashort pulses

Each subpulse in the output sequence of Fig. 20.3(a) has its own central frequency, which is slightly different from those of other subpulses. This makes it possible to use such sequences for data coding by removing spectral peaks associated with certain subpulses.

In Fig. 20.5(a) we present a chain with the removed subpulse in the time domain marked by the dashed line. This signal can be obtained by cutting out the respective spectral peak [Fig. 20.5(b)] from the output quasi-discrete supercontinuum of Fig. 20.3 ($z = 1.5$ mm). Figure 20.6 gives a more meaningful example of coding the bit sequence 11011001111 using the same supercontinuum source from Fig. 20.3.

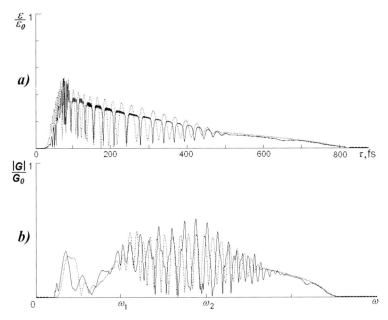

Figure 20.4 Temporal (a) and spectral (b) profiles of the chain of ultrashort pulses. The solid line is for interacting and the dashed one is for interfering pulses.

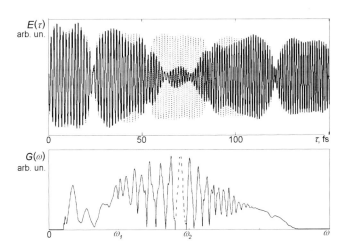

Figure 20.5 The chain of ultrashort pulses with a removed subpulse marked by the dashed line in the temporal (a) and spectral (b) domains.

20.2.4 Interaction of femtosecond pulses with supercontinuum spectra in bulk optical media

For the analysis of optical field dynamics in bulk media, Eq. (20.1) should be extended to the form[5,6,12]

$$\frac{\partial E}{\partial z} + \frac{N_0}{c}\frac{\partial E}{\partial t} - a\frac{\partial^3 E}{\partial t^3} + b\int_{-\infty}^{t} E d\tau + gE^2 \frac{\partial E}{\partial t} = \frac{c}{2N_0}\Delta_\perp \int_{-\infty}^{t} E d\tau, \qquad (20.7)$$

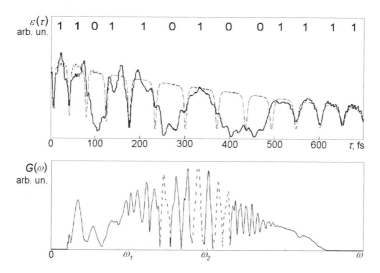

Figure 20.6 Coding of a bit sequence by quasi-discrete spectral supercontinuum in the temporal (a) and spectral (b) domains.

with the new term on the right-hand side, which describes diffraction; $\Delta_\perp$ is the transverse Laplacian.

Similar to the fiber case, we will consider the interaction of two pulses as

$$E(0,r,t) = E_1 \exp\left[-2\ln(2)\left(\frac{t}{\Delta t_1}\right)^2\right]\exp\left[-2\ln(2)\left(\frac{r}{\Delta r_1}\right)^2\right]\sin(\omega_1 t)$$
$$+E_2 \exp\left[-2\ln(2)\left(\frac{t+\Delta\tau}{\Delta t_2}\right)^2\right]\exp\left[-2\ln(2)\left(\frac{r}{\Delta r_2}\right)^2\right]\sin[\omega_2(t+\Delta\tau)], \qquad (20.8)$$

with the central frequencies $\lambda_1 = 780$ nm and $\lambda_2 = 390$ nm, but with finite transverse sizes Δr_1, Δr_2 and equal durations $\Delta t_1 = \Delta t_2 = 22$ fs as they propagate in a fused silica bulk with the same characteristics as in Section 20.2.2. for the silica fiber. Parameters of the approximation (20.2) remain the same because the waveguide dispersion of standard telecommunication-type fibers made from fused silica is negligibly small in the considered spectral range and the constants are defined by the material dispersion.

Figures 20.7(a)–(d) give a comparison of the nonlinear interaction of pulses with the initial transverse sizes $\Delta r_1 = \Delta r_2 = 33$ μm, the delay time $\Delta\tau = 240$ fs, and the intensities $I_1 = I_2 = 10^9$ W/cm^2 (left column) and the linear interference of the same pulses propagating independently (right column). The computations were made for the pulse electric field using the model (20.7). The input distribution for the interaction case was taken in the form of Eq. (20.8). For the independent propagation, we simulated the dynamics of each single pulse from Eq. (20.8) and then superposed the fields to obtain the interference pattern.

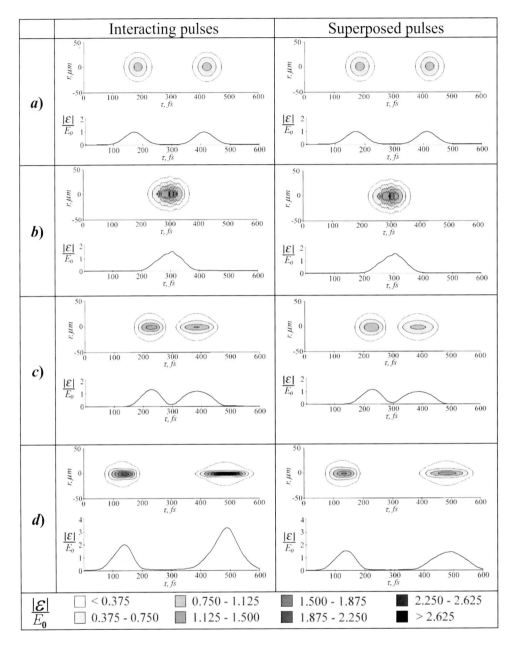

Figure 20.7 The dynamics of the field envelope of interacting (left column) and superposed (right column) 22 fs pulses with the initial intensities $I_1 = I_2 = 10^{12}$ W/cm^2 and transverse sizes $\Delta r_1 = \Delta r_2 = 33$ μm in the fused silica bulk.

The scale of Fig. 20.7 is insufficient to depict each field cycle. Thus, we present contour plots of spatiotemporal distributions of pulse envelopes obtained artificially from the field values via analysis of local maximums in the postprocessing stage. In addition, we inset axial envelope cuts for each layer.

Figure 20.7(a) introduces the input spatiotemporal field distribution. The pulses of fundamental and second harmonics have maximums near 420 fs and 170 fs, respectively, i.e., the former is initially delayed from the latter. The higher-frequency pulse has a lower group velocity due to normal dispersion of fused silica, and by $z \sim 0.6$ mm the pulse centers appear to be coincident in time and space [Fig. 20.7(b)]. It is seen that the distributions in the left and right columns of Fig. 20.7(b) are practically identical, which means that the interaction does not play a role within this distance. The complicated envelope structure with mutual beatings comes from the interference of the two pulses with ultrabroad, but essentially separated spectra. Figure 20.7(b) is not detailed enough to show the interference pattern, so we depict the axial distribution of the squared electric field in Fig. 20.8. Particularly, this demonstrates that any mathematical definition of the envelope will be ambiguous for the case considered.

The pulses pass through each other by the distance $z \sim 1.2$ mm [Fig. 20.7(c)]. Here it is seen that the interaction leads to the additional focusing and axial intensity increase [Fig. 20.7(c), left and right columns]. Peak field values of the interacting pulses are higher than those of the independently propagating pulses by no more than 10%, but the exceeding axial energy substantially affects further propagation. Therefore, by the medium output ($z \sim 1.8$ mm), the peak field values of interacting pulses appear to be twice as high as those of the noninteracting pulses [Fig. 20.7(d)]. The total electric energy is conserved in both cases, so the axial intensity rise occurs at the expense of the peripheral energy. Because of the nonlinear focusing, the beams are narrowing in the transverse dimension. This effect is demonstrated in greater detail in Fig. 20.9, where transverse envelope distributions on the medium output are cut at the time of maximum intensities. The transverse width of the high-frequency pulse at half-maximum intensity is only 5 µm in the interaction case (solid line), but 22 µm after the independent propagation and following interference (dashed line).

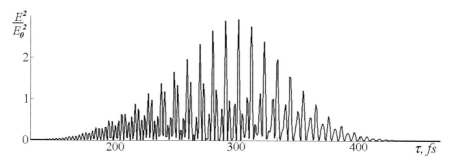

Figure 20.8 Axial distribution of the squared electric field of interacting pulses at the distance z = 0.6 mm from the input of the silica bulk.

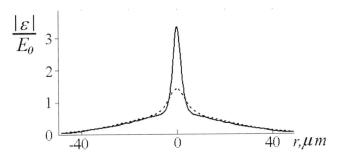

Figure 20.9 Transverse distribution of the field envelope in the nonlinear focus. The full beam half-maximum width is 5 and 22 µm for interacting (solid line) and superposed (dashed line) pulses, respectively.

Figure 20.10 represents the axial spectral dynamics of interacting (solid line) and noninteracting pulses (dashed line). The spectra are broadening due to the nonlinearity of fused silica in both cases, but the spectral width appears to be larger when the pulses interact. It should be noted that the pulse spectra remain separated by a considerable frequency range. The pulses are also separated spatially [Fig. 20.7(d)], so after the collision they propagate independently.

The interaction of pulses with the higher intensities $I_1 = I_2 = 7.5 \times 10^{12}$ W/cm^2 but the same central wavelength and durations is shown in Figs. 11(a)–(d). The initial delay time and transverse sizes are $\Delta\tau = 120$ fs and $\Delta r_1 = \Delta r_2 = 130$ µm. The distribution of the artificial field envelope is visualized as a three-dimensional surface $\varepsilon(r, t)$ normalized to its maximum on the medium input ε_0. In Fig. 20.12 we illustrate the axial spectral dynamics of interacting pulses.

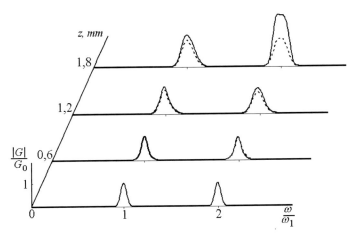

Figure 20.10 The axial spectral dynamics of interacting (solid line) and superposed (dashed line) 22 fs pulses with the initial intensities $I_1 = I_2 = 10^{12}$ W/cm^2 and transverse sizes $\Delta r_1 = \Delta r_2 = 33$ µm in the fused silica bulk. The values are normalized to the input spectral amplitude G_0.

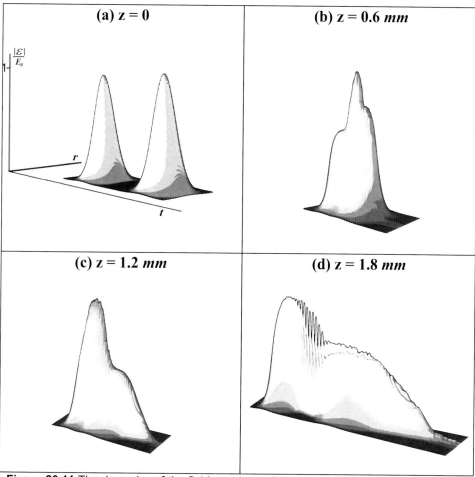

Figure 20.11 The dynamics of the field envelope of interacting 22 fs pulses with the initial intensities $I_1 = I_2 = 7.5 \times 10^{12}$ W/cm² and transverse sizes $\Delta r_1 = \Delta r_2 = 130$ μm in the fused silica bulk.

As before, the pulse centers coincide by the distance $z \sim 0.6$ mm. The superposition of the pulse fields produces the axial intensity peak near the centers, which self-focuses stronger than other areas. Therefore, its intensity is further growing while the transverse size diminishes. The spectrum of the whole spatiotemporal structure broadens dramatically (Fig. 20.12). By the distance $z \sim$ 1.2 mm, the greater part of the low-frequency pulse [left in Fig. 20.11(c)] passes through the higher-frequency pulse [right in Fig. 20.11(c)]. Between the maximums of pulse envelopes, an area is formed where tail frequencies of the leading pulse are practically the same as the front frequencies of the lagging pulse (see also Fig. 20.12). Close frequencies produce an interference pattern, visible as envelope beatings in the form of a sequence of few-cycle subpulses similarly to the fiber case. Because the interacting pulses are superposed in space and their spectra overlap, they form an inseparable field structure.

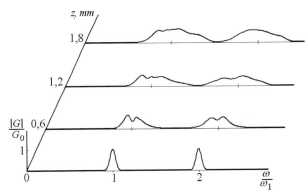

Figure 20.12 The axial spectral dynamics of interacting 22 fs pulses with the initial intensities $I_1 = I_2 = 7.5 \times 10^{12}$ W/cm^2 and transverse sizes $\Delta r_1 = \Delta r_2 = 130$ μm in the fused silica bulk. The values are normalized to the input spectral amplitude G_0.

Each subpulse in the sequence further undergoes self-focusing. The result of interaction is shown in Fig. 20.11(d). Nonlinear effects continue enriching the spectrum and the field structure significantly stretches in time due to the group dispersion. The chain of few-cycle subpulses is finally formed in the center of the structure. Each subpulse has a duration of ~10 fs and transverse size of ~30 wavelengths of the Ti:sapphire oscillator.

Figure 20.13 depicts the axial distribution of the squared field of the final structure. The chain of few-cycle pulses is visible in more detail here. As it has been mentioned, the structure is formed by high-frequency components of the low-frequency pulse and low-frequency components of the high-frequency pulse. The opportunity for generation of sequences of few-cycle pulses as a result of interaction of pulses with ultrabroad spectra in bulk optical media can be important for the development of systems with information transmission rates higher than 10 Tb/s (see Section 20.2.3).

Let us note that merging of interacting pulses into an inseparable field structure occurs in all cases when the spectra overlap after ultrabroadening in a nonlinear medium with a smooth dispersion dependence. However, the selection of conditions for a favorable generation of chains of ultrashort pulses in bulk

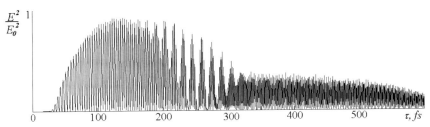

Figure 20.13 Axial distribution of the squared electric field of interacting 22 fs pulses with the initial intensities $I_1 = I_2 = 7.5 \times 10^{12}$ W/cm^2 and transverse sizes $\Delta r_1 = \Delta r_2 = 130$ μm on the output of the silica bulk ($z = 1.8$ mm).

media is much more sensible to the parameters of input femtosecond pulses and medium characteristics and, consequently, much more labor consuming than in optical waveguides.

20.3 Experiment

Our experimental goal was to confirm the hypothesis that the interaction of time-delayed ultrashort pulses centered on the fundamental and second harmonics of a Ti:sapphire laser produces a much broader and more modulated supercontinuum than after the propagation of isolated pulses.

A schematic view of our experimental setup is shown in Fig. 20.14. The femtosecond laser source was a 50 Hz repetition rate Ti:sapphire multipass laser system (MPA-30, Avesta Project Ltd.) with the FEMOS-2 master oscillator (Novosibirsk, Russia). The duration of output pulses on the fundamental wavelength of 800 nm was 30 fs with an energy of 1 mJ/pulse. After the frequency doubling in the 200 μm thick BBO crystal, we obtained 40 fs, 0.2 mJ pulses centered on 400 nm.

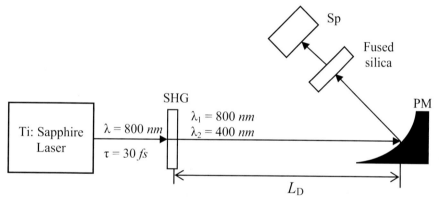

Figure 20.14 The experimental setup: PM, parabolic mirror; Sp, spectrometer.

To generate supercontinuum, we focused the fundamental and second-harmonic beams into the 10 mm thick fused-silica plate using a 250 mm focal-length parabolic mirror (Fig. 20.14). The peak pulse intensity in the focus under these conditions was estimated to be in the range between 2×10^{14} and 8×10^{14} W/cm^2. To measure the output spectrum we used the ASP-100 spectrometer (Avesta Project Ltd.) with a spectral coverage extending from 190 to 1100 nm and a spectral resolution of 0.05 nm. In our observations, both the fundamental and second-harmonic beams produced a supercontinuum in the silica plate (Fig. 20.15). These ultrabroad spectra could interact in the range of their overlapping. We registered the output spectrum and analyzed the dependence of the interference structure on the time delay between the pulses. The delay originated from the difference in the pulses' group velocities in air and could be controlled by the length of the air line L_D (see Fig. 20.14) as

$$\Delta\tau = L_D \Delta\beta, \quad (20.9)$$

where $\Delta\beta$ is the group velocity mismatch of 0.27 fs/cm in the case considered.

We changed the delay from 3 to 100 fs. The observed spectra were ultrabroad and characterized by a quasi-discrete spectral structure in the range from 500 to 700 nm with strong interference for the delays up to 80 fs. The best results were obtained with small delays and agree qualitatively with theoretical predictions. We did not observe quasi-discrete spectral features when considering independent propagation of either fundamental or second-harmonic pulses.

An example of the interference spectrum registered in the visible region is illustrated in Fig. 15 for $\Delta\tau = 20$ fs by curve 1. Such a delay provided partial temporal overlapping and interaction of pulses in the silica bulk and they both contributed to the supercontinuum generation. The quasi-discrete spectral structure is clearly seen in the figure. The highest spectral modulation presented here was observed at ~3 mm off the beam axis. For reference, we show supercontinuum spectra generated separately by the fundamental (curve 2) and second-harmonic pulses (curve 3).

20.4 Conclusion

In conclusion, we have shown theoretically that two few-cycle pulses with different spectral contents interacting in a bulk or waveguiding nonlinear media can produce chains of ultrashort pulses with central frequencies varying from one pulse to another. The chain is characterized by a quasi-discrete supercontinuum spectrum and each spectral peak corresponds to a certain pulse in the chain, which can be used efficiently for data coding.

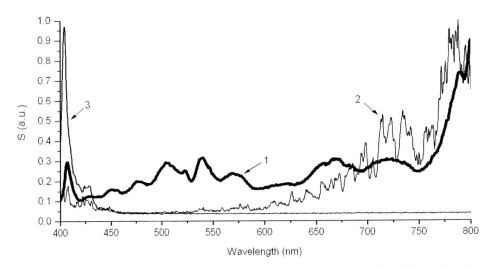

Figure 20.15 Supercontinuum signal registered on the output of the silica plate: the interaction of fundamental and second-harmonic pulses with the initial time delay of 20 fs (curve 1) and the self-action of the single fundamental (curve 2) or the second-harmonic pulse (curve 3).

We presented experimental measurements of supercontinuum spectra excited by 30–40 fs pulses centered on the fundamental and second harmonic of a Ti:sapphire laser with energy up to 1 mJ. It is demonstrated that the interaction produces a much broader spectrum than the propagation of isolated pulses and gives rise to a strong spectral modulation in the wavelength range from 500 to 700 nm.

Acknowledgment

This research is supported by the program "Development of Scientific Potential of Higher Education," project RNP.2.1.1.6877, and grant N05-02-16556a of the Russian Foundation for Basic Research (RFBR). Yuri A. Shpolyanskiy thanks the private noncommercial foundation Dynasty, Moscow, Russia, by Prof. Dmitry Zimin for the postdoc fellowship.

References

1. Cerullo, G., De Silvestri, S., Nisoli, M., Sartania, S., Stagira, S., and Svelto, O., "Few-optical cycle laser pulses: From high peak power to frequency tenability," *IEEE J. Sel. Top. Quant. Electron.*, **6**, 948–958, 2000.

2. Steinmeyer, G., Sutter, D.H., Gallman, L., Matuschek, N., and Keller, U., "Frontiers in ultrashort pulse generation: Pushing the limits in linear and nonlinear optics," *Science*, **286**, 1507–1512, 1999.

3. Brabec, Th., and Krausz, F. "Intense few-cycle laser fields: Frontiers of nonlinear optics," *Rev. Mod. Phys.*, **72**, N2, 545-591, 2000.

4. Sudrie, L., Couairon, A., Franco, M., Lamouroux, B., Prade, B., Tzortzakis, S., and Mysyrowich, A., "Femtosecond laser-induced damage and filamentary propagation in fused silica," *Phys. Rev. Lett.*, **89**, N18, 186601, 2002.

5. Bespalov, V.G., Kozlov, S.A., Shpolyansky, Yu.A., and Walmsley, I.A., "Simplified field wave equations for nonlinear propagation of extremely short light pulses," *Phys. Rev. A.*, **66**, 013811, 2002.

6. Berkovsky, A.N., Kozlov, S.A., and Shpolyanskiy, Yu.A., "Self-focusing of few-cycle light pulses in dielectric media," *Phys. Rev. A*, **72**, 043821, 2005.

7. Aközbek, N., Trushin, S.A., Baltuška, A., Fuß W., Goulielmakis E., Kosma K., Krausz F., Panja S., Uiberacker M., Schmid W.E., Becker A., Scalora M. and M Bloemer. "Extending the supercontinuum spectrum down to 200nm with few-cycle pulses," *New J. Phys.*, **8**, 177–188, 2006.

8. Karasawa, N., Morita, R., Xu, L., Shigekawa, H., and Yamashita, M., "Theory of ultrabroadband optical pulse generation by induced phase modulation in a gas-filled hollow waveguide," *J. Opt. Soc. Am. B*, **16**, 662–668, 1999.

9. Karasawa, N., Morita, R., Shigekawa, H., and Yamashita, M., "Generation of intense ultrabroadband optical pulses by induced-phase modulation in an argon-filled single-mode hollow waveguide," *Opt. Lett.*, **25**, 183–185, 2000.

10. Akhmanov, S.A., Vysloukh, V.A., and Chirkin, A.S., *Optics of Femtosecond Laser Pulses*, American Institute of Physics, New York (1992).

11. Agrawal, G.P., *Nonlinear Fiber Optics*, Academic, San Diego (1995).

12. Kozlov, S.A., and Sazonov, S.V., "Nonlinear propagation of optical pulses of a few oscillations duration in dielectric media," *JETP*, **84**, 221–228, 1997.

13. Bakhtin, M. A., and Kozlov, S. A., "Formation of a sequence of ultrashort signals in a collision of pulses consisting of a small number of oscillations of the light field in nonlinear optical media," *Opt. Spectrosc.*, **98**, 425–430, 2005.

14. Bakhtin, M.A., Kolesnikova, S.Yu., and Shpolyanskiy, Yu.A., "The approximation accuracies of fused-silica dispersion in the methods of slowly-varying envelope and slowly-varying profile," *Modern Technologies* (in Russian), S.A. Kozlov, Ed., University ITMO, St.Petersburg, pp. 196–203 (2001).

15. Corsi, C., Tortora, A., and Bellini, M., "Mutual coherence of supercontinuum pulses collinearly generated in bulk media," *Appl. Phys. B*, **77**, 285–290, 2003.

Chapter 21
Noise in Classical and Quantum Photon-Correlation Imaging

Bahaa E. A. Saleh and Malvin Carl Teich
Boston University

21.1 Introduction
21.2 Classical Photon-Correlation Imaging
    21.2.1 Ghost imaging
    21.2.2 Van Cittert–Zernike theorem
    21.2.3 Hanbury-Brown-Twiss interferometer
21.3 Quantum Photon-Correlation Imaging
    21.3.1 Ghost imaging
    21.3.2 Van Cittert–Zernike theorem
    21.3.3 Quantum microscopy and lithography
21.4 Noise in Photon-Correlation Imaging
21.5 Conclusion
Acknowledgments
References

21.1 Introduction

Imaging is the estimation of the spatial distribution of a physical object by measuring the optical radiation it emits, or by making use of an optical wave that interacts with the object, via reflection or transmission, for example, before being measured by a detector.[1] The resolution of an imaging system is limited by the inability to localize the optical field at points of the object. Under otherwise ideal conditions, resolution is limited by diffraction. The sensitivity of an imaging system is limited by the uncertainty in the measurement. Under ideal conditions, this is determined by photon noise, which depends on the statistical fluctuations of the light.

In conventional imaging systems, an extended detector, such as a CCD camera or an array detector, measures the spatial distribution of the optical intensity, which is proportional to the photon flux density. In interferometric systems, the spatial distribution of the optical field is inferred from measurements of the optical intensity.[2]

With the emergence of coherence theory,[2-5] imaging systems based on measurements of the second-order coherence function at pairs of points in the detection plane were developed. An example is the imaging of an incoherent object based on the van Cittert-Zernike theorem. Imaging systems based on measurement of intensity correlation, or the photon coincidence rate, at pairs of points, were developed in the 1960s. A classic example of the photon-correlation imaging of an object emitting thermal light is stellar imaging using a Hanbury–Brown–Twiss intensity-correlation interferometer.[4-7]

More recently, two-photon light, which may be generated via spontaneous parametric downconversion in a second-order nonlinear optical crystal,[8] has been used for imaging.[9-17] This type of two-photon (or biphoton) imaging, which has come to be called quantum imaging, is also based on the measurement of photon coincidence by the use of photon-counting array detectors or by scanning two photon-counting detectors at pairs of points.

To compare the resolution and sensitivity of imaging systems based on the aforementioned types of measurements, it is necessary to derive expressions for the measured quantities in terms of the object distribution. The point-spread functions based on such expressions can be used to assess the resolution. One measure of the sensitivity of the imaging process is the signal-to-noise ratio (SNR) of the measured variables. The statistical nature of the light source must be known in order to determine the SNR.[1,5]

The purpose of this chapter is to compare the resolution and sensitivity of photon-correlation imaging systems that make use of thermal light and two-photon light. We will henceforth refer to these two imaging modalities as classical and quantum photon-correlation imaging, or simply classical and quantum imaging, respectively. Clearly, light in other quantum states can also be used for imaging.

21.2 Classical Photon-Correlation Imaging

Consider the imaging system shown schematically in Fig. 21.1. The source emits quasi-monochromatic, spatially incoherent light with intensity $I_s(\mathbf{x})$, where $\mathbf{x} = (x, y)$. The emitted light reaches the two detectors via two linear systems with impulse response functions $h_1(\mathbf{x}_1, \mathbf{x})$ and $h_2(\mathbf{x}_1, \mathbf{x})$. The object may reside in either of these systems, or in the source itself. The two systems may also be combined as one system in which the object resides.

Based on coherence theory and a systems description of the imaging process,[2,18] the second-order coherence function $G^{(1)}(\mathbf{x}_1, \mathbf{x}_2) = \langle E^*(\mathbf{x}_1) E(\mathbf{x}_2) \rangle$ is related to the source intensity and the impulse response functions by the integral,

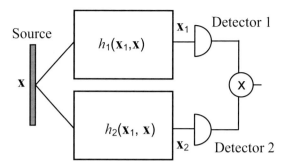

Figure 21.1 Imaging system.

$$G^{(1)}(\mathbf{x}_1,\mathbf{x}_2) = \int \acute{Z}_s(\mathbf{x}) h_1^*(\mathbf{x}_1,\mathbf{x}) h_2(\mathbf{x}_2,\mathbf{x}) \acute{Z} d\mathbf{x} \ . \qquad (21.1)$$

The range of the integrals is [-∞, ∞] throughout, unless otherwise indicated.

21.2.1 Ghost imaging

If the object resides in the first system and has a complex amplitude transmittance (or reflectance) $O(\mathbf{x})$, then $h_1(\mathbf{x}_1,\mathbf{x})$ is linearly related to $O(\mathbf{x})$. Consequently, if $G^{(1)}(x_1,x_2)$ is measured as a function of $\mathbf{x}_1$ with $\mathbf{x}_2$ fixed, the measurement would be linearly related to $O(\mathbf{x})$ so that the imaging system is coherent [i.e., it measures the complex amplitude transmittance (or reflectance)]. For example, consider an object placed in one branch of the system at a distance d_1 from the source, with a lens used to collect the transmitted light and focus it onto detector 1. The other branch of the system contains a lens of focal length f placed in the space between the source and detector 2 at distances d_2 and d_3, as shown in Fig. 21.2. For such a system, Eq. (21.1) leads to diffraction-limited imaging if the focusing condition

$$\frac{1}{d_2-d_1}+\frac{1}{d_3}=\frac{1}{f} \qquad (21.2)$$

is satisfied. This type of imaging is peculiar since the light field that is transmitted through the object is collected and observed with a point detector at the fixed point $\mathbf{x}_2$. The image is acquired by scanning detector 1 at all points $\mathbf{x}_1$ and by observing the light that has not interacted with the object. This type of image may be dubbed ghost imaging, although that appellation historically originated in the context of quantum imaging.

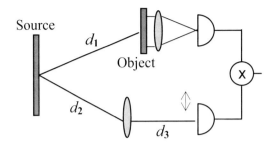

Figure 21.2 Ghost-imaging system.

21.2.2 Van Cittert–Zernike theorem

As another example, assume that the object itself is the source, i.e., $O(\mathbf{x}) = I_s(\mathbf{x})$, and assume that the two systems are combined into a single 2-f optical system, as shown in Fig. 21.3, whereupon

$$h_1(\mathbf{x}_1, \mathbf{x}) = h_2(\mathbf{x}_1, \mathbf{x}) \propto \exp\left(-i\frac{2\pi}{\lambda f} \mathbf{x}_1 \cdot \acute{\mathbf{x}}\right), \qquad (21.3)$$

Where λ is the wavelength of the light and f is the focal length of the lens. In this case, Eq. (21.1) provides

$$G^{(1)}(\mathbf{x}_1, \mathbf{x}_2) \propto \int O(\mathbf{x}) \exp\left[-i\frac{2\pi}{\lambda f}(\mathbf{x}_2 - \mathbf{x}_1) \cdot \acute{\mathbf{x}}\right] \acute{Z} d\mathbf{x} . \qquad (21.4)$$

When measured as a function of the position difference $\mathbf{x}_1 - \mathbf{x}_2$, the coherence function $G^{(1)}$ is proportional to the Fourier transform of $O(\mathbf{x})$. Equation (21.4) is the van Cittert–Zernike theorem and is the basis of a well-known technique for measuring the angular diameter of stars.[2–6]

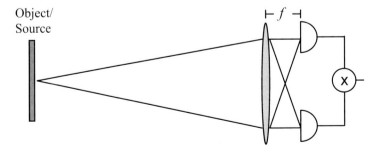

Figure 21.3 Photon-correlation imaging of the source.

21.2.3 Hanbury-Brown-Twiss interferometer

The intensity-correlation function, which is a special fourth-order coherence function,[18] at the positions $\mathbf{x}_1$ and $\mathbf{x}_2$ is $G^{(2)}(\mathbf{x}_1,\mathbf{x}_2) = \langle I_1(\mathbf{x}_1) I_2(\mathbf{x}_2) \rangle$. If the incoherent light is thermal (which at one time was called chaotic), the rate of photon coincidence is related to the second-order coherence function by the Siegert relation,[5]

$$G^{(2)}(\mathbf{x}_1,\mathbf{x}_2) = G^{(1)}(\mathbf{x}_1,\mathbf{x}_1) G^{(1)}(\mathbf{x}_2,\mathbf{x}_2) + \frac{1}{M} \left| G^{(1)}(\mathbf{x}_1,\mathbf{x}_2) \right|^2, \quad (21.5)$$

where $M \geq 1$ is a factor (known as the degrees of freedom) that increases with increase of the ratio of the detector response time to the optical coherence time. This equation is the basis of the Hanbury-Brown-Twiss effect.[4–7] Thus, the measurement of the intensity-correlation function $G^{(2)}(\mathbf{x}_1,\mathbf{x}_2)$ and the intensity at $G^{(1)}(\mathbf{x},\mathbf{x})$ at all points permit the magnitude of the second-order coherence function to be determined. This principle has served as the basis of a number of schemes for ghost imaging with thermal light.[19–22] A fundamental difficulty in such systems is the small relative magnitude of the second term on the right-hand side of Eq. (21.5) when the coherence time of the detected field is much smaller than the detector's response time.[5]

21.3 Quantum Photon-Correlation Imaging

We now consider a light source that emits photons in pairs. The coherence properties of such light must be described using quantum coherence theory.[18] For each pair, the quantum state is

$$|\Psi\rangle = \iint d\mathbf{x}\, d\mathbf{x}' \, \psi_s(\mathbf{x},\mathbf{x}') |1_\mathbf{x}, 1_{\mathbf{x}'}\rangle, \quad (21.6)$$

where $\psi(\mathbf{x},\mathbf{x}')$ is the two-photon wave function, i.e., the probability that the photons are emitted from positions $\mathbf{x}$ and $\mathbf{x}'$ is $|\psi_s(\mathbf{x},\mathbf{x}')|^2$. In an entangled state, $\psi_s(\mathbf{x},\mathbf{x}') = \xi_s(\mathbf{x}) \delta(\mathbf{x}-\mathbf{x}')$ so that the photons are emitted from a common position, although that position is random with probability density $|\xi_s(\mathbf{x})|^2$.

Such light may be generated by spontaneous parametric downconversion (SPDC) from a thin planar nonlinear crystal. For each annihilated pump photon, a pair of photons, the signal and idler, are generated in a two-photon, or biphoton, state. In this case, the complex function $\xi_s(\mathbf{x})$ in Eq. (21.6) is proportional to the conjugate of the optical field of the pump. This spatially entangled state emerges as a result of conservation of momentum, which makes the directions of the photons anticorrelated.

If the two photons are directed through systems with impulse response functions $h_1(\mathbf{x}_1,\mathbf{x})$ and $h_2(\mathbf{x}_1,\mathbf{x})$, and directed to two detectors, as in Fig. 21.1, then it can be shown that the probability of detecting photons simultaneously with detectors at positions $\mathbf{x}_1$ and $\mathbf{x}_2$ is

$$G^{(2)}(\mathbf{x}_1,\mathbf{x}_2) = \langle :\hat{I}(\mathbf{x}_1)\hat{I}(\mathbf{x}_2): \rangle = |\psi(\mathbf{x}_1,\mathbf{x}_2)|^2, \qquad (21.7)$$

i.e., is simply the square magnitude of the two-photon wave function

$$\psi(\mathbf{x}_1,\mathbf{x}_2) = \int \xi_s(\mathbf{x}) h_1(\mathbf{x}_1,\mathbf{x}) h_2(\mathbf{x}_2,\mathbf{x}) d\mathbf{x}. \qquad (21.8)$$

The similarity between Eq. (21.8) for the two-photon wave function $\psi(\mathbf{x}_1,\mathbf{x}_2)$ and Eq. (21.1) for the second-order coherence function $G^{(1)}(\mathbf{x}_1,\mathbf{x}_2)$ is remarkable. The source function $\xi_s(x)$ plays the role of the intensity of the incoherent source $I_s(x)$, and, except for a conjugation operation in the incoherent case [Eq. (21.1)], the impulse response functions of the optical systems play similar roles. The origin of this similarity may be attributed to the fact that both the second-order wave function and the two-photon wave function satisfy the Wolf equations.[23]

For thermal light, in accordance with the Siegert relation set forth in Eq. (21.5), the two-photon coincidence rate is proportional to $|G^{(1)}(\mathbf{x}_1,\mathbf{x}_2)|^2$, to which a background term is added, whereas in the two-photon case, the photon coincidence rate is simply $|\psi(\mathbf{x}_1,\mathbf{x}_2)|^2$. The background term, which typically dominates Eq. (21.5), as discussed earlier, is absent in the two-photon case, as was recognized by Belinskii and Klyshko.[9]

21.3.1 Ghost imaging

If the imaging configuration depicted in Fig. 21.2 is used with a two-photon source, then diffraction-limited imaging is attained if the condition

$$\frac{1}{d_2+d_1} + \frac{1}{d_3} = \frac{1}{f} \qquad (21.9)$$

is satisfied.[12] The sign change in Eq. (21.9), in comparison to Eq. (21.2), is attributed to the conjugation in the imaging equation.

21.3.2 Van Cittert–Zernike theorem

If the imaging configuration depicted in Fig. 21.3 is used with a two-photon source, which is itself the object, then the imaging equation becomes

$$G^{(1)}(\mathbf{x}_1,\mathbf{x}_2) \propto \int O(\mathbf{x})\exp\left[-i\frac{2\pi}{\lambda f}(\mathbf{x}_2+\mathbf{x}_1).\mathbf{x}\right]d\mathbf{x}. \qquad (21.10)$$

This equation is identical to Eq. (21.4) except for a change in sign. When measured as a function of $\mathbf{x}_1+\mathbf{x}_2$, the coherence function $G^{(1)}$ is proportional to the Fourier transform of $O(\mathbf{x})$. For example, if $O(\mathbf{x})$ is uniform, then $G^{(1)}(\mathbf{x}_1,\mathbf{x}_2) \propto \delta(\mathbf{x}_2+\mathbf{x}_1)$, i.e., if a photon is observed at $\mathbf{x}_1$, then another must be detected at $-\mathbf{x}_1$.

21.3.3 Quantum microscopy and lithography

The intensity correlation may be measured at the same position by making use of a single detector responsive to the rate of two-photon absorption. For two-photon light in an entangled state, $G^{(2)}(\mathbf{x}_1,\mathbf{x}_1) = |\psi(\mathbf{x}_1,\mathbf{x}_1)|^2$, with

$$\psi(\mathbf{x}_1,\mathbf{x}_1) = \int \xi_s(\mathbf{x})h^2(\mathbf{x}_1,\mathbf{x})d\mathbf{x}, \qquad (21.11)$$

in the special case when the two optical systems h_1 and h_2 collapse into a single imaging system h, as can be shown by use of Eq. (21.8). The imaging system is therefore linear with impulse response function $h^2(\mathbf{x}_1,\mathbf{x})$. For example, for a 2-$f$ Fourier optics system, $h(\mathbf{x}_1,\mathbf{x})$ is given by Eq. (21.3). The squaring operation increases the frequency by a factor of two. For example, for a two-slit object, the system creates a sinusoidal pattern at twice the spatial frequency. This feature plays a role in quantum (entangled-photon) microscopy[24,25] and photoemission,[26] and is, in essence, the principle behind quantum lithography,[27] which exploits entanglement to enhance resolution. This enhancement cannot be attained by making use of a classical thermal-imaging system.

21.4 Noise in Photon-Correlation Imaging

In this section we compare the accuracy of classical and quantum imaging systems by determining the error of measurement of the photon-correlation functions in each case. It is convenient to define the normalized photon-correlation function $g^{(2)} = G^{(2)}/(I_1 I_2)$, the normalized second-order coherence function $g^{(1)} = G^{(1)}/(I_1 I_2)^{1/2}$, and the normalized two-photon wave function $\varphi = \psi/(I_1 I_2)^{1/2}$, where I_1 and I_2 are the intensities (or the mean number of photons) at the two detectors. For classical imaging, Eq. (21.5) becomes

$$g^{(2)} - 1 = \frac{1}{M}|g^{(1)}|^2, \qquad (21.12)$$

while for quantum case, Eq. (21.7) becomes

$$g^{(2)} = |\varphi|^2. \tag{21.13}$$

The detectors measure estimates of the functions $G^{(2)}$ and I, which we label $\hat{G}^{(2)}$ and $\hat{I}$, respectively. These estimates are used to calculate estimates of $g^{(2)}$ via the relation $\hat{g}^{(2)} = \hat{G}^{(2)}/(\hat{I}_1 \hat{I}_2)$. We now proceed to determine the errors in the estimate $\hat{g}^{(2)}$ in the classical and quantum cases.

The uncertainty in the measurement of the intensity-correlation function $G^{(2)}(\mathbf{x}_1, \mathbf{x}_2) = \langle : \hat{I}(\mathbf{x}_1)\hat{I}(\mathbf{x}_2) : \rangle$ stems from the finite time available to measure the average intensity products. This function is usually measured by registering the number of photon counts n_1 and n_2 detected in each of the two detectors during a sequence of short time intervals, each of duration T, and averaging the product of the counts,[5]

$$\hat{G}^{(2)} = \frac{1}{N}\sum_{m=1}^{N} n_1(m) n_2(m). \tag{21.14}$$

Here, the index m refers to the mth time interval and N is the total number of intervals observed. The total duration of the measurement is NT. Likewise, an estimate of the intensity is measured at each detector by computing the averages

$$\hat{I}_j = \frac{1}{N}\sum_{m=1}^{N} n_i(m), \quad i = 1, 2. \tag{21.15}$$

Clearly, if $N \to \infty$, the measured functions $\hat{G}^{(2)}$ and $\hat{I}_i$ equal exactly the true functions $G^{(2)}$ and I_i, respectively.

The normalized measurement errors e in the classical (C) and quantum (Q) cases are defined by the following normalized variances:

$$e_C^2 = \frac{\mathrm{Var}\{\hat{g}^{(2)}\}}{\left[g^{(2)} - 1\right]^2}, \quad e_Q^2 = \frac{\mathrm{Var}\{\hat{g}^{(2)}\}}{\left[g^{(2)}\right]^2}. \tag{21.16}$$

For comparison, we define the ratio

$$R = \frac{e_C^2}{e_Q^2}. \tag{21.17}$$

If $R > 1$, then quantum imaging offers a statistical advantage, and vice versa.

Computation of the variances in Eq. (21.16) is a lengthy process, particularly in the classical case for which the photon counts obey Bose-Einstein statistics. We assume that the photon counts $n_i(m)$ are statistically independent for the different counting intervals. Since it is difficult to determine the variance of the ratio $\hat{g}^{(2)} = \hat{G}^{(2)}/(\hat{I}_1 \hat{I}_2)$, we assume that the errors are sufficiently small so that we can use the relation $1/\hat{I} \approx (1/I)[1-(\hat{I}-I)/I]$ to simplify the computation in terms of statistical moments of n_i. To further simplify the computation, we have also assumed that the thermal light in the classical case has a Lorentzian spectrum, i.e., an exponential coherence function, with coherence time τ_c. The following expressions result:

$$e_C^2 = \frac{1}{N}\frac{M^2}{\gamma^2}\left[\frac{1+g^2}{g^4}\frac{1}{\bar{n}_C^2} + \frac{2+4g^2-2g^4}{g^4}\frac{1}{\bar{n}_C} + \frac{5+10g^2-11g^4}{2g^4}\gamma\right], \quad (21.18)$$

$$e_Q^2 = \frac{1}{N}\frac{1}{g^2}\frac{1}{\bar{n}_Q^2}. \quad (21.19)$$

Here, the symbol g is used to denote $|g^{(1)}|$ or $|\varphi|$, for the classical and quantum cases, respectively, and $\bar{n}_C$ and $\bar{n}_Q$ are the mean number of photon detected in each time interval in the classical and quantum cases, respectively. The quantity M is the degrees-of-freedom parameter, which is a function of the ratio $\gamma = T/\tau_c$.

As expected, the errors e_C and e_Q depend on $\sqrt{N}$ for both cases, but the dependence on the mean number of photons, the ratio $\gamma = T/\tau_c$, and the quantity g, which is to be ultimately estimated, are different. It is useful to take the following two limiting cases:

Case 1. If the mean number of counts in both cases are equal and small, i.e., $\bar{n}_C = \bar{n}_Q \ll 1$, then

$$R = \frac{M^2}{\gamma^2}f_1(g), \quad f_1(g) = \frac{1+g^2}{g^2}. \quad (21.20)$$

The quantity R can then be substantially greater than unity, as can be seen from the plots of $f_1(g)$ and M/γ in Fig. 21.4, for thermal light with Lorenzian spectrum, for which $M = 2\gamma^2/(e^{-2\gamma}+2\gamma-1)$. It follows that quantum imaging can offer a significant statistical advantage under these conditions.[28,29]

Case 2. In reality, the mean number of photons in the quantum case is typically small, i.e., $\bar{n}_Q \ll 1$, since the generation of a high flux of biphotons is generally difficult. Assuming strong thermal light, i.e., $\bar{n}_C \gg 1$, we obtain the ratio

$$R = \frac{\bar{n}_Q^2}{\bar{n}_C^2} \frac{M^2}{\gamma} f_2(g), \quad f_2(g) = \frac{5 + 10g^2 - 11g^4}{2g^2}. \quad (21.21)$$

As shown in Fig. 21.4, the factor $f_2(g)$ can be large. Also, M and M/γ are greater than unity and can be large for large values of γ. These factors favor quantum imaging. However, it is the ratio of the mean counts $\bar{n}_Q/\bar{n}_C$ that can be sufficiently small, allowing classical imaging to outdo quantum imaging.

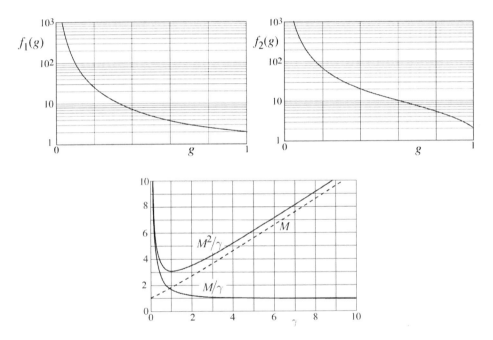

Figure 21.4 Factors affecting the error ratio R.

21.5 Conclusion

Imaging based on photon-correlation measurement with thermal light exploits the photon-bunching effect, which is accompanied by a large background. Two-photon light, on the other hand, comprises complete bunching since the photons arrive simultaneously, and it therefore offers the same possibilities for imaging without the background attendant to thermal light. Both systems offer possibilities for coherent imaging, including ghost imaging, which is *not* a unique feature of quantum imaging.[30] Because the only difference in the imaging equations is a conjugation factor, quantum imaging offers no advantage in resolution in a configuration using two detectors. In the degenerate case for which the two detectors become one, i.e., if the detector is a two-photon absorber, the quantum paradigm offers a factor of two advantage in resolution.

The sensitivity of the classical and quantum imaging systems are, of course, different because the noises associated with the sources are different. Quantum imaging has a significantly greater signal-to-noise ratio—if the same mean number of photons are used. Because the generation of a high two-photon flux is difficult, this advantage has not yet been exploited in real imaging systems.

Acknowledgments

This work was supported by a U.S. Army Research Office (ARO) Multidisciplinary University Research Initiative (MURI) Grant and by the Bernard M. Gordon Center for Subsurface Sensing and Imaging Systems (CenSSIS), an NSF Engineering Research Center.

References

1. Barrett, H.H., and Myers, K., *Foundations of Image Science*, Wiley, Hoboken, NJ (2003).
2. Saleh, B.E.A., and Teich, M.C., *Fundamentals of Photonics*, 2nd ed., Wiley, Hoboken, NJ (2007).
3. Wolf, E., *Introduction to the Theory of Coherence and Polarization of Light*, Cambridge University Press, New York (2007).
4. Mandel, L., and Wolf, E., *Optical Coherence and Quantum Optics*, Chap. 22, Cambridge University Press, New York (1995).
5. Saleh, B.E.A., *Photoelectron Statistics*, Springer, New York (1978).
6. Hanbury-Brown, R., *The Intensity Interferometer*, Taylor & Francis, London (1974).
7. Goldberger, M.L., Lewis, H.W., and Watson, K.M., "Use of intensity correlation to determine the phase of a scattering amplitude," *Phys. Rev.*, **132**, 2764–2787, 1963.
8. Klyshko, D.N., *Photons and Nonlinear Optics*, Nauka, Moscow (1980) [Translation: Gordon and Breach, New York (1988)].
9. Belinskii, A.V., and Klyshko, D.N., "Two-photon optics: diffraction, holography, and transformation of two-dimensional signals," *Zh. Eksp. Teor. Fiz.*, **105**, 487–493, 1994 [Translation: *Sov. Phys. JETP*, **78**, 259–262, 1994].
10. Strekalov, D.V., Sergienko, A.V., Klyshko, D.N., and Shih, Y.H., "Observation of two-photon 'ghost' interference and diffraction," *Phys. Rev. Lett.*, **74**, 3600–3603, 1995.
11. Pittman, T.B., Shih, Y.H., Strekalov, D.V., and Sergienko, A.V., "Optical imaging by means of two-photon quantum entanglement," *Phys. Rev. A*, **52**, R3429–R3432, 1995.

12. Pittman, T.B., Strekalov, D.V., Klyshko, D.N., Rubin, M.H., Sergienko, A.V., and Shih, Y.H., "Two-photon geometric optics," *Phys. Rev. A*, **53**, 2804–2815, 1996.

13. Saleh, B.E.A., Popescu, S., and Teich, M.C., "Generalized entangled-photon imaging," *Proc. Ninth Ann. Mtg. IEEE Lasers and Electro-Optics Society*, Vol. 1, pp. 362–363, IEEE, Piscataway, NJ (1996).

14. Abouraddy, A.F., Saleh, B.E.A., Sergienko, A.V., and Teich, M.C., "Role of entanglement in two-photon imaging." *Phys. Rev. Lett.* **87**, 123602, 2001.

15. Abouraddy, A.F., Saleh, B.E.A., Sergienko, A.V., and Teich, M.C., "Quantum holography," *Opt. Expr.*, **9**, 498–505, 2001.

16. Abouraddy, A.F., Saleh, B.E.A., Sergienko, A.V., and Teich, M.C., "Entangled-photon Fourier optics," *J. Opt. Soc. Am. B*, **19**, 1174–1184, 2002.

17. Goodman, J.W., *Introduction to Fourier Optics*, 2nd ed., Roberts and Company, Englewood (2004).

18. Glauber, R.J., "The quantum theory of optical coherence," *Phys. Rev.*, **130**, 2529–2539, 1963.

19. Saleh, B.E.A., Abouraddy, A.F., Sergienko, A.V., and Teich, M.C., "Duality between partial coherence and partial entanglement," *Phys. Rev. A*, **62**, 043816, 2000.

20. Gatti, A., Brambilla, E., Bache, M., and Lugiato, L.A., "Ghost imaging with thermal light: comparing entanglement and classical correlation," *Phys. Rev. Lett.*, **93**, 093602, 2004.

21. Gatti, A., Brambilla, E., Bache, M., and Lugiato, L.A., "Correlated imaging, quantum and classical," *Phys. Rev. A*, **70**, 013802, 2004.

22. Ferri, F., Magatti, D., Gatti, A., Bache, M., Brambilla, E., and Lugiato, L.A., "High-resolution ghost image and ghost diffraction experiments with thermal light," *Phys. Rev. Lett.*, **94**, 183602, 2005.

23. Saleh, B.E.A., Teich, M.C., and Sergienko, A.V., "Wolf equations for two photon light," *Phys. Rev. Lett.*, **94**, 223601, 2005.

24. Teich, M.C., and Saleh, B.E.A., "Entangled-photon microscopy." *Československý časopis pro fyziku* (Prague), **47**, 3–8, 1997; Teich, M.C., and Saleh, B.E.A., "Entangled-Photon Microscopy, Spectroscopy, and Display," U.S. Patent No. 5,796,477 (issued August 18, 1998).

25. Nasr, M.B., Abouraddy, A.F., Booth, M., Saleh, B.E.A., Sergienko, A.V., Teich, M.C., Kempe, M., and Wolleschensky, R., "Biphoton focusing for two-photon excitation," *Phys. Rev. A*, **65**, 023816, 2002.

26. Lissandrin, F., Saleh, B.E.A., Sergienko, A.V., and Teich, M.C., "Quantum theory of entangled-photon photoemission," *Phys. Rev. B*, **69**, 165317, 2004.

27. Boto, A.N., Kok, P., Abrams, D.S., Braunstein, S.L., Williams, C.P., and Dowling, J.P., "Quantum interferometric optical lithography: Exploiting

entanglement to beat the diffraction limit," *Phys. Rev. Lett.*, **85**, 2733–2736, 2000.

28. Jakeman, E., and Rarity, J.G., "The use of pair production processes to reduce quantum noise in transmission measurements," *Opt. Commun.*, **59**, 219–223, 1986.

29. Hayat, M.M., Joobeur, A., and Saleh, B.E.A., "Reduction of quantum noise in transmittance estimation using photon-correlated beams," *J. Opt. Soc. Am. A*, **16**, 348–358, 1999.

30. Gatti, A., Brambilla, E., and Lugiato, L.A., "Quantum imaging," *Progress in Optics*, Vol. 51, E. Wolf, Ed., Elsevier, New York (2008).

Chapter 22
Spectral and Correlation Properties of Two-Photon Light

Maria V. Chekhova
M.V. Lomonosov Moscow State University, Russia

22.1 Introduction
22.2 Characterization of Two-Photon Light
    22.2.1 Spectrum and first-order correlation function
    22.2.2 Second-order correlation function
    22.2.3 Anticorrelation ("dip") effect
    22.2.4 Propagation of biphotons through a dispersive medium
22.3 Pulsed Two-Photon Light
    22.3.1 Two-photon spectral amplitude
    22.3.2 Anticorrelation effect
    22.3.3 Frequency-correlated, anticorrelated, and noncorrelated two-photon light
    22.3.4 Measurement of intensity correlation functions
22.4 Interference Effects
    22.4.1 Interference of biphotons generated from time-separated pump pulses
    22.4.2 Different entangled states inside the SPDC bandwidth
22.5 Conclusion
References

22.1 Introduction

Although light quanta were known since the works by Planck and Einstein, for several decades after these works, there was no need to treat light quantum mechanically, and all experiments could be well explained using the quantum description of matter and classical description of light (the semiclassical approach). The situation changed with the development of sources of so-called nonclassical light, i.e., light whose properties cannot be understood in the framework of the semiclassical approach. Two simple examples can be given here. The first one is the suppression of shot noise in optical detectors. Shot noise, i.e., fluctuations of photocurrent scaling as the square root of the photocurrent mean value, can be, in principle, attributed to the discrete nature of the charge generated by the detector. At the same time, the effect of shot-noise suppression, which is observed with certain light sources (squeezed light) is clearly related to the statistics of light, and quantum mechanics is completely necessary for its description. Another example, which brings us closer to the subject of the present paper, is two-photon light. This type of light contains only pairs of photons (often called biphotons), and hence, a pair of properly aligned photon-counting detectors registering two-photon light, in an ideal case, will produce only simultaneous photocounts. This behavior, again, can only be explained in terms of a quantum-mechanical description for the light.

Two-photon light is one of the most important types of nonclassical light. Its role in the development of quantum optics can hardly be overestimated. First of all, it enabled experimental tests of Bell's inequalities, which proved the validity of the "Copenhagen" approach versus hidden-variable theories.[1] It is remarkable that the foundations of quantum mechanics, initially a subject of purely theoretical or even philosophical discussions, due to two-photon light became directly testable by experiment. Later, two-photon light was used in quantum photometry for the absolute measurement of the quantum efficiency of photodetectors.[2] Finally, recent development of quantum information showed that two-photon light can be considered as a fundamental resource for quantum communications and quantum computation.[1]

At present, three main sources of two-photon light are known: cascaded transitions in atoms,[3] spontaneous parametric down-conversion (SPDC),[4] and, since recently, spontaneous four-wave mixing.[5] Although the properties of two-photon light generated via all three effects are very similar, below we will only consider SPDC, so far the simplest and most common source of photon pairs.

Speaking of photon pairs, one can point out various types of pair-wise correlations. Correlation in time means that although the birth moments of photons are random, they are almost simultaneous for two photons belonging to a single pair. Similarly, from the spectral viewpoint, each photon of a pair can have any frequency within a rather broad range, but at the same time there is a strict relation between its frequency and the frequency of its match. (Traditionally, the photon with a larger frequency is called the signal photon while the other one, the idler photon.) This property is called entanglement and can be mathematically formulated as the impossibility to represent the wave function of a pair as a product of the wave functions of separate photons[*]. In experiment, two-photon light can be recognized and characterized by observing coincidences of photocounts for two detectors that register photon pairs.

In a similar way, one can speak about pairwise correlations in space (near-field correlations) and wave vector direction (far-field correlations). These spatial and angular properties of two-photon light are described and observed similarly to its temporal and frequency properties; below we will only consider the

---

[*] This is the definition of entangled pure states. Entangled mixed states are defined in a more complicated way, see Ref. 1.

frequency representation of the two-photon state (and its Fourier conjugate, time representation) and the corresponding effects.

In addition to entanglement in time, frequency, space, and wave-vector direction, two-photon light can manifest polarization entanglement, which is based on its polarization properties. From this viewpoint, one distinguishes between type I and type II two-photon light: in the first case, both photons of a pair have the same polarization while in the second case, they have orthogonal polarizations.[6]

22.2 Characterization of Two-Photon Light

Let us consider two-photon light with the wave vector direction fixed for both photons of a pair (in experiment, this can be done by placing small pinholes at a large distance from the nonlinear crystal generating biphotons). Then the quantum state of two-photon light can be written as

$$|\Psi\rangle = |vac\rangle + C \int d\omega d\omega' F(\omega, \omega') a_1^+(\omega) a_2^+(\omega') |vac\rangle, \quad (22.1)$$

where C is a constant depending on the properties of the pump (mainly the amplitude) and the parameters of the nonlinear crystal (mainly quadratic susceptibility), $a_{1,2}^+$ are photon creation operators in the two polarization or angular modes where the photons are born; the modes coincide in the case of type I collinear SPDC and are different in the case of type II or noncollinear SPDC. The amplitude $F(\omega, \omega')$, usually termed the two-photon spectral amplitude, is the main characteristics of two-photon light since it fully determines its spectral and correlation properties. In the simplest case, where two-photon light is generated from a continuous-wave (CW) pump, the two-photon spectral amplitude contains a δ function,[4]

$$F(\omega, \omega') \propto \delta(\omega + \omega' - \omega_p), \quad (22.2)$$

ω_p being the pump frequency. The case of a pulsed pump will be considered separately, in Section 22.3. Meanwhile, assuming the pump to be continuous-wave and taking into account Eq. (22.2), it is convenient to represent the state as

$$|\Psi\rangle = |vac\rangle + C \int d\Omega F(\Omega) a_1^+(\omega_0 + \Omega) a_2^+(\omega_0 - \Omega) |vac\rangle, \quad (22.3)$$

with $\omega_0 = \omega_p / 2$.

22.2.1 Spectrum and first-order correlation function

The spectrum of two-photon light, which can be observed by means of any spectral device with a sufficient resolution (typically, the spectral width of SPDC radiation at a fixed angle is between 1 and 100 nm), is given by

$$S(\Omega) = |F(\Omega)|^2. \qquad (22.4)$$

Usually, the spectrum for SPDC has the shape of a "sinc-square" function,

$$S(\Omega) = \left[\frac{\sin(D\Omega L/2)}{D\Omega L/2} \right]^2, \qquad (22.5)$$

where L is the length of the crystal and D is the inverse group velocity difference for the photons of a pair, $D = 1/u_1 - 1/u_2$. This difference is due to the crystal birefringence in the type II case and to the crystal dispersion in the case where signal and idler photons are considerably separated in frequency. For the frequency-degenerate type I SPDC, the argument of the sinc function has a different form, and

$$S_I(\Omega) = \left[\frac{\sin(\Omega^2 L k''/2)}{\Omega^2 L k''/2} \right]^2, \qquad (22.6)$$

with $k'' \equiv d^2k/d\omega^2$ given by the group velocity dispersion (GVD). Usually, the bandwidth of type I frequency-degenerate two-photon light is larger than in the nondegenerate (or type II) case.

Less commonly discussed but still worth mentioning is the first-order correlation function (CF) of two-photon light, defined as $G^{(1)}_{1,2}(\tau) \equiv \langle E^{(-)}_{1,2}(t) E^{(+)}_{1,2}(t+\tau) \rangle$, where $E^{(\pm)}_{1,2}(t)$ are positive- and negative-frequency field operators for the signal and idler modes. This CF can be measured, for instance, by feeding signal or idler radiation into a Michelson interferometer. According to the Wiener-Khinchin theorem, the first-order CF is related to the spectrum through the Fourier transformation, and hence, up to a normalization constant,

$$G^{(1)}_{1,2}(\tau) \propto \int d\Omega e^{i\Omega\tau} |F(\Omega)|^2. \qquad (22.7)$$

For instance, in the case of type II SPDC, the first-order CF has a triangular shape, which was shown experimentally in Ref. 7 (Fig. 22.1).

Spectral and Correlation Properties of Two-Photon Light 441

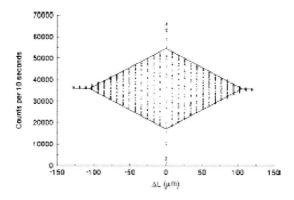

Figure 22.1 First-order correlation function of two-photon light (the envelope of the interference pattern) generated via type II SPDC in a 3 mm BBO crystal (from Ref. 7).

22.2.2 Second-order correlation function

Much more informative for the description of two-photon light is second-order Glauber's CF, a direct characteristic of pairwise correlations,

$$G_{12}^{(2)}(\tau) \equiv \left\langle E_1^{(-)}(t) E_2^{(-)}(t+\tau) E_1^{(+)}(t) E_2^{(+)}(t+\tau) \right\rangle. \tag{22.8}$$

It is convenient to consider normalized CF,

$$g_{12}^{(2)}(\tau) \equiv \frac{G_{12}^{(2)}(\tau)}{G_1^{(1)}(0) G_2^{(1)}(0)}. \tag{22.9}$$

Using Eq. (22.3), one can express the second-order CF in terms of the two-photon spectral amplitude,

$$G_{12}^{(2)}(\tau) \propto \left| \int d\Omega e^{i\Omega\tau} F(\Omega) \right|^2. \tag{22.10}$$

In other words, the second-order CF, similarly to the first-order one, is related to the two-photon spectral amplitude. However, its dependence on $F(\Omega)$ is different. In particular, while the first-order CF is not sensitive to the phase of the two-photon spectral amplitude, the second-order CF depends on this phase; this leads to some interesting effects, which will be considered below.

It is important that the expression (22.3) is derived in the first order of the perturbation theory, where only two-photon emissions are considered.[4] If multiphoton emissions are taken into account, one immediately obtains that the second-order CF has a background, which, after the normalization, is equal to unity. The exact expression for the normalized CF can be written as[8]

$$g_{tot}^{(2)}(\tau) = 1 + g_{12}^{(2)}(\tau), \qquad (22.11)$$

and, since $g_{12}^{(2)}(\tau)$ at its maximum scales as the inverse number of photons per mode, the background of $g_{tot}^{(2)}(\tau)$ is more pronounced for intense two-photon light.

The second-order CF can be measured in a simple experiment [Fig. 22.2(a)]. Signal and idler photons are registered by a pair of photon-counting detectors, photo counts of the detectors are sent to a coincidence circuit, and the coincidence counting rate R_c is measured as a function of the electronic delay τ. This dependence has the shape of a peak, whose width, in an ideal case, would be given by the width of $G_{12}^{(2)}(\tau)$; according to Eq. (22.10), it is on the order of the coherence time $\tau_{coh} = 1/\Delta\Omega$, where $\Delta\Omega$ is the spectral width of two-photon light. However, since the time resolution is usually much worse than that (mainly because of the large response time of photon-counting detectors), the peak width is determined by the time resolution of the measurement setup T_c. The coincidence counting rate as a function of τ is a convolution of the second-order CF and the point-spread function of the measurement setup.[9] For $T_c \gg \tau_{coh}$, the coincidence counting rate at the maximum of the peak is

$$R_c^{max} \approx g_{12}^{(2)}(0) R_1 R_2 \tau_{coh}, \qquad (22.12)$$

where $R_{1,2}$ are the counting rates of the two detectors. At the same time, the coincidence counting rate at the background is

$$R_c^b = R_1 R_2 T_c. \qquad (22.13)$$

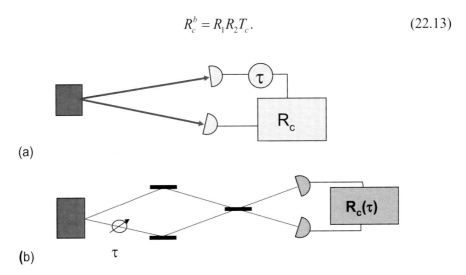

Figure 22.2 Schemes for the measurement of (a) second-order correlation function and (b) anticorrelation dip.

Hence, because of the limited resolution of the measurement setup, one cannot measure the width of the second-order CF and, moreover, the observed ratio of the maximum to the background is reduced $m \equiv T_c / \tau_{coh}$ times [see Eq. (22.11)]. One can say that in the measurement of the second-order CF for CW SPDC, detection is always multimode.[10] Furthermore, we will show that this drawback is eliminated when SPDC is excited with a pulsed pump.

22.2.3 Anticorrelation ("dip") effect

Often, a different scheme is used for the study of two-photon light [Fig. 22.2(b)]. This scheme is based on the effect that is usually called anticorrelation, or the "dip" effect. To observe the dip effect, signal and idler photons are directed to different input ports of a 50% beamsplitter, and one measures the coincidence counting rate of two detectors placed in the output ports.[9] The coincidence counting rate is measured as a function of the delay τ, which in this case is introduced optically before the beamsplitter. When the optical paths for the signal and idler photons before the beamsplitter are exactly balanced, one observes a dip in the coincidence counting rate, which is otherwise high due to two-photon correlations. The dip is caused by the destructive quantum interference of the two alternative ways for a coincidence to occur: when both photons of a pair are reflected from the beamsplitter and when they are both transmitted through it. The width of the dip is given by the coherence time of two-photon light.

At first sight, this setup allows one to study two-photon light with a much better time resolution than in the case of $G_{12}^{(2)}(\tau)$ measurement. However, one can show[11] that the shape of the dip is given not by $G_{12}^{(2)}(\tau)$, but by the first-order normalized CF, $g_{12}^{(1)}(\tau) \equiv G_{12}^{(1)}(\tau) / G_{12}^{(1)}(0)$,

$$R_c(\tau) \propto 1 - g_{12}^{(1)}(2\tau). \qquad (22.14)$$

For instance, in the case of type II SPDC, one observes a triangular dip (Fig. 22.3), repeating the shape of the first-order CF; the dip shape becomes Gaussian if narrowband Gaussian filters are used.[6]

22.2.4 Propagation of biphotons through a dispersive medium

The difference between the first-order and second-order correlation functions of two-photon light is clearly seen from the effects arising when biphotons propagate through a medium with group velocity dispersion (GVD)[12]. If the medium is transparent and linear, the spectrum, and hence the first-order CF, does not change. At the same time, propagation through a GVD medium changes the second-order CF, since the two-photon spectral amplitude $F(\Omega)$ acquires a

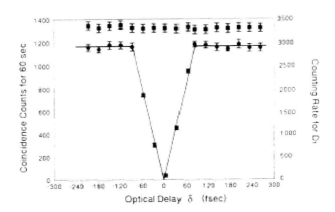

Figure 22.3 The "dip" shape for type II SPDC in a BBO crystal without filtration (from Ref. [6]). Upper points show the single counts, which are independent of the optical delay δ.

frequency-dependent phase, proportional to the GVD of the medium k'', its length Ω, and the square of the frequency Ω. As a result, the second-order CF is broadened and, if the medium is sufficiently long, takes the shape of the spectrum of two-photon light, $S(\Omega)$ as

$$G_{12}^{(2)}(\tau,z) \propto \left|F(\Omega)\right|^2\Big|_{\Omega=(\tau/2k''z)}, \qquad (22.15)$$

(here, GVD is assumed to be the same for the signal and idler photons).

The situation is similar to dispersive spreading of short pulses: while the spectrum of such a pulse does not change, its shape broadens and at a sufficiently large distance it repeats the pulse spectrum.

The broadened second-order CF can be measured in experiment, which was carried out in Ref. 12 for both type I and type II two-photon light transmitted though a 500 m optical fiber. The result for the case of type II is shown in Fig. 22.4.

Since the shape of the anticorrelation dip is directly related to the shape of the first-order CF, it should not change if biphotons propagate through a GVD medium. This effect was observed in Ref. 13 and called "dispersion cancellation." The effect of the second-order CF dispersive spreading can be used for the study of the two-photon spectral amplitude (see Section 22.4.2).

22.3 Pulsed Two-Photon Light

Several applications of two-photon light, such as quantum cryptography and quantum computation, require pairs of photons produced within given short time intervals. This is provided by generating two-photon light from pulsed pump sources. In the case of sufficiently short pulses, whose spectral width is comparable with the spectral width $\Delta\Omega$ or larger, Eq. (22.2) is not valid. The theory of two-photon light generation from a short-pulsed pump was developed in Refs. 14 and 15.

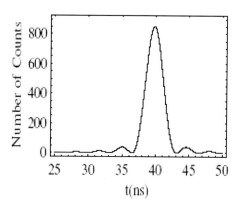

Figure 22.4 Second-order CF of type-II SPDC in a 0.4 mm BBO crystal, broadened due to the propagation through a 500m optical fiber (from Ref. 12).

22.3.1 Two-photon spectral amplitude

The state of two-photon light generated from a pulsed pump has the form (22.1) with[15]

$$F(\omega,\omega') = F_c(\omega,\omega')E_p(\omega+\omega'-\omega_p), \qquad (22.16)$$

where $F_c(\omega,\omega')$ is the two-photon spectral amplitude determined by phase matching, and $E_p(\omega)$ is the spectral amplitude of the pump. It is convenient to denote $\omega \equiv \omega_p/2+\Omega, \omega' \equiv \omega_p/2+\Omega'$. If only collinear directions of wave vectors are selected, the phase matching part of the two-photon spectral amplitude can be written as

$$F_c(\Omega,\Omega') = \frac{\sin[(L/2)(\gamma\Omega+\gamma'\Omega')]}{(L/2)(\gamma\Omega+\gamma'\Omega')}, \gamma \equiv \frac{1}{u_p}-\frac{1}{u}, \gamma' \equiv \frac{1}{u_p}-\frac{1}{u'}. \qquad (22.17)$$

Here, u_p, u, u' are the group velocities of the pump, signal, and idler photons, respectively. The total two-photon spectral amplitude is then the product

$$F(\Omega,\Omega') = E_p(\Omega+\Omega')F_c(\Omega,\Omega'). \qquad (22.18)$$

To calculate the spectrum of, say, the signal radiation, one should integrate the square modulo of the two-photon amplitude with regard to the idler frequency. In the most interesting case of very short pump pulses, the resulting spectrum will have the shape of the pump spectrum but may be stretched or compressed, depending on the group velocities of the pump, signal, and idler radiation,

$$S(\Omega) = |E_p(\xi\Omega)|^2, \quad \xi \equiv \frac{-\gamma}{\gamma'}. \tag{22.19}$$

It is interesting that the second-order correlation function, which can be calculated according to Eq. (22.8), can contain, depending on the value of ξ, several peaks instead of a single one.

22.3.2 Anticorrelation effect

Due to the asymmetry of the two-photon spectral amplitude (22.16) with respect to the interchange of signal and idler frequencies,[1] the anticorrelation dip for type II SPDC has very low visibility. A typical dip shape[16] is shown in Fig. 22.5. The width of the dip is determined by the crystal length, as in the CW case, while its depth depends on the ratio between the coherence time of CW SPDC and the pulse width. The dip is shallow because it originates from the destructive interference of the two probability amplitudes: corresponding to both photons transmitted through the beamsplitter and reflected from it [Fig. 22.2(b)]. In the case of type II SPDC, this corresponds to the events when the ordinarily polarized photon hits detector 1 and the extraordinarily polarized photon, detector 2 (the o1-e2 event), and vice versa (the e1-o2 event). Because, due to the group velocity difference, one of the photons is always ahead of the other one, an e-o delay should be introduced before the beamsplitter. In the CW case, this solves the problem, and the two interfering events become indistinguishable. But if the pump has a short pulse, the o1-e2 and e1-o2 events can still be distinguished using the delay of signal and idler pulses from the pump pulse.[15]

A shallow dip for type II SPDC means the absence of polarization entanglement. At the same time, polarization entanglement is necessary for most quantum optics and quantum information applications. This problem is overcome in experiments by using either narrowband filters[17] or interferometric schemes involving several crystals or several passages of the pump pulse through the same

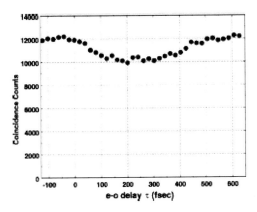

Figure 22.5 Shape of the anticorrelation "dip" obtained with a 2 mm type II BBO crystal pumped by 80 fs pulses (from Ref. 15).

crystal (see, for instance, Ref. 18). Another possibility is a special choice of the nonlinear crystal, whose optical parameters (group velocities of the signal, idler, and pump photons) should satisfy a certain condition, which will be considered in the next section.

22.3.3 Frequency-correlated, anticorrelated, and noncorrelated two-photon light

From Eq. (22.18), we see that unlike in the CW case, where the frequencies of signal and idler photons obey a strict relation, $\omega + \omega' = \omega_p$ (*frequency anticorrelation*), pulsed two-photon light has more "relaxed" conditions on the signal and idler frequencies. Indeed, let us plot the two-photon frequency distribution $|F(\omega,\omega')|^2$.[15] This distribution [Fig. 22.6(c)] is the overlap of the pump spectrum, $|E_p(\omega+\omega'-\omega_p)|^2$ [Fig. 22.6(a)], and the phase-matching function, $|F_c(\omega,\omega')|^2$ [Fig. 22.6(b)]. For most nonlinear crystals, typical values of group velocities u_p, u, u' are such that the phase-matching function has a negative tilt on the ω,ω' diagram and hence the state is still frequency anticorrelated. However, in certain frequency ranges and in certain crystals, it is possible to satisfy the condition

$$\gamma = -\gamma', \tag{22.20}$$

[see Eq. (22.17)], under which the phase-matching function shape has a unity positive tilt,[19] as shown in Fig. 22.6(d). The resulting joint distribution [Fig. 22.6(e)] has frequency correlation, $\omega - \omega' \approx 0$, which is strong if the crystal is long enough.

In some experiments (such as, for instance, preparation of pure single-photon states from two-photon light), it is necessary to completely eliminate the frequency correlation between photons in pairs.[20] Mathematically, this is achieved when the two-photon amplitude can be factorized as

$$F(\omega,\omega') = f(\omega)g(\omega'). \tag{22.21}$$

Condition (22.21) means the absence of frequency entanglement. Note that two-photon light satisfying Eq. (22.21) can still be entangled in other parameters, such as polarization and, in any case, photon number: a single pump pulse may create photons with only a small probability, but signal and idler photons are always created in pairs.

A frequency-uncorrelated state with the spectral amplitude of the form (22.21) can be generated, for instance, by picking a crystal with parameters satisfying condition (22.20) and with a length short enough so that the width of the phase-matching function is equal to the width of the pump spectrum. The resulting two-photon frequency distribution is shown in Fig. 22.6(f).

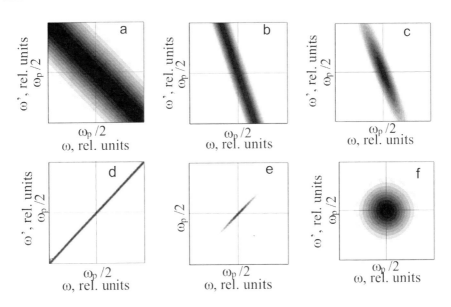

Figure 22.6 Typical pump spectrum (a); typical shape of the phase-matching function (b); resulting frequency distribution (c); the phase-matching function in the case of $\gamma = -\gamma'$ and a very long crystal (d); joint frequency distribution in this case (e); frequency-uncorrelated state (f).

Another important advantage of satisfying condition (22.20) is that it provides symmetry to the two-photon spectral amplitude with respect to the interchange $\omega \leftrightarrow \omega'$ [see Figs. 22.6(e) and (f)] and hence, a high interference visibility in the "anticorrelation dip" experiments. Similarly, a deep dip can be obtained by choosing a crystal with $\gamma = \gamma'$.

22.3.4 Measurement of intensity correlation functions

As mentioned in Section 22.2.2, because the second-order CF of two-photon light is very narrow (on the order of hundreds of femtoseconds), its measurement in experiment has insufficient time resolution. The ratio of the peak value to the background is reduced m times, where m is the number of detected longitudinal modes. For higher-order intensity correlation functions, the situation is even worse: the contrast is reduced m^{n-1} times for the nth-order CF measurement.[10]

If the SPDC is excited by short pump pulses, this reduction of contrast can be avoided.[10] Indeed, consider the second-order CF of SPDC (Fig. 22.7). In the CW case [Fig. 22.7(a)], let a CF of width τ_0, height g, and a unity background be measured with a time resolution T_c (shown by a shaded rectangle). The measured peak height becomes on the order of $g\tau_0$ and the measured background level, T_c; hence, the peak-to-background ratio becomes $g/m \ll g$. In the pulsed case [Fig. 22.7(b)], the CF (dashed line) is modulated by the CF of the pulse sequence,

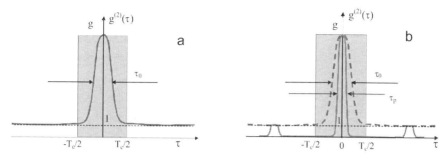

Figure 22.7 Measurement of second-order intensity CF for SPDC excited by (a) CW pump and (b) pulsed pump.

which is also a sequence of pulses with the width τ_p. The resulting CF is shown by a solid line; its measured peak value will be $g\tau_p$ and the measured background will consist of separate peaks of height τ_p. Thus, the measured peak-to-background ratio will remain g, as for the original CF. As a result, the second-order CF of pulsed two-photon light can be measured without contrast reduction.

22.4 Interference Effects

22.4.1 Interference of biphotons generated from time-separated pump pulses

Quantum interference, one of the central phenomena of the quantum world, occurs whenever a certain event may happen via several indistinguishable paths. A typical example is interference of two-photon light, which has been observed in many experiments with biphotons created in several space-time domains or propagating through several arms of an interferometer. Below we consider two-photon interference of biphotons generated by time-separated pump pulses.

There are two main types of interference effects for two-photon light: first-order interference, observed in the intensity distribution, and second-order interference (also called two-photon interference), observed in coincidences of two detectors. Second-order interference of biphotons generated from two nonoverlapping pump pulses was reported in Ref. 21. The main parameters of the experiment were the spectral width $\Delta\omega$ of the interference filters placed before the detectors, and the distance T_p between the pump pulses. Interference was observed by measuring the coincidence counting rate as a function of the phase delay between the pulses. By varying T_p [Figs. 22.8(a)–(c)], one could change the second-order interference visibility: at $T_p\Delta\omega \ll 2\pi$, a pair born from the first pulse is indistinguishable from a pair born from the second one, and the interference visibility is high; in the opposite case, $T_p\Delta\omega \gg 2\pi$, pairs born from different pulses are distinguishable and the interference visibility is zero.

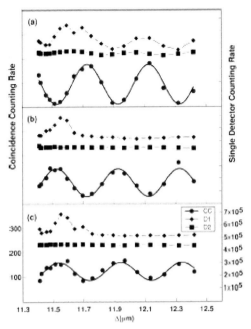

Figure 22.8 Second-order interference of biphotons generated from two pump pulses separated by (a) 236 fs, (b) 420 fs, and (c) 701 fs. Interference filters have a bandwidth of 1 nm. Squares and diamonds are single-photon counts; circles are coincidence counts. (From Ref. 21.)

In the cases shown in Figs. 22.8(a)–(c), the ratio $T_p \Delta\omega / 2\pi$ takes the values 0.118, 0.21, and 0.351, respectively. The corresponding visibility values are 87%, 65%, and 31%.

First-order interference in a similar scheme was observed in Ref. 22. It was shown that there are two necessary conditions for the first-order interference of the signal radiation:

$$T_p \Delta\omega \ll 2\pi, \quad (22.22)$$

which is the same as for second-order interference, and

$$T_p \ll L\gamma'. \quad (22.23)$$

Note that (a usual situation for first-order interference of biphotons) for the interference to be observed in the signal radiation, a condition is imposed on the idler radiation. The physical meaning of condition (22.23) is as follows. For the signal photons born from two different pump pulses to be indistinguishable, overlapping of the signal wave packets is not sufficient [Eq. (22.22)]: one can still distinguish the signal photons by registering the delay of their idler matches with respect to the pump pulses. However, condition (22.23) means that the delay between the idler photon and the pump pulse has an uncertainty (due to the finite length of the crystal) much larger than the distance between the pulses, which "erases" the distinguishing information.

Similarly, a necessary condition for the first-order interference of idler radiation is

$$T_p \ll L\gamma. \tag{22.24}$$

These first-order and second-order interference effects are essentially related to the nonstationarity of the pump. However, even for CW-pumped SPDC, one can observe interference effects in the two-photon spectral amplitude. An example of such an effect is presented in the next section.

22.4.2 Different entangled states inside the SPDC bandwidth

An interesting example of two-photon interference is the "fine structure" of the type II SPDC two-photon amplitude. One can show[23] that with an account for the phase of the two-photon spectral amplitude, the state (22.3) of two-photon light generated from a CW pump has the form

$$|\Psi\rangle = |vac\rangle + C \int_{\Omega \geq 0} d\Omega |F(\Omega)| \{ e^{i\Omega DL/2} a_e^+(\omega_0 + \Omega) a_o^+(\omega_0 - \Omega) \\ + e^{-i\Omega DL/2} a_o^+(\omega_0 + \Omega) a_e^+(\omega_0 - \Omega) \} |vac\rangle, \tag{22.25}$$

with $|F(\Omega)|^2$ given by Eq. (22.5). The indices of the creation operators denote the polarization directions of the two photons; the frequency-dependent phase originates from the delay accumulated between the orthogonally polarized photons of a pair.

It follows that the line shape of type II SPDC contains a set of various maximally entangled states. For instance, in the middle of the SPDC spectral band, the frequency mismatch $\Omega = 0$, and the two-photon part of the state is $|\Psi\rangle = a_o^+ a_e^+ |vac\rangle$, with $a_{o,e}^+ \equiv a_{o,e}^+(\omega_0)$. After being split on a beamsplitter, it becomes $|\Psi^{(+)}\rangle = (1/\sqrt{2})(a_{1o}^+ a_{2e}^+ + a_{1e}^+ a_{2o}^+)|vac\rangle$, one of the triplet Bell states,[1] with the two spatial modes labeled with the indices 1,2. At the same time, at the frequency mismatch $\Omega = 1/DL$, i.e., at approximately half-maximum of the line, the two-photon state generated is the singlet $\Psi^{(-)}$ state, $|\Psi^{(-)}\rangle = (1/\sqrt{2})(a_{1o}^+ a_{2e}^+ - a_{1e}^+ a_{2o}^+)|vac\rangle$. The states generated at intermediate frequencies are maximally entangled states of the form $|\Psi^{(-)}\rangle = (1/\sqrt{2})(e^{i\varphi} a_{1o}^+ a_{2e}^+ + e^{-i\varphi} a_{1e}^+ a_{2o}^+)|vac\rangle$. Note that for the two-photon amplitude to have a frequency-dependent phase, the o-e delay should not be compensated after the crystal. If the e-o delay is compensated, the resulting state

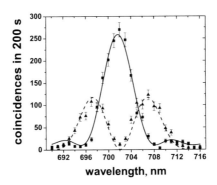

Figure 22.9 Spectral dependence of the coincidence counting rate between the two detectors registering (45 deg, 45 deg) polarizations (squares) and (45 deg, –45 deg) polarizations (triangles). The input state is collinear type II two-photon state. Points: experiment. Lines: theory.

becomes $|\Psi^{(+)}\rangle$ for all frequency mismatches. In the opposite case, when an additional e-o delay introduced after the crystal adds to the internal one instead of compensating for it, the period of the "fine structure" reduces.

The "fine structure" was observed[23] in a setup with the two-photon light beam generated by a 0.5 mm type II BBO crystal, split by a nonpolarizing beamsplitter, a Glan prism and a photon-counting detector inserted into each output port, and the coincidence counting rate of the two detectors measured as a function of the prism orientations as well as the wavelength selected by a spectral device in one of the arms. For instance, Fig. 22.9 shows the spectral distribution of coincidences for the two sets of the Glan prisms orientations: (45 deg, 45 deg) (solid line, squares) and (45 deg, –45 deg) (dashed line, triangles). In the middle of the spectrum (wavelength 702 nm), there is a peak (constructive interference) for the (45 deg, 45 deg) case and a dip (destructive interference) for the (45 deg, –45 deg) case. This indicates that the state is $|\Psi^{(+)}\rangle$. At the same time, at wavelengths 695.5 nm and 708.5 nm, the situation is reversed, which is a signature of the $|\Psi^{(-)}\rangle$ state.

The results shown in Fig. 22.9 were obtained using a monochromator in the signal arm. Due to the effect of two-photon amplitude dispersive spreading, similar dependencies can be measured by transmitting two-photon light through a sufficiently long optical fiber and then time selecting the coincidence distribution.[24]

22.5 Conclusion

In conclusion, we have considered correlation properties of two-photon light in the time and frequency domains. The main characteristic of two-photon light is the two-photon spectral amplitude. It was shown how this characteristic is related to most commonly measured properties of two-photon light, such as its spectrum, first- and second-order correlation functions, and the shape of the anticorrelation dip. It was also shown that spreading and Fourier transformation of the second-order correlation function in a GVD medium can be used as additional means for studying the spectral properties of two-photon light.

Generation of two-photon light from femtosecond pulses provides additional possibilities for "tailoring" the spectral properties of two-photon light. It enables one to produce not only frequency-anticorrelated photon pairs, but also frequency-correlated and frequency-uncorrelated pairs, which are useful for certain applications. Another advantage of pulsed two-photon light is that its intensity correlation functions can be measured without contrast reduction, even by "slow" detection systems. Finally, examples of two-photon interference in the spectral domain were presented, both for the pulsed case and for the continuous wave case.

References

1. Bouwmeester, D., Ekert, A., and Zeilinger, A., *The Physics of Quantum Information*, Springer-Verlag, Berlin (2000).
2. Klyshko, D.N., and Penin, A.N., "The prospects of quantum photometry," *Sov. Phys. Usp.*, **30**, 716–723, 1987; Migdall, A., "Correlated-photon metrology without absolute standards," *Phys. Today*, **52**, 41–46. 1999.
3. Kocher, C.A., and E.D. Commins, "Polarization Correlation of Photons Emitted in an Atomic Cascade," *Phys. Rev. Lett.*, **18**, 575–577, 1967 [doi: 10.1103/PhysRevLett.18.575].
4. Klyshko, D.N., *Photons and Nonlinear Optics*, Gordon and Breach Science, New York (1988).
5. Li, X., et al., "Optical-fiber source of polarization-entangled photons in the 1550 nm Telecom Band," *Phys. Rev. Lett.*, **94**, 053601, 2005; Fulconis, J., et al., "Photonic crystal fiber source of correlated photon pairs," *Opt. Expr.*, **13**, 19757219, 2005; Takesue, H., and Inoue, K., "Generation of polarization-entangled photon pairs and violation of Bell's inequality using spontaneous four-wave mixing in a fiber loop," *Phys. Rev. A.* **70**, 031802(R) (2004); Fan, J., Migdall, A., and Wang, L.J., "Efficient generation of correlated photon pairs in a microstructure fiber," *Opt. Lett.*, **30**, 3368, 2005.
6. Rubin, M.H., Klyshko, D.N., Shih, Y.H., and Sergienko, A.V., "Theory of two-photon entanglement in type-II optical parametric down-conversion," *Phys. Rev. A*. **50**, 5122 (1994).
7. Strekalov, D.V., Kim, Y.H., and Shih, Y.H., "Experimental study of a subsystem in an entangled two-photon state," *Phys. Rev. A,* **60**, 2685, 1999 [doi: 10.1103/PhysRevA.60.2685].
8. Klyshko, D.N., "Transverse photon bunching and two-photon processes in the field of parametrically scattered light," *Sov. Phys. JETP*, **56**, 753–758, 1982.
9. Mandel, L., and Wolf, E., *Optical Coherence and Quantum Optics*, Cambridge University Press, Cambridge (1995).

10. Ivanova, O.A., et al., "Multiphoton correlations in parametric down-conversion and their measurement in the pulsed regime," *Quantum Electron.,* **36**, 951–956, 2006.

11. Burlakov, A.V., et al., "Collinear two-photon state with spectral properties of type-I and polarization properties of type-II spontaneous parametric down-conversion: Preparation and testing," *Phys. Rev. A,* **64**, 041803, 2001 [doi: 10.1103/PhysRevA.64.041803].

12. Valencia, A., et al., "Entangled two-photon wave packet in a dispersive medium," *Phys. Rev. Lett.,* **88**, 183601, 2002 [doi: 10.1103/PhysRevLett.88.183601].

13. Steinberg, A.M., Kwiat, P.G., and Chao, R.Y., "Dispersion cancellation in a measurement of the single-photon propagation velocity in glass," *Phys. Rev. Lett.,* **68**, 2421–2424, 1992.

14. Keller, T.E., and Rubin, M.H., "Theory of two-photon entanglement for spontaneous parametric down-conversion driven by a narrow pump pulse," *Phys. Rev. A,* **56**, 1534–1541, 1997.

15. Grice, W.P., and I.A.Walmsley, "Spectral information and distinguishability in type-II down-conversion with a broadband pump," *Phys. Rev. A* **56**, 1627–1634 (1997).

16. Kim, Y.H., et al., "Anticorrelation effect in femtosecond-pulse pumped type-II spontaneous parametric down-conversion," *Phys. Rev. A,* **64**, 011801(R), 2001 [doi: 10.1103/PhysRevA.64.011801].

17. Di Giuseppe, G., et al., "Quantum interference and indistinguishability with femtosecond pulses,"De Martini, Sergienko *Phys. Rev. A,* **56**, R21–R24, 1997.

18. Kim, Y.H., et al., "Interferometric Bell-state preparation using femtosecond-pulse-pumped spontaneous parametric down-conversion," *Phys. Rev. A,* **63**, 062301, 2001 [doi: 10.1103/PhysRevA.63.062301].

19. Giovannetti, V., et al., "Extended phase-matching conditions for improved entanglement generation," *Phys. Rev. A,* **66**, 043813, 2002 [doi: 10.1103/PhysRevA.66.043813].

20. Grice, W.P., U'Ren, A.B., and Walmsley, I.A., "Eliminating frequency and space-time correlations in multiphoton states," *Phys. Rev. A* **64**, 063815, 2001 [doi: 10.1103/PhysRevA.63.063815].

21. Kim, Y.H., et al, "Temporal indistinguishability and quantum interference," *Phys. Rev. A* **62**, 043820, 2000.

22. Kim, Y.H., et al, "First-order interference of nonclassical light emitted spontaneously at different times," *Phys. Rev. A,* **61**, 051803, 2000.

23. Brida, G., et al., "Generation of different Bell states within the spontaneous parametric down-conversion phase-matching bandwidth," *Phys. Rev. A,* **76**, 053807, 2007.

24. Brida, G., et al., "Interference structure of two-photon amplitude revealed by dispersion spreading," *Phys. Rev. A*, **75**, 015801, 2007 [doi: 10.1103/PhysRevA.75.015801].

Maria V. Chekhova is a senior researcher at M.V. Lomonosov Moscow State University. She received her MS, PhD, and Habilitation degrees from M.V. Lomonosov Moscow State University in 1986, 1989, and 2004, respectively. She is the author of more than 70 journal papers. Her current research interests include generation, study, and applications of nonclassical light, in particular, multiphoton light and various polarization-squeezed states of light. She is teaching a course in quantum optics at Moscow State University.

Chapter 23
Entanglement-Based Quantum Communication

Alexios Beveratos
Alcatel de Marcoussis, France

Sébastien Tanzilli
Université de Nice Sophia-Antipolis, France

23.1 Preface
23.2 Entanglement: From the Basics to Recent Developments
    23.2.1 From the fundamental side
    23.2.2 Nonlinear optics for producing pairs of photons
    23.2.3 Different kinds of entanglement
23.3 From Quantum Teleportation to Quantum Relays
    23.3.1 Basics of quantum relays
    23.3.2 The Bell-state measurement
    23.3.3 Quantum teleportation
    23.3.4 Entanglement swapping using highly coherent CW sources
23.4 A Photonic Quantum Information Interface
23.5 Conclusion
References

23.1 Preface

The aim of this chapter is not to focus one more time on quantum cryptography, which is extensively discussed in a collection of books and review papers, but rather to provide the reader with recent advances and challenges that concern experimental, entanglement-based, guided-wave quantum communication.

In our everyday world, almost all the information exchanged, stored, and processed is encoded using elementary entities called bits, conventionally represented by the discrete values 0 or 1. In today's fiber-based telecommunications systems, these classical bits are carried by light pulses, corresponding to macroscopic packets of photons, allowing a classical description of their behavior and propagation. To draw a simple picture, each light pulse consists of at least hundreds, of photons to encode the bit value 1, or of no photons to encode the bit value 0.

In the past 20 years, physicists have realized that individual quantum objects, for instance photons, could also be employed to deal with another kind of information. Here information is no longer encoded on the number of involved photons, but individual photons merely serve as carriers and quantum information is encoded on their quantum properties, such as polarization or time-bins of arrival.[1] Indeed, by selecting two orthogonal states spanning the Hilbert space, $|0\rangle$ and $|1\rangle$ now encode the 0 and 1 values of the quantum bit (qubit), and quantum superposition makes it possible to create states of the form

$$|\psi\rangle = \alpha|0\rangle + e^{i\phi}\beta|1\rangle, \qquad (23.1)$$

provided α and β follow the normalization rule $|\alpha|^2 + |\beta|^2 = 1$. Quantum superposition lies at the heart of quantum physics and is one of the resources for quantum communication protocols. Let us emphasize that $\{|0\rangle, |1\rangle\}$ can represent any observable related to the considered quantum system. Moreover, from an abstract point of view, the nature of the carrying particle is irrelevant because only amplitudes and relative phases are exploited in Eq. (23.1) to encode the qubits. It is clear however that photons are the natural "flying qubit carriers" for quantum communication, and the existence of telecommunication optical fibers makes the wavelengths of 1310 and 1550 nm particularly suitable for distribution over long distances.

Among quantum communication protocols, quantum key distribution (QKD), also known as quantum cryptography, is probably seen as the most striking application because it offers a provably secure way to establish a confidential key between distant partners, commonly called Alice and Bob, by detection eavesdropping. Such quantum communication links are already commercially available. In spite of progress in photonics and telecommunications technologies, losses in optical fibers, and especially dark counts in the detectors, limit the maximum achievable distance for QKD to about ≈ 100 km.

Any communication link will eventually suffer from losses leading to signal attenuation while traveling from Alice to Bob. When the attenuation is too strong, Bob cannot extract any useful information out of the received signal due to low signal-to-noise ratio. Classical telecommunication amplifiers have been well developed for a long time in order to compensate for losses. Quantum communication does suffer from exactly the same issues because quantum information cannot travel over long distances without being lost. But in this case, because qubits of information are generally encoded on single photons, the loss of the latter is irreversible, dropping the communication bit rate below any useful level or raising the error rate above any tolerable value. In spite of progress in photonics and telecommunication technologies, losses in optical fibers, and especially dark counts in the detectors (getting a click without any incoming photon), both limit the maximum achievable distance for QKD to about ≈ 100 km. Moreover, classical amplifiers cannot serve in quantum communication for reamplification because an unknown qubit cannot be perfectly cloned without introducing errors.[2] The implementation of quantum relays allows overcoming optical losses by relaying quantum information and has the same important role as classical amplifiers. The basic resource of quantum relays is entanglement, and we will see that such an implementation, based on quantum teleportation,[3] allows increasing the quantum communication signal-to-noise ratio and thus extending the maximum achievable distance.[4]

23.2 Entanglement: From the Basics to Recent Developments

23.2.1 From the fundamental side

A profound way in which quantum information differs from classical information lies in the properties, implications, and uses of quantum entanglement when two or more correlated particles are involved in the considered quantum system. More precisely, most of the information contained in such a system is stored in the form of quantum correlations between its subsystems that have no classical analog. Entanglement can be seen as a generalization of the superposition principle [Eq. (23.1)] to multi particle systems. In this case however, the entangled state describing the whole system cannot be factorized, (i.e., written as a tensor product of the properties associated with each subsystem). Entangled pairs of qubits, implying two particles, can be described by a state of the form

$$|\Phi\rangle_{A,B} = \alpha|0\rangle_A|0\rangle_B + e^{i\phi}\beta|1\rangle_A|1\rangle_B, \quad |\alpha|^2 + |\beta|^2 = 1, \tag{23.2}$$

where indices A and B label the two qubits. Then, measuring, for instance, qubit A, provides a random result, but, if we find it to be $|0\rangle$ (with probability $|\alpha|^2$), then we learn from state (23.2) that qubit B will be found in the same state for a similar measurement. A convenient basis in the Hilbert space $\mathcal{H}_A \otimes \mathcal{H}_B$ is the so-called Bell basis, made of the four mutually orthogonal maximally entangled states

$$|\Phi^+\rangle_{A,B} = \tfrac{1}{\sqrt{2}}\left(|0\rangle_A|0\rangle_B + |1\rangle_A|1\rangle_B\right), \tag{23.3}$$

$$|\Phi^-\rangle_{A,B} = \tfrac{1}{\sqrt{2}}\left(|0\rangle_A|0\rangle_B - |1\rangle_A|1\rangle_B\right), \tag{23.4}$$

$$|\Psi^+\rangle_{A,B} = \tfrac{1}{\sqrt{2}}\left(|0\rangle_A|1\rangle_B + |1\rangle_A|0\rangle_B\right), \tag{23.5}$$

$$|\Psi^-\rangle_{A,B} = \tfrac{1}{\sqrt{2}}\left(|0\rangle_A|1\rangle_B - |1\rangle_A|0\rangle_B\right), \tag{23.6}$$

where amplitudes α and β are equal to $1/\sqrt{2}$. Note that the phase ϕ is set to 0 for states $|\Psi^+\rangle$ and $|\Phi^+\rangle$, and to π for states $|\Psi^-\rangle$ and $|\Phi^-\rangle$, respectively. As usual with quantum formalism when an orthogonal basis is given, any two-qubit state can be described using the Bell basis as a linear combination of these four states.

The important point to be noted lies in the fact that an entangled state intrinsically contains strong correlations that cannot be reproduced by any classical probability distribution. These correlations are exclusively quantum because no information about the entangled state can be acquired by any local measurement on any one of the two qubits. In other words, if the qubits are spatially separated, qubit A and B sent to Alice and Bob, respectively, performing local measurements on both without any additional classical communication will provide no information about the complete state; thus no discrimination between the four Bell states would be possible. Information is therefore not associated with the single qubits themselves but rather shared by them. Entangled qubits therefore have to be considered as a whole (i.e., a unique quantum object), from the moment they were created to the moment they are measured. Moreover, trying to describe the single qubits is meaningless because their individual states are not defined.[i]

Since the discovery of quantum physics, this point has been extensively discussed over the years and different approaches have been suggested to oppose these counter intuitive features and description. For some physicists, single qubits, or particles out of an entangled pair, should have some elements of reality (i.e., their individual properties should not depend on any measurement process). Moreover, these properties should be defined at the source, when the qubit pair is prepared. Thus, hidden variables, if we could have access to them, could help to predict all the joint measurement results. Regarding this, the major contributions are certainly that of Einstein, Podolsky, and Rosen (EPR) in 1935,[5] on the one hand, and that of Bell in 1964,[6] on the other hand. In the first paper,[5] the completeness of quantum physics is discussed as a description of physical reality. The validity of quantum physics was not questionned but the interpretation was for the authors not satisfying. From that paper,[5] lots of physical and philosophical discussions arose and people reached a point where it seemed that one could make either choice. An experimental test or proof was missing, but even the way to perform it was not known. Then Bell's paper[6] came out, in which he demonstrated the existence of an inequality able to discriminate between quantum theory and classical theories based on local hidden variables. This inequality deals with an experimental parameter S, corresponding to a combination of different measurement settings applied on the two individual qubits. These settings are local and theoretically independent from each other. Then, if classical local hidden variable are true, S should follow the inequality rule $-2 < S < 2$. Otherwise, no local description of the individual qubits is allowed, and something more is needed (i.e., quantum non-locality). Bell's idea, extended by Clauser et al.[7] provided a way to test this. Fifteen more years were necessary to overcome technical limitations that prevented one from making the test experimentally. In the beginning of the 1980's, Aspect et al. performed in Orsay their famous experiments on the violation of the Bell inequality[6,7] using

---

[i] Here "not defined" means that the qubits are in a mixture of states, which is the opposite of a coherent superposition.

a quite complicated polarization entangled photon-pair source based on a double atomic cascade. transition[8,9] Although these experimental results were not really the first ones, they were unambiguous in the sense that a clear violation had been obtained.[8,9] These demonstrations brought a kind of final point to decades of discussions. Finally, a pair of entangled particles had to be considered as a whole and non separable quantum system. From that point on, more and more accurate tests have been carried out, all leading to conclusions in favor of quantum physics. Photons from a pair have to be considered as a whole quantum object, even if physically separated by kilometers of fiber[10–12] or free space.[13,14] This amazing feature actually provided a way to understand the quantum world more deeply. A short but intersting review on that story is given in Ref. 15, and references therein.

23.2.2 Nonlinear optics for producing pairs of photons

As it occurs sometimes in science, this fundamental achievement and the related works[15] are admitted as a starting point of an entire new field of research, that of quantum communication, and broadly speaking, of quantum information science. Nowadays, sources creating entangled photon pairs are an essential tool for a variety of quantum communication experiments such as quantum teleportation,[3,16–19] entanglement swapping,[4,20] quantum relays,[4,21] as well as for fiber-based or free space quantum key distribution.[13,14,22–27] The Orsay group's source has been advantageously replaced by much simpler and compact sources based on spontaneous parametric downconversion (SPDC), a process that occurs under some conditions in nonlinear crystals (i.e., that show a second order nonlinear susceptibility[28] (χ^2)). In such a crystal, photons from a pump field can be, with a small probability, converted into so-called signal and idler photons provided energies and k-vectors are conserved,

$$\omega_p = \omega_s + \omega_i, \tag{23.7}$$
$$\vec{k_p} = \vec{k_s} + \vec{k_i}, \tag{23.8}$$

where the indices label the frequency and k-vector of the pump, signal and idler fields, respectively. The latter equation is also known as the phase-matching condition and can be achieved, despite chromatic dispersion in the crystal, by employing different field polarizations (birefringent phase-matching) or by a periodic inversion of the sign of the crystal nonlinear coefficient along the propagation [quasi phase-matching (QPM)]. Dispersion is canceled in the latter case by an additional grating-type k-vector, $\left|\vec{K}_{QPM}\right| = 2\pi/\Lambda$, where Λ is the step of the periodic structure. Equation (23.8) therefore, becomes $\vec{k_p} = \vec{k_s} + \vec{k_i} + \vec{K}_{QPM}$. Thus by an appropriate choice of Λ, QPM can be engineered to phase match practically any desired interaction within the crystal transparency window. In addition, QPM also offers the possibility to integrate a waveguiding structure on the considered periodic substrate, structures that are known for enhancing nonlinear optical interactions.

The SPDC efficiency (i.e., the probability to downconvert a pump photon) is generally quite low, ranging from 10^{-6} for waveguiding structures integrated on QPM crystals[29,30] to 10^{-10} for the best bulk. crystals[10,31,32] Note that downconversion efficiency is found to be much higher in the case of waveguide-based sources because the waveguiding structure offers a strong confinement of the interacting fields over much smaller sections (with a cross surface of tens of square microns) and over much longer lengths (up to a few centimeters) compared to bulk configurations.[ii] In addition, QPM technique permits using the highest nonlinear element of the considered material. Regarding such sources, widely employed devices are waveguides integrated on periodically poled lithium niobate also known as PPLN waveguides.[30,33] In the following, the described experiments are based on LBO bulk crystals (see Section 23.3.3) and PPLN waveguides (see Sections 23.3.4 to 23.4).

Eventually, the three past years have also seen an intensive development of photon-pair sources based on four-wave mixing in microstructured fibers taking advantage of third-order nonlinearities.[34–36] Sources relying on quantum dots are also being actively developed and seem to be very promising.[37,38]

23.2.3 Different kinds of entanglement

Now that the physical process to generate pairs of photons has been presented, we must focus on the different quantum information encoding schemes and what they are made for. This corresponds to defining to which type of quantum observable "0" and "1" of Eq. (23.2) refer to. Depending on the type of entanglement, polarization, energy-time or time-bin, different experimental conformations are adopted.

Polarization is probably the most widespread observable for quantum entanglement. Two main strategies have been carried out during the past 20 years. On the one hand, the two photons can have the same polarization which leads to states corresponding to those of Eqs. (23.3) and Eq. (23.4). This can be achieved using two cascaded type-I nonlinear bulk BBO crystals as described in Ref. 32 and depicted in Fig. 23.1(a). In addition to an increased overall efficiency, the particular feature here lies in the fact that the two crystals are oriented at 90 deg with respect to each other (i.e., the optic axis of the first (second) crystal lies in the vertical (horizontal) plane). Then, using a pump beam polarized at 45 deg and showing a coherence length greater than the combined size of the two crystals, it is impossible to know, when a photon-pair is generated, whether it comes from crystal 1 or 2. Because of type-I phase matching, the former leads to two horizontally polar-

---

[ii]Note that SPDC efficiency increases linearly with the interaction length, but also depends on the field intensities per cross section of propagation. The confinement is then much higher in waveguides (a cross surface of tens of square microns over a few centimeters) than in bulk crystals for which the pump intensity is the highest only for a few mm around the focal point of the front lens.

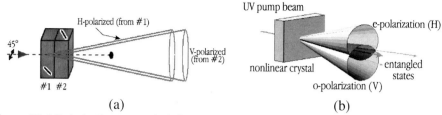

Figure 23.1 Polarization entangled photons obtained by two different means. Two cascaded type-I BBO crystals are used to produce identically polarized photons[32] (a), and a type-II BBO crystal is employed, together with a selection of the particular spatial modes, to get orthogonally polarized photons[31] (b). The cascade BBO source has initially been developed in Los Alamos and then extensively used in Urbana-Champaign, whereas the type-II source has been developed and used by the Vienna group. Because of their practical interest, both sources have been replicated many times by a variety of groups around the world.

ized photons [vertical (V) component of the pump] and the latter to two vertically polarized photons [horizontal (H) component of the pump]. This lack of information is responsible for a coherent superposition of two quantum states of the form $|\Psi\rangle = 1/\sqrt{2}\,(|0\rangle_s|0\rangle_i \pm |1\rangle_s|1\rangle_i)$, after normalization, and where 0 and 1 replace H and V polarization, respectively.

On the other hand, photons can have orthogonal polarizations, leading to states of the form of Eqs. (23.5) and (23.6). In such a situation, one can take advantage of type-II bulk crystals, again, such as BBO, and phase matching implies that ordinary and extraordinary polarized SPDC photons are emitted along two light cones with no common axis. Here, the idea lies in the fact that, at the intersection of these two axes, one cannot differentiate between an ordinary and an extraordinary polarized photon. It is however sure that the emitted two photons will have orthogonal polarizations. Hence, by only selecting the two spatial modes right at these intersections, the emitted two photons carry an entangled polarization state of the form $|\Psi\rangle = 1/\sqrt{2}\,(|0\rangle_s|1\rangle_i + |1\rangle_s|0\rangle_i)$. Note that such a source has very recently been used for a demonstration of a entanglement-based free-space quantum communication over 144 km.[14]

Other types of entanglement, energy-time and time-bin, have been largely studied, especially when the purpose is transmit photons over long distances in optical telecommunication fibers.

The simplest sort of entanglement that one can work with in this case appears to be energy-time entanglement because it only requires a SPDC generator and a CW pump laser. Indeed, Eq. (23.7) indicates that signal and idler photons are always energy correlated within their bandwidths, which essentially depends on the phase matching. This means that, within these bandwidths, the sum of two single frequencies $\omega_s + \omega_i$ will always give the same sum (i.e., the pump frequency ω_p that can be considered monochromatic). In addition, because SPDC is a spontaneous effect, the emission time of a pair remains unknown, but because paired signal and idler photons are always emitted at the same time, they are time correlated. This energy-time entanglement can be revealed and quantified using the so-called

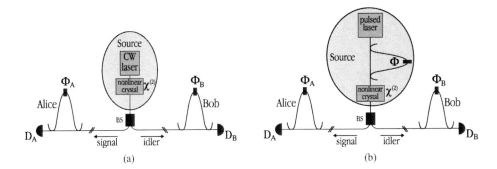

Figure 23.2 Schematics of fiber optic Franson conformations to measure two-photon interferences with energy-time (a) and time-bin (b) entangled photons. The sources are made of a nonlinear crystal operated in the downconversion regime, a fiber optic beam splitter (BS) to separate the twins, and of a CW or a pulsed laser, respectively. Note that in the time-bin case, a third interferometer, also called preparation interferometer is used. In principle, signal (Alice) and idler (Bob) photons can be spatially separated by a large number of kilometers without destroying entanglement. Both types of entanglement sources have been extensively developed and used by the Geneva group especially for long-distance quantum communication ranging from 10 to 50 km.[10–12,40]

Franson interferometer,[39] as depicted in Fig. 23.2(a). In such a setup, the photons, after generation and separation at a beam splitter are sent to Alice and Bob, respectively, where they encounter two analyzers corresponding to equally unbalanced Mach-Zehnder type interferometers. These devices have a path length difference D_L several orders of magnitude larger than the coherence length of the incoming photons, preventing single-photon interference. The single count rates S_1 and S_2, respectively, are therefore independent of the phases ϕ_A and ϕ_B applied in the long arms of the interferometers. However, provided the coherence length of the pump laser is larger than D_L, an "optical-path" entangled state of the form (after normalization) $|\Phi\rangle_{s,i} = 1/\sqrt{2} \left(|0\rangle_s|0\rangle_i + e^{i(\phi_A+\phi_B)}|1\rangle_s|1\rangle_i\right)$ can be post selected using a high-resolution two-fold coincidence technique,[10,11,30] such as an AND-gate. Indices "s" and "i" label signal and idler photons, and 0 and 1 represent the short and long arms of the interferometers, respectively. Such a state corresponds to the cases where both photons pass either through the short or long arms of their interferometers. From the quantum side, as the two photons are emitted at the same time, these two cases cannot be distinguished, and, provided the non interfering possibilities (the photons pass through different arms) are discarded by the coincidence measurement, quantum interference with an ideal visibility of 100% is expected. We measure this interference in the coincidence rate, which evolves as a function of the sum of the phases $\phi_A + \phi_B$. Here, the quantum coherence or information responsible for the two-photon interference is hidden in the coherence length of the pump photon that is transferred to the pair during the SPDC process. In other words, we have to consider these photon pairs as single quantum objects.

Extension to time-bin entanglement[41] is relatively easy to understand now. As already mentioned for the energy-time case, the coherence length of the pump photons plays an important role. Obviously, if one has to pump the non-linear crystal with a pulsed laser for experimental reasons, this coherence will be partially lost. The way to give quantum coherence back to the system is to place a third interferometer in the path of the pump photons, as shown in Fig. 23.2(b), and works as follows. A short light pulse emitted at a time t_0 enters the preparation interferometer having a path-length difference greater than the pulse duration, again avoiding single-photon interference at this stage. The pump pulse is then split into two pulses of smaller amplitudes following each other with a fixed phase relation. At one output port of the interferometer, a pump photon can be described by a coherent superposition of states, i.e., the pump pulse taking the short ($|0\rangle_p$) or the long arm ($|1\rangle_p$) of the interferometer, respectively. It comes, after normalization, $|\Psi\rangle_p = 1/\sqrt{2}\left(|0\rangle_p + e^{i\phi}|1\rangle_p\right)$, which is a qubit state. Then, after the SPDC stage, the two-photon state takes the form of Eq. (23.3), $|\Psi\rangle_p = 1/\sqrt{2}\left(|0\rangle_p + e^{i\phi}|1\rangle_p\right) \stackrel{SPDC}{\longmapsto} |\Psi\rangle_{s,i} = 1/\sqrt{2}\left(|0\rangle_s|0\rangle_i + e^{i\phi}|1\rangle_s|1\rangle_i\right)$, where ϕ is the phase acquired by the pump photon in its interferometer, and p, s, and i label the pump, signal, and idler photons, respectively. The previously described analyzers are used here for two-photon entanglement analysis. Thanks to a three-fold coincidence technique (pump laser trigger + the two photons) interference fringes of 100% visibility are expected at the output of the setup.[11,12,30,40] Again here, two indistinguishable paths can be identified, the pump photon having travelled along the short and both photons of the pair along the long arms of their respective interferometer, and conversely.

More details on experimental setups implemented for long-distance quantum communication based on both energy-time and time-bin entanglement will be given in the following sections.

Even if the above sources rely on a probabilistic process and are not optimal[42] for the generation of entanglement, they are the key elements of quantum communication experiments. On the other hand, it turns out that entanglement is precisely the quantum resource that enables such protocols by offering what is now called a quantum channel. In particular, one of the main goals is the realization of a quantum network able to link people with the distribution of quantum information over longer distances than would be possible without the use of entanglement.[21,43] This is what we are going to discuss in the next section.

23.3 From Quantum Teleportation to Quantum Relays

As introduced in the previous sections, entanglement lies at the heart of quantum physic, and has no classical counterpart. It is also quite easy to implement experimentally using "off the shelf equipment". This novel resource can be used to implement quantum relays to extend the maximum achievable distance in quantum communication.

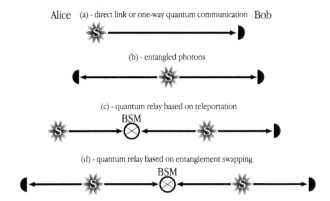

Figure 23.3 Different scenarios for quantum communication links. (a) A direct link between Alice and Bob. (b) A source of entangled photons between Alice and Bob. (c) Quantum relay based on quantum teleportation between Alice and Bob, a third party (Charlie) makes a Bell-Sstate measurement (BSM). (d) Quantum relay based on entanglement swapping, where two sources of entangled photons and a measurement are used.

23.3.1 Basics of quantum relays

Figure 23.3 depicts different links that Alice and Bob can implement for quantum communication purposes. The first and simplest is called the "one-way" link [Fig. 23.3(a)] where Alice encodes the qubits on single photons and sends them to Bob who performs measurements and detection of these qubits. This scheme is probably the most largely used, because, only a single photon source and an encoding system are necessary on Alice's side. The single-photon source can be a simple pulsed laser attenuated down to the single photon-level. When working with photons at telecom wavelengths, such a protocol is however limited in terms of distance essentially because of the fiber losses combined with the intrinsic noise level of commercially available single-photon counting modules based on InGaAs avalanche photodiode technology. It can easily be demonstrated that after a 100 kilometers of fiber, the probability to detect a photon at Bob's location is close to the probability of getting a dark count in his detector. Other new detection schemes based on nonlinear wavelength upconversion processes[44–46] or on superconductors[47–49] will be discussed in Section 23.3.4.1.

What can be done to increase the maximum achievable distance between Alice and Bob? The simplest quantum relay [Fig. 23.3(b)] is when the single photon-source initially at Alice's is replaced by an entangled photon source placed at mid-distance between Alice and Bob.[22] In this way, one can either increase the overall bit rate at fixed distance, since each photon travels half of the total distance, or increase the communication link distance, at the fixed bit rate.[4] Experimentally, the distribution of quantum information now depends on a coincidence measurement leading to a greater reduction of the overall noise compared to the loss in the bit rate (true coincidence events). Then, following this idea, more sophisticated quantum relays can be implemented using quantum teleportation[3] [Fig. 23.3(c)] or entanglement swapping [Fig. 23.3(d)]. In such situations, the number of joint detection is increased, allowing for a fine selection of the true events that contribute to the bit rate. Another way to understand how these configurations can impact the maximum distance is to realize that the greater the number of places where quantum information is present along the channel (increasing the number of detectors), the shorter the distance this quantum information has to travel.[21] More precisely, from case (a) to case (d) in Fig. 23.3, the number of sources and detectors changed from one (emitting single photon) to two (emitting entangled photons), and from 1 to 4,[iii] respectively.

Before we discuss teleportation and entanglement swapping, let us focus on Bell-state measurements which are needed for both cases.

23.3.2 The Bell-state measurement

The realization of quantum relays is fully conditioned on the ability to implement a Bell-state measurement (BSM). A BSM is the operation that projects two incoming photons on one of the four Bell atates [Eqs. (23.3–23.6)]. More precisely, quantum mechanics treats a BSM as a self-adjoint operator whose eigenvectors are entangled states. Such a measurement can be obtained by using only linear optics, such as a simple beamsplitter, or some more complicated linear-optics schemes.[50] Despite the simplicity of these realizations, they have a success rate of only 50% and only two (or three[50]) Bell-states can be resolved[51], meaning that in half of the cases, the output of the Bell-state measurement does not project on one of the Bell-states and the obtained result has to be discarded.

Let us consider the simple case of a 50/50 beam splitter acting as a BSM. If a photon enters the beam splitter at one port (see Fig. 23.4), its evolution, using the formalism of creation operator, can be written as

$$a^\dagger \to \tfrac{1}{\sqrt{2}}\left(ia^\dagger + b^\dagger\right), \quad b^\dagger \to \tfrac{1}{\sqrt{2}}\left(a^\dagger + ib^\dagger\right), \qquad (23.9)$$

where $a^\dagger$ and $b^\dagger$ are the operators associated with the input and output ports of

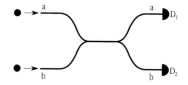

Figure 23.4 Bell-state analyzer apparatus based on a simple fiber optic beam splitter. a and b denote the upper and lower modes of the device, and D_1 and D_2 two detectors.

[iii]Two are not represented and they are part of the BSM. Four detectors are needed because in this case, four photons are involved.

Table 23.1 Two-qubit detection patterns when a simple beam splitter is used. A result value of the type "XY", with X,Y∈0,1, in front of a given D_i means that the two qubits were measured to be in the state X or Y, and that the two associated photons were detected by the same detector D_i. Having results in separated detectors means that the two qubits were measured to be in the state X or Y, but that the two associated photons were detected in two different detectors. Then, one can read for each detection pattern the probability with which each Bell-state contributes to this pattern.

D_1	00	11	01		0	0	1	1
D_2		00	11	01	0	1	0	1
$\|\Phi^+\rangle$	1/4	1/4	1/4	1/4				
$\|\Phi^-\rangle$	1/4	1/4	1/4	1/4				
$\|\Psi^+\rangle$					1/2	1/2		
$\|\Psi^-\rangle$							1/2	1/2

the considered[iv] device, respectively. For instance, $a^\dagger$ means that one photon is in the spatial mode a, regardless of whether it is at the input or the output. Suppose now that two unknown photonic qubits enter each port of the beam splitter, but are completely indistinguishable otherwise and detected by two perfect detectors (D_1 and D_2) at the output ports. The above scheme will lead to the following detection patterns for the two unknown qubits.

By measuring the qubit state of each photon on D_1 and D_2 one projects the two photons on one of the four Bell-states. From Table 23.1, one can deduce that only the $|\Psi^\pm\rangle$ states are unambiguously detected, whereas $|\Phi^\pm\rangle$ can never be detected. From the overall probabilities, it is simple to conclude that the above BSM has an overall success rate of 50%. Note that the detection patterns are independent of the encoding scheme since 0 and 1 can be made of any quantum observable as previously mentionned.

In all the above cases, photons entering the BSM apparatus ought to be completely indistinguishable with respect to all their degrees of freedom, but the qubit value itself. Hence the two photons must have the same spatial, spectral, and temporal modes.

In the case of time-bin qubits encoded on photons at telecom wavelength, obtaining equal spatial and polarization modes for photons is straightforward, especially in single-mode optical fibers. In addition, the same spectral properties can be obtained using commercially available custom fiber Bragg gratings. Temporal indistinguishability, however, is much more complex to achieve. Consider a non-linear crystal operated in the SPDC regime pumped under CW excitation. Photon pairs are emitted at random times with a spectral width of $\Delta\lambda$ corresponding to a coherence time of $\delta\tau$. In a quantum relay, such two photons generated from two independent sources must arrive at the same time on the BSM apparatus (commonly made of a simple beamsplitter) for the protocol to be successful. Hence the two

---

[iv] The phase i on the reflected port is necessary for energy conservation.

Figure 23.5 Arrival time of each photon on detectors 1 and 2. Time has been divided into equal parts, each corresponding to the detector's resolution (i.e., timing jitter) δT. The points represent single photons and the envelop, their coherence time (i.e., temporal extention), given by the Fourier transform of their spectrum $\Delta\lambda$. The left part of the figure represents two photons detected at the same time, because the detectors resolution is sub coherence. Conversely, the right part depicts two photons detected at the same time but having a coherence greater than the resolution of the detectors. In the latter case, they can, in principle, be distinguished.

detectors placed on each output port must click exactly at the same time, meaning that the detectors time resolution (δT) has to be shorter than the photons coherence time in order to single out a precise detection time [Fig. 23.5(a)]. On the other hand, when $\delta T > \delta\tau$ [Fig. 23.5(b)], two simultaneous clicks do not imply that the photons arrived exactly at the same time on the beam splitter.

State-of-the-art detectors suitable for the third telecom window ($\sim$1550 nm) offer time resolutions ΔT of 50 ps, implying a maximum bandwith $\Delta\lambda$ for the downconverted photons on the order of 30 pm. Almost all entangled photon sources based on parametric downconversion have an spectrum ranging from 1 to 40 nm, two to three orders of magnitude larger than desired. A way to circumvent the problem consists of transferring the timing information from the detectors to the excitation source. Hence, using a pulsed excitation laser, as is the case for time-bins, with a repetition rate T greater than $\delta\tau$, one can emit the photon pairs at exactly the same time. The pulse duration (δT) takes over the role of the detectors for defining a single temporal mode. Again, as for the case of photon detectors, the coherence time of the photons has to be greater than the pulse duration as shown on Fig. 23.6. By filtering the photons down to approximatively $\Delta\lambda = 5$ nm, their coherence time is approximately $\delta\tau = 700$ fs and, hence, femtosecond excitation pulses of about $\delta T = 150$ fs can be used to achieve a suitable matching of the two photons at the beam splitter.

Figure 23.6 The boxes represent the temporal extension of the excitation pulse. Whenever the coherence time of the downconverted photons is smaller than the excitation pulse, photons are emitted anywhere inside the time window defined by the optical pulse (left). On the other hand, if the coherence time is greater, then the excitation pulse singles out a single temporal mode (right).

As already mentionned, experimental BSMs most often rely on a simple beam splitter, although only two of the four Bell-states can, in principle, be detected. Performing a complete BSM is much more complicated since it requires the use of controlled-NOT gates[v] based on either non linear interactions or auxiliary photons. Such schemes are quite inefficient in terms of probability of success and demand a large amount of resources. However, as has been demonstrated in Ref. 51, it is the only way to measure all four Bell-states.

23.3.3 Quantum teleportation

Quantum teleportation is the art of transferring information from Alice's qubit to Bob's qubit without these two qubits ever having interacted[3] (Fig. 23.3). This may sound trivial, because Alice could measure her qubit and communicate the information to Bob, who could reconstruct the state. But because α and β in Eq. (23.1) can take any complex value (provided $|\alpha|^2 + |\beta|^2 = 1$), a single measurement on a single qubit does not provide enough information to allow deducing $|\alpha|$. In addition to Alice's qubit, fulfilling such a protocol requires other resources, one quantum, an entangled pair of qubits, and one technical, a BSM apparatus discussed in the previous section. The latter is operated by Charlie who only makes the necessary measurement and communicates the related result to Bob. Consider that Bob and Charlie initially share an entangled state of the form $|\Phi^+\rangle_{BC} = 1/\sqrt{2}\,(|0\rangle_B|0\rangle_C + |1\rangle_B|1\rangle_C)$, i.e., they have each one qubit of the pair. Moreover, suppose that Alice encodes a single qubit of the form $\alpha|0\rangle_A + \beta|1\rangle_A$, which is also sent to Charlie. At Charlie's, the whole state, corresponding to to description of the three involved qubits can be rewritten as

$$\begin{aligned}|\Xi\rangle_{ABC} &= (\alpha|0\rangle_A + \beta|1\rangle_A) \otimes |\Phi^+\rangle_{BC} \\ &= \frac{1}{\sqrt{2}}\big[\alpha|0\rangle_A|0\rangle_B|0\rangle_C + \beta|1\rangle_A|1\rangle_B|1\rangle_C \\ &\quad + \alpha|0\rangle_A|1\rangle_B|1\rangle_C + \beta|1\rangle_A|0\rangle_B|0\rangle_C\big] \\ &= \frac{1}{2}\big[|\Phi^+\rangle_{AC} \otimes (\alpha|0\rangle_B + \beta|1\rangle_B) \\ &\quad\ |\Phi^-\rangle_{AC} \otimes (\alpha|0\rangle_B - \beta|1\rangle_B) \\ &\quad\ |\Psi^+\rangle_{AC} \otimes (\beta|0\rangle_B + \alpha|1\rangle_B) \\ &\quad\ |\Psi^-\rangle_{AC} \otimes (-\beta|0\rangle_B + \alpha|1\rangle_B)\big]\end{aligned}$$

(23.10)

in which the four Bell-states of Eqs. (23.3)–(23.6), relative to qubit A and C, have been introduced thanks to easy algebraic combinations. What we can understand from Eq. (23.10) is that projecting qubits A and C on one of the four Bell-states (each has a probability of 1/4 to be found) will project qubit B on a state similar to that of initial qubit A at the price of a unitary transformation, a bit flip, a phase

---

[v] A CNOT gate is the quantum equivalent of the well known exclusive OR gate (XOR) acting on classical bits in digital electronics.

flip, or both. The information of the BSM (i.e., which Bell-state has been obtained) has to be communicated classically (two bits are necessary) in order to perform the required unitary transformation, otherwise, qubit B remains in a mixed state and no information has traveled from Alice to Bob. From the fundamental side, the latter statement implies that no faster than light communication exists.

A pioneering experiment of quantum teleportation was performed back in 1997 at the University of Innsbruck with polarization entangled photons,[52] and since than repeated a number of times with either polarization or time-bin entanglement. In all aforementioned experiments, entanglement had not been distributed in advance. In the following experiment depicted in Fig. 23.7, this issue is overcome. To do so, fiber spools have been inserted in order to delay the BSM. The entangled photon pair is created by a laser pulse. One of the photons is sent to Bob using the Swiss-Com network situated at 500 m (800 m of fiber) at a SwissCom substation. The other photon is sent to Charlie who keeps it in a 200 m fiber spool. Hence, Bob's photon is already 200 m away when the BSM takes place. Meanwhile Alice's qubit is prepared using a photon created by another nonlinear crystal and sent in a fiber spool before being measured by Charlie. Because Alice's qubit is created by SPDC, detecting the paired photon leads to a heralded single photon source.[53] Charlie performs the BSM, and in the case of a successful result (detection of the $|\Psi^-\rangle$ state) sends a confirmation pulse over to Bob. Bob, on his side, delays the incoming photon long enough, waiting for the classical information from Charlie. He then measures his photon. In order to measure the fidelity of the teleportation, Alice keeps her phase fixed (always sending the same qubit) and Bob varies it. All the units (Alice, Bob, Charlie, and creation of entanglement) are completely independent, and each interferometer is stabilized with respect to its own local stabilization laser. A TCP/IP link allows one to remotely control Bob for data acquisition purposes only.

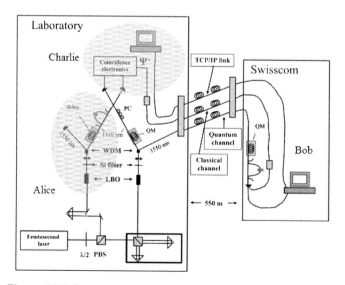

Figure 23.7 Quantum teleportation over the SwissCom network.

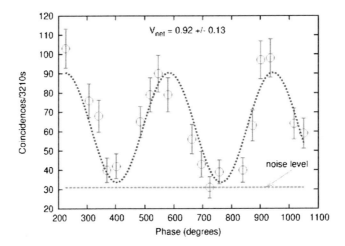

Figure 23.8 Experimental result of the Geneva group "out-of-the-lab" quantum teleportation.

Figure 23.8 shows the obtained result. The visibility of the interference fringe is 92%, and no correction has been applied. More important than the interference visibility value is the fidelity of the teleported state. The fidelity F, which can be understood as describing how closely the teleported and initial qubit states are (in an ideal experiment, $F = 1$), is defined by

$$F = \int \langle \Psi_{Alice} | \rho_{out} | \Psi_{Alice} \rangle d\Psi_{Alice}, \qquad (23.11)$$

in which ρ_{out} is written as

$$\rho_{out} = V |\Psi_{Alice}\rangle\langle\Psi_{Alice}| + (1 - V)I. \qquad (23.12)$$

According to Eq. (23.11), the fidelity equals 1 with probability V and equals 1/2 with probability $1 - V$. Hence the overall fidelity is given by

$$F = \frac{1 + V}{2} \qquad (23.13)$$

Hence, a fringe visibility of 92% leads to a raw fidelity of the teleported state of 96%, this being the highest obtained fidelity for a quantum teleportation protocol[54] ever reported.

23.3.4 Entanglement swapping using highly coherent CW sources

Quantum relays based on entanglement swapping [Fig. 23.3(d)] can increase the communication distances. Polarization and time-bin-based quantum relays have both been demonstrated[4,55] while using the same excitation source, but similar experiments with independent sources have been carried out in the last two years.[55,56] In the former experiment, synchronization of the excitation pulses was obtained by sophisticated electronics for frequency control. In the latter experiment, an entanglement swapping was performed using two pump lasers, sharing a common Kerr medium. Hence, synchronization of the two lasers was easily controllable and adjusted for. In all the above-mentioned experiments, the use of independent, or quasi-independent sources is a first step toward real-world quantum relays. Nevertheless, synchronization of remote laser sources in both frequency and phase[vi] with a precision of 1–2 ps seems unfeasible in practice for lasers separated by more than some meters. Moreover, because photons have to arrive at exactly the same time on the beam splitter, any change in fiber length due to temperature fluctuations has to be compensated for with a precision of 1 ps.

The main reason for using pulsed excitation sources is the fact that one has to be able to select a single temporal mode in order to achieve a successful BSM. A single temporal mode can also be singled out just by measuring the arrival time of the incident photon very precisely. Just as with creation jitter under pulsed excitation, detection jitter plays the same role. If the coherence time of the photon (τ_c) is smaller than the detectors resolution (T_d) then a single temporal mode cannot be post selected. On the other hand, if $\tau_c > T_d$ the photon detection will project it to a single temporal mode. Hence, quantum relays using CW excitation sources can be implemented. The advantage of such a setup lies in the fact that no synchronization of pulsed sources is required, allowing for real-world quantum relays. To do so, however, as previously discussed, there is a crucial need for high-resolution detectors, i.e., showing timing jitters down to 50 ps. This is precisely the aim of the next section.

23.3.4.1 High resolution photodetectors: state of the art

Current commercial InGaAs photon-counting avalanche photodiodes (APD) have a resolution of 300 ps and can only be operated in gated mode, meaning that the arrival time of the photon has to be known in advance. Under CW excitation, the photons creation time remain unknown, implying that free-running detectors have to be employed. On the other hand, silicon-based detectors (Si APD) can be run in continuous regime and, depending on the technology, some of them show a jitter as short as 50 ps. Unfortunately, the associated detection window ranges from 500 to 800 nm, meaning that the signal-photon wavelength has to be adapted to this visible detection band. This can be done by mixing the signal single photons with a high power source at a different wavelength in a nonlinear crystal operated in the upconversion regime, thus providing resulting single photons at a wavelength right in the Si APD detection window detection. For single photons at 1550 nm, this has been achieved several times using different types of crystals and pump wavelength

---

[vi] By "phase" we do not mean the optical phase of the pulse, but the temporal delay of one pulsed laser with respect to the other one. Both lasers can have exactly the same frequency, but their temporal delay can vary over time, especially after a small current surge.

configurations.[44–46] Although the overall quantum efficiency is at most 10%, the noise level is of the order of 100k-counts/s, three orders of magnitude higher than the Si APD's intrinsic dark-count level.

Another solution consists of using single-photon superconducting bolometers.[47–49] A superconducting NbN stripe is brought below its critical temperature and polarized with a current just below the critical current density (J_c). Once a photon is absorbed, a hot spot is formed. The current is forced to flow around the hot spot, hence, rising locally the current density (J). If $J > J_c$, then the stripe turns resistive and a voltage drop can be observed. Once the hot spot has been absorbed, the current density falls below the critical point and the detector is able to detect a new photon. Despite their huge size and limited handability, these detectors offer characteristics close to that obtained with Si APDs. In free-running mode, the quantum efficiency is 5%, dark counts below 100 s^{-1} and detectors can reach a maximal count rate on the order of 100 MHz. Moreover, their temporal resolution is <50 ps. Let us mention that the two following experiments would not have been possible if such new detectors had not been developed and commercially available.

23.3.4.2 Highly coherent photon-pair source

Most of the SPDC sources emit entangled photon pairs with a coherence time of < 1 ps (bandwidth > 5 nm at 1550 nm), which is two to three orders of magnitude shorter than the resolution of the best available photon detectors. Hence, a new highly coherent source has to be developed for this special purpose (i.e., a quantum relay based on CW excitation). Because the best available detectors have a timing resolution (jitter) of 50 ps, the coherence time of the downconverted photons has to be at least four times greater, leading to a bandwidth that ought to be <10 pm. Because existing sources emit photons with a spectrum of >5 nm, downconverted photons have to be filtered. Special filters (described in Fig. 23.9), made of phase-shifted fiber Bragg gratings, have been developed and are now commercially available. They offer 45–50 dB extinction rate of outside the bandwith of interest and over the whole remaining downconversion spectrum, 400 pm spectral tuning, and a drift of <1 pm over several days. The filtered bandwidth of 10 pm provides photons with a coherence time τ_c of 350 ps.

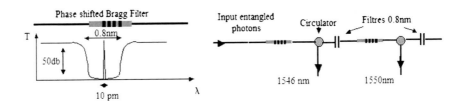

Figure 23.9 Spectral response of fiber Bragg grating filters (left). Two of them are used for the narrow filtering of downconverted photons (right).

The nonlinear crystal is a PPLN waveguide with a conversion efficiency of 10^{-6}, allowing for very low pump powers.[30] The crystal is pumped with 2 mW coming from a 780 nm diode laser that is stabilized against the D2 transition of Rubidium for long-term stability (linewidth on the order of 1 MHz). After collection and filtering, we obtain a photon-pair production rate of 120 k-pairs/second.

23.3.4.3 Hong-Ou-Mandel dip with completely autonomous CW sources

The simplest way to prove that detecting photons with sub coherence length resolution enables projecting on a single temporal mode consists of measuring the Hong-Ou-Mandel (HOM) dip.[57] Figure 23.10 shows the experimental setup. From all the possible events, we choose to keep only those with $\delta t = \Delta t \pm 100$ ps, thus enabling four-fold coincidences arising from two photon pairs. If the time delay δt is smaller than the coherence time, then the two photons interfere at the beam splitter and take the same output port together. This can be understood from Eqs. (23.9); in this case, the input states reads $a_{in}^\dagger b_{in}^\dagger$ and evolves, through the beam splitter, as

$$a^\dagger b^\dagger \xrightarrow{BS} \frac{1}{2}\left(ia^{\dagger 2} + b^{\dagger 2} + a^\dagger b^\dagger - b^\dagger a^\dagger\right) = \frac{i}{2}\left(a^{\dagger 2} + b^{\dagger 2}\right),$$

where the power 2 indicates that there are two photons in the same output port, a or b, provided they cannot be distinghuised. At the output of the beam splitter, a dip in the coincidence rate is therefore expected. Let us emphasize that such an interference effect is purely quantum. One should note that under CW excitation, all different time delays are recorded continually, and no fiber length adjustment is needed. There will always be photons arriving at the same time and photons

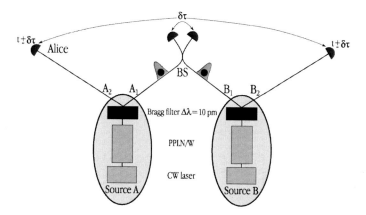

Figure 23.10 Experimental setup for an CW Hong-Ou-Mandel dip.[57] Photons A_1 and A_2 from source A, respectively, are sent at the two inputs of a beam splitter. They take the same output port of the device provided they are indistinghuisable. More details on the detection process are given in the next section.

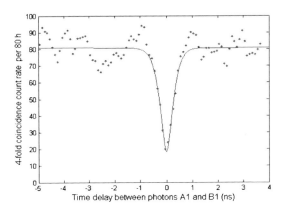

Figure 23.11 HOM dip with two independent CW sources of highly coherent photon pairs.

arriving with a time difference. Figure 23.11 shows the obtained curve where a dip clearly appears when photons arrive at the same time on the beam splitter.[33] The width of the dip is given by the photon's coherence time (350 ps) convolved with the detectors jitter and the obtained visibility is $V_{dip} = 77\%$. Experimentally speaking, having such a dip is seen as the key point toward the achievement of entanglement swapping, which lies at the heart of extended quantum relays.

23.3.4.4 Entanglement swapping with independent CW sources

Let us consider the downconversion of a single pump photon. This effect can occur at any time t and all the possible events are coherent within the kilometer-long coherence length of the laser. Hence, the downconverted state can be written as

$$|\Psi\rangle_A \propto \sum_t |t,t\rangle_A, \qquad (23.14)$$

describing a pair of signal and idler photons emitted by source A at any time t.

Thus, the state produced by two such sources A and B, as described in Fig. 23.12, is given by

$$|\Psi\rangle_A \otimes |\Psi\rangle_B \propto \sum_t \left(|t,t\rangle_A |t,t\rangle_B + \sum_{\tau>0} |t,t\rangle_A |t+\tau, t+\tau\rangle_B \right.$$
$$\left. + |t+\tau, t+\tau\rangle_A |t,t\rangle_B \right) \qquad (23.15)$$

The first term in the above sum describes four photons created at the same time t. In this case, two of them [i.e., the inner photons A_1 and B_1 (see Fig. 23.12)] arrive at the same time at a beam splitter, corresponding to the Hong-Ou-Mandel experiment[57] discussed in the previous section. Moreover, the second term of the sum describes two photon pairs created with a time difference $\tau \neq 0$. Hence, if

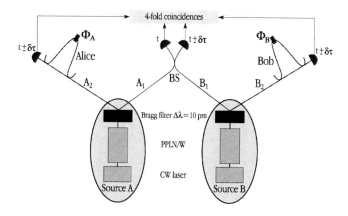

Figure 23.12 Entanglement swapping setup based on two CW independent photon-pair sources A and B. Both sources are made of a CW laser, a PPLN waveguide, and of a Bragg grating filters having a bandwidth of 10 pm. In this experiment, BSM is achieved using a high resolution superconductor detector (see Section 23.3.4.1). Final entanglement shared by photons A_2 and B_2 is analyzed using two interferometers.

the inner photons A_1 and B_1 are sent to the beam splitter for BSM purposes, it is possible to detect one photon at time t and the other one at $t + \tau$. This means that they are, in this case, projected onto the $|\Psi\rangle^-$ Bell-state. As a consequence, the two remaining photons, A_2 and B_2, are also projected onto the $|\Psi\rangle^-$ Bell-state of the form

$$|\Psi\rangle^-_{A_2,B_2} \propto |t\rangle_{A_2}|t+\tau\rangle_{B_2} - |t+\tau\rangle_{A_2}|t\rangle_{B_2} \qquad (23.16)$$

which is a time-bin entangled state with a time difference τ. This state is the same as the one defined in Eq. (23.6) in Section 23.2.1. This coherent projection, also called teleportation of entanglement, is directly related to the photons indistinguishability at the beam splitter where the BSM occurs.

As in Section 23.2.3, quantum correlations shared by these two newly entangled photons are analyzed using two unbalanced interferometers. Here, this corresponds to measuring the fidelity of the final state $|\Psi\rangle^-_{A_2,B_2}$ compared to the state onto which the two inner photons have been projected.

The results are depicted in Fig. 23.13 in which a total of 13 phase points has been measured eight times. The overall integration time was 104 hours (4.3 days). The experiment could be stopped and restarted at any time since no alignment is required. The obtained visibility is equal to V = 63±2%, high enough to demonstrate that the obtained photons are entangled but too low to violate Bell's inequalities.[6,15] Several experimental imperfections account for the loss of visibility, such as finite detector's resolution, imperfect filters, polarization alignment, and, most importantly, multiple photon-pair creation at each source. The latter contribution can be reduced by decreasing the pump power but would lower the count rates.

This proof of principle experiment demonstrates that quantum relays can be developed and implemented in an out of the lab environment because drawbacks

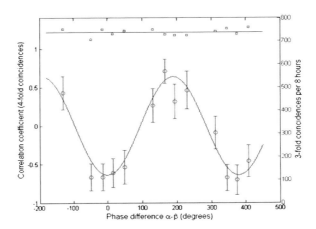

Figure 23.13 Results of the entanglement swapping under CW excitation with independent sources. The evolution in the coincidence rate depends on the phases settings Alice and Bob apply in their interferometers.

due to laser synchronization and fiber alignment have been bypassed. The two parties only need to agree on the filters to be used. The clock can be recovered by agreeing on coincidence measurements. One could make an analog with clock recovery in classical communications allowing for long-distance communications without any clock sharing.

23.4 A Photonic Quantum Information Interface

As we already mentioned, quantum entanglement is widely recognized to be at the heart of quantum physics, with all its counter intuitive features. Especially, the new science of quantum information treats entanglement as a resource for quantum teleportation and relays, which is essential for the coming age of quantum technology. Yet, to be of any practical value, entanglement has to be proven robust enough, at least under some circumstances, i.e., that one can really manipulate it. Accordingly, a growing community of experimental physicists are working with entanglement, in view of finding optimal conditions to master it, especially when the purpose is the connection of different users on a quantum network made of quantum channels along which photonic qubits travel[1] and atomic ensembles are used to store and process the qubits.[58,59] It is thus timely to develop quantum information interfaces that allow transferring the qubits from one type of carrier to another (i.e., in the case of photonic carriers), able to provide a wavelength adaptation while preserving quantum coherence[60,61] of the initial state.

In a recent experiment, the coherent transfer of a qubit from a photon at the telecom wavelength of 1312 nm to another photon at 712 nm, a wavelength close to that of alkaline atomic transitions,[58,59] has been demonstrated. This operation has been performed in the most general way because the initial information-carrying photon was actually entangled in energy and time with a third one at 1555 nm. We could then verify that the transfer does indeed preserve the quantum coherence by testing the entanglement between the receiving photon at 712 nm and the third photon. The entanglement is found to by large enough to violate Bell's inequalities,[6,15,39] despite the fact that these two photons do not satisfy the energy conservation rule of Eq. 23.7 anymore.[60] In other words, a coherent transfer from one photon to another at a different wavelength do not alter the original entangled state.

The transfer of quantum information has been achieved using sum frequency generation, also called up conversion, i.e., by mixing the initial 1312 nm single photon with a highly coherent pump laser at 1560 nm in a PPLN waveguide, as depicted in Fig. 23.14. The probability of a successful up conversion was 5%, including the losses due to phase-matching in the nonlinear waveguide and the addition of an interference filter centred on the resulting wavelength. The receiving single photon at 712 nm, filtered out of the huge flow of pump photons, and the remaining 1555 nm photon were then analyzed using two unbalanced Michelson interferometers in the Franson conformation[39] (i.e., the usual setup for testing energy-time or time-bin

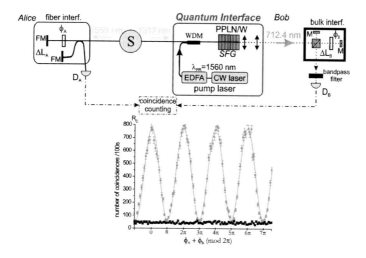

Figure 23.14 A photonic quantum interface. First, the source (S) produces, by parametric down conversion in a PPLN waveguide (not represented), pairs of energy-time entangled photons whose wavelengths are centered at 1555 and 1312 nm, respectively. By means of telecommunications optical fibers, these photons are sent to Alice and Bob, respectively. Next, the qubit transfer is performed at Bob's location from a photon at 1312 nm to a photon at 712 nm using sum frequency generation (SFG) in a PPLN waveguide (PPLN/W). This crystal is pumped by a CW, 700 mW, and high coherence laser working at 1560 nm. The success of the transfer is tested by measuring the quality of the final entanglement between the newly created 712 nm photon and Alice's 1555 nm photon using two unbalanced Michelson interferometers and a coincidence counting technique between detectors D_A and D_B. The graph shows the interference pattern obtained for the coincidence rate as a function of the sum of the phases acquired in the interferometers ($\phi_A + \phi_B$). The corresponding visibility is >97%. Compared to the 98% obtained with the initial entanglement (curve not represented), the quantum fidelity of this transfer >99%.[60]

entanglement). From the high quality of the resulting two-photon interference, we conclude that the fidelity of this quantum information transfer is found to be excellent, higher than 99%. Here, the fidelity can be understood as the comparison between the visibilities of the interference pattens obtained with and without the quantum transfer (more details are given in the caption of Fig. 23.14). Let us emphasize that this experimental result is very encouraging for quantum information science and that mastering of entanglement is making serious progress.

Note that the quantum interface discussed above is not limited to the specific wavelengths chosen. Indeed, suitable modifications of phase-matching conditions and pump wavelengths enable tuning the result of the up-conversion process to any desired wavelengths. More precisely, periodically poled waveguiding structures, as described previously, provide an easy tuning over a broad range of wavelengths by changing the grating period and the pump wavelength.[30] For instance, having an original qubit carried by a photon at 1550 nm and a pump laser at 1570 nm would lead to a transfered qubit carried by a photon at the wavelength of 780 nm (i.e., that of alkaline atomic transitions).

Second, these components yield very high up and down conversion efficiencies.[29,30,45,46] This permits the use of a modest reservoir power to achieve a reasonable qubit transfer probability. In the case of applications requiring very narrow photon bandwidths, for instance, when transferring quantum information onto atoms,[43,58,59,62] bright down converters make very narrow spectral filtering possible while maintaining high photon-pair creation rates with reasonable pump powers as described in Sections 23.11 and 23.12 and in Ref. 33.

However, although current technology of Bragg grating filters allows having photon bandwidths as narrow as 10 pm, this is still a limiting factor when the purpose is to map these photons to atomic transitions. Depending on the nature of the considered atomic ensembles, the bandwidths are off by one to three orders of magnitude. Even if the past ten years have seen the development of single and entangled photon sources of higher and higher brilliance,[vii] it is very important to note that efforts are also coming from the atomic side since physicists are now trying to develop broadband mapping of single photons as described in a very recent proposal.[63]

To conclude this part, note that this photonic and atomic qubit mapping is of great interest since it would allow actual building of quantum networks and computers. And the latter is not a dream anymore since researchers have already been able, thanks to an ion-based experiment,[64] to implement Shor's quantum algorithm able to provide the prime factors of a large number in a polynomial time.[65] In addition two very recent demonstrations of qubit storage and manipulation have been made in experiments based on superconducting Josephson junctions and resonant cavities[66,67].

[vii]This is the number of photon pairs created per spectral and per miliWatt of pump power.

23.5 Conclusion

Quantum entanglement is probably the most counter intuitive feature in quantum physics. It allows for correlations between two distant photons with a degree of correlation that cannot be reproduced by any classical theories. Nonetheless, these correlations are the building blocks of future quantum relays, which play an important role in quantum communication. They allow one to overcome the maximal distance over which quantum communication links can be established and enable quantum networking. Such relays have been implemented in the laboratory since pioneering experiments back in 1997. In this chapter, we have shown that such a relay has been implemented in out of the lab experiments, using dark fibers of the telecommunication network. Also, by using state-of-the-art detectors and new sources, one is able to create quantum relays under continuous wave excitation allowing for the implementation of asynchronous relays, offering more realistic implementations. Finally, together with the development of quantum relays, quantum memories will play an important role in future quantum communication systems. It is therefore necessary to match telecom photons to the operation wavelengths of atomic-ensemble based quantum memories, typically in the visible range. Such a quantum interface preserves, as demonstrated, the intrinsic coherence of entanglement.

Although several technical limits still exist, important steps toward real-world quantum relays and networks have been accomplished.

Acknowledgments

The authors acknowledge Prof. N. Gisin, Dr. H. Zbinden, and Prof. D.B. Ostrowsky for fruitful discussions. Most of the experimental work presented here was obtained at the Geneva Group of Applied Physics. Financial support by the European project QAP and the Swiss NCCR Quantum Photonics are acknowledged.

References

1. Weihs, G., and Tittel, W., "Photonic entanglement for fundamental tests and quantum communication," *Quant. Inf. Comp.*, **1**, 3–56, 2001.

2. Scarani, V., Iblisdir, S., Gisin, N., and Acìn, A., "Quantum cloning," *Rev. Mod. Phys.*, **77**, 1225–1256, 2005.

3. Bennett, C.H., Brassard, G., and Crepeau, C., "Teleporting an unknown quantum state via dual classical and Einstein-Podolsky-Rosen channels," *Phys. Rev. Lett.*, **70**, 1895–1899, 1993.

4. de Riedmatten, H., Marcikic, I., Tittel, W., Zbinden, H., and Gisin, N., "Teleportation in the quantum relay configuration," *Phys. Rev. Lett.* **92**, 047904, 2004.

5. Einstein, A., Podolsky, B., and Rosen, N., "Can quantum mechanical description of physical reality be considered complete ?" *Phys. Rev.*, **47**, 777–780, 1935.

6. Bell, J.S., "On the Einstein-Podolsky-Rosen Paradox," *Physics (Long Island City, N.Y.)*, **1**, 195–200, 1964.

7. Clauser, J.F., Horne, M.A., Shimony, A., and Holt, R.A., "Proposed experiment to test local hidden-variable theories," *Phys. Rev. Lett.*, **23**, 880–884, 1969.

8. Aspect, A., Grangier, P., and Roger, G., "Experimental test of realistic theories via Bell's inequality," *Phys. Rev. Lett.*, **47**, 460–463, 1981.

9. Aspect, A., Dalibard, J., and Roger, G., "Experimental test of Bell's inequalities using time-varying analyzers," *Phys. Rev. Lett.*, **49**, 1804–1807, 1982.

10. Tittel, W., Brendel, J., Zbinden, H., and Gisin, N., "Violation of Bell inequalities by photons more than 10 km apart," *Phys. Rev. Lett.*, **81**, 3563–3566, 1998.

11. Thew, R.T., Tanzilli, S., Tittel, W., Zbinden, H., and Gisin, N., "Experimental investigation of the robustness of partially entangled photons over 11 km," *Phys. Rev. A*, **66**, 062304, 1999.

12. Marcikic, I., de Riedmatten, H., Tittel, W., Zbinden, H., Legré, M., and Gisin, N., "Distribution of time-bin entangled qubits over 50 km of optical fiber," *Phys. Rev. Lett.*, **93**, 180502, 2004.

13. Peng, C.-Z., Yang, T., Bao, X.-H., Zhang, J., Jin, X.-M., Feng, F.-Y., Yang, B., Yang, J., Yin, J., Zhang, Q., Li, N., Tian, B.-L., and J.-W.P., "Experimental free-space distribution of entangled photon pairs over 13 km: towards satellite-based global quantum communication," *Phys. Rev. Lett.*, **94**, 150501, 2005.

14. Ursin, R., Tiefenbacher, F., Schmitt-Manderbach, T., Weier, H., Scheidl, T., Lindenthal, M., Blauensteiner, B., Jennewein, T., Perdigues, J., Trojek, P., Ömer, B., Fürst, M., Meyenburg, M., Rarity, J., Sodnik, Z., Barbieri, C., Weinfurter, H., and Zeilinger, A., "Entanglement-based quantum communication over 144 km," *Nature Phys.*, **3**, 481–486, 2007.

15. Aspect, A., "Bell's inequality test: more ideal than ever," *Nature*, **398**, 189–190, 1999.

16. Bouwmeester, D., Pan, J.-W., Mattle, K., Eibl, M., Weinfurter, H., and Zeilinger, A., "Experimental quantum teleportation," *Nature*, **390**, 575–579, 1997.

17. Boschi, D., Branca, S., Martini, F.D., Hardy, L., and Popescu, S., "Experimental realization of teleporting an unknown pure quantum state via dual classi-

cal and Einstein-Podolsky-Rosen channels," *Phys. Rev. Lett.*, **80**, 1121–1124, 1998.

18. Kim, Y.H., Kulik, S.P., and Shih, Y., "Quantum teleportation of a polarization state with a complete Bell state measurement," *Phys. Rev. Lett.*, **86**, 1370–1373, 2001.

19. Marcikic, I., de Riedmatten, H., Tittel, W., Zbinden, H., and Gisin, N., "Long-distance teleportation of qubits at telecommunication wavelengths," *Nature*, **421**, 509–513, 2003.

20. Pan, J.-W., Bouwmeester, D., and Weinfurter, H., "Experimental entanglement swapping: entangling photons that never interacted," *Phys. Rev. Lett.* **80**, 3891–3894, 1998.

21. Collins, D., Gisin, N., and de Riedmatten, H., "Quantum relays for long distance quantum cryptography," *J. Mod. Opt.*, **52**, 735–753, 2005.

22. Ekert, A.K., "Quantum cryptography based on Bell's theorem," *Phys. Rev. Lett.*, **97**, 661–664, 1991.

23. Gisin, N., Ribordy, G., Tittel, W., and Zbinden, H., "Quantum key distribution," *J. Mod. Phys.*, **74**, 145–195, 2002.

24. Naik, D.S., Peterson, C.G., White, A.G., Berglund, A.J., and Kwiat, P.G., "Entangled state quantum cryptography: eavesdropping on the Ekert protocol," *Phys. Rev. Lett.*, **84**, 4733–4736, 2000.

25. Tittel, W., Brendel, J., Gisin, N., and Zbinden, H., "Quantum cryptography using entangled photons in energy-time Bell states," *Phys. Rev. Lett.*, **84**, 4737–4740, 2000.

26. Jennewein, T., Simon, C., Weihs, G., Weinfurter, H., and Zeilinger, A., "Quantum cryptography with entangled photons," *Phys. Rev. Lett.*, **84**, 4729–4732, 2000.

27. Fasel, S., Gisin, N., Ribordy, G., and Zbinden, H., "Quantum key distribution over 30 km of standard fiber using energy-time entangled photon pairs: a comparison of two chromatic dispersion reduction methods," *Eur. J. Phys. D*, **30**, 143–148, 2004.

28. Boyd, R.W., *Nonlinear Optics*, Elsevier Science USA, 2nd ed. (2003).

29. Banaszek, K., U'Ren, A.B., and Walmsley, I.A., "Generation of correlated photons in controlled spatial modes by downconversion in nonlinear waveguides," *Opt. Lett.*, **26**, 1367–1369, 2001.

30. Tanzilli, S., Tittel, W., de Riedmatten, H., Zbinden, H., Baldi, P., Micheli, M.D., Ostrowsky, D., and Gisin, N., "PPLN waveguide for quantum communication," *Eur. Phys. J. D*, **18**, 155–160, 2002.

31. Kwiat, P.G., Mattle, K., Weinfurter, H., Zeilinger, A., Sergienko, A.V., and Shih, Y., "New high-intensity source of polarization-entangled photon pairs," *Phys. Rev. Lett.*, **75**, 4337–4341, 1995.

32. Kwiat, P.G., Waks, E., White, A.G., Appelbaum, I., and Eberhard, P.H., "Ultrabright source of polarization-entangled photons," *Phys. Rev. A* **60**, R773–R776, 1999.

33. Halder, M., Beveratos, A., Gisin, N., Scarani, V., Simon, C., and Zbinden, H., "Entangling independent photons by time measurement," *Nature Phys.* **3**, 692, 2007.

34. Takesue, H., and Inoue, K., "Generation of polarization-entangled photon pairs and violation of Bell's inequality using spontaneous four-wave mixing in a fiber loop," *Phys. Rev. A*, **70**, 031802, 2004.

35. Li, X., Voss, P.L., Sharping, J.E., and Kumar, P., "Optical-fiber source of polarization-entangled photon pairs in the 1550 nm telecom band," *Phys. Rev. Lett.*, **94**, 053601, 2005.

36. Fulconis, J., Alibart, O., O'Brien, J.L., Wadsworth, W.J., and Rarity, J.G., "Nonclassical interference and entanglement generation using a photonic crystal fiber pair photon source," *Phys. Rev. Lett.*, **99**, 120501, 2007.

37. Simon, C., and Poizat, J.-P., "Creating single time-bin-entangled photon pairs," *Phys. Rev. Lett.*, **94**, 030502, 2005.

38. Stevenson, R.M., Young, R.J., Atkinson, P., Cooper, K., Ritchie, D.A., and Shields, A.J., "A semiconductor source of triggered entangled photon pairs," *Nature*, **439**, 179–182, 2006.

39. Franson, J.D., "Bell inequality for position and time," *Phys. Rev. Lett.*, **62**, 2205–2208, 1989.

40. Marcikic, I., de Riedmatten, H., Tittel, W., Scarani, V., Zbinden, H., and Gisin, N., "Time-bin entangled qubits for quantum communication created by femtosecond pulses," *Phys. Rev. A*, **66**, 062308, 2002.

41. Brendel, J., Gisin, N., Tittel, W., and Zbinden, H., "Pulsed energy-time entangled twin-photon source for quantum communication," *Phys. Rev. Lett.*, **82**, 2594–2597, 1999.

42. Scarani, H., de Riedmatten, H., Marcikic, I., Zbinden, H., and Gisin, N., "Four-photon correction in two-photon Bell experiments," *Eur. Phys. J. D*, **32**, 129–138, 2005.

43. Lloyd, S., Shahriar, M.S., Shapiro, J.H., and Hemmer, P.R., "Long distance, unconditional teleportation of atomic states via complete Bell state measurements," *Phys. Rev. Lett.*, **87**, 167903, 2001.

44. Vandevender, A.P., and Kwiat, P.G., "High efficiency single photon detection via frequency up-conversion," *J. Mod. Opt.*, **51**, 1433–1445, 2004.

45. Langrock, C., Diamanti, E., Roussev, R.V., Yamamoto, Y., Fejer, M.M., and Takesue, H., "Highly efficient single-photon detection at communication wavelengths by use of upconversion in reverse-proton-exchanged periodically poled LiNbO$_3$ waveguides," *Opt. Lett.*, **30**, 1725–1727, 2005.

46. Thew, R.T., Tanzilli, S., Krainer, L., Zeller, S.C., Rochas, A., Rech, I., Cova, S., Zbinden, H., and Gisin, N., "Low jitter up-conversion detectors for telecom wavelength GHz QKD," *New J. Phys.*, **8**, 32, 2006.

47. Gol'tsman, G.N., Okunev, O., Chulkova, G., Lipatov, A., Semenov, A., Smirnov, K., Voronov, B., Dzardanov, A., Williams, C., and Sobolewski, R., "Picosecond superconducting single-photon optical detector," *Appl. Phys. Lett.* **79**, 705–707, 2001.

48. Verevkin, A., Pearlman, A., Slstrokysz, W., Zhang, J., Currie, M., Korneev, A., Chulkova, A., Okunev, O., Kouminov, P., Smirnov, K., Voronov, B., Gol'tsman, G.N., and Sobolewski, R., "Ultrafast superconducting single-photon detectors for near-infrared-wavelength quantum communications," *J. Mod. Opt.*, **51**, 1447–1458, 2004.

49. Engel, A., Semenov, A., Hübers, H.-W., Il'in, K., and Siegel, M., "Superconducting single-photon detector for the visible and infrared spectral range," *J. Mod. Opt.*, **51**, 1459–1466, 2004.

50. van Houwelingen, J.A.W., Beveratos, A., Brunner, N., Gisin, N., and Zbinden, H., "Experimental quantum teleportation with a three-Bell-state analyzer," *Phys. Rev. A*, **74**, 022303, 2006.

51. Calsamiglia, J., and Lütkenhaus, N., "Maximum efficiency of a linear-optical Bell-state analyzer," *Appl Phys. B: Lasers Opt.*, **72**, 67–71, 2001.

52. Ursin, R., Jennewein, T., Aspelmeyer, M., Kaltenbaek, R., Lindenthal, M., and A. Zeilinger, "Quantum teleportation link across the Danube," *Nature*, **430**, 849–849, brief com., 2004.

53. Fasel, S., Alibart, O., Tanzilli, S., Baldi, P., Beveratos, A., Gisin, N., and Zbinden, H., "High-quality asynchronous heralded single-photon source at telecom wavelength," *New J. Phys.*, **16**, 163, 2004.

54. Landry, O., van Houwelingen, J.A.W., Beveratos, A., Zbinden, H., and Gisin, N., "Quantum teleportation over the Swisscom telecommunication network," *JOSA B*, **24**, 398–403, 2007.

55. Kaltenbaek, R., Blauensteiner, B., Zukowski, M., Aspelmeyer, M., and Zeilinger, A., "Experimental interference of independent photons," *Phys. Rev. Lett.* **96**, 240502, 2006.

56. Yang, T., Zhang, Q., Chen, T.-Y., Lu, S., Yin, J., Pan, J.-W., Wei, Z.-Y., Tian, J.-R., and Zhang, J., "Experimental synchronization of independent entangled photon sources," *Phys. Rev. Lett.*, **96**, 110501, 2006.

57. Hong, C.K., Ou, Z.Y., and Mandel, L., "Measurement of subpicosecond time intervals between two photons by interference," *Phys. Rev. Lett.*, **59**, 2044–2046, 1987.

58. Julsgard, B., Sherson, J., Cirac, J.I., Fiuràsek, J., and Polzik, E.S., "Experimental demonstration of quantum memory for light," *Nature*, **432**, 482–486, 2004.

59. Langer, C., Ozeri, R., Jost, J.D., Chiaverini, J., Marco, B.D., Ben-Kish, A., Blakestad, R.B., Britton, J., Hume, D.B., Itano, W.M., Leibfried, D., Reichle, R., Rosenband, T., Schaetz, T., Schmidt, P.O., and Wineland, D.J., "Long-lived qubit memory using atomic ions," *Phys. Rev. Lett.*, **060502**, 2005.

60. Tanzilli, S., Halder, M., Tittel, W., Alibart, O., Baldi, P., Gisin, N., and Zbinden, H., "A photonic quantum information interface," *Nature*, **437**, 116–120, 2005.

61. Giorgi, G., Mataloni, P., and Martini, F.D., "Frequency hopping in quantum interferometry: efficient up-down conversion for qubits and ebits," *Phys. Rev. Lett.*, **90**, 027902, 2003.

62. Chanelière, T., Matsukevich, D.N., Jenkins, S.D., Kennedy, T.A.B., Chapman, M.S., and Kuzmich, A., "Quantum telecommunication based on atomic cascade transitions," *Phys. Rev. Lett.*, **96**, 093604, 2006.

63. Nunn, J., Walmsley, I.A., Raymer, M.G., Surmacz, K., Waldermann, F.C., Wang, Z., and D. Jaksch, "Mapping broadband single-photon wave packets into an atomic memory," *Phys. Rev. A*, **75**, 011401(R), 2007.

64. Vandersypen, L.M.K., Steffen, M., Breyta, G., Yannoni, C.S., Sherwood, M.H., and Chuang, I., "Experimental realization of Shor's quantum factoring algorithm using nuclear magnetic resonance," *Nature*, **414**, 883–887, 2001.

65. Shor, P.W., "Polynomial-time algorithms for prime factorization and discrete logarithms on a quantum computer," *Proc. of the 35th Symposium on Foundations of Computer Science*, Ed. S. Goldwasser, IEEE Computer Society, Los Alamitos, 124–134 (1994).

66. Sillanpää, M.A., Park, J.I., and Simmonds, R.W., "Coherent quantum state storage and transfer between two phase qubits via a resonant cavity," *Nature*, **449**, 438–442, 2007.

67. Majer, J., Chow, J.M., Gambetta, J.M., Koch, J., Johnson, B.R., Schreier, J.A., Frunzio, L., Schuster, D.I., Houck, A.A., Wallraff, A., Blais, A., Devoret, M.H., Girvin, S.M., and Schoelkopf, R.J., "Coupling superconducting qubits via a cavity bus," *Nature*, **449**, 443–447, 2007.

Alexios Beveratos obtained his PhD in optics and photonics in 2002 at the University of Paris XI, where he worked on single photon sources for quantum cryptography based on color center in diamond nanocrystals. He then developed a complete quantum cryptography setup allowing one to demonstrate, for the first time, a quantum key distribution link with single photons. He then joined the Group of Applied Physics at the University of Geneva, where he lead the quantum communication team developing out-of-the-lab teleportation and continuous wave quantum relays. In 2006, he joined the Laboratory of Photonics and Nanostructures to develop, among others, single-photon sources and deterministic entangled photon sources at telecom wavelengths.

Sébastien Tanzilli obtained his PhD in physics in 2002 at the University of Nice where he worked on entangled photon sources for quantum communication based on integrated optics. In collaboration with the Group of Applied Physics (GAP) at the University of Geneva, he demonstrated the most efficient energy-time entangled photon source ever reported. This type of source is commonly used in many quantum communication experiments as a key element. Then he joined the GAP where he aimed at developing the first photonic quantum interface based on coherent upconversion in order to map photonic and atomic qubit carriers as well as fast and high repetition rate quantum key distribution at telecom wavelength based on up conversion detectors. In 2005, he joined the Laboratoire de Physique de la Matière Condensée in Nice, where he leads the quantum communication group that aims at applying the technology of integrated and guided-wave optics to quantum communication and information science.

Chapter 24
Exploiting Optomechanical Interactions in Quantum Information

Claudiu Genes, David Vitali, and Paolo Tombesi
Università di Camerino, Italy

24.1 Introduction
24.2 Ground-State Cooling of a Micromirror
    24.2.1 Hamiltonian description of the system
    24.2.2 Quantum Langevin equations
    24.2.3 Cooling techniques
    24.2.4 Spectrum of quadrature fluctuations
    24.2.5 Realistic considerations on cooling limits
    24.2.6 Summary
24.3 Light-Light and Light-Mirror Entanglement Generation
    24.3.1 Heterodyne detection: Detection quadratures
    24.3.2 Theory of entanglement in bipartite Gaussian states
    24.3.3 Readout CM derivation
    24.3.4 Analysis of achievable entanglement
    24.3.5 Limitations of the model
    24.3.6 Summary
References

24.1 Introduction

A multipartite quantum system is entangled when its state cannot be factorized into a product of states or a mixture of such products. Quantum information schemes that exploit this purely quantum mechanical feature have been devised initially for systems with two discrete quantum levels (i.e., qubits), and afterward, extended to continuous-variable (CV) systems characterized by observables with continuous spectra, such as the position and momentum of a particle, or the quadratures of an electromagnetic field. Optomechanical systems, such as a driven optical cavity with vibrating end-mirrors or a classical coherent field scattering sidebands off a freely suspended mirror, hold a promise of being candidates for CV quantum information processing applications. As in most other proposals, light is the information carrier; the difference lies in the choice of a micromirror as a quantum storage device. Desired quantum states can be encoded in acoustic modes. The optomechanical coupling, owing to which quantum state transfer between radiation and mirror is possible, is realized via the radiation pressure effect,[1,2] which consists in momentum transfer between field and mirror during the scattering process.

The use of the vibrational modes of a micromirror as a quantum state encoder system is somehow unexpected because this is a macroscopic ensemble of atoms expected to behave rather classically. However, there are proposals[3-6] that show that, in principle, the vibrational occupation number of even such a large collection of atoms in a solid can be reduced close to zero, which signifies a clear transition from the classical to the quantum regime. Remarkable progress has been made to experimentally cool the resonator by a few orders of magnitude to quite low occupation numbers.[7-15] However, in order to approach the quantum limit, one needs to integrate small, light, and flexible micromechanical elements into high-finesse cavities (which are typically much more rigid and massive), which is technically challenging. Furthermore, the mechanical quality factor has to be quite high, but present microfabrication techniques yield a mechanical quality factor scaling down for smaller and smaller microresonators.

This chapter starts with developing a general framework for treating the dynamics of a driven optical cavity with a vibrating end-mirror system in which two cooling techniques have been recently demonstrated: (i) cold-damping quantum feedback,[3,4,16] where the oscillator position is measured through a phase-sensitive detection of the cavity output and the resulting photocurrent is used for a real-time correction of the dynamics,[7,10,11,13] and (ii) backaction or self-cooling,[18] in which the off-resonant operation of the cavity results in a retarded back action on the mechanical system, and hence, in a "self"-modification of its dynamics.[9,12,14,15,17] The two situations are illustrated in Fig. 24.1. Analytical results for the final phonon number are obtained for both schemes and their applicability within today's state-of-the-art laboratory setups is analyzed.

As a consequence of these theoretical predictions on the possibility of entering a purely quantum regime with a meso- or macroscopic mirror, one can proceed to design setups in which optomechanical entanglement can be generated and exploited. Starting from an observation (which will be further detailed in Section 24.2.3.2) that backaction cooling can be viewed as sideband cooling, we proceed with analyzing a three-mode problem in which phonon-photon interactions occurring at the scattering of the laser off the mirror are responsible for producing intermode coupling. More precisely, laser photons combine with vibrational phonons either constructively leading to more energetic quanta (anti-Stokes scattering) or destruc-

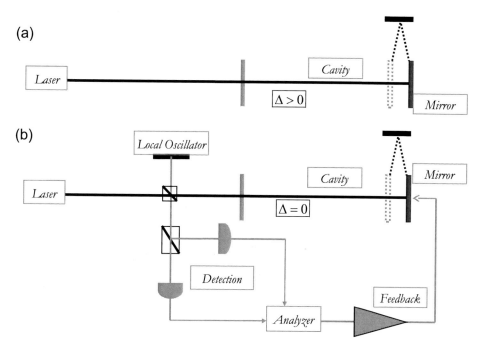

Figure 24.1 (a) Schematic description of a driven, off-resonant ($\Delta > 0$) optical cavity, which is needed for the implementation of the backaction cooling technique: the retarded backaction of the driving laser results in effective cooling of the microresonator. (b) Alternatively, the cavity output is mixed with a local oscillator signal and fed into a feedback loop, which adds an additional viscous force cooling the mirror. This scheme works best at resonance ($\Delta = 0$).

tively to produce lower-frequency photons (Stokes scattering). The dynamics of this light-light-vibration system is phenomenologically similar to another system already analyzed in the literature[19] and that is proven to lead to tripartite entanglement. In Section 24.3.3, a quantum Langevin equation (QLE) approach, including all dissipative and noise effects, is followed with the purpose of obtaining an analytical expression for the 6×6 correlation matrix (CM) of the system; the analysis of the degree of produced entanglement is done by using the logarithmic negativity[20] as an entanglement measure.

24.2 Ground-State Cooling of a Micromirror

We consider a driven optical cavity, which is coupled by radiation pressure with a micromechanical oscillator. The typical experimental configuration is a Fabry-Perot cavity with one mirror much lighter than the other (see, e.g., Refs. 7, 9–11, 21 and 12), but our treatment applies to other configurations, such as the silica toroidal microcavity of Refs. 14, 22. Radiation pressure typically excites several mechanical degrees of freedom of the system with different resonant frequencies. However, a single mechanical mode can be considered when a bandpass filter in the detection scheme is used[23] and coupling between the different vibrational modes of the movable mirror can be neglected. Therefore, we consider a single mechanical mode only, modeled as an harmonic oscillator with frequency ω_m, representing the mechanical resonator we want to cool.

24.2.1 Hamiltonian description of the system

The dynamics of the optical cavity-mirror system with no external driving besides the pump laser and ignoring the coupling to dissipative baths is described by the following Hamiltonian:

$$H = \hbar\omega_c a^\dagger a + \frac{1}{2}\hbar\omega_m(p^2 + q^2) - \hbar G_0 a^\dagger a q + i\hbar E(a^\dagger e^{-i\omega_0 t} - a e^{i\omega_0 t}). \quad (24.1)$$

The first term describes the energy of the cavity mode, with lowering operator a ($[a, a^\dagger] = 1$) and cavity frequency ω_c. The second term gives the energy of the considered mechanical mode of frequency ω_m, described by dimensionless position and momentum operators q and p ($[q, p] = i$). The third term is the radiation-pressure term mediated by a coupling constant

$$G_0 = \frac{\omega_c}{L}\sqrt{\frac{\hbar}{m\omega_m}}, \quad (24.2)$$

where m is the effective mass of the mechanical mode,[23] and L is an effective length that depends on the cavity geometry: it coincides with the cavity length in the Fabry-Perot case and with the toroid radius in the case of Refs. 22 and 14. The last term describes the input driving by a laser at frequency ω_0, where E is related to the input laser power P by

$$|E| = \sqrt{\frac{2P\kappa}{\hbar\omega_0}}, \quad (24.3)$$

where κ is the cavity decay rate. One can adopt the single-cavity mode description of Eq. (24.1) as long as one drives only one cavity mode and the mechanical frequency is much smaller than the cavity free spectral range $\sim c/2L$. In this case, in fact, scattering of photons from the driven mode into other cavity modes is negligible.[2]

24.2.2 Quantum Langevin equations

The dynamics are also determined by the fluctuation-dissipation processes affecting both the optical and the mechanical mode. They can be taken into account in a fully consistent way[24] by considering the following set of nonlinear quantum Langevin equations (QLEs), written in the interaction picture with respect to $\hbar\omega_0 a^\dagger a$

$$\dot{q} = \omega_m p, \quad (24.4a)$$

$$\dot{p} = -\omega_m q - \gamma_m p + G_0 a^\dagger a + \xi, \quad (24.4b)$$

$$\dot{a} = -(\kappa + i\Delta_0)a + iG_0 a q + E + \sqrt{2\kappa}a^{in}, \quad (24.4c)$$

where $\Delta_0 = \omega_c - \omega_0$ is the intrinsic detuning between the laser and cavity frequencies, γ_m is the rate at which the oscillator's state decays and ξ and a^{in} are thermal and optical noise operators.

24.2.2.1 Description of noise terms

The mechanical mode Brownian stochastic force, with zero mean value ξ, possesses the time correlation function[24,25]

$$\langle \xi(t)\xi(t') \rangle = \frac{\gamma_m}{\omega_m} \int \frac{d\omega}{2\pi} e^{-i\omega(t-t')} \omega \left[\coth\left(\frac{\hbar\omega}{2k_B T}\right) + 1 \right], \quad (24.5)$$

where k_B is the Boltzmann constant and T is the temperature of the reservoir of the micromechanical oscillator. The Brownian noise $\xi(t)$ is a Gaussian quantum stochastic process, and its non-Markovian nature (neither its correlation function nor its commutator are proportional to a Dirac delta) guarantees that the QLEs of Eqs. (24.4) preserve the correct commutation relations between operators during the time evolution.[24] The cavity mode amplitude instead decays at the rate κ and is affected by the vacuum radiation input noise a^{in}, whose only nonvanishing correlation function is given by

$$\langle a^{in}(t) a^{in,\dagger}(t') \rangle = \delta(t - t'). \quad (24.6)$$

under the assumption that at optical frequencies $\hbar\omega_c/k_B T \gg 1$ and the equilibrium mean thermal photon number is effectively zero.

24.2.2.2 Steady state

By taking an average over the Langevin equations, one finds that the system is characterized by a semiclassical steady state with the cavity mode in a coherent state with amplitude α_s ($|\alpha_s| \gg 1$), a vanishing momentum $p_s = 0$, and a new equilibrium position for the oscillator, displaced by q_s. The parameters α_s and q_s are the solutions of the nonlinear algebraic equations obtained by factorizing Eqs. (24.4) and setting the time derivatives to zero. They are given by

$$q_s = \frac{G_0 |\alpha_s|^2}{\omega_m}, \quad (24.7a)$$

$$\alpha_s = \frac{E}{\kappa + i\Delta}, \quad (24.7b)$$

where the latter equation is, in fact, the nonlinear equation determining α_s, because the effective cavity detuning Δ, including radiation pressure effects, is given by

$$\Delta = \Delta_0 - \frac{G_0^2 |\alpha_s|^2}{\omega_m}. \quad (24.8)$$

In the following, without any loss of generality, α_s will be chosen real valued, which can always be realized in practice by properly adjusting the phase of the driving laser.

24.2.2.3 Linearization

Rewriting each Heisenberg operator of Eqs. (24.4) as the c-number steady state value plus an additional fluctuation operator with zero mean value

$$a = \alpha_s + \delta a, \quad (24.9a)$$
$$q = q_s + \delta q, \quad (24.9b)$$
$$p = p_s + \delta p, \quad (24.9c)$$

and neglecting nonlinear terms like $\delta a^\dagger \delta a$ and $\delta a \delta q$ based on the assumption that $\alpha_s \gg 1$, one arrives at a linearized QLEs system

$$\delta \dot{q} = \omega_m \delta p, \quad (24.10a)$$
$$\delta \dot{p} = -\omega_m \delta q - \gamma_m \delta p + G_0 \alpha_s \left(\delta a^\dagger + \delta a \right) + \xi, \quad (24.10b)$$
$$\delta \dot{a} = -(\kappa + i\Delta)\delta a + i G_0 \alpha_s \delta q + \sqrt{2\kappa} a^{\text{in}}. \quad (24.10c)$$

The linearized QLEs show that the mechanical mode is coupled to the cavity mode quadrature fluctuations by the effective optomechanical coupling

$$G = G_0 \alpha_s \sqrt{2} = \frac{2\omega_c}{L} \sqrt{\frac{P\kappa}{m\omega_m \omega_0 (\kappa^2 + \Delta^2)}}, \quad (24.11)$$

which can be made very large by increasing the intracavity amplitude α_s. Note that, together with the condition $\omega_m \ll c/L$, which is required for the single-cavity mode description, $\alpha_s \gg 1$ is the *only* assumption required by the present approach.

24.2.3 Cooling techniques

The two situations used for cooling and described in the following are depicted in Figs. 24.1(a) and (b). In Fig. 24.1(a), the setup is quite simple, involving a laser field appropriately detuned with respect to the cavity frequency, a situation that gives rise to an increase in the damping force experienced by the mirror. However, we start the treatment of cooling with the more complex situation of Fig. 24.1(b), which adds a feedback loop for the control of the mirror state read by a homodyne measurement, and which is more physically obvious being simply understood as the addition of an externally controlled damping force.

24.2.3.1 Feedback cooling

This technique is based on the application of a negative derivative feedback, which increases the damping of the system without increasing the thermal noise [see Fig. 24.1(b)]. The oscillator's position is measured by means of a phase-sensitive detection of the cavity output, which is then fed back to the oscillator by applying a force whose intensity is proportional to the time derivative of the output signal, and therefore, to the oscillator velocity.[4] One measures the phase quadrature δY defined as $\delta Y \equiv (\delta a - \delta a^\dagger)/i\sqrt{2}$, whose Fourier transform, according to Eqs. (24.10), is given by

$$\delta Y(\omega) = \frac{G(\kappa - i\omega)}{(\kappa - i\omega)^2 + \Delta^2} \delta q(\omega) + \text{noise terms}. \quad (24.12)$$

This shows that the highest sensitivity for position measurements is achieved for a resonant cavity, $\Delta = 0$ and in the large cavity bandwidth limit $\kappa \gg \omega_m, \gamma_m$, when the cavity mode adiabatically follows the oscillator dynamics $\delta Y(\omega) \simeq (G/\kappa)\delta q(\omega)$.

An additional feedback force is, as a consequence, applied to the vibrating mirror and Eq. (24.10b) now reads

$$\delta \dot{p} = -\omega_m \delta q - \gamma_m \delta p + \frac{G}{\sqrt{2}}\left(\delta a^\dagger + \delta a\right) - \int_{-\infty}^{t} ds\, g(t-s)\delta Y^{\text{est}}(s) + \xi, \quad (24.13)$$

where $g(t)$ is a causal kernel, proportional to a derivative of a Dirac delta in the ideal derivative feedback limit, and the estimated phase quadrature $\delta Y^{\text{est}}(s)$ is given by the input-output relation

$$\delta Y^{\text{est}}(t) = \frac{Y^{\text{out}}(t)}{\eta\sqrt{2\kappa}} = \delta Y(t) - \frac{Y^{\text{in}}(t) + \sqrt{\eta^{-1}-1}\, Y^v(t)}{\sqrt{2\kappa}}. \quad (24.14)$$

We have considered a nonunit detection efficiency, in general, and modeled a detector with quantum efficiency η with an ideal detector preceded by a beam splitter with transmissivity $\sqrt{\eta}$, mixing the incident field with an uncorrelated vacuum-field $Y^v(t)$.[27]

Solving Eqs. (24.10) in the Fourier domain and looking for the modification of the mechanical susceptibility of the microresonator owing to the presence of the cavity and of the feedback loop, we obtain

$$\chi_{\text{eff}}^{\text{cd}}(\omega) = \omega_m \left[\omega_m^2 - \omega^2 - i\omega\gamma_m + \frac{g(\omega)G\omega_m}{\kappa - i\omega}\right]^{-1}. \quad (24.15)$$

The simplest choice for the Fourier transform of the feedback gain function is

$$g(\omega) = \frac{-i\omega g_{\text{cd}}}{1 - i\omega/\omega_{\text{fb}}}, \quad (24.16)$$

corresponding to a standard derivative high-pass filter

$$g(t) = g_{\text{cd}} \frac{d}{dt}\left[\theta(t)\omega_{\text{fb}} e^{-\omega_{\text{fb}} t}\right]. \quad (24.17)$$

The inverse of the feedback bandwidth ω_{fb}^{-1} plays the role of the time delay of the feedback loop, and $g_{\text{cd}} > 0$ is the feedback gain. The ideal derivative limit is

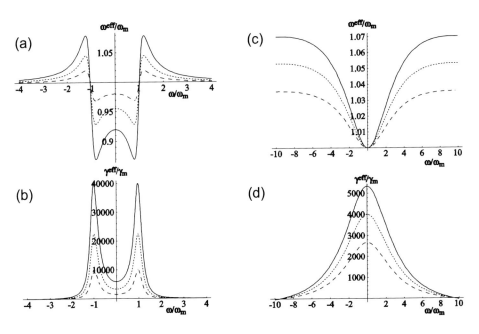

Figure 24.2 Plot of the normalized effective mechanical frequency ($\omega^{\text{eff}}/\omega_m$) and damping ($\gamma^{\text{eff}}/\gamma_m$) versus normalized frequency (ω/ω_m) for both backaction (plots (a) and (b)) and feedback (plots (c) and (d)) cooling. Common parameter values for the two schemes are $\omega_m/2\pi = 10$ MHz, $\gamma_m/2\pi = 100$ Hz, $G = 0.2\,\omega_m$ (dashed line), $G = 0.3\,\omega_m$ (dotted line) and $G = 0.4\,\omega_m$ (full line). For feedback cooling we chose $g_{cd} = 4$, $\omega_{fb} = 3\omega_m$ and $\kappa = 30\,\omega_m$ while for self-cooling $\Delta = \omega_m$ and $\kappa = 0.2\,\omega_m$.

obtained for $\omega_{fb} \to \infty$, implying $g(\omega) = -i\omega g_{cd}$ and therefore $g(t) = g_{cd}\delta'(t)$. The susceptibility of Eq. (24.15) can be read as that of a mechanical oscillator with the following frequency-dependent frequency and damping, respectively,

$$\omega_m^{\text{eff,cd}}(\omega) = \left[\omega_m^2 + \frac{g_{cd}G\omega_m\omega_{fb}\omega^2(\kappa+\omega_{fb})}{(\kappa^2+\omega^2)(\omega_{fb}^2+\omega^2)}\right]^{\frac{1}{2}}, \qquad (24.18)$$

$$\gamma_m^{\text{eff,cd}}(\omega) = \gamma_m + \frac{g_{cd}G\omega_m\omega_{fb}(\kappa\omega_{fb}-\omega^2)}{(\kappa^2+\omega^2)(\omega_{fb}^2+\omega^2)}. \qquad (24.19)$$

Equation (24.19) shows that an increase in the feedback gain can increase the effective damping rate affecting the oscillator and therefore can lead to effective cooling.

24.2.3.2 Backaction cooling

Climbing down to the ground state can be pursued alternatively by making use of a self-cooling effect in which, by properly detuning the laser with respect to the cavity frequency, an extra channel of dissipation from the resonator to the field can be added. Before providing an explanation for this effect, note that if we linearize the radiation pressure Hamiltonian of Eq. (24.1) in the same way we proceeded with the QLEs, one recovers an interaction term $G\delta a \delta q + G\delta a^\dagger \delta q$, which is responsible for the creation or annihilation of photons and phonons. In fact, laser photons are scattered by the moving mirror and two different processes are described by this interaction: (i) the creation of an upshifted photon (anti-Stokes scattering) simultaneously with the destruction of a vibrational quantum and ii) the creation of a downshifted photon (Stokes scattering) simultaneously with the creation of a vibrational phonon.

Applying energy conservation arguments, one sees that energy may be transferred between the resonator and the sideband modes of the field. In fact, an imbalance between the two sideband scattering rates yields *cooling* when the anti-Stokes scattering predominates over the Stokes one, or heating in the opposite scenario. Such an imbalance can be easily produced by simply shifting the sideband one desires to amplify on resonance with the cavity. In order to cool, one needs only to shift the laser to a frequency about one mechanical frequency lower than the cavity frequency ($\Delta \simeq \omega_m$) to make the scattering of anti-Stokes photons ($\omega_L + \omega_m \simeq \omega_c$) more likely.

The accompanying mathematical explanation for this effect can be obtained by solving Eqs. (24.10) in the Fourier domain and looking for the modification of the mechanical effective susceptibility of the mirror owing to the presence of the driving laser

$$\chi_{\text{eff}}^{\Delta}(\omega) = \omega_m \left[\omega_m^2 - \omega^2 - i\omega\gamma_m - \frac{G^2 \Delta \omega_m}{(\kappa - i\omega)^2 + \Delta^2} \right]^{-1}. \qquad (24.20)$$

The susceptibility of Eq. (24.20) can be read as that of a mechanical oscillator with the following frequency-dependent frequency and damping, respectively,

$$\omega_m^{\text{eff},\Delta}(\omega) = \left\{ \omega_m^2 - \frac{G^2 \Delta \omega_m (\kappa^2 - \omega^2 + \Delta^2)}{\left[\kappa + (\Delta - \omega)^2\right]\left[\kappa + (\Delta + \omega)^2\right]} \right\}^{\frac{1}{2}}, \qquad (24.21)$$

$$\gamma_m^{\text{eff},\Delta}(\omega) = \gamma_m + \frac{2G^2 \Delta \omega_m \kappa}{\left[\kappa + (\Delta - \omega)^2\right]\left[\kappa + (\Delta + \omega)^2\right]}. \qquad (24.22)$$

The conclusion reached via physical arguments is validated now by the expression of $\gamma_m^{\text{eff},\Delta}(\omega)$. If the laser is red detuned with respect to the cavity mode ($\Delta > 0$), one achieves an increase in damping and, therefore, cooling, while when the laser is blue detuned with respect to the cavity mode, the system heats up eventually to the point of becoming unstable [when $\gamma_m^{\text{eff},\Delta}(\omega)$ becomes zero].

24.2.4 Spectrum of quadrature fluctuations

We have to evaluate the mean energy of the oscillator in the steady state

$$U = \frac{\hbar\omega_m}{2}\left[\langle \delta q^2 \rangle + \langle \delta p^2 \rangle\right] \equiv \hbar\omega_m \left(n_{\text{eff}} + \frac{1}{2}\right) \qquad (24.23)$$

and see under what conditions it approaches the ground state value of $\hbar\omega_m/2$. The two oscillator variances $\langle\delta q^2\rangle$ and $\langle\delta p^2\rangle$ can be obtained by solving Eqs. (24.10) in the frequency domain and integrating the corresponding fluctuation spectrum. One gets

$$\langle\delta q^2\rangle = \int_{-\infty}^{\infty}\frac{d\omega}{2\pi}S_q^{\Delta,\mathrm{cd}}(\omega), \quad \langle\delta p^2\rangle = \int_{-\infty}^{\infty}\frac{d\omega}{2\pi}\frac{\omega^2}{\omega_m^2}S_q^{\Delta,\mathrm{cd}}(\omega), \quad (24.24)$$

containing the position noise spectrum $S_q^{\Delta,\mathrm{cd}}(\omega)$ which will be particularized for the two distinct cooling mechanisms in the following.

24.2.4.1 Feedback cooling

The position spectrum is given by

$$S_q^{\mathrm{cd}}(\omega) = |\chi_{\mathrm{eff}}^{\mathrm{cd}}(\omega)|^2[S_{\mathrm{th}}(\omega) + S_{\mathrm{rp}}(\omega) + S_{\mathrm{fb}}(\omega)], \quad (24.25)$$

where the first two contributions: the thermal noise

$$S_{\mathrm{th}}(\omega) = \frac{\gamma_m\omega}{\omega_m}\coth\left(\frac{\hbar\omega}{2k_B T}\right), \quad (24.26)$$

and the radiation pressure noise

$$S_{\mathrm{rp}}(\omega) = \frac{G^2\kappa}{\kappa^2+\omega^2}, \quad (24.27)$$

are intrinsic to the cavity-mirror dynamics, while the last term represents an additional noise source due to feedback

$$S_{\mathrm{fb}}(\omega) = \frac{|g(\omega)|^2}{4\kappa\eta}. \quad (24.28)$$

24.2.4.2 Backaction cooling

In this case, the noise is a sum of only two contributions

$$S_q^{\Delta}(\omega) = |\chi_{\mathrm{eff}}^{\Delta}(\omega)|^2[S_{\mathrm{th}}(\omega) + S_{\mathrm{rp}}^{\Delta}(\omega)], \quad (24.29)$$

where $S_{\mathrm{th}}(\omega)$ is the same as in Eq. (24.26), whereas the off-resonant radiation pressure noise assumes a slightly more complicated expression

$$S_{\mathrm{rp}}^{\Delta}(\omega) = \frac{G^2\kappa\left[\Delta^2+\kappa^2+\omega^2\right]}{|(\kappa-i\omega)^2+\Delta^2|^2}. \quad (24.30)$$

24.2.5 Realistic considerations on cooling limits

The exact expressions for the position and momentum variances can be obtained by performing the integrals in Eq. (24.24). In this section, the results are particularized for the two distinct cooling techniques and analyzed for a range of realistic parameters.

24.2.5.1 Feedback cooling

In this case the exact expressions are cumbersome; therefore, we discuss the results of $\langle \delta q^2 \rangle$ and $\langle \delta p^2 \rangle$ only in the adiabatic limit (where $\kappa \gg \omega_m$), by distinguishing two situations, depending on the value of the feedback bandwidth ω_{fb}: (i) almost instantaneous feedback, $\omega_{fb} \gg \kappa \gg \omega_m, \gamma_m$ and (ii) $\kappa \gg \omega_{fb} \sim \omega_m \gg \gamma_m$. In the first case, one has

$$\langle \delta q^2 \rangle \simeq \frac{\bar{n} + 1/2 + \zeta/4 + g_2^2/(4\eta\zeta)\left[1 + g_2\gamma_m/\kappa\right]}{1 + g_2}, \qquad (24.31)$$

$$\langle \delta p^2 \rangle \simeq \frac{(\bar{n} + 1/2)\left[1 + g_2\gamma_m/\kappa\right] + \zeta/4}{1 + g_2} + \frac{g_2^2}{4\eta\zeta}\frac{\omega_{fb}\gamma_m}{\omega_m^2}, \qquad (24.32)$$

where we have defined the scaled dimensionless input power $\zeta = 2G^2/\kappa\gamma_m$ and feedback gain $g_2 = g_{cd}G\omega_m/\kappa\gamma_m$. We have also approximated the thermal noise spectral contribution as

$$\frac{\gamma_m \omega}{\omega_m} \coth\left(\frac{\hbar\omega}{2k_B T}\right) \simeq \gamma_m \frac{2k_B T}{\hbar\omega_m} \simeq \gamma_m (2\bar{n} + 1), \qquad (24.33)$$

where $\bar{n} = (\exp\{\hbar\omega_m/k_B T\} - 1)^{-1}$ is the mean thermal excitation number. This is always justified because $k_B T/\hbar \simeq 10^{11}$ s^{-1} even at cryogenic temperatures. The above results provide the generalization of the results of Refs. 3, 4, where the quantum limits of cold damping have been already discussed within the adiabatic limit. In fact, Eqs. (24.31) and (24.32) reproduce the results of [4] in the large feedback bandwidth limit except for the addition of the nonadiabatic correction term $g_2\gamma_m/\kappa$ for both $\langle \delta q^2 \rangle$ and $\langle \delta p^2 \rangle$. The almost instantaneous feedback regime $\omega_{fb} \gg \kappa \gg \omega_m, \gamma_m$ is not convenient for cooling because of the last contribution to $\langle \delta p^2 \rangle$, which is very large because it diverges linearly with ω_{fb}. This is because the derivative feedback injects a large amount of shot noise when its bandwidth is very large.

In the other limit where the feedback delay time is comparable to the oscillator timescales, that is, $\kappa \gg \omega_{fb} \sim \omega_m \gg \gamma_m$, one has

$$\langle \delta q^2 \rangle \simeq \frac{g_2^2/4\eta\zeta + (\bar{n} + 1/2 + \zeta/4)\left(1 + \omega_{fb}^2/\omega_m^2\right)}{1 + g_2 + \omega_{fb}^2/\omega_m^2}, \qquad (24.34)$$

$$\langle \delta p^2 \rangle \simeq \left[1 + g_2 + \omega_m^2/\omega_{fb}^2\right]^{-1} \left[\frac{g_2^2}{4\eta\zeta}\left(1 + \frac{g_2\gamma_m\omega_{fb}}{\omega_m^2}\right)\right.$$
$$\left. + \left(\bar{n} + \frac{1}{2} + \frac{\zeta}{4}\right)\left(1 + \frac{\omega_m^2}{\omega_{fb}^2} + \frac{g_2\gamma_m}{\omega_{fb}}\right)\right]. \qquad (24.35)$$

These results allow one to modify the conclusions of Refs. 3, 4 about the quantum limits of cold damping. These papers concluded that in the ideal limit $\eta = 1$,

$g_2 \simeq \zeta \to \infty$, that is, very large input power and feedback gain, one achieves ground-state cooling independently from the Q of the oscillator. The general treatment illustrated here puts some limits to this statement. First of all, at finite values of κ and ω_{fb}, the feedback gain g_2 cannot be taken arbitrarily large because otherwise the system becomes unstable. This fact, together with the fact that the nonadiabatic additional factors $g_2\gamma_m/\omega_{fb}$ and $g_2\gamma_m\omega_{fb}/\omega_m^2$, in Eq. (24.35) makes $\langle \delta p^2 \rangle$ to increase too fast for increasing g_2 implies that the minimum oscillator energy is now achieved at a finite, intermediate, optimal value of g_2, whose explicit expression is however involved even in the simplified regime of Eqs. (24.34) and (24.35). Nonetheless, we can still provide a simple recipe for determining the best conditions for ground-state cooling with cold damping. For $\eta = 1$ and $g_2 \simeq \zeta \gg 1, \omega_{fb}^2/\omega_m^2$, one can approximate Eq. (24.34) with the simpler expression $\langle \delta q^2 \rangle \simeq 1/2 + \omega_m^2/4\omega_{fb}^2 + \bar{n}/g_2\left(1 + \omega_m^2/\omega_{fb}^2\right)$. Assuming that the initial reservoir temperature is not too large so that $\bar{n}/g_2 \ll 1$, this approximate expression suggests that the ground state is approached in the limit $\omega_m/\omega_{fb} \ll 1$. However, in the same regime of large feedback gain and input power, the momentum variance of Eq. (24.35) can be approximated as $\langle \delta p^2 \rangle \simeq 1/2 + (\omega_{fb}^2/4\omega_m^2)(g_2\gamma_m/\omega_{fb}) + \bar{n}/g_2(1 + \omega_m^2/\omega_{fb}^2 + g_2\gamma_m/\omega_{fb})$, clearly showing that the feedback bandwidth ω_{fb} cannot be too large; otherwise, $\langle \delta p^2 \rangle$ becomes too large due to the presence of the factor $(\omega_{fb}^2/4\omega_m^2)(g_2\gamma_m/\omega_{fb})$. One finds that the best cooling regime is achieved for $g_2 \simeq \xi$, i.e., $g_{cd} \simeq 2G/\omega_m$, and $\omega_{fb} \sim 3\omega_m$. This is illustrated in Fig. 24.3, where the effective mean excitation number is plotted as a function of g_{cd} and input power P for fixed κ and ω_{fb} [Fig. 24.3(a)], and as a function of κ/ω_m and ω_{fb}/ω_m for fixed values of P and g_{cd} [Fig. 24.3(b)]. In this parameter region, we find the minimum value $n_{eff} \simeq 0.2$, which corresponds to an effective temperature $T_{eff} \simeq 0.27$ mK. Lower values of n_{eff} can be obtained only if the mechanical quality factor Q is further increased.

24.2.5.2 Backaction cooling

It is convenient to introduce the notations[5]

$$A_\pm = \frac{G^2 \kappa}{2\left[\kappa^2 + (\Delta \pm \omega_m)^2\right]}, \quad (24.36)$$

which define the rates at which laser photons are scattered by the moving oscillator, simultaneously with the absorption (Stokes, A_+) or emission (anti-Stokes, A_-) of the oscillator vibrational phonons. For $\Delta > 0$, one has $A_- > A_+$ and a net laser cooling rate $\Gamma = A_- - A_+ > 0$ can be defined, giving the rate at which mechanical energy is taken away by the leaking cavity. As a consequence, the total mechanical damping rate is given by $\gamma_m + \Gamma$ and this is consistent with the expression of the effective (frequency-dependent) damping rate of Eq. (24.22). In fact, it is easy to check that $\gamma_m^{eff}(\omega = \omega_m) = \gamma_m + \Gamma$. For small γ_m, the simplified expressions for

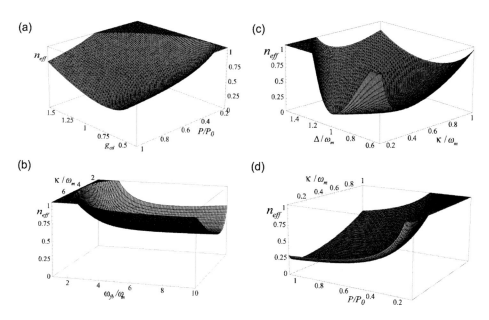

Figure 24.3 (a) Plot of n_{eff} versus the feedback gain g_{cd} and the scaled input power P/P_0 ($P_0 = 100$ mW), around the optimal cooling regime for $\omega_m/2\pi = 10$ MHz, $\gamma_m/2\pi = 100$ Hz, $m = 250$ ng, a cavity of length $L = 0.5$ mm driven by a laser with wavelength 1064 nm and bandwidth $\kappa = 3\omega_m$. The feedback bandwidth is $\omega_{\text{fb}} = 3.5\omega_m$. The oscillator reservoir temperature is $T = 0.6$ K, corresponding to $\bar{n} \simeq 1250$. The minimum value $n_{\text{eff}} \simeq 0.2$ corresponds to an effective temperature $T_{\text{eff}} \simeq 0.27$ mK. (b) Plot of n_{eff} versus κ/ω_m and $\omega_{\text{fb}}/\omega_m$ at fixed input power $P_0 = 100$ mW and feedback gain $g_{cd} = 0.8$. The other parameters are as in (a). (c) Plot of n_{eff} versus Δ/ω_m and κ/ω_m around the optimal ground state cooling regime for $\omega_m/2\pi = 10$ MHz, $\gamma_m/2\pi = 100$ Hz, $m = 250$ ng, a cavity of length $L = 0.5$ mm driven by a laser with power $P = 50$ mW, and wavelength 1064 nm. The oscillator reservoir temperature is $T = 0.6$ K, corresponding to $\bar{n} \simeq 1250$. The minimum value $n_{\text{eff}} \simeq 0.1$ corresponds to an effective temperature $T_{\text{eff}} \simeq 0.2$ mK. (d) Plot of n_{eff} versus κ/ω_m and the normalized power P/P_0, ($P_0 = 50$ mW) at the fixed, optimal value for the detuning, $\Delta = \omega_m$. The other parameters are the same as in (c).

the quadrature variances are

$$\langle \delta p^2 \rangle = \frac{1}{\gamma_m + \Gamma} \left\{ \frac{A_+ + A_-}{2} + \gamma_m \bar{n} \left(1 + \frac{\Gamma}{2\kappa}\right) \right\}, \quad (24.37)$$

$$\langle \delta q^2 \rangle = \frac{1}{\gamma_m + \Gamma} \left\{ a \frac{A_+ + A_-}{2} + \frac{\gamma_m \bar{n}}{\eta_\Delta} \left(1 + \frac{\Gamma}{2\kappa} b\right) \right\}, \quad (24.38)$$

where we have defined the additional coefficients

$$a = \frac{\kappa^2 + \Delta^2 + \eta_\Delta \omega_m^2}{\eta_\Delta (\kappa^2 + \Delta^2 + \omega_m^2)}, \quad (24.39)$$

$$b = \frac{2(\Delta^2 - \kappa^2) - \omega_m^2}{\kappa^2 + \Delta^2}, \quad (24.40)$$

$$\eta_\Delta = 1 - \frac{G^2 \Delta}{\omega_m(\kappa^2 + \Delta^2)}, \quad (24.41)$$

with $0 < \eta_\Delta < 1$ (due to a stability condition not explicitly discussed here). Equations (24.37) and (24.38) show that equipartition $\langle \delta q^2 \rangle = \langle \delta p^2 \rangle$ is not generally verified in the steady state, even though it is often assumed. In fact, η_Δ, a and b are generally different from one.

Another important fact is that Eqs. (24.37) and (24.38) provide a generalization of the results of Refs. 5, 6, which are reproduced if we take $\omega_m \gg \bar{n}\gamma_m, G$ and $\kappa \gg \gamma_m, G$ (assumed in Refs. 5, 6) in Eqs. (24.37) and (24.38). In these limits $a, \eta_\Delta \to 1$, $\Gamma/\kappa \to 0$ and, therefore,

$$\langle \delta p^2 \rangle \simeq \langle \delta q^2 \rangle = n_{\text{eff}} + 1/2, \tag{24.42}$$

where

$$n_{\text{eff}} = \frac{\gamma_m \bar{n} + A_+}{\gamma_m + \Gamma} \tag{24.43}$$

is the basic result of Ref. 6 [see Eq. (9)] and Ref. 5 [see Eq. (5) and its derivation].

In order to approach ground-state cooling ($n_{\text{eff}} < 1$) efficiently, one first of all needs a large value of the effective cooling rate Γ while keeping the heating rate A_+ limited. For a fixed value of G, this is done by matching the detuning to the mechanical frequency, $\Delta = \omega_m$, thus leading to an optimization of energy transfer from the mirror to the anti-Stokes sideband [see Fig. 24.3(c)]. Further optimization requires a quite large value of G, which however cannot be increased indefinitely, because otherwise one approaches the threshold for the violation of the stability condition meaning that η_Δ tends to zero, and $\langle \delta q^2 \rangle$ becomes too large. An upper bound for G is found by requiring that $\eta_\Delta \sim 1$. To this purpose, one has to adjust the ratio κ/ω_m so that also the values of a and b are optimized. We find that the best cooling regime, within a set of experimentally achievable parameters, for which we get $n_{\text{eff}} \simeq 0.1$, is obtained in the good cavity limit condition $\kappa/\omega_m \simeq 0.2$ [see Fig. 24.3(d)], which is close to the value of $1/\sqrt{32}$ suggested in Ref. 5.

24.2.6 Summary

Ground-state cooling can be achieved in today's state-of-the-art laboratories by using either passive and active setups as described in this section. Both techniques require high mechanical quality factors and are limited by the noise introduced by the same interaction that leads to cooling (laser shot noise in the case of passive cooling and feedback noise in the active case). Thus, the final choice of experimental technique lies within the experimentalist who can decide to use a passive setup when in possession of a high-finesse optical cavity or an active setup when a lower finesse cavity is used and the stability of working on resonance is preferred.

24.3 Light-Light and Light-Mirror Entanglement Generation

After concluding in Section 24.2 that a microresonator coupled to a quantum field can be treated as a quantum object, one can proceed with analyzing possibilities for creating multipartite optomechanical entanglement. The existence of bipartite intracavity acousto-optical entanglement has already been established;[26] however, here we provide a more physical interpretation of this entanglement and analyze it in terms of quantities that can be measured at the cavity exit ports. This comes also as an extension of a well-understood entanglement generation scheme[19] (exploitable as a resource for quantum communications), which assumes a three mode sideband-sideband-mirror quantum system resulting from the scattering of a classical laser pulse off an oscillating mirror in free space. Realistic physical limitations, such as the intrinsic quantum nature of the pump field, the decay of the mirror state owing to the inherent thermal noise, and also the leaking of the cavity field to the electromagnetic vacuum, are taken into account in the present setup.

24.3.1 Heterodyne detection: Detection quadratures

Our analysis starts from the observation of the previous section that backaction cooling is sideband-induced cooling. The cavity outgoing noise spectrum shows two peaks at frequencies $\omega_L \pm \omega_m$ corresponding to Stokes and anti-Stokes sidebands, which contain information about the quantum state of the mirror [see Fig. 24.4(a)]. To the purpose of describing the two sidebands as independent optical modes, we introduce a heterodyne detection scheme [shown in Fig. 24.4(b)] in which the scattered field is mixed with the signal from local oscillators shifted up and down a mechanical frequency ω_m from ω_L. Starting from the Fourier components of the intracavity field $\delta a(\omega)$, we write the input-output relation

$$\sqrt{2\kappa}\delta a(\omega) = \delta a_{\text{out}}(\omega) + \delta a_{\text{in}}(\omega), \qquad (24.44)$$

where $\delta a_{\text{out}}(\omega)$ is the operator corresponding to the fluctuations of the field exiting the cavity. The detected quadratures are defined in terms of this latter operator (by integration over a detection interval $I_{\omega_m,D}$ centered at $\omega_L + \omega_m$ and with bandwidth D) as

$$\delta X_s \equiv \frac{1}{\sqrt{2D}} \int_{I_{\omega_m,D}} d\omega \left[\delta a_{\text{out}}(-\omega) e^{i\omega t} + \delta a_{\text{out}}^\dagger(\omega) e^{-i\omega t} \right], \qquad (24.45a)$$

$$\delta Y_s \equiv \frac{1}{i\sqrt{2D}} \int_{I_{\omega_m,D}} d\omega \left[\delta a_{\text{out}}(-\omega) e^{i\omega t} - \delta a_{\text{out}}^\dagger(\omega) e^{-i\omega t} \right], \qquad (24.45b)$$

for the Stokes sideband and

$$\delta X_{as} \equiv \frac{1}{\sqrt{2D}} \int_{I_{\omega_m,D}} d\omega \left[\delta a_{\text{out}}(\omega) e^{-i\omega t} + \delta a_{\text{out}}^\dagger(-\omega) e^{i\omega t} \right], \qquad (24.46a)$$

$$\delta Y_{as} \equiv \frac{1}{i\sqrt{2D}} \int_{I_{\omega_m,D}} d\omega \left[\delta a_{\text{out}}(\omega) e^{-i\omega t} - \delta a_{\text{out}}^\dagger(-\omega) e^{i\omega t} \right], \qquad (24.46b)$$

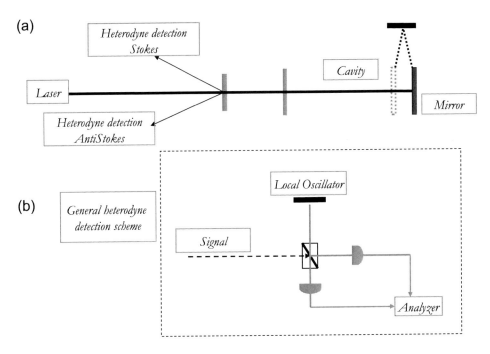

Figure 24.4 Illustration of the entanglement generation-detection scheme in (a). The cavity output field shows noise at the sideband frequencies $\omega_L \pm \omega_m$. Pairs of field quadratures at these frequencies are detected via heterodyne detection (represented in (b)), where the local oscillator is tuned at $\omega_L - \omega_m$ (to detect Stokes photons) and at $\omega_L + \omega_m$ (for anti-Stokes photons detection).

for the anti-Stokes sideband. The mirror state detection is assumed to be performed by a detector with a bandwidth much larger than all the typical oscillator frequencies. In such a case, the frequency integration simply becomes the inverse Fourier transform operation

$$\delta Q \equiv \frac{1}{\sqrt{2\pi}} \int_{-\infty}^{\infty} d\omega dq(\omega) e^{-i\omega t}, \qquad (24.47a)$$

$$\delta P \equiv \frac{1}{\sqrt{2\pi}} \int_{-\infty}^{\infty} d\omega dp(\omega) e^{-i\omega t}. \qquad (24.47b)$$

The tripartite system Stokes-anti-Stokes-mirror can be described by a 6×6 readout correlation matrix (CM)

$$V_{\text{read}}^{ij} = \frac{1}{2} \langle \delta u_i \delta u_j + \delta u_j \delta u_i \rangle, \qquad (24.48)$$

defined in term of components of the quadrature fluctuations vector

$$\delta u \equiv [\delta Q, \delta P, \delta X_s, \delta Y_s, \delta X_{\text{as}}, \delta Y_{\text{as}}]^\mathsf{T}. \qquad (24.49)$$

24.3.2 Theory of entanglement in bipartite Gaussian states

Because we are interested in quantum-state transfer between light and mirror and in mechanically mediated sideband entanglement, an analysis of bipartite entanglement suffices. We use a bimodal Gaussian state description, whose CM can be written as

$$V \equiv \begin{pmatrix} A & C \\ C^\intercal & B \end{pmatrix}, \tag{24.50}$$

in terms of the 2×2 matrices A, B, and C. Its negative symplectic eigenvalue is

$$\eta^-(V) \equiv \frac{1}{\sqrt{2}} \left(\Sigma(V) - \sqrt{\Sigma(V)^2 - 4 \det V} \right)^{1/2}, \tag{24.51}$$

with $\Sigma(V) \equiv \det A + \det B - 2 \det C$. The bipartite state is entangled if and only if $\eta^-(V) < 1/2$, which is equivalent with Simon's necessary and sufficient entanglement positive partial transpose (PPT) criterion for Gaussian states, which can be written as $4 \det V < \Sigma(V) - 1/4$. In the following, a quantity which has already been proposed as a measure of entanglement, the logarithmic negativity, defined as

$$E_N(V) = \max\{0, -\ln 2\eta^-(V)\} \tag{24.52}$$

will be used for the characterization of entanglement properties. Positive nonzero values of $E_N(V)$ indicate entanglement.

24.3.3 Readout CM derivation

A derivation of the readout CM is sketched in the following. One starts from the Fourier transform of Eqs. (24.10) and using Eq. (24.44), the following set of linear algebraic equations is obtained

$$\delta a_{\text{out}}(\omega) = id(\omega)\delta q(\omega) + \frac{d(\omega)}{d^*(\omega)} a_{\text{in}}(\omega), \tag{24.53a}$$

$$\delta a_{\text{out}}^\dagger(\omega) = -id^*(-\omega)\delta q(\omega) + \frac{d^*(-\omega)}{d(-\omega)} a_{\text{in}}^\dagger(\omega), \tag{24.53b}$$

$$\delta q(\omega) = \chi_{\text{eff}}^\Delta(\omega)) \left[\xi(\omega) + d(\omega) a_{\text{in}}(\omega) + d^*(-\omega) a_{\text{in}}^\dagger(\omega) \right], \tag{24.53c}$$

$$\delta p(\omega) = -i\frac{\omega}{\omega_m}\delta q(\omega), \tag{24.53d}$$

where

$$d(\omega) = \frac{G\sqrt{\kappa}}{\kappa + i(\Delta - \omega)}, \tag{24.54}$$

is defined, in terms of which the Stokes and anti-Stokes frequency-dependent scattering rates are simply expressed as

$$A_+(\omega) = \frac{G^2 k/2}{\kappa^2 + (\Delta + \omega)^2} = \frac{1}{2}|d(-\omega)|^2, \quad (24.55a)$$

$$A_-(\omega) = \frac{G^2 k/2}{\kappa^2 + (\Delta - \omega)^2} = \frac{1}{2}|d(\omega)|^2. \quad (24.55b)$$

Making use of the correlations of the mechanical and optical noise operators [Eqs. (24.5), (24.6), and of the approximation of Eq. (24.33)], we have

$$\langle \xi(\omega)\xi(\omega') \rangle = \gamma_m \left[(2n+1) + \frac{\omega}{\omega_m} \right] \delta(\omega + \omega'), \quad (24.56a)$$

$$\langle a_{\text{in}}(\omega) a_{\text{in}}^\dagger(\omega') \rangle = \delta(\omega + \omega'), \quad (24.56b)$$

the frequency-dependent, delta correlated, 4×4 output CM, $V_{\text{out}}(\omega, \omega') = V_{\text{out}}(\omega)\delta(\omega + \omega')$, of the output fluctuation vector

$$\left[\delta q(\omega), \delta p(\omega), \delta a_{\text{out}}(\omega), \delta a_{\text{out}}^\dagger(\omega) \right]^\mathsf{T}. \quad (24.57)$$

can be computed. The elements of V_{read}, in which the expressions of Eqs. (24.45) and (24.46) are introduced, are frequency integrals over elements of the $V_{\text{out}}(\omega)$. The integration is not a simple task since the denominators include high powers of ω and the integral is performed over a finite interval $I_{\omega_m,D}$.

A simplification that allows one to obtain fairly simple analytical expressions for the elements of V_{read} is introduced by the assumption of high Q quality factor mechanical oscillator. In other words, we assume that in spite of a serious increase in the effective microoscillator damping rate due to the action of the laser, the mechanical frequency is always much larger than the damping rate. This allows one to consider that the effective mechanical susceptibility $\chi_{\text{eff}}^\Delta(\omega)$ is always the sharpest function in ω present inside the integrals and to approximate any other slowly varying term with its value at the peak of $\chi_{\text{eff}}^\Delta(\omega)$ (i.e., at ω_m).

24.3.4 Analysis of achievable entanglement

Under the assumption described above, the 6×6 CM V_{read} is a Gaussian matrix. The Gaussian nature of the system is also preserved after a partial trace operation in which one of the modes is traced out and one can analyze bipartite entanglement of the three separate systems. In the following, we present analytical results for the bipartite sideband-sideband and sideband-mirror entanglement.

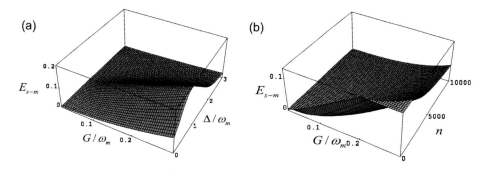

Figure 24.5 (a) Logarithmic negativity of the mirror-Stokes sideband system ($E_\text{s-m}$) as a function of the coupling G and the detuning Δ. $E_\text{s-m}$ is peaked at $\Delta = \omega_m$ and increases with larger values of the optomechanical coupling G. (b) One can note the appreciable robustness of the entanglement with temperature increase (entanglement is present still at $n = 10^5$).

24.3.4.1 Stokes mirror

The Stokes-mirror CM can be put in the form of Eq. (24.50) with

$$V_\text{s-m} = \begin{pmatrix} \mathcal{B}_m + 1/2 & 0 & \mathcal{C}_\text{s-m} & \mathcal{D}_\text{s-m} \\ 0 & \mathcal{B}_m + 1/2 & \mathcal{D}_\text{s-m} & -\mathcal{C}_\text{s-m} \\ \mathcal{C}_\text{s-m} & \mathcal{D}_\text{s-m} & \mathcal{B}_s + 1/2 & 0 \\ \mathcal{D}_\text{s-m} & -\mathcal{C}_\text{s-m} & 0 & \mathcal{B}_s + 1/2 \end{pmatrix}, \quad (24.58)$$

where the matrix elements are

$$\mathcal{B}_m \simeq \frac{1}{\gamma_m + \Gamma} [\gamma_m n + A_+], \quad (24.59\text{a})$$

$$\mathcal{B}_s \simeq \frac{2\pi}{D} \frac{A_+}{\gamma_m + \Gamma} [\gamma_m (n+1) + A_-], \quad (24.59\text{b})$$

$$\mathcal{C}_\text{s-m} \simeq \sqrt{\frac{\pi}{D} \frac{2(\omega_m + \Delta)}{G\sqrt{\kappa}} A_+ \left(\mathcal{B}_m + \frac{3}{4}\right)}, \quad (24.59\text{c})$$

$$\mathcal{D}_\text{s-m} \simeq \sqrt{\frac{\pi}{D} \frac{2\kappa}{G\sqrt{\kappa}} A_+ \left(\mathcal{B}_m + \frac{3}{4}\right)}, \quad (24.59\text{d})$$

and the identifications $A_\pm = A_\pm(\omega_m)$ have been made. The behavior of the logarithmic negativity, computed as detailed in Section 24.3.2 is illustrated in Fig. 24.5(a) versus G and Δ and in Fig. 24.5(b) versus G and n. Figure 24.5(a) shows that there is a quite wide region around what it was concluded in Section 24.2.5.2 to be the optimal cooling regime (around $\Delta = \omega_m$) and entanglement to be present even around zero detuning. In Fig. 24.5(b), the dependence on temperature, or equivalently on the initial environment-induced mirror vibrational level (n) is plotted. Entanglement shows signs of robustness, still showing its presence at a value of $n = 10^5$, which, for the numbers considered in Section 24.2.5.2, corresponds to a temperature of 6 K (a temperature available by means of cryogenic techniques).

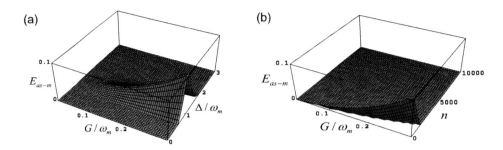

Figure 24.6 (a) The entanglement between mirror and antiStokes sideband ($E_{\text{as-m}}$) is peaked at $\Delta = \omega_m$ but it falls off fast with variations from this optimal value. The optimal value is also smaller than in the Stokes-mirror case. (b) As opposed to the Stokes-mirror case, the generated entanglement is fragile at nonzero temperature (for n around 10^5 it completely vanishes).

24.3.4.2 Anti-Stokes mirror

The anti-Stokes mirror CM is given by

$$V_{\text{as-m}} = \begin{pmatrix} \mathcal{B}_m + 1/2 & 0 & \mathcal{C}_{\text{as-m}} & \mathcal{D}_{\text{as-m}} \\ 0 & \mathcal{B}_m + 1/2 & -\mathcal{D}_{\text{as-m}} & \mathcal{C}_{\text{as-m}} \\ \mathcal{C}_{\text{as-m}} & -\mathcal{D}_{\text{as-m}} & \mathcal{B}_{\text{as}} + 1/2 & 0 \\ \mathcal{D}_{\text{as-m}} & \mathcal{C}_{\text{as-m}} & 0 & \mathcal{B}_{\text{as}} + 1/2 \end{pmatrix}, \quad (24.60)$$

with elements ($\mathcal{B}_m$ is given in Eq. (24.59a)

$$\mathcal{B}_{\text{as}} \simeq \frac{2\pi}{D} \frac{A_-}{\gamma_m + \Gamma} [\gamma_m n + A_+] = \frac{2\pi}{D} A_- \mathcal{B}_m, \quad (24.61\text{a})$$

$$\mathcal{C}_{\text{as-m}} \simeq \sqrt{\frac{\pi}{D}} \frac{2(\Delta - \omega_m)}{G\sqrt{\kappa}} A_- \left(\mathcal{B}_m + \frac{1}{4} \right), \quad (24.61\text{b})$$

$$\mathcal{D}_{\text{as-m}} \simeq \sqrt{\frac{\pi}{D}} \frac{2\kappa}{G\sqrt{\kappa}} A_- \left(\mathcal{B}_m + \frac{1}{4} \right). \quad (24.61\text{c})$$

Similarly to the Stokes-mirror case, the optimal entanglement for this bipartite state is obtained at the cooling limit; however, it reaches to zero very fast as Δ is swept around ω_m [see Fig. 24.6(a)]. At difference with the Stokes case, it is very fragile as temperature increases [Fig. 24.6(b)].

24.3.4.3 Stokes-AntiStokes

The Stokes antiStokes CM is given by

$$V_{\text{s-as}} = \begin{pmatrix} \mathcal{B}_s + 1/2 & 0 & \mathcal{C}_{\text{s-as}} & \mathcal{D}_{\text{s-as}} \\ 0 & \mathcal{B}_s + 1/2 & \mathcal{D}_{\text{s-as}} & -\mathcal{C}_{\text{s-as}} \\ \mathcal{C}_{\text{s-as}} & \mathcal{D}_{\text{s-as}} & \mathcal{B}_{\text{as}} + 1/2 & 0 \\ \mathcal{D}_{\text{s-as}} & -\mathcal{C}_{\text{s-as}} & 0 & \mathcal{B}_{\text{as}} + 1/2 \end{pmatrix}, \quad (24.62)$$

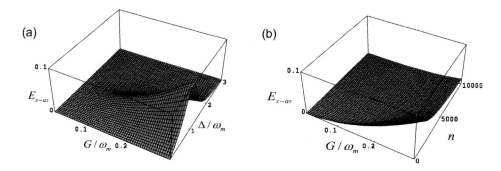

Figure 24.7 (a) The entanglement between the two sideband modes ($E_{\text{s-as}}$) is peaked at $\Delta = \omega_m$ and increases with larger values of the optomechanical coupling G. (b) The robustness at temperature increase is similar to the one shown in the Stokes-Mirror case.

with elements

$$\mathcal{B}_s \simeq \frac{2\pi}{D(\gamma_m + \Gamma)}\left[A_+\gamma_m(n+1) + A_+A_-\right], \tag{24.63a}$$

$$\mathcal{B}_{as} \simeq \frac{2\pi}{D(\gamma_m + \Gamma)}\left[A_-\gamma_m n + A_+A_-\right], \tag{24.63b}$$

$$\mathcal{C}_{\text{s-as}} \simeq \frac{-\pi\Gamma}{D(\gamma_m + \Gamma)}\frac{\kappa^2 + \Delta^2 - \omega_m^2}{2\omega_m\Delta}\left[\gamma_m\left(n+\frac{1}{2}\right) + \frac{A_+ + A_-}{2}\right], \tag{24.63c}$$

$$\mathcal{D}_{\text{s-as}} \simeq \frac{\pi\Gamma}{D(\gamma_m + \Gamma)}\frac{\kappa}{2\omega_m}\left[\gamma_m\left(n+\frac{1}{2}\right) + \frac{A_+ + A_-}{2}\right]. \tag{24.63d}$$

24.3.5 Limitations of the model

There are two limitations that restrict the values of the parameters that one can consider for optimal sideband-mirror entanglement: (i) stability conditions and (ii) good quality factor approximation. The stability conditions insure that the steady state reached by the system is unique independently of the initial conditions. They can be derived by applying the Routh-Hurwitz criterion[28] to Eqs. (24.10). For positive Δ, which is the cooling regime where anti-Stokes scattering dominates Stokes scattering, the stability condition requires that

$$\omega_m\left(\kappa^2 + \Delta^2\right) - G^2\Delta > 0. \tag{24.64}$$

For Δ around ω_m, this simply reads

$$G < \sqrt{\kappa^2 + \omega_m^2}. \tag{24.65}$$

For negative Δ, where Stokes scattering dominates and where one would like to look for entanglement, the stability condition can be approximated by the condition that the effective mechanical damping is positive $\gamma_m + \Gamma > 0$. Using the expression of $\Gamma = A_- - A_+$ combined with the definitions in the Eq. (24.36), in the limit of G and κ being of the order of ω_m the condition simply reads $\Delta > -\gamma_m$. This drastically restricts the negative detuning regime to a very narrow one; thus making it clearly not the preferred regime to work in.

The approximation made in Section 24.3.3 that made possible obtaining analytical results for the readout CM comes with its own limitation. A large quality factor means that the effective damping should be small compared to ω_m. Using again $\Gamma = A_- - A_+$ combined with the definitions in the Eq. (24.36) for $\Delta = \omega_m$ it results that this condition simply imposes that $G \ll \omega_m$, thus making it the more restrictive limitation.

24.3.6 Summary

The emergence of stationary, temperature robust, tripartite entanglement among a micromirror and the sidebands responsible with its cooling in a passive, detuned regime has been analyzed in this section. Based on this result, optomechanical schemes, to teleport unknown quantum states onto the vibrational degree of freedom of the mirror, can be designed. Moreover, a way of measuring this generated entanglement has been analyzed where heterodyne detection can distinguish between the optical sidebands scattered off the mirror.

References

1. Shampire, P., Loudon, R., and Babiker, M., "Quantum theory of radiation-pressure fluctuations on a mirror," *Phys. Rev. A*, **51**, 2726–2737, 1995.

2. Law, C.K., "Interaction between a moving mirror and radiation pressure: A Hamiltonian formulation," *Phys. Rev. A*, **51**, 2537–2541, 1995.

3. Courty, J.-M., Heidmann, A., and Pinard, M., "Quantum limits of cold damping with optomechanical coupling," *Eur. Phys. J. D*, **17**, 399–408, 2001.

4. Vitali, D., Mancini, S., Ribichini, L., and Tombesi, P, "Mirror quiescence and high-sensitivity position measurements with feedback," *Phys. Rev. A*, **65**, 063803, 2002; **69**, 029901(E), 2004; "Macroscopic mechanical oscillators at the quantum limit through optomechanical cooling," *J. Opt. Soc. Am. B*, **20**, 1054–1065, 2003.

5. Wilson-Rae, I., Nooshi, N., Zwerger, W., and Kippenberg, T.J., "Theory of ground gtate cooling of a mechanical oscillator using dynamical backaction," *Phys. Rev. Lett.*, **99**, 093901, 2007.

6. Marquardt, F., Chen, J.P., Clerk, A.A., and Girvin, S.M., "Quantum theory of cavity-assisted sideband cooling of mechanical motion," *Phys. Rev. Lett.*, **99**, 093902, 2007.

7. Cohadon, P.F., Heidmann, A., and Pinard, M., "Cooling of a mirror by radiation pressure," *Phys. Rev. Lett.*, **83**, 3174–3177, 1999.

8. LaHaye, M.D., Buu, O., Camarota, B., and Schwab, K.C., "Approaching the quantum limit of a nanomechanical resonator," *Science*, **304**, 74–77, 2004.

9. Gigan, S., Böhm H., Paternostro, M., Blaser, F., Langer, G., Hertzberg, F., Schwab, K.C., Bäuerle, D., Aspelmeyer, M., and Zeilinger, A., "Self-cooling of a micromirror by radiation pressure," *Nature, (London)*, **444**, 67–70, 2006.

10. Arcizet, O., Cohadon, P.F., Briant, T., Pinard, M., Heidmann, A., Mackowski, J.M., Michel, C., Pinard, L., Francais, O., and Rousseau, L., "Radiation-pressure cooling and optomechanical instability of a micromirror, "High-sensitivity optical monitoring of a micromechanical resonator with a quantum-limited optomechanical sensor," *Phys. Rev. Lett.*, **97**, 133601, 2006.

11. Kleckner, D., and Bouwmeester, D., "Sub-kelvin optical cooling of a micromechanical resonator," *Nature, (London)*, **444**, 75–78, 2006.

12. Arcizet, O., Cohadon, P.F., Briant, T., Pinard, M., and Heidmann, A., "Radiation-pressure cooling and optomechanical instability of a micromirror," *Nature, (London)*, **444**, 71–74, 2006.

13. Poggio, M., Degen, C.L., Mamin, H.J., and Rugar, D., "Feedback cooling of a cantilever's fundamental mode below 5 mK," *Phys. Rev. Lett.*, **99**, 017201, 2007.

14. Schliesser, A., Del'Haye, P., Nooshi, N., Vahala, K.J., and Kippenberg, T.J., "Radiation pressure cooling of a micromechanical oscillator using dynamical backaction," *Phys. Rev. Lett.*, **97**, 243905, 2006.

15. Corbitt, T., Chen, Y., Innerhofer, E., Müller-Ebhardt, H., Ottaway, D., Rebhein, H., Sigg, D., Whitcomb, S., Wipf, C., and Mavalvala, N., "An all-optical trap for a gram-scale mirror," *Phys. Rev. Lett.*, **98**, 150802, 2007.

16. Mancini, S., Vitali, D., and Tombesi, P., "Optomechanical cooling of a macroscopic oscillator by homodyne feedback," *Phys. Rev. Lett.*, **80**, 688–691, 1998.

17. Bhattacharya, M., and Meystre, P., "Trapping and cooling a mirror to its quantum mechanical ground state," *Phys. Rev. Lett.*, **99**, 073601, 2007.

18. Braginsky, V.B, Strigin, S.E., and Vyatchanin, S.P., "Parametric oscillatory instability in Fabry–Perot interferometer," *Phys. Rev. A*, **287**, 331–338, 2001.

19. Pirandola, S., Mancini, S., Vitali, D., and Tombesi, P., "Continuous-variable entanglement and quantum-state teleportation between optical and macroscopic vibrational modes through radiation pressure," *Phys. Rev. A*, **68**, 062317, 2003.

20. Vidal, G., and Werner, R.F., "Computable measure of entanglement," *Phys. Rev. A*, **65**, 032314, 2002.

21. Metzger, C.H., and Karrai, K., "Cavity cooling of a microlever," *Nature, (London)*, **432**, 1002–1005, 2004.

22. Kippenberg, T.J., Rokhsari, H., Carmon, T., Scherer, A., and Vahala, K.J., "Analysis of radiation-pressure induced mechanical oscillation of an optical microcavity," *Phys. Rev. Lett.*, **95**, 033901, 2005.

23. Pinard, M., Hadjar, Y., and Heidmann, A., "Effective mass in quantum effects of radiation pressure," *Eur. Phys. J. D*, **7**, 107–116, 1999.

24. Giovannetti, V., and Vitali, D., "Phase-noise measurement in a cavity with a movable mirror undergoing quantum Brownian motion," *Phys. Rev. A*, **63**, 023812, 2001.

25. Landau, L., and Lifshitz, E., *Statistical Physics*, Pergamon, New York (1958).

26. Vitali, D., Gigan, S., Ferreira, A., Böhm, H.R., Tombesi, P., Guerreiro, A., Vedral, V., Zeilinger, A., and Aspelmeyer, M., "Optomechanical entanglement between a movable mirror and a cavity field," *Phys. Rev. Lett.*, **98**, 030405, 2007.

27. Gardiner, C.W., and Zoller, P., *Quantum Noise*, Springer, Berlin (2000).

28. Gradshteyn, I.S. and Ryzhik, I.M., *Table of Integrals, Series and Products*, Academic Press, New York, 1119 (1980).

Chapter 25
Optimal Approximation of Nonphysical Maps via Maximum Likelihood Estimation

Vladimír Bužek, Mário Ziman, and Martin Plesch
Research Center for Quantum Information and QUNIVERSE, Slovakia

25.1 Introduction
25.2 Quantum States and Quantum Channels
    25.2.1 States of quantum systems and density operators
    25.2.2 State tomography
    25.2.3 Quantum processes
    25.2.4 Processes tomography
25.3 Structure of Qubit Channels
25.4 Method of Maximum Likelihood
25.5 U-NOT gate
25.6 Nonlinear Transformations
    25.6.1 Nonlinear polarization rotation
    25.6.2 Powers of density operator: $\rho \longrightarrow \rho^2$
25.7 Analysis of Results
25.8 Conclusions
References

25.1 Introduction

Any quantum dynamics,[1,2] i.e., the process that is described by a completely positive (CP) map of a quantum-mechanical system, can be probed in two different ways. Either we use a single entangled state of a bipartite system,[3] or we use a collection of linearly independent single-particle test states[4,5] (forming a basis of

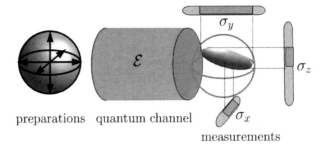

Figure 25.1 Schematic representation of a reconstruction of a single-qubit channel. Input (test) states ρ_k of the single-qubit channels are represented by the Bloch sphere (the state space of a single qubit). At the output of the single-qubit channel (modeled as a transformation of the Bloch sphere into an ellipsoid, i.e., the Bloch sphere is "deformed" by the action of the channel), a complete measurement of test states is performed. The complete measurement is performed via the projective measurement of σ operators. Based on correlations between input and output states of the test qubits, the action of the quantum channel (a CP map) is determined (reconstructed).

the vector space of all Hermitian operators). Given the fragility of entangled states, in this chapter we will focus our attention on the process reconstruction using only single-particle states.

The task of a process reconstruction is to determine an unknown quantum channel (a "black box") using correlations between known input states and results of measurements performed on states that have been transformed by the channel (see Fig. 25.1). The linearity of quantum dynamics implies that the channel $\mathcal{E}$ is exhaustively described by its action $\rho_j \to \rho'_j = \mathcal{E}[\rho_j]$ on a set of basis states (i.e., a collection of linearly independent states ρ_j) that play a role of *test states*. Therefore, to perform a reconstruction of the channel $\mathcal{E}$, we have to perform a complete state tomography[1,6,7] of ρ'_j. The number of test states equals d^2, where $d = \dim \mathcal{H}$ is the dimension of the Hilbert space $\mathcal{H}$ associated with the system. Consequently, in order to reconstruct a channel, we have to determine $d^2(d^2-1)$ real parameters [i.e., 12 numbers in the case of qubit ($d=2$)].

In what follows, we will assume that test states can be prepared on demand, perfectly. Nevertheless, the reconstruction of the channel $\mathcal{E}$ can be affected by the lack of required information, because each test state is represented by a finite ensemble of identically prepared test particles (e.g., qubits). Correspondingly, measurements performed at the output can result in an approximate estimation of transformed test states. This situation is typical for experiments—one cannot prepare an infinite ensemble of identically prepared particles; thus, frequencies of the measured outcomes are only approximations of probability distributions. Consequently, a straightforward reconstruction of output states $\rho_j^{(\text{out})}$ might lead to nonphysical conclusions about the action of the quantum channel. As a result, we can find a negative operator $\rho_j^{(\text{out})}$, or a channel $\mathcal{E}$, which is not CP.[1,2]

We will recall the method of maximum likelihood (MML) to perform an estimation of an unknown channel.[8] We will use this method to perform a reconstruction based on numerical simulation of the antiunitary universal-NOT gate [(U-NOT), the spin-flip operation].[9–14] This is a linear, not a CP map, and we will show how our estimation will result in the optimal physical approximations of the U-NOT operation. In order to demonstrate the power of this approach, we will also apply it to obtain an approximation of nonlinear quantum-mechanical maps: the so-called nonlinear polarization rotation (NPR)[15] and a highly nonlinear transformation $\rho \longrightarrow \rho^2$.

Having the result of the MML method, one may always analyze the question of how reliable this result is. As with any numerical method, MML may fail by falling to (or rather climbing up to) a nonlocal maxima. On the other hand, data provided by experiment may lead to nonphysical results. In this case, one shall not wonder that any physical approximation based on inconsistent data will result in an inconsistent map, e.g., a map that does not reproduce the experimental data perfectly. To deal with such cases, we will examine the resulting value of the MML functional. We will show that in certain cases it may be used as an indicator of whether the whole reconstruction scheme is consistent (e.g., whether we have a proper a priori knowledge about input test states, etc.).

This chapter is organized as follows: In Section 25.2, we present some basic facts about quantum states and quantum channels. This should help a reader who is not familiar with the problem of quantum state and process estimation (reconstruction) to understand some technical details that are presented later in the chapter. Then, in Section 25.3, we describe properties of single-qubit channels. In Section 25.4, we briefly introduce the method of maximum likelihood. In Section 25.5, we apply the MML for an estimation of the U-NOT gate, and in Section 25.6, we will present two physical approximations of nonlinear maps, namely, the nonlinear polarization rotation and the map $\rho \longrightarrow \rho^2$. In Section 25.7, we will examine how to use the value of the MML functional to verify consistency of the reconstruction scheme, namely, whether the prior knowledge about the input states is correct. In Section 25.8, we will summarize our results.

25.2 Quantum States and Quantum Channels

25.2.1 States of quantum systems and density operators

In quantum theory, a state of a physical system is described by a state vector or a density operator ρ that acts on a Hilbert space $\mathcal{H}$ (of a specific dimension d) associated with the system. Density operators are mathematical objects that allow us to describe states of physical systems that have not been prepared in a unique way. Specifically, a state of a quantum system that has been prepared in a unique way is described by a state vector $|\Psi\rangle$. One can associate with the state vector $|\Psi\rangle$ a density operator (projector) $\rho = |\Psi\rangle\langle\Psi|$. In many situations though, the preparation is not under full control and the system is prepared in one state from a specific set of states $\{|\Psi_j\rangle\langle\Psi_j|\}$ with corresponding probabilities p_j ($p_j \geq 0$ and $\sum_j p_j = 1$).

Such a "mixed" state of a quantum system is described by a density operator

$$\rho = \sum_j p_j |\Psi_j\rangle\langle\Psi_j|. \tag{25.1}$$

The density operator has to have nonnegative eigenvalues, and its trace has to be equal to unity, i.e., $\mathrm{Tr}\rho = 1$. The purity of the density operator can be quantified in terms of a von Neumann entropy

$$S = -\mathrm{Tr}\left(\rho \ln \rho\right). \tag{25.2}$$

If we assume that the state vectors $|\Psi_j\rangle$ are mutually orthogonal ($|\Psi_j\rangle$ are basis vectors of the Hilbert space $\mathcal{H}$ such that $\langle\Psi_k|\Psi_j\rangle = \delta_{j,k}$), then the eigenvalues of the density operator (25.1) are equal to probabilities p_j and the von Neumann entropy (25.2) reads

$$S = -\sum_{j=0}^{d} p_j \ln p_j. \tag{25.3}$$

It is clear now that if the system has been prepared in a unique (pure) state $|\Psi_k\rangle$, i.e., $p_k = \delta_{j,k}$, then the entropy S is equal to zero. On the other hand, if the system is prepared in a mixture of mutually orthogonal states $|\Psi_j\rangle$ (basis vectors of d-dimensional Hilbert space $\mathcal{H}$) with the equal probability, i.e., $p_j = 1/d$, then the von Neumann entropy of the corresponding (total) mixture achieves its maximum value $S = \ln d$.

Alternatively, a concept of a mixed state can appear when our quantum system is considered to be quantum-mechanically correlated with another quantum system. To be more specific, let us consider two quantum systems A and B with Hilbert spaces $\mathcal{H}_A$ ($\dim\mathcal{H}_A = d_A$) and $\mathcal{H}_B$ ($\dim\mathcal{H}_B = d_B$), respectively. The two Hilbert spaces are spanned by basis vectors $|\Psi_k^{(A)}\rangle$ and $|\Psi_k^{(B)}\rangle$, respectively. For simplicity, let us assume that $d_A = d_B$. Now we consider a pure bipartite state $|\Phi\rangle_{AB}$ ($S_{AB} = 0$). If this pure state can be expressed in a factorized form, i.e., as a product of two state vectors,

$$|\Phi\rangle_{AB} = |\phi^{(A)}\rangle_A \otimes |\phi^{(B)}\rangle_B, \tag{25.4}$$

then the two systems under consideration are each in a pure state. The two systems are not correlated, which can be seen by performing a partial trace over one of the subsystem (say, B) and from the state vector $|\Phi\rangle_{AB}$, we obtain a pure state $|\phi^{(A)}\rangle_A$ ($S_A = 0$). On the other hand, if the bipartite state $|\Phi\rangle_{AB}$ cannot be expressed in a factorized form (25.4) then we say, the two subsystems are mutually entangled (quantum-mechanically correlated). Any such state can be expressed in a form of the so-call Schmidt decomposition

$$|\Phi\rangle_{AB} = \sum_j \lambda_j |\xi_j^{(A)}\rangle_A |\xi_j^{(B)}\rangle_B, \tag{25.5}$$

where λ_j are nonnegative real numbers satisfying the normalization condition $\sum_j \lambda_j^2 = 1$ and $|\xi_j^{(A)}\rangle_A$ and $|\xi_j^{(B)}\rangle_B$ are orthonormal states from $\mathcal{H}_A$ and $\mathcal{H}_B$, respectively. From the Schmidt decomposition, it follows that states of the two subsystems are described the density operators

$$\rho_{(A)} = \sum_j \lambda_j^2 |\xi_j^{(A)}\rangle\langle\xi_j^{(A)}|, \qquad \rho_{(B)} = \sum_j \lambda_j^2 |\xi_j^{(B)}\rangle\langle\xi_j^{(B)}|, \qquad (25.6)$$

from where we see that eigenvalues of the density operators $\rho_{(A)}$ and $\rho_{(B)}$ are equal to λ_j^2. Consequently, von Neumann entropies of both subsystems are identical and equal to $S_A = S_B = -\sum_j \lambda_j^2 \ln \lambda_j^2$.

25.2.2 State tomography

A state of a quantum system can be completely reconstructed when mean values of all system observables (the so-called quorum) are known from a measurement. A typical example would be a state of a qubit—a spin-1/2 particle. This systems has a two-dimensional Hilbert space, and in general, its state can be described by a density operator

$$\rho = \frac{1}{2}\left(I + \vec{n}\cdot\vec{\sigma}\right), \qquad (25.7)$$

where $\vec{\sigma} = \{\sigma_x, \sigma_y, \sigma_z\}$, with $\sigma_{x,y,z}$ being Pauli matrices. Components of the vector $\vec{n} = \{n_x, n_y, n_z\}$ are the mean values of the Pauli operators in the state ρ, e.g., $n_x = \text{Tr}(\rho\sigma_x)$. These mean values can be determined when probabilities $p_\pm$ on the eigenvectors $|\pm x\rangle$ of the operator σ_x (we have $\sigma_x|\pm x\rangle = \pm|\pm x\rangle$) are obtained from a measurement. In order to obtain a probability distributions $p_\pm = \langle \pm x|\rho|\pm x\rangle$, a measurement on a sufficiently large (infinite) ensemble of identically prepared qubits has to be performed. Once the measurement of all observables belonging to the quorum is done, the complete tomography (reconstruction) of the state of a qubit can be performed.

It may happen that, for one or an other reason, mean values of just a subset of observables from the quorum are known from the measurement. In this case, a complete tomography of the state cannot be performed because of missing information. Nevertheless, a partial reconstruction (an estimation) of the measured state can be performed. One of possible approaches in this case is the state reconstruction based on the principle of maximum entropy, as introduced by Jaynes (see, e.g., Ref. 7, and references therein). This method works as follows: Let us assume that a subset $\{G_\nu\}$ of system operators has been measured in the experiment and that the corresponding mean values are

$$\langle G_\nu \rangle = \text{Tr}(\rho G_\nu). \qquad (25.8)$$

The question is how to select, from a set of all density operators ρ_G that satisfy the condition (25.8), the one that is considered to be the "best" estimation of the true density operator. Jaynes has suggested that among all density operators ρ_G that satisfy the condition (25.8), one should select the operator ρ_{est} that maximizes the von Neumann entropy.

In a real physical situation, it happens that infinite ensembles of identically prepared particles are not available. Therefore, strictly speaking, the reconstruction methods that are based on the "perfect" knowledge of probability distributions are not applicable. In these cases, when only a finite set of registered "clicks" are obtained from measurements, one should use more appropriate methods. In particular, the method based on Bayesian inversion (for details, see Ref. 16, and references therein) results in a reliable estimation of states of quantum systems. Alternatively, the method of maximum likelihood can be employed (see, e.g., Ref. 17, and references therein).

25.2.3 Quantum processes

If a quantum system is isolated from its environment, then it evolves unitarily, its dynamics is described by a unitary operator U, and the density operator is transformed in a usual way $\rho^{(out)} = U\rho U^\dagger$. On the other hand, if the system interacts with its environment, then its dynamics is described by a transformation

$$\rho^{(out)} = \mathcal{E}[\rho], \qquad (25.9)$$

where $\mathcal{E}$ is a map that fulfills three conditions:

(1) The output density operator $\rho^{(out)}$ has to describe a physical state. $\text{Tr}(\mathcal{E}[\rho])$ is the probability that the process represented by $\mathcal{E}$ occurs when the system is originally in the state ρ. Therefore, $0 \leq \text{Tr}(\mathcal{E}[\rho]) \leq 1$ for all input states ρ. That is, the map $\mathcal{E}$ is positive.

(2) $\mathcal{E}$ is a convex-linear map on a set of density operators. Therefore, if the initial state ρ is expressed as a mixture $\rho = \sum_j p_j \rho_j$ (such that $p_j \geq 0$ and $\sum_j p_j = 1$), then

$$\mathcal{E}\left[\sum_j p_j \rho_j\right] = \sum_j p_j \mathcal{E}[\rho_j]. \qquad (25.10)$$

(3) The map $\mathcal{E}$ has to complete positive.

In order to understand the condition of complete positivity, let us consider two systems A and B. The state of this composite system is described by a density operator ρ_{AB}. Let us assume that this state is transformed in the following way: the subsystem A undergoes a transformation described by the map $\mathcal{E}_A$ while the subsystem B is left unchanged; that is, we have

$$\rho_{AB}^{(out)} = (\mathcal{E}_A \otimes I_B)[\rho_{AB}], \qquad (25.11)$$

where I_B is the identity operator on system B. The condition of complete positivity requires that the operator $\rho_{AB}^{(out)}$ is positive for any input state ρ_{AB}.

We note, that not all positive maps $\mathcal{E}$ are also completely positive. A typical example of a positive but not completely positive operator is the so-called U-NOT gate $\mathcal{E}_{\text{U-NOT}}$, which is an antiunitary transformation, that "flips" an input state of a qubit $|\Psi\rangle = \alpha|0\rangle + \beta|1\rangle$ into its antipode $|\Psi^\perp\rangle = \beta^*|0\rangle - \alpha^*|1\rangle$, such that $\langle\Psi|\Psi^\perp\rangle = 0$ for all possible input states $|\Psi\rangle$. It is easy to show the operation $\mathcal{E}_{\text{U-NOT}}$, which is supposed to act on a single qubit state as

$$|\Psi\rangle\langle\Psi| \to \mathcal{E}_{\text{U-NOT}}[|\Psi\rangle\langle\Psi|] = |\Psi^\perp\rangle\langle\Psi^\perp|, \qquad (25.12)$$

is positive because both the input and output of the channel (25.12) are positive density operators. In order to show that the operation is not completely positive, let us consider the action of the transformation $\mathcal{E}_{\text{U-NOT}} \otimes I$ on the singlet state $|\Xi\rangle_{AB}$, which we define as

$$|\Xi\rangle_{AB} = \frac{1}{\sqrt{2}}\left(|\Psi\rangle_A|\Psi^\perp\rangle_B - |\Psi^\perp\rangle_A|\Psi\rangle_B\right). \qquad (25.13)$$

This is a maximally entangled state that is $SU(2) \otimes SU(2)$ invariant and violates Bell inequalities (for more details see Ref. 1). The operator $\rho_{AB} = |\Xi\rangle\langle\Xi|$ has eigenvalues $\lambda = \{1,0,0,0\}$. The output state $\rho_{AB}^{(out)}$ of the action of the operator $\mathcal{E}_{\text{U-NOT}} \otimes I$ on the singlet state (25.13) has the eigenvalues $\lambda = \{-1/2, 1/2, 1/2, 1/2\}$. From here it follows that the operator $\rho_{AB}^{(out)}$ has the trace equal to unity but does not describe a physical state because its eigenvalues are negative. From above, it follows that the map $\mathcal{E}_{\text{U-NOT}}$ is positive but not complety positive (see also Section 25.5).

Every map $\mathcal{E}$ satisfies the three conditions specified above if and only if it can be expressed as

$$\mathcal{E}[\rho] = \sum_j E_j \rho E_j^\dagger \qquad (25.14)$$

via the set of operators $\{E_j\}$, which map the input Hilbert space to the output Hilbert space and satisfy the condition

$$\sum_j E_j^\dagger E_j = I. \qquad (25.15)$$

25.2.4 Processes tomography

In order to determine the channel $\mathcal{E}$, we have to find out the operators E_j that define this map via the decomposition [Eq. (25.14)]. In order to determine operators E_j, it is useful to express them via a set of "fixed" operators $\tilde{E}_k$, which form a basis for the set of all operators on the state space. Therefore, one can express arbitrary E_j

as a superposition of basis operators $\tilde{E}_k$

$$E_j = \sum_{k=0}^{d^2-1} e_{jk} \tilde{E}_k \qquad (25.16)$$

with e_{jk} being complex numbers. Now $\mathcal{E}$ can be expressed as

$$\mathcal{E}[\rho] = \sum_{mn=0}^{d^2-1} \chi_{mn} \tilde{E}_m \rho \tilde{E}_n^\dagger, \qquad (25.17)$$

where by construction the matrix $\chi_{mn} = \sum_j e_{jm} e_{nj}^*$ is a positive Hermitian. This matrix completely describes $\mathcal{E}$ (once the set of basis operators $\tilde{E}_j$ is fixed). In general, the matrix χ contains $d^2(d^2-1)$-independent real parameters. To see this, we remind ourselves that $\mathcal{E}$ implements transformation between $d \times d$ complex matrices (input and output states of the quantum system under consideration). Therefore, χ is a $d^2 \times d^2$ matrix. If we take into account that in and out states are described by Hermitian operators with the trace equal to unity [which results into d^2 constraints that are expressed by the completeness relation (25.15)], then we see that the matrix χ is specified by $d^2(d^2-1)$ real parameters. Complete reconstruction of a quantum channel is equivalent to determination of these parameters. For this, we need to have a capacity to prepare input test states and, after the action of the channel, to perform complete reconstruction of the output states.

From above, it follows that in order to perform a complete reconstruction of an unknown CP map $\mathcal{E}$, we need a set d^2 linearly independent test states $\{\rho_k\}_{k=0}^{d^2-1}$ that are transformed by the channel. The states $\rho_k^{(\text{out})} = \mathcal{E}[\rho_k]$ are supposed to be reconstructed (completely) and out of the correlations between the in and out states, ρ_k and $\rho_k^{(\text{out})}$, respectively, the map (channel) can be reconstructed.

Because the map $\mathcal{E}$ is linear, it can be reconstructed by the following inversion procedure: Let the set of operators $\{\rho_k\}_{k=0}^{d^2-1}$ form a linearly independent basis for the space of $d \times d$ matrices (density operators). This means that any $d \times d$ matrix can be written in a form of a linear combination of the basis operators ρ_k. Usually (see, e.g., Ref. 1), these basis operators are taken in the form $\rho_k = |k_1\rangle\langle k_2|$, where $\{k_j\}_{j=0}^{d-1}$ is an orthonormal basis set in the Hilbert space $\mathcal{H}$ of the system under consideration.* We note that the transformation of the operator $\rho_k = |k_1\rangle\langle k_2|$ can be determined via performing preparation and measurement of four projectors, namely,

$$|k_1\rangle\langle k_2| = |+\rangle\langle+| + i|-\rangle\langle-| - \frac{1+i}{2}\left(|k_1\rangle\langle k_1| + |k_2\rangle\langle k_2|\right), \qquad (25.18)$$

---

*We assume that k is represented by a pair $\{k_1, k_2\}$ such that $k = 0, \ldots, d^2 - 1$. We can think of k being expressed in a d-nary form via $k_1 k_2$.

where $|+\rangle = (|k_1\rangle + |k_2\rangle)/\sqrt{2}$ and $|-\rangle = (|k_1\rangle + i|k_2\rangle)/\sqrt{2}$, respectively. The linearity of the map $\mathcal{E}$ implies that it is sufficient to determine the output states $\mathcal{E}[|+\rangle\langle+|]$, $\mathcal{E}[|-\rangle\langle-|]$, $\mathcal{E}[|k_1\rangle\langle k_1|]$, and $\mathcal{E}[|k_2\rangle\langle k_2|]$ in order to determine the transformation on the basis operators ρ_k, i.e.,

$$\mathcal{E}[|k_1\rangle\langle k_2|] = \mathcal{E}[|+\rangle\langle+|] + i\mathcal{E}[|-\rangle\langle-|] - \frac{1+i}{2}\left(\mathcal{E}[|k_1\rangle\langle k_2|] + \mathcal{E}[|k_1\rangle\langle k_2|]\right). \tag{25.19}$$

Moreover, because the operators ρ_k form a basis, we have

$$\mathcal{E}[\rho_k] = \sum_{l=0}^{d^2-1} \lambda_{kl}\rho_l, \tag{25.20}$$

where the parameters λ_{kl} are determined by the measurement results in the operator basis, i.e.,

$$\lambda_{kl} = \mathrm{Tr}\left(\tilde{E}_k\, \mathcal{E}[\rho_l]\right), \tag{25.21}$$

where $\tilde{E}_k = \rho_k$. This is a usual choice of the basis operators $\tilde{E}_k$ because ρ_k are Hermitian operators; thus, they can play a role of genuine observables. To proceed, we use the expression

$$\tilde{E}_m \rho_k \tilde{E}_n^\dagger = \sum \beta_{mn,kl}\rho_l, \tag{25.22}$$

where $\beta_{mn,kl}$ are complex numbers that are determined from a simple algebraic relation between $\tilde{E}_m$ and ρ_k. If $\tilde{E}_m = \rho_m$, then given the fact that $\rho_x = |x_1\rangle\langle x_2|$ ($x = l, m, n, k$ with $0 \le x \le d^2 - 1$), we find

$$\beta_{mn,kl} = \delta_{l_1,m_1}\delta_{n_1,l_2}\delta_{m_2,k_1}\delta_{k_2,n_2}. \tag{25.23}$$

When we combine Eqs. (25.21) and (25.22), we obtain

$$\sum_{k=0}^{d^2-1}\sum_{m,n=0}^{d^2-1} \chi_{mn}\beta_{mn,kl}\rho_k = \sum_{j=0}^{d^2-1} \lambda_{lj}\rho_j. \tag{25.24}$$

From the linear independence of the operators ρ_k, it follows that for each k

$$\sum_{m,n=0}^{d^2-1} \chi_{mn}\beta_{mn,kl} = \lambda_{kl}. \tag{25.25}$$

This relation is a necessary and sufficient condition for the matrix χ to give the correct quantum channel $\mathcal{E}$. Formally, we can express χ as

$$\chi_{mn} = \sum_{j,k=0}^{d^2-1} \kappa_{mn,kl}\lambda_{kl}, \tag{25.26}$$

where the matrix κ is the inverse of the $d^4 \times d^4$ matrix β. From here, one obtains the reconstructed expression for the map $\mathcal{E}$ (for technical details, see, e.g., Ref. 1, and references therein).

25.3 Structure of Qubit Channels

As discussed in the previous section, (physical) quantum channels are described by linear trace-preserving CP maps $\mathcal{E}$ defined on a set of density operators.[1,2,18] Any qubit channel $\mathcal{E}$ can be imagined as an affine transformation of a three-dimensional Bloch vector $\vec{r}$ (representing a qubit state), i.e., $\vec{r} \to \vec{r}' = T\vec{r} + \vec{t}$, where T is a real 3×3 matrix and $\vec{t}$ is a translation.[18] This form guarantees that the transformation $\mathcal{E}$ is Hermitian and trace preserving. The CP condition defines (nontrivial) constraints on possible values of the involved parameters. In fact, the set of all CP trace-preserving maps forms a specific convex subset of all affine transformations. Representing the qubit states by four-dimensional vectors $\vec{v}_\rho = (1, \vec{r})$, where the first element corresponds to the normalization of the state ($\mathrm{Tr}\rho = 1$), one can express the action of the channel $\mathcal{E}$ in a more compact matrix form

$$\mathcal{E}[\rho] = \begin{pmatrix} 1 & \vec{0} \\ \vec{t} & T \end{pmatrix} \begin{pmatrix} 1 \\ \vec{r} \end{pmatrix} = \begin{pmatrix} 1 \\ \vec{t} + T\vec{r} \end{pmatrix}. \qquad (25.27)$$

In other words, the qubit channels form 4×4 matrices of the affine form.

The matrix T can be written in the so-called singular-value decomposition, i.e., $T = R_U D R_V$ with R_U, R_V corresponding to orthogonal rotations and $D = \mathrm{diag}\{\lambda_1, \lambda_2, \lambda_3\}$ being diagonal where λ_k are the singular values of T. This means that any map $\mathcal{E}$ is a member of a less parametric family of maps of the "diagonal form" $\Phi_\mathcal{E}$, i.e., $\mathcal{E}[\rho] = U \Phi_\mathcal{E}[V \rho V^\dagger] U^\dagger$, where U, V are unitary operators. The reduction of parameters is very helpful, and most of the properties (including complete positivity) of $\mathcal{E}$ are reflected by the properties of $\Phi_\mathcal{E}$. The map $\mathcal{E}$ is CP only if $\Phi_\mathcal{E}$ is. Let us note that $\Phi_\mathcal{E}$ is determined not only by the matrix D, but also by a new translation vector $\vec{\tau} = R_U \vec{t}$, i.e., under the action of the map $\Phi_\mathcal{E}$ the Bloch sphere transforms as follows: $r_j \to r'_j = \lambda_j r_j + \tau_j$.

A special class of CP maps are the unital maps, which transform the total mixture ($\rho = I/2$) into itself, $\mathcal{E}[I/2] = I/2$. In this case, $\vec{t} = \vec{\tau} = \vec{0}$, and the

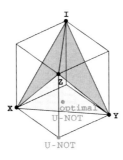

Figure 25.2 Unital CP maps are embeded in the set of all positive unital maps (cube). The CP maps form a tetrahedron with four unitary transformations in its vertices (extremal points) I, x, y, z corresponding to the Pauli σ matrices. The nonphysical U-NOT operation ($\lambda_1 = \lambda_2 = \lambda_3 = -1$) and its optimal completely positive approximation quantum U-NOT gate ($\lambda_1 = \lambda_2 = \lambda_3 = -1/3$) are shown (for more details, see Section 25.5).

corresponding map $\Phi_{\mathcal{E}}$, is uniquely specified by just three real parameters. The positivity of the transformation $\Phi_{\mathcal{E}}$ results into conditions $|\lambda_k| \leq 1$, while to fulfill the CP condition, we need the four inequalities $|\lambda_1 \pm \lambda_2| \leq |1 \pm \lambda_3|$ to be satisfied. These conditions specify a tetrahedron lying inside a cube of all positive unital maps. In this case, the extreme points represent four unitary transformations $I, \sigma_x, \sigma_y, \sigma_z$ (see Fig. 25.2).

25.4 Method of Maximum Likelihood

The MML is a general estimation scheme[17,19] that has already been considered for a reconstruction of quantum operations from incomplete data. It has been studied by Fiurášek and Hradil[20] and by Sacchi[21] (critically reviewed in Ref. 22). The task of the maximum likelihood in the process reconstruction is to find a map $\mathcal{E}$, for which the likelihood is maximal. By definition, we assume that the estimated map has to be CP. Let us now briefly describe the main idea of this method in more detail.

Given the measured data represented by couples ρ_k, F_k (ρ_k is one of the test states and F_k is a positive operator corresponding to the outcome of an individual measurement used in the kth run of the experiment), the likelihood functional is defined by

$$L(\mathcal{E}) = -\log \prod_{k=1}^{N} p(k|k) = -\sum_{k=1}^{N} \log \mathrm{Tr}\mathcal{E}[\rho_k] F_k , \qquad (25.28)$$

where N is the total number of measurement events (clicks) and $p(j|k) = \mathrm{Tr}\mathcal{E}[\rho_k] F_j$ is the conditional probability of using the test state ρ_k and observing the outcome F_j. The aim is to find a physical map $\mathcal{E}_{\mathrm{est}}$ that maximizes this function, i.e., $L(\mathcal{E}_{\mathrm{est}}) = \max_{\mathcal{E}} L(\mathcal{E})$. This variational task is usually performed numerically.

Our approach (see also Ref. 8) is different from those described in Refs. 20–22 in the way that we find the maximum of the functional defined in Eq. (25.28). The parametrization of $\mathcal{E}$, itself, guarantees the trace-preserving condition. Hence, only the CP condition must be checked separately during the numerical maximization. Instead of using the Lagrange multipliers (and increasing thereby the number of parameters for the numerical procedure), we introduce the CP condition as an external boundary for a Nelder-Mead simplex algorithm. The maximization is performed by the Mathematica 5.0 built-in function with following parameters:

(1) **Method = Nelder-Mead.** We chose the Simplex algorithm because it gives the most stable results with the smallest memory requirements.

(2) **Shrink ratio and contract ratio = 0.95.** These parameters are normally taken somewhere around 0.5. Their values close to unity induce a rather slow "cooling" of the process and prevents from falling into a local maxima. Thus, the global minimum can be determined reliably. The price to pay is, as usual, a longer time for a numerical search.

(3) **Reflect ratio** = **1.5.** This parameter is bigger than the standard choice but helps us to enhance the probability of finding the global maximum.

In what follows, we will analyze different examples of nonphysical operations and reconstructions that were obtained via the MML method.

25.5 U-NOT gate

Let us assume the U-NOT (spin-flip) operation discussed in Section 25.2. This operation corresponds to the inversion of the Bloch sphere (see Fig. 25.3). It is well known that this inversion preserves angles (which is related to the scalar product $|\langle\Phi,\Psi\rangle|$ of rays). Therefore, by the arguments of the Wigner theorem, the ideal spin-flip operation must be implemented either by a unitary or by an antiunitary operation. Unitary operations correspond to proper rotations of the Bloch sphere, whereas antiunitary operations correspond to orthogonal transformations with determinant -1. The spin-flip operation is an antiunitary operation, i.e., it is not completely positive (see Section 25.2).

Because the tensor product of an antilinear and a linear operator is not correctly defined, the U-NOT gate cannot be applied to a qubit while the rest of the world is governed by unitary evolution.[†] Therefore, the ideal (perfect) U-NOT gate that would flip a qubit initially prepared in an arbitrary state does not exist.

Obviously, if the state of the qubit is known, then we can always perform a flip operation. In this situation, the classical and quantum operations share many similar features, because the knowledge of the state is classical information, which can be manipulated according to the rules of classical information processing (e.g., known states can be copied, flipped, etc.). But, the universality of the operation is lost. That is, the gate that would flip the state $|0\rangle \to |1\rangle$ is not able to perform a flip $|(0\rangle + |1\rangle)/\sqrt{2} \to (|0\rangle - |1\rangle)/\sqrt{2}$.

Because it is not possible to realize a perfect U-NOT gate[9] that would flip an arbitrary (unknown) qubit state, it is of interest to study what *is* the best approximation to the perfect U-NOT gate. Here one can consider two possible scenarios. The first one is based on the measurement of input qubit(s)—using the results of an optimal measurement, one can manufacture an orthogonal qubit or any desired number of them. Obviously, the fidelity of the U-NOT operation in this case is equal to the fidelity of estimation of the state of the input qubit(s). The second scenario would be to approximate an antiunitary transformation on a Hilbert space of the input qubit(s) by a unitary transformation on a larger Hilbert space, which describes the input qubit(s) and ancillas.

It has been shown recently, that the best achievable fidelity of both flipping scenarios is the same.[10–12] That is, the fidelity of the optimal U-NOT gate is equal to the fidelity of the best state estimation performed on input qubits[25–27] (one might

---

[†]In fact, exactly this property makes the spin-flip operation so important in all criteria of inseparability for two-qubit systems.[23,24]

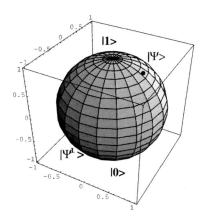

Figure 25.3 The state space of a qubit is the Bloch sphere. Pure states are represented by points on the sphere, while statistical mixtures are points inside the sphere. The U-NOT operation corresponds to the inversion of the sphere, since the states $|\Psi\rangle$ and $|\Psi^\perp\rangle$ are antipodes.

say, that in order to flip a qubit, we have to transform it into a bit). A more detailed description of the unitary transformation realizing the quantum scenario for the spin-flip operation can be found in Refs. 10–12. The experimental realization of the optimal U-NOT gate has been reported in Ref. 14. In this experiment, qubits were encoded in polarization states of photons.

As we have said, the U-NOT gate ($\mathcal{E}_{\text{U-NOT}} : |\Psi\rangle \to |\Psi^\perp\rangle$) is associated with the inversion of the Bloch sphere, i.e., $\vec{r} \to -\vec{r}$, which is not a CP map. It represents an nonphysical transformation specified by $\lambda_1 = \lambda_2 = \lambda_3 = -1$. The distance (see Fig. 25.2) between this map and the tetrahedron of completely positive maps is extreme, i.e., it is the most nonphysical map among linear transformations of a single-qubit and can be performed only approximatively. A quantum "machine" that optimally implements an approximation of the U-NOT is represented by the map $\tilde{\mathcal{E}}_{\text{U-NOT}} = \text{diag}\{1, -1/3, -1/3, -1/3\}$. The distance[28] between the U-NOT gate and its optimal physical approximation reads

$$d(\tilde{\mathcal{E}}_{\text{U-NOT}}, \mathcal{E}_{\text{U-NOT}}) = \int_{\text{states}} d\rho \, \text{Tr}|(\mathcal{E}_{\text{U-NOT}} - \tilde{\mathcal{E}}_{\text{U-NOT}})[\rho]| = \frac{1}{3}. \quad (25.29)$$

This channel corresponds to the best CP approximation of the U-NOT gate (the spin-flip operation).

For the estimation of the U-NOT gate via our numerical Gedanken experiment, we chose as inputs six test states—the eigenstates of $\sigma_x, \sigma_y, \sigma_z$. The data are generated as (random) results of three projective measurements $\sigma_x, \sigma_y, \sigma_z$ applied in order to perform the output-state reconstruction. In order to analyze the convergence of the method, we have performed the reconstruction for a different number of detected events (clicks) and compared the distance between the original map $\mathcal{E}_{\text{U-NOT}}$ and the estimated map $\mathcal{E}_{\text{est}}$. The result is plotted in Fig. 25.4, where we can see

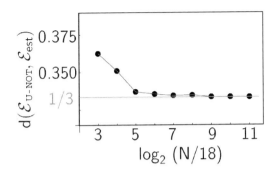

Figure 25.4 The distance $d(\mathcal{E}_{\text{U-NOT}}, \mathcal{E}_{\text{est}})$ as a function of the number of measured outcomes N in a logarithmic scale. We used six input states (eigenvectors of $\sigma_x, \sigma_y, \sigma_z$) and measured $\sigma_x, \sigma_y, \sigma_z$. The distance converges to the theoretical value $1/3$ that corresponds to the optimal U-NOT.

that the distance converges to $1/3$, as calculated in Eq. (25.29). For $N = 100 \times 18$ clicks (i.e., each measurement is performed 100 times per particular input state), the algorithm leads us to the map

$$\mathcal{E}_{\text{est}} = \begin{pmatrix} 1 & 0 & 0 & 0 \\ -0.0002 & -0.3316 & -0.0074 & 0.0203 \\ 0.0138 & -0.0031 & -0.3334 & 0.0488 \\ -0.0137 & 0.0298 & -0.0117 & -0.3336 \end{pmatrix}, \quad (25.30)$$

which is very close $[d(\mathcal{E}_{\text{est}}, \tilde{\mathcal{E}}_{\text{U-NOT}}) = 0.0065]$ to the best approximation of the U-NOT operation, i.e., $\tilde{\mathcal{E}}_{\text{U-NOT}} = \text{diag}\{1, -1/3, -1/3, -1/3\}$.

We conclude that, for sufficiently large N, the MML reconstruction gives us the same result as a theoretical prediction derived in Ref. 10. In order to illustrate the power of this approach, we will find approximations of nonlinear quantum mechanical transformations.

25.6 Nonlinear Transformations

Quantum mechanics is intrinsically linear theory, and therefore, nonlinear transformations cannot be considered as legitimate quantum maps. Nevertheless, one can consider a toy model in which one looks for optimal physical approximations of nonlinear quantum transformations. Such maps are sometimes used as an "effective" description of specific processes. In what follows, we will consider two specific examples: The nonlinear polarization rotation and the transformation-generating powers of an input density operator.

25.6.1 Nonlinear polarization rotation

Let us consider a nonlinear transformation of a qubit defined[15]

$$\mathcal{E}_\theta[\rho] = e^{i\frac{\theta}{2}\langle\sigma_z\rangle_\rho \sigma_z} \rho\, e^{-i\frac{\theta}{2}\langle\sigma_z\rangle_\rho \sigma_z}. \quad (25.31)$$

Unlike the U-NOT gate, this map is nonlinear. Four test states are not sufficient to allow us to determine the action of nonlinear maps. Therefore, in our Gedanken experiment, we have to consider all possible input states (that cover the whole Bloch sphere) as test states, but still we use only three different measurements performed on the outcomes of the channel. These measurement data are sufficient for the channel reconstruction.

First, we present an analytic derivation of a physical approximation of $\mathcal{E}_\theta$. This approximation is the closest physical map $\tilde{\mathcal{E}}_\theta$, i.e., $d(\tilde{\mathcal{E}}_\theta, \mathcal{E}_\theta) = \min$. The map $\mathcal{E}_\theta$ exhibits two symmetries: the continuous U(1) symmetry (rotations around the z-axis) and the discrete σ_x symmetry (rotation around the x-axis by π). The physical approximation $\tilde{\mathcal{E}}_\theta$ should possess these properties as well. Exploiting the two symmetries, possible transformations of the Bloch vector are restricted as follows $x \to \lambda x, y \to \lambda y, z \to pz$. In the process of minimization, the parameter p behaves trivially and is equal to unity. This means that $\tilde{\mathcal{E}}_\theta$ is of the form $\mathcal{E}_\lambda = \text{diag}\{1, \lambda, \lambda, 1\}$. Our task is to minimize the distance $d(\mathcal{E}_\theta, \mathcal{E}_\lambda) = \int d\rho |\mathcal{E}_\theta[\rho] - \mathcal{E}_\lambda[\rho]|$ in order to find the physical approximation $\tilde{\mathcal{E}}_\theta$ (i.e., the functional dependence of λ on θ).

We plot the parameter λ that specifies the best physical approximation of the NPR map in Fig. 25.5. Also in Fig. 25.5, we also present a result of the maximum likelihood estimation of the NPR map based on a finite number of measurements. Here, for every point (θ), the nonlinear operation was applied to 1800 input states, which have been chosen randomly (via a Monte Carlo method so they uniformly cover the whole Bloch sphere). These input states have been transformed according to the nonlinear transformation (25.31). Subsequently, simulations of random projective measurements have been performed. With these experimental data, a

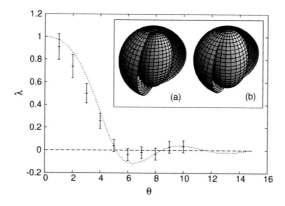

Figure 25.5 We present analytical as well as numerical results of an approximation of a nonlinear map $\mathcal{E}_\theta$ for different values of the parameter θ (measured in radians). The numerical ("experimental") results shown in the graph in terms of a set of discrete points with error bars are obtained via the MML. The theoretical approximation $\tilde{\mathcal{E}}_\theta$ of the nonlinear NPR map is characterized by the parameter λ that is plotted (solid line) in the figure as a function of the parameter θ. In the inset (a) is the Bloch sphere transformation for $\theta = 3$ obtained by MML and in (b) the same transformation obtained analytically.

maximization procedure was performed as described in the previous section. The resulting approximation specified by a value of λ (error bars shown in the graph represent the variance in outcomes for subsequent runs with different test states, but the same procedure parameters) transforms the original Bloch sphere, as it is shown in an inset for particular value θ = 3. Figure 25.5(a) corresponds to the result obtained by MML, and Fig. 25.5(b) has been obtained via analytic calculations. We see that the original Bloch sphere is transformed into an ellipsoid, one axis of which is significantly longer than the remaining two axes, which are of a comparable length. The mean of these two lengths corresponds to the parameter λ that specifies the map. We conclude that the MML is in excellent agreement with our analytical calculations.

25.6.2 Powers of density operator: $\rho \longrightarrow \rho^2$

Similar to the previous example, the map $\rho \longrightarrow \rho^2$ is intrinsically nonlinear and we have to define the map for all possible inputs. The most general state of a qubit can be written as $\rho(\vec{r}) = 1/2(I + \vec{r}\cdot\vec{\sigma})$, where $\vec{\sigma} = (\sigma_x, \sigma_y, \sigma_z)$ are Pauli matrices and $\vec{r}$ is a real vector and $\vec{r} \leq 1$. In such a notation, we can define the map under consideration as follows:

$$\mathcal{E}[\rho] = \frac{1}{2}\left(I + |\vec{r}|\ \vec{r}\cdot\vec{\sigma}\right) = |\vec{r}|\rho + (1 - |\vec{r}|)\frac{I}{2}. \tag{25.32}$$

Because the action of the map cannot be written as a matrix independent of ρ times ρ, the map is not linear. Hence, it is a very interesting example to study reconstruction schemes, because it is does not change the typical test states (pure states and complete mixture state).

Because of the high symmetry of the map, one would expect that the best physical approximation is a contraction of the whole Bloch sphere

$$\mathcal{E}[\rho]_{\text{physical}} = k\rho + (1-k)\frac{I}{2} \tag{25.33}$$

with $k \leq 1$. Indeed, the result of the MML method shows within the precision given by the finite number of test states, a result of the expected form

$$\mathcal{E}[\rho]_{\text{physical}} = 0.85\,\rho + (1.00 - 0.85)\frac{I}{2}. \tag{25.34}$$

Even though here we have considered only the second power of the density operator ρ, using the same arguments one can approximate channels that generate an arbitrary power of the original density operator. In addition, taking into account all symmetries associated with the transformation $\rho \longrightarrow \rho^k$, one can derive an analytical expression for its physical (CP) approximation.

25.7 Analysis of Results

With the three examples presented in previous section, we have demonstrated the power of the MML method to find physical approximations of operations that are truly nonphysical. In reality (when analyzing data of real experiments), one always expects to have physical operations. However, if the data indicate a nonphysical operation, this may be a consequence of errors in experiment, wrong interpretation of data, but also a failure of the reconstruction method.

To rule out the last possibility, one has always to precisely analyze the outcome of the procedure. By performing the maximization numerically, the functional itself is a rather simple function of the input data. The only challenges are boundary conditions imposed by the CP condition. These may cause the search engine to "stick" in a point, which is not a local maxima of the functional, but is confined by the boundary conditions. This case is easy to detect by calculating the CP condition of the resulting operation and to check to see if the result is on the boundary of the CP maps. If so, it is worthwhile to run the maximization procedure again with different starting conditions. However, for parameters of the maximization procedure specified in Section 25.3 for every testing case, the MML resulted in the proper maxima at the first attempt.

If the result of the MML method is correct and the resulting operation is yet on the boundary of CP maps, there is strong evidence that the incoming data were biased by some kind of errors. To analyze this problem more closely, we must take into account, not only the resulting operations, but also the value of L in Eq. (25.28). This value defines the logarithm of the probability to obtain, for specified input states (used in the MML method as test states), the same results as the experimental results. For proper data, this probability should be comparable to the probability of a sequence of measured data produced directly by the reconstructed operation. However, for improper data (in our examples, these data are produced by nonphysical operations), the reconstructed operations may reproduce these data with a much smaller probability.

We define the $L_{\text{data}}(\mathcal{E})$ to be the value of the functional (25.28) for the resulting approximation. For the same set of test states as used in the original experiments (denoted by ρ_k), we perform a Gedanken experiment. We apply the reconstructed operation on every such state and then apply the measurement in the same direction as in the original experiment (the resulting positive operators we denote as $\bar{F}_k$). Then, we define

$$L_{\text{test}}(\mathcal{E}) = -\sum_{k=1}^{N} \log \text{Tr} \mathcal{E}[\rho_k] \bar{F}_k. \qquad (25.35)$$

The same procedure may be repeated sufficiently many times to obtain a typical value of the functional [calculated as the mean of all runs and denoted by $\bar{L}_{\text{test}}(\mathcal{E})$] and the typical variance of this value $\sigma[L_{\text{test}}(\mathcal{E})]$.

Table 25.1 Values of the functional L for different examples of nonphysical operations. For the U-NOT gate, it is clear that the probability to obtain a sequence similar to the sequence given by the experimental data is much lower than to obtain a typical sequence of data produced by the reconstructed operation. For the other two examples, the difference is not significant

	$L_{\text{data}}(\mathcal{E})$	$\bar{L}_{\text{test}}(\mathcal{E})$	$\sigma[L_{\text{test}}(\mathcal{E})]$
U-NOT	-1371	-1472	6.79
NPR	-1423	-1426	12.64
$\rho \longrightarrow \rho^2$	-1343	-1335	17.44

In Table 25.1, results of the calculations for three examples presented in this chapter are shown. As one can clearly see, for the physical approximation of the U-NOT gate, the difference between $L_{\text{data}}(\mathcal{E})$ and $\bar{L}_{\text{test}}(\mathcal{E})$ is rather large, showing clear evidence that the input data originated in a nonphysical operation. This is, however, not the case for the nonlinear polarization rotation and for the transformation $\rho \longrightarrow \rho^2$. In these cases, the typical sequence (as a whole) of the experimental data has a comparable probability to appear as any other sequence produced by the reconstructed operation. Thus, we may conclude that this method gives us a partial tool (a necessary condition) to identify data that would lead to nonphysical operations.

25.8 Conclusions

In this chapter, we have shown that the method of maximum likelihood can be efficiently used for derivation of physical approximations of nonphysical maps (both non-CP linear maps as well as nonlinear quantum-mechanical transformations). We have applied this method for approximating qubit transformations (the U-NOT gate, the nonlinear polarization rotation, and the map $\rho \longrightarrow \rho^2$). We have analyzed the resulting operations and provided a tool to detect the quality of the input data.

Finally, we note that in this chapter we have considered an idealized situation when the input states of test particles can be perfectly prepared (i.e., the action of the initial-state preparator is totally known). Certainly, this is an approximation of a real situation, when test states are prepared with a finite precision. This additional source of uncertainty has to be taken into account in realistic estimation procedures of quantum channels.

Acknowledgments

This work was supported, in part, by the European Union projects QAP, HIP, and QUROPE, by the Slovak Academy of Sciences via the project CE-PI and the APVV project.

References

1. Nielsen, M.A., and Chuang, I.L., *Quantum Computation and Quantum Information*, (Cambridge University Press, Cambridge, 2000).

2. Preskill, J., *Lecture Notes on Physics 229: Quantum Theory of Information and Computation* (1998), available at http://www.theory.caltech.edu/people/preskill/.

3. D'Ariano, G.M., and Lo Presti, P., "Quantum tomography for measuring experimentally the matrix elements of an arbitrary quantum operation," *Phys. Rev. Lett.*, **86**, 4195–4198, 2001.

4. Poyatos, J.F., Cirac, J.I., and Zoller, P., "Complete characterization of a quantum process: The two-bit quantum gate," *Phys. Rev. Lett.*, **78**, 390–393 (1997).

5. Bužek, V., "Reconstruction of Liouvillian superoperators," *Phys. Rev. A*, **58**, 1723–1727, 1998.

6. Bužek, V., Adam, G., and Drobný, G., "Reconstruction of Wigner functions on different observation levels," *Ann. Phys. (N.Y.)*, **245**, 37–97, 1996.

7. Bužek, V., "Quantum tomography from incomplete data via MaxEnt principle," *Quantum Estimations: Theory and Experiment*, pp. 189–234, Springer-Verlag, Berlin (2004).

8. Ziman, M., Plesch, M., Bužek, V., and Štelmachovič, P., "Process reconstruction: From unphysical to physical maps via maximum likelihood," *Phys. Rev. A*, **72**, 022106-1-5, 2005.

9. Bechmann-Pasquinucci, H., and Gisin, N., "Incoherent and coherent eavesdropping in the six-state protocol of quantum cryptography," *Phys. Rev A*, **59**, 4238–4248, 1999.

10. Bužek, V., Hilery, M., and Werner, R.F., "Optimal manipulations with qubits: Universal-NOT gate," *Phys. Rev. A*, **60**, R2626–R2629, 1999.

11. Bužek, V., Hillery, M., and Werner, R.F., "Universal-NOT gate," *J. Mod. Opt.*, **47**, 211–232, 2000.

12. Rungta, P., Bužek, V., Caves, C.M., Hillery, M., and Milburn, G.J., "Universal state inversion and concurrence in arbitrary dimensions," *Phys. Rev. A*, **64**, 042315-1-13, 2001.

13. Gisin, N., and Popescu, S., "Spin flips and quantum information for antiparallel spins," *Phys. Rev. Lett.*, **83**, 432–435, 1999.

14. DeMartini, F., Bužek, V., Sciarrino, F., and Sias, C., "Experimental realization of the quantum universal NOT gate," *Nature*, **419** (6909) 815–818, 2002.

15. Vinegoni, C., Wegmuller, M., Huttner, B., and Gisin, N., "Measurement of nonlinear polarization rotation in a highly birefringent optical fibre using a Faraday mirror," *J. Opt. A*, **2**, 314–318, 2000.

16. Bužek, V., and Derka, R., "Quantum observations," *Coherence and Statistics of Photons and Atoms*, J. Peřina, Ed., pp. 198–261, Wiley, New York (2001).

17. Hradil, Z., Řeháček, J., Fiurášek, J., and Ježek, M., "Maximum likelihood methods in quantum mechanics," *Quantum Estimations: Theory and Experiment*, p. 63, Springer-Verlag, Berlin (2004).

18. Ruskai, M.B., Szarek, S., and Werner, E., "An analysis of completely positive trace-preserving maps on $M-2$," *Lin. Alg. Appl.* **347**, 159–187, 2002.

19. Fisher, R.A., "Theory of statistical estimation," *Proc. Cambridge Phil. Soc.*, **22**, 700–725, 1925.

20. Fiurášek, J., and Hradil, Z., "Maximum-likelihood estimation of quantum processes," *Phys. Rev. A*, **63**, 020101(R)–1-4, 2001.

21. Sacchi, M.F., "Maximum-likelihood reconstruction of completely positive maps," *Phys. Rev. A*, **63**, 054104–1-4, 2001.

22. Fiurášek, J., and Hradil, Z., "Comment on "Maximum likelihood reconstruction of CP maps", quant-ph/0009104," Los Alamos e-print arXiv quant-ph/0101048.

23. Peres, A., "Separability criterion for density matrices," *Phys. Rev. Lett.*, **77**, 1413–1415, 1996.

24. Horodecki, P., "Separability criterion and inseparable mixed states with positive partial transposition," *Phys. Lett. A*, **223**, 333–339, 1997;
Horodecki, M., Horodecki, P., and Horodecki, R., "Inseparable two spin-1/2 density matrices can be distilled to a singlet form," *Phys. Rev. Lett.*, **78**, 574–577, 1997.

25. Holevo, A., *Probabilistic and Statistical Aspects of Quantum Theory*, North Holland, Amsterdam (1982).

26. Massar, S., and Popescu, S., "Optimal extraction of information from finite quantum ensembles," *Phys. Rev. Lett.*, **74**, 1259–1263, 1995.

27. Derka, R., Bužek, V., and Ekert, A., "Universal algorithm for optimal estimation of quantum states from finite ensembles via realizable generalized measurement," *Phys. Rev. Lett.*, **80**, 1571–1575, 1998, and references therein.

28. Ziman, M., Plesch, M., and Bužek, V., "Reconstruction of superoperators from incomplete measurements," *Foundations of Physics*, **36**, 127–156, 2006.

Chapter 26
Quantum Processing Photonic States in Optical Lattices

Christine A. Muschik, Inés de Vega, Diego Porras,
and J. Ignacio Cirac
Max-Planck-Institut für Quantenoptik, Germany

26.1 Introduction
    26.1.1 Quantum information processing with light
    26.1.2 Entanglement of light states using cold atoms in optical lattices
26.2 Important Concepts and Procedures
    26.2.1 Atom-light interface schemes
    26.2.2 Atom lattices
26.3 Entangling Gate for Photons
    26.3.1 Processing of atomic states
    26.3.2 Quantum gate protocol
    26.3.3 Performance of the quantum gate in the presence of noise
References

26.1 Introduction

In this chapter, we present a proposal for the realization of an entangling gate for photons. The protocol works deterministically and is experimentally feasible under realistic conditions. We start outlining the general problem of quantum information processing with photons and describe the current status in this field. Then we introduce the proposed scheme and present its main features. This part is followed by a review of the concepts and procedures employed in our proposal. Finally, we explain the quantum gate, in detail, in the last section of this chapter, where we also briefly discuss the performance of the gate in the presence of imperfections.

26.1.1 Quantum information processing with light

Photons play a key role in applications of quantum information because they are ideally suited to transmit quantum states between distant sites. This feature makes them indispensable for the realization of quantum communication protocols and the construction of quantum networks.

While being a good flying carrier of information, photons are naturally less adequate for storage than the long-lived matter degrees of freedom. For this reason, matter systems are employed as quantum memories. In order to combine both elements in a quantum network (i.e., photons as flying qubits and atoms as memory devices), light-matter interface schemes have been developed to transfer quantum states of light onto an atomic system. Some of these schemes are based on quantum-nondemolition interactions,[1,2] electromagnetically induced transparency,[3,4] and Raman processes.[5] Moreover, Raman processes have been used to entangle two distant atomic ensembles,[5] which is an important step to realize quantum repeaters[6,7] and, thus to solve the problem of losses and decoherence that exists in a photonic channel.

Apart from the transmission and storage of information, in a quantum network it is necessary to process quantum states. However, manipulation of quantum states of light is still challenging, since it requires the ability to create entanglement between photons. This task is difficult because photons are noninteracting particles, in principle. One possibility to entangle photons is to employ materials that possess optical nonlinearities, but so far, there are no materials available whose nonlinearities are strong enough to allow for short gate times. An alternative approach was put forward by Knill et al.,[8] which requires only linear optical operations and measurements. However, this scheme is probabilistic and not very efficient in practice.

In this chapter, we present a scheme for the realization of a deterministic entangling gate for photons by using an atom lattice. An atom lattice consists of an ensemble of cold atoms loaded in the periodic optical potential created by a standing wave. In order to use the lattice as a quantum register for quantum computation, it is necessary to prepare it in a Mott insulating phase, in which the number of atoms in each site of the potential (i.e., in each of its minima) is approximately constant, and set equal to 1.[9,10]

In the scheme we describe here, the photonic input state is mapped to a collective atomic state following light-matter interface schemes.[1-5] Once the input state is transferred to a collective state of the atomic system, the desired gate operation is performed by means of controlled atomic interactions. These controlled interactions or collisions between atoms are basic elements for quantum processing within a lattice and have been already used experimentally to create and manipulate highly entangled states, such as, for instance, Cluster states.[11-15] As will be explained later, performing collisions between atoms in a controlled way requires the ability to transport them within the lattice, depending on their internal state. The resulting entangled atomic state is then released back to the photonic channel, and the protocol results in a deterministic two-qubit gate for photons. In that way, the scheme combines the advantages of two successful experimental techniques that have been recently demonstrated, namely, the techniques of quantum memories and repeater schemes on the one hand, and the ability to manipulate neutral atoms in a very clean and controllable environment on the other hand. Moreover, it shows that atoms in optical lattices are not only suited to store quantum information, but also to process it at the same time.

26.1.2 Entanglement of light states using cold atoms in optical lattices

The gate operation realized in our scheme transforms the photonic input state

$$|\Psi^{\text{in}}\rangle_L = \alpha|0\rangle_L + \beta|1\rangle_L + \gamma|2\rangle_L,$$

consisting of a a superposition of 0-, 1-, or 2-photon Fock states with corresponding coefficients α, β, and γ, into the output state

$$|\Psi^{\text{out}}\rangle_L = \alpha|0\rangle_L + i\beta|1\rangle_L + \gamma|2\rangle_L.$$

A phase i is applied in case the parity of the photon number n is odd ($n = 1$), while no phase is applied in case of even parity ($n = 0$ or $n = 2$). Together with one-qubit rotations, this entangling operation is sufficient for universal quantum computation.[16]

As described above, the quantum gate operation is implemented by mapping the input light state to an atomic ensemble, manipulating the resulting atomic state, and reconverting it back to a light state. More specifically, we consider an atomic ensemble of N identical atoms possessing two internal states $|a\rangle$ and $|b\rangle$, which may refer to two different hyperfine ground states. All atoms are assumed to be initially prepared in state $|a\rangle$. The mapping of the light state to the ensemble can be done by employing a quantum memory protocol that maps each incoming photon to a collective atomic excitation. In this way, the one-photon Fock state $|1\rangle_L$ results in the collective atomic state $|1\rangle_A = \sum_{j=1}^{N} f_j |a\rangle_1 \ldots |b\rangle_j \ldots |a\rangle_N$ with $\sum_j^N |f_j|^2 = 1$. This state is a superposition of all possible N particle product states containing 1 atom in $|b\rangle$ and $N - 1$ atoms in $|a\rangle$ and represents one atomic excitation that is delocalized over the whole ensemble. The state containing two atomic excitations is defined in an analogous fashion. More generally, photonic Fock states $|n\rangle_L$ are mapped to collective atomic states with n excitations $|n\rangle_A$. The initial atomic state is therefore given by $|\Psi^{\text{in}}\rangle_A = \alpha|0\rangle_A + \beta|1\rangle_A + \gamma|2\rangle_A$.

In the next step, the quantum gate operation has to be applied to this state. As will be explained below, a CNOT operation between two particles can be implemented by inducing a controlled collision between them. However, when employing this tool, we face the challenge of efficiently implementing a nonlinear operation between collective states, having only local interactions at hand. The following lien of thought illustrates this difficulty. Using a naive approach, every atom has to interact with all other atoms in the ensemble, which would require $\mathcal{O}(N^2)$ operations. In a more sophisticated approach, one could choose a messenger atom that interacts with all the other atoms in the sample. This procedure would still require $\mathcal{O}(N)$ controlled collisions, which is not experimentally feasible either. In contrast, our scheme requires only $\mathcal{O}(N^{1/3})$ interactions. This striking reduction is achieved by projecting the input state, which is shared by all N atoms in the ensemble, to a single qubit, which can be accessed and manipulated directly and easily.

In order to perform the entangling gate, four kinds of basic operations are needed, which are all within the experimental state of the art and do not require addressability of individual atoms. The first is the application of external laser pulses, in order to transfer population between the two internal atomic states in which the information is encoded. Second, external magnetic fields have to be applied that produce a state-dependent single particle rotation. In other words, the magnetic field produces a phase shift that depends on the particular internal state. The other two operations, namely, state-dependent transport and cold collisions, are discussed in Section 26.2.2. A detailed presentation of the set of operational steps that have to be performed in order to process the desired entangling gate is given in Sections 26.3.1 and 26.3.2.

26.2 Important Concepts and Procedures

This section provides an introduction to the quantum optical procedures adopted in our scheme. We explain how light states can be mapped to collective states of atomic ensembles and review some of the main aspects concerning atom lattices as well as their use for quantum computation.

26.2.1 Atom-light interface schemes

Any device realizing a light-matter quantum interface relies on a coherent state transfer in which, ideally, the quantum state of light and the atomic state are swapped. Such a device cannot rely on any classical strategy, because any approach that involves the measurement of a quantum state and the subsequent reconstruction from the obtained classical data is fundamentally limited by the no-cloning theorem.

Typically, the faithful transfer of a quantum state of light to an atomic system requires strong coupling between these systems. One possibility to achieve this condition is to use a high-finesse optical cavity. Another possibility is to make use of the collective enhancement to be found in atomic ensembles. This effect originates from the fact that the initial light state excites a collective mode in the atomic ensemble, which is shared by all particles in the sample. The mapping of the incoming light field to collective atomic excitations can be achieved by adding a strong classical field to the signal field, which is used to manipulate and control the propagation of the signal field and its interaction with the atoms. The original photonic state can be retrieved later on by a similar procedure.

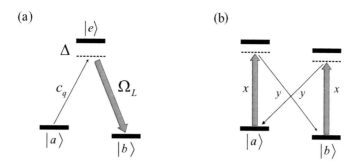

Figure 26.1 Level schemes used in quantum memory protocols. Thick arrows represent classical fields. Quantum fields are indicated by thin arrows. (a) Λ configuration. Details are given in the text. (b) Atomic structure employed in light-matter interfaces based on a quantum nondemolition interaction.

In recent years, significant progress has been made toward the realization of an efficient light-matter quantum interface, including atomic ensembles at room temperature as well as cold samples. There are two main directions: methods based on a Λ configuration [Fig. 26.1(a)], and techniques using a quantum nondemolition interaction [Fig. 26.1(b)]. In the following, we explain the basic principles of these schemes by outlining the mapping of a quantum state of light to an atomic ensemble employing a Λ configuration. The quantun nondemolition scheme is summarized later in this section.

Let us consider a uniform optically thick cloud of N identical atoms with an internal level structure as shown in Fig. 26.1(a). $|e\rangle$ denotes an excited state, while $|a\rangle$ and $|b\rangle$ are metastable ground states and may, for instance, correspond to hyperfine states of Alkali atoms. All atoms in the ensemble are assumed to be initially prepared in state $|a\rangle$. The light field corresponding to the input state to be mapped onto the atomic ensemble has wave vector $\mathbf{q}$ and is described by the destruction operator $c_\mathbf{q}$. It couples the levels $|a\rangle$ and $|e\rangle$, while the transition $|e\rangle \to |b\rangle$ is assisted by a strong classical field with wave vector $\mathbf{k}_L$ and Rabi frequency Ω_L. We assume that selection rules prohibit transitions between the levels $|a\rangle$ and $|b\rangle$. If a photon of the quantum field is absorbed by an atom and subsequently reemitted into the classical field, then a transition between the two atomic ground states is induced thus creating an atomic excitation $\sigma^\dagger = |b\rangle\langle a|$ in the ensemble.

For large detunings Δ the excited state $|e\rangle$ can be adiabatically eliminated, yielding an effective coupling between $\sigma^\dagger$ and the quantum light field. In the interaction picture, the Hamiltonian is given by

$$H = \sum_{j,\mathbf{k}} g_\mathbf{k} \left[\sigma_j^\dagger c_\mathbf{k} e^{i(\mathbf{k}-\mathbf{k}_L)\mathbf{r}_j + i(\omega-\omega_L)t} + \text{h.c.}\right],$$

$$g_k = \frac{\Omega_L}{2\Delta}\sqrt{\frac{\hbar\omega_k}{2\epsilon_0 V}}(\boldsymbol{\epsilon}_\mathbf{k} \cdot \mathbf{d}_{ae}),$$

(26.1)

where $j = 1, \ldots, N$ labels the atoms. ω_L is the frequency of the classical field, while ϵ_k and d_{ae} are the polarization vectors and the dipole matrix element for the $|a\rangle - |e\rangle$ transition, respectively. We will refer to this dynamic as coherent interaction. Other radiative processes are omitted in the description above and will be referred to as incoherent interactions.

Because the incident quantum field has wavevector $\mathbf{q}$, only terms describing the coupling to this light mode need to be considered. For simplicity, we consider the case $\mathbf{q} = \mathbf{k}_L$. Therefore, the Hamiltonian reduces to

$$H = \sum_j g_{\mathbf{k}_L} \left(\sigma_j^\dagger c_{\mathbf{k}_L} + \text{h.c.} \right). \tag{26.2}$$

We can now define a collective mode

$$S_{\mathbf{k}_L}^\dagger = \frac{1}{\sqrt{N}} \sum_j \sigma_j^\dagger. \tag{26.3}$$

As stated above $\sigma_j^\dagger$ excites the jth atom from $|a\rangle$ to $|b\rangle$. As the summation is taken over all atoms, $S_{\mathbf{k}_L}^\dagger$ describes an atomic excitation that is delocalized over the whole ensemble. By expressing the Hamiltonian in terms of the collective mode, we obtain

$$H = g_{\mathbf{k}_L} \sqrt{N} \left(S_{\mathbf{k}_L}^\dagger c_{\mathbf{k}_L} + \text{h.c.} \right). \tag{26.4}$$

Note that the effective coupling is given by $g_{\mathbf{k}_L} \sqrt{N}$, which can be high for large particle numbers. This enhancement is not present in the incoherent part of the interaction, which scales on the single-particle level. This leads to a collectively enhanced signal-to-noise ratio. A rigorous treatment of this effect can be found elsewhere.[17] The Hamiltonian $H \sim S_{\mathbf{k}_L}^\dagger c_{\mathbf{k}_L} + \text{h.c.}$ describes an interspecies beam-splitter-like interaction. For simplicity we assume that propagation effects do not have to be taken into account. Given that only a very small fraction of the atomic population in state $|a\rangle$ is transferred to state $|b\rangle$, atomic operators fulfill bosonic commutation relations (within the Holstein-Primakoff approximation), such that $[S_{\mathbf{k}_L}, S_{\mathbf{k}_L}^\dagger] = \sum_i (|a\rangle_i \langle a| - |b\rangle_j \langle b|)/N \simeq 1$. Under these assumptions, the solution to the Heisenberg equations is given by the unitary evolution,

$$\begin{aligned} c_{\mathbf{k}_L}(t) &= \cos(g_{\mathbf{k}_L}\sqrt{N}t) c_{\mathbf{k}_L}(0) + \sin(g_{\mathbf{k}_L}\sqrt{N}t) S_{\mathbf{k}_L}(0), \\ S_{\mathbf{k}_L}(t) &= \cos(g_{\mathbf{k}_L}\sqrt{N}t) S_{\mathbf{k}_L}(0) - \sin(g_{\mathbf{k}_L}\sqrt{N}t) c_{\mathbf{k}_L}(0), \end{aligned} \tag{26.5}$$

where $c_{\mathbf{k}_L}(0)$ and $S_{\mathbf{k}_L}(0)$ refer to the initial photonic and atomic state, respectively, and the phase of the laser was chosen such that $i\Omega^* g_{\mathbf{k}_L}^* = |\Omega g_{\mathbf{k}_L}|$. These equations show that the quantum state of light can be transferred to the atomic state. In the ideal case, both states can be swapped.

Now we describe a memory scheme using the level structure shown in Fig. 26.1(b). It can be understood as two superimposed Λ schemes. However, we will outline the basic mapping mechanism from a slightly different perspective. We assume an atomic ensemble interacting with a pulse of light propagating along $\hat{z}$. The atomic ensemble is assumed to be spin polarized along $\hat{x}$, while the pulse of light consists of a strong coherent $\hat{x}$-polarized component and a copropagating quantum field in $\hat{y}$ polarization. The relevant atomic property used for the mutual exchange of quantum information with the field is the collective spin of the atomic ensemble J. The $\hat{x}$ component of the collective spin J is a macroscopic quantity, because the atomic ensemble is strongly polarized along $\hat{x}$. Therefore, it is described by a C number and commuting observables can be assigned to the deviation of the collective atomic spin in the $\hat{y}$ and $\hat{z}$ direction by identifying $x_A = J_y/J_x$ and $p_A = J_z/J_x$. As signal field, we consider a coherent light state, which is described by its quadratures x_L and p_L. The off-resonant scattering interaction between atoms and light is assumed to be well within the dispersive regime such that absorptive processes can be neglected. Moreover, it is assumed that the collective spin experiences only small deviations from its alignment along $\hat{x}$. In this limit, the interaction can be described by a Hamiltonian of the form $H \sim p_A p_L$. The light field experiences a Faraday rotation depending on the alignment of the collective spin, while the dipole interaction imposes a Stark-shift on the atomic magnetic sublevels, which causes a rotation of the collective spin depending on the helicity of the light field. This mutual influence leads to the creation of entanglement between atoms and light field. The corresponding input-output relations for atomic and photonic quadratures are given by[1,2]

$$\begin{aligned} x_A^{\text{out}} &= x_A^{\text{in}} + \kappa p_L^{\text{in}}, \\ p_A^{\text{out}} &= p_A^{\text{in}}, \\ x_L^{\text{out}} &= x_L^{\text{in}} + \kappa p_A^{\text{in}}, \\ p_L^{\text{out}} &= p_L^{\text{in}}, \end{aligned} \quad (26.6)$$

where κ is a constant describing the coupling between atoms and light. These equations show that p_L is mapped onto the atomic x quadrature. Mapping of the photonic x quadrature to the atomic system can be achieved by measuring x_L via homodyne detection on the transmitted light and correspondingly applying a feedback on the atomic p quadrature. In this way, every coherent state can be transferred to the collective state of an atomic ensemble. Because the set of coherent states is complete, arbitrary quantum states can be stored using this method.

26.2.2 Atom lattices

Once the photonic input state is transferred to an atomic ensemble in an optical lattice, it needs to be processed in a controlled way, before it is released back to the radiation field. As noted in the introduction, the processing of quantum states in a lattice is based mainly on two types of operations, namely, state-dependent transport and controlled collisions between the atoms. In the following, we briefly explain the basic principles of an atom lattice, as well as these two basic operations that will later be used to perform a deterministic entangling gate on collective atomic states.

26.2.2.1 Preliminaries

A one-dimensional atom lattice is created by loading a cold gas of neutral atoms into the field of a standing wave. The addition of standing waves along the other two directions of space gives rise to higher dimensional atom lattices.

The interaction of a standing wave with the atoms can be of a dissipative type, arising due to the absorption of photons by the atoms and subsequent spontaneous emission, and of a conservative type, produced by the interaction of the light field with the induced dipole moment of the atom. The dissipative part of the interaction can be used to cool, whereas the conservative is used to trap. In order to create an atom lattice, it is necessary to maximize the conservative interaction with respect to dissipation, which is done by choosing the detuning of the lasers that produce the standing wave, δ, large enough.

The conservative interaction produces an energy level shift of the atom (the Stark shift), which depends on the amplitude of the standing wave field at each point of space. The spatial dependency of the shift in the potential energy produces the so-called dipolar force, which in the case of detuning $\delta > 0$ leads the atoms to the minima of the standing wave.[18] In other words, the interaction between the standing waves and the atoms gives rise to an optical potential, and the atoms tend to stay in its minima. For a three-dimensional lattice created by six beams of wave vector $k = 2\pi/\lambda$, this potential has the form

$$V(x, y, z) = V_{0x} \sin^2(kx) + V_{0y} \sin^2(ky) + V_{0z} \sin^2(kz), \qquad (26.7)$$

with V_0 proportional to the dynamic polarizability and the strength of the lasers. Note that the separation between atoms in different lattice sites is $\lambda/2$, which precludes the possibility of individually addressing single atoms with optical lasers. Atoms in an optical lattice are usually described within the second quantized form by the so-called Bose-Hubbard Hamiltonian.[9,10] This Hamiltonian has two important terms, a term proportional to the parameter U, which describes the repulsion between two atoms at the same lattice site due to contact collisions, and a kinetic term proportional to the so-called tunneling matrix element J, which describes the possibility for the atoms to move from one lattice site to the next one by tunneling through the optical potential. One important thing about atom lattices is that they can be externally manipulated. For instance, the ratio J/U can be increased by decreasing the lattice depth V_0, thus allowing for more tunneling between different lattice sites. In a similar way, the scattering length a_s corresponding to a collision between two atoms can be increased with an external magnetic field which is tuned to a Feshbach resonance. Since $U \sim a_s$, the ratio J/U can be made arbitrarily small near this resonance. An adiabatic change between the regimes $J/U \gg 1$ and $J/U \ll 1$ gives rise to a quantum phase transition between a superfluid state, with atoms moving almost freely through the lattice, and fluctuating atom number at each lattice site, and a Mott insulator state, in which atoms are frozen to each lattice site.

A final remark in this section concerns the effects of the periodicity of the optical potential. Just as in the case for electrons in a solid, periodicity gives rise to a band structure for the atomic vibrational levels. At low-temperature conditions, and with a low filling factor (which means that the average number of atoms in each site is small), we can assume that atoms are living mainly in the lowest of these Bloch bands. Nevertheless, care should be taken when moving the atoms through the lattice by shifting the optical potential. One possibility is to make this shift adiabatically, such that atoms do not change their vibrational state moving to a different Bloch band. However, for computational purposes, it is sufficient to ensure that both the initial and final states after the lattice shift are in the ground state. In this respect, the intermediate vibrational states for the atom during the process do not matter.

26.2.2.2 State-dependent transport

A Mott insulator is a very interesting state to perform simulations of many solid state systems, such as for instance, magnetic Hamiltonians, as well as for performing quantum computation. In particular, a Mott state with one atom per site can be used to realize a quantum register. (For this reason, protocols have been put forward, which allow for the defect-free realization of such a quantum state.[19]) More specifically, this quantum register can be used to perform quantum computation by producing collisions between atoms in a coherent and controlled way.[11,12] The basic element to induce controlled collisions in a lattice is the ability to perform state-dependent transport. This can be done by considering two counterpropagating linearly polarized laser beams with an angle 2θ between their polarization axes. The resulting light field can be decomposed into two cirularly polarized standing waves σ^+ and σ^- with potentials $V_\pm \cos^2(kx \pm \theta)$. As noted above, the strength of the optical potential depends on the dynamical polarizability, which is proportional to the atomic dipole moment. Hence, selection rules for the optical transitions induced by the standing waves $\sigma^\pm$ can be used to create different trap potentials for different atomic levels. For instance, for alkali atoms ^{23}Na and ^{87}Rb which are the most commonly used,[13] the standing wave σ^+ couples the ground level $S_{1/2}$ $m_s = -1/2$ to two excited levels $P_{1/2}$ and $P_{3/2}$ with $m_s = 1/2$ [see Fig. 26.2(a)]. Each of these transitions has a detuning of a different sign, and therefore, their contributions to the Stark shift of the ground level with $m_s = -1/2$ also have a different sign. In fact, for a certain frequency, these contributions cancel out and, therefore, the standing wave σ^+ does not produce an optical potential for the ground

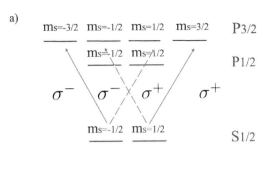

Figure 26.2 Diagram (a) represents the fine structure of alkali atoms, while (b) represents the hyperfine structure corresponding to the ground level $S_{1/2}$ for the alkali atoms ^{23}Na and ^{87}Rb, which have nuclear spin $I = 3/2$.

level $S_{1/2}$ with $m_s = -1/2$. Hence, this level is only affected by the potential V_-. A similar argument leads to the conclusion that the level of $S_{1/2}$ with $m_s = 1/2$ is only affected by V_+. Thus, if the potentials $V_\pm$ are shifted with respect to each other (which can be done by changing their relative phase θ), then the atomic levels $m_s = -1/2$ and $m_s = 1/2$ will be also displaced through the lattice in opposite directions.

It is important to note that quantum information is encoded in the hyperfine levels corresponding to the ground state $S_{1/2}$, because these states do not suffer much from decoherence. For the alkali atoms considered above, the hyperfine structure of the ground state consists of two manifolds with $F = 2$ and $F = 1$, respectively, each one consisting of $2F + 1$ levels with quantum numbers $m_F = -F, \ldots, F$ [see Fig. 26.2(b)]. Quantum information can be encoded, for instance, by identifying $|a\rangle = |F = 2, m_F = 2\rangle$ and $|b\rangle = |F = 1, m_F = -1\rangle$.[13] By representing $|a\rangle$ and $|b\rangle$ as a combination of the two ground state fine levels with $m_s = -1/2$ and $m_s = 1/2$, their corresponding level shifts can be written in terms of $V^\pm$ as $V_a = V_+$ and $V_b = V_+/4 + 3V_-/4$, where the factors follow from the corresponding Clebsch-Gordan coefficients. Hence, the hyperfine levels $|a\rangle$ and $|b\rangle$ are affected by different optical potentials and can therefore be shifted with respect to each other.

26.2.2.3 Collisional phases

State-dependent transport can be used to produce controlled collisions and, therefore entanglement between atoms in the lattice. Let us consider, for instance, two adjacent atoms placed at positions $\mathbf{r}_j$, with $j = 1, 2$, prepared in the state

$$\psi_j(\mathbf{r}_j, t) = \frac{1}{\sqrt{2}} \left[\psi_j^a(\mathbf{r}_j, t) + \psi_j^b(\mathbf{r}_j, t) \right] \tag{26.8}$$

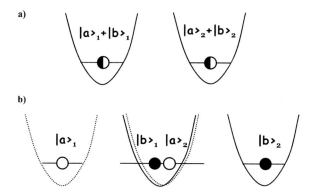

Figure 26.3 Moving potentials. (a) describes the situation at times $t = \pm\tau$, while (b) shows the situation at $-\tau < t < \tau$. The solid (dashed) line corresponds to the optical potential affecting the atomic level $|b\rangle$ ($|a\rangle$).

at some initial time $t = -\tau$. $\psi_j^\beta(\mathbf{r}_j, t)$ denotes the wave function of atom j in state $\beta = a, b$. Now a state-dependent shift of the potentials is produced by changing their relative phase θ. We assume here that θ is changed such that the potential V_b moves to the right while V_a moves to the left within a certain spatial direction, so that after a certain time the wave functions ψ_1^b and ψ_2^a start to overlap. Their interaction is described by a contact potential $V^{ba}(\mathbf{r}_1 - \mathbf{r}_2) = (4\pi a_s^{ba} \hbar^2/m)\delta^3(\mathbf{r}_1 - \mathbf{r}_2)$, where m is the atomic mass and a_s^{ba} is the scattering length corresponding to the collision. Subsequently the lattices are shifted back so that the atoms reach their original positions at $t = \tau$ (see Fig. 26.3 for a schematic representation). The phase accumulated in the time interval $[-\tau, \tau]$ is then given by

$$\phi^{ba} = \frac{1}{\hbar} \int_{-\tau}^{\tau} dt \Delta E^{ba}(t), \qquad (26.9)$$

where $\Delta E^{ba}(t)$ is the energy correction due to the collision, which can be calculated in perturbation theory. This energy is proportional to the scattering length a_s^{ba} and the overlap of the wave functions at each time.

In addition to the collisional phase, moving the atomic wave function ψ_i^β during a time interval $[-\tau, \tau]$ gives rise to a kinetic phase ϕ^β. Within the adiabatic limit, in which atoms remain in the motional ground state during the shift, the kinetic phase can be written as

$$\phi^\beta = \frac{m}{2\hbar} \int_{-\tau}^{\tau} dt (\dot{\mathbf{r}}^\beta)^2, \qquad (26.10)$$

where $\dot{\mathbf{r}}^\beta$ denotes the spatial trajectory followed by the wave function ψ^β of the atom. Adiabaticity requires the condition $|\ddot{\mathbf{r}}^\beta| \ll v_{osc}/\tau$, where v_{osc} is the mean velocity of atoms in the vibrational ground state.

Considering both collisional and kinetic phases, the internal states of the atoms after the controlled interaction are

$$|a\rangle_1|a\rangle_2 \to e^{-2i\phi^a}|a\rangle_1|a\rangle_2,$$
$$|a\rangle_1|b\rangle_2 \to e^{-i(\phi^a+\phi^b)}|a\rangle_1|b\rangle_2,$$
$$|b\rangle_1|a\rangle_2 \to e^{-i(\phi^a+\phi^b+\phi^{ba})}|b\rangle_1|a\rangle_2, \quad (26.11)$$
$$|b\rangle_1|b\rangle_2 \to e^{-2i\phi^b}|b\rangle_1|b\rangle_2.$$

Note that from now on atomic states are denoted in terms of their internal levels, while the representation in the spatial basis is omitted. This scheme realizes a fundamental two-qubit gate (a sign gate) by choosing the interaction time τ such that $\phi_{\rm col} = \phi^{ba} = \pi$.

26.3 Entangling Gate for Photons

In this section, the quantum gate protocol is presented. First, we explain how atomic states are processed in the lattice. This is done by introducing two methods, which are based on the collisional gate operation introduced in the previous section. Then, we explain the actual protocol and consider several sources of noise.

26.3.1 Processing of atomic states

As was shown above, controlled collisions can be used to manipulate a two-qubit state such that a phase π is introduced if the first qubit is in state $|b\rangle$ and the second atom is in state $|a\rangle$. This operation can be used to implement a CNOT gate between two qubits. The key procedure is to move control atoms in $|b\rangle$ with respect to a set of target atoms in $|a\rangle$, thus inducing collisions with the target qubits and transferring the atoms along its path to state $|b\rangle$. This tool is employed in two related methods, which lie at the heart of the proposed scheme and are introduced now.

Creation of a d dimensional structure from a $(d-1)$-dimensional one. Starting from a control atom in $|b\rangle$ and an ensemble of target atoms in $|a\rangle$, a line of atoms in $|b\rangle$ can be produced. To this end many CNOT operations are run in series, such that the control qubit in $|b\rangle$ acts successively on several target atoms in a row, which are accordingly transferred to state $|b\rangle$. The operational sequence, which has to be carried out is shown in Fig. 26.4. In Fig. 26.4(a) the atomic ensemble, which is initially prepared in state $|a\rangle$, is shown as grey sphere as the atomic sample occupies a spherical volume under typical experimental conditions. The black dot outside the sphere represents an isolated atom in state $|b\rangle$. We will explain later how such a configuration can be obtained. The atoms in the ensemble are the target qubits, while the isolated particle is the control qubit. First a $\pi/2$ pulse is applied to the atomic ensemble transferring all target atoms to the superposition state $(|a\rangle+|b\rangle)/\sqrt{2}$ as shown in Fig. 26.4(b). Then, the $|b\rangle$ lattice is displaced along

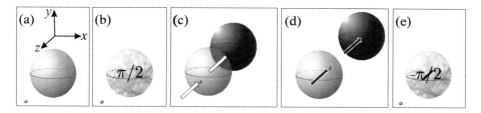

Figure 26.4 Creation of a line of atoms in $|b\rangle$. (a) A control atom in $|b\rangle$ is located outside an ensemble of atoms in $|a\rangle$. (b) A $\pi/2$ pulse is applied to the target atoms. (c) $|a\rangle$ and $|b\rangle$ components of the target qubits are separated spatially by a $|b\rangle$ lattice shift along $-\hat{z}$. (d) The $|b\rangle$ lattice is further displaced along $-\hat{z}$, such that the control atom in $|b\rangle$ interacts successively with the the $|a\rangle$ part of the target atoms along its path, each time leading to a collisional phase π. (e) The lattice is shifted such that all atoms are taken back to in their original positions and no collisional phases are accumulated. Then a $-\pi/2$ pulse is applied to the target qubits. Atoms that have interacted with the control atom are transferred to $|b\rangle$, while all other atoms in the ensemble are transferred back to $|a\rangle$.

$-\hat{z}$, such that $|a\rangle$ and $|b\rangle$ components of the target qubits are separated spatially. This separation step is shown in Fig. 26.4(c) and has to be performed such that no collisional phases are accumulated. This can be done by displacing the lattices first by half a lattice spacing along $\hat{x}$ (or $\hat{y}$) and then by a distance exceeding the length of the ensemble along $-\hat{z}$*. By displacing the $|b\rangle$ lattice further along $-\hat{z}$ in the next step [compare Fig. 26.4(d)], collisions between the control atom and target atoms encountered on its path through the ensemble are induced. In this way, the control atom interacts successively with all atoms along the indicated line, each time leading to a collisional phase $\phi_{\text{col}} = \pi$. The affected target atoms are therefore left in the state $(-|a\rangle + |b\rangle)/\sqrt{2}$. The last step is shown in Fig. 26.4(e). The initial positions are restored by shifting the $|b\rangle$ lattice back along $\hat{z}$, such that no collisional phases are accumulated. Then a $-\pi/2$ pulse is applied to the target atoms. Atoms, which are located on the path of the control qubit, have acquired a change of sign in the wave function. Accordingly these atoms are transferred to state $|b\rangle$, while all other atoms in the ensemble are transferred back to state $|a\rangle$.

In an analogous fashion, a plane of atoms in state $|b\rangle$ can be produced using a line of control atoms intead of a single qubit. To this end a line of control atoms in state $|b\rangle$ has to be swept through an ensemble of target atoms in $|a\rangle$. Following the steps described above, each control qubit in the line interacts with a line of atoms along its path, and finally all target atoms located in this plane are transferred to $|b\rangle$.

Mapping of collective excitations from an atomic ensemble of dimensionality d to a sample of dimensionality $d-1$. Figure 26.5 illustrates how excitations in a three-dimensional ensemble can be mapped to a plane of particles. Figure 26.5(a) shows

---

*Alternatively the lattice can be moved fast along $-\hat{z}$. In this way, the interaction time is very short and no appreciable collisional phase is accumulated. The displacement has to be done such that the atoms start and end up in their motional ground state.[14, 20]

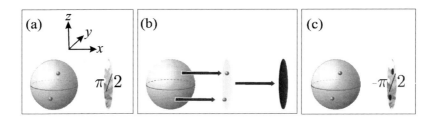

Figure 26.5 Mapping of excitations in the bulk to the plane. (a) A $\pi/2$ pulse is applied to the plane transferring all these t'qubits to the state $(|a\rangle + |b\rangle)/\sqrt{2}$. (b) The $|b\rangle$ lattice is shifted along $\hat{x}$ such that atoms in $|b\rangle$ in the bulk interact with the $|a\rangle$ part of the plane. The time spent after each single site displacement is chosen such that a phase π is accumulated if a collision occurs. Then the initial atomic positions are restored. Each target atom that is located on the path of a control atom is in state $(-|a\rangle + |b\rangle)/\sqrt{2}$. All other target atoms are still in state $(|a\rangle + |b\rangle)/\sqrt{2}$. (c) Finally the plane is subjected to a $-\pi/2$ pulse, which transfers most of the atoms back to $|a\rangle$. Only atoms, that suffered a collision are transferred to $|b\rangle$.

an ensemble of atoms, where two atoms are in state $|b\rangle$, indicated by black dots. Note that collective excitations are delocalized. The state $|2\rangle_A$ corresponds to a superposition of all N-particle states with two atoms (at specific locations) in state $|b\rangle$. In this sense Fig. 26.5 shows only one term of the superposition. Located next to the ensemble, a plane of particles in state $|a\rangle$ is shown. Both structures are separated in space but confined by the same optical lattice. Excitations in the ensemble can be mapped to excitations in the plane by using the atoms in the ensemble as control qubits acting on the target qubits in the plane, as described in the figure caption. Most atoms in the ensemble are in state $|a\rangle$ and do not induce any changes. However, in this example, two atoms in the ensemble are in state $|b\rangle$. Thus, collisions are induced if the $|b\rangle$ lattice is shifted along $\hat{x}$ and the target atoms hit by these control atoms along their paths through the plane are transformed to $|b\rangle$. In this way, atoms in $|b\rangle$ are projected from the bulk to the plane. More precisely, the procedure maps a state with n atoms in $|b\rangle$ the bulk to a state with n atoms in $|b\rangle$ in the plane, except if two atoms in $|b\rangle$ in the bulk are located in a line along $\hat{x}$, leaving the corresponding target atom in $|a\rangle$ (CNOT$^2 = 1$). In any case, an even/odd number of excitations is mapped to an even/odd number of excitations in the target object. This method allows us to reduce, stepwise, the dimensionality of the problem. Collective excitations can be mapped from a plane to a lane in an analogous fashion. Finally, excitations are mapped from a line to a single site ($d = 1$) and an odd number of excitations in the line transfers the target atom to state $|b\rangle$; whereas, in case of an even number of excitations, this atom is left in state $|a\rangle$. Thus, the parity information is encoded in the state of a single atom.

26.3.2 Quantum gate protocol

This subsection starts with the main idea behind the proposed protocol and gives a summary of the steps that have to be carried out in order to perform the desired

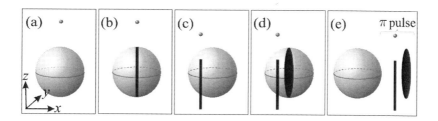

Figure 26.6 Initialization of the lattice. (a) A control atom in $|b\rangle$ is placed outside the ensemble. (b) The control qubit interacts successively with a row of target atoms in the ensemble, thus transferring them to state $|b\rangle$, as explained in Fig. 26.4. We obtain a line of atoms in $|b\rangle$, that is aligned along $\hat{z}$. (c) The line is separated from the ensemble along $-\hat{y}$. (d) The line of control qubits is now used to create a plane of atoms in $|b\rangle$. For this purpose, a $\pi/2$ pulse is applied to the ensemble, collisions are induced by a $|b\rangle$ lattice shift along $\hat{y}$ and a $-\pi/2$ pulse is applied to the bulk. Because each control atom in the line leads to a line of atoms in $|b\rangle$, that is aligned along $\hat{y}$, we obtain a plane in the $\hat{y}\hat{z}$ plane. (e) The plane is separated from the ensemble by a $|b\rangle$ lattice shift along $\hat{x}$, and a π pulse is applied, which leaves the atoms in the plane, the line, and the dot in state $|a\rangle$.

quantum gate by combining the methods introduced above. Furthermore the gate operation realized by this scheme is explained on the basis of its truth table.

26.3.2.1 Operation sequence

As mentioned above, the light input state is transferred to an atomic ensemble. The gate operation is performed on the collective atomic input state, which is thereafter converted back to light. The main idea of processing the delocalized atomic state is to map the collective excitations stepwise to structures of lower dimension. More specifically, collective excitations are first mapped from the three-dimensional Mott insulator to a plane of particles, then to a line, and finally, to a single atom, which can be directly manipulated. The whole scheme comprises two stages. During an initialization phase, the atoms are divided into four sets, namely, the bulk, a plane, a line, and a dot, which are confined by the same optical lattice but spatially separated. This setup has to be prepared once and can afterward be used many times to perform gates. In the second phase, the quantum gate protocol itself is performed. In the following, we explain first the preparation of the setup and then how the processing part is performed.

Preparation of the lattice. The initialization protocol is summed up in Fig. 26.6. First, a single collective excitation is created in the ensemble, for example, by mapping a one-photon Fock state to the atomic sample by means of a light-atom interface scheme.[1-5][†] To be more specific, after the absorption of the single photon, the

---

[†]This can, for instance, be done employing heralded single photons from an EPR source. An atomic excitation can also be created by means of a weak coherent field together with a postselecting photon detection.[3-5]

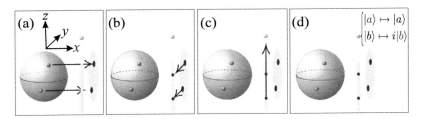

Figure 26.7 Quantum gate protocol transforming the input state $|\Psi^{\text{in}}\rangle_A = \alpha|0\rangle_A + \beta|1\rangle_A + \gamma|2\rangle_A$ into $|\Psi^{\text{out}}\rangle_A = \alpha|0\rangle_A + i\beta|1\rangle_A + \gamma|2\rangle_A$. (a) to (c) Excitations in the Mott insulator are successively mapped to structures of lower dimensionality resulting in a single atom being in state $|a\rangle/|b\rangle$ in case of an even/odd number of excitations in the Mott insulator. (d) A state-dependent phase is applied to the isolated particle such that $|1\rangle_A \mapsto i|1\rangle_A$. Subsequently steps (a) to (c) have to be reversed.

atomic ensemble is in a superposition of all possible N particle product states with one atom in $|b\rangle$ and all other atoms in $|a\rangle$. Now, a $|b\rangle$ lattice shift is applied such that the (single) atom in state $|b\rangle$ is separated from the bulk of atoms in $|a\rangle$. This is a global operation acting on all atoms at the same time but affecting only one atom (being in state $|b\rangle$). In the next step, method (I) is applied to create a line of atoms in $|b\rangle$, using the isolated atom as control qubit[‡]. Subsequently, this line is separated from the bulk and utilized to produce a plane of atoms in $|b\rangle$ employing the same method. Finally, the isolated atom, the line, and the plane are displaced by another global $|b\rangle$ lattice shift such that the constellation shown in Fig. 26.6(e) is obtained. Finally, a π pulse is applied to the plane, the line, and the dot transferring these atoms to state $|a\rangle$.

Quantum gate protocol. The quantum gate protocol is summarized in Fig. 26.7. Once the light state is transferred to the bulk, the resulting collective atomic excitations are mapped to the plane of atoms by means of method (II). These excitations are mapped from the plane to the line in the next step in an analogous fashion. The projection sequence is completed by applying method (II) a third time and finally, mapping the excitations from the line to the solitary particle. In this way, the parity of the number of excitations contained in the bulk is mapped to the dot, transferring the isolated atom to state $|b\rangle$ in case of one excitation, while it remains in state $|a\rangle$ otherwise. Now a state-dependent phase shift is applied to the dot such that a phase $\pi/2$ is introduced if the atom is in state $|b\rangle$. In this way, the phase i is only applied if an odd number of collective excitations has been initially present in the bulk and atomic states are transformed according to $|0\rangle_A \mapsto |0\rangle_A$, $|1\rangle_A \mapsto i|1\rangle_A$, $|2\rangle_A \mapsto |2\rangle_A$.

[‡]If the path of the control atom is located at the edge of the cloud, short lines are produced. This can be avoided by focusing the light field, which is used in the first step of the preparation, to the center of the Mott insulator along x and y. In this way, the light field acts on a cylindrical volume centered at the center of the Mott insulator, restricting the deviation in length. Note that this procedure does not require individual addressing.

The described mapping sequence results in an entangled state of the bulk, the plane, the line, and the dot. In order to apply a light-matter interface protocol, which transfers the atomic state back to a photonic channel, the initial atomic state (up to the introduced phase) has to be restored. To this end, the previous steps have to be reversed. After converting the final atomic state to light, the setup is left in the original state and can be used again to perform a quantum gate operation.

26.3.2.2 Truth table

Now the truth table corresponding to the protocol is considered. After mapping of the photonic state to the atomic ensemble, the bulk is in a state containing $n = 0$, 1, or 2 collective excitations. Remember that a collective atomic state $|n\rangle_A$ is a superposition of atomic states with n atoms in $|b\rangle$ and $N - n$ atoms in $|a\rangle$. All possible configurations are included in this superposition. We denote these possible initial atomic states of the bulk with n atoms in $|b\rangle$ located at certain lattice sites, corresponding to a configuration k by $|B_k^n\rangle_b$. $|P\rangle_p$ and $|L\rangle_l$ refer to the initial state of the plane and the line, respectively, where all atoms are in state $|a\rangle$. The three-step projection sequence of the protocol described above produces the map

$$|B_k^n\rangle_b |P\rangle_p |L\rangle_l |a\rangle_d \mapsto |B_k^n\rangle_b |P_k^{n'}\rangle_p |L_k^{n''}\rangle_l |a_n\rangle_d. \tag{26.12}$$

$|P_k^{n'}\rangle_p$ and $|L_k^{n''}\rangle_l$ denote the states of the plane and line after the excitations have been mapped and $|a_n\rangle_d$ describes the state of the dot with $a_0 = a_2 = a$ and $a_1 = b$. After applying the phase i state dependently and reversing all the previous steps, the protocol results in

$$|B_k^n\rangle_b |P\rangle_p |L\rangle_l |a\rangle_d \mapsto i^{n \bmod 2} |B_k^n\rangle_b |P\rangle_p |L\rangle_l |a\rangle_d. \tag{26.13}$$

Note that the initial atomic state under consideration features not only a superposition of atoms in $|b\rangle$ at different positions in the ensemble, but also a superposition of different positions of the plane, the line, and the dot. This is because the excitation created at the beginning of the initialization is delocalized and these positions depend on the initial position of the control atom in $|b\rangle$. However, for any term in the superposition, the final state differs only in a phase from the initial state. By adding the terms in Eq. (26.13) with respect to the positions of the excitations, k, and the positions of the plane, the line, and the dot, we obtain the desired quantum gate transformation. We remark that the protocol does not rely on a coherent superposition of the different positions of the dot, the line, and the plane. It can also be run using a mixed state.

26.3.3 Performance of the quantum gate in the presence of noise

The specific design of the protocol minimizes decoherence and is optimized such that imperfections have only little effect on the performance of the quantum gate despite the large number of particles involved. Now, we briefly estimate limitations on the gate fidelity, whereas detailed quantitative results will be presented elsewhere.[21]

Decoherence is suppressed, first of all, by providing short runtimes, such that mechanisms leading to decoherence do not have much time to act on and corrupt the quantum states, which are manipulated. The time required to perform the scheme is essentially given by the time needed to run the collisional steps, because population transfers and separations can be done much faster. Each collisonal step has to be performed along a whole ensemble length and therefore requires a time $t_{\text{int}} N^{1/3}$, where t_{int} is the time spent in a single collision. Remarkably, the three-dimensional problem scales like a one-dimensional one in time. This is because the task of scanning N particles in a three-dimensional lattice is accomplished by a one-dimensional projection scheme.

Moreover, cat states in the internal basis states are avoided completely, because they are extremely susceptible to decoherence. The collective states used in our scheme, in contrast, are very robust.

An important source of errors are transitions from $|a\rangle$ to $|b\rangle$ or to another trapped state affected by $|b\rangle$ lattice shifts. As noted before, hyperfine levels, in principle, do not suffer strong decoherence because their energy separation is very small, on the order of microwaves. Nevertheless, uncontrolled transitions between them can occur indirectly, via two-photon processes induced by the standing wave lasers. If such an event happens, for instance, during the initialization, it can lead to the creation of several lines and planes. In the following, we give an example how a judicious choice of atomic levels allows us to suppress this effect, while still being able to perform the operations that are necessary for the quantum gate. We consider an atom with nuclear spin $I = 1/2$. In this case, the hyperfine structure consists of two manifolds, one with $F = 1$, comprising three sublevels with quantum numbers $m_F = 1, 0, -1$, and one with $F = 0$ and $m_F = 0$. If we identify $|a\rangle = |F = 1, m_F = -1\rangle$ and $|b\rangle = |F = 1, m_F = 1\rangle$, then indirect transitions between those levels will be forbidden by selection rules. Although there will still be some transitions to $|F = 1, m_F = 0\rangle$ and $|F = 0, m_F = 0\rangle$, the atomic population in those levels is being depleted continuously when shifting the lattice. To be more specific, within a lattice shift, there is a certain value for the phase θ at which the optical potential for these levels vanishes, so that atoms in $|F = 1, m_F = 0\rangle$ and $|F = 0, m_F = 0\rangle$ will become untrapped and escape from the lattice.

Among the remaining noise mechanisms, the most important ones are imperfect population transfer and dephasing of quantum states[§] between two $\pi/2$ pulses. The

---

[§]Inhomogeneous background fields are an important source of dephasing because they lead to uncontrolled relative phases in $(|a\rangle + e^{i\beta}|b\rangle)/\sqrt{2}$ during the state-dependent transport. This effect can be avoided by reversing each lattice shift involving atomic collisions. More precisely, each shift involving collisions is first performed along a certain direction such that a phase $\pi/2$ is accumulated in each collision. Then, this lattice movement is reversed. Thus, each target atom that is located on the path of a control atom in

corresponding probability of error is proportional to the number of target atoms in the mapping steps $N^{2/3}$. This failure probability can be reduced by using an elongated atomic ensemble having a spatial extend L along the direction of the first lattice shift in the quantum gate protocol and a length $l < L$ along the other directions. In this case the probability of obtaining a wrong result is proportional to l^2.

The probability of error due to the remaining noise mechanism scales at worst like $N^{1/3}$ (i.e., proportional to the runtime of the protocol). First, we consider imperfections in the π pulse, which is performed at the end of the initialization of the lattice. Because an imperfect population transfer leaves atoms in a superposition state, the $|b\rangle$ lattice should be emptied as an additional step of the initialization after the π pulse. Occupation number defects are another source of errors. We only have to deal with empty lattice sites, because double occupied sites can be avoided by choosing low filling factors. Holes in the plane and the line lead to a wrong result, if they are located at specific sites that interact with an atom in $|b\rangle$ in the course of the processing protocol. The failure probability due to defects that are initially present in the Mott insulator are given by the probability for a single site to be unoccupied and does not depend on the size of the system. Holes can also be created as consequence of atomic transitions into untrapped states. This dynamical particle loss induces an error that scales like the duration of the gate, $N^{1/3}$. Another limiting factor is imperfect collision. The phase acquired in each lattice shift during the collisional steps may differ from $\phi_{col} = \pi$. However, as in the case of unoccupied lattice sites, the probability of obtaining a wrong result due to such an event is given by the probability on the single-site level. The fidelity of the scheme is also decreased by undesired collisional phases. The corresponding failure probability is proportional to $N^{1/3}$, because these phases are accumulated in one-dimensional operations each covering one ensemble length. Finally, kinetic phases acquired by the atoms during lattice shifts do not play a role in the proposed scheme. Employing the common technique for state-dependent transport, the nodes of two optical potentials forming standing waves are moved in opposite directions $V_{\pm}(x) = \cos^2(kx \pm \phi)$ for some wave vector k, spatial variable x, and angle ϕ. Lattice shifts therefore affect both atomic species in the same way and lead only to global phases of the resulting quantum states. In conclusion, we have shown that cold atoms in an optical lattice in a Mott insulator state can be used to perform a deterministic entangling operation on a photon state. The time required for the proposed scheme scales like the length of the Mott insulator and is robust against imperfections and decoherence during the course of the quantum gate.

$|b\rangle$ experienced two collisions and picked up a total phase of π. Because all lattice shifts are reversed, uncontrolled relative phases can be suppressed by applying a single π pulse that swaps levels $|a\rangle$ and $|b\rangle$ after each shift, before the atoms are transported back. In this way, both components acquire the same phase. A π pulse has to be applied once each time the motion of the atoms is reversed, that is only six times for the whole quantum gate protocol.

Acknowledgments

We thank Eugene Polzik for discussions and acknowledge support from the Elite Network of Bavaria (ENB) project QCCC, the EU projects SCALA and COVAQUIAL, the DFG-Forschungsgruppe 635, and Ministerio de Educacion y Ciencia EX-2006-0295.

References

1. Julsgaard, B., Sherson, J., Cirac, J.I., Fiurasek, J., and Polzik, E.S., "Experimental demonstration of quantum memory of light," *Nature*, **432**, 482–486, 2004.

2. Sherson, J., Julsgaard, B., and Polzik, E.S., "Determinsitic atom-light quantum interface," in *Adv. in Atomic, Molecular and Optical Phys.* **54**, Academic Press, New York (2006).

3. Chanelière, T., Matsukevich, D.N., Jenkins, S.D., Lan, S.-Y., Kennedy, T.A.B., and Kuzmich, A., "Storage and retrieval of single photons transmitted between remote quantummemories," *Nature*, **438**, 833–836, 2005.

4. Eisaman, M.D., Andre, A., Massou, F., Fleischhauer, M., Zibrov, A.S., and Lukin, M.D., "Electromagnetically induced transparency with tunable single-photon pulses," *Nature*, **438**, 837–841, 2005.

5. Chou, C.W., de Riedmatten, H., Felinto, D., Polyakov, S.V., van Enk, S.J., and Kimble, H.J.,"Measurement induced entanglement for excitation stored in remote atomic ensembles," *Nature* **438**, 828–832, 2005.

6. Dür, W., Briegel, H.-J., Cirac, J.I., and Zoller, P., "Quantum repeaters based on entanglement purification," *Phys. Rev. A*, **59**, 169–181, 1999.

7. Duan, L.-M., Cirac, J.I., Lukin, M., and Zoller, P., "Long distance quantum communication with atomic ensembles and linear optics," *Nature*, **414**, 413–418, 2001.

8. Knill, E., Laflamme, R., and Milburn, G.J., "A scheme for efficient quantum computation with linear optics," *Nature*, **409**, 46–52, 2001.

9. Jaksch, D., Bruder, C., Cirac, J.I., Gardiner, C.W., and Zoller, P., "Cold bosonic atoms in optical lattices," *Phys. Rev. Lett.*, **81**, 3108–3111, 1998.

10. Greiner, M., Mandel, O., Esslinger, T., Hänsch, T.W., and Bloch, I., "Quantum phase transition from a superfluid to a Mott insulator in a gas of ultracold atoms," *Nature*, **415**, 39–44, 2002.

11. Mandel, O., Greiner, M., Widera, A., Rom, T., Hänsch, T.W., and Bloch, I., "Controlled collisions for multi-particle entanglement of optically trapped atoms," *Nature*, **425**, 937–940, 2003.

12. Jaksch, D., Briegel, H.-J., Cirac, J.I., Gardiner, C.W., and Zoller, P., "Entanglement of atoms via cold controlled collisions," *Phys. Rev. Lett.*, **82**, 1975–1978, 1999.

13. Jaksch, D., and Zoller, P., "The cold atom Hubbard toolbox," *Ann. of Phys.*, **315**, 52–79, 2005.

14. Briegel, H.-J., Calarco, T., Jaksch, D., Cirac, J.I., and Zoller, P., "Quantum computing with neutral atoms," *J. Mod. Opt.*, **47**, 415–451, 2000.

15. Brennen, G.K., Caves, C.M., Jessen, P.S., and Deutsch, I.H., "Quantum logic gates in optical lattices," *Phys. Rev. Lett.*, **82**, 1060–1063, 1999.

16. Barenco, A., Bennett, C.H., Cleve, R., DiVincenzo, D.P., Margolus, N., Shor, P., Sleator, T., Smolin, J.A., and Weinfurter, H., "Elementary gates for quantum computation," *Phys. Rev. A*, **52**, 3457–3467, 1995.

17. Duan, L.M., Cirac, J.I., and Zoller, P., "Three-dimensional theory for interaction between atomic ensembles and free-space light," *Phys. Rev. A*, **66**, 023818, 2000.

18. Metcalf, H.J., and van der Straten, P., *Laser Cooling and Trapping*, Springer-Verlag, Berlin (1999).

19. Popp, M., García-Ripoll, J.-J., Vollbrecht, K.G.H., and Cirac, J.I., "Cooling toolbox for atoms in optical lattices," *New. J. Phys.*, **8**, 164, 2006.

20. Dorner, U., Calarco, T., Zoller, P., Browaeys, A., and Grangier, P., "Quantum logic via optimal control in holographoc dipole traps," *J. Opt. B*, **7**, 341–346, 2005.

21. Muschik, C.A., de Vega, I., Porras, D., and Cirac, J.I., in preparation.

Chapter 27
Strongly Correlated Quantum Phases of Ultracold Atoms in Optical Lattices

Immanuel Bloch
Johannes Gutenberg-Universität, Germany

27.1 Introduction
27.2 Optical Lattices
    27.2.1 Optical dipole force
    27.2.2 Optical lattice potentials
27.3 Bose-Hubbard Model of Interacting Bosons in Optical Lattices
    27.3.1 Ground states of the Bose-Hubbard Hamiltonian
    27.3.2 Double-well case
    27.3.3 Multiple-well case
    27.3.4 Superfluid to Mott insulator-transition
27.4 Quantum Noise Correlations
27.5 Outlook
References

27.1 Introduction

Ultracold quantum gases in optical lattices form almost ideal conditions to analyze the physics of strongly correlated quantum phases in periodic potentials. Such strongly correlated quantum phases are of fundamental interest in condensed matter physics, because they lie at the heart of topical quantum materials, such as high-T_c superconductors and quantum magnets, which pose a challenge to our basic understanding of interacting many-body systems. Quite generally, such strongly interacting quantum phases arise when the interaction energy between two particles dominates over the kinetic energy of the two particles. Such a regime can either be achieved by increasing the interaction strength between the atoms via Feshbach resonances or by decreasing the kinetic energy, such that eventually the interaction energy is the largest energy scale in the system. The latter can, for example, simply be achieved by increasing the optical lattice depth.

This chapter tries gives an introduction into the field of optical lattices and the physics of strongly interacting quantum phases. A prominent example hereof is the superfluid-to-Mott insulator transition,[1-5] which transforms a weakly interacting quantum gas into a strongly correlated many body system. Dominating interactions between the particles are in fact crucial for the Mott insulator transition and also for the realization of controlled interaction-based quantum gates,[6-8] of which several have been successfully realized experimentally.[9-11]

27.2 Optical Lattices

27.2.1 Optical dipole force

In the interaction of atoms with coherent light fields, two fundamental forces arise.[12,13] The so-called Doppler force is dissipative in nature and can be used to efficiently laser cool a gas of atoms and relies on the radiation pressure together with spontaneous emission. The so-called dipole force, on the other hand, creates a purely conservative potential in which the atoms can move. No cooling can be realized with this dipole force; however, if the atoms are cold enough initially, then they may be trapped in such a purely optical potential.[14,15]

How does this dipole force arise? We may grasp the essential points through a simple classical model in which we view the electron as harmonically bound to the nucleus with oscillation frequency ω_0. An external oscillating electric field of a laser **E** with frequency ω_L can now induce an oscillation of the electron resulting in an oscillating dipole moment **d** of the atom. Such an oscillating dipole moment will be in phase with the driving oscillating electric field, for frequencies much lower than an atomic resonance frequency and 180 deg out of phase for frequencies much larger than the atomic resonance frequency. The induced dipole moment again interacts with the external oscillating electric field, resulting in a dipole potential V_{dip} experienced by the atom[15-19]

$$V_{dip} = -\frac{1}{2}\langle \mathbf{dE}\rangle, \qquad (27.1)$$

where $\langle \cdot \rangle$ denotes a time average over fast oscillating terms at optical frequencies. From Eq. (27.1) it becomes immediately clear that for a red detuning ($\omega_L < \omega_0$), where **d** is in phase with **E**, the potential is attractive, whereas for a blue detuning ($\omega_L > \omega_0$), where **d** is 180 deg out of phase with **E**, the potential is repulsive. By relating the dipole moment to the polarizability $\alpha(\omega_L)$ of an atom and expressing the electric field amplitude E_0 via the intensity of the laser field I, one obtains for the dipole potential

$$V_{dip}(\mathbf{r}) = -\frac{1}{2\epsilon_0 c}\text{Re}(\alpha)I(\mathbf{r}). \qquad (27.2)$$

A spatially dependent intensity profile $I(\mathbf{r})$ can therefore create a trapping potential for neutral atoms.

For a two-level atom a more useful form of the dipole potential may be derived within the rotating wave approximation, which is a reasonable approximation provided that the detuning $\Delta = \omega_L - \omega_0$ of the laser field ω_L from an atomic transition frequency ω_0 is small compared to the transitions frequency itself, $\Delta \ll \omega_0$. Here, one obtains[15]

$$V_{dip}(\mathbf{r}) = \frac{3\pi c^2}{2\omega_0^3} \frac{\Gamma}{\Delta} I(\mathbf{r}), \qquad (27.3)$$

with Γ being the decay rate of the excited state. Here, a red- detuned laser beam ($\omega_L < \omega_0$) leads to an attractive dipole potential and a blue-detuned laser beam ($\omega_L > \omega_0$) leads to a repulsive dipole potential. By simply focusing a Gaussian laser beam, this can be used to attract or repel atoms from an intensity maximum in space (see Fig. 27.1).

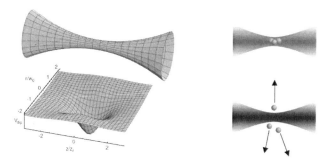

Figure 27.1 (a) Gaussian laser beam together with corresponding trapping potential for a red-detuned laser beam. (b) A red- detuned laser beams leads to an attractive dipole potential, whereas a blue detuned laser beam leads to a repulsive potential (c).

For such a focused gaussian laser beam, the intensity profile $I(r, z)$ is given by

$$I(r, z) = \frac{2P}{\pi w^2(z)} e^{-2r^2/w^2(z)}, \qquad (27.4)$$

where $w(z) = w_0(1 + z^2/z_R^2)$ is the $1/e^2$ radius, depending on the z coordinate, $z_R = \pi w^2/\lambda$ is the Rayleigh length and P is the total power of the laser beam.[20] Around the intensity maximum a potential depth minimum occurs for a red-detuned laser beam, leading to an approximately harmonic potential of the form

$$V_{dip}(r, z) \approx -V_0 \left[1 - 2\left(\frac{r}{w_0}\right)^2 - \left(\frac{z}{z_R}\right)^2\right]. \qquad (27.5)$$

This harmonic confinement is characterized by radial ω_r and axial ω_{ax} trapping frequencies $\omega_r = (4V_0/mw_0^2)^{1/2}$ and $\omega_z = (2V_0/mz_R^2)$.

Great care must be taken to minimize spontaneous scattering events, because they lead to heating and decoherence of the trapped ultracold atom samples. For a two-level atom, the scattering rate $\Gamma_{sc}(\mathbf{r})$ can be estimated[15] through

$$\Gamma_{sc}(\mathbf{r}) = \frac{3\pi c^2}{2\hbar\omega_0^3}\left(\frac{\Gamma}{\Delta}\right)^2 I(\mathbf{r}) \qquad (27.6)$$

From Eqs. (27.3) and (27.6) it can be seen that the ratio of scattering rate to optical potential depth can always be minimized by increasing the detuning of the laser field. In practice however, such an approach is limited by the maximum available laser power. For experiments with ultracold quantum gases of alkali atoms, the detuning is typically chosen to be large compared to the excited-state hyperfine structure splitting and in most cases even large compared to the fine structure splitting in order to sufficiently suppress spontaneous scattering events. Typical detunings range from several tens of nanometers to optical trapping in CO_2 laser fields. A laser trap formed by a CO_2 laser field can be considered as a quasi-electrostatic trap, where the detuning is much larger than the optical resonance frequency of an atom.[21-23]

One final comment should be made about state-dependent optical potentials. For a typical multi-level alkali atom, the dipole potential will both depend on the internal magnetic substate m_F of a hyperfine ground state with angular momentum F, as well as on the polarization of the light field $P = +1, -1, 0$ (circular $\sigma^\pm$ and linear polarization). One can then express the lattice potential depth through[15,24]

$$V_{dip}(\mathbf{r}) = \frac{\pi c^2 \Gamma}{2\omega_0^3}\left(\frac{2 + Pg_F m_F}{\Delta_{2,F}} + \frac{1 - Pg_F m_F}{\Delta_{1,F}}\right) I(\mathbf{r}). \qquad (27.7)$$

Here, g_F is the Landé factor and $\Delta_{2,F}, \Delta_{1,F}$ refer to the detuning relative to the transition between the ground state with hyperfine angular momentum F and the center of the excited state hyperfine manifold on the D_2 and D_1 transition, respectively. For large detunings relative to the fine structure splitting Δ_{FS}, the optical potentials become almost spin independent again. For detunings of the laser frequency in between the fine structure splitting, very special spin dependent optical potentials can be created that will be discussed below.

27.2.2 Optical lattice potentials

A periodic potential can simply be formed by overlapping two counterpropagating laser beams. Because of the interference between the two laser beams an optical standing wave with period $\lambda/2$ is formed in which the atoms can be trapped. By interfering more laser beams, one can obtain one-, two, and three-dimensional (1D,2D,3D) periodic potentials,[25] which, in their simplest form, will be discussed below. Note that by choosing to let two laser beams interfere under an angle of $< 180\,\text{deg}$, one can also realize periodic potentials with a larger period.

27.2.2.1 One-dimensional lattice potentials

The simplest possible periodic optical potential is formed by overlapping two counterpropagating focused gaussian laser beams, which results in a trapping potential of the form

$$V(r, z) = -V_{lat} \cdot e^{-2r^2/w^2(z)} \sin^2(kx) \approx -V_{lat} \left(1 - 2\frac{r^2}{w^2(z)}\right) \sin^2(kz), \quad (27.8)$$

where w_0 denotes the beam waist, $k = 2\pi/\lambda$ is the wave vector of the laser light, and V_{lat} is the maximum depth of the lattice potential. Note that due to the interference of the two laser beams, V_{lat} is four times larger than V_0 if the laser power and beam parameters of the two interfering lasers are equal.

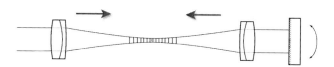

Figure 27.2 One-dimensional optical lattice potential. By interfering two counterpropagating Gaussian laser beams, a periodic intensity profile is created due to the interference of the two laser fields.

27.2.2.2 Two-dimensional lattice potentials

Periodic potentials in two dimensions can be formed by overlapping two optical standing waves along different directions. In the simplest form one chooses two orthogonal directions and obtains at the center of the trap an optical potential of the form (neglecting the harmonic confinement due to the gaussian beam profile of the laser beams):

$$V(y, z) = -V_{lat} \left[\cos^2(kx) + \cos^2(ky) + 2\mathbf{e_1} \cdot \mathbf{e_2} \cos\phi \cos(kx)\cos(ky)\right]. \quad (27.9)$$

Here, $\mathbf{e}_{1,2}$ denote the polarization vectors of the laser fields, each forming one standing wave and ϕ is the time between them. If the polarization vectors are chosen not to be orthogonal to each other, then the resulting potential will not only be the sum of the potentials created by each standing wave, but will be modified according to the time phase ϕ used (see Fig. 27.3). In such a case, it is absolutely essential to stabilize the time phase between the two standing waves,[26] because small vibrations will usually lead to fluctuations of the time phase, resulting in severe heating and decoherence effects of the ultracold atom samples.

In such a two-dimensional optical lattice potential, the atoms are confined to arrays of tightly confining one-dimensional tubes [see Fig. 27.4(a)]. For typical experimental parameters, the harmonic trapping frequencies along the tube are very

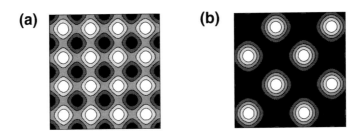

Figure 27.3 Two-dimensional optical lattice potentials for a lattice with (a) orthogonal polarization vectors and (b) with parallel polarization vectors and a time phase of $\phi = 0$.

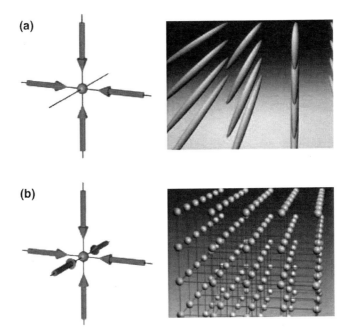

Figure 27.4 Two-dimensional (a) and three-dimensional (b) optical lattice potentials formed by superimposing two or three orthogonal standing waves. For a two-dimensional optical lattice, the atoms are confined to an array of tightly confining one-dimensional potential tubes whereas, in the three-dimensional case, the optical lattice can be approximated by a three-dimensional simple cubic array of tightly confining harmonic oscillator potentials at each lattice site.

weak and on the order of 10–200 Hz, whereas in the radial direction the trapping frequencies can become as high as up to 100 kHz, thus allowing the atoms to effectively move only along the tube for deep lattice depths.[27–31]

27.2.2.3 Three-dimensional lattice potentials

For the creation of a three-dimensional lattice potential, three orthogonal optical standing waves have to be overlapped. Here we only consider the case of independent standing waves, with no cross interference between laser beams of different standing waves. This can for example be realized by choosing orthogonal polarization vectors between different standing wave light fields and also by using different wavelengths for the three standing waves. In this case, the resulting optical potential is simply given by the sum of three standing waves

$$V(\mathbf{r}) = -V_x e^{-2(y^2+z^2)/w_x^2} \sin^2(kx) - V_y e^{-2(x^2+z^2)/w_y^2} \sin^2(ky) \\ -V_z e^{-2(x^2+y^2)/w_z^2} \sin^2(kz). \quad (27.10)$$

Here, $V_{x,y,z}$ are the potential depths of the individual standing waves along the different directions. In the center of the trap, for distances much smaller than the beam waist, the trapping potential can be approximated as the sum of a homogeneous periodic lattice potential and an additional external harmonic confinement due to the gaussian laser beam profiles

$$V(\mathbf{r}) \approx V_x \sin^2(kx) + V_y \sin^2(ky) + V_z \sin^2(kz) \\ + \frac{m}{2}\left(\omega_x^2 x^2 + \omega_y^2 y^2 + \omega_z^2 z^2\right), \quad (27.11)$$

where $\omega_{x,y,z}$ are the effective trapping frequencies of the external harmonic confinement. They can again be approximated by

$$\omega_x^2 = \frac{4}{m}\left(\frac{V_y}{w_y^2} + \frac{V_z}{w_z^2}\right); \quad \omega_{y,z}^2 = \text{(cycl. perm.)}. \quad (27.12)$$

In addition to this harmonic confinement due to the Gaussian laser beam profiles, a confinement due to a magnetic trapping typically exists, which has to be taken into account as well for the total harmonic confinement of the atom cloud.

For sufficiently deep optical lattice potentials, the confinement on a single lattice site is also approximately harmonic. Here the atoms are very tightly confined with typical trapping frequencies ω_{lat} of up to 100 kHz. One can estimate the trapping frequencies at a single lattice site through a Taylor expansion of the sinusoidally varying lattice potential at a lattice site and obtains

$$\omega_{lat} \approx \sqrt{\frac{V_{lat}}{E_r}\frac{\hbar^2 k^4}{m^2}}. \quad (27.13)$$

Here, $E_r = \hbar^2 k^2/2m$ is the so-called recoil energy, which is a natural measure of energy scales in optical lattice potentials.

27.3 Bose-Hubbard Model of Interacting Bosons in Optical Lattices

The behavior of bosonic atoms with repulsive interactions in a periodic potential is fully captured by the Bose-Hubbard Hamiltonian of solid state physics,[1,2] which in the homogeneous case can be expressed through

$$H = -J \sum_{\langle i,j \rangle} \hat{a}_i^\dagger \hat{a}_j + \frac{1}{2} U \sum_i \hat{n}_i(\hat{n}_i - 1). \tag{27.14}$$

Here $\hat{a}_i^\dagger$ and $\hat{a}_i$ describe the creation and annihilation operators for a boson on the ith lattice site and $\hat{n}_i$ counts the number of bosons on the ith lattice site. The tunnel coupling between neighboring potential wells is characterized by the tunnel matrix element

$$J = -\int d^3x \, w(\mathbf{x} - \mathbf{x}_i) \left[-\frac{\hbar^2}{2m} \nabla^2 + V_{lat}(\mathbf{x}) \right] w(\mathbf{x} - \mathbf{x}_j), \tag{27.15}$$

where $w(\mathbf{x} - \mathbf{x}_i)$ is a single-particle Wannier function localized to the ith lattice site and $V_{lat}(\mathbf{x})$ indicates the optical lattice potential. The repulsion between two atoms on a single lattice site is quantified by the on-site matrix element U

$$U = \frac{4\pi \hbar^2 a_s}{m} \int |w(\mathbf{x})|^4 d^3x, \tag{27.16}$$

with a_s being the scattering length of an atom. Because of the short range of the interactions compared to the lattice spacing, the interaction energy is well described by the second term of Eq. (27.14) which characterizes a purely on-site interaction.

Both the tunneling matrix element J and the onsite interaction matrix element can be calculated from a band-structure calculation. The tunnel matrix element J is related to the width of the lowest Bloch band through

$$4J = \left| E_0(q = \frac{\pi}{a}) - E_0(q = 0) \right|, \tag{27.17}$$

where a is the lattice period, such that $q = \pi/a$ corresponds to the quasimomentum at the border of the first Brillouin zone. Care has to be taken to evaluate the tunnel matrix element through Eq. (27.15), when the Wannier function is approximated by the Gaussian ground-state wave function on a single lattice site. This usually results in a severe underestimation of the tunnnel coupling between lattice sites.

The interaction matrix element can be evaluated through the Wannier function with the help of Eq. (27.16). In this case, however, the approximation of the Wannier function through the Gaussian ground-state wave function yields a very good approximation.

Recently, Zwerger[32] carried out a more sophisticated approximation of the tunnel matrix element and the onsite interaction matrix element. He found (in units of the recoil energy)

$$J \approx \sqrt{\frac{8}{\pi}} \left(\frac{V_{lat}}{E_r} \right)^{3/4} e^{-2\sqrt{V_{lat}/E_r}} \tag{27.18}$$

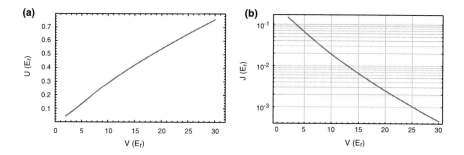

Figure 27.5 Onsite interaction matrix element U for ^{87}Rb (a) and tunnel matrix element J (b) vs lattice depth. All values are given in units of the recoil energy E_r.

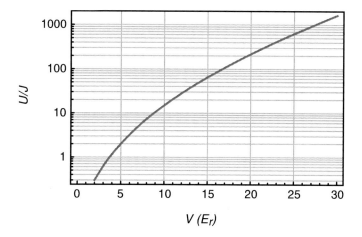

Figure 27.6 U/J vs optical lattice potential depth for ^{87}Rb. By increasing the lattice depth one can tune the ratio U/J, which determines whether the system is strongly or weakly interacting.

and for the interaction matrix element

$$U \approx 4\sqrt{2\pi}\frac{a}{\lambda}\frac{V_{lat}}{E_r}. \tag{27.19}$$

The ratio U/J is crucial for determining, whether one is an a strongly interacting or a weakly interacting regime. It can tuned continuously by simply changing the lattice potential depth (see Fig. 27.6).

27.3.1 Ground states of the Bose-Hubbard Hamiltonian

The Bose-Hubbard Hamiltonian of Eq. (27.14) has two distinct ground states, depending on the strength of the interactions U relative to the tunnel-coupling J. In order to gain insight into the two limiting ground states, let us first consider the case of a double-well system with only two interacting neutral atoms.

27.3.2 Double-well case

In the double-well system, the two lowest lying states for noninteracting particles are the symmetric $|\varphi_S\rangle = 1/\sqrt{2}(|\varphi_L\rangle + |\varphi_R\rangle)$ and the antisymmetric $|\varphi_A\rangle = 1/\sqrt{2}(|\varphi_L\rangle - |\varphi_R\rangle)$ states, where $|\varphi_L\rangle$ and $|\varphi_R\rangle$ are the ground states of the left- and right-hand sides of the double-well potential. The energy difference between $|\varphi_S\rangle$ and $|\varphi_A\rangle$ will be named $2J$, which characterizes the tunnel coupling between the two wells and depends strongly on the barrier height between the two potentials.

In the case of no interactions, the ground state of the two-body system is realized when each atom is in the symmetric ground state of the double-well system [see Fig. 27.7(a)]. Such a situation yields an average occupation of one atom per site with the single-site many-body state actually being in a superposition of zero, one and two atoms.

Let us now consider the effects due to a repulsive interaction between the atoms. If both atoms are again in the symmetric ground state of the double well, then the total energy of such a state will increase due to the repulsive interactions between the atoms. This higher energy cost is a direct consequence of having contributions where both atoms occupy the same site of the double well. This leads to an interaction energy of $1/2U$ for this state.

If this energy cost is much greater than the splitting $2J$ between the symmetric and antisymmetric ground states of the noninteracting system, then the system can minimize its energy when each atom is in a superposition of the symmetric and antisymmetric ground state of the double well $1/\sqrt{2}(|\varphi_S\rangle \pm |\varphi_A\rangle)$. The resulting many body state can then be written as $|\Psi\rangle = 1/\sqrt{2}(|\varphi_L\rangle \otimes |\varphi_R\rangle + |\varphi_R\rangle \otimes |\varphi_L\rangle)$. Here exactly one atom occupies the left and right site of the double well. Now the interaction energy vanishes because both atoms never occupy the same lattice site. The system will choose this new "Mott-insulating" ground state when the energy costs of populating the antisymmetric state of the double-well system are

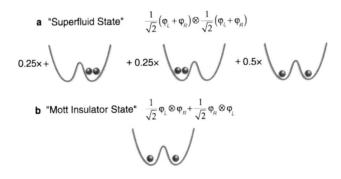

Figure 27.7 Ground state of two interacting particles in a double well. For interaction energies U smaller than the tunnel coupling J, the ground state of the two-body system is realized by the "superfluid" state (a). If on the other hand, U is much larger than J, then the ground state of the two-body system is the Mott-insulating state (b).

outweighed by the energy reduction in the interaction energy. It is important to note that, precisely the atom number fluctuations due to the delocalized single-particle wave functions make the "superfluid" state unfavorable for large U.

Such a change can be induced by adiabatically increasing the barrier height in the double-well system, such that J decreases exponentially and the energy cost for populating the antisymmetric state becomes smaller and smaller. Eventually it will then be favorable for the system to change from the "superfluid" ground state, where for $U/J \to \infty$ each atom is delocalized over the two wells, to the "Mott-insulating" state, where each atom is localized to a single lattice site.

27.3.3 Multiple-well case

The above ideas can be readily extended to the multiple well case of the periodic potential of an optical lattice. For $U/J \ll 1$, the tunneling term dominates the Hamiltonian, and the ground-state of the many-body system with N atoms is given by a product of identical single particle Bloch waves, where each atom is spread out over the entire lattice with M lattice sites

$$|\Psi_{SF}\rangle_{U/J\approx 0} \propto \left(\sum_{i=1}^{M} \hat{a}_i^\dagger\right)^N |0\rangle. \tag{27.20}$$

Because the many-body state is a product over identical single-particle states, a macroscopic wave function can be used to describe the system. Here the single site many-body wave function $|\phi\rangle_i$ is almost equivalent to a coherent state. The atom number per lattice site then remains uncertain and follows a Poissonian distribution with a variance given by the average number of atoms on this lattice site $\mathrm{Var}(n_i) = \langle \hat{n}_i \rangle$. The nonvanishing expectation value of $\psi_i = \langle \phi_i | \hat{a}_i | \phi_i \rangle$ then characterizes the coherent matter wave field on the ith lattice site. This matter wave field has a fixed phase relative to all other coherent matter wave fields on different lattice sites.

If, on the other hand, interactions dominate the behavior of the Hamiltonian, such that $U/J \gg 1$, then fluctuations in the atom number on a single lattice site become energetically costly and the ground state of the system will instead consist of localized atomic wave functions that minimize the interaction energy. The many-body ground state is then a product of local Fock states for each lattice site. In this limit the ground state of the many-body system for a commensurate filling of n atoms per lattice site is given by

$$|\Psi_{MI}\rangle_{J\approx 0} \propto \prod_{i=1}^{M} (\hat{a}_i^\dagger)^n |0\rangle. \tag{27.21}$$

Under such a situation the atom number on each lattice site is exactly determined but the phase of the coherent matter wave field on a lattice site has obtained a maximum uncertainty. This is characterized by a vanishing of the matter wave field on the ith lattice site $\psi_i = \langle \phi_i | \hat{a}_i | \phi_i \rangle \approx 0$.

In this regime of strong correlations, the interactions between the atoms dominate the behavior of the system and the many body state is not amenable anymore to a description as a macroscopic matter wave, nor can the system be treated by the theories for a weakly interacting Bose gas of Gross, Pitaevskii, and Bogoliubov.[33,34]

For a 3D system, the transition to a Mott insulator occurs around $U/J \approx z\, 5.6$,[2,32,35,36] where z is the number of next neighbors to a lattice site (for a simple cubic crystal $z = 6$).

27.3.4 Superfluid to Mott insulator-transition

In the experiment, the crucial parameter U/J that characterizes the strength of the interactions relative to the tunnel coupling between neighboring sites can be varied by simply changing the potential depth of the optical lattice potential. By increasing the lattice potential depth, U increases almost linearly due to the tighter localization of the atomic wave packets on each lattice site and J decreases exponentially due to the decreasing tunnel coupling. The ratio U/J can therefore be varied over a large range from $U/J \approx 0$ up to values, in our case, of $U/J \approx 2000$.

In the superfluid regime,[37] phase coherence of the matter wave field across the lattice characterizes the many body state. This can be observed by suddenly turning off all trapping fields, such that the individual matter wave fields on different lattice sites expand and interfere with each other. After a fixed time of flight period the atomic density distribution can then be measured by absorption imaging. Such an image directly reveals the momentum distribution of the trapped atoms. In Fig. 27.8(b), an interference pattern can be seen after releasing the atoms from a three-dimensional lattice potential.

If on the other hand the optical lattice potential depth is increased such that the system is very deep in the Mott-insulating regime ($U/J \to \infty$), phase coherence is lost between the matter wave fields on neighboring lattice sites due to the formation of Fock states.[1,2,32,38] In this case, no interference pattern can be seen in the

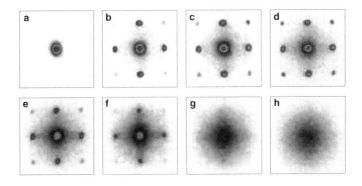

Figure 27.8 Absorption images of multiple matter wave interference patterns after releasing the atoms from an optical lattice potential with a potential depth of (a) $7\,E_r$ and (b) $20\,E_r$. The ballistic expansion time was 15 ms.

time-of-flight images [see Fig. 27.8(h)].[3-5] For a Mott insulator at finite U/J, one expects a residual visibility in the interference pattern,[39,40] which can be caused, on the one hand, by the residual superfluid shells and, on the other hand, through an admixture of coherent particle hole pairs to the ideal MI ground state.

A more detailed picture for the residual short-range coherence features beyond the SF-MI transition is obtained by considering perturbations deep in the Mott-insulating regime at $J = 0$. There, the first-order correlation function $G^{(1)}(\mathbf{R})$ describing the coherence properties vanishes beyond $\mathbf{R} = 0$ and the momentum distribution is a structureless Gaussian, reflecting the Fourier transform of the Wannier wave function. With increasing tunneling J, the Mott state at $J/U \to 0$ is modified by a coherent admixture of particle-hole pairs. However, due to the presence of a gapped excitation spectrum, such particle hole pairs cannot spread out and are rather tightly bound to close distances. They do, however, give rise to a significant degree of short-range coherence (see Fig. 27.9). Using first-order perturbation theory with the tunneling operator as a perturbation on the dominating interaction term, one finds that the amplitude of the coherent particle hole admixtures in a Mott-insulating state is proportional to J/U

$$|\Psi\rangle_{U/J} \approx |\Psi\rangle_{U/J \to \infty} + \frac{J}{U} \sum_{\langle i,j \rangle} \hat{a}_i^\dagger \hat{a}_j |\Psi\rangle_{U/J \to \infty}. \quad (27.22)$$

Within a local density approximation, the inhomogeneous situation in a harmonic trap is described by a spatially varying chemical potential $\mu_\mathbf{R} = \mu(0) - \epsilon_\mathbf{R}$

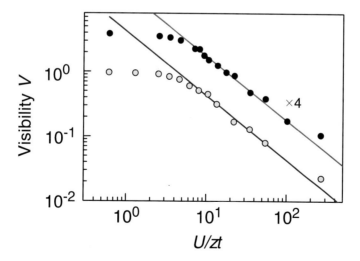

Figure 27.9 Visibility of the interference pattern versus U/zt, the characteristic ratio of interaction to kinetic energy. The data are shown for two atom numbers 5.9×10^5 atoms (black circles), and 3.6×10^5 atoms (gray circles). The former curve has been offset vertically for clarity. The lines are fits to the data in the range 14–25 E_r, assuming a coherent particle hole admixture, as in Eq. (27.22) (see Ref. 41).

with $\epsilon_R = 0$ at the trap center. Assuming, for example, that the chemical potential $\mu(0)$ at trap center falls into the $\bar{n} = 2$ 'Mott-lobe', one obtains a series of MI domains separated by a SF by moving to the boundary of the trap where μ_R vanishes. In this manner, all the different phases that exist for a given J/U below $\mu(0)$ are present simultaneously. The SF phase has a finite compressiblity $\kappa = \partial n/\partial \mu$ and a gapless excitation spectrum of the form $\omega(q) = cq$ because there is a finite superfluid density n_s. By contrast, in the MI phase both n_s and κ vanish identically. As predicted by Ref. 2, the incompressibility of the MI phase allows one to distinguish it from the SF by observing the local density distribution in a trap. Because $\kappa = \partial n/\partial \mu = 0$ in the MI, the density stays constant in the Mott phases, even though the external trapping potential is rising.

The existence of such wedding-cake-like density profiles of a Mott insulator has been supported by accurate Monte Carlo[42,43] and Density Matrix Renormalization Group (DMRG)[44] calculations in one, two, and three dimensions. Very recently in-trap density profiles have been detected experimentally by Campbell et al.[45] and Fölling et al.[46] In the latter case, it has been possible to directly observe the wedding-cake density profiles and thus confirm the incompressibiltiy of the Mott-insulating regions of the atomic gas in the trapping potential (see Fig. 27.10). It should be noted that the in-trap density profiles can be used as a sensitive thermometer for the strongly interacting quantum gas. Already for small temperatures around $T \approx 0.2U/k_B$, the wedding-cake profiles become completely washed out.

Close to the transition point, higher-order perturbation theory or a Green's function analysis can account for coherence beyond nearest neighbors and the complete liberation of the particle-hole pairs, which eventually leads to the formation of long-range coherence in the superfluid regime. The coherent particle hole admixture and its consequence on the short-range coherence of the system have been investigated theoretically and experimentally in Refs. 41, 47, 48. It has been demonstrated that through a quantitative analysis of the interference pattern, one can even observe traces of the shell structure formation in the Mott-insulating regime.

In addition to the fundamentally different momentum distributions in the superfluid and Mott-insulating regime, the excitation spectrum is markedly different as well in both cases. Although the excitation spectrum in the superfluid regime is gapless, it is gapped in the Mott-insulating regime. This energy gap of order U (deep in the MI regime) can be attributed to the now localized atomic wave functions of the atoms.[1,2,32,36]

Let us consider for example a Mott insulating state with exactly one atom per lattice site. The lowest lying excitation to such a state is determined by removing an atom from a lattice site and placing it into the neighboring lattice site [see Fig. 27.11(a)]. Because of the onsite repulsion between the atoms, however, such an excitation costs energy U, which is usually not available to the system. Therefore these are only allowed in virtual processes and an atom in general has to remain im-

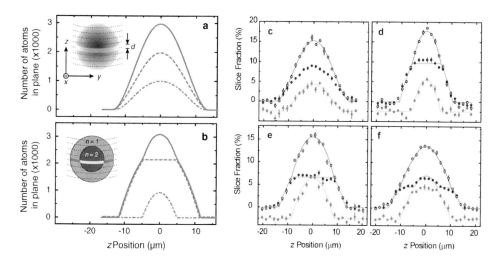

Figure 27.10 Integrated distribution of a superfluid (a) and a Mott- insulating state (b) calculated for a lattice with harmonic confinement. Gray solid lines denote the total density profiles, blue (red) lines the density profiles from singly (doubly) occupied sites. A vertical magnetic field gradient is applied that creates almost horizontal surfaces of equal Zeeman shift over the cloud (dashed lines in inset). A slice of atoms can be transferred to a different hyperfine state by using microwave radiation only resonant on one specific surface (colored areas in insets). Spin-changing collisions can then be used to separate singly (blue) and doubly occupied sites (red) in that plane into different hyperfine states. Experimental data: (c) 1.0×10^5 atoms in the superfluid regime ($V_0 = 3E_r$), (d) 1.0×10^5 atoms in the Mott regime ($V_0 = 22E_r$), (e) 2.0×10^5 atoms, and (f) 3.5×10^5 atoms. The gray data points denote the total density distribution and the red points the distribution of doubly occupied sites. The blue points show the distribution of sites with occupations other than $n = 2$. The solid lines are fits to an integrated Thomas-Fermi distribution in (c) and an integrated shell distribution for (d)–(f). The $n = 2$ data points are offset vertically for clarity.

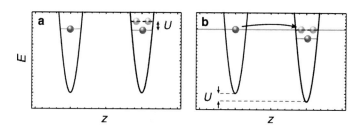

Figure 27.11 (a) Excitation gap in the Mott-insulator phase with exactly $n = 1$ atom on each lattice site. (b) If a correct potential gradient is added, then atoms can tunnel again.

mobile at its original position. If one adds a potential gradient such that the energy difference between neighboring lattice sites ΔE exactly matches the onsite energy cost U, then such an excitation becomes energetically possible and one is able to resonantly perturb the system [see Figs. 27.11(b) and 27.12]. It has been possible to measure this change in the excitation spectrum by applying varying magnetic

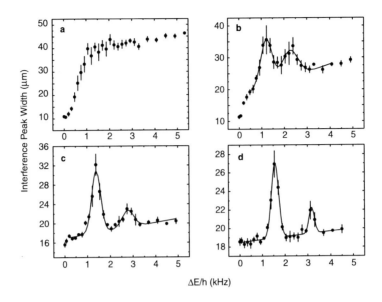

Figure 27.12 Probing the excitation probability vs an applied vertical potential gradient. Width of interference peaks after adiabatic rampdown vs. the energy difference between neighboring lattice sites E, due to the potential gradient at different lattice depths (a) $V_{max} = 10E_r$, (b) $V_{max} = 13E_r$, (c) $V_{max} = 16E_r$ and (d) $V_{max} = 20E_r$. The emergence of an energy gapped can be observed as the system is converted into a Mott insulator (adapted from Ref. 3).

field gradients to the system for different lattice potential depths and detecting the response of the system to such perturbations.[3,49,50]

Recently, the excitation spectrum has also been probed via an intensity modulation of the lattice potential, which is effectively a Bragg spectroscopy for momentum transfer $q = 0$.[5] Here a similar behavior was predicted and observed: in the superfluid a broad excitation spectrum was detected,[5,51] whereas the Mott-insulating state displays sharp resonances at integer multiples of the onsite interaction matrix element U.[5] Using Bragg spectroscopy, it should also be possible to probe the excitation spectrum of a Mott insulator in more detail at various momentum transfers, which has been suggested and evaluated recently by.[52]

27.4 Quantum Noise Correlations

For almost 10 years, absorption imaging of released ultracold quantum gases has been a standard detection method for revealing information on the macroscopic quantum state of the atoms in the trapping potential. For strongly correlated quantum states in optical lattices, however, the average signal in the momentum distribution that one usually observes, e.g., for a Mott-insulating state of matter, is a featureless Gaussian wave packet. From this Gaussian wave packet, one cannot deduce anything about the strongly correlated quantum states in the lattice potential apart from the fact that phase coherence has been lost. Recently, however, the widespread interest in strongly correlated quantum gases in optical lattices as quantum simulators has lead to the prediction of fascinating new quantum phases for ultracold atoms, e.g., with antiferromagnetic structure, spin waves or charge density waves. So far it has not been clear as to how one could detect those states. Recently a theoretical proposal by Altman et al.[53] has shown that noise correlation interferometry could be a powerful tool to directly visualize such quantum states. Noise correlation in expanding ultracold atom clouds can, in fact, be seen as a powerful way to read out the quantum states of an optical lattice-based quantum simulator.

The basic effect relies on fundamental Hanbury-Brown and Twiss correlations[54–57] in the fluctuation signal of an atomic cloud. For bosons, a bunching effect of the fluctuations is predicted to occur at special momenta of the expanding cloud, which directly reflect the ordering of the atoms in the lattice. Such bunching effects in momentum space can be directly revealed as spatial correlations in the expanding atom cloud. Our goal, therefore, is to reveal correlations in the fluctuations of the expanding atomic gas after it has been released from the trap. Such correlations in the expanding cloud at at distance $\mathbf{d}$ can be quantified through the second-order correlation function

$$C(\mathbf{d}) = \frac{\int \langle n(\mathbf{x}+\mathbf{d}/2)\, n(\mathbf{x}-\mathbf{d}/2)\rangle d^2\mathbf{x}}{\int \langle n(\mathbf{x}+\mathbf{d}/2)\rangle \langle n(\mathbf{x}-\mathbf{d}/2)\rangle d^2\mathbf{x}}. \tag{27.23}$$

Here $n(\mathbf{x})$ is the density distribution of a single expanding atom cloud and the angular brackets $\langle \cdot \rangle$ denote a statistical averaging over several individual images taken for different experimental runs.

Spatial correlations in the noise of expanding atom clouds arise here due to a fundamental indistinguishability of the particles, as is well known from the foundational experiments in quantum optics of Hanbury-Brown and Twiss.[54–57] Let us for simplicity consider two detectors spaced at a distance d below our trapped atoms and furthermore restrict the discussion two only two atoms trapped in the lattice potential (see Fig. 27.13). As the trapping potential is removed and the particles propagate to the detectors, there are two possibilities for the particles to reach these detectors, such that one particle is detected at each detector. First, the particles can propagate along path A in Fig. 27.13 to achieve this. However, another propagation path exists, which is equally probable, path B in Fig. 27.13. If we fundamentally cannot distinguish which way the particles have been propagating to our detectors, we have to form the sum for bosons or difference for fermions of the two propagation amplitudes and square the resulting value to obtain the two particle detection probability at the detectors. As one increases the separation between the detectors, the phase difference between the two propagation paths increases, leading to constructive and destructive interference effects in the two-particle detection probability. The length scale of this modulation in the two particle detection probability of the expanding atom clouds depends on the original separation

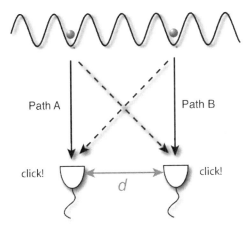

Figure 27.13 Hanbury-Brown and Twiss correlations in expanding quantum gases from an optical lattice. For bosonic particles that are detected at distances d (e.g. on a CCD camera), an enhanced detection probability exists due to the two indistinguishable paths the particles can take to the detector. This leads to enhanced fluctuations at special detection distances d, depending on the ordering of the atoms in the lattice. Detection of the noise correlation can therefore yield novel information on the quantum phases in an optical lattice.

of the trapped particles at a distance a and is given by the characteristic length scale

$$l = \frac{h}{ma}t, \qquad (27.24)$$

where t is time of flight.

Such Hanbury-Brown and Twiss correlations in the shot noise of an expanding atom cloud from a Mott-insulating state of matter have recently been observed for 3D and 2D Mott insulating states.[58,59] Similar pair correlations at the shot noise level have also been obtained in the group of Jin (see Ref. 60) for dissociated molecular fragments. Here, the correlations arise, however, due to the dissociation event and not due to indistinguishable pathways of the particles.

Very recently, our group has been able to analyze the quantum noise correlation of a band-insulating single-component Fermi gas in an optical lattice.[61] Although the density distribution there is almost identical to the one of a bosonic Mott-insulating cloud for $J \to 0$, a profound difference can be observed in both cases. Whereas the bosonic particles exhibit an HBT-type bunching, the fermionic particles exhibit an antibunching, which can be seen as a decreased detection probability for finding two particles at integer multiples of l (see Fig. 27.15). Such a quantum statistical antibunching effect for Fermions has been observed for electrons in semiconductors,[62,63] as well as for free electrons[64] and neutrons.[65] Together with the experiments performed by Jeltes et al.,[66] our experiments are the first demonstration of antibunching with atoms.[61]

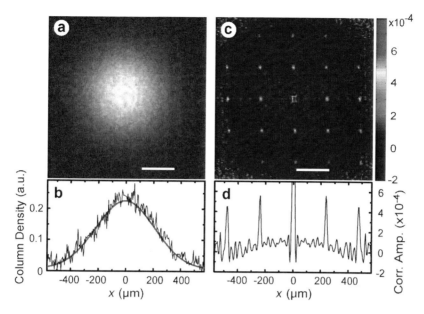

Figure 27.14 Single-shot absorption image, including quantum fluctuations and associated spatial correlation function. (a) 2D Column density distribution of a Mott-insulating atomic cloud containing 6×10^5 atoms, released from a 3D optical lattice potential with a lattice depth of 50 E_r. The white bars indicate the reciprocal lattice scale l defined in Eq. (27.24). (b) Horizontal cut (black line) through the center of the image in a and Gaussian fit (red line) to the average over 43 independent images each one similar to (a). (c) Spatial noise correlation function obtained by analyzing the same set of images, which shows a regular pattern revealing the lattice order of the particles in the trap. (d) Horizontal profile through centre of pattern, containing the peaks separated by integer multiples of l. The width of the individual peaks is determined by the optical resolution of our imaging system.

27.5 Outlook

Ultracold atoms in optical lattices have proven to be a versatile model system for the investigation of strongly correlated quantum physics with ultracold atoms. For the future, the focus of research will shift toward the realization of quantum magnetism with ultracold atoms. Although no long-range direct interaction exists between the particles on different lattice sites, superexchange processes can mediate effective long-range spin-spin interaction between the particles.[67] Such superexchange processes fundamentally rely on second-order tunneling events, which lead to effective spin-spin coupling strengths on the order of J^2/U. To observe these phases, the temperature of the many-body system will have to be below this superexchange-interaction energy scale, which presents a challenge to current experiments, however, does not seem out of reach.[68] If such systems for fermionic particles could eventually be doped or analyzed at noninteger filling, then one could hope to solve one of the long-standing problems of condensed matter physics, namely, whether the fermionic Hubbard model with repulsive interactions contains a superconducting phase[69,70] and what the mechanism behind such a superconducting phase could be.

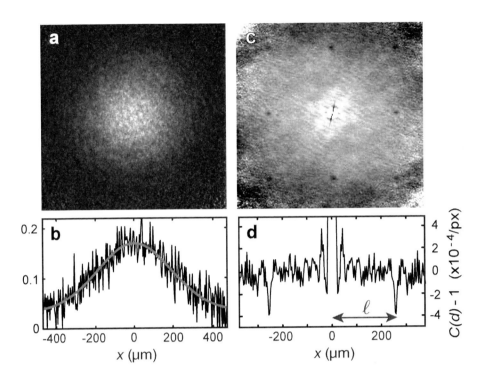

Figure 27.15 Noise correlations of a band-insulating Fermi gas. Instead of the correlation "bunching" peaks observed in Fig. 27.14 the fermionic quantum gas shows an HBT-type antibunching effect, with dips in the observed correlation function.[61]

Acknowledgments

I.B. acknowledges funding through a EU Marie-Curie-Excellence grant (QUASI-COMBS), EU-STREP (OLAQUI), the DFG and AFOSR.

References

1. Fisher, M.P.A., Weichman, P.B. Grinstein, G., and Fisher, D.S., "Boson localization and the superfluid-insulator transition," *Phys. Rev. B*, **40**, 546–570, 1989.

2. Jaksch, D., Bruder, C., Cirac, J.I., Gardiner, C.W., and Zoller, P., "Cold bosonic atoms in optical lattices," *Phys. Rev. Lett.*, **81**, 3108–3111, 1998.

3. Greiner, M., Mandel, O., Esslinger, T., Hänsch, T.W., and Bloch, I., "Quantum phase transition from a superfluid to a mott insulator in a gas of ultracold atoms," *Nature*, **415**, 39–44, 2002.

4. Porto, J., Rolston, S., Tolra, B., Williams, C., and Phillips, W., "Quantum information with neutral atoms as qubits," *Phil. Trans. Roy. Soc. A*, **361**, 1417–1427, 2003.

5. Stöferle, T., Moritz, H., Schori, C., Köhl, M., and Esslinger, T., "Transition from a strongly interacting 1d superfluid to a mott insulator," *Phys. Rev. Lett.*, **92**, 130403, 2004.

6. Jaksch, D., Briegel, H.J., Cirac, J.I., Gardiner, C.W., and Zoller, P., "Entanglement of atoms via cold controlled collisions," *Phys. Rev. Lett.*, **82**, 1975–1978, 1999.

7. Brennen, G., Caves, C., Jessen, P., and Deutsch, I.H., "Quantum logic gates in optical lattices," *Phys. Rev. Lett.*, **82**, 1060–1063, 1999.

8. Brennen, G., and Deutsch, I., "Entangling dipole-dipole interactions for quantum logic with neutral atoms," *Phys. Rev. A*, **61**, 062309, 2000.

9. Mandel, O., Greiner, M., Widera, A., Rom, T., Hänsch, T.W., and I. Bloch, "Controlled collisions for multiparticle entanglement of optically trapped atoms," *Nature*, **425**, 937, 2003.

10. Widera, A., Mandel, O., Greiner, M., Kreim, S., Hänsch, T.W., and Bloch, I., "Entanglement interferometry for precision measurement of atomic scattering properties," *Phys. Rev. Lett.* **92**, 160406–1–4, 2004.

11. Widera, A., Gerbier, F., Fölling, S., Gericke, T., Mandel, O., and Bloch, I., "Coherent collisional spin dynamics in optical lattices," *Phys. Rev. Lett.*, **95**, 190405, 2005.

12. Cohen-Tannoudji, C., Dupont-Roc, J., and Grynberg, G., *Atom–Photon Interactions*, Wiley-VCH, Berlin (1992).

13. Metcalf, H., and van der Straten, P., *Laser Cooling and Trapping*, Springer, New York (1999).

14. Chu, S., Björkholm, J., Ashkin, A., and Cable, A., "Experimental observation of optically trapped atoms," *Phys. Rev. Lett.* **57**, 314, 1986.

15. Grimm, R., Weidemüller, M., and Ovchinnikov, Y., "Optical dipole traps for neutral atoms," *Adv. At. Mol. Opt. Phys.*, **42**, 95, 2000.

16. Askar'yan, G., "Effects of the gradient of a strong electromagnetic beam on electrons and atoms," *Sov. Phys. JETP*, **15**, 1088, 1962.

17. Kazantsev, A., "Acceleration of atoms by a resonance field," *Sov. Phys. JETP*, **36**, 861, 1973.

18. Cook, R., "Atomic motion in resonant radiation: An application of ehrenfest's theorem," *Phys. Rev. A*, **20**, 224, 1979.

19. Gordon, J., and Ashkin, A., "Motion of atoms in a radiation trap," *Phys. Rev. A*, **21**, 1606, 1980.

20. Saleh, B., and Teich, M., *Fundamentals of Photonics*, Wiley, New York (1991).

21. Takekoshi, T., Yeh, J., and Knize, R., "Quasi-electrostatic trap for neutral atoms," *Opt. Comm.*, **114**, 421, 1995.

22. Friebel, S., D'Andrea, C., Walz, J., and Weitz, M., "Co2-laser optical lattice with cold rubidium atoms," *Phys. Rev. A*, **57**, R20, 1998.

23. Barrett, M., Sauer, J., and Chapman, M., "All-optical formation of an atomic bose-einstein condensate," *Phys. Rev. Lett.*, **87**, 010404, 2001.

24. Jessen, P., and Deutsch, I.H., "Optical lattices," *Adv. At. Mol. Opt. Phys.*, **37**, 95–139, 1996.

25. Petsas, K., Coates, A., and Grynberg, G., "Crystallography of optical lattices," *Phys. Rev. A*, **50**, 5173–5189, 1994.

26. Hemmerich, A., Schropp, D., Esslinger, T., and Hänsch, T., "Elastic scattering of rubidium atoms by two crossed standing waves," *Europhys. Lett.* **18**, 391, 1992.

27. Greiner, M., Bloch, I., Mandel, O., Hänsch, T.W., and Esslinger, T., "Exploring phase coherence in a 2d lattice of bose-einstein condensates," *Phys. Rev. Lett.*, **87**, 160405, 2001.

28. Moritz, H., Stöferle, T., Köhl, M., and Esslinger, T., "Exciting collective oscillations in a trapped 1d gas," *Phys. Rev. Lett.*, **91**, 250402, 2003.

29. Laburthe-Tolra, B., O'Hara, K., Huckans, J., Phillips, W., Rolston, S., and Porto, J., "Observation of reduced three-body recombination in a correlated 1d degenerate bose gas," *Phys. Rev. Lett.*, **92**, 190401, 2004.

30. Paredes, B., Widera, A., Murg, V., Mandel, O., Fölling, S., Cirac, J.I., Shlyapnikov, G., Hänsch, T.W., and Bloch, I., "Tonks-girardeau gas of ultracold atoms in an optical lattice," *Nature*, **429**, 277–281, 2004.

31. Kinoshita, T., Wenger, T., and Weiss, D., "Observation of a one-dimensional tonks-girardeau gas," *Science*, **305**, 1125–1128, 2004.

32. Zwerger, W., "Mott-hubbard transition of cold gases in an optical lattice," *J. Opt. B*, **5**, S9–S16, 2003.

33. Pitaevskii L., and Stringari, S., *Bose-Einstein Condensation*, International Series of Monographs on Physics, Oxford University Press, Oxford (2003).

34. Pethick, C., and Smith, H., *Bose-Einstein Condensation in Dilute Gases*, Cambridge University Press, Cambridge (2001).

35. Sheshadri, K., Krishnamurthy, H.R., Pandit, R., and Ramakrishnan, T.V., "Superfluid and insulating phases in an interacting-boson model: mean-field theory and the rpa," *Europhys. Lett.*, **22**, 257–263, 1993.

36. van Oosten, D., van der Straten, P., and Stoof, H., "Quantum phases in an optical lattice," *Phys. Rev. A*, **63**, 053601, 2001.

37. Cataliotti, F.S., Burger, S., Fort, C., Maddaloni, P., Minardi, F., Trombettoni, A., Smerzi, A., and Inguscio, M., "Josephson junction arrays with bose-einstein condensates," *Science*, **293**, 843–846, 2001.

38. Sachdev, S., *Quantum Phase Transitions*, Cambridge University Press, Cambridge (1999).

39. Kashurnikov, V.A., Prokof'ev, N.V., and Svistunov, B., "Revealing the superfluid-mott insulator transition in an optical lattice," *Phys. Rev. A*, **66**, 031601(R), 2002.

40. Roth, R., and Burnett, K., "Quantum phases of atomic boson-fermion mixtures in optical lattices," *Phys. Rev. A*, **69**, 021601(R), 2004.

41. Gerbier, F., Widera, A., Fölling, S., Mandel, O., Gericke, T., and Bloch, I., "Phase coherence of an atomic mott insulator," *Phys. Rev. Lett.*, **95**, 050404, 2005.

42. Kashurnikov, V.A., Prokof'ev, N.V., and Svistunov, B.V., "Revealing the superfluid to mott-insulator transition in an optical lattice," *Phys. Rev. A*, **66**, 031601, 2002.

43. Wessel, S., Alet, F., Troyer, M., and Batrouni, G.G., "Quantum monte carlo simulations of confined bosonic atoms in optical lattices," *Phys. Rev. A*, **70**, 053615, 2004.

44. Kollath, C., Schollwöck, U., von Delft, J., and Zwerger, W., "Spatial correlations of trapped one-dimensional bosons in an optical lattice," *Phys. Rev. A*, **69**, 031601(R), 2004.

45. Campbell, G., Mun, J., Boyd, M., Medley, P., Leanhardt, A.E., Marcassa, L.G., Pritchard, D.E., and Ketterle, W., "Imaging the mott insulator shells by using atomic clock shifts," *Science*, **313**, 5787, 2006.

46. Fölling, S., Widera, A., Müller, T., Gerbier, F., and Bloch, I., "Formation of spatial shell structures in the superfluid to mott insulator transition," *Phys. Rev. Lett.*, **97**, 060403, 2006.

47. Gerbier, F., Widera, A., Fölling, S., Mandel, O., Gericke, T., and Bloch, I., "Interference pattern and visibility of a mott insulator," *Phys. Rev. A*, **72**, 053606, 2005.

48. Sengupta, P., Rigol, M., Batrouni, G., Denteneer, P., and Scalettar, R.T., "Phase coherence, visbility, and the superfluid-mott-insulator transition on one-dimensional optical lattices," *Phys. Rev. Lett.*, **95**, 220402, 2005.

49. Sachdev, S., Sengupta, K., and Girvin, S., "Mott insulators in strong electric fields," *Phys. Rev. B*, **66**, 075128, 2002.

50. Braun-Munzinger, K., Dunningham, J.A., and Burnett, K., "Excitations of bose-einstein condensates in optical lattices," *Phys. Rev. A*, **69**, 053613, 2004.

51. Schori, C., Stöferle, T., Moritz, H., Köhl, M., and Esslinger, T., "Excitations of a superfluid in a three-dimensional optical lattice," *Phys. Rev. Lett.*, **93**, 240402, 2004.

52. van Oosten, D., Dickerscheid, D., Farid, B., van der Straten, P., and Stoof, H., "Inelastic light scattering from a mott insulator," *Phys. Rev. A*, **71**, 021601, 2005.

53. Altman, E., Demler, E., and Lukin, M.D., "Probing many-body states of ultracold atoms via noise correlations," *Phys. Rev. A*, **70**, 013603, 2004.

54. Hanbury Brown, R., and Twiss, R.Q., "Correlation between photons in two coherent beams of light," *Nature*, **177**, 27–29, 1956.

55. Hanbury Brown, R., and Twiss, R.Q., "The question of correlation between photons in coherent light rays," *Nature*, **178**, 1447–1448, 1956.

56. Hanbury Brown, R., and Twiss, R.Q., "A test of a new type of stellar interferometer on sirius," *Nature*, **178**, 1046–1448, 1956.

57. Baym, G., "The physics of hanbury brown-twiss intensity interferometry: From stars to nuclear collisions," *Act. Phys. Pol. B*, **29**, 1839–1884, 1998.

58. Fölling, S., Gerbier, F., Widera, A., Mandel, O., Gericke, T., and Bloch, I., "Spatial quantum noise interferometry in expanding ultracold atomic gases," *Nature*, **434**, 481–484, 2005.

59. Spielman, I.B., Phillips, W.D., and Porto, J.V., "The mott insulator transition in two dimensions," *Phys. Rev. Lett.*, **98**, 080404, 2007.

60. Greiner, M., Regal, C., Stewart, J., and Jin, D., "Probing pair-correlated fermionic atoms through correlations in atom shot noise," *Phys. Rev. Lett.*, **94**, 110401, 2005.

61. Rom, T., Best, T., van Oosten, D., Schneider, U., Fölling, S., Paredes, B., and Bloch, I., "Free fermion antibunching in a degenerate fermi gas released from an optical lattice," *Nature*, **444**, 733–736, 2006.

62. Henny, M., Oberholzer, S., Strunk, C., Heinzel, T., Ensslin, K., Holland, M., and Schonenberger, C., "The fermionic hanbury brown and twiss experiment," *Science*, **284**, 296–298, 1999.

63. Oliver, W.D., Kim, J., Liu, R.C., and Yamamoto, Y., "Hanbury brown and twiss-type experiment with electrons," *Science*, **284**, 299–301, 1999.

64. Kiesel, H., Renz, A., and Hasselbach, F., "Observation of hanbury brown-twiss anticorrelations for free electrons," *Nature*, **418**, 392–394, 2002.

65. Iannuzzi, M., Orecchini, A., Sacchetti, F., Facchi, P., and Pascazio, S., "Direct experimental evidence of free-fermion antibunching," *Phys. Rev. Lett.*, **96**, 080402, 2006.

66. Jeltes, T., McNamara, J., Hogervorst, W., Vassen, W., Krachmalnicoff, V., Schellekens, M., Perrin, A., Chang, H., Boiron, D., Aspect, A., and Westbrook, C., "Hanbury brown twiss effect for bosons versus fermions," *Nature*, **445**, 402–405, 2007.

67. Auerbach, A., *Interacting Electrons and Quantum Magnetism*, Springer, New York (2006).

68. Werner, F., Parcollet, O., Georges, A., and Hassan, S.R., "Interaction-induced adiabatic cooling and antiferromagnetism of cold fermions in optical lattices," *Phys. Rev. Lett.*, **95**, 056401, 2005.

69. Hofstetter, W., Cirac, J.I., Zoller, P., Demler, E., and Lukin, M.D., "High-temperature superfluidity of fermionic atoms in optical lattices," *Phys. Rev. Lett.*, **89**, 220407, 2002.

70. Lee, P., Nagaosa, N., and Wen, X.-G., "Doping a mott insulator: Physics of high-temperature superconductivity," *Rev. Mod. Phys.*, **78**, 17–85, 2006.

Chapter 28
The Intimate Integration of Photonics and Electronics

Ashok V. Krishnamoorthy
Sun Microsystems

28.1 Background
28.2 History of Optical Data Links in Communication
28.3 The Opportunity for Photonics in Ultrashort-Reach Interconnect
28.4 The Convergence in Optical and Electrical Packaging
28.5 Photonics-on-CMOS: Some Examples
    28.5.1 Multiple-quantum-well modulators on CMOS
    28.5.2 VCSELs and detectors on CMOS
28.6 Silicon Photonics
28.7 Commercialization of Optoelectronics-on-VLSI Technologies
    28.7.1 VCSELs and PIN photodetectors on bulk CMOS
    28.7.2 VCSELs and PIN photodetectors on SOS CMOS
    28.7.3 Silicon photonic active cable
28.8 Discussion
References

28.1 Background

It is evident that computing and switching performance must continue to scale to meet the growing proliferation of computing and communications capability and the reduced price per performance expected from systems manufacturers. Indeed, we have already witnessed over five orders of magnitude improvement in compute performance-per-dollar over the last three decades, well in excess of the improvements delivered by technology scaling alone. In fact, the demand from applications is underserved by Moore's law, and requires continued scaling and improvements in efficiencies at the system level. Examples of such applications include content servers and high-performance computing.

For reasons that include hiding memory and wire latency, reducing power dissipation, and managing design complexity, it is now evident that multi-core, multi-threaded processors are necessary for continued scaling of microprocessor performance. But this architectural scaling of processors alone is insufficient to meet all application requirements. System limitations in interconnect power, bandwidth, and density threaten to mask the benefits of improved chip performance, and hinder our ability to create large, power-efficient machines with optimal bisection bandwidth and adequate performance on critical performance benchmarks. For instance, high-performance computing (HPC) systems require sophisticated architectures for optimizing an increased number of processor-memory units, lower latency signaling between chips, and larger system bisection bandwidths for communication. Currently, the top HPC systems have bisection bandwidths between 1 and 10Tbps and next-generation systems are looking for 10-100× improvements in bandwidth.

Relying on current and soon-to-be-available optical interconnect modules, one can design enough system bandwidth to enable optical interconnection of modest arrays of processor/memory units. Simply scaling such system architectures would necessitate vast numbers of optical modules and fiber optic cables. Among the issues of concern in such systems are ways to increase the link data rate while maintaining low cost and high overall system reliability. We believe that techniques that increase the number of effective channels using dense integration of photonic devices with silicon, combined with WDM and multi-level encoding to increase the number of multiplexed optical data channels per optical waveguide will be key. By virtue of its abilities to achieve dense integration, promote electrical-optical symbiosis,* utilize wafer-scale silicon manufacturing, achieve WDM and other encoding techniques, photonic interconnects appear to be a scalable interconnect candidate. In this article, we will review some technology examples and opportunities for high-density optical interconnects.

28.2 History of Optical Data Links in Communication

It is useful to examine the history of penetration optical data links into communication over the last 30 years. Figure 28.1 shows an approximate perspective of the rate of penetration of optics versus the link distance and the bandwidth. The lower horizontal axis represents the first commercial introduction of the optical link. The vertical axis represents the minimum range of the link, and the upper horizontal axis represents the bandwidth per fiber connector. The approximate trend suggests that over the last three decades, optical links have achieved an order-of-magnitude deeper penetration into the interconnection

---

* There are many opportunities for electrical-optical symbiosis; for instance, we (and other groups) have shown that it is possible to significantly extend (by over 2x) the modulation bandwidth of silicon optical modulators by using electronic pre-emphasis and equalization circuits. As a complementary example, it is easier for the optical channels to transport 10-20Gbps modulated signals globally on-chip; to do so would otherwise require either electrical transmission lines or power-hungry repeaters that are not convenient or efficient to implement in CMOS.

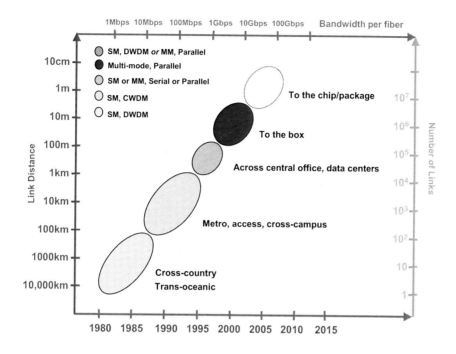

Figure 28.1 Penetration of optical links into communications.

hierarchy every five years. Continuing this trend suggests that during the next five years, optical links can be expected to reach right to the chip-scale package on a printed circuit board, representing link distances as short as one meter. Although this prediction might be viewed by some as a radical departure from current technology solutions, in the remainder of this article, we will support the view that the technologies that could achieve this in a cost-effective manner are under development. In this article we will concentrate on the active components, although there also has been strong progress made in fiber and waveguide platforms to support this.

28.3 The Opportunity for Photonics in Ultrashort-Reach Interconnect

In the past two decades, ultrashort reach interconnects (i.e. distances of <30meters) have been dominated by electrical links. This has been true because of the relative ease of designing and building many electrical interconnect drivers and receivers on a silicon CMOS VLSI chip. Another factor that has supported this has been that the electrical channel has typically not limited the signal bandwidth up to 1-2GHz. Hence the continued scaling of VLSI technologies to finer lithography and larger silicon wafers has driven the density and bit-rates of such high-speed electrical signaling up and simultaneously driven down the cost per Gigabit per second and the power consumption of such solutions.

Unfortunately, the conditions that surround the continued scaling of VLSI interconnects in the 21st century no longer support these simple and powerful assumptions. As data rates reach and exceed 10Gbps, the off-chip signaling channel becomes bandwidth-limited due to the skin effect in the wires and the dielectric losses in the transmission media, causing inter-symbol interference and associated power penalties. In addition, the necessity of multiple vias and connectors in a typical backplane link configuration inevitably cause nonideal impedance discontinuities resulting in potentially severe reflections that further limit bandwidth (Fig. 28.2). Furthermore, at the chip level, voltages are no longer scaling down proportionally with feature size, resulting in fixed power for a given electrical link. Another difficulty related to the shortest interconnect scales (on-chip or between chips few centimeters apart) is that the high resistance of the wires and the hence the RC delay of the wires is rapidly increasing. Although this does not immediately affect the bandwidth of the link (because it can be mitigated by using optimized electrical links with repeaters), it has resulted in signal transmission speeds on-chip to be reduced to a small fraction (e.g. less than 10%) of the speed of light.

At the board and backplane level, electrical techniques that use transmitter pre-emphasis and receiver equalization (to reduce inter-symbol interference), and transmission of multiple bits/Hz (to improve coding efficiency and effective bit-rate) have been proposed to solve some of the bandwidth limitations of the off-chip signaling environment and thus to continue to allow electrical interconnects to scale to support ultrashort reach interconnects for VLSI systems. However these solutions are coming at substantially increased complexity, volume, weight, and most importantly, power dissipation. Furthermore, the cost of such electrical solutions, particularly when taking into account the improvements needed of the electrical signaling medium to support the increased bandwidth (e.g. better dielectrics, back-drilled vias, more expensive cables) is not always advantageous when compared to an optical waveguide or fiber.

Because the optical channel has substantially greater bandwidth than the electrical channel, and because the availability of electrical-to-optical conversion technologies are now quite well understood, the use of optical interconnects in ultrashort-reach applications is becoming more plausible and represents an important opportunity for the optics industry. In this context, a useful optical interconnect can be viewed as simply an electrical interconnect (when viewing the endpoints) with efficient optical-to-electrical converters (and vice-versa) designed to minimize the additional cost, weight, and size of the link. To be able to seize this opportunity in its broadest terms, it is also incumbent upon the designer to intimately integrate the photonic components onto the foundry VLSI circuits.

28.4 The Convergence in Optical and Electrical Packaging

To impact electrical interconnects in a meaningful manner, the progress in the integration of optical devices with VLSI electronics is a necessary but not

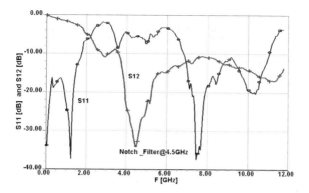

Figure 28.2 S11 and S12 measurement result of single through-hole via of a typical 20+ layer PCB. Signal degradation due to vias become a major limitation for further performance improvements. The increasing need for more off-chip I/Os drives the number of layer counts of PCB and hence the thickness of PCB. For a typical 20+ layer PCB the thickness can exceed 4mm. For signal routing on the top layers, the electrical length of the through-hole via stub is close to the quarter wavelength of a 4-5GHz signal and the via itself becomes a 4-5GHz notch filter. This notch filter will significantly deteriorate signal propagation.[1]

sufficient condition. Progress is also needed at the package level to enable optical interconnects to be truly low-cost, to be compatible with electrical packaging, and to be consistent with electrical system assembly and manufacturing techniques. Fortunately, the electrical and optical packaging industries have seen substantial convergence after the turn of the century as optical packages have begun using silicon carriers to host the photonic components and silicon flip-chip techniques to provide alignment, electrical contacts, and more efficient heat sinking. Optical packages have also experienced the inclusion of intelligent electrical functionality including electrical monitoring and integration of electrical multiplexing (serialization–deserialization), and clock recovery. The more advanced higher-performance photonic modules have also used flip-chip techniques for integration of the photonic and optical components onto the driver circuits (Fig. 28.3).

On the electrical packaging front, flip-chip and ball-grid-array interconnects, once reserved only for the highest performance applications have gradually become mainstream and now represent a significant fraction of the electrical packaging industry. In addition, electrical circuits have increasingly seen clock and data recovery circuits and advanced modulation techniques (including transmitter pre-emphasis, receiver feedback equalization, and multi-level signaling) be integrated into the chips to help mitigate signal integrity problems caused by 2nd level and 3rd level packaging (Figure 28.3).

Looking to the future, one can deduce that for the higher performance optical and electrical packages, the method of delivering data to the optical and electrical components and the need to break electrical bottlenecks, particularly in the 2nd level package, becomes a universal problem. In fact this becomes one of the most critical issues for optical transceivers, particularly if the photonics and electronics are not tightly integrated. It is our opinion that this electrical bottleneck will fuel the further convergence of optical and electrical packaging and will drive further efforts to intimately integrate optical transceivers into electrical packages (such as electrical packages with optical I/O and also optical waveguide-based printed circuit boards) and finally to integrate optical I/O directly with electronic circuits. There is another driver: power. In the past, the electrical power delivery and the thermal and cooling aspects of system design were looked upon as the last step in the design cycle, with electrical power and thermal engineers forced to find acceptable power delivery and cooling solutions in a highly constrained environment with little ability to alter the logical design of the system or to influence the design of the microchips. This is changing. With the advent of 100W+ super-processors that may require over 100Amps of current with low resistive losses and low inductance to be delivered to a single chip socket in a printed circuit board, it is becoming critical to include power delivery and thermal considerations early in the design of the chips.

In the future, we will see power delivery and thermal issues begin to heavily influence the systems design approach and, especially in high-end systems, to dictate the design of the processing and interconnections within the system.

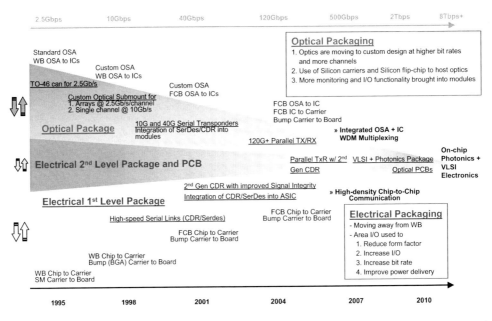

Figure 28.3 Historical trends in packaging of high-performance photonic components and VLSI electronics. Lower horizontal axis shows time and upper horizontal axis show total data-rate from an optical package. Vertical zones show electrical 1st level package (chip and chip carrier), electrical 2nd-level carrier and printed circuit board, and optical package. Progress, convergence, and typical configurations for each zone is shown as a function of time. At the nexus where electrical and optical packaging converge, a unified optical-electronic package that provides both optical and electrical connectivity as well as power delivery and heat removal will be needed to support a new design era that exploits co-existing VLSI photonics and electronics.

Indeed the ability to deliver optical power with low overhead to modulators on a VLSI circuit may be a true, hidden, benefit of optical signaling. In this context, the cost of the interconnect solution will then dictated by the transmitter cost (per Gbps), the cabling cost, and also the power dissipation of the link and the associated cost for cooling the system.

28.5 Photonics-on-CMOS: Some Examples

In the following, we briefly review some of the progress made in integrating optical devices to foundry VLSI CMOS. This article will not attempt to provide a comprehensive and exhaustive review of the literature, but instead offers a sampling of the technology developments, and highlights a few of the successful product development efforts. For a general review, readers may refer to Ref. 2.

We note that tremendous progress has recently being made on monolithic techniques for optoelectronic-VLSI in materials systems that include InP and silicon.[3-4] Because we view computing as being dominated by CMOS-based electronics, we will limit our discussion of monolithic photonics interconnect technologies to silicon. The first work discussed here is based on the use of a hybrid technique via flip-chip bonding.[5] We further note that there are other military and mass consumer applications of hybrid photonics-on-silicon VLSI such as liquid-crystal-on-silicon and epitaxial-liftoff that are being widely deployed such as in display and sensor products. We will focus here on the high-speed optical device and interconnect applications.

28.5.1 Multiple-Quantum Well Modulators on CMOS

The highest-density hybrid optical interconnects to silicon VLSI circuits have been prototyped with flip-chip bonded multiple quantum-well (MQW) modulators-on-CMOS (Fig. 28.4). These devices have high-yield and have low leakage, making them suitable for both light modulation and detection. Over 5,700 thousand high-speed detector/modulator devices have been simultaneously flip-chip integrated onto CMOS chips with a device yield exceeding 99.95%. Each bonded device had a load capacitance of approximately 50fF (65fF including a 15µm × 15µm bond pad) and could be driven by a CMOS inverter at 2.5Gbps (using 0.5micron CMOS circuits) to accomplish the electrical-to-optical interface. This technology produced several switching system demonstrators and was the basis of two multi-project foundry shuttles of optoelectronics that resulted in CMOS-based optoelectronic chips being distributed to research groups around the world.[6] We note that even larger number of such devices have recently been bonded to circuits for infrared imaging applications, firmly establishing the potential of this technology for density and yield.[†]

[†] Up to 1 million quantum well devices have recently been bump-bonded to electronic readout circuits to create high-density quantum-well infrared photodetector arrays for infrared imaging applications with pixel pitches as small as 25microns and fill factors as high as 85% [7].

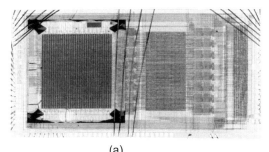
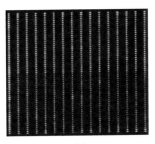

(a)   (b)

Figure 28.4 (a) Micrograph of a mixed-signal CMOS-MQW dense-WDM switch chip with 2000 multiple quantum well modulators attached to the left side of the chip. Each device was approximately 50microns x 20microns. Center-to-center spacing of modulators was 70microns along the horizontal direction and 35microns along the vertical direction. (b) half the array (alternate columns) consisting of 1000 devices were forward-biased as LEDs to show that all devices were functional and connected.

Because the MQW absorption modulators have a relatively broad spectral range (tunable by current and temperature), one may employ wavelength-to-space-division multiplexing via dispersive optics to access multiple devices from a single input and output fiber. Examples of a free-space grating-based WDM data transmission link and switching system for computing applications are provided in Ref. 8 and Ref. 9 respectively.

28.5.2 VCSELs and detectors on CMOS

Single-channel and parallel optical interconnects based on vertical-cavity surface-emitting lasers (VCSELs) are being widely deployed today in switching and routing systems, local area networks, central offices, and data centers. This penetration is a result of many well-known performance and cost advantages of the VCSEL device: its low drive current, favorable high-speed modulation characteristics, high wall-plug efficiency, the ability to complete manufacture and testing at the wafer level, and the ability to tailor the light output to improve coupling to optical fiber. Its low cost is also manifest in its choice as the preferred laser for use in optical mice.

One advantage of the VCSEL device that is currently seeing much interest in the industry today is its ability, with certain modifications, to be directly connected to electrical circuits at the chip and wafer levels.[10] Arrays of VCSELs have been bonded directly to CMOS VLSI chips, with each VCSEL capable of multi-Gigabit/s modulation by the CMOS circuits. Initial approaches used 980nm lasers so that the substrate was transparent to the light output from the bottom-emitting VCSELs.[11] These approaches were extended to 850nm VCSELS bonded to CMOS followed by removal of the substrate.

CMOS chips with interleaved VCSELs and detectors have also been developed to create circuits with optical input and output. Two bonding and substrate-removal steps were applied to accomplish this: the first to bond the PIN detectors, and the second for the VCSELs. The resulting chip after the second

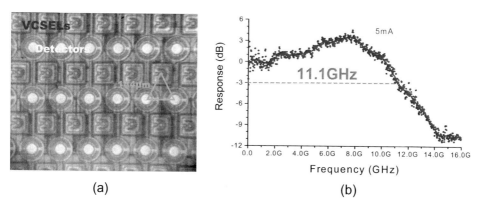

Figure 28.5 (a) micrograph of a CMOS chip with dual-bonded 16 x 16 array of 850nm VCSELs and detectors spaced at 144microns center to center; (b) S12 measurement result of small signal bandwidth of an early flip-chip bonded, substrate-removed, bottom-emitting, 850nm VCSEL in the array when bonded to a silicon tester die.

bond and substrate removal is shown in Fig. 28.5(a). Figure 28.5(b) shows the small signal response of a flip-chip bonded VCSEL in the array, when bonded to an appropriate test chip with high-speed traces. Compact detectors and CMOS transimpedance receiver circuits have been developed to execute the optical-to-electrical conversion. Operation of single-ended receivers (one diode per optical input) fabricated in a 0.25µm linewidth CMOS technology, were been demonstrated over 3Gigabit/s, limited by the driver and receiver circuits. Subsequently, 10Gbps VCSEL driver circuits have been achieved using silicon germanium BiCMOS. Total power dissipation per channel was held to approximately 50mW per transmitter/receiver. More recently, array VCSEL modules have been developed with line rates up to 10Gbps and with array bandwidths of up to 160Gbps.[12]

Another notable development is the integration of coarse WDM VCSELs with CMOS using multiple linear arrays of VCSELs grown separately, but aligned and bonded in a single step.[13] Combined with a space-to-wavelength multiplexer, this allows the fiber connectorization and packaging costs to be amortized over multiple data channels. This technique has been used to create terabits-per-second density interconnections to a single chip.

For commercialization purposes, an important requirement is the need to monitor laser output. Direct optical power monitoring for edge-emitting lasers is relatively straightforward due to the fact that these diode lasers emit light from both front and back facets. This allows the laser to be placed in an assembly where one facet is designed to be the output coupler, while the other facet provides a convenient means of monitoring the output power.

For VCSELs, power monitoring is more complex, because the device does not generally emit light in the rear direction, i.e. through the substrate wafer. A common solution in the industry is to place the completed VCSEL die in an TO-can package fitted with a partially reflective window above the VCSEL aperture and a photodiode also placed in package to collect a portion of the light from the

VCSEL. Such an arrangement is commonly known as a back-reflection monitor. A limitation of back-reflection monitors is the fact that the detector does not uniformly measure the VCSEL output. This is because the optical power emanating from the VCSEL is concentrated into multiple modes, each mode having unique angular and radial optical power distributions that may be selectively captured by the mirror and detector. Hence, VCSELs with different mode structures and divergence angles will result in vastly different monitor photocurrents, even though the output power and packaging of the VCSELs may be identical. This results in considerable complexity during the packaging and calibration of the VCSELs, and is further complicated by the mechanical and optical stability and uniformity of the reflective surface in the TO-can. Monitoring is particularly difficult for flip-chip bonded 850nm wavelength VCSELs since the substrate is absorbing at this wavelength. Because of these and other issues, it is impractical to extend this technique to VCSEL arrays.

A solution often used for optical monitoring of arrays is to build extra VCSELs into the array and use these extra VCSELs solely for optical power monitoring purposes. This technique does not account for the fact that all lasers, even when fabricated closely together on a common substrate, may have different characteristics, and hence is generally not useful when there is variation in power levels across arrays. It is also not valuable as a means of predicting failure or monitoring wearout in the individual VCSELs. Another solution is to monitor side-emissions from the VCSEL by building photodiodes surrounding the VCSEL structure. This usually adds complexity to the VCSEL fabrication process and also does not monitor the emission modes of interest, i.e. the modes that are actually used for data communications. A fourth possibility is to include photodiodes integrally built in the VCSEL structure close to the active layer. Although this approach provides, in principle, the most compact solution, it typically reduces the efficiency and yield of the VCSELs. To circumvent these issues, integrated silicon monitor photodetectors were exploited to continuously capture front-side emissions from a flip-chip bonded VCSEL array.[14]

Note that the relative intensities of the modes emitted by a VCSEL will change with temperature and as the device ages, which means that the conditions observed initially may not exist over the entire operating lifetime of the VCSEL, and significant distortions of the power feedback signal can result over time. One of the advantages of having continuous monitoring of the VCSEL on an individual basis, is that one may use such data to help predict wearout or failure of an individual device, and effectively use sparing to boost the lifetime of the array.[15]

28.6 Silicon Photonics

The field of silicon-based optoelectronics has experienced renewed interest recently and has seen the development of on-chip waveguides, optical couplers, and high-speed optical modulators.[16-17] The development of the waveguide transmission medium and the optical output modulator devices is significant, because together with silicon CMOS-compatible optical detectors, a complete silicon CMOS tool suite is now available enabling us to consider, for the first time, widespread deployment of custom silicon photonics modules.[4]

There are several advantages in considering such an optical interconnect system. Most previous planar lightwave circuits were constructed of lightly doped silicon dioxide waveguides with very weak index contrasts (Δn) on the order of ~0.01. With the use of silicon waveguides and SiO_2 cladding, a much higher index contrast (greater than 2) can be achieved. Compared to previous technologies that relied on on-chip waveguide dimensions ~5λ-10λ (similar to an optical fiber), the use of silicon as the waveguide material allows the confinement of light in a much smaller waveguide having sub-wavelength width (<1micron). This in turn enables faster and more efficient electrical modulators and detectors.

Along with these advantages of high-index contrast waveguides are some concerns. As the index contrast in increased, the scattering and transmission losses of the index guided modes due to the roughness of the waveguide sidewalls and other manufacturing imperfections are increased. Another interpretation of this is that the constraints or requirements of roughness and manufacturing uniformity are increased if a particular loss target is required. These issues have fortunately been automatically solved as the feature sizes of silicon CMOS have been reduced. Indeed the enormous investments into the electronics industry have, in fact, paved the way for sub-wavelength width, low-loss waveguides, and high integration density.

There has also been strong progress made in active devices such as silicon modulators.[18-19] Because of the weak electro-optic effect in silicon, early modulators were based on Mach-Zehnder interferometers and were relatively large (several square millimeters) and not suitable for large-scale integration. More recent results have shown that the use of resonant structures can enhance the electro-optic effect and can be controlled to produce fast modulation in compact structures.[20]

28.7 Commercialization of Optoelectronics-on-VLSI Technologies

Several efforts have been launched to commercialize various photonics-on-CMOS technologies. In the following, we briefly describe three examples.

28.7.1 VCSELs and PIN photodetectors on bulk CMOS

The first two-dimensional array parallel optical interconnect module having 36 channels, with each channel operating up to 3.3Gigabit/s, is described in Ref. 21. The transmitter/receiver module pair eliminated wire-bonds between the driver/receiver chips and the corresponding VCSEL/detector chips thereby reducing crosstalk and power dissipation, simplifying packaging and alignment, and simultaneously improving bandwidth and total link jitter performance (Fig. 28.6).

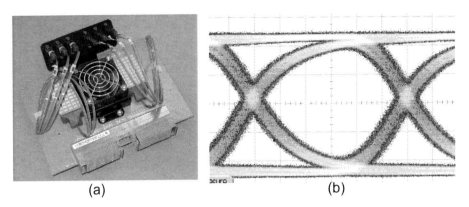

Figure 28.6 (a) photograph of an optical mezzanine test card assembled and ready to be mated to a system line card. The mezzanine card held two 36-channel modules: one 36-channel receiver module based on P-I-N photodetectors flip-chip bonded onto CMOS, and one 36-channel transmitter module based on 850nm VCSELs flip-chip bonded onto CMOS. Each module was pigtailed to three standard 12-channel multi-mode fiber connectors (b) Typical eye diagram for complete transmitter-receiver link at 3.3Gbit/s using a $2^{23}-1$ pseudo-random bit sequence.[21]

28.7.1 VCSELs and PIN photodetectors on bulk CMOS

The first two-dimensional array parallel optical interconnect module having 36 channels, with each channel operating up to 3.3Gigabit/s, is described in Ref. 21. The transmitter/receiver module pair eliminated wire-bonds between the driver/receiver chips and the corresponding VCSEL/detector chips thereby reducing crosstalk and power dissipation, simplifying packaging and alignment, and simultaneously improving bandwidth and total link jitter performance (Fig. 28.6).

The module pair had a transmission distance range of 1m to 1km (the latter achieved with 2000MHz-km 50micron core fiber) and was designed to work with both 50micron core and 62.5micron core graded-index fiber. The module had a remateable electrical connector and a built-in fiber management system as well as many built-in testability and system-level monitoring tools, including loss-of-signal on a per-channel basis, temperature sensors and VCSEL bias current and voltage monitors on a per-channel basis. Two key issues in a dense optoelectronic module of this kind are crosstalk and reliability. Solutions to these issues are respectively shown in Figures 28.7(a) and 28.7(b).

28.7.2 VCSELs and PIN photodetectors on SOS-CMOS technology

Another option is to flip-chip bond Gallium Arsenide VCSEL and PIN arrays onto CMOS-on-Sapphire integrated circuits. The transparent Sapphire substrate provides the ability to flip-chip bond top-emitting VCSELs so that the light emits from the substrate side of the assembly. Because insulating substrates such as Sapphire are often used for radiation hardening, SoS-based optoelectronics can

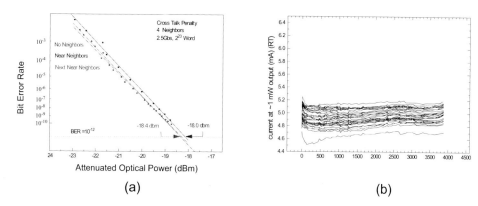

Figure 28.7 (a) Crosstalk measurement of parallel optical module, showing minimal penalty to channel under test as the four neighboring and all other channels are operated under stressed conditions (6dB excess power); (b) reliability of VCSEL arrays used in 36-channel parallel optical module. This is measured by plotting the electrical current required to maintain 1mW of optical output as a function of time (in hours) when operated under in a stressed thermal environment (80C case, 120C junction).

address several technical hurdles in building parallel optical transceivers for harsh space environments and military-aerospace applications. The first product to be deployed based on this technology was a 4 + 4 transceiver with each channel operating at 2.5 gigabits per second.

The integrated transceiver circuit shown in Fig. 28.8 was fabricated in Peregrine Semiconductor's silicon-on-sapphire (SoS) 0.5 micron CMOS process.[22] This process uses a ~100 nm epitaxial silicon layer to form metal-oxide semiconductor (MOS) transistors on a sapphire substrate. The active regions of individual transistors are fully isolated by sapphire and silicon-dioxide. Source/drain junction capacitance and interconnect-to-substrate capacitance are eliminated, as is substrate crosstalk. Flip-chip bonding was used to attach four-element commercially available top-emitting VCSEL and PIN detector arrays to the IC. The VCSEL array was mounted so that the optical signals propagate through the transparent sapphire substrate. The SoS transceiver was constructed by mounting a corner-turn microlens array on the sapphire substrate side of the transceiver IC. This microlens array couples the VCSEL light into 50 micron core fiber and couples light from the fiber to the PIN detectors. The lens and reflecting surfaces are environmentally sealed. The heat from the IC is coupled to the host PCB with a contoured thermal plug. A lid is attached to the substrate to add structural support and the MT ferrule connection. Lateral integrated photodiodes implemented in the IC process were used to monitor optical output power using the front-side emissions, and embedded controllers were developed to regulate power on a per-channel basis. The resulting module offered optical power monitoring on each VCSEL to maintain constant power and extinction ratio over the -40 to 100 degrees Celsius temperature range and VCSEL aging.

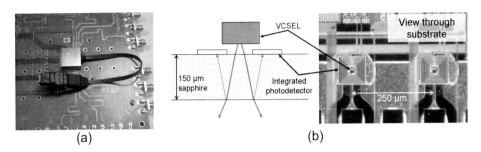

Figure 28.8 (a) Four-channel SoS transceiver chip with flip-chip attached VCSEL and photodetector array; (b) Transmitter optical path showing with top-emitting VCSEL on SoS with optical signal monitoring.[22]

We note that power monitoring and other intelligent functionality on an individual channel-by-channel basis is an important general advantage of the intimate integration of VCSELs with VLSI circuits and has also been proven for bottom-emitting VCSEL geometries.[14] Since the transceivers described above monitors the average power per channel at both ends of the fiber link, the network can calculate the cable plant link loss, and VCSEL threshold and slope efficiency can be continually monitored for signs of aging.

28.7.3 Silicon photonic active cable

Active optical cables were recently introduced as a new form of transceivers that included the optoelectronics components into the end-points of the cable, so the user could plug in the optical active cable into an existing socket intended for an electrical transceiver. Power for the optical transceiver is provided from the connection to the socket. This eliminates the need to redesign the system for the optical part, and also enables new optical technologies to compete with existing optical technologies in a manner transparent to the user. The first transceiver product based on silicon photonics is an active optical cable. The first 40 Gigabit active optical cable, developed by Luxtera is depicted in Fig. 28.9. The cable is based on a four-channel, 10Gbps per channel, silicon photonic transceiver chip with integrated CMOS-compatible optical couplers, silicon modulator-based transmitters, and integrated Si-Ge based photodetectors. Light from a single laser (inset) is coupled into the chip using a grating coupler, is split equally into four waveguides, and is used to power four separate channels. Each transmit channel includes a Mach-Zenhder modulator with a distributed driver and is routed to an output coupler. Closed loop control using a monitor detector and a variable optical attenuator maintains the correct bias points and output signal. On the receiver side, each channel includes an input coupler, and a photodetector whose output is fed to a transimpedance amplifier built in 130nm CMOS.[23]

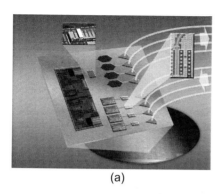 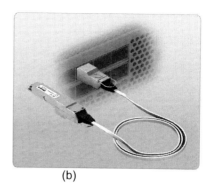

(a) (b)

Figure 28.9 (a) A four-channel silicon photonic transceiver chip with integrated CMOS-compatible optical couplers, silicon modulator-based transmitters, and integrated Si-Ge based photodetectors. A single flip-chip attached laser (inset) is used to power all four channels; (b) Active optical cable plugs into a socket intended for a Quad-SFP transceiver.[23]

28.8 Discussion

We have reviewed the motivations, challenges, and potential for achieving "optical-interconnects-to-the-chip" via the intimate integration of photonics components such as lasers, detectors, and modulators with silicon VLSI electronics. We also reviewed the progress made towards commercializing this technology for high-density optical transceivers. We believe that one of the primary arguments for optical I/O is that electrical I/O alone will not be able to efficiently meet the continually advancing needs of systems based on silicon VLSI electronics. For serious consideration of any photonics-on-VLSI technology as a viable replacement for electrical I/O, it is important that the quality of the electronics should not be compromised. Therefore, any integration scheme must be able to use state-of-the-art electronics without any significant degradation, preferably allowing three-dimensional integration so that the density of the electronic integration is not reduced. Hence a strong preference for processes which are as non-invasive as possible to the VLSI electronics manufacturing processes. Here, we have described one such hybrid technique that allows attachment of optical devices after the fabrication of the electronic devices is completed. Thousands of microsolder bumps can be produced on VLSI wafers by photolithography and evaporation. Flip-chip bonding optionally followed by substrate removal is then used to attach many devices at once. The temperatures involved are below $200^{\circ}C$ and so do not affect the silicon. Modulators, detectors, and lasers were bonded to metal interconnect layers on the Si chip directly above transistor layers, creating a three-dimensional optoelectronic/VLSI circuit. LEDs could also be a useful device for several applications but were not explicitly discussed here. Many other techniques have been proposed, each with its own merits. Perhaps the ultimate integration will allow photonics-"in"-VLSI, so that the design and manufacture of the photonic components will share an identical wafer-scale fabrication facility and the integration will be seamless and wafer-scale. Such concepts are being investigated in a silicon-based material system.

Another guiding principle for any integration technique also derives from the argument that electrical I/O will be unable to keep pace with the needs of silicon VLSI electronics. From this argument, it is apparent that an optical I/O technology will not be useful in the broadest sense unless the number of optical channels per chip can be measured in hundreds or even thousands, since it is not until this number is reached by pin-out count that there will be a serious density or power dissipation crisis. Therefore, any integration process must be capable of producing such numbers of optoelectronic devices, preferably in a single step. The area and power of the optical devices must be appropriately tailored to meet this density goal. The ability of the optical interconnects to provide a complex interconnection topology also becomes an advantage at these large numbers.

As we look forward, the efficient delivery of power to the chip and the effective dissipation of the heat created on the circuits will be critical. The need for the most energy efficient optical transceivers will dominate over other design considerations. We can expect to see a hundred-fold reduction in energy to communicate an optical bit of information, thereby enabling optical interconnects to transition to the intra-chip stage and provide systems level benefits exceeding that offered by scaled electrical technologies. To meet this aggressive challenge, a completely new class of disruptive components not previously explored with photonic-enhanced silicon VLSI circuits will be needed. Developing both a broad spectrum of components as well as a comprehensive toolbox of components and design tools will greatly help. Future research programs will explore changing the physical dimensionality of component designs, exploit newly discovered material science, and develop novel fabrication methods for 3-D bulk effects, 2-D quantum wells, 1-D nanowires, and zero-dimensional quantum-dot structures.

Further innovation will come from the electronics side. Innovative low-voltage, low-energy circuit architectures and families can result in dramatically reduced energy per bit for end-to-end optical links. For instance low-area and low-power TIAs and sense-amplifier-based receivers using simple CMOS front-end circuits can scale to electrical energies of 100fJ with photon energies of 1fJ. In addition to physical improvements, we can also expect architecture and system innovations to play a large role in heralding optical interconnects deeper into the system interconnect hierarchy.

Based on relatively conservative assumptions on how these components will evolve, a general conclusion is that it appears this integrated photonics-on-VLSI technology has substantial room for continued scaling to large numbers of higher-speed interconnects. Indeed, future optoelectronic-VLSI technologies can be expected to provide an I/O bandwidth to a chip that is commensurate with the processing and switching power of the chip, even in the finest linewidth silicon: a task that cannot be expected from conventional electrical interconnect technologies. Prototype technology demonstrations using hybrid integration techniques have already shown that multiterabit capacities are achievable. We have also seen the advent of CMOS-compatible silicon photonics technologies that allow some of the functions of detection and modulation to be implemented in the VLSI electronics, and hence presumably be more cost-effective and efficient. With emerging technologies such as those we have briefly glimpsed in this paper, one might readily imagine 100Tbps of I/O from a single electronic chip package.

Acknowledgements

The author gratefully acknowledges exciting and productive collaborations with John E. Cunningham, Keith W. Goossen, William Y. Jan, James Walker, and Richard Rozier, without whom the work described above would not have been possible. The author is also grateful to Dawei Huang, Charles Kuznia and Cary Gunn for valuable input and many helpful discussions.

References

1. D. Huang et al., "The Chip Multithreading Architecture and Parallel Optical Interconnects," *Proc. IEEE LEOS Summer Topical Meeting on Optical Interconnects*, San Diego, pp. 75-76 (2004).

2. D. A. B. Miller, "Rationale and challenges for optical interconnects to electronic chips," *IEEE Proceedings,* Vol. 88, No. 6, pp. 728-749 (2000).

3. R. Nagarajan, M. Kato, et al., "Single-chip 40-channel InP transmitter photonic integrated circuits capable of aggregate data rate of 1.6Tbps," *Electronic Letters,* Volume 42, No. 13, pp. 347-349 (2006).

4. C. Gunn, "CMOS Photonics for High-Speed Interconnects," *IEEE Micro*, Vol. 26, No. 2, pp. 58-66 (2006).

5. K. W. Goossen and A. V. Krishnamoorthy., "Optoelectronics-in-VLSI" in *Wiley Encyclopedia of Electrical and Electronic Engineering,* Vol. 15, pp. 380-395 (1999).

6. A. V. Krishnamoorthy and K. W. Goossen., "Optoelectronic-VLSI: photonics integrated with VLSI circuits," *IEEE Journal. of Select. Top. in Quantum Electronics,* Vol. 4, No. 6, pp. 899-912 (1998).

7. M. Jhabvala, K. K. Choi, C. Monroy, and A. La, "Development of a 1K · 1K, 8–12 micron QWIP array," *Infrared Physics & Technology,* Vol. 50, pp. 234–239 (2007).

8. B. E. Nelson, G. A. Keeler, et al., "Wavelength Division Multiplexed Optical Interconnect Using Short Pulses," *IEEE Journal. of Select. Top. in Quantum Electronics,* Vol. 9, pp. 486-491 (2003).

9. A. V. Krishnamoorthy et al., "The Amoeba switch: an optoelectronic switch for multiprocessor networking using dense-WDM" *IEEE Journal. of Select. Top. in Quantum Electronics,* Vol. 5, No. 2, pp. 261-275 (1999).

10. L. M. F. Chirovsky et al., "Vertical-cavity surface-emitting lasers specifically designed for integration with electronic circuits," Chap. 2 in *Heterogeneous*

Optoelectronics Integration **CR76**, SPIE Press, Bellingham, WA pp. 49-74 (2000).

11. A. V. Krishnamoorthy et al., "Vertical-cavity surface emitting lasers flip-chip bonded to gigabit/s CMOS circuits," *Photonics Technology Letters,* Volume 11, pp. 128-130 (1999).

12. F. E. Doany, et al., "160-Gb/s Bidirectional Parallel Optical Transceiver Module for Board-Level Interconnects Using a Single-Chip CMOS IC," *Proc. Electronic Components and Technology Conference, 2007. ECTC apos;07.* Vol. 57, pp. 1256 – 1261 (2007).

13. B. E. Lemoff et al., "MAUI: enabling fiber-to-the-Processor with parallel multiwavelength optical interconnects," *IEEE J. Lightwave Technology,* Volume 22, No. 9, pp. 2043 – 2054 (2004).

14. K. W. Goossen et al., "1x12 VCSEL array with optical monitoring via flip-chip bonding," *IEEE Photonics Technology Letters,* Vol. 18, No. 18, pp. 1219-1221 (2006).

15. J. E. Cunningham, D. K. McElfresh, et al., "Scaling VCSEL reliability for petascale systems," *Applied Optics,* Vol. 45, No. 25, pp. 6343-6348 (2006).

16. R. Soref, "Silicon based Optoelectronics," *Proceedings of the IEEE,* Vol. 81, No. 12, pp. 1687-1706 (1993).

17. B. Jalali, and S. Fathpour "Silicon photonics," *IEEE Journal of Lightwave Technology,* Vol. 24, No. 12, pp. 4600-4615 (2006).

18. A. Liu, D. Samara-Rubio, L. Liao, and M. Paniccia, "Scaling the modulation bandwidth and phase efficiency of a silicon optical modulator," *IEEE JSTQE,* Vol 11, No. 2, pp. 367-372 (2005).

19. P. Dainesi, A. Küng, et al., "CMOS compatible fully integrated Mach-Zehnder interferometer in SOI technology," *IEEE Photonics Technology Letters,* Vol. 12, No. 6, pp. 660-662 (2000).

20. Q. Xu, B. Schmidt, S. Pradhan, M. Lipson, "Micrometre-scale silicon electro-optic modulator," *Nature,* Vol 435, No. 19, pp. 325-327 (2005).

21. C. Cook, J. Cunningham et al., "A 36-channel transceiver parallel optical interconnect module based on optoelectronics-on-VLSI technology," *IEEE Journal of Selected Topics in Quantum Electronics,* Vol. 9, No. 2, pp. 387-399 (2003).

22. C. B. Kuznia, J. F. Ahadian, R. J. Pommer, and R. Hagan, "Integration of fiber optical cable diagnostics within aerospace transceivers," *Proc. Avionics Fiber-Optics and Photonics Conf,* pp. 58-59 (2006).

23. A. Narasimha, B. Analui, et al., "A 40Gbps QSFP optoelectronic transceiver in a 0.13micron CMOS silicon-on-insulator technology," *Proc. Optical Fiber Communications (OFC/NFOEC) Conf.,* (2008).

Chapter 29
Echelle and Arrayed Waveguide Gratings for WDM and Spectral Analysis

Pavel Cheben, André Delâge, Siegfried Janz, and Dan-Xia Xu
National Research Council Canada

29.1 From Rittenhouse's Discovery to Planar Waveguide Gratings
29.2 Waveguide Echelle Gratings
    29.2.1   Grating geometry
    29.2.2   Grating equation
    29.2.3   Stigmatization
    29.2.4   Passband
    29.2.5   Waveguide platform
    29.2.6   Slab waveguide birefringence and polarization-dependent wavelength shift
    29.2.7   Polarization dependence of grating diffraction efficiency
    29.2.8   Grating loss
    29.2.9   Errors in facet positions and the influence on crosstalk
    29.2.10 State-of-the-art waveguide echelle grating devices
29.3 Arrayed Waveguide Gratings
    29.3.1   Fundamentals
    29.3.2   Phase error influence on crosstalk and chromatic dispersion
    29.3.3   Passband
    29.3.4   Loss
    29.3.5   Polarization-dependent wavelength shift
    29.3.6   Temperature dependence
    29.3.7   AWGs in glass waveguides
    29.3.8   AWGs in InGaAsP/InP
    29.3.9   AWGs in silicon-on-insulator
    29.3.10 Fourier transform AWGs and interferometer arrays
Acknowledgments
References

29.1 From Rittenhouse's Discovery to Planar Waveguide Gratings

The diffraction grating story is one of a remarkable success. When in 1789 American astronomer David Rittenhouse made the first diffraction grating by wrapping a wire around the threads of two fine-pitch screws and used it to measure the wavelength of light,[1] no one could foresee the impact this device would have on our lives. Diffraction gratings have been, for more than a century, an essential tool for uncovering the world at both microscopic and cosmological scales by means of spectroscopy. Also, the profound recent changes in the way our society communicates, including the Internet, have been made possible thanks to various diffraction grating-based devices that are key elements in modern telecommunication optical networks.

The foundations of diffraction grating technology were laid in the first half of the nineteenth century by pioneering work of Joseph von Fraunhofer.[2] To make his gratings, he built the first ruling engine. With such ruled gratings he discovered dark lines in light emitted by several substances and observed similar spectral lines in light from the sun and other stars. Later, he used these absorption lines, now known as Fraunhofer's lines, as markers for precise measurements of the refractive index of glass used in achromatic lenses. He also derived and experimentally verified the grating equation. In the 1870s, Rayleigh predicted that gratings could outperform even the most powerful prisms in terms of spectral resolution, and several such gratings were soon made. Toward the end of the nineteenth century, Henry A. Rowland at John Hopkins University was producing gratings of unprecedented high spectral resolution and accuracy. The availability of new and accurate spectroscopic data obtained with such gratings was the key for fundamental discoveries on the nature of spectral lines by Rydberg, Balmer, Runge, Zeeman, and others. These spectral studies were of singular importance paving the road toward the era of atomic physics and quantum mechanics.

However, the grating alone does not suffice for making accurate spectroscopic measurements. It needs to be specifically positioned[i] in an optomechanical setup capable of collecting the input light and focusing the spectrally dispersed light by the grating on an output port. The latter can be either a simple slit aperture in the case of a monochromator, or a photographic film (in modern devices often a photodetector array) in the case of a spectrometer. Essential for the development of spectroscopic instruments was a discovery, made by Rowland, that diffraction gratings can be formed on a concave (rather than plane) substrate. This way, the light is simultaneously dispersed and focused, obviating the need for additional focusing optics (lenses). Rowland also proposed a grating configuration that improves focusing, eliminates the primary coma, and minimizes spherical aberration. In this mounting arrangement, both the input and its spectrally dispersed image are formed on an arc that lies on a

---

[i] This grating positioning in an instrument is often referred to as "grating mount."

circle[ii] tangent to the concave grating, but with the radius of half the grating focal length. The Rowland circle geometry is used in various configurations of modern spectroscopic instruments,[3] and also, as we will see later in this chapter, in waveguide echelle grating[4-8] and arrayed waveguide grating (AWG) devices.[9]

In modern optical fiber networks, wavelength division multiplexing (WDM)[10] is used to increase the amount of data transmitted in an optical fiber by employing multiple wavelengths. These different wavelengths, each modulated with a different stream of data, need to be combined (multiplexed) and coupled into the input end of the optical fiber, and then at the output end, separated from each other (demultiplexed). These two functions are often performed by arrayed waveguide grating optical (de)multiplexers, and more recently also by waveguide echelle grating (de)multiplexers. The applications of these devices are not limited to WDM. They can also be used as compact spectrometer chips for applications in spectroscopy, metrology, and chemical and biological sensing, among others.

The aim of this chapter is to provide the reader with an overview on waveguide echelle grating (EG) and arrayed waveguide grating (AWG) technologies, from the basic principles to the most recent developments, including examples of advanced devices. For recent reviews, also see Refs. 11-14.

Before proceeding to the next two sections on waveguide echelle grating[iii] and arrayed waveguide grating devices, a general unifying statement can be made: Despite obvious differences in EG and AWG layouts, they both rely on the same fundamental principle, namely, multiple path optical interference. We will point out this and other parallels, and also distinctions, between EG and AWG devices at various points of this chapter.

29.2 Waveguide Echelle Gratings

An obvious question that one may ask is how small a spectrometer can be made and using what kind of fabrication technologies. A straightforward approach is to assemble separate miniature optical elements such as lenses and diffraction gratings in a miniature mount with input and output optical fibers attached.

Such devices have been developed and are commercially available for WDM[11] and spectroscopic[12] applications. Device size can be further reduced using optical waveguides. The compact optical fiber-based spectrometer was first demonstrated in the late 1970s. It used a gradient index (GRIN) lens as a collimator attached on one end to the input and output fibers and, on the other end, to a miniature diffraction grating.[16]

The next step in miniaturization is to use the planar waveguide technology. The first such devices were demonstrated in the early 1980s[17-21] and were further

---
[ii] The Rowland circle (see Section 29.2.1, Fig. 29.2).
[iii] Since planar waveguide gratings for WDM applications are typically designed to work in high diffraction orders, i.e., as *echelle* gratings, in this chapter we will preferentially use the term "waveguide echelle grating" rather than the term "planar waveguide grating."

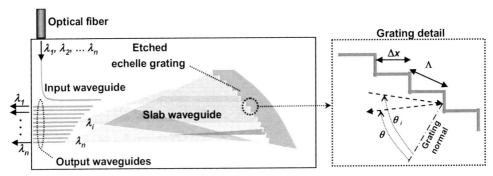

Figure 29.1 Schematics of an echelle grating planar waveguide demultiplexer.

developed in the 1990s at Siemens AG,[4,22] NRC Canada[6] and Bellcor.[5,23] In planar waveguide grating devices, a conventional bulk optics design is, essentially, "projected" onto a waveguide plane, as is schematically shown in Fig. 29.1.

We will now discuss the fundamentals of waveguide echelle grating devices, starting from the basic principles and finalizing with a few examples of state-of-the-art devices.

29.2.1 Grating geometry

In planar waveguide grating (de)multiplexers, the configuration shown schematically in Fig. 29.1, and including more device details in Fig. 29.2, is often used.[4-8] The input and output channel waveguides join the slab waveguide at the Rowland circle. The light from the input waveguide is coupled into the slab waveguide where it diverges in the waveguide plane, illuminating the grating. The light is diffracted backward from the concave (focusing) grating. Since the grating is used in the reflection geometry, the grating facets are often metalized to increase the efficiency.

Different wavelengths are focused at different positions along the Rowland circle where they are intercepted by different output waveguides. The physical separation between the adjacent output waveguides at the Rowland circle is determined, for a given wavelength channel separation $\Delta\lambda$, by the grating focal length[iv] and dispersion. The latter can be controlled by varying the grating order (m) or pitch (Λ), as will be discussed in the next section.

29.2.2 Grating equation

In a waveguide grating spectrometer, the light propagates as an optical mode guided in the slab waveguide of effective index n_{eff}. The grating is etched through the waveguide layers in the direction z perpendicular to the waveguide plane xy (Fig. 29.2). The grating facets are often metal coated to improve reflectivity so that each facet acts as a small mirror reflecting the incident light. The light

---

[iv] The grating focal length $f = 2r$, where r is the Rowland circle radius.

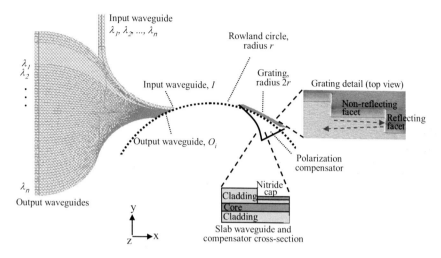

Figure 29.2 Demultiplexer layout similar to that reported in Ref. [7].

reflected by different facets will add in phase, hence producing interference maxima in those directions θ^v for which the phase difference between the light diffracted by two adjacent facets is $2\pi m$. This is expressed by the scalar grating equation,

$$\sin\theta + \sin\theta_i = \frac{m\lambda}{n_{\text{eff}}\Lambda}, \qquad (29.1)$$

where θ_i is the angle of incidence, λ is the wavelength measured in vacuum, Λ is the grating period (pitch), and $m = 0, \pm 1, \pm 2, \ldots$ is the diffraction order (see Fig. 9.1, grating detail). For gratings with small ratios λ/Λ, many diffraction orders may exist. Such gratings are called echelle gratings[vi] and their important characteristics are the large dispersion and resolution.

By differentiating Eq. (29.1), the grating angular dispersion is obtained,

$$\frac{d\theta}{d\lambda} = \frac{m}{n_{\text{eff}}\Lambda\cos\theta}. \qquad (29.2)$$

The grating resolution, $R = \lambda/\Delta\lambda$, also called resolving power,[vii] is the ability of a grating to separate adjacent wavelengths,[3]

---

[v] The angles measured from the grating normal and positive in the sense of the arrow shown in Fig. 29.1, grating detail.

[vi] From French *échelle*, ladder, because of the staircaselike arrangement of the grating facets as shown in Figs. 29.1 and 29.2, grating detail.

[vii] The definition of grating resolving power by Eq. (29.3) is obtained using Rayleigh criterion: the two peaks of wavelengths λ and $\lambda+\Delta\lambda$ are considered resolved if their far fields are angularly separated by at least the half-width $\Delta\theta$ of the far-field diffraction pattern, where $\Delta\theta \sim \lambda/(Ln_{\text{eff}}\cos\theta)$. In (de)multiplexers, the adjacent channels need to be separated more than this Rayleigh limit in order to minimize the interchannel crosstalk.

$$R = mN = \frac{L(\sin\theta_i + \sin\theta)n_{eff}}{\lambda}, \qquad (29.3)$$

where N is the number of grating facets and $L = N\Lambda$ is the grating length.

For a concave grating in a Rowland circle configuration in the coordinates of Fig. 29.2 and including the effect of waveguide effective index dispersion $dn_{eff}/d\lambda$, the following relation for the grating linear dispersion is obtained (assuming $\sin\theta \sim \sin\theta_i$)[24]

$$D = \frac{dy}{d\lambda} = \frac{4r\sin\theta}{\lambda}\left(1 - \frac{\lambda}{n_{eff}}\frac{dn_{eff}}{d\lambda}\right), \qquad (29.4)$$

where r is the radius of the Rowland circle.

From the above equations it follows that the resolution and dispersion increase with the diffraction order m at which the grating operates. Both waveguide echelle grating and arrayed waveguide grating devices use comparatively large diffraction orders. For WDM applications, $m \sim 20$ is typically sufficient for (de)multiplexing of wavelength channels separated at 100 GHz. Working in high orders, however, comes at a cost of a reduced free spectral range (FSR). The FSR is the difference between two wavelengths that are diffracted into the same direction in successive orders. From Eq. (29.1), $\theta(m,\lambda) = \theta(m\pm1, \lambda \mp \text{FSR})$, and $\text{FSR} = \lambda/m$. In some applications, for example, in broadband spectroscopy, a large free spectral range is preferred, which demands working in low diffraction orders.

29.2.3 Stigmatization

In well-designed waveguide EG devices, the position of the grating teeth are slightly shifted from the exact Rowland geometry in order to reduce residual aberrations in the Rowland circle configuration. This procedure is called stigmatization.[25,13] Designs with two stigmatic, i.e., aberration-free, wavelengths λ_i and λ_k are often used.[25] Such stigmatic correction provides aberration-free (stigmatic) imaging of the input waveguide I onto the outputs O_i (for λ_i) and O_k (for λ_k) (see Fig. 29.2). The stigmatic imaging is guaranteed if the light that originates from arbitrary adjacent grating teeth arrives at these two focal points in phase (modulo 2π), which can be achieved by introducing small displacements in the grating teeth positions. This procedure eliminates phase errors at the wavelengths λ_i and λ_k. The residual phase errors at other wavelengths can be minimized by a further stigmatization step in which output waveguide positions are slightly displaced from the Rowland circle. This stigmatization procedure would obviously be extremely difficult to implement with bulk optics diffraction gratings. In waveguide gratings, though, the optimized positions of the grating facets and the output channels can easily be produced on the lithographic mask and then transferred on the chip.

29.2.4 Passband

For a concave (focusing) grating, the light spatial distribution in the focal region is the image of the input profile, both being typically Gaussian. As the wavelength changes, the focal spot moves across the output waveguide aperture. The focal field is coupled to the fundamental mode of the output waveguide with a coupling efficiency given by the overlap integral of the focal field and the output waveguide mode. The channel passband, which is the spectral dependence of the coupling efficiency for a given wavelength channel, is thus determined by the convolution of the focal field and the output waveguide mode spatial distributions. Since both distributions are nearly Gaussian, the channel passband is also, to a good approximation, Gaussian.[viii]

In practical devices though, a flattened and broadened passband is preferred. Compared to a Gaussian passband, the latter provides better tolerance to laser wavelength drifts and variations of the passband center wavelength due to manufacturing inaccuracies, temperature changes, or polarization-dependent wavelength shifts. This also helps widen the cumulative passband with cascading multiple filters. The channel passband can be widened and flattened by several techniques originally developed for AWG devices (see Section 29.3.3). A technique successfully implemented in EGs uses the small and controllable changes (dithering) of the grating teeth positions,[26] similar to the stigmatization process discussed in Section 29.2.3. This way, a specific phase and intensity distribution can be created along the focal line. For example, a top-hat or a double peak intensity distribution can be created. Such focal fields, after convolution with the output waveguide mode profile, result in a flatter and a wider passband compared to the original Gaussian profile.

29.2.5 Waveguide platforms

Waveguide EG devices can be implemented on different waveguide platforms; for example, silica (SiO_2) glass on silicon (Si) substrate, also called silica-on-silicon [Fig. 29.3(a)], silicon-on-insulator [Fig. 29.3(b)], and InGaAsP/InP [Fig. 29.3(c)].

Silica glass waveguides are the most conventional and widely used. In these waveguides, the light is guided in the waveguide core made of SiO_2 glass, which is slightly doped (e.g., with Ge or P) to increase its refractive index compared to the surrounding silica claddings. The core and cladding layers are deposited on a silicon wafer. The principal advantages of silica glass waveguides, making them currently the predominant commercial waveguide technology, include a very low waveguide propagation loss (< 0.05 dB/cm), optimal modal field matching with standard (SMF-28) optical fibers yielding a high fiber-chip coupling efficiency, and superbly developed glass deposition technologies. The latter include plasma-enhanced chemical vapor deposition (PECVD), a common technique in

---

[viii] For a typical Gaussian channel passband, see Section 29.3.2, Fig. 29.6.

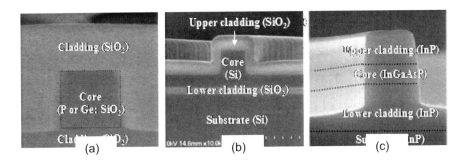

Figure 29.3 Examples of planar optical waveguides. Cross sections of (a) glass buried channel waveguide (silica-on-silicon), (b) silicon-on-insulator (SOI) ridge waveguide, and (c) deep ridge waveguide in InGaAsP/InP.

microelectronic industry, and flame hydrolysis (FH), which was originally developed for making fiber optics preforms. For planar waveguide fabrication technologies, see the recent review by Lamontagne.[27]

We will now inspect some aspects of the waveguide echelle grating device that are of critical importance for its performance in WDM networks.

29.2.6 Slab waveguide birefringence and polarization-dependent wavelength shift

The slab waveguide is an essential part of a waveguide echelle grating device. It is where the grating is formed and where the input light propagates to illuminate the grating and is diffracted toward the focal region, with each wavelength λ propagating at a specific angle $\theta(\lambda)$. In order to guarantee well-defined conditions required for these functions, the slab waveguide must support only a fundamental mode. However, even in a single-mode waveguide, a complication arises since the fundamental mode can exist in two polarization states, known as the transverse electric (TE) and transverse magnetic (TM) modes.[ix] The TE and TM modes have the electric field and the magnetic field polarized in the plane of the waveguide (Fig. 29.2, plane xy), respectively. These two polarization modes generally have distinct propagation constants (β_{TM}, β_{TE}), thus different mode effective indices $n_{eff,TM,TE} = k/\beta_{TM,TE}$, where $k = 2\pi/\lambda$ is the wave vector. This results in an effective index birefringence of $\Delta n_{eff} = n_{eff,TE} - n_{eff,TM}$. This also means that the TE and TM modes have different effective wavelengths in the waveguide, $\lambda_{eff} = \lambda/n_{eff}$. Since the very nature of a diffraction grating is its diffraction direction dependence on the wavelength [Eq. (29.1)], the TE and TM modes will be diffracted by the grating at slightly different angles $\theta_{TE} \neq \theta_{TM}$, and

---

[ix] Since optical fibers deployed in telecommunication networks do not preserve light polarization, the light coupled from the fiber to a (de)multiplexer device does not have a defined polarization state. One needs to assume that both TE and TM modes can be excited in the device at any time.

thus will arrive at different focal positions. If at a given vacuum wavelength λ the light of one polarization is focused at a certain output waveguide, the other polarization would arrive shifted along the focal line and may reach another output waveguide. In such a situation, a specific output waveguide receives TE and TM signals with different vacuum wavelengths, λ_{TE} and $\lambda_{TM} = \lambda_{TE} + \Delta\lambda$, respectively. This polarization-dependent wavelength shift $\Delta\lambda$ in demultiplexer spectral response would seriously degrade the device performance, particularly the interchannel crosstalk. The specifications for practical devices demand the wavelength shifts $\Delta\lambda$ of less than 0.01 nm, which requires the waveguide birefringence[x] Δn be reduced to a level of 10^{-5}. However, this is about 50–100 times less than the birefringence level typically found in state-of-the-art glass waveguides. We will now inspect the origins of birefringence in glass waveguides and show how the subsequent problems may be overcome in EG devices.

An important source of the birefringence in glass waveguides on silicon substrate is the residual stress that originates in the high-temperature anneals required to form silica films. Both plasma-enhanced chemical vapor deposition (PECVD) and flame hydrolysis deposition (FHD) require high temperature anneals. On cooling the wafer down to the room temperature, the mismatch in the thermal expansion coefficients of the Si wafer ($3.6 \times 10^{-6}\,K^{-1}$) and of the SiO_2 glass ($5.4 \times 10^{-7}\,K^{-1}$) layers makes the silicon wafer contract more than the glass, resulting in a large anisotropic (in-plane) compressive stress in the glass. The stress produces, via photo-elastic effect, anisotropic changes in the glass refractive index, thus a waveguide birefringence.

Various techniques have been developed for reducing the polarization sensitivity in AWGs (see Section 29.3.5), but many of them are not applicable for EG devices. An effective solution for reducing the polarization sensitivity of EGs is the polarization compensator[7] shown in Fig. 29.2. The compensator comprises a prism-shaped region where the slab waveguide birefringence is modified. The TE and TM polarized light is refracted at the compensator prism boundary at slightly different angles to compensate for $\Delta\theta = \theta_{TE} - \theta_{TM}$ existing in the uncompensated device. This assures that both polarizations eventually arrive at the same focal position. Waveguide birefringence can be modified by partially etching the waveguide layers.[28] However, to fully compensate for birefringence in glass waveguides, all of the cladding, and even part of the core, may need to be etched away. This introduces a loss penalty due to mode mismatch between the compensator and the unmodified slab waveguide regions. To avoid this problem, only a part of the cladding can be etched and capped by a few hundred nanometers thick overlayer with a high refractive index (silicon nitride, $n \sim 2$; see Fig. 29.2, compensator cross section). Since the TE mode overlap with the high index layer is slightly larger compared to the TM mode, the birefringence is increased in the compensator region. Using this method, the polarization dependent wavelength shift was reduced to < 0.01 nm.[7]

---

[x] $\Delta n/n \sim \Delta\lambda/\lambda$, by arguments similar to those in Section 29.3.5.

29.2.7 Polarization dependence of the grating diffraction efficiency

In metal-coated diffraction gratings, the diffraction efficiency depends on the light polarization. This means that the device loss varies as the polarization fluctuates. This so-called polarization-dependent loss (PDL) is one of the critical parameters for WDM components, which typically require that the PDL is reduced to 0.3 dB or less. The grating polarization dependency and also comparatively large grating loss observed in early EG MUX/deMUX devices[4-6] made many experts skeptical about the usefulness of EGs for WDM applications. Not surprisingly, when the alternative technology, the arrayed waveguide grating (AWG), was proposed in 1988,[29] interesting developments in waveguide EGs were overshadowed by impressive advances in AWGs.

The situation has changed. It has been shown that the waveguide echelle grating polarization dependency can be significantly reduced.[7] The polarization dependency of a metalized grating originates in different boundary conditions for the TE and TM polarizations at the reflecting facets and nonreflecting facets (see Fig. 29.2, grating detail). At the reflecting facet, both TE and TM polarizations have electric field E parallel to the facet, so that the same boundary condition applies to both. The boundary conditions for TE and TM are different at the nonreflecting facet, since the electric field polarized parallel to the nonreflecting facet (TM polarization) must be continuous at the interface, while the perpendicular component (TE polarization) must satisfy the requirement that the displacement $D = \varepsilon E$ is continuous across the boundary. The corresponding difference in field distributions near the facet means that the power coupled into higher diffraction orders is larger for TM than for TE polarized light incident on the grating. This is the main cause of PDL in metalized echelle grating devices and it can be alleviated by avoiding metallization on nonreflecting facets.[xi,7]

29.2.8 Grating loss

Deviation from the verticality of a grating sidewall is a critical factor affecting the grating loss. A loss incurred on reflection on a grating etched in the waveguide at an angle of 90 deg + α can be obtained from the overlap integral of the fundamental waveguide mode with the same mode tilted by 2α on reflection, as shown by Delâge et al.[24] In glass waveguides with an index step of $\Delta n \sim 0.01$, a grating wall offset of 1 deg from the vertical produces a coupling loss of about 0.5 dB, as the incident mode couples to the reflected mode. The verticality error can also result in coupling to the higher-order waveguide modes. The ability to etch nearly vertical sidewalls is thus essential for making waveguide EG devices. Planar waveguide gratings are typically fabricated by reactive ion etching (RIE). The etch profile, including the verticality, depends on the chemistry, pressure, temperature, ion density, and ion energy.[30] For the relatively deep etches as needed in glass waveguide EG devices (~10 μm), high etch rates are preferred.

[xi] The metallization is not needed anyway on the nonreflective facets since only the reflective (illuminated) facets are required for the EG function.

ICP (inductively coupled plasma) RIE systems can provide low and controllable ion energy, high ion density that is adjustable independently from the ion energy, and operate at low pressures. These are regarded as desirable conditions for deep vertical etching.[27]

Other sources of loss are reflectivity reduction caused by mirror imperfections such as metal absorption and sidewall roughness. With controlled etching and metal[xii] deposition these can be minimized, and metalized waveguide mirrors with reflectivities higher than $R = 0.9$ have been reported.[7] Corner rounding arising from limitations in the lithographic resolution, and the mask pattern transfer by etching to the waveguide also contributes to the loss. The rounding effect can be reduced by using specifically shaped small compensatory patches (called serifs) near the grating corners, and also by reducing the ratio between the rounded and straight areas of the grating teeth by increasing the length of the facets.

29.2.9 Errors in facet positions and the influence on crosstalk

Interchannel crosstalk is one of the most important parameters in optical multiplexing devices. Crosstalk refers to an undesired presence in a given wavelength channel of a signal of another wavelength or wavelengths. In an EG device, the crosstalk mainly originates in the errors in the grating facet positions from their ideal locations. The main source of these errors is lithography mask pixelation. The facet positioning errors alter the phase relationship between the light reflected from different facets. Resulting distortions of the optical phase and amplitude distributions along the focal line will increase undesired light coupling (crosstalk) between different spectral channels. Since the phase error at the ith facet is $\delta\phi_i = 4\pi n_{\text{eff}} \delta x_i /\lambda$, where δx_i is the facet position error, waveguides in materials with lower refractive index (e.g., silica glass) can tolerate a larger pixelation error. From the phase error condition above it is observed that the error tolerance is improved by a factor of ~2.4 for EGs in glass ($n \sim 1.45$) compared to silicon-on-insulator ($n \sim 3.5$) waveguides.

It should be noted that the main source of the phase errors in echelle gratings is the geometrical fluctuations in the positions of the facets. This is in contrast to AWGs, where the light propagates in different waveguides of the phased array and the phase errors are influenced by nonuniformities in waveguide parameters that are difficult to control over large areas.[xiii] The influence of these imperfections obviously increases with the size of the waveguide array, which in turn increases with decreasing channel spacing. This gives an important advantage to echelle gratings over AWGs for WDM networks with a large number of densely spaced channels.

[xii] Aluminum is typically used.
[xiii] A simple argument can also be made: since an EG layout is smaller compared to AWG, the waveguide uniformity requirements are less critical.

29.2.10 State-of-the-art waveguide echelle grating devices

Since the 1980s, important advances were achieved in the waveguide echelle grating mux/dmux technology.[4-7,17-21,31] In the early 2000s, planar waveguide echelle grating devices were developed in glass waveguides by Optenia Inc.[7,31] with performance comparable to arrayed waveguide gratings, but of a significantly smaller size. Optenia's demultiplexer device (40 channels, 100 GHz) has a layout similar to that shown in Fig. 29.2. This is arguably the first device in which the grating loss, the polarization-dependent loss (PDL), and the polarization-dependent spectral shift were all reduced to levels acceptable for WDM applications. The devices were fabricated in silica waveguides deposited on a silicon substrate, with a refractive index contrast of $\Delta n = 0.012$. The slab waveguide core layer thickness is 5 µm and the input and output waveguides have a square (5 × 5 µm^2) cross section. The grating was formed by etching a deep vertical trench through the waveguide layers using an anisotropic etching process with a verticality error of $< \pm 1$ deg. The spectra of a 40-channel 100 GHz demultiplexer is shown in Fig. 29.4.

The device has a Gaussian passband spectrum, the adjacent channel crosstalk is better then -35 dB, the polarization-dependent loss is < 0.2 dB across the C-band, and the polarization-dependent wavelength shift is less than 10 pm. These parameters are comparable to AWG devices, but the device size is significantly smaller, only 18 × 17 mm^2. The insertion loss was ~ 4 dB, thus about 2 dB larger than that for the best AWGs at that time.[32] This higher loss was mainly due to a comparatively large propagation loss of the specific waveguide platform used in the fabrication. A 256-channel 25 GHz spaced demultiplexer was also demonstrated by increasing the grating focal length from 7.5 mm (48-channel device) to 20 mm (256-channel device).[7] A comparable AWG on a waveguide platform with the same index step would occupy about 10 times the area of the EG device.

Several other EG devices were also reported. MetroPhotonics Inc. developed various InP-based EG devices, including an optical power monitor,[33] a dynamic channel equalizer,[34] a data receiver,[35] and a triplexer.[36] An echelle grating comb filter working in a very high order of $m \sim 2000$ and with a free spectral range of 100 GHz has been reported by LNL Optenia.[37] The chip size was only 0.57 cm^2 for the comb filter with a Gaussian passband. This is a remarkably small size, given the device was made in low index contrast glass waveguides ($\Delta n = 0.75\%$). Recently, very compact EG-based transceiver modules were reported by Bidnyk et al. at Enablence Inc. for fiber to the home (FTTH) application.[38]

Concluding this brief overview of waveguide echelle gratings, we would like to emphasize again the remarkable size advantage of EG compared to AWG devices. This allows for higher integration densities, more functions per chip, and a larger number of spectral channels with a denser frequency grid. Though at the present time EGs are still overshadowed by the remarkably successful AWG technology, their obvious miniaturization advantages can be utilized in various applications, including WDM, spectroscopy, sensing, and optical interconnects.

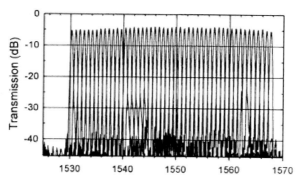

Figure 29.4 Measured spectra of a 40-channel 100 GHz waveguide echelle grating demultiplexer (Ref. 7).

29.3 Arrayed Waveguide Gratings

29.3.1 Fundamentals

The arrayed waveguide grating (AWG)[xiv] is an optical analogue of a microwave phased array antenna: by controlling the phase relationship between radiating sources forming the array output, the radiated beam direction can be changed without mechanical movement. The AWG was invented by Smit[29] in 1988. Along with the semiconductor lasers, the AWG is arguably one of the most complex, superbly developed, and commercially successful planar waveguide devices. The AWG is a key device in WDM optical communication systems where it performs functions such as wavelength multiplexing and demultiplexing, wavelength filtering, signal routing, and optical cross-connects. AWGs are also increasingly used in areas other than WDM, such as signal processing, spectral analysis, and sensing. In the limited space of this section, we will briefly introduce the AWG fundamentals followed by some recent developments in AWG technologies, including devices in high index contrast platforms and Fourier-transform waveguide arrays. For comprehensive reviews on AWGs, see Refs. 13 and 39-43.

The operation of an AWG as a demultiplexer, or a spectrometer, can be visualized as follows (see Fig. 29.5): polychromatic light is coupled from an optical fiber via the input channel waveguide to the input slab waveguide combiner. In the slab waveguide, the light is confined in a direction normal to the waveguide plane, while in the in-plane direction it diverges to illuminate the waveguide array.

At the end of the input combiner, the light couples into the array of waveguides each starting at an arc with a radius equal to the length of the combiner (focal length) and centered at the end of the input waveguide. Ideally, this arc should follow the wavefront curvature of the field illuminating the arrayed waveguides, so that the light from the central input waveguide is launched into each array waveguide with the same phase.

[xiv] AWGs are also known as the phasars (phased arrays) or waveguide grating routers (WGRs).

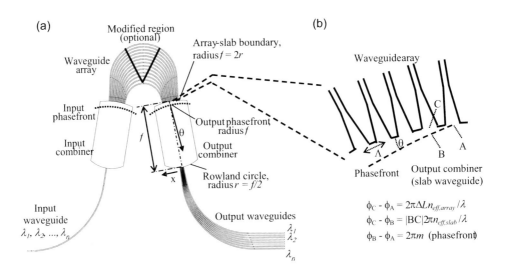

Figure 29.5 (a) Schematic of an arrayed waveguide grating. (b) Detail of the phased array output aperture showing the arrayed waveguides joining the slab waveguide output combiner. Included are phase conditions at different points (A, B, and C), from which the AWG dispersion formula [Eq. (29.8)] is derived.

Light then propagates in the waveguide array and it arrives at the end of the array with a phase difference $\Delta\varphi_{AWG}$ between the adjacent waveguides,

$$\Delta\varphi_{AWG} = \beta\Delta L = \frac{2\pi\Delta L n_{eff,a}}{\lambda}, \quad (29.5)$$

where β is the propagation constant, ΔL is the length difference (constant) between the adjacent waveguides, and $n_{eff,a}$ is the effective index of an arrayed waveguide. It is observed that the phase shift is $\Delta\varphi_{AWG} = 2\pi m$ for $\lambda = \lambda_c$ (the central wavelength of a demultiplexer),

$$\lambda_c = \frac{n_{eff,a}\Delta L}{m}. \quad (29.6)$$

This results in the constructive interference of order m of the light emerging from the respective waveguides of the phased array to the output combiner (slab waveguide) at $\theta=0$. The waveguide array output aperture is curved along an arc centered at the focal point where the central output channel waveguide joins the output combiner. This assures that the phase front emerging from the waveguide array bears the same curvature and hence converges toward the focal region, as it would for an equivalent lens of a focal length f. The light of central wavelength λ_c converges toward the central receiver (output) waveguide. For a wavelength λ, the constructive interference condition is satisfied in direction θ with respect to the combiner axis if the phase difference at positions A and B [located at the phase front, in proximity of the adjacent waveguides, see Fig. 29.5(b)] is

$$\varphi_B - \varphi_A = 2\pi m. \quad (29.7)$$

We recall that the phase difference between the fields measured directly at the adjacent waveguide outputs (positioned at A and C, Fig. 29.5) is, by the AWG definition [Eq. (29.5)], $\varphi_C - \varphi_A = 2\pi\Delta L n_{\text{eff},a}/\lambda$. Since the phase difference at positions C and B is $\varphi_C - \varphi_B = |CB| 2\pi n_{\text{eff},s}/\lambda$, and distance $|CB| = \Lambda \sin\theta$, Eq. (29.7) can be written as

$$n_{\text{eff},s}\Lambda \sin\theta + n_{\text{eff},a}\Delta L = m\lambda, \tag{29.8}$$

where Λ is the waveguide pitch at the output of the phased array and $n_{\text{eff},s}$ is the slab waveguide effective index. Equation (29.8) is the AWG equivalent of the grating equation (29.1). It is convenient to express it in a form explicitly including the effect of the change in waveguide effective index with wavelength,

$$\sin\theta = \frac{(\lambda - \lambda_c)M}{n_{\text{eff},s}\Lambda}, \tag{29.9}$$

where $M = m(n_{g,a}/n_{\text{eff},a})$ is the modified interference order, and $n_{g,a} = n_{\text{eff},a} - \lambda dn_{\text{eff},a}/d\lambda$ is the group index of the arrayed waveguides. The AWG linear dispersion in the focal region can be obtained by differentiating Eq. (29.9),

$$\frac{dx}{d\lambda} = f\frac{d\theta}{d\lambda} = f\frac{M}{n_{\text{eff},s}\Lambda}, \tag{29.10}$$

where f is the coupler (focal) length and x is the focal coordinate [see Fig. 29.5(a)].

It is thus obvious that the operating principles of the AWG and the waveguide echelle grating are identical in the sense that both use interference of multiple beams, resulting in a wavelength-dependent light propagation direction with different wavelengths arriving at different positions along the focal curve. We recall that in an echelle grating, the light reflected by the adjacent facets, the positions of which are spatially shifted by Δx (Fig. 29.1, grating detail), is phase shifted $\Delta\varphi(\lambda) = 2\beta\Delta x$, where $\beta(\lambda)$ is the propagation constant. In an AWG, the required wavelength-dependent phase difference is obtained by allowing the light to propagate in N waveguides, each of a length ΔL longer than the adjacent one, resulting in the wavelength-dependent phase shift $\Delta\varphi(\lambda) = \beta\Delta L$ at the waveguide ends. The design factors ΔL in an AWG and $2\Delta x$ in an EG device are thus equivalent.

Many parallels with echelle grating construction discussed in the previous section apply to AWGs. For example: The geometries of the phased array output aperture and of the echelle grating curve are identical (in terms of arc radii and lengths), both typically use the Rowland circle design;[xv] the effect of changing

---

[xv] The focal curve in AWG lies on the Rowland circle (radius $r = f/2$) tangent to the arc (radius $f = 2r$) of the phased array output aperture, see Fig. 29.5a.

the grating facet width is equivalent to changing the width of an array waveguide at the phased array output aperture,[xvi] etc.

AWGs have been successfully implemented in different waveguide platforms, both low- and high-index contrast, including glass (silica-on-silicon), polymer, silicon-on-insulator, and InP waveguides. Glass waveguides are the most conventional and widely used, for the same reasons that were discussed in Section 29.2.5.

We will now review some important properties of AWGs, including phase errors, crosstalk, chromatic dispersion, passband, loss, and polarization and temperature sensitivities.

29.3.2 Phase error influence on crosstalk and chromatic dispersion

The main source of crosstalk in AWGs is phase errors arising from variations in the effective index of the array waveguides, caused by fluctuations of parameters such as the waveguide width and thickness, material composition, and stress. Since the influence of these errors increases with the size of the waveguide array, their effect on crosstalk can be severe for large AWGs, such as those with dense channel spacing and large numbers of channels. The crosstalk degradation with increasing size of the waveguide array is illustrated in Fig. 29.6, where simulated demultiplexer channel spectra are presented for AWGs with 100, 50, and 25 GHz channel spacing, respectively, for a sinusoidal phase fluctuation of $\delta n = 10^{-6}$. Empirically, it is known that the crosstalk degrades approximately by 5 dB when the channel spacing is reduced to half, or by 10 dB when δn is increased $10^{1/2}$ (= 3.2) times.[43] The crosstalk can be reduced by correcting the phase errors in the individual arrayed waveguides, but this is a rather tedious procedure involving measurements of phase delays for each waveguide of the array[44] and a subsequent modification of the effective index for each waveguide.[45]

The phase errors are not only a source of crosstalk but also of chromatic dispersion, $D = d\tau/d\lambda$. The latter, typically measured in ps/nm, is determined by the group delay

$$\tau = -\frac{d\varphi}{d\omega} = \frac{\lambda^2}{2\pi c}\frac{d\varphi}{d\lambda}, \quad (29.11)$$

where $\varphi(\omega)$ is the device phase response for light of angular frequency ω, and c is the speed of light in vacuum. A typical AWG has a symmetric intensity distribution across the phased array and it belongs to a category of finite-impulse response filters. In such filters, the group delay is nearly constant within the channel passband, thus $D \sim 0$. However, in a practical AWG, this symmetry is disturbed by the phase and amplitude errors in the arrayed waveguides, and this

---

[xvi] By changing the waveguide width, loss inhomogeneity (roll-off) experienced by the peripheral output channels can be controlled. The roll-off is determined by the far field envelope of the field corresponding to diffraction by an individual waveguide aperture.

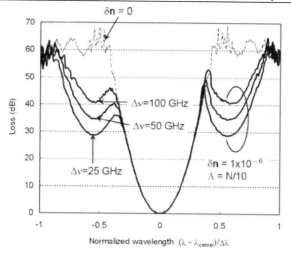

Figure 29.6 Phase error influence on crosstalk. Simulated channel spectra for AWGs with 100, 50, and 25 GHz channel spacing, respectively, with a sinusoidal phase error $\delta n = 10^{-6}$ (after Ref. 43).

increases the chromatic dispersion. As the errors increase with the size of the waveguide array, the chromatic dispersion increases alike. For a typical 100 GHz AWG with Gaussian passband, chromatic dispersion is small, usually a few picoseconds per nanometer. However, D can easily exceed user specifications[xvii] for AWGs with 50 GHz and smaller channel spacing.

29.3.3 Passband

The principles that were discussed in Section 29.2.4, including the motivations for passband flattening, also apply to AWGs. Since the channel passband is determined by the convolution of the focal field and the output waveguide mode distributions, both being almost Gaussian (assuming Gaussian input field), the AWG passband is also nearly Gaussian. Various passband flattening techniques exist,[xviii] we describe two of them here. A popular technique, invented by Okamoto and Sugita at NTT, uses a parabolic waveguide taper at the end of the input waveguide.[46] This creates a slightly double-peaked electric field distribution where the input waveguide joins the input combiner. Given that an AWG is an imaging device, the same field is reproduced in the focal region of the output combiner. After the convolution with the output waveguide modal field, this double-peaked focal field results in a flattened passband. In the second technique, the light intensity and phase across the phase array are modified such that an approximately sinc-function like field distribution is achieved at the output of the waveguide array.[47] Since the field distribution at the waveguide array output aperture is related with the field at the focal region by the Fourier transform, the sinclike field at the former will produce a quasi-rectangular (top-hat) distribution at the latter, thus flatten the passband.

[xvii] The specifications typically demand $D < 10$ ps/nm.
[xviii] See, for example, Ref. 43 and Refs. 12–19 therein.

29.3.4 Loss

Apart from waveguide propagation losses, the most significant source of excess loss in an AWG is the field mismatch between the slab waveguide and the waveguide array. At the slab-array interface, the continuous field in the slab waveguide couples to a segmented field of an array of waveguides.[xix] In a simple picture, a part of the light that falls into the gaps between the array waveguides is lost. The excess loss can be reduced by using an adiabatic (slowly varying) transition between the slab waveguide and the waveguide array. This can be achieved, for example, by using waveguide segmentation near the slab-array interface[48] or vertical tapering. The latter has been shown to be effective in both low index contrast (glass)[49] and high index contrast (InP)[50] waveguides.

29.3.5 Polarization dependent wavelength shift

As we discussed in Section 29.2.6, it is important to assure that the demultiplexer spectral response does not change as the polarization of incoming light fluctuates. In an AWG, effective index birefringence $\Delta n_{\text{eff},a}$ in the arrayed waveguides produces a polarization-dependent wavelength shift $\Delta\lambda$ in the device spectral response. From Eq. 29.6, the polarization-dependent wavelength shift in the demultiplexer central wavelength is $\Delta\lambda_c = \Delta n_{\text{eff},a}\Delta L/m$, and using Eq. 29.6 again, $\Delta\lambda_c/\lambda_c = \Delta n_{\text{eff},a}/n_{\text{eff},a}$. In contrast to echelle grating devices where $\Delta\lambda$ originates from the slab waveguide birefringence, the latter does not significantly contribute to $\Delta\lambda$ in AWGs.[xx]

Various techniques have been developed to mitigate this polarization-dependent wavelength shift. A large group of techniques relies on making the arrayed waveguides birefringence free. Since the stress is the main source of birefringence in glass waveguides, stress compensation is often used. These techniques include stress balancing by applying stress in the direction parallel to the buried channel vertical sidewalls,[51] using waveguide layers with a thermal expansion coefficient (TEC) matched to the Si substrate,[52,53] eliminating TEC mismatch by using silica instead of a silicon substrate,[54] making stress-releasing grooves beside the arrayed waveguides,[55] and using raised-strip waveguides.[56,57]

However, making the arrayed waveguide birefringence free is not the only possible method for achieving $\Delta\lambda = 0$. For example, $\Delta\lambda$ can be compensated by modifying the waveguide birefringence over a triangular region of the array [Fig. 29.5(a), modified region] that acts as a prism, producing a compensatory polarization-dependent wavefront tilt in the phased array, similar to the prism slab compensator we discussed in Section 29.2.6. Another popular technique is inserting a half-wave plate in the middle of the phased array to convert TE

[xix] This source of loss is obviously absent in waveguide echelle grating devices.

[xx] $\Delta n_{\text{eff},a}$ is the main source of $\Delta\lambda$ in an AWG. However, $\Delta n_{\text{eff},a} = 0$ yields $\Delta\lambda = 0$ only at the central wavelength. At $\lambda \neq \lambda_c$ there is a small residual $\Delta\lambda$, increasing with $|\lambda - \lambda_c|$. Complete $\Delta\lambda$ compensation over the full AWG spectral range would demand making the factor $M/n_{\text{eff},s} = m(n_{g,a}/(n_{\text{eff},a} n_{\text{eff},s}))$ polarization independent [see Eq. (29.9)].

polarization to TM, and vice versa.[58] This way, the overall phase accumulation through the waveguide array and the AWG spectral response are made polarization independent.

29.3.6 Temperature dependence

Since the environment temperature at which an AWG operates may vary considerably, athermal devices, that is, devices with substantially reduced temperature sensitivity, are desirable. Several athermal designs use mechanically moving parts attached to the chip to compensate for temperature-induced changes in the position of the focal spot.[59-61] Static athermalization (i.e., with no moving parts) can also be achieved. By differentiating Eq. (29.6), it follows that the central wavelength change with the temperature is[xxi]

$$\frac{d\lambda_c}{dT} = \lambda_c \left[\frac{dn}{ndT} + \alpha \right], \tag{29.12}$$

where α is the linear thermal expansion coefficient. An obvious athermal solution of the equation above is: $(1/n)dn/dT = -\alpha$. This can be achieved, for example, by using a TiO_2 waveguide core and SiO_2 cladding on a silicon substrate.[62]

A popular athermalization technique is based on creating a temperature-compensating triangular region in the waveguide array[63] or in the combiner slab waveguide.[64] This compensating region is filled with a material with a different dn/dT [xxii] compared to the material outside the region. In a certain analogy to the polarization compensating prism discussed in Sections 29.2.6 and 29.3.5, the temperature-compensating prism provides a compensatory temperature-dependent wavefront tilt as a consequence of the varying refraction of the prism with temperature. The compensator is designed to give a temperature-dependent wavefront tilt with the opposite sign compared to the tilt arising from the remainder of the AWG device.

29.3.7 AWGs in glass waveguides

Glass AWGs have been commercially produced since 1994.[xxiii] They typically have up to 40 channels spaced at 100 (or 50) GHz and use waveguides with index contrast $\Delta n \sim 0.75\%$ (core size 6×6 μm, minimum bend radius 5 mm). Channel glass waveguides [Fig. 29.3(a)] can be fabricated with very low losses[xxiv] and they also provide a good mode matching with standard optical fibers. Insertion losses as low as 0.75 dB have been achieved in glass AWGs with

---

[xxi] In glass AWGs, $d\lambda_c/dT \sim 0.01$ nm/°C.
[xxii] Silicone with large negative $dn/dT \sim -3.7 \times 10^{-4}$/°C is often used.
[xxiii] Current commercial suppliers are JDS Uniphase, Gemfire, ANDevices, Wavesplitter, Hitachi Cable, NEC, NEL, NKT, and PPI Technology.
[xxiv] Losses as low as 0.035 dB/cm have been achieved for glass waveguides of $\Delta n = 0.75\%$.

vertically tapered waveguides near the slab-array interface and a spot-size converter to reduce fiber-to-chip coupling loss.[49]

Waveguide platforms with larger Δn and still comparatively low losses of 0.05 dB/cm for 1.5% Δn (bend radius 2 mm) and 0.09 dB/cm for 2.5% Δn (bend radius 1 mm), have been developed for large-scale AWGs.[41] A demultiplexer with 400 channels spaced at 25 GHz covering the full C and L bands (1530–1610 nm), a loss from 3.8 to 6.4 dB, adjacent crosstalk of –20 dB, and a device size of 12 × 6 cm was fabricated on a 1.5% Δn platform.[65] Methods have been developed to reduce the fiber-chip coupling loss due to the mode mismatch between optical fibers and these high Δn glass waveguides.

It has been demonstrated that dense channel spacing and large numbers of channels can be achieved using a two-stage (tandem) AWG configuration. For example, a two-stage 1080-channel 25 GHz AWG covering the full range of S, C, and L bands has been reported.[66] The first stage is a 10-channel 2.5 THz AWG producing 10 wavelength bands, each being further demultiplexed at the second stage by 10 secondary 200-channel 25 GHz AWGs. Two-stage demultiplexers with channel spacing as small as 5 GHz (4200 channels) have been reported.[67]

AWGs with such massive numbers of densely spaced channels are large in size and difficult to fabricate. An obvious solution to reduce AWG sizes is to abandon the glass waveguide technology and to use high index contrast (HIC) waveguide platforms, such as InP, silicon-on-insulator, and silicon oxynitride. In HIC waveguides, the waveguide modes of submicrometer dimensions and waveguide bend radii of a few micrometers can be achieved, thus very compact devices can be built.

29.3.8 AWGs in InGaAsP/InP

An InP waveguide platform not only allows making ultracompact devices, but also provides (unlike glass waveguides) a unique potential for monolithic integration with electronics and active optical elements.[68,69] InGaAsP/InP AWGs typically use deep-ridge waveguides [Fig. 29.3(c)] formed by plasma etching. Deep etching provides a strong lateral confinement (bend radius ~100 μm or less). Waveguide losses are on the order of ~0.5 dB/cm, which is larger compared to silica-on-silicon waveguides, but this is compensated by a smaller device size. Making birefringence free InP waveguides requires rather strict control of the waveguide composition and width, but AWG polarization-dependent wavelength shift can be compensated by inserting in the phased array a triangular waveguide section with modified birefringence. Due to the large refractive index of InGaAsP/InP, phase errors are increased compared to glass waveguides. Nevertheless, crosstalk levels of 30 dB have been achieved and temperature-insensitive designs have also been demonstrated.[68]

AWGs with up to 64 channels[70] and ultracompact size[71] (< 0.1 mm^2) were achieved. Complex AWG-based photonic circuits such as WDM receivers, channel monitors, optical cross-connects, add-drop multiplexers, channel selectors, equalizers, and multiwavelength lasers were developed. An example of a very advanced device is a multiwavelength transmitter module by Infinera.[72] It integrates a wavelength multiplexer with a tunable DFB laser, a power monitor, an electro-absorption modulator, and an attenuator, for each of the multiplexed channels. For reviews on these complex circuits see Refs. 42, 68, and 69.

29.3.9 AWGs in silicon-on-insulator

Since the discovery in 1958 by Kilby that resistors, capacitors, and transistors can be made in single-crystal silicon, we have witnessed an unprecedented growth in electronic device integration density,[xxv] with silicon being the dominant platform for the microelectronics industry. Furthermore, the new field of silicon photonics has emerged since the 1980s.[73,74]

Among the optical properties of silicon, particularly attractive is the high refractive index ($n \sim 3.5$ at 1.55 μm).[xxvi] When silicon is used as the waveguide core surounded by silica cladding (n ~ 1.5), an index step of $\Delta n \sim 2$ is obtained. Such high index contrast (HIC) waveguides can be made with cross sections as small as ~250 × 250 nm^2 and bending radii of a few micrometers. Thus, very compact waveguide devices can be made in Si.

Silicon photonic devices are typically fabricated in the silicon-on-insulator (SOI) platform. SOI consists of a top single-crystal Si layer separated from the Si substrate by an insulating SiO$_2$ layer, called the buried oxide (or box). When SOI is used as a waveguide platform, the top Si layer acts as the waveguide core and the buried oxide as the bottom cladding, thus providing the vertical confinement for light. Lateral confinement is typically achieved by using a ridge geometry [Fig. 29.3(b)]. An important advantage of a Si ridge waveguides[xxvii] is that the single-mode condition can be obtained even for comparatively large cross sections (a few μm^2) by choosing a specific ratio between the ridge width and thickness, and etch depth, first explained by Soref et al.[75]

Large waveguide propagation loss, waveguide birefringence, and fiber-chip coupling loss have been the main obstacles to practical applications of Si waveguides, but various effective solutions have recently been demonstrated. The waveguide sidewall roughness and the resulting loss can be reduced by using a thermal oxidation step to smoothen the sidewalls after the reactive ion etching (ridge-forming) step.[76] Also, low-loss Si waveguides can be formed using a deep UV lithography and reactive ion etching.[77]

The silicon waveguide birefringence is primarily caused by differences in boundary conditions for TM- and TE-like polarizations due to different

---

[xxv] Device density has been doubling every 24 months, close to what was predicted by G. Moore, the Intel co-founder, in 1962.

[xxvi] Intrinsic silicon is transparent in the near-infrared range of 1.2–5μm, thus including the 1.3 and 1.55 μm telecommunication windows.

[xxvii] Instead of ridge waveguides, it is possible to use channel waveguides with a square Si core buried in SiO$_2$ cladding. Nevertheless, in such waveguides the single-mode condition demands a cross section of ~0.32 × 0.32 μm. With present fabrication technologies, the loss in such photonic wire waveguides due to the light scattering at the sidewall roughness is still large for practical AWG devices to be built.

waveguide cross-sectional symmetries along the vertical and horizontal directions.[xxviii] As shown by Xu et al.,[78,79] this so-called waveguide geometrical birefringence can be eliminated by using stress engineering for a wide range of waveguide geometries. This technique relies on the deposition of a stressed cladding film over the Si waveguide core. The stressed film not only optically confines the mode (acts as the upper cladding) but also exerts a force on the Si ridge [Fig. 29.7(a)] that modifies the refractive index and the birefringence of the waveguide via the photoelastic effect. By adjusting the SiO_2 cladding thickness or by modifying the stress in the oxide film [Fig. 29.7(b)], birefringence free silicon waveguides and polarization-insensitive SOI devices, including AWGs[79] [Fig. 29.7(c)], can be fabricated.

Fiber-chip coupling loss mainly arises from a large mode size disparity between an optical fiber and an SOI waveguide.[xxix] Various spot-size converters have been proposed to reduce the coupling loss. These include the inversely tapered waveguide (the waveguide is narrowed to ~100 nm, causing the mode to expand and to eventually match that of the fiber),[80,81] the GRIN coupler (graded-index layer deposited over the Si core making the optical fiber mode to converge into the Si waveguide core),[82] and the subwavelength grating (SWG) coupler.[83] The principle of the latter is shown in Fig. 29.8(a). The grating is made in a Si core such that the grating duty ratio, thus average refractive index of the core region, is gradually varied by replacing the Si core with SiO_2 cladding. In this way, the effective index, which matches that of the optical fiber, is obtained at the coupler input. Since the grating is subwavelength, diffraction effects are suppressed and the coupler function is based on the effective index effect.[xxx] Finite difference time domain (FDTD) simulations [Fig. 29.8(b)] predict a fiber-chip coupling loss of < 1.2 dB for coupling from an SMF-28 fiber to a 0.3 µm SOI waveguide. Schmid et al. have recently extended the SWG principle to control waveguide facet reflectivity over a wide range of values.[84] An SOI waveguide terminated with a triangular SWG facet is shown in Fig. 29.8(c). Such triangular SWG acts as a GRIN medium, thereby suppressing Fresnel reflection at the silicon-air boundary.

The first silicon AWG was demonstrated in 1997.[85] An SOI with a thick (5 µm) Si layer was used. SOI AWG devices were later commercialized by Bookham Technology.[86] Since a thick Si waveguide core layer was used, these devices were comparatively large. The first miniature SOI AWG was developed by Pearson et al.[87] It is an eight-channel 200 GHz demultiplexer with an overall

---

[xxviii] SOI ridge [Fig. 29.3(b)] is a typical example of an asymmetric waveguide.

[xxix] The mode area of a standard (SMF-28) single-mode fiber (mode diameter 10.4 µm) is about two orders of magnitude larger compared to photonic wire waveguides.

[xxx] The frustration of diffraction effect demands the grating period to be smaller than the first-order Bragg period $\Lambda_{Bragg} = \lambda/(2n_{eff})$, where n_{eff} is the mode effective index. Thus the name "subwavelength."

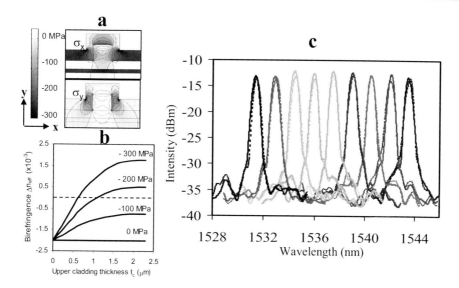

Figure 29.7 (a) Anisotropic stress distributions in an SOI ridge waveguide. (b) Waveguide birefringence as a function of the oxide cladding thickness for different levels of film stress. (c) Polarization-compensated AWG spectra. Polarization-dependent wavelength shift is reduced to < 0.02 nm by depositing a 0.8 μm PECVD cladding oxide with an in-plane compressive stress of $\sigma_x \sim -320$ MPa. TM (dashed) and TE (solid) polarizations.

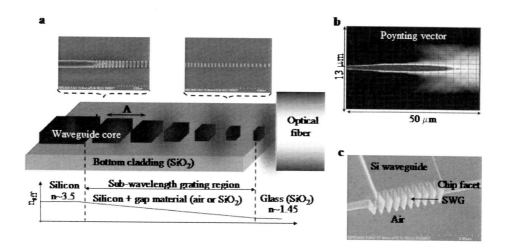

Figure 29.8 Subwavelength grating fiber-chip coupler. (a) Coupler schematics, including SEM micrographs (top) with details of a SWG coupler fabricated in SOI. (b) 2D-FDTD simulation of coupling from an SMF-28 fiber to a 0.3 μm silicon wire waveguide. (c) Triangular SWG etched in an SOI waveguide facet, acting as a GRIN antireflective boundary.

chip size of 5 × 5 mm². Other compact SOI AWGs were subsequently developed,[88] including polarization compensated devices. Currently, the smallest reported AWG[89] uses Si photonic wire waveguides and has a size of 70 × 60 μm.

Various solutions have been proposed to mitigate polarization-dependent wavelength shift in SOI AWGs.[79,90,91] We have already discussed a technique based on waveguide stress engineering to mitigate waveguide birefringence.[78,79] The technique was experimentally demonstrated on a compact nine-channel 200 GHz AWG demultiplexer.[79] The measured TE and TM spectra of the polarization compensated AWG are shown in Fig. 29.7(c).

New applications of AWGs are emerging, for example, in optical interconnects, spectroscopy, metrology, chemical and biological sensing, medical instrumentation, and space-based (satellite) sensing. Compact waveguide spectrometer chips can be made in SOI with many channels and a high spectral resolution. The wavelength resolution of a bulk optics grating spectrometer (or monochromator) can be maximized, for a given grating, by reducing the input and output slit widths. In an AWG spectrometer, the input and output waveguide mode size plays an analogous role to the slit width. This property was exploited in a recently reported SOI AWG microspectrometer, which is shown in Fig. 29.9.[92] This device uses deeply etched rectangular (1.5×0.6 μm^2) waveguides [Fig. 29.9(b)] separated by narrow (0.4 μm) air gaps at the spectrometer input and focal regions [Fig. 29.9(c)]. In this way, the width of the input mode launched into the input combiner, and its corresponding image at the focal region (the Rowland circle) are minimized. The spectrometer has 50 output waveguides, each receiving a different wavelength channel. The output waveguides are densely arrayed (1 μm pitch) along the Rowland circle [Fig. 29.9(c)]. Adiabatic tapers are used to join these narrow rectangular waveguides with the single-mode ridge waveguides; the latter are nominally used in all other regions of the AWG. The measured microspectrometer spectra are show in Fig. 29.9(d). The channel spacing is 0.2 nm (25 GHz at 1550 nm), which allows to resolve spectral lines separated by ~0.1 nm. This is the highest resolution and the largest number of channels reported to date for an SOI AWG. The size advantage of this device is obvious. State-of-the-art 25 GHz glass AWGs occupy a four-inch wafer, compared to the 8×8 mm footprint of this SOI AWG microspectrometer.

Recently, an interesting possibility to increase the AWG resolution without increasing the size of the waveguide array has been proposed by Martinez et al.[93] Dispersion of a conventional AWG [Eq. (29.9)] can be increased by forming a strongly dispersive region in the waveguide array [Fig. 29.5(a), modified region] comprising waveguides with a large group index. Very large group indices and corresponding dispersion enhancement can be obtained over a limited wavelength range near the edge of the stop band of gratings and photonic crystals, or the near-resonance condition for optical resonators.

For broadband spectroscopic applications in near infrared, an SOI AWG microspectrometer has been fabricated operating over a wide wavelength range (1250–1550 nm). The device has 40 wavelength channels separated at 10 nm and the overall size of this device is 1 cm $\times$ 2 cm.[94]

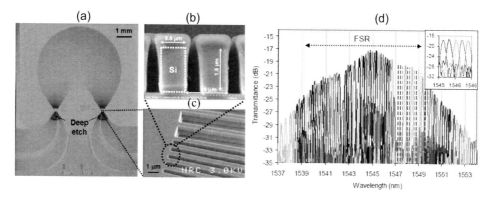

Figure 29.9 (a) An optical microphotograph of a high-resolution SOI AWG microspectrometer. (b) and (c) Narrow rectangular waveguides at the input and focal regions. (d) Measured microspectrometer spectra (Source: Opt. Express, Vol. 15, p. 2299, 2007).

29.3.10 Fourier transform AWGs and interferometer arrays

The spectroscopic and sensing applications that detect low-power optical signals present additional challenges to high resolution or a broad spectral range spectrometer. In such devices, often a figure of merit to be maximized is the optical throughput, or étendue. High étendue is an intrinsic property of a Michelson interferometer, as it was first noticed by Jacquinot in 1954.[95] This so-called Jacquinot (or étendue) advantage is a reason why Fourier transform Michelson interferometers are currently dominating the field of infrared spectroscopy.

The AWG devices that we have discussed up to this point are generalized multi-path Mach-Zehnder interferometers. Recently though, the first Michelson-type AWG has been proposed.[96] It is a Fourier transform waveguide spectrometer that exploits the étendue benefit of a Michelson interferometer. The principle of this device is as follows [see Fig. 29.10(a)]. Light emerging from an input waveguide is coupled through a slab waveguide combiner into two interleaved reflective arrayed waveguide gratings, AWG_1 and AWG_2, of interference orders m and $-m$, respectively. The light propagates through each individual waveguide toward the mirror that reverses the light propagation direction. As the wavelength changes, the two wavefronts originating from the waveguide arrays AWG_1 and AWG_2 tilt in the slab waveguide combiner region in opposite directions, each according to the AWG dispersion relation [Eq. (29.9)]. This results in Fizeau interference fringes with a wavelength-dependent period in the slab waveguide combiner where the two wavefronts overlap.

The imaging lens brings the interference fringes to a convenient location (image curve), where the interferogram $I(x)$ is sampled by an array of receiver waveguides. Monochromatic light of wavelength λ produces a sinusoidal interferogram, whereas for arbitrary input spectral distribution $B(\lambda)$, the spectrum $B(\lambda)$ can be retrieved by Fourier transformation of the interferogram $I(x)$. An important advantage of this device is that, unlike a conventional Fourier transform Michelson spectrometer, which requires moving parts (a scanning mirror), the FT AWG is a static device with no scanning elements.

In contrast to conventional AWG devices where an image of the input waveguide is formed dispersively in the focal region (i.e., operating in the Fraunhoffer regime), interference fringes in an FT AWG are formed in the proximity of the waveguide array (Fresnel diffraction regime). Rodrigo et al.[97] has shown that in an interleaved FT AWG, each AWG produces in the combiner region a distinct spatial Talbot effect [Fig. 29.10(b)], whereas the superposition of the Talbot patterns yields a spatial Moiré pattern [Fig. 29.10(c)]. This results in a unique interference pattern referred to as the Moiré-Talbot effect. The influence of this effect on spectral retrieval has been explained.[97]

Florjańczyk et al. have recently generalized the FT waveguide spectrometer concept into a multiaperture configuration.[98] The key advantage of this configuration is that the optical throughput is largely increased by using multiple inputs. An example of such device is shown in Fig. 29.11(a). It is a waveguide array formed by N unbalanced Mach-Zehnder interferometers (MZIs), with a linearly increasing path difference between the two arms of each MZI. As the

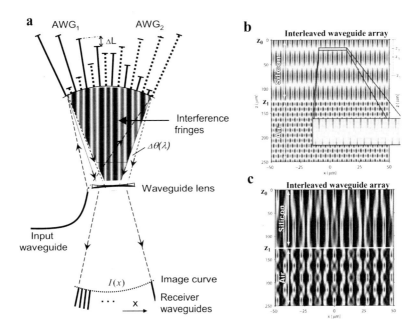

Figure 29.10 (a) Schematic of an FT AWG in reflective configuration. (b) Talbot effect in the slab waveguide combiner (silicon) and free space (air) at Littrow wavelength λ_L = 1550 nm. The inset shows the spatial frequency doubling effect at the fractional Talbot plane, as expected from Talbot effect theory. (c) Moiré-Talbot effect in the slab waveguide combiner (silicon) and free space (air) at λ = 1502 nm. In (b) and (c), the interface between the interleaved waveguide arrays and the slab waveguide combiner is at $z = z_0$, and the slab waveguide combiner terminates at the chip output edge located at $z = z_1$.

wavelength changes, the output signal of each MZI oscillates at a frequency increasing with the degree of unbalancing, that is, with the path difference ΔL between the two MZI arms. Figure 29.11(b) shows power P_i measured by photodetectors at the outputs x_i of different MZIs for different wavelengths. For a given monochromatic wavelength, constant (for Littrow wavelength, $\lambda_{Littrow}$) or sinusoidal (for $\lambda \neq \lambda_{Littrow}$) power distributions P_i are obtained at the output ports. The spatial power distribution and the input spectrum are the Fourier transform pair, and the polychromatic input light spectrum produces a complex power distribution from which the spectrum can be calculated using Fourier transformation.

In addition to the increased light throughput, an important advantage of this device compared to conventional AWGs is its fabrication robustness. In AWGs, there is no direct physical access to the waveguide array output aperture, and measuring and correcting the AWG phase errors is a formidable task (see Section 29.3.2). Unlike an AWG, an MZI array provides physical access to each MZI output, where both the phase and amplitude errors can be measured as a part of the spectrometer calibration procedure. Once the errors are measured, their influence can readily be subtracted in the spectral calculus with no need for costly physical modification of the optical path lengths in each MZI.

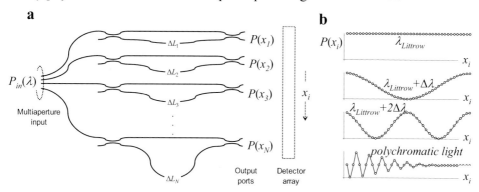

Figure 29.11 Multiaperture Mach-Zehnder array spectrometer. (a) Device schematic. (b) Spatial fringes for different input wavelengths.

Acknowledgments

We would like to thank Dr. Boris Lamontagne, Dr. Jens Schmid, Dr. Jean Lapointe, Dr. Mirosław Florjańczyk, Edith Post, Dr. Adam Densmore, Dr. Philip Waldron, Przemek Bock, Prof. Maria L. Calvo, Prof. Tatiana Alieva, Jose Rodrigo, Prof. Oscar Martinez Matos, Prof. Brian Solheim, and Dr. Alan Scott for many insightful discussions, and the Institute for Microstructural Sciences, National Research Council Canada, for continuous support. The support from the Canadian Space Agency, York University, ComDev Ltd., Complutense University of Madrid, and the NRC Genomics and Health Initiative, is also gratefully acknowledged.

References

1. Rittenhouse, D., "An optical problem proposed by F. Hopkinson and solved." *J. Am. Phil. Soc.,* **201**, 202–206, 1786.
2. Fraunhofer, J., "Über die Brechbarkeit des electrishen Lichts." *K. Acad. d. Wiss. zu München,* 61–62, 1824.
3. Hutley, M.C., *Diffraction Gratings*, Academic Press, London, 1982.
4. Cremer, C., et al., "Grating spectrometer in InGaAsP/InP for dense wavelength division multiplexing." *Appl. Phys. Lett.,* **59**, 627–629, 1991.
5. Soole, J.B.D., et al., "Monolithic InP/InGaAsP/InP grating spectrometer for the 1.48-1.56 µm wavelength range." *Appl. Phys. Lett.,* **58**, 1949–1951, 1991.
6. Fallahi, M., et al., "Demonstration of grating demultiplexer in GaAs/AlGaAs suitable for integration." *Electron. Lett.,* **28**, 2217–2218, 1992.
7. Janz, S., et al., "Planar waveguide echelle gratings in silica-on-silicon." *Photon. Technol. Lett.,* **16**, 503–505, 2004.
8. Tolstikhin, V.I., et al., "Monolithically integrated optical channel monitor for DWDM transmission systems." *J. Lightwave Technol.,* **22**, 146–153, 2004.
9. Smit, M.K., and van Dam, C., "Phasar-based WDM devices: Principles, design and applications." *IEEE J. Select. Topics Quantum Electron.,* **2**, 236–250, 1996.
10. Keiser, G.E., "A review of WDM technology and applications." *Opt. Fiber Technol.,* **5**, 3–39, 1999.
11. Laude, J.-P., "Diffraction gratings WDM components." in: *Wavelength Filters in Fibre Optics*, H. Venghaus, Ed., Springer, New York, 2006.
12. Doerr, C.R., and Okamoto, K., "Advances in silica planar lightwave circuits." *J. Ligtwave Technol.* **24**, 4763–4789, 2006.
13. Cheben, P., "Wavelength dispersive planar waveguide devices: echelle gratings and arrayed waveguide gratings." in *Optical Waveguides: from Theory to Applied Technologies*, M. L. Calvo and V. Lakshminarayanan, Eds., CRC Press, London, 2007.
14. Leijtens, X.J.M., Kuhlow, B., and Smit, M., "Arrayed waveguide gratings." in *Wavelength Filters in Fibre Optics*, H. Venghaus, Ed., Springer, New York, 2006.
15. For example: http://www.oceanoptics.com/.
16. Tomlinson, W., "Wavelength multiplexing in multimode optical fibers." *Appl. Opt.,* **16**, 2180–2185, 1977.
17. Tangonan, G.L., et al., "Planar multimode devices for fiber optics." *Proc. Int. Conf. Integrated Opt. and Optical Fiber Commun.,* 21-5, Amsterdam, The Netherlands, Sept. 17–19, 1979.

18. Watanabe, R., and Nosu, K., "Slab waveguide demultiplexer for multimode optical transmission in the 1.0-1.4 µm wavelength region." *Appl. Opt.,* **19**, 3588–3590, 1980.

19. Fujii, Y., and Minowa, J., "Optical demultiplexer using a silicon concave diffraction grating." *Appl. Opt.,* **22**, 974–978, 1983.

20. Yen, H.W., et al., "Planar Rowland spectrometer for fiber-optic wavelength demutiplexing." *Opt. Lett.,* **6**, 639–641, 1981.

21. Suhara, T., et al., "Integrated-optic wavelength multi- and demultiplexers using a chirped grating and a ion-exchanged waveguide." *Appl. Opt.,* **21**, 2195–2198, 1982.

22. Clemens, P.C., et al., "8-Channel optical demultiplexer realized as SiO_2/Si flat-field spectrograph." *IEEE Photon. Technol. Lett.,* **6**, 1109–1111, 1994.

23. Soole, J.B.D., et al., "Integrated grating demultiplexer and pin array for high-density wavelength division multiplexed detection at 1.5 µm." *Electron. Lett.,* **29**, 558–560, 1993.

24. Delâge, A., et al., "Recent developments in integrated spectrometers." *Proc. IEEE ICTON'04*, 78–83, 2004.

25. März, R., *Integrated optics Design and Modeling*, Artech House, London, 1994.

26. He, J.-J., "Phase-dithered waveguide grating with flat passband and sharp transitions." *IEEE J. Select. Top. Quant. Electron.,* **8**, 1186–1193, 2002.

27. Lamontagne, B., "Enabling fabrication technologies for planar waveguide devices." in *Optical Waveguides: from Theory to Applied Technologies*, M. L. Calvo and V. Lakshminarayanan, eds., CRC Press, London, 2007.

28. He, J.-J., et al., "Integrated polarization compensator for WDM waveguide demultiplexers." *IEEE Photon. Technol. Lett.,* **11**, 224–226, 1999.

29. Smit, M.K., "New focusing and dispersive component based on optical phased array." *Electron. Lett.,* **24**, 385–386, 1988.

30. Humpreys, B., and Koteles, E., "Fabrication challenges for enabling metropolitan WDM network technologies." *Compound Semiconduct.*, **7**, 87–94, 2001.

31. Janz, S., et al., "The scalable planar waveguide component technology: 40 and 256-channel echelle grating demultiplexers." *Tech. Dig. Integrated Photonic Research*, Vancouver, Canada, July 2002, p. IFE1-1.

32. Dixon, M., et al., "Performance improvements in arrayed waveguide-grating modules." *Proc. SPIE,* **4640**, 79–92, 2002.

33. Tolstikhin, V., et al., "Monolithically integrated optical channel monitor for DWDM transmission systems." *J. Lightwave Technol.,* **22**, 146–153, 2004.

34. Tolstikhin, V., et al., "Monolithically integrated InP-based dynamic channel equalizer using waveguide electroabsorptive attenuators-photodetectors."

European Conference on Optical Communications (ECOC 2002), paper No. P2.21, Copenhagen, Sept. 2002.

35. Densmore, A., et al., "DWDM data receiver based on monolithic integration of an echelle grating demultiplexer and waveguide photodiodes." *Electron. Lett.,* **41**, 766–767, 2005.

36. Tolstikhin, V., "InP-based photonic integrated circuit triplexer for FTTP applications." *31st European Conference on Optical Communications*, Glasgow, UK, 25–29 September 2005.

37. Bidnyk, S., et al., "Planar comb filters based on aberration-free elliptical grating facets." *J. Lightwave Technol.,* **23**, 1239–1243, 2005.

38. Bidnyk, S., et al., "Recent progress in design and hybridization of planar grating-based transceivers." *Proc. SPIE,* **6796**, 67963H-1, 2007.

39. Special issue on Arrayed grating routers/WDM mux/demuxs and related applications/uses, *IEEE J. Select. Top. Quantum Electron.,* **8**, 1087–1214, 2002.

40. Okamoto, K., *Fundamentals of Optical Waveguides*, Academic Press, London, 2000.

41. Hibino, Y., "Recent advances in high density and large scale AWG multi-demultiplexers with higher index contrast silica based PLCs." *IEEE J. Select. Topics Quantum Electron.,* **8**, 1090-1101, 2002.

42. Leijtens, X.J.M., Kuhlow, B., and Smit, M.K., "Arrayed waveguide gratings." in *Wavelength Filters in Fibre Optics*, H. Venghaus, Ed., Springer, Berlin, 2006.

43. Doerr, C.R., and Okamoto, K., "Advances in silica planar lightwave circuits." *J. Lighwave Technol.,* **24**, 4763–4789, 2006.

44. Takada, K., et al., "Measurement of phase error distributions in silica-based arrayed-waveguide grating multiplexers by using Fourier transform spectroscopy." *Electron. Lett.,* **30**, 1671–1672, 1994.

45. Takada, K., et al., "Beam-adjustement-free crosstalk reduction in 10 GHz-spaced arrayed-waveguide grating via photosensitivity under UV laser irradiation through metal mask." *Electron. Lett.,* **36**, 60–61, 2000.

46. Okamoto, K., and Sugita, A., "Flat spectral response arrayed waveguide grating multiplexer with parabolic waveguide horns." *Electron. Lett.,* **32**, 1661–1662, 1996.

47. Okamoto, K., and Yamada, H., "Arrayed waveguide grating multiplexer with flat spectral response." *Opt. Lett.,* **20**, 43–45, 1995.

48. Li, Y.P, "Optical device having low insertion loss." U.S. Patent 5,745,618, April 28, 1998.

49. Sugita, A., et al., "Very low insertion loss arrayed-waveguide grating with vertically taoered waveguides." *J. Lightwave Technol.* **12**, 1180–1182, 2000.

50. Herben, C.G.P., et al., "Low-loss and compact phased array demultiplexer using a double etch process." *Proc. ECIO'99*, 211–214, 1999.

51. Ojha, S.M., et al., "Simple method of fabricating polarization insensitive and very low crosstalk AWG grating devices." *Electron. Lett.*, **34**, 78–79, 1998.

52. Chun, Y.Y., et al., "Birefringence reduction in a high boron-doped core silica-on-silicon planar optical waveguides." *J. Korean. Phys. Soc.*, **29**, 140–142, 1996.

53. Suzuki, S., "Polarization insensitive arrayed waveguide gratings using dopant-rich silica-based glass with thermal expansion adjusted to Si substrate." *Electron. Lett.*, **33**, 1173–1174, 1997.

54. Suzuki, S., Innoue, Y., and Ohmori, Y., "Polarization-insensitive arrayed-waveguide grating multiplexer with SiO_2-on-SiO_2 structure." *Electron. Lett.*, **30**, 642–643, 1994.

55. Nadler, C.K., et al., "Polarization insensitive, low-loss, low-crosstalk wavelength multiplexer modules." *IEEE J. Sel. Top. Quantum Electron.*, **5**, 1407–1412, 1999.

56. Bissessur, H., et al., "Polarization-independent phased-array demultiplexer on InP with high fabrication tolerance." *Electron. Lett.*, **31**, 1372–1373, 1995.

57. Kasahara, R., "Birefringence compensated silica-based waveguide with undercladding ridge." *Electron. Lett.*, **38**, 1178–1179, 2002.

58. Takahashi, H., Hibino, Y., and Nishi, I., "Polarization-insensitive arrayed waveguide grating wavelength demultiplexer on silicon." *Opt. Lett.*, **17**, 499–501, 1992.

59. Heise, G., Shneider, P.C., and Clemens, P.C., "Optical phased array filter module with passively compensated temperature dependence." *Proc. 24th European Conf. Optical Commun.* (ECOC'98), Madrid, Spain, 1998.

60. Ooba, N. et al., "Athermal silica-based arrayed-waveguide grating multiplexer using bimetal plate temperature compensator." *Electron. Lett.*, **36**, 1800-1801, 2000.

61. Soole, J.B.D., et al., "Athermalised monolithic VMUX employing silica arrayed waveguide grating multiplexer." *Electron. Lett.*, **39**, 1318–1319, 2003.

62. Hirota, H., et al., "Athermal arrayed-waveguide grating multi/demultiplexers composed of TiO_2-SiO_2 waveguides on Si." *IEEE Photon. Technol. Lett.*, **17**, 375–377, 2005.

63. Inoue, Y., et al., "Athermal silica-based arrayed-waveguide grating multiplexer." *Electron. Lett.*, **33**, 1945–1946, 1997.

64. Maru, K., et al., "Super-high-Δ athermal arrayed waveguide grating with resin-filled trenches in slab region." *Electron. Lett.*, **40**, 374–375, 2004.

65. Hida, Y., et al., "400-channel 25-GHz spacing arrayed-waveguide grating covering a full range of C- and L-bands." *Proc. OFC2001*, paper WB2, 2001.

66. Takada, K., et al., "A 25-GHz-spaced 1080-channel tandem multi/demultiplexer covering the S-, C-, and L-bands using arrayed waveguide grating with Gaussian passband as primary filter." *IEEE Photon. Technol. Lett.*, **14**, 648–650, 2002.

67. Takada, K., Abe, M., Shibata, T., and Okamoto, K., "5Ghz-spaced 4200-channel two-stage tandem demultiplexer for ultra-multi-wavelength light source using supercontinuum generation." *Electron. Lett.*, **38**, 572–573, 2002.

68. Yoshikuni, Y., "Semiconductor arrayed waveguide gratings for photonic integrated devices." *IEEE J. Select. Top. Quantum Electron.*, **8**, 1102–1114, 2002.

69. Leijtens, X., "Developments in photonic integrated circuits for WDM applications." *Proc. SPIE*, **5247**, 19–25, 2003.

70. Kohtoku, M., et al., "InP-based 64-channel arrayed waveguide grating with 50 GHz channel spacing and up to -20 dB crosstalk." *Electron. Lett.*, **33**, 1786–1787, 1997.

71. Barbarin, Y., et al., "Extremely small AWG demultiplexer fabricated on InP by using a double-etch process." *IEEE Photon. Technol. Lett.*, **16**, 2478–2480, 2004.

72. Nagarajan, R., et al., "Large-scale photonic integrated circuits." *IEEE J. Select. Topics Quantum Electron.*, **11**, 50–65, 2005.

73. Reed, G., and Knights, A.P., *Silicon Photonics—An Introduction*, Wiley, Chichester, 2004.

74. Pavesi, L., and Lockwood, D.J., Eds., *Silicon Photonics*, Springer, Berlin, 2004.

75. Soref, R.A., Schmidtchen, J., and Petermann, K., "Large single-mode rib waveguides in GeSi-Si and Si-on-SiO$_2$." *IEEE J. Quantum Electron.*, **27**, 1971–1974, 1991.

76. Lee, K.K., et al., "Fabrication of ultra-low loss Si/SiO$_2$ waveguides by roughness reduction." *Opt. Lett.*, **26**, 1888–1890, 2001.

77. Dumon, P., et al., "Low loss photonic wires and ring resonators fabricated with deep UV lithography." *IEEE Photon. Technol. Lett.*, **16**, 1328–1330, 2004.

78. Xu, D.-X., et al., "Eliminating the birefringence in silicon-on-insulator ridge waveguides by use of cladding stress." *Opt. Lett.* **29**, 2384–2386, 2004.

79. Xu, D.-X., et al., "Stress engineering for the control of birefringence in SOI waveguide components." *Proc. SPIE.*, **5730**, 158–172, 2005.

80. Shoji, T., et al., "Low loss mode size converter from 0.3 µm square Si wire waveguides to single mode fibers." *Electron. Lett.*, **38**, 1669–1670, 2002.

81. Almeida, V.R., Panepucci, R.R., and Lipson, M., "Nanotaper for compact mode conversion." *Opt. Lett.,* **28**, 1302–1304, 2003.

82. Delâge, A., et al., "Monolithically integrated symmetric graded and step-index couplers for microphotonic waveguides." *Opt. Expr.,* **14**, 148–161, 2006.

83. Cheben. P., Xu, D.-X., Janz, S., and Densmore, A., "Subwavelength waveguide grating for mode conversion and light coupling in integrated optics." *Opt. Expr.,* **14**, 4695–4702, 2006.

84. Schmid, J.H., et al., "Gradient index antireflective subwavelength structures for planar waveguide facets." *Opt. Lett.,* **32**, 1794–1796, 2007.

85. Trinh, P.D., et al., "Silicon-on-insulator (SOI) phased-array wavelength multi/demultiplexer with extremely low-polarization sensitivity." *IEEE Photon. Technol. Lett.,* **9**, 940–941, 1997.

86. Bozeat, R.J., et al., "Silicon based waveguides." in *Silicon photonics*, L. Pavesi and Lockwood, D. J., Eds., Springer, Berlin, 2004.

87. Pearson, M.R.T., et al., "Arrayed waveguide grating demultiplexer in silicon-on-insulator." *Proc. SPIE,* **3953**, 11–18, 2000.

88. Cheben, P., et al., "Scaling down photonic waveguide devices on the SOI platform." *Proc.SPIE,* **5117**, 147–156, 2003.

89. Sasaki, K., Ohno, F., Motegi, A., and Baba, T., "Arrayed waveguide grating of 70×60 µm^2 size based on Si photonic wire waveguides." *Electron. Lett.,* **41**, 801–802, 2005.

90. Cheben, P., et al., "Birefringence compensation in silicon-on-insulator arrayed waveguide grating devices." *Proc. SPIE,* **3953**, 11–18, 2000.

91. Cheben, P., et al., "Birefringence compensation in silicon-on-insulator planar waveguide demultiplexers using a buried oxide layer." *Proc. SPIE,* **4997**, 181–197, 2003.

92. Cheben, P., et al., "A high-resolution silicon-on-insulator arrayed waveguide grating microspectrometer with sub-micrometer aperture waveguides." *Opt. Expr.,* **15**, 2299–2306, 2007.

93. Martínez, O.M. et al., "Arrayed waveguide grating based on group index modification." *J. Lightwave Technol.,* **24**, 1551–1557, 2006.

94. Cheben, P. et al., "Silicon-based arrayed waveguide grating microspectrometers and subwavelength structures." *5th Workshop in Fibers and Optical Passive Components*, Taipei, Taiwan, 2007.

95. Jacquinot, P., "The luminosity of spectrometers with prisms, gratings, or Fabry-Perot etalons," *J. Opt. Soc. Am.,* **44**, 761–765, 1954.

96. Cheben, P., Powell, I., Janz, S., and Xu, D.-X., "Wavelength-dispersive device based on a Fourier-transform Michelson-type arrayed waveguide grating." *Opt. Lett.,* **30**, 1824–1826, 2005.

97. Rodrigo, J.A., et al., "Fresnel diffraction effects in Fourier-transform arrayed waveguide grating spectrometer." *Opt. Expr.*, **15**, 16431-16441, 2007.

98. Florjańczyk, M., Cheben, P., Janz, S., Scott, A., Solheim, B., and Xu, D.-X., "Multi-aperture planar waveguide spectrometer formed by arrayed Mach-Zehnder interferometers." *Opt. Expr.*, **15**, 18176-18189, 2007.

Chapter 30
Silicon Photonics— Recent Advances in Device Development

Andrew P. Knights and J. K. Doylend
McMaster University, Canada

31.1 Silicon Photonics Fundamentals
    31.1.1 A historical perspective
    30.1.2 Silicon waveguide
30.2 Coupling Light to Silicon Waveguides
    30.2.1 Problem of external coupling
    30.2.2 Nonvertical taper (NVT)
    30.2.3 Grating coupler
    30.2.4 Nanotaper
30.3 Resonant Structures
    30.3.1 Principle of directional coupling
    30.3.2 Mach-Zehnder interferometers
    30.3.3 Ring Resonators
30.4 Modulation of Optical Signals
30.5. Detection
    30.5.1 Incompatibility of optical propagation and detection
    30.5.2 Silicon-germanium for detection
    30.5.3 Defect-engineered detectors
30.6 Integrated Optical Source Development
30.7 Conclusion

30.1. Silicon Photonics Fundamentals

30.1.1 A historical perspective

In 1958 while working at Texas Instruments, Jack Kilby demonstrated that it was possible to fabricate a resistor, capacitor, and transistor using single-crystal silicon.[1] This technological landmark led directly to the first truly integrated circuit and its importance was recognized by the award of a Nobel Prize to Kilby in 2000. In the subsequent almost five decades, the microelectronics industry witnessed a miraculous reduction in individual device size, and hence increases in chip functionality. This trend has seen the doubling of device density approximately every 24 months, roughly in line with the prediction of Gordon Moore in 1962. Moore's law has, more or less, remained relevant to the present day, forming the motivation for the International Technology Roadmap for Semiconductors (ITRS) in 1993. The roadmap is a needs-driven document that assumes that the industry will be dominated by complementary metal oxide semiconductor (CMOS) silicon technology. In fact, the MOSFET transistor forms the basic element of many standard products such as high-speed MPU, DRAM, and SRAM. Of some significance is the fact that no single material possesses the optimum properties for each individual device found in an integrated electronic circuit; however, silicon provides a base material from which all the required devices can be fabricated to an acceptable performance specification.

Borrowing many of the design and manufacturing principles from the microelectronics industry, several researchers began projects in the 1980s on the adoption of silicon as the base material for the fabrication of photonic circuits, i.e., those circuits that have light as the carrier of information as opposed to electrical charge. Of note at that time was the work of Richard Soref at the Rome Air Development Center in Maine[2] and Graham Reed at the University of Surrey, UK.[3] The Surrey work was of particular importance for the future commercialization of silicon photonic technology because Reed's group showed that very low-loss propagation was possible in silicon-on-insulator (SOI) rib waveguides, a structure in which light could be confined and manipulated.

Many of the optical properties of silicon would suggest it to be an ideal material for planar lightwave circuit (PLC) fabrication (not least the availability of waveguide structures in the form of SOI as shown by Reed). Silicon is virtually transparent to wavelengths > 1100 nm, while silicon dioxide (SiO_2) shares its chemical composition with glass fiber, providing a degree of compatibility with long-haul, fiber-optic technology. Silicon has a relatively high refractive index around 3.5 (compared to that for glass fiber, for example, which is around 1.5), which allows the fabrication of waveguides on the nanometer scale. However, there remains an outstanding limitation of silicon in the photonics arena, in the form of the size and nature of its indirect bandgap, which prevents the straightforward formation of efficient optical sources (maybe the greatest challenge to silicon photonics researchers), and detectors compatible with subbandgap wavelengths.

Prior to 2000, the primary application for integrated silicon photonics was viewed to lay in telecommunications. So-called first-generation (earlier than 2004) silicon photonics was dominated by the development of relatively large waveguides (cross sections of ~10–100 μm^2), which were suitable for use in fiber-optic networks performing roles such as wavelength division multiplexing and optical switching. This telecomcentric motivation was further fueled by the telecommunications boom of the late 1990s, which saw a plethora of technologies touted as the preferred platform for integrated photonic circuit fabrication. At that time, Bookham Technology of the UK (founded by Andrew Rickman, a graduate of Graham Reed's Surrey group) championed silicon photonics and showed that the methods used in the microelectronics industry to fabricate large volumes of devices in a cost-effective manner could be applied to photonic circuits. The dramatic increase in engineers and scientists dedicated to silicon photonic device design in the 1990s led directly to the spawning of a vast array of novel devices together with a dawning realization that silicon photonics had a role to play well beyond the telecommunications arena.

Second-generation silicon photonics arrived in February 2004 when the group led by Mario Paniccia at Intel Corp. announced the demonstration of an optical device, fabricated wholly in silicon with the same procedures and protocols as those used for transistor fabrication, which was able to modulate an embedded optical signal at speeds greater than 1 GHz.[4] The potential for integration of photonic and electronic functionality as a method for reducing the excessive power dissipation in microelectronic circuits was thus demonstrated. The year 2004 also saw the publication of the first textbooks to deal specifically with the subject of silicon photonics, a further sign of its acceptance as a mainstream technology.[5,6] In the relatively short period since this watershed year, the field has expanded rapidly. Waveguide dimensions are now measured in square nanometers rather than square microns, and modulation speeds in excess of 20 GHz have been demonstrated.[7] Subbandgap detection[8] has been shown to be possible at speeds compatible with the fast modulation speeds reported in Ref. 7, and perhaps the most significant development is the integration of a laser technology with a silicon circuit.[9]

The purpose of this report is to convey the excitement that currently surrounds the field of highly integrated silicon photonics. We have chosen just a few highlights of the many breakthroughs that have been reported over the last few years. We hope these go some way to show that silicon photonics is at the forefront of the next technological revolution—one that will impact our lives in a manner similar to that of the microelectronics revolution initiated by Kilby's work in 1958.

30.1.2 Silicon waveguide

The silicon waveguide forms the basic building block of all silicon photonic circuits. It is therefore necessary to briefly describe its structure.

Silicon has a bandgap of 1.12 eV, which places its optical absorption band edge at a wavelength of 1100 nm. For wavelengths shorter than this, silicon is highly absorbing and is an important photonic material for photodetectors and for CCD and CMOS imaging. For wavelengths longer than 1100 nm, including the most important optical communications bands centered at 1300 and 1550 nm, high-purity silicon is transparent, suitable for use as an optical waveguide material. Highly integrated silicon photonic circuits are thus designed for use with these longer wavelengths, most commonly those around 1550 nm. There exists a significant compatibility with the long-haul communications technologies associated with fiber optics.

The confinement of an optical signal requires a material system with appropriate variation in refractive index, such that it may support low-order optical mode propagation (or usually single-mode propagation). Such a system can be fabricated using silicon in a number of ways. For instance, the reduction in refractive index due to free carriers is sufficient to allow highly doped silicon to act as a cladding for a low-doped silicon waveguide layer,[10] while silicon-germanium alloy layers (having a refractive index greater than pure silicon) have been used as optical waveguides, with pure silicon serving as the substrate material.[11] In the last decade, however, silicon-on-insulator (SOI) has been shown to be the most suitable platform for silicon photonic device fabrication. There is a strong and growing demand for high-quality SOI material (specifically silicon-on-SiO_2) from the microelectronics industry and hence the supply of substrates for silicon photonic fabrication is guaranteed. The dimensions of the Si and SiO_2 layers in SOI may be varied in a straightforward manner from 100 nm to a few microns, providing flexibility for the array of devices required in a silicon PLC. The variation in refractive index between Si and SiO_2 provides strong vertical confinement for light traveling in the silicon overlayer of SOI. Lateral confinement may be achieved by the fabrication of a rib structure through which a variation in the effective index of the overlayer may be induced. Furthermore, the dimensions of the waveguide may be controlled such that only the fundamental mode will propagate with low loss (< 1 dBcm^{-1}).[12] Figure 30.1 shows a schematic representation of an SOI waveguide with electron microscope images of waveguides fabricated in SOI. Figure 30.1(b) resulted from the masked plasma etch of a 2.5 μm overlayer. While this is the most convenient method for the fabrication of waveguides with dimensions >1 μm, any surface roughness induced by the etch process leads to a significant propagation loss for small devices. The waveguide may be smoothed subsequent to etch via a thermal oxidation, which has been shown to reduce scattering loss.[13] Figure 30.1(c) shows a structure that was fabricated through the local oxidation of silicon (LOCOS) process, without recourse to any silicon etching.[14] Although the overlayer thickness in this case was in excess of 4 μm, the process is suitable for the fabrication of waveguides in material of only a few 100 nanometers, with a resulting propagation loss of < 0.2 dBcm^{-1} for the TE mode.[15]

30.2 Coupling Light to Silicon Waveguides

30.2.1 Problem of external coupling

The preferred method for transferring signals between optical components in a system is standard single-mode fiber (SMF). Coupling between fiber and silicon waveguides is therefore critical for practical device operation in most applications.

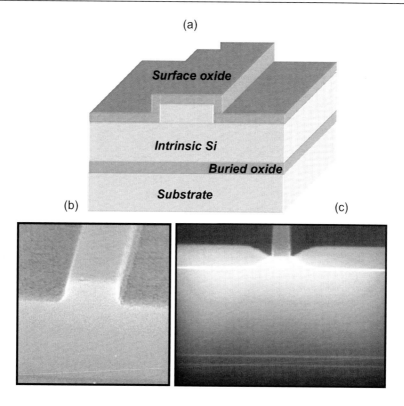

Figure 30.1 (a) Schematic representation of a SOI rib waveguide. (b) SOI waveguide formed by masked etch, and (c) by the LOCOS process.

Due to the high index of silicon compared to the silica core of SMF, however, a single-mode waveguide in silicon has s few % the area cross section of the SMF core. Therefore, a simple butt-coupling approach entails significant excess loss from mode mismatch both at the input (fiber to waveguide) and output (waveguide to fiber) of the chip. As shown in Fig. 30.2, mode mismatch loss is roughly 12 dB per facet for SMF to a typical waveguide in SOI. Fresnel reflection losses and scattering from interface imperfections add to this total, resulting in tens of dB lost from the fiber-waveguide interfaces alone.

Mode mismatch can be mitigated by the use of lensed or tapered fiber, a commercially available product in which the fiber is drawn to a narrower tip than the fiber core. Spot sizes smaller than 4 μm allow a significant reduction in mode mismatch loss, but lensed fiber is relatively costly and entails additional fiber alignment, packaging, and handling challenges due to the size and fragility of the tip.

As discussed in the previous section, the device fabrication process must be CMOS compatible in order to benefit from the cost and integration advantages of fabrication in silicon. Implementation of a simple taper from the fiber to the waveguide cross section, although effective,[17] is problematic because the vertical

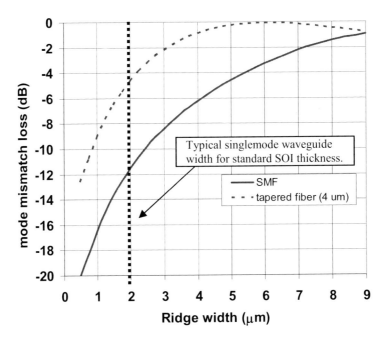

Figure 30.2 Calculated mode mismatch loss for coupling from SMF and tapered fiber to a silicon ridge waveguide (adapted from Ref. 16).

as well as the horizontal dimensions must be varied. This is difficult to achieve with standard deposition and etch techniques.

There are several approaches to reduce the coupling loss between the rib mode and fiber. We will briefly examine three of the most popular methods.

30.2.2 Nonvertical taper (NVT)

In this approach, a wedge-shaped taper is fabricated atop the waveguide such that the mode initially occupies both the rib and the wedge when launched from the butt-coupled fiber. As the "wedge" width tapers to a narrow point, the mode evolves adiabatically (i.e., without exciting other modes) into a standard rib mode, which then propagates within the waveguide. The overall structure is shown in Fig. 30.3. Coupling losses of less than 0.5 dB per facet have been demonstrated.[18] The "wedge" as shown must have a height of several microns in order to be compatible with the waveguide. Using a thick SOI overlayer, patterning the "wedge," and etching to the top of the rib is a simple fabrication approach, but introduces considerable waveguide loss due to the unavoidable etch nonuniformity across the length of the waveguide for such a deep vertical profile. An alternate process is to mask the top of a waveguide and grow the "wedge" using selective epitaxy; however, this can be awkward due to growing a thick overlayer.

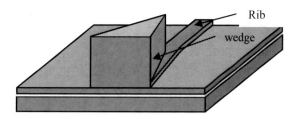

Figure 30.3 Nonvertical taper coupler.

30.2.3 Grating coupler

Use of a grating to diffract the launch signal into a waveguide mode has been demonstrated with coupling loss of 0.75 dB.[19] Gratings can also be stacked along separate deposited films to improve coupling efficiency by interposing an intermediate-index waveguide (e.g., Si_3N_4) between an outer low-index layer (e.g., SiON) and the SOI rib.[20] A 2-D grating coupler as illustrated in Fig. 30.4 has also been demonstrated[21] with higher loss (several dB) but with the added advantages that (1) fiber can be coupled at normal incidence to the wafer surface, and (2) both polarizations can be captured from the fiber as TE polarized light in both output rib waveguides. Normal incidence coupling allows a signal to be launched into the wafer prior to dicing, thus providing a means of testing devices during fabrication. Capturing both polarizations from the fiber eliminates the need for a polarization-maintaining launch fiber and significantly reduces alignment complexity; the conversion of both to TE polarized signal on chip is an added bonus since most on-chip functions are polarization dependent. Polarization-dependent coupling loss of less than 0.7 dB has been reported.[22] Grating fabrication can be accomplished using either e-beam or deep-UV lithography since the period is typically only several hundred nanometers.

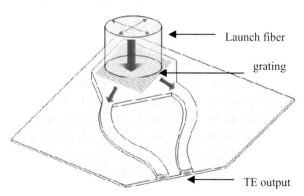

Figure 30.4 Polarization splitting 2-D grating coupler.

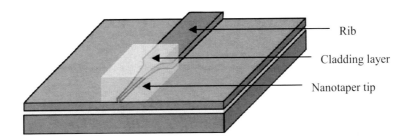

Figure 30.5 Nanotaper coupler.

30.2.4 Nanotaper

In a standard taper, the width (and height) of the rib waveguide is increased until it matches the area cross section of the launch fiber at the interface between the two. If the waveguide width is instead decreased to a narrow tip, mode confinement is significantly reduced; the mode expands and can be matched to the fiber mode, as shown in Fig. 30.5. Conversion loss of less than 1 dB has been reported[23] for tip widths of 100 nm fabricated by e-beam lithography. The device can be very compact relative to standard tapers (tens of microns long versus hundreds of microns or even several millimeters). Performance can be enhanced by means of an intermediate cladding layer deposited on top of the tip to further expand the mode by reducing the index contrast between the core (rib) and cladding.

30.3 Resonant Structures

30.3.1 Principle of directional coupling

When two waveguides are routed alongside each other such that the evanescent tails of their guided modes overlap, the signal in one waveguide will excite the overlapping mode in the other. Power can be transferred between waveguides in this manner; such a device is called a directional coupler.

Relative field amplitudes within the arms of a directional coupler are described by[24]

$$a_1(z) = e^{-j2\Delta\beta z}\left[\cos(\beta_c z) + j\frac{2\Delta\beta}{\beta_c}\sin(\beta_c z)\right],$$

$$a_2(z) = e^{j2\Delta\beta z}\left[-j\frac{\kappa_c}{\beta_c}\sin(\beta_c z)\right],$$

(30.1)

where $\Delta\beta$ is the effective index difference between the two waveguides, κ_c is the coupling coefficient (determined by mode overlap), and $\beta_c = (\kappa_c^2 + 4\Delta\beta^2)^{1/2}$. Relative power transfer from arm 1 to arm 2 is given by

$$\left|\frac{a_2(z)}{a_1(z)}\right|^2 = \frac{1}{1+(2\Delta\beta/\kappa_c)^2} \sin^2(\beta_c z). \tag{30.2}$$

From Eq. (30.2) we see that complete power transfer is only possible for waveguides with identical effective indices. This is generally achievable by design since directional coupler arms are close together and are therefore affected equally by any process variation. Assuming the arms are identical, Eqs. (30.1) simplify to

$$\begin{aligned} a_1(z) &= \cos(\kappa_c z), \\ a_2(z) &= -j\sin(\kappa_c z). \end{aligned} \tag{30.3}$$

Overall coupling κ from one arm to the other is therefore determined by the mode overlap between the arms and the length of the coupler. For a rib waveguide in SOI, overlap is increased by reducing any one of (1) the gap between waveguides, (2) etch depth, (3) rib width, or (4) sidewall verticality.

Noting that there is a $-\pi/2$ phase shift for power coupled to the opposite arm, and that power at either output will be comprised both of power coupled from the opposite arm and power transmitted through the same arm, we arrive at the transfer function for a directional coupler,

$$\begin{bmatrix} b_1 \\ b_2 \end{bmatrix} = \begin{bmatrix} t & -j\kappa \\ -j\kappa & t \end{bmatrix} \begin{bmatrix} a_1 \\ a_2 \end{bmatrix}, \tag{30.4}$$

where $\kappa = \sin(\kappa_c z)$ and $t^2 = 1 - \kappa^2$.

30.3.2 Mach-Zehnder interferometers

A directional coupler can be used to split power between two arms and then recombine them at an output to form a Mach-Zehnder interferometer (MZI) as illustrated in Fig. 30.6. Using the transfer function (30.4) derived above and assuming a phase difference of θ between the two arms, the MZI coupled output is given (for a single input and identical directional couplers) by

$$\left|\frac{b_2}{a_1}\right|^2 = 4\kappa^2(1-\kappa^2)\sin^2\left(\frac{1}{2}\theta\right). \tag{30.5}$$

It is worth noting that the maximum possible coupling decreases as coupler ratio κ^2 deviates from 50%. Apart from this restriction, MZI output can be arbitrarily tuned by changing the relative phase of the interferometer arms. MZI devices are therefore useful as modulators,[25] dispersion compensators,[26] wavelength filters,[27] and tunable couplers. Tuning can be accomplished via the thermo-optic effect

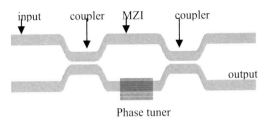

Figure 30.6 Mach-Zehnder interferometer.

using a resistive heater near the rib (see Fig. 30.7), or by carrier injection/depletion within the rib. In the latter case, additional loss due to free carrier absorption in the tuned arm can degrade the extinction ratio unless it is balanced by a similar loss in the other arm. Modulation speeds in excess of 30 Gb/s have been demonstrated.[25]

30.3.3 Ring Resonators

A directional coupler (or MZI) with one arm routed back to the input acts as a feedback loop. Substituting feedback into the directional coupler transfer function (30.4) and solving for relative output power yields

$$\left|\frac{b_1}{a_1}\right|^2 = \frac{t^2 + \sigma^2 - 2\sigma t \cos(\theta)}{1 + \sigma^2 t^2 - 2\sigma t \cos(\theta)}, \quad (30.6)$$

where σ is the round-trip ring transmission, $t^2 = 1 - \kappa^2$ for the coupler, and θ is the round-trip phase incurred in the feedback loop.

From (30.6) we see that for θ an even multiple of π (resonance) and $t = \sigma$, output power will be zero. In this case there is perfect equilibrium between power injected into the ring and power lost while circulating. The ring is said to be "critically coupled". For $t > \sigma$, more power is transmitted past the ring than is lost circulating, and the resonator is said to be undercoupled. For $t < \sigma$, more

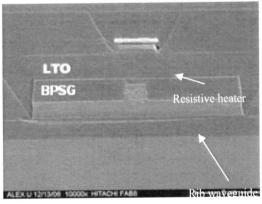

Figure 30.7 SEM cross section of a rib waveguide in silicon with a metal heater positioned directly above the waveguide for tuning.[26]

power is lost than transmitted; the resonator is overcoupled. These three cases are plotted in Fig. 30.8 for an arbitrary silicon ring resonator. It is worth noting that near resonance, the phase delay is also highly wavelength dependent, i.e., there is significant dispersion.

The sensitivity of the ring resonator to small phase changes within the feedback loop has been used for modulation by tuning and detuning the ring from resonance (modulation is dealt with more generally in the following section). Modulation speeds of 1.5 Gb/s have been demonstrated with a 3 V *pk-pk* drive.[28]

Silicon ring resonators have also been used for Raman amplification[29]. In this case the resonator is designed to be critically coupled at the pump wavelength and overcoupled at the signal wavelength. By tuning the ring to resonance, pump power within the ring is maximized, thus optimizing amplification. However, the signal makes a single amplification pass and has a broad resonance peak such that high data rates are possible without imposing dispersion or bandwidth limitations on the signal.

30.4 Modulation of Optical Signals

The ability to modulate an optical signal is one of the key functional building blocks of any photonic circuit. Modulation implies an induced change in the optical field; for example, amplitude or phase. This is achieved via a change in the complex refractive index (n). For elemental semiconductors, such as silicon, demonstration of the Pockels effect is not possible. In contrast, the refractive index of silicon does exhibit a change in response to an applied electric field, which is of quadratic form. However, the so-called Kerr effect for silicon is extremely weak. It was quantified by Soref and Bennett[2] for a wavelength of 1300 nm to be approximately $\Delta n = 10^{-4}$, for an applied field of 10^6 Vcm^{-1} (a value above that corresponding to the electrical breakdown of silicon).

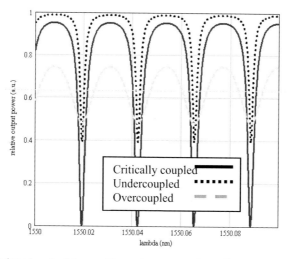

Figure 30.8 Calculated output for a silicon ring resonator with round-trip loss of 1 dB and coupling to the ring of 35% (critical), 10% (under), and 95% (over).

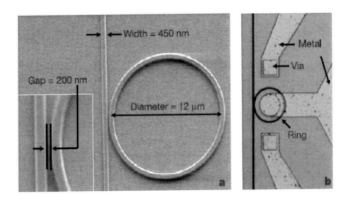

Figure 30.9 Ring resonator modulator in silicon.[28]

For materials that are centrosymmetric and for which only a small Kerr response is observed, alternative methods are required to achieve significant changes to the refractive index, and hence allow the formation of monolithically integrated modulators. The optical properties of silicon are strongly affected by the presence of free charge. In the same work in which they considered the effect of applied electric field, Soref and Bennett also determined the change in the refractive index of silicon as a function of free carrier concentration. In a rigorous treatment, they extracted a large range of experimental values of optical absorption from the research literature. Using the Kramers-Kronig relationship, they subsequently calculated values for Δn versus carrier concentration. These were compared to theoretical relationships obtained from the classical Drude model where

$$\Delta n = -\left(\frac{e^2 \lambda^2}{8\pi c^2 \varepsilon_0 n}\right)\left(\frac{\Delta N_e}{m_{ce}^*} + \frac{\Delta N_h}{m_{ch}^*}\right), \qquad (30.7)$$

$$\Delta \alpha = -\left(\frac{e^3 \lambda^2}{4\pi c^3 \varepsilon_0 n}\right)\left(\frac{\Delta N_e}{m_{ce}^{*2}} + \frac{\Delta N_h}{m_{ch}^{*2}}\right), \qquad (30.8)$$

where e is the electronic charge; α is the absorption coefficient; ε_0 is the permittivity constant; λ is the wavelength; n is the unperturbed, real part of the refractive index; m_{ce}^* is the effective electron mass; m_{ch}^* is the effective hole mass; ΔN_e is the change in electron concentration; and ΔN_h is the change in hole concentration.

Following early work on the integration of p-i-n diodes with silicon waveguides, most notably by the Reed group at the University of Surrey,[30] the free carrier effect has been exploited with remarkable success in depletion type devices that have bandwidths well in excess of 1 GHz. In 2004, the first efficient depletion modulator capable of operation in the GHz regime was reported.[4] A

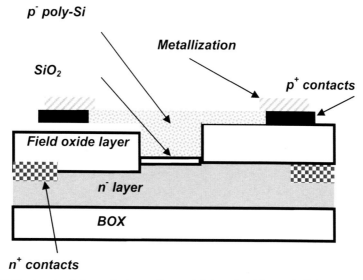

Figure 30.10 Carrier depletion, electro-optic modulator after Liu et al. (following Ref. 4).

schematic representation of this device is reproduced in Fig. 30.10. The waveguide structure consists of a lightly doped *n*-type slab and a lightly doped *p*-type *poly-Si* region, separated by a thin oxide, which acts as the insulating gate during modulation. Simulation and measurement confirmed that the device propagated a single optical mode at a wavelength of 1550 nm. It was noted, however, that the gate oxide induced a strong polarization effect on the waveguide and hence all results were reported for TE polarization only. Several modulators were measured, varying in length from 1 to 8 mm. When the *poly-Si* was biased positively, a small accumulation layer of free charge was induced on either side of the gate oxide. It is this charge that induces a change in the waveguide refractive index. By placing the modulator in one arm of a MZI structure the authors were able to quantify the phase change of the waveguide versus the applied bias.

The significant result of the work reported by the Intel group was the modulation bandwidth of the device. For the first time, an all-silicon optical waveguide, fabricated using standard monolithic processing technology, could be modulated at a rate greater than 1 GHz. In fact, the 3 dB bandwidth of the devices was greater than 3 GHz. The same Intel group reported recently on methods to increase the bandwidth of their depletion device beyond 20 GHz.[31]

30.5 Detection

30.5.1 Incompatibility of optical propagation and detection

The development of power monitoring presents one of the key challenges for integrated optical communication technology. The facilitation of signal interrogation requires the capability for efficient optical to electrical conversion. Although optical detectors fabricated using silicon device technology have been available since the concept of silicon integrated circuits was conceived, the silicon bandgap of 1.12 eV ensures that they are most commonly marketed with sensitivity around 700 nm, and are still suitable for the short-haul telecommunications wavelength of 850 nm. They are, though, incompatible with wavelengths in the infrared. This presents a challenge for those wishing to integrate detection with other photonic functionality on a single silicon chip. Clearly, it is desirable that the signal is carried at a wavelength in the infrared (> 1100 nm) to avoid significant, on-chip attenuation; however, this would imply virtually zero responsivity for monolithically integrated detectors or optical monitors. Research currently in progress then attempts to reconcile this contradiction in performance specification.

There are several approaches for long wavelength integration functionality. The most straightforward (in concept at least) is the addition of III-V semiconductor detector material onto a silicon platform—so-called hybridization. Compound semiconductors may be fabricated with a direct bandgap of the appropriate size providing both high speed and efficiency. Whereas the most elegant form of hybridization would involve the direct growth or deposition of the III-V material onto a silicon substrate, the disparity in lattice parameters makes this approach extremely challenging to implement. Recently, the silicon photonics group at Intel Corp. successfully demonstrated the direct bonding of III-V material to silicon, which has been shown to produce integrated devices that possess the required functionality of both emission and detection.[32] This important work is dealt with in more detail in the section on optical emission. Although heterobonding has allowed this globally leading team to progress with their aim of photonic integration for applications in transistor technology, an "all-silicon" (or at least CMOS compatible approach) would be preferable for the development of monolithic, all-optical communication at the chip level. The following two sections outline two such approaches. The first deals with the integration of germanium with silicon photonic devices, while the second describes detection via midgap states introduced through defect engineering.

30.5.2 Silicon-germanium for detection

The addition of Ge to the silicon matrix (i.e., the formation of $Si_{1-x}Ge_x$ alloy) shifts the absorption edge from 1100 nm, deeper into the infrared. For $x > 0.3$, absorption (and, hence, detection) of 1300 nm is possible; and for $x > 0.85$, even 1550 nm wavelengths can no longer traverse a SiGe sample unattenuated. Of some importance, there is a concomitant increase in refractive index with increasing x; a property that suggests the fabrication of optical detector integration with silicon waveguides via evanescent coupling.

The growth of germanium or high-concentration silicon-germanium, on a silicon substrate presents a number of problems related to the introduction of crystal defects. This is a result of the lattice mismatch of 4.2% between Si and Ge, which leads to significant stain in the grown epilayer. In general, there exists a critical thickness specific to Ge concentration,[33] beyond which the growth of the epilayer cannot proceed without the introduction of large concentrations of dislocations. The impact of such dislocations on the fabrication of detectors may manifest as an unacceptable dark current, even for the lowest detector bias. The values of critical thickness given by Ref. 33 are too thin for efficient detection for devices fabricated in a planar geometry, providing insufficient absorption of wavelengths around 1550 nm. This has prompted work that seeks to overcome the equilibrium constraints of epilayer relaxation and the subsequent introduction of defects, thus combining a large absorption coefficient with acceptable responsivity at infrared wavelengths. The decade of work on the material issues of integration of high-concentration germanium with silicon for photonic applications that precedes this article is too vast to review. The reader is referred to several key papers.[34–39] Instead, we here describe significant, recent results that owe much to the academic research into the direct growth of Ge on Si.

Perhaps the strongest evidence for the potential of SiGe in silicon PLCs is its reported adoption by two of the most influential industrial research groups. Recently Morse et al. of Intel Corp. described the fabrication process for a Ge-on-Si photodetector.[40] Epitaxial Ge films were grown on a p-type silicon substrate in a commercial CVD reactor. The growth included an initial seed layer of 0.1 μm Ge deposited at a temperature between 350 and 400°C, followed by a thicker Ge film grown between 670 and 725°C. Postgrowth, circular mesas were etched through the Ge film down to the silicon substrate. The films were passivated with amorphous silicon and Si_3N_4 and then annealed at 900°C for 100 minutes before contacts were added via ion implantation and aluminum metallization. The dislocation density in the devices was determined to be ~1 × $10^7 cm^{-2}$. The reported optical characterization of the fabricated devices was performed at 850 nm, with emphasis being placed on the detector performance relative to commercially available GaAs structures. The leakage current for 50 μm diameter mesa structures was ~1 μA for a reverse bias of 3 V, whereas the responsivity was found to saturate for a Ge film thickness of 1.5 μm at 0.6 A/W. The bandwidth was determined to be ~9 GHz.

Koester et al. have also reported an update on Ge/SOI detector technology under development at IBM.[41] Previously, the IBM group had demonstrated the successful fabrication of lateral p-i-n Ge-on-SOI photodetectors with bandwidths as high as 29 GHz, while also showing that these device geometries, combined with a CMOS IC, could produce error-free operation at 19 Gb/s. In Ref. 41 they addressed one of the major issues of concern to those wishing to integrate Ge detectors on Si substrates, namely, temperature sensitivity, particularly as it relates to the issue of dark current. Fig. 30.11 shows the measured variation of dark current as a function of reverse bias for measurement temperatures ranging from 179 to 359 K. Analysis of these results showed that the dark current generation mechanism has a distinctive activation energy close to half that of the bandgap, confirming the dominant role of trap (defect) assisted carrier generation. The authors acknowledged the consistency of this result with the relatively high concentration of defects (~$10^8 cm^{-2}$) in the devices, but proceeded

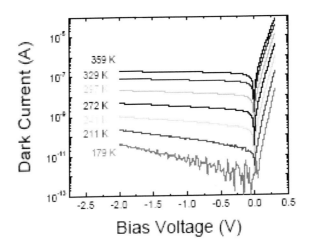

Figure 30.11 Dark current versus reverse bias for Ge on SOI photodetector (reproduced from Ref. 41, © 2006 IEEE).

to show that this did not impact the 10 Gbs^{-1} performance of the detectors at an elevated temperature of 85°C.

The Intel group has led attempts to integrate Ge detectors with SOI waveguides.[42] Their recent report of a detector with a bandwidth of 31 GHz is of some significance. The Ge layer was grown on top of a SOI waveguide using a procedure similar to that outlined in Ref. 40, with the addition of a planarization step. The detector was shown to have a repsonsivity of 0.89 AW^{-1} at a wavelength of 1550 nm, while the dark current was limited to 169 nA. An electron micrograph of the device is shown in Fig. 30.12. Although the current detector design appears limited to terminal detection, one might imagine how these devices could be used to couple fractions of an optical signal for monitoring purposes.

30.5.3 Defect-engineered detectors

A relatively straightforward solution to integrated detection has been pioneered by the McMaster University Silicon Photonics research group. Response to subband radiation is achieved via the introduction of perturbations to the silicon lattice through ion implantation and subsequent thermal annealing. The resultant band structure is modified such that midgap states are introduced at a concentration large enough to create measurable charge separation, but not so large as to prevent carrier drift to the electrical contacts adjacent to the rib waveguide. Recent work has led to the development of a waveguide detector suited to tapping a small fraction of an optical signal while allowing the vast majority of the signal to pass unaffected—this device then forms the basis of an almost perfect tap monitor providing signals that permit a health check on an

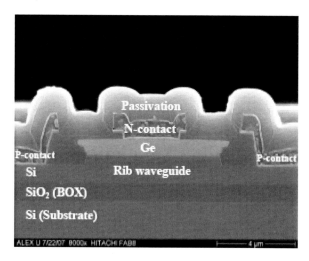

Figure 30.12 Cross sectional SEM of the Intel germanium/SOI detector (from Ref. 42).

optical circuit, or allowing the electrical signal to be used elsewhere in the same integrated circuit.[43] A schematic of this device is shown in Fig. 30.13. The device design is extremely flexible, allowing for performance to be controlled through the geometry of the detector, or the type of defect introduced during the implantation and annealing stage of the process.

More recently, work from the Lincoln Labs at MIT has scaled this type of device to the submicron level, showing an increase in responsivity concomitant with the reduction in size.[44] A 3 mm long detector was reported to absorb 99% of the incident light at a wavelength of 1545 nm, coupled from a lensed fiber. At a reverse bias of 25 V, the responsivity exceeds unity quantum efficiency, indicating carrier multiplication in the strong electric field. The authors

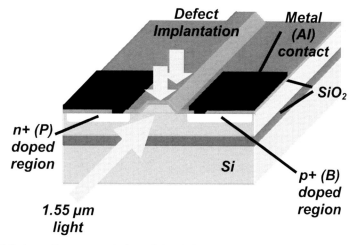

Figure 30.13 Schematic representation of the defect engineered waveguide detectors.[43]

demonstrated the thermal stability of these devices (tentatively associated with the presence of oxygen), showing that annealing at 300°C increased the detector responsivity. Perhaps the most important result from Ref. 44 is that associated with bandwidth. For a detector of length 250 µm, the frequency response was measured using a vector network analyzer and an optical modulator capable of 50 GHz operation. The half-power point of the detector frequency response after correcting for the frequency response of the modulator was approximately 20 GHz.

30.6 Integrated Optical Source Development

The previous sections have described several of the essential elements required of a highly integrated optoelectronic system and how they could be fabricated using a silicon substrate. In this final section we draw attention to the greatest challenge that faces those wishing to fabricate all-optical circuits in silicon—the development of an optical source.

Silicon is not well suited to optical emission because of its indirect bandgap. Band-edge luminescence in bulk Si is a three-body process involving an electron, hole, and phonon. The low probability of such an event means that the luminescence lifetime is very long—on the order of milliseconds. As the electron and hole move through the sample during this time period, they typically come into contact with a defect or trapping center within a few nanoseconds and recombine nonradiatively, releasing their energy as phonons. The room-temperature internal quantum efficiency of Si is thus on the order of 10^{-6}. In terms of achieving optical amplification in Si, there are two further mechanisms that tend to limit population inversion. The first is a nonradiative, three-body process in which an electron and hole recombine; but instead of creating a photon in the process, the recombination energy is instead transferred to another free carrier, exciting it to a higher energy—the so-called Auger process. Doping, current injection, and increased temperature all increase the probability of such an event because they promote population of the conduction band. The second nonradiative process is free-carrier absorption. As with the Auger process, the absorption probability increases with the Si free-carrier density. Further complicating the specifications required from a silicon optical source is the preference for emission at a wavelength for which silicon is transparent.

There are several approaches currently under investigation as routes toward efficient silicon optical sources: bulk Si systems, e.g., dislocation loops[45] and stimulated Raman scattering in Si waveguides[46] band structure engineering via alloying with Ge,[47] quantum confined structures,[48] and impurity centers (e.g., rare earth doping).[49] Of particular note is the vast array of work devoted to emission via quantum confinement, most often through the creation of silicon nanocrystals. Indeed, nanocrystals have been integrated into a low-loss waveguide system,[50] although electrical pumping of structures embedded in a dielectric matrix remains challenging.

Recently, work that emerged from UC Santa Barbara has considerably simplified the problems associated with hybridization.[51] Their unique approach utilized a silicon waveguide mode evanescently coupled to III-V semiconductor multiple quantum wells, thus combining the advantages of high-gain III-V materials and the integration capability of silicon. Moreover, the difficulty of coupling to silicon-based passive optical devices was overcome by confining most of the optical mode to the silicon. This approach restricted laser operation to the region defined by the silicon waveguide, relaxing the requirement for high-precision pick and place of the III-V device on the silicon substrate. Concerns over processing compatibility of the disparate components were minimized because the bonding procedure used to attach the III-V device and the silicon is positioned at the back end of the process flow. The fabrication thus consists of standard CMOS-compatible processing of the silicon waveguides and a low-temperature oxide-mediated wafer bonding process for heterogeneous integration. The authors reported the first demonstration of a silicon evanescently coupled laser operating at a wavelength of 1538 nm with an optically pumped threshold of 30m W and a maximum power output of 1.4 mW. The calculated and observed optical modes for this remarkable device are shown in Fig. 30.14. A further and significant development related to the UCSB/Intel project was announced in 2006. Whereas the device reported in Ref. 51 was optically pumped, the group had proceeded to fabricate an electrically pumped hybrid laser on a silicon waveguide.[52] This holds great promise as a method for the introduction of virtually any optical functionality (including high-performance detection) using a CMOS compatible approach.

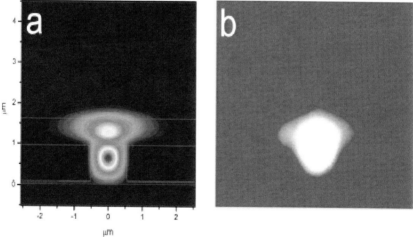

Figure 30.14 (a) Calculated fundamental TE mode. (b) Observed lasing mode of the UCSB hybrid laser (reproduced from Ref. 51).

30.7 Conclusion

In this brief report we have attempted to provide an insight into the development and vast potential application of silicon-based waveguides and devices derived therefrom. The recurring message of those working in this field is the ease with which one may integrate photonic functionality with electronic functionality on the same substrate in one seamless process flow. Furthermore, this process flow is compatible with fabrication technologies already in place in the highly developed silicon microelectronics industry. Although not yet dominant as a material for optoelectronic fabrication, it is difficult to imagine highly integrated devices of any kind not based on a silicon technology. The next few decades will then likely witness the increased migration of silicon photonics from the research lab to the manufacturing facility.

References

1. Kilby, J.S., "Invention of the integrated-circuit," *IEEE Trans. Electron Devices*, **ED-23**, 648–654, 1976.
2. Soref, R.A., and Bennett, B.R., "Electrooptical effects in silicon," *IEEE J. Quant. Electron.*, **QE-23**, 123–129, 1987.
3. Tang, C.K., Kewell, A.K., Reed, G.T., Rickman, A.G., and Namavar, F., "Development of a library of low-los silicon-on-insulatoroptoelectronic devices," *IEE Proc. Optoelectron.*, **143**, 312–315, 1996.
4. Liu, A., Jones, R., Liao, L., Samara-Rubio, D., Rubin, D., Cohen, O., Nicolaescu, R., and Paniccia, M., "A high-speed silicon optical modulator based on a metal-oxide-semiconductor capacitor," *Nature*, **427**, 615–618, 2004.
5. Reed G., and Knights, A.P., *Silicon Photonics: An Introduction*, Wiley, New York, (2004).
6. Pavesi, L., and Lockwood, D.J., *Silicon Photonics*, Springer-Verlag, Berlin, (2004).
7. Analui, B., Guckenberger, D., Kucharski, D., and Narasimba, A., "A fully integrated 20Gb/s optoelectronic transceiver implemented in a standard 0.13 micron CMOS SOI technology," *IEEE J. Solid-State Circuits*, **41**, 2945–2955, 2006.
8. Yin, T., Cohen, R., Morse, M.M., Sarid, G., Chetrit, Y., Rubin, D., and Paniccia, M.J., "31GHz Ge n-i-p waveguide photodetectors on silicon-on-insulator substrate," *Opt. Expr.*, **15**, 13965–13971, 2007.
9. Fang, A.W., Park, H., Cohen, O., Jones, R., Paniccia, M.J., and Bowers, J.E., "Electrically pumped hybrid AlGaInAs-silicon evanescent laser," *Opt. Expr.*, **14**, 9203–9210, .2006
10. Cocorullo, G., Della Corte, F.G., Iodice, M., Rendina, I., and Sarro, P.M., "Silicon-on-silicon rib waveguides with a high-confining ion-implanted lower cladding," *IEEE J. Sel. Top. Quantum Electron.*, **4**, 983–989, 1998.

11. Janz, S., Baribeau, J.-M., Delage, A., Lafontaine, H., Mailhot, S., Williams, R.L., Xu, D.-X., Bruce, D.M., Jessop, P.E., and Robillard, M., "Optical properties of pseudomorphic $Si_{1-x}Ge_x$ for Si-based waveguides at the λ = 1300nm and 1550 nm telecommunications wavelength bands," *IEEE J. Sel. Top. Quantum Electron.*, **4**, 990–996, 1998.

12. Pogossian, S.P., Vescan, L., and Vonsovici, A., "The single-mode condition for semiconductor rib waveguides with large cross-section," *J. Lightwave Technol.*, **16**, 1851–1853, 1998.

13. Lee, K.K., Lim, D.R., Kimerling, L.C., Shin, J., and Cerrina, F., "Fabrication of ultra-low loss Si/SiO_2 waveguides by roughness reduction," *Opt. Lett.* **26**, 1888–1890, 2001.

14. Rowe, L.K., Elsey, M., Tarr, N.G., Knights, A.P., and Post, E., "CMOS-compatible optical rib waveguides defined by local oxidation of silicon," *Electronics Letters*, **43**, 392–393, 2007.

15. Gardes, F.Y., and Reed, G.T., private communication, (2007).

16. Doylend, J.K., and Knights, A.P., "Design and simulation of an integrated fiber-to-chip coupler for silicon-on-insulator waveguides," *IEEE J. Sel. Top. Quantum Electron.*, **12**, 1363–1370, 2006.

17. Fijol, J.J., Fike, E.E., Keating, P.B., Gilbody, D., LeBlanc, J.J., Jacobson, S.A., Kessler, W.J., and Frish,M.B., Fabrication of silicon-on-insulator adiabatic tapers for low-loss optical interconnection of photonic devices, *Proc. SPIE*, **4997**, 157–170, 2003.

18. Day, I., Evans, I., Knights, A., Hopper, F., Roberts, S., Johnston, J., Day, S., Luff, J., Tsang, H., and Asghari, M., "Tapered silicon waveguides for lowinsertion-loss highly efficient high-speed electronic variable attenuators," *Proc. IEEE Opt. Fiber Commun. Conf.*, Vol. 1, pp. 249–251, IEEE, Piscataway, NJ (2003).

19. Ang, T.W., Reed, G.T., Vonsovici, A., Evans, G.R., Routley, P.R., and Josey, M.R., "Highly efficient unibond silicon-on-insulator blazed grating couplers," *Appl. Phys. Lett.*, **77**, 4214–4216, 2000.

20. Masanovic, G.Z., Reed, G.T., Headley, W., Timotijevic, B., Passaro, V.M.N., Atta, R., Ensell, G., and Evans, A.G.R., "A high efficiency input/output coupler for small silicon photonic devices," *Opt. Expr.*, **13**, 7374–7376, 2005.

21. Taillaert, D., Chong, H., Borel, P.I., Frandsen, L.H., De La Rue, R.M., and Baets, R., "A compact two-dimensional grating coupler as a polarization splitter," *IEEE-Phon.Tech.Letts.*, **15**, 1249–1251, 2003.

22. Bogaerts, W., Taillaert, D., Dumon, P., Pluk, E., Van Thourhout, D., Baets, R., A Compact Polarization-Independent Wavelength Duplexer Using a Polarization-Diversity SOI Photonic Wire Circuit, OFC/NFOEC (2007).

23. Almeida, V.R., Panepucci, R.R., and Lipson, M., "Nanotaper for compact mode conversion," *Opt. Lett.*, **28**, 1302–1304, 2003.

24. Nishihara, H., Haruna, M., and Suhara, T., *Optical Integrated Circuits*, pp. 46–61, McGraw-Hill, New York (1989).

25. Liu, A., Liao, L., Rubin, D., Nguyen, H., Ciftcioglu, B., Chetrit, Y., Izhaky, N., and Paniccia, M., "High-speed optical modulation based on carrier depletion in a silicon waveguide," *Opt. Expr.*, **15**, 660–668, 2007.

26. Jones, R., Doylend, J., Ebrahimi, P., Ayotte, S., Raday, O., and Cohen, O., "Silicon photonic tunable optical dispersion compensator," *Opt. Expr.*, **15**, 15846–15841, 2007.

27. Wooten, E.L., Stone, R.L., Miles, E.W., and Bradley, E.M., "Rapidly tunable narrowband wavelength filter using LiNbO3 unbalanced Mach-Zehnder interferometers," *J. Lightwave Technol.*, **14**, 2530–2536, 1996.

28. Xu, Q., Schmidt, B., Pradhan, S., and Lipson, M.L., "Micrometre-scale silicon electro-optic modulator," *Nature Lett.*, **435**, 325–327, 2005.

29. Kuo, Y., Rong, H. and Paniccia, M., "High bandwidth silicon ring resonator Raman amplifier," *Group IV Photonics*, Ottawa 2006. 228-230, 2006.

30. Hewitt, P.D., and Reed, G.T., "Improving the response of optical phase modulator in SOI by computer simulation," *J. Lightwave Tech.*, **18**, 443–450, 2000.

31. Liu, A., Liao, L., Rubin, D., Nguyen, H., Ciftcioglu, B., Chetrit, Y., Izhaky, N., and Paniccia, M., "High-speed optical modulation based on carrier depletion in a silicon waveguide," *Opt. Expr.*, **15**, 660–668, 2007.

32. Park, H., Kuo, Y., Fang, A.W., Jones, R., Cohen, O., Paniccia, M.J., and Bowers, J.E., "Hybrid AlGaInAs-silicon evanescent preamplifier and photodetector," *Opt. Expr.*, **15**, 13539–13546, 2007.

33. Bean, J.C., "Silicon-based semiconductor heterostructures- column-IV bandgap engineering," *Proc. IEEE*, **80**, 571–587, 1992.

34. Colace, L., Masini, G., Galluzi, F., Assanto, G., Capellini, G., Di Gaspare, L., Palange, E., and Evangelisti, F., "Metal-semiconductor-metal near infrared light detector based on epitaxial Ge/Si," *Appl. Phys. Lett.*, **72**, 3175–3178, 1998.

35. Luan, H.-C., Lim, D.R., Lee, K.K., Chen, K.M., Sandland, J.S., Wada, K., and Kimerling, L.C., "High quality Ge epilayers on Si with low threading-dislocation densities," *Appl. Phys. Lett.*, **75**, 2909–2911, 1999.

36. Fama, S., Colace, L., Masini, G., Assanto, G., and Luan, H.-C., "High performance germanium-on-silicon detectors for optical communications," *Appl. Phys. Lett.*, **81**, 586–588, 2002.

37. Ishikawa, Y., Wada, K., Cannon, D.D., Liu, J., Luan, H.-C., and Kimerling, L.C., "Strain-induced band gap shrinkage in Ge grown on Si substrate," *Appl. Phys. Lett.*, **82**, 2044–2046, 2003.

38. Dosunmu, O.I., Cannon, D.D., Emsley, M.K., Ghyselen, B., Liu, J., Kimerling, L.C., and Selim Ünlü, M., "Resonant cavity enhanced Ge

photodetectors for 1550nm operation on reflecting Si substrates," *IEEE J. Selected Top Quant Electron*, **10**, 694–701, 2004.

39. Masini, G., Colace, L., and Assanto, G., "2.5 Gbit/s polycrystalline germanium-on-silicon photodetector operating from 1.3 to 1.55 microns," *Appl. Phys. Letts.*, **82**, 2524–2526, 2003.

40. Morse, M., Dosunmu, F., Ginsburg, E., Chetrit, Y., and Sarid, G., *Proceedings of IEEE Group IV Photonics Conference,* Ottawa, pp. 170–171, IEEE, Piscataway, NJ (2006).

41. Koester, S.J., Schares, L., Schow, C.L., Dehlinger, G., and John, R.A., *Proceedings of IEEE Group IV photonics conference,* Ottawa, pp. 179–181, IEEE, Piscataway, NJ (2006).

42. Yin, T., Cohen, R., Morse, M.M., Sarid, G., Chetrit, Y., Rubin, D., and Paniccia, M.J., "31GHz Ge *n-i-p* waveguide photodetectors on silicon-on-insulator," *Optics Express*, **15**, 13965-13971, 2007.

43. Bradley, J.D.B., Jessop, P.E., and Knights, A.P., "Silicon waveguide-integrated optical power monitor with enhanced sensitivity at 1550nm," *Appl. Phys. Lett.*, **86**, 241103, 2005.

44. Geis, M.W.. Spector, S.J., Grein, M.E., RSchulein, .T., Yoon, J.U., Lennon, D.M., Deneault, S., Gan, F., Kaertner,F.X., and Lyszczarz, T.M., "CMOS compatible all-Si high-speed waveguide photodiodes with high responsivity in near-infrared communication band," *IEEE Photon. Technol. Lett.*, **19**, 152–154, 2007.

45. Ng, W.L., Lourenço, M.A., Gwilliam, R.M., Ledian, S., Shao, G., Homewood, K.P., "An efficient room-temperature silicon-based light-emitting diode," *Nature*, **410**, 192–194, 2001.

46. Rong, H., Jones, R., Liu, A., Cohen, O., Hak, D., Fang, A., Paniccia, M., *A continuous wave Raman silicon laser, Nature*, **433**, 725–728, 2005.

47. Presting, H., Zinke, T., Splett, A., Kibbel, H., Jaros, M., "Room-temperature electroluminescence from Si/Ge/SiGe quantum-well diodes grown by molecular beam epitaxy," *Appl. Phys. Lett.*, **69**, 2376–2378, 1996.

48. Pavesi, L., Dal Negro, L., Mazzoleni, C., Franzo, G., Priolo, F., "Optical gain in silicon nanocrystals," *Nature*, **408**, 440–444, 2000.

49. Franzo, G., Coffa, S., Priolo, F., and Spinella, C., "Mechanism and performance of forward and reverse bias electroluminescence at 1.54 microns from Er-doped Si diodes," *J. Appl. Phys.,* **81**, 2784–2793, 1997.

50. J.N. Milgram, J. Wojcik, P. Mascher, and A.P. Knights, "Optically pumped Si nanocrystal emitter integrated with low loss silicon nitride waveguides," *Opt. Expr.*, **15**, 14679–14688, 2007.

51. Park, H., Fang, A.W. Kodama, S. and Bowers, J.E. "Hybrid silicon evanescent laser fabricated with a silicon waveguide and III-V offset quantum wells," *Opt. Expr.*, **13**. 9460–9464, 2005.

52. Fang, A.W., Park, H., Cohen, O., Jones, R., Paniccia, M.J., and Bowers, J.E., "Electrically pumped hybrid AlGaInAs-silicon evanescent laser," *Opt. Expr.*, **14**, 9203–9210, 2006.

Chapter 31
Toward Photonic Integrated Circuit All-Optical Signal Processing Based on Kerr Nonlinearities

David J. Moss and Benjamin J. Eggleton
University of Sydney, Australia

31.1 Introduction
31.2 Nonlinear Materials
31.3 Optical Regeneration
    31.3.1 2R regeneration in integrated waveguides
31.4 Wavelength Conversion
31.5 Future Prospects
31.6 Conclusions
References

31.1 Introduction

New technologies and services such as voice-over-Internet protocol and streaming video are driving global bandwidth and traffic demand, which in turn is driving research and development on ultra-high-bandwidth optical transmission capacities (see Fig. 31.1). The clusters of points in Fig. 31.1 represent different generations of lightwave communication systems, from the original 0.8 µm sources and multimoded fiber to today's erbium-doped fiber amplifier (EDFA) WDM systems and onward. The resulting "optical Moore's law" corresponds to a 10× increase in "capacity × distance" every four years, making it faster than the original Moore's law for integrated circuits! This drive to higher bandwidths is being realized on many fronts—by opening up new wavelength bands (S, L, etc.), to ever-higher WDM channel counts, density, and spectral efficiency,[1] to higher bit rates via optical and/or electrical time division multiplexing (OTDM, ETDM). In parallel with this is the drive toward

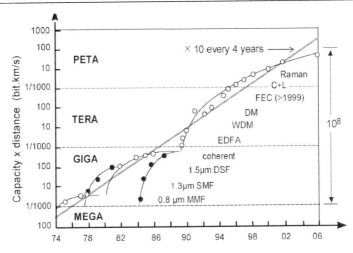

Figure 31.1 Growth in optical transmission link capacity–bandwidth product.

increasingly optically transparent and agile networks, toward full "photonic networks," in which ultrafast optical signals—independent of bit rate and modulation format—will be transmitted and processed from end to end without costly, slow, and bulky optical-electrical-optical conversion. All of these factors will result in a critical future demand for high-performance, cost effective, ultra-high-speed, all-optical signal processing devices.

A key part of this evolution will be the development of full-scale photonic integrated circuits (PICs) for all-optical signal processing that are capable of operating from 40 Gb/s to potentially > 1 Tb/s. The motivation for this is clear, and involves many factors such as reducing cost and achieving high performance in the same manner as electronic integrated circuits (ICs) have revolutionized our world in the past 40 years. Nonlinear optical materials underpin nonlinear all-optical signal processing devices, and the recent explosion in interest in this area arguably was ignited by the observation in 2003 of stimulated Raman gain in silicon.[2] In parallel with this has been the recent realization of nanophotonic devices—primarily in silicon.[3] The ultimate solution for low-cost high-performance all-optical signal processing PICs will likely be based on nanophotonic structures circuits that greatly enhance nonlinear optical efficiencies through the drastic reduction in mode field area. Finally, the development of high-quality fabrication processes for other novel nonlinear materials, such as chalcogenide glasses,[4] that traditionally have not lent themselves to semiconductor industry standard processing, raises the promise of realizing nanowire waveguides in these new materials as well.

Although there have been notable recent achievements in all-optical PICs,[5–7] this field is still arguably in its infancy. Ultimately, the full range of both passive (e.g., filtering functions) and active (switching, regeneration, wavelength conversion, demultiplexing etc.) will need to be integrated in a single chip. Figure 31.2 shows an artist's impression of what a possible embodiment of such a chip might look like, incorporating all of the latest breakthrough devices such

Figure 31.2 Artist's impression of a future embodiment of a fully integrated all optical signal processing chip (www.cudos.org.au).

as nanowires, photonic crystals, and other structures to achieve these functionalities. Regardless of the specific embodiment or material system, however, it is clear that the combination of greatly reducing cost and improving performance both in terms of speed as well as reducing peak power requirements to below the current levels of a few watts[5,8] to milliwatts or even lower, will go a long way to ensure that all-optical nonlinear devices will play an integral part in future telecommunications systems.

The full scope of all-optical signal processing devices is too large to comprehensively review here and so this chapter focuses on recent progress on PICs for all-optical 2R regeneration and wavelength conversion based on Kerr nonlinearities—particularly in chalcogenide glass—where the intrinsic device speed is in theory almost unlimited (> 1 Tb/s). We note that there has been significant activity in other areas, such as semiconductor optical amplifiers (SOAs)[31–33] or electroabsorption modulators,[34] with recent results even reaching 640 Gb/s.[35] In contrast to devices based on pure Kerr nonlinearities, these devices are based on real carrier densities and exploit ever-increasingly subtle ultra-high-speed carrier dynamics to improve the nonlinear optical response speed. These devices, while showing significant promise, are outside the scope of this chapter.

Although this chapter focuses on integrated all-optical signal processing PICs, much of the groundwork for this has been laid through work based on novel nonlinear optical fiber devices such as bismuth oxide[9] or chalcogenide glass,[10] and so a brief summary of some of this work is included here. Chalcogenide fiber, in particular, has exhibited the largest material nonlinearitiy of any fiber to date used for all-optical signal processing,[11] and so in addition to paving the way toward the development of all-optical PICs it is also attractive in its own right, with the potential to advance all-optical device performance through novel methods of mode field area reduction such as chalcogenide fiber nanotapers,[12] as well as highly nonlinear glass photonic crystal fibers.[13]

Whether in fiber or PICs, however, all-optical signal-processing devices are universally based on optical nonlinearities (either third order or, to a lesser degree, second order). It is interesting and perhaps somewhat paradoxical to note that these nonlinear processes have historically posed significant challenges and limitations for long-haul wavelength division multiplexed (WDM) systems[14] in fiber. By "taming" these nonlinearities to perform desired functions, they offer the promise of addressing the bandwidth bottleneck for signal processing for future ultra-high-speed optical networks as they evolve beyond 40 Gb/s to 160 Gb/s and ultimately to 1 Tb/s and beyond.[1] Key attributes of nonlinear optical signal processing include huge potential benefits in cost, speed, simplicity, and footprint size. In particular, all-optical devices based on third-order $\chi^{(3)}$ (nonresonant) optical nonlinearity are nearly instantaneous—relying as it does on virtual electron or hole transitions (rather than generating real carrier densities) with a response time typically < 10 fs. Also, being a third-order nonlinearity, $\chi^{(3)}$ is responsible for a wide range of both phase-matched and non-phase-matched phenomena. Non-phase-matched processes do not rely on critical matching of waveguide refractive indices between incoming and outgoing signals, and include cross-phase and self-modulation (XPM, SPM) based on the nonlinear (Kerr) refractive index, two-photon absorption (TPA), Raman gain, and others. These have been exploited to demonstrate a wide range of novel all-optical functions such as optical performance monitoring,[15,16] 2R and 3R optical regeneration,[7,11,17,18] wavelength conversion,[5,19,20] optical buffering and delay,[3] demultiplexing,[21] optical performance monitoring,[15,16] and others. Phase-matched processes, on the other hand, are probably best represented by optical parametric signal processing techniques that have had a long and diversified history spanning more than two decades.[22-30] While each of these approaches has its advantages and disadvantages, the bulk of the work presented here is based on non-phase-matched processes that do not require sophisticated design of waveguides and/or device structures in order to achieve phase matching.

We begin with a brief look at recent developments in material and platform technology. This is followed by a review of progress on one of the most important applications of all-optical signal processing, namely, all-optical regeneration based on Kerr nonlinearities, with a focus on our work based on chalcogenide glass fiber and waveguide devices. In Section 31.4 we review all-optical wavelength conversion in chalcogenide glass waveguides. Finally, we end with a discussion of the immediate future directions for all-optical PICs, focusing on a particular potential embodiment of an all-optical device based on nanowaveguides.

31.2 Nonlinear Materials

Nonlinear materials have been of significant interest for almost as long as lasers have existed and have been the subject of many books in their own right. It is not the intention of this chapter to comprehensively review nonlinear materials as a whole, but to briefly highlight a few of the key materials that relate most directly to nonlinear all-optical devices with an emphasis on the material system that has been the focus of the work reviewed in this chapter, namely, chalcogenide glasses.

The benchmark platform for novel materials and devices is arguably highly nonlinear silica-based fiber (HNLF), which is generally recognized as the most mature platform due to its exceptional characteristics of low loss (< 0.5 dB/km), high nonlinearity (~30 W^{-1} km^{-1}), and precise dispersion slope control (~0.03 ps/km nm^2). Any novel material or devices based on alternative platforms, such as photonic crystal fibers (PCFs) or nonsilica single-mode fiber, need to not only compete with HNLF, but exceed its performance in terms of effective interaction length, nonlinearity, and possibly phase-matching (chromatic dispersion) control.

Chalcogenide glasses (ChGs) have attracted significant attention in recent years as a promising nonlinear material for all-optical devices.[36-38] They are amorphous materials containing at least one of the nonoxide group VI element (S, Se, and Te)—a chalcogen—combined with other elements such as Ge, As, Sb, and Ga.[38,39] Some of the crucial properties that have made them attractive for all-optical devices include the following:

1. Transparency in the near IR as well as into the far IR due to the relatively high atomic masses and weak bond strengths resulting in low phonon energies.[38,40]
2. Photosensitivity to above bandgap light. The weak bonding arrangements and inherent structural flexibility result in many distinct photoinduced phenomena.[38] Photo darkening and photo bleaching, which shift the absorption edge and induce changes in the refractive index,[38] are particularly noteworthy and have been used to write both channel waveguides[39,41-45] and periodic structures.[46-52]
3. High refractive index ($n \approx 2-3$), providing strong mode field confinement[4] and tight bending radii, and allowing complete bandgaps to be engineered in photonic crystal structures.[53]
4. Large third-order nonlinearities, with reported values for the nonlinear refractive index (n_2) up to three orders of magnitude larger than silica, and even more for the Raman gain coefficient (g_R).[3,38,54] Because these nonlinearities are based on nonresonant virtual electronic transitions, the intrinsic response time is in the tens of femtoseconds, making devices based on these materials potentially much faster than those based on real carrier dynamics such as semiconductor optical amplifiers.[36,54]
5. Low two-photon absorption (TPA),[36] of particular importance since it has been known since 1989[55] that TPA poses a fundamental limitation for nonlinear switching, manifesting through nonlinear figure of merit (FOM); FOM = $n_2/\beta\lambda$, where β is the TPA coefficient and n_2 is the Kerr coefficient. Since then it has come to be accepted that, for nonlinear all-optical switching at least, FOM > 1 is required for efficient operation.[56] It is important to note that FOM, essentially the ratio of the real to imaginary parts of $\chi^{(3)}$, is a fundamental property of a material, varying only with wavelength.[57]

Table 31.1

Material	n^2 (× silica)	β (TPA) (cm/GW)	FOM
Chalcogenide Glasses	100 to 400	< 0.03 to 0.25	From 2 to 15 and higher
Silicon	270	0.9	0.5

Probably the most significant development in the past few years for chalcogenide glasses, in terms of their use for practical all-optical signal processing, has been the realization of low-loss single-mode chalcogenide fibers;[58] and for waveguides, the development of deposition techniques for high-quality low-loss thin films, such as pulsed laser deposition,[59] as well as high-quality fabrication processes such as dry etching.[4] Together with the demonstration of sophisticated writing techniques to produce Bragg grating filters in chalcogenide waveguides of high enough quality to be useful for signal-processing applications,[60] the door has been opened for chalcogenide glass fiber and waveguide devices to join mainstream materials such as silicon as the basis for potential solutions for integrated all-optical signal processing.

Although this review does not focus explicitly on silicon, it is useful to contrast chalcogenide devices against the backdrop of intense activity that has taken place the past few years on integrated silicon all-optical devices.[61-67] Silicon has, in many ways, become the benchmark material platform with which to compare new materials and devices in a wide range of areas, and nonlinear optics is perhaps no exception. Since the observation of Raman gain in 2003,[2] there has been intense interest worldwide in developing silicon as a material platform for all-optical devices. This has been highlighted by the achievement of ultrasmall nanowires[68] to greatly reduce operating powers.[69] In addition, together with techniques such as ion implantation[70] and the use of p-n junctions to sweep out carriers,[71] nanowires have helped in solving the problem of two-photon-induced free carrier absorption that has been known since 2004[71,72] to pose a significant obstacle to realizing net Raman gain in silicon. A critical point, however, is that even if free carrier effects are completely eliminated, one is still left with the intrinsic material two-photon absorption coefficient, which determines the nonlinear FOM, and which is non-negligible in silicon. Table 31.1 lists the key parameters for silicon as well as typical chalcogenide glasses where we see that, purely from the nonlinear FOM viewpoint, silicon is rather poor with a FOM of ~ 0.3 to 0.5 (depending on which sets of experimental results one takes).[73-76] Purely from their nonlinear material characteristics, chalcogenide glasses arguably represent a more promising platform, with nonlinearities ranging from being comparable to, to several times larger than silicon, along with a FOM ranging from ~ 2 to > 15. Having said this, however, it has recently been realized that for some other applications such as 2R regeneration[77] a modest

Table 31.2

Parameter	Units	SiO_2	Bi_2O_3 fiber	As_2Se_3 fiber
Nonlinear index (n_2)	n_2 of silica*	1	42	470
Effective core area (A_{eff})	Mm_2	60	3.3	37
Nonlinearity coefficient (γ)	$W^{-1}km^{-1}$	1.9	1360	1200
Length (L)	M	5000	1	1
Dispersion (D)	Ps/nm/km	−0.69	−260	−560
Figure of merit (FOM)	-	High†	High†	2.3

*n_2 of silica=$2.6\times10^{-20} m^2/W$
†No direct measurement found. Literature suggests a FOM greatly in excess of unity.

degree of TPA can actually enhance performance, in sharp contrast with all-optical switching. For silicon, however, it is not clear if its FOM is adequate even under these circumstances, and it will be interesting to watch continuing developments following the recent groundbreaking report of 2R regeneration in silicon-on-insulator (SOI) nanowires.[8]

For novel nonlinear fibers, probably the two most promising contenders are single-mode selenide-based chalcogenide fiber[54] and bismuth oxide fiber.[9,78] Table 31.2 shows some of the key material properties of both of these, along with highly nonlinear silica fiber. The key contrasts between the bismuth and chalcogenide fibers are that, whereas the chalcogenide fiber has an intrinsically higher material nonlinearity n_2 (up to 10 times bismuth oxide), because the core area of SM chalcogenide fiber[58] is still quite large, the effective nonlinearity, $\gamma = n_2\omega/cA_{eff}$, for both fibers is comparable at this point in time. For chalcogenide fiber, however, the hope is that further improvement in fabrication methods or the introduction of novel geometries, such as fiber tapers or photonic crystal chalcogenide fibers, will yield mode field areas comparable to (or even smaller than) current bismuth fiber, which would yield $\gamma > 10{,}000$ $W^{-1} m^{-1}$, well and truly in the regime of practical device requirements.

31.3 Optical Regeneration

Noise and interference sources in optical communication systems take many forms, the most obvious of which being the amplified spontaneous emission (ASE) noise from amplifiers, which grows proportionally with the number of cascaded amplifiers in the link. If not compensated for, ASE noise is one of the most significant impairments in communication systems reaching thousands of kilometers, by interfering with the signal. In order to limit signal degradation, all-optical signal regenerators including 2R (reamplification and reshaping) and 3R (reamplification, reshaping, and retiming) regenerators[79] are expected to play a major role both in ultralong signal transmission and in terrestrial photonic networks. The advantages of all optical regeneration over conventional techniques include huge potential savings in cost (by avoiding O/E/O conversion and demultiplexing), speed, and footprint.

A wide range of 2R and 3R optical regeneration techniques[80] have been studied that rely on optical nonlinear effects in order to replace the widespread optoelectronic regenerators currently used primarily for long-haul and ultra-long-haul systems. The aim of this section is not to review all of this work, but to focus on our research, which has centered around all-optical 2R and 3R regeneration methods that exploit self-phase modulation (SPM) and cross phase modulation (XPM) processes.[11,18,77,81,82]

Typical 2R regenerators achieve their functionality from a nonlinear power transfer function, governing the instantaneous output versus input power of the regenerator (Fig. 31.3). Such transfer functions have been realized through nonlinear phenomena such as SPM,[83–86] interference within an optical loop mirror,[87–89] four-wave mixing,[90] parametric amplification,[17] saturable absorption,[91] or gain dynamics in a semiconductor optical amplifiers.[92] Numerous papers have experimentally and theoretically demonstrated a reduction in bit error rate (BER) degradation, or a decrease in Q factor degradation, at the end of systems with cascaded erbium-doped fiber amplifiers (EDFAs), by the addition of optical regenerators between the EDFAs.[85,86] These regenerators prevent the BER degradation by subsequent noise sources (e.g., EDFAs) in an optical link by periodically compressing the noise fluctuations on the logical ones and zeros, which prevents a sudden noise buildup with its associated BER degradation. These 2R optical regenerators that operate on both noise and signal equally do not directly reduce the BER of a noisy signal, however. That is, the BER just before and after these regenerators is theoretically identical.[81]

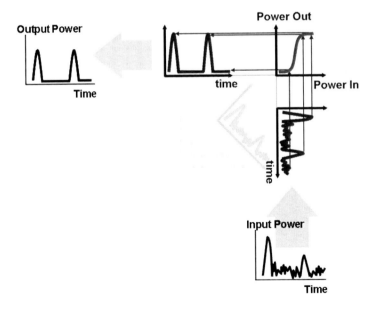

Figure 31.3 Principle of noise compression by a 2R optical regenerator that operates via an instantaneous nonlinear power transfer curve.

Contrasting this are 2R regenerators based on SPM followed by filtering, as first proposed by Mamyshev.[83] These regenerators have been the focus of significant interest and, among other things, have been shown to be capable of improving not only the Q factor but directly improving the bit error rate (BER) of input data.[81,82] Other advantages include simplicity and a bandwidth limited only by the intrinsic material nonlinear (Kerr) response. Mamyshev regenerators have been demonstrated in highly nonlinear silica-based fiber[81,84] and have even been used to achieve a million kilometers of error-free transmission without electrical conversion.[86] Recently,[11] we investigated 2R regeneration via the Mamyshev technique in highly nonlinear chalcogenide fiber, achieving operation below 10 W peak power and demonstrating that high material dispersion as well as moderate two-photon absorption (TPA)—in contrast to previous expectations—can actually enhance the device performance by improving the transfer function. For the case of TPA (which determines the nonlinear FOM), this is rather remarkable since it has been known for 20 years[55] that for nonlinear switching, it can only degrade, not improve, device performance.

The principle of operation of a Mamyshev regenerator is shown in Fig. 31.4. A noisy input signal (return-to-zero in this case) is passed through a nonlinear medium (which can contain dispersion) producing SPM-induced spectral broadening. Noise on the input signal experiences low-SPM spectral broadening, and so is filtered out by the bandpass filter (offset from the input center wavelength), whereas signal pulses experience significant broadening and are consequently partially transmitted through the bandpass filter. This results in a nonlinear power transfer curve, which improves[84] the OSNR and Q factor for modulated signals at 40 Gb/s. Furthermore, as discussed above, it has recently been shown[81,82] that these regenerators attenuate noise (relative to signal pulses) even more than what the nonlinear power transfer function would imply, the result being a direct improvement in bit error rate rather than merely a prevention of degradation in signal to noise.

Figure 31.5 shows the experimental configuration used to demonstrate the device performance. The device was based on a 2.8 m long As_2Se_3 single-mode fiber with a mode field diameter (MFD) ~ 6.4 μm at 1550 nm, a core/cladding refractive index of ~ 2.7 and a numerical aperture of 0.18. Pulses from a mode-locked laser near 1550m with a repetition rate of 9.04Hz and width of 5.8s were

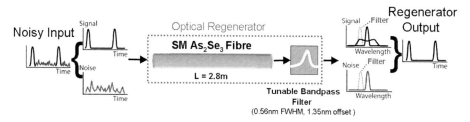

Figure 31.4 Principle of operation of self-phase modulation (SPM)–based ("Mamyshev") regenerator.

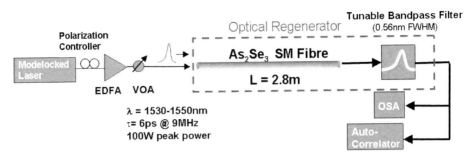

Figure 31.5 Experimental setup for demonstrating 2R regeneration in chalcogenide fiber.

passed through a polarization controller and then through a custom-built amplifier designed to keep extraneous SPM-induced spectral broadening to a minimum. The signal was then butt-coupled into the chalcogenide fiber from SMF fiber, with typical coupling losses being 2.5–3.5 dB per facet (− 0.8 dB Fresnel loss, < 1dB mode overlap loss, and the rest due to facet cleave quality). Propagation loss in the fiber was 0.9 dB/m. The output of the fiber was filtered [FWHM = 0.56 nm (70GHz), offset by 1.3 nm] and then measured by an optical spectrum analyzer or optical autocorrelator. The peak power levels of 10 W (100 MW/cm^2) was much less than the photo-darkening threshold reported for this material.[36]

Figure 31.6 shows the output pulse spectra of the chalcogenide fiber with no filter present for different peak input powers, and clearly illustrates the principle of operation of this device—as the peak input power increases, the SPM broadens the signal spectrum so that it overlaps more with the bandpass filter, creating a nonlinear power transfer function (PTF), a key benchmark of the performance of a 2R regenerator. The theoretical results (red) show good agreement with experiment and were calculated using a split-step Fourier transform method with a value of n_2 of ~ 1.1 × 10^{-14} cm^2/W, or ~ 400 times silica, in line with earlier measurements of n_2 in this Se-based chalcogenide glass fiber.[36] Also shown in Fig. 31.6 is the tunable bandpass (0.56 nm FWHM) filter transmission spectrum, offset by 1.35 nm from the input center frequency. Note that the device performance was not exhaustively optimized in terms of filter offset.

Figure 31.7 shows the resulting experimental and theoretical nonlinear power transfer curves as a function of peak in-fiber input power of the full device with the filter in place. The curves show a clear output power limiting function at ~ 5 W peak input power, as well as a clear delayed "turn-on." The former is effective in suppressing the noise in the logical "1s" while the latter contributes to suppressing noise in the "0s". There is still a slight oscillation at high peak power levels but this is much smaller than what would result[84] without the presence of the large dispersion of −504 ps/nm/km at 1550 nm of the 2.8 m long As$_2$Se$_3$ fiber, with an average dispersion slope of +3 ps/nm^2/km. This is not only much larger than standard SMF-28 fiber (+17 ps/nm/km) but of opposite sign—normal—as opposed to anomalous in SMF-28. This is ideal for SPM-based optical regenerators and played a critical role in the success of our device. SPM-induced spectral broadening, without the presence of linear dispersion, produces an oscillating spectrum at very high peak powers corresponding to nonlinear phase shifts $\gg 2\pi$, which results in an oscillating power transfer curve. The large

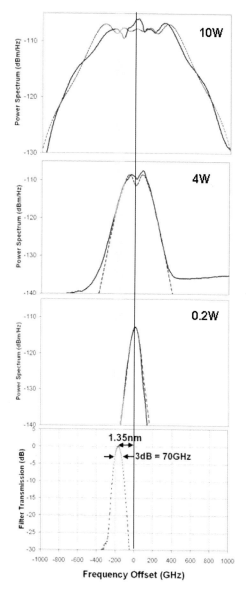

Figure 31.6 Experimental output pulse spectra for different input pulse peak powers, theory (dashed) and experiment (solid). Bottom: transmission bandpass filter transmission spectrum, offset by 1.35 nm from the input centre wavelength, and with a 3 dB bandwidth of 0.56 nm (70 GHz).

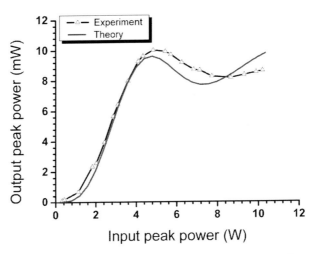

Figure 31.7 Experimental (data points) and theoretical (solid) output nonlinear power transfer curves of fiber-based 2R regenerator).

linear normal dispersion in As_2Se_3 linearizes the induced chirp, which averages out the spectral oscillations, thus smoothing out the nonlinear transfer curve. Note that even though the total dispersion of 1 ps/nm to 2 ps/nm present in the device is comparable with that of typical lengths of silica-based HNLF (one to several kilometers) used in regeneration experiments,[84] the key is that for chalcogenide glass, this level of dispersion is present in very short lengths of a few meters.

The overall success of this device was largely a result of optimizing the parameters such as length and input pulse widths. It has been shown[84] that the optimum device length is $L_{opt} = 2.4\, L_D/N$, where $N = \sqrt{L_D/L_{NL}} \propto \sqrt{P}$ is the soliton number (although clearly with normal dispersion solitons not present), P is the input launch power, and L_D and L_{NL} are the dispersion and nonlinear lengths, respectively.[84] For this As_2Se_3 fiber, L_D is ~13 m and N is ~ 13, yielding an optimum length of $L_{opt} = 2.7$ m, very close to the length used (2.8 m).

This initial proof-of-concept demonstration employed low duty cycle pulses. However, Fig. 31.7 indicates that this device already operates at power levels commensurate with typical powers achievable in a 33% RZ signal at 40 Gb/s of 5 W, corresponding to ~800 mW of average power. This is within about a factor of 2 of typical peak powers used in highly nonlinear silica fiber.[84] We recently demonstrated operation at full data bit rates at 10 Gb/s in both fiber[93] and waveguides[94] for devices operating via cross-phase modulation, rather than self-phase modulation.

One of the more remarkable findings of our work on these 2R regenerators was that a modest amount of nonlinear, or two-photon absorption can actually enhance the device performance. This is in stark contrast with nonlinear

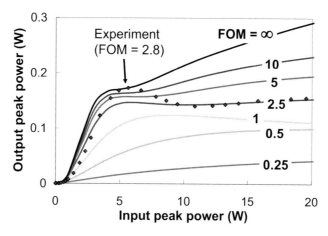

Figure 31.8 Calculated power transfer curves for 2R regenerator for different FOM with fixed n_2, including experimental data for a device with a FOM of 2.8 (diamonds).

switching applications where it has been known for almost 20 years[55] that TPA can only degrade device performance and that it represents a fundamental limit that in fact inspired the definition of the nonlinear FOM ($= n_2/\beta\lambda$) in the first place. Figure 31.8 shows theoretical power transfer curves for this device obtained by varying the FOM of the material from infinite (no TPA) to 0.25 (very high TPA). Materials with an FOM in the range of 1 to 5 produce a flatter PTF than those with no TPA. This saturation of output power versus input power (or "step-like" function) will result in a reduction in noise on the logical "1s" and on the logical "0s"—an important requirement of 2R regenerators. For FOM < 1, the transfer curves lose their sharp edge characteristics and round-off. The experimental data points in Fig. 31.8 agree well with the theoretical curve for FOM = 2.5. This value is close to the experimental value of FOM = 2.8 for As_2Se_3.[37,54,57,95]

To better understand the performance of this 2R regenerator in a system, simulations of a pseudo-random bit sequence (PRBS) at 40 Gb/s with an RZ signal made of 33% duty cycle pulses, including ASE noise to a level of OSNR = 15 dB, were performed. A series of results were calculated by varying the material FOM, keeping other parameters constant. Figure 31.9 shows the calculated Q factor at the output as a function of peak input power, for various FOMs, where it is clear that the improvement in PTF characteristics for the moderate FOM of 2.5 seen in Fig. 31.8 also translates into improved eye diagrams and Q factors. It is also evident that the performance of the regenerators depends strongly not only on the FOM, but on the absolute input peak power as well: 7 W for a device with no TPA and 14 W for an FOM of 2.5. The optimum input peak power occurs above the step-threshold power in the PTF and so the lower power pulses remain in the higher output power plateau. It is also clear from Fig. 31.9 that while a device with an FOM of less than unity does not effectively increase the Q factor, an infinite FOM is clearly not optimum either. The best device performance is obtained with an intermediate FOM and with

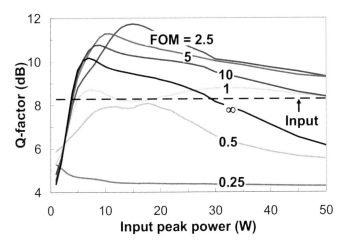

Figure 31.9 Calculated Q factor for 40 Gb/s RZ signal with OSNR of 15 dB, for different figures of merit (FOMs) versus input signal power. The horizontal dashed line represents the input Q factor.

slightly higher input power, although still well within practical power levels. Figure 31.10 plots the maximum optimized (by adjusting the input peak power) output Q factor as a function of FOM and shows that there is clearly an optimum range of FOM, from 2 to 4, where the improvement in Q factor is maximized and noticeably better than a device with no TPA (infinite FOM). It is interesting to note that As_2Se_3 chalcogenide glass has a figure of merit of 2.8, which is near the optimum value.

Finally, a key requirement for all-optical regenerators is to keep the signal pulse distortion by the regenerator to a minimum. The spectral and temporal response of the output pulses were measured[96] by frequency-resolved optical gating, showing explicitly that the spectral and temporal pulse distortion, in both phase and amplitude, is very small—less than a few percent.

In summary, these results show promising performance from As_2Se_3 chalcogenide-fiber- based 2R all-optical regenerators, already with advantages over silica-based HNLF devices such as a reduction in fiber length from kilometers to meters, and an intrinsic nonlinear figure of merit—a fundamental quality that can intrinsically limit device performance—that is nearly ideal for 2R regeneration. There is significant room for future improvements in, for example, reducing operating power—currently comparable with the best silica-based devices. A reduction in operating power by over an order of magnitude would be achieved by reducing the mode field diameter as well as possibly increasing the intrinsic nonlinearity of the chalcogenide glass by optimizing stoichiometry. In addition, because our fiber lengths are so short, there is also scope to further reduce the required operating power further by using slightly longer lengths (up to 5 to 10 m). The passive propagation loss of this fiber, currently at 0.9 dB/m (resulting in a total loss for 2.8 m of fiber comparable with that for 2 to 3 km long silica HNLF devices) is also expected to decrease significantly with

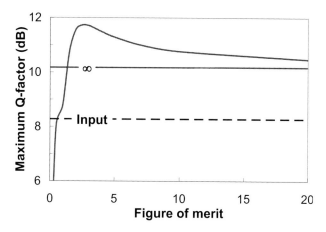

Figure 31.10 Optimized Q factor improvement versus figure of merit, showing that modest TPA can enhance device performance.

improvement in fabrication processes. Losses due to spectral filtering (in this case ~13dB) together with losses from additional components (tunable filter, VOA, coupling loss at exit face, etc.; in this case ~12 dB) can also be significantly reduced by optimizing filter design, improving fiber coupling (cleave quality and AR coatings), and reducing the insertion loss of all of the components. Ultimately, however, the goal is to move on beyond fiber-based devices to fully integrated all-optical photonic integrated circuits, and we turn to this topic next.

31.3.1 2R regeneration in integrated waveguides

As discussed in the Introduction, there is a compelling motivation to develop fully integrated all-optical photonic integrated circuits (PICs) capable of operating from 40 Gb/s to potentially > 1 Tb/s. Recently,[6,7] we reported the first fully integrated all-optical 2R regenerator based on Kerr nonlinearities, and even more recently[8] a similar approach has been reported in silicon, where the use of nanowire waveguides enabled an order of magnitude reduction in peak operating power.

Figure 31.11 shows a schematic of the chalcogenide glass-based device. It consists of a 5 cm long rib waveguide in (As_2S_3) chalcogenide glass integrated with a 5-mm long double grating that acts as a bandpass filter. This device is a result of recent advances in low-loss rib waveguides[4] and high-quality Bragg gratings in chalcogenide glass (As_2S_3).[60,97] The waveguides were fabricated by pulsed laser deposition (PLD)[59] of a 2.4-μm thick As_2S_3 film, with a refractive index of 2.38, followed by photolithography and reactive-ion etching to form a 5-cm long, 4-μm wide, rib waveguide with a rib height of 1.1 μm, and then overcoated with a polymer film transparent in the visible. The fiber-waveguide-fiber insertion loss (with high NA fiber, 4-μm mode field diameter) was 10.5 dB, of which ~2.5 dB was propagation loss, equating to a coupling efficiency of ~4.0

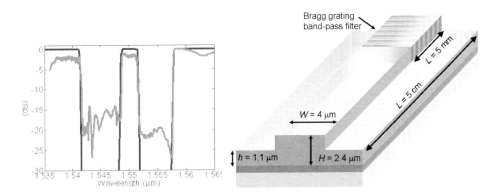

Figure 31.11 Integrated all-optical 2R regenerator in As$_2$S$_3$ chalcogenide glass: a 5 cm long ridge waveguide integrated with a 5 mm long double grating bandpass filter (inset—theory and experiment).

dB per facet (including about 0.8 dB of Fresnel reflection loss from the material refractive index mismatch).

Waveguide gratings were written[60,97] near the exit facet of the waveguide using a Sagnac interferometer along with a CW 532 nm doubled Nd:YAG laser having a coherence length of 4 mm, which provided a naturally compensated refractive index apodization. The sample was exposed with 10 mW for 60 s, resulting in gratings with an estimated index change $\Delta n = 0.004$–0.005. The extremely high quality of the gratings in terms of both width (>v6 nm each), depth (> 25 dB), and very low out-of-band sidelobes was critical for the successful performance of this device. Figure 31.11 (inset) shows the transmission spectrum of the waveguide Bragg grating for TE polarization. The grating in fact consisted of two gratings offset from each other to produce an overall rejection bandwidth of 16.3 nm with a 2.8 nm wide pass band in the middle. The TM polarized grating was separated by ~8 nm from the TE grating due to waveguide birefringence, and so these experiments were carried out with TE polarized light. TE polarized pulses could be launched to better than 20 dB extinction ratio by adjusting the polarization controller before the amplifier.

The experimental setup to demonstrate device performance is similar to that for fiber (Fig. 31.5). Pulses from a mode-locked "figure of 8" laser (1.5 ps, full-width, half-maximum) were passed through a polarization controller and then a custom optical amplifier (designed to minimize excess spectral broadening), resulting in 8.75 MHz repetition rate pulses at peak powers up to 1.2 kW, nearly transform limited with a spectral width of 1.9 nm, tunable from 1530 to 1560 nm. The pulse spectra were tuned to slightly offset wavelengths relative to the grating transmission pass band. The output of the amplifier was butt-coupled into the waveguide using a short (to minimize spectral broadening within the fiber) 20 cm length of standard single-mode telecommunications fiber (SMF) followed by 5 mm of high-NA fiber spliced on the end to improve coupling efficiency to the waveguide. The output of the waveguide was then directed either into a power meter, an optical spectrum analyzer, or an optical autocorrelator.

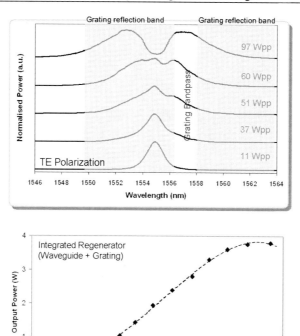

Figure 31.12 Top: Unfiltered pulse output spectra of waveguide for different input peak powers. Bottom: resulting nonlinear power transfer curve of integrated 2R regenerator.

Figure 31.12 shows the pulse spectra for 2 ps pulses after passing through a bare waveguide with no grating present, showing significant broadening with increasing input peak power, due to SPM equivalent to a nonlinear optical phaseshift (ϕ_{NL}) of approximately $3\pi/2$, which was spectrally symmetric. The maximum peak power in the waveguide was 97 W, corresponding to a maximum intensity of ~1.8 GW/cm^2. We verified that SPM-induced spectral broadening by the amplifier was negligible and that no photo-darkening occurred. Figure 31.12 also shows the resulting power transfer curve of the device integrated with the waveguide Bragg grating, obtained by tuning the pulse center wavelength to within the rejection band of the Bragg grating to the longer wavelength side of the pass band, for TE polarization. As seen in Fig. 31.12 this resulted in low transmission at low input power since the filter rejection bandwidth was much wider than the 3 dB spectral bandwidth of the input pulse. As the input pulse power was increased, SPM broadened the spectrum so that power was transmitted through the pass band of the grating, resulting in a clear nonlinear "S" shape, required for optical regeneration.

The temporal transfer characteristics of this device were measured via autocorrelation at high power[7] and showed that the output pulses were broadened slightly to ~3 ps. This broadening is not due to either waveguide or material dispersion (both of which are negligible on this length scale), but rather a combination of grating dispersion near the edge of the stop band and a comparatively wide pass band (3 nm square filter) relative to the input pulse spectrum (1.9 nm FWHM). Optimizing the grating pass band shape and width is expected[98] to reduce this.

Although this device operates at peak powers from 10 to 76 W, and was tested with low duty cycle optical pulses, practical devices would need to operate at subwatt power levels and with high duty cycle (e.g., 33% return-to-zero) optical pulses. By increasing the device length to 50 cm through the use of serpentine or spiral structures, increasing the material nonlinearity (by using As_2Se_3 rather than As_2S_3, for example), and by decreasing the waveguide area by a factor of 10 (to 1µm^2), a reduction in operating power by two orders of magnitude should be achievable, resulting in subwatt power level operation. Furthermore, the increased (normal) dispersion of longer waveguides would linearize the frequency chirp of the output pulses, reducing pulse distortion and improving transfer function characteristics.[11]

31.4 Wavelength Conversion

Of the many approaches to wavelength conversion, all-optical solutions based on the ultrafast Kerr nonlinearity arguably have the greatest potential for both speed and simplicity. We focus here on our work based on wavelength conversion in chalcogenide glass planar waveguides, using a technique based on non-phase-matched wavelength conversion via cross-phase modulation (XPM). This technique was first demonstrated by Ohlsen[99] in highly nonlinear fiber, and was subsequently employed in bismuth fiber[100] and chalcogenide fiber,[93] followed by our recent work in chalcogenide waveguides.[5] The primary advantage of this technique is that because it is non-phase-matched, it doesn't require dispersion engineered fiber or waveguides. Pump-probe walk-off is certainly a concern,[93] but this has not prevented demonstrations of up to 25 nm of conversion range, even in devices with quite high material dispersion.[5,93]

Figure 31.13 shows the principle of operation of a wavelength converter based on XPM.[99] A continuous-wave (CW) probe experiences XPM from copropagating signal pump pulses, which generate optical sidebands. This is converted to amplitude modulation by using a band-pass filter to select a single sideband. Importantly, the interaction length between the signal pump and the CW probe, i.e., pulse walk-off, is the phase-matching condition. Recently,[93] this approach was used to demonstrate wavelength conversion in As_2Se_3 chalcogenide glass fiber—the fiber with the highest Kerr nonlinearity to date (n_2 = 400 × silica) for which any nonlinear signal processing has been reported. An important aspect of this work was demonstrating device operation with a full data bit stream rather than just low duty cycle optical pulses where achieving very high peak powers (for typical system average signal powers) is much easier.

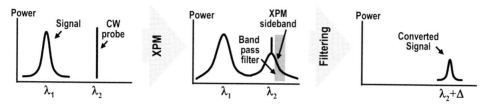

Figure 31.13 Principle of XPM wavelength conversion. Amplified pulsed pump signal (at λ_1) imposes a nonlinear frequency chirp onto a copropagating wavelength tunable CW probe (at λ_2) through the nonlinear refractive index. Filtering one of the XPM-generated sidebands results in wavelength conversion (to $\lambda_2 + \Delta$).

Error-free conversion showing full system bit error rate (BER) measurements was demonstrated at 10 Gb/s return-to-zero (RZ) near 1550 nm over a 10 nm wavelength range with an excess system penalty of 1.4 dB in only 1 m of fiber and 2.1 W of peak optical pump power.

This approach to λ conversion has also been reported[100] in bismuth oxide fiber over a 15 nm wavelength range at 160 Gb/s. Both fiber types offer advantages and disadvantages, and the fiber parameters are compared in Table 31.1. The significantly larger n_2 for As_2Se_3—an order of magnitude greater than Bi_2O_3—is a key advantage, although because of a much larger effective core area of 37 um^2 [and hence similar nonlinear coefficients $\gamma = n_2\omega/(cA_{eff})$], the fiber length in both demonstrations was the same (1 meter), which is still excessive for future integrated devices. Because of the higher material nonlinearity and larger mode area of the As_2Se_3 fiber, there arguably is much more potential for future improvement through the reduction in mode area by exploiting the much stronger confinement achievable by the large linear refractive index (2.7) of As_2Se_3. This would result in a greatly reduced core area and consequently device length, which would in turn reduce both fiber losses and dispersion-related impairments.

Although dispersion is not as critical a factor as in phase-matched devices (see next section), the conversion bandwidth is still limited by the difference in pump and probe group velocity, commonly referred to as walk-off length.[93] In fact, this is also an issue for many pump-probe devices such as signal demultiplexing and Raman modulation.[101] While the dispersion parameter of As_2Se_3 is high relative to Bi_2O_3, the potentially shorter device lengths just described make significant conversion bandwidth a possibility. For example, an effective core area of the As_2Se_3 equivalent to the Bi_2O_3 fiber would result in $\gamma = 11,100$ W^{-1}km^{-1}, requiring only 12 cm of As_2Se_3 fiber, which would yield > 40 nm of conversion bandwidth compared to 10 nm for 1 m of Bi_2O_3 (for 8 ps pulses). Both fibers (As_2Se_3, and Bi_2O_3) have relatively constant dispersion D over the entire communications band,[11] in contrast with dispersion-shifted silica fiber.

The ultimate route to realizing the full potential of chalcogenide glass (or any material) all-optical devices lies in exploiting their high refractive index (2.4–2.8) and high n_2 by reducing the waveguide mode area, whether by the use of microstructured optical fiber,[104] fiber tapering,[103] or planar waveguides,[7] to which we turn next.

To date, all-optical signal processing has been mainly limited to fiber-based devices—only very recently have there been reports on integrated devices based on pure Kerr nonlinearities (2R regeneration).[7,8] Furthermore, there have been even fewer reports of system bit error rate (BER) measurements for integrated all-optical devices operating at full data rates of 10 Gb/s or higher, primarily because of peak power requirements. These include (non-phase-matched) XPM-based wavelength conversion[5] in As_2S_3 chalcogenide glass waveguides and wavelength conversion via FWM in silicon nanowires.[20,105] The lowest peak operating power for non-phase-matched integrated all-optical devices is comparable in silicon (2R regeneration) and chalcogenide glasses at ~6W. Ultimately, for practical operation, devices with peak power requirements well below 1W would be needed. Here, we focus on a recent report[94] of system penalty measurements of a waveguide all-optical wavelength converter based on XPM that achieved error-free wavelength conversion at 10 Gb/s in a 5 cm long As_2S_3 chalcogenide glass planar waveguide, over a 25 nm wavelength range with a Q-factor penalty of 2.3 dB.

The experimental configuration (Fig. 31.14) consists of the input pump provided by a 10 GHz clock signal (FWHM 2.1 ps), modulated with a $2^{29} - 1$ long PRBS, and then amplified and filtered to remove out-of-band ASE noise. This was combined through a 50/50 coupler with a CW probe (also amplified and filtered for out-of-band ASE). The polarizations of both the pump and probe were aligned to the lower-loss TE mode of the waveguide to maximize XPM. The combined signals are butt-coupled to the waveguide via SMF fiber pigtails (with index-matching fluid) with a short length of UHNA-4 spliced onto the end for better mode matching, resulting in a loss of ~2.2 dB per facet. The 5 cm long, 3 μm wide As_2S_3 chalcogenide glass ridge waveguide was fabricated using pulsed laser deposition and dry etching, resulting in a low propagation loss of ~0.3 dB/cm.[4] The pump was centered at 1541 nm with an average power of 115 mW (~5.8 W peak powers in the waveguide). The CW probe was 25 mW coupled in and set to three different wavelengths: 1555.4, 1559.5, and 1563.6 nm.

Figure 31.15 shows the XPM spectral broadening of the probe at the output of the waveguide, which is independent of the signal-probe wavelength offset due to the short device length, which ensures the dispersive pulse walk-off between the signal and probe is negligible. Note that a significant portion of the CW probe remains due to the comparatively low signal duty cycle. The bit error rates (BERs) of the converted pulses were measured at all three probe wavelengths and the corresponding Q factors were calculated[106] (Fig. 31.16). The converted pulses were centered at 1557.0, 1560.6, and 1565.0 nm with Q factors of 20.2, 20.7, and 20.7 dB, respectively. Comparing these Q factors to that of the back-to-back signal at 1541 nm, we obtain a Q-factor penalty of 2.3 dB, achieved over a range of 25 nm. Due to the short device length, group velocity

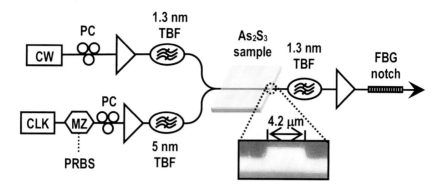

Figure 31.14 Experimental setup for demonstrating cross-phase modulation-based wavelength conversion in chalcogenide ridge waveguides.

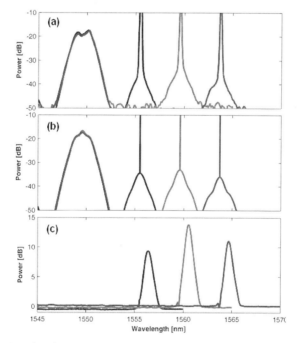

Figure 31.15 Spectra showing cross-phase modulation sidebands with the probe set to three different wavelengths; (a) input spectra, (b) theory, and (c) filtered output spectra.

mismatch will potentially allow a conversion range of up to ±45 nm. As mentioned, this is one of very few demonstrations of error-free wavelength conversion at telecommunication bit rates (10 Gb/s and above) based on pure Kerr (n_2) nonlinearities—the others being in highly nonlinear fiber[99,100] and via FWM in silicon.[20]

31.5 Future Prospects

Clearly, there are still challenges to be met by all-optical signal processing devices in order to be fully viable in high-bit-rate systems. A key issue is the required operating peak power. The lowest peak powers that have been achieved in integrated devices for non-phase-matched processes[5,8] are in the range of 5W—probably two orders of magnitude larger than the ideal. Phase-matched wavelength conversion has been achieved with roughly 100 mW of pump power,[20] but even this ideally should be reduced. Ultimate solutions will require

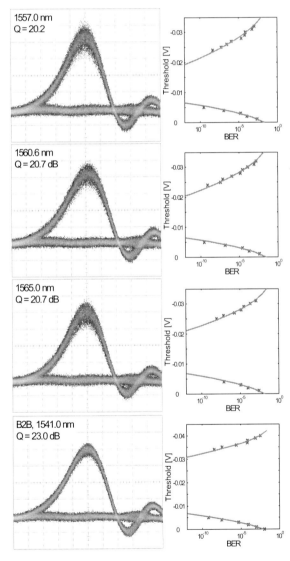

Figure 31.16 Eye diagrams of the data converted to three different wavelengths and the back-to-back pulse for comparison.

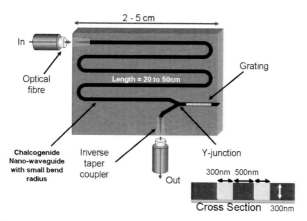

Figure 31.17 Schematic layout for a future nanowire-based integrated all-optical regenerator in chalcogenide glass.

greatly reduced waveguide cross-sectional areas (to enhance the nonlinear parameter γ), and increased device lengths, while maintaining low linear and nonlinear losses. Figure 31.17 shows a layout of a future concept for a device that could potentially meet all of these requirements—based in chalcogenide glass. The key features include the use of a very small mode area (300 × 300 nm) waveguides—commonly referred to as nanowires—which are achievable because of the large refractive index of chalcogenide glasses, as well as long waveguide lengths (and low propagation loss requirements). With nonlinear induced phaseshifts of $3\pi/2$ (a rough benchmark for nonlinear device performance), then usable phase shifts can be achieved for most chalcogenide glass waveguides with lengths of a few 10s of centimeters at 100 MW/cm^2 as long as losses are <0.1dB/cm. These are certainly challenging requirements, but they are not beyond the realm of feasibility given the great strides that have been made in nonlinear glass waveguide fabrication the past few years.[4]

31.6 Conclusions

We have reviewed some of the recent progress on all-optical nonlinear signal processing devices over the past few years based primarily on chalcogenide glass. Initial proof-of-concept demonstrations were achieved in fiber, followed by a subsequent demonstration in integrated waveguides. It is clear to many of us working in the field that all-optical signal processing devices are steadily coming of age, and the prospect for reducing peak power requirements of PIC-based all-optical circuits below the current levels of a few watts[8,15] to tens of milliwatts, or even lower, through the use of nanowire structures and other novel approaches will go a long way to ensure that all-optical nonlinear devices will play an integral part in future telecommunications systems.

References

1. Weber, H.G., et al., "Ultrahigh-speed OTDM-transmission technology," *J. Lightwave Technol.*, **24**, 4616–4627, 2006.
2. Claps, R., et al., "Observation of stimulated Raman amplification in silicon waveguides," *Opt. Expr.*, **11**(15), 1731–1739, 2003.
3. Mok, J.T., and Eggleton, B.J., "Photonics - Expect more delays," *Nature*, **433**(7028), 811–812, 2005.
4. Ruan, Y.L., et al., "Fabrication and characterization of low loss rib chalcogenide waveguides made by dry etching," *Opt. Expr.*, **12**(21), 5140–5145, 2004.
5. Ta'eed, V.G., et al., "All optical wavelength conversion via cross phase modulation in chalcogenide glass rib waveguides," *Opt. Expr.*, **14**(23), 11242–11247, 2006.
6. Ta'eed, V.G., et al., "Self-phase modulation-based integrated optical regeneration in chalcogenide waveguides," *IEEE J. Sel. Top. Quant. Electron.*, **12**, 360–370, 2006.
7. Ta'eed, V.G., et al., "Integrated all-optical pulse regenerator in chalcogenide waveguides," *Opt. Lett.*, **30**(21), 2900–2902, 2005.
8. Reza Salem, M.A.F., Turner, A.C., Geraghty, D.F., Lipson, M., and Gaeta, a.A.L., "All-optical regeneration on a silicon chip," *Opt. Expr.*, **15**, 7802-7809, 2007.
9. Sugimoto, N., et al. "Bismuth-based optical fiber with nonlinear coefficient of 1360 $W^{-1}.km^{-1}$," IEEE *Proceedings of Optical Fiber Communications Conference (OFC)*, Los Angeles, Postdeadline paper 26, 2004.
10. Aggarwal, I.D., and Sanghera, J.S., "Development and applications of chalcogenide glass optical fibers at NRL," *J. Optoelectron. Adv. Mater.*, **4**, 665–678, 2002.
11. Fu, L.B., et al., "Investigation of self-phase modulation based optical regeneration in single mode As2Se3 chalcogenide glass fiber," *Opt. Expr.*, **13**(19), 7637–7644, 2005.
12. Fu, L. B., Pelusi, M. D., Magi, E. C., Ta'eed, V. G., Eggleton, B. J., "Broadband all-optical wavelength conversion of 40 Gbit/s signals in nonlinearity enhanced tapered chalcogenide fibre", Electronics Letters **44**, 44-45, 2008.
13. Le Person, J., et al., "Light guidance in new chalcogenide holey fibres from GeGaSbS glass," *Mater. Res. Bull.*, **41**, 1303–1309, 2006.
14. Bergano, N.S., "Wavelength division multiplexing in long-haul transoceanic transmission systems," *J. Lightwave Technol.*, **23**, 4125–4139, 2005.
15. Ng, T.T., et al., "In-band OSNR and chromatic dispersion monitoring using a fibre optical parametric amplifier," *Opt. Expr.*, **13**(14), 5542–5552, 2005.

16. Luo, T., et al., "All-optical chromatic dispersion monitoring of a 40-Gb/s RZ signal by measuring the XPM-generated optical tone power in a highly nonlinear fiber," *IEEE Photon. Technol. Lett.*, **18**, 430–432, 2006.
17. Radic, S., et al., "All-optical regeneration in one- and two-pump parametric amplifiers using highly nonlinear optical fiber," *IEEE Photon. Technol. Lett.*, **15**, 957–959, 2003.
18. Rochette, M., Blows, J.L., and Eggleton, B.J., "3R optical regeneration: An all-optical solution with BER improvement," *Opt. Expr.*, **14**(14), 6414–6427, 2006.
19. Jiang, R., et al., "Continuous-wave band translation between the near-infrared and visible spectral ranges," *J. Lightwave Technol.*, **25**, 58–66, 2007.
20. Yamada, K., et al., "All-optical efficient wavelength conversion using silicon photonic wire waveguide," *IEEE Photon. Technol. Lett.*, **18**, 1046–1048, 2006.
21. Lee, J.H., and Kikuchi, K., "All fiber-based 160-Gbit/s add/drop multiplexer incorporating a 1-m-long Bismuth Oxide-based ultra-high nonlinearity fiber," *Opt. Expr.*, **13**(18), 6864–6869, 2005.
22. Radic, S., and McKinstrie, C.J., "Optical amplification and signal processing in highly nonlinear optical fiber," *IEICE Trans. Electron.*, **E88C**(5), 859–869, 2005.
23. Hansryd, J., and Andrekson, P.A., "Broad-band continuous-wave-pumped fiber optical parametric amplifier with 49-dB gain and wavelength-conversion efficiency," *IEEE Photon. Technol. Lett.*, **13**, 194–196, 2001.
24. Inoue, K., "Polarization-independent wavelength conversion using fiber 4-wave-mixing with 2 orthogonal pump lights of different frequencies," *J. Lightwave Technol.*, **12**, 1916–1920, 1994.
25. Stolen, R.H., and Bjorkholm, J.E., "Parametric amplification and frequency-conversion in optical fibers," *IEEE J. Quant. Electron.*, **18**, 1062–1072, 1982.
26. Jopson, R.M., and Tench, R.E., "Polarization-independent phase-conjugation of lightwave signals," *Electron. Lett.*, **29**(25), 2216–2217, 1993.
27. Kumar, P., and Shapiro, J.H., "Squeezed-state generation via forward degenerate 4-wave mixing," *Phys. Rev. A*, **30**, 1568–1571, 1984.
28. Marhic, M.E., et al., "High-nonlinearity fiber optical parametric amplifier with periodic dispersion compensation," *J. Lightwave Technol.*, **17**(2), 210–215, 1999.
29. Radic, S., et al., "Continuous-wave parametric gain synthesis using nondegenerate pump four-wave mixing," *IEEE Photon. Technol. Lett.*, **14**(10), 1406–1408, 2002.
30. Li, R.D., Kumar, P., and Kath, W.L., "Dispersion compensation with phase-sensitive optical amplifiers," *J. Lightwave Technol.*, **12**, 541–549, 1994.

31. Monroy, I.T., et al., "Monolithically integrated reflective SOA-EA carrier remodulator for broadband access nodes," *Opt. Expr.*, **14**(18), 8060–8064, 2006.
32. van der Poel, M., et al., "Ultrafast gain and index dynamics of quantum dash structures emitting at 1.55 mu m," *Appl. Phys. Lett.*, **89**, 081102 , 2006.
33. Ohman, F., et al., "Noise and regeneration in semiconductor waveguides with saturable gain and absorption," *IEEE J. Quant. Electron.*, **40**, 245–255, 2004.
34. Murai, H., et al., "EA-modulator-based optical time division multiplexing/demultiplexing techniques for 160-Gb/s optical signal transmission," *IEEE J. Sel. Top. Quant. Electron.*, **13**, 70–78, 2007.
35. Tangdiongga, E., et al., "All-optical demultiplexing of 640 to 40 Gbits/s using filtered chirp of a semiconductor optical amplifier," *Opt. Lett.*, **32**, 835–837, 2007.
36. Slusher, R.E., et al., "Large Raman gain and nonlinear phase shifts in high-purity As_2Se_3 chalcogenide fibers," *J. Opt. Soc. Am. B*, **21**, 1146–1155, 2004.
37. Asobe, M., Kanamori, T., and Kubodera, K., "Applications of Highly Nonlinear Chalcogenide Glass-Fibers in Ultrafast All-Optical Switches," *IEEE J. Quant. Electron.*, **29**, 2325–2333, 1993.
38. Zakery, A., and Elliott, S.R., "Optical properties and applications of chalcogenide glasses: A review," *J. Non-Cryst. Solids*, **330**, 1–12, 2003.
39. Ramachandran, S., and Bishop, S.G., "Photoinduced integrated-optic devices in rapid thermally annealed chalcogenide glasses," *IEEE J. Sel. Top. Quant. Electron.*, **11**, 260–270, 2005.
40. DeCorby, R.G., et al., "High index contrast waveguides in chalcogenide glass and polymer," *IEEE J. Sel. Top. Quant. Electron.*, **11**, 539–546, 2005.
41. Ramachandran, S., and Bishop, S.G., "Low loss photoinduced waveguides in rapid thermally annealed films of chalcogenide glasses," *Appl. Phys. Lett.*, **74**, 13–15, 1999.
42. Rode, A.V., et al., "Laser-deposited As_2S_3 chalcogenide films for waveguide applications," *App. Surf. Sci.*, **197**, 481–485, 2002.
43. Zakery, A., "Low loss waveguides in pulsed laser deposited arsenic sulfide chalcogenide films," *J. Phys. D*, **35**(22), 2909–2913, 2002.
44. Zoubir, A., et al., "Direct femtosecond laser writing of waveguides in As_2S_3 thin films," *Opt. Lett.*, **29**(7), 748–750, 2004.
45. Viens, J.F., et al., "Fabrication and characterization of integrated optical waveguides in sulfide chalcogenide glasses," *J. Lightwave Technol.*, **17**(7), 1184–1191, 1999.
46. Tanaka, K., Toyosawa, N., and Hisakuni, H., "Photoinduced Bragg gratings in As_2s_3 optical fibers," *Opt. Lett.*, **20**(19), 1976–1978, 1995.

47. Asobe, M., et al., "Fabrication of Bragg grating in chalcogenide glass fibre using the transverse holographic method," *Electron. Lett.*, **32**(17), 1611–1613, 1996.
48. Saliminia, A., et al., "Photoinduced Bragg reflectors in As-S-Se/As-S based chalcogenide glass multilayer channel waveguides," *Fiber Integr. Opt.*, **20**, 151–158, 2001.
49. Ramachandran, S., et al., "Fabrication of holographic gratings in As2S3 glass by photoexpansion and photodarkening," *IEEE Photon. Technol. Lett.*, **8**, 1041–1043, 1996.
50. Robinson, T.G., et al., "Strong Bragg gratings photoinduced by 633-nm illumination in evaporated AS(2)Se(3) thin films," *Opt. Lett.*, **28**, 459–461, 2003.
51. Ponnampalam, N., et al., "Small core rib waveguides with embedded gratings in As2Se3 glass," *Opt. Expr.*, **12**(25), 6270–6277, 2004.
52. Vallee, R., et al., "Real-time observation of Bragg grating formation in As2S3 chalcogenide ridge waveguides," *Opt. Commun.*, **230**, 301–307, 2004.
53. Freeman, D., Madden, S., and Luther-Davies, B., "Fabrication of planar photonic crystals in a chalcogenide glass using a focused ion beam," *Opt. Expr.*, **13**, 3079–3086, 2005.
54. Asobe, M., "Nonlinear optical properties of chalcogenide glass fibers and their application to all-optical switching," *Opt. Fiber Technol.*, **3**, 142–148, 1997.
55. Mizrahi, V., et al., "2-Photon absorption as a limitation to all-optical switching," *Opt. Lett.*, **14**(20), 1140–1142, 1989.
56. Lenz, G., et al., "Large Kerr effect in bulk Se-based chalcogenide glasses," *Opt. Lett.*, **25**, 254–256, 2000.
57. Nguyen, H.C., et al., "Dispersion in nonlinear figure of merit of As2Se3 chalcogenide fibre," *Electron. Lett.*, **42**, 571–572, 2006.
58. Sanghera, J.S., et al., "Nonlinear properties of chalcogenide glass fibers," *J. Optoelectron. Adv. Mater.*, **8**, 2148–2155, 2006.
59. Zakery, A., et al., "Low-loss waveguides in ultrafast laser-deposited As2S3 chalcogenide films," *J. Opt. Soc. Am. B*, **20**, 1844–1852, 2003.
60. Shokooh-Saremi, M., et al., "High-performance Bragg gratings in chalcogenide rib waveguides written with a modified Sagnac interferometer," *J. Opt. Soc. Am. B*, **23**, 1323–1331, 2006.
61. Xu, Q.F., and Lipson, M., "All-optical logic based on silicon micro-ring resonators," *Opt. Expr.*, **15**, 924–929, 2007.
62. Preble, S.R., et al., "Ultrafast all-optical modulation on a silicon chip," *Opt. Lett.*, **30**(21), 2891–2893, 2005.
63. Almeida, V.R., et al., "All-optical switching on a silicon chip," *Opt. Lett.*, **29**(24), 2867–2869, 2004.

64. Almeida, V.R., et al., "All-optical control of light on a silicon chip," *Nature*, **431**(7012), 1081–1084, 2004.
65. Okawachi, Y., et al., "All-optical slow-light on a photonic chip," *Opt. Expr.*, **14**, 2317–2322, 2006.
66. Manolatou, C., and Lipson, M., "All-optical silicon modulators based on carrier injection by two-photon absorption," *J. Lightwave Technol.*, **24**, 1433–1439, 2006.
67. Liang, T.K., et al., "High speed logic gate using two-photon absorption in silicon waveguides," *Opt. Commun.*, **265**, 171–174, 2006.
68. Vlasov, Y.A., and McNab, S.J., "Losses in single-mode silicon-on-insulator strip waveguides and bends," *Opt. Expr.*, **12**, 1622–1631, 2004.
69. Espinola, R.L., et al., "Raman amplification in ultrasmall silicon-on-insulator wire waveguides," *Opt. Expr.*, **12**(16), 3713–3718, 2004.
70. Liu, Y., and Tsang, H.K., "Nonlinear absorption and Raman gain in helium-ion-implanted silicon waveguides," *Opt. Lett.*, **31**, 1714–1716, 2006.
71. Liang, T.K., and Tsang, H.K., "On Raman gain in silicon waveguides: Limitations from two-photon-absorption generated carriers," Paper No. CThT48, Conference on Lasers and Electro-Optics, 2004 (CLEO), San Francisco, 2004.
72. Liang, T.K., and Tsang, H.K., "Role of free carriers from two-photon absorption in Raman amplification in silicon-on-insulator waveguides," *Appl. Phys. Lett.*, **84**(15), 2745–2747, 2004.
73. Tsang, H.K., et al., "Optical dispersion, two-photon absorption and self-phase modulation in silicon waveguides at 1.5 mu m wavelength," *Appl. Phys. Lett.*, **80**, 416–418, 2002.
74. Dinu, M., Quochi, F., and Garcia, H., "Third-order nonlinearities in silicon at telecom wavelengths," *Appl. Phys. Lett.*, **82**(18), 2954–2956, 2003.
75. Cowan, A.R., Rieger, G.W., and Young, J.F., "Nonlinear transmission of 1.5 mu m pulses through single-mode silicon-on-insulator waveguide structures,", *Opt. Expr.*, **12**, 1611–1621, 2004.
76. Rieger, G.W., Virk, K.S., and Young, J.F., "Nonlinear propagation of ultrafast 1.5 mu m pulses in high-index-contrast silicon-on-insulator waveguides," *Appl. Phys. Lett.*, **84**, 900–902, 2004.
77. Lamont, M.R.E., et al., "Two-photon absorption effects on self-phase-modulation-based 2R optical regeneration," *IEEE Photon. Technol. Lett.*, **18**, 1185–1187, 2006.
78. Sugimoto, N., et al., "Third-order optical nonlinearities and their ultrafast response in Bi2O3-B2O3-SiO2 glasses," *J. Opt. Soc. Am. B*, **16**, 1904–1908, 1999.
79. Boscolo, S., Turitsyn, S.K., and Mezentsev, V.K., "Performance comparison of 2R and 3R optical regeneration schemes at 40 Gb/s for application to all-optical networks," *J. Lightwave Technol.*, **23**, 304–309, 2005.

80. Leclerc, O., et al., "Optical regeneration at 40 Gb/s and beyond," *J. Lightwave Technol.*, **21**, 2779–2790, 2003.

81. Rochette, M., et al., "2R optical regeneration: An all-optical solution for BER improvement," *IEEE J. Sel. Top. Quant. Electron.*, **12**, 736–744, 2006.

82. Rochette, M., et al., "Bit-error-ratio improvement with 2R optical regenerators," *IEEE Photon. Technol. Lett.*, **17**, 908–910, 2005.

83. Mamyshev, P.V. "All-optical data regeneration based on self-phase modulation effect," *Proceedings of European Conference on Optical Communication (ECOC)*, p475 Madrid, (1998).

84. Her, T.H., Raybon, G., and Headley, C., "Optimization of pulse regeneration at 40 Gb/s based on spectral filtering of self-phase modulation in fiber," *IEEE Photon. Technol. Lett.*, **16**, 200–202, 2004.

85. Yoshikane, N., et al., "Benefit of SPM-based all-optical reshaper in receiver for long-haul DWDM transmission systems," *IEEE J. Sel. Top. Quant. Electron.*, **10**, 412–420, 2004.

86. Raybon, G., et al. "40 Gbit/s Pseudo-linear transmission over one million kilometers," Technical Digest of Optical Fiber Communications Conference, OFC 2002, Anaheim, 2002.

87. Huang, Z.J., et al., "10-Gb/s transmission over 100 mm of standard fiber using 2R regeneration in an optical loop mirror," *IEEE Photon. Technol. Lett.*, **16**(11), 2526–2528, 2004.

88. Meissner, M., et al., "3.9-dB OSNR gain by an NOLM-based 2-R regenerator," *IEEE Photon. Technol. Lett.*, **16**(9), 2105–2107, 2004.

89. Ludwig, R., et al. "Experimental verification of noise squeezing by an optical intensity filter in high-speed transmission," *Proceedings of the European Conference on Optical Communication*, Paper Tu B.2.7, Amsterdam, (2001).

90. Ciaramella, E., "A new scheme for all-optical signal reshaping based on wavelength conversion in optical fibers," IEEE *Proceedings of the Optical Fiber Communication Conference (OFC)*, Paper WM37 (2000).

91. Rouvillain, D., et al. "40 Gbit/s optical 2R regenerator based on passive saturable absorber for WDM long-haul transmission," IEEE *Proceedings of the Optical Fiber Communication Conference (OFC)*, Anaheim, Paper FD11 CA, (2002).

92. Ohman, F., et al. "Semiconductor devices for all-optical regeneration," IEEE *Proceedings of the International Conference on Transparent Optical Networks*, (ICTON) Warsaw, paper WE.B.4, p41–46, (2003).

93. Ta'eed, V.G., et al., "Error free all optical wavelength conversion in highly nonlinear As-Se chalcogenide glass fiber," *Opt. Expr.*, **14**(22), 10371–10376, 2006.

94. Lamont, M., "Error-free 2R regeneration in chalcogenide rib waveguides at10Gb/s," *Electron. Lett.* (submitted).

95. Asobe, M., et al., "Nonlinear Refractive-Index Measurement in Chalcogenide-Glass Fibers by Self-Phase Modulation," *Appl. Phys. Lett.*, **60**, 1153–1154, 1992.
96. Lamont, M.R.E., et al., "2R optical regenerator in AS(2)Se(3) chalcogenide fiber characterized by a frequency-resolved optical gating analysis," *Appl. Opt.*, **45**(30), 7904–7907, 2006.
97. Shokooh-Saremi, M., et al., "Ultra-strong, well-apodised Bragg gratings in chalcogenide rib waveguides," *Electron. Lett.*, **41**(13), 738–739, 2005.
98. Lenz, G., et al., "Dispersive properties of optical filters for WDM systems," *IEEE J. Quant. Electron.*, **34**, 1390–1402, 1998.
99. Olsson, B.E., et al., "A simple and robust 40-Gb/s wavelength converter using fiber cross-phase modulation and optical filtering," *IEEE Photon. Technol. Lett.*, **12**, 846–848, 2000.
100. Lee, J.H., et al., "Wavelength conversion of 160 Gbit/s OTDM signal using bismuth oxide-based ultra-high nonlinearity fibre," *Electron. Lett.*, **41**(16), 918–919, 2005.
101. Burdge, G., et al., "Ultrafast intensity modulation by Raman gain for all-optical in-fiber processing," *Opt. Lett.*, **23**, 606–608, 1998.
102. Lamont, M., et al., "Influence of Two Photon Absorption on 2R Optical Regeneration," *Photon. Technol. Lett.* (in press).
103. Lize, Y.K., et al., "Microstructured optical fiber photonic wires with subwavelength core diameter," *Opt. Expr.*, **12**(14), 3209–3217, 2004.
104. Monro, T.M., et al., "Chalcogenide holey fibres," *Electron. Lett.* **36**(24), 1998–2000, 2000,..
105. Fukuda, H., et al., "Four-wave mixing in silicon wire waveguides," *Opt. Expr.*, 2005. **13**(12), 4629–4637.
106. Bergano, N.S., Kerfoot, F.W., and Davidson, C.R., "Margin measurements in optical amplifier systems," *IEEE Photonics Technol. Lett.*, **5**, 304–306, 1993.

Chapter 32
Ultrafast Photonic Processing Applied to Photonic Networks

Hideyuki Sotobayashi
Aoyama Gakuin University
　and
National Institute of Information and Communications Technology, Japan

32.1 Introduction
32.2 Wavelength-Band Generation
32.3 Tunable Wavelength Conversion
32.4 Multiplexing Format Conversion
32.5 Wavelength-Band Conversion
32.6 OCDM Applied for Improved Spectral Efficiency
　　32.6.1 Key technologies
　　32.6.2 Experiments and discussions
32.7 Transparent Virtual Optical Code/Wavelength Path Network
　　32.7.1 Virtual optical code/wavelength path (VOCP/VWP) network
　　32.7.2 Experiments and discussions
32.8 Conclusion
References

32.1 Introduction

Ultrafast photonic processing is expected to play a major role in photonic networks and photonic sensing systems. At the bandwidth of 40 GHz or above, electronics imposes severe technology and economic constraints, which ultrafast photonic processing could advantageously remove.

The quasi-instantaneous response of Kerr nonlinearity in fibers makes it the most attractive effect to overcome bandwidth limitations. For simplicity, consider two optical beams of different wavelength copropagating in the same optical fiber; the intensity dependence of the refractive index leads to a large number of interesting nonlinear effects, namely, self-phase modulation (SPM), cross-phase modulation (XPM), and four-wave mixing (FWM). These ultrafast phenomena could be favorably applied to photonic processing. One application is optical multiplexing, which is used as optical switches, multiplexers, and demultiplexers. Another possible application is wavelength conversion, which is to be used in wavelength division multiplexing (WDM) networks. The other application is supercontinuum (SC) generation, which is a promising technique for various applications in photonic networks and sensing systems.

In present photonic networks, there are two primary techniques for multiplexing data signals: optical time division multiplexing (OTDM) and WDM. Optical code division multiplexing (OCDM) is an alternative method for future options. Optical code division multiple access (OCDMA), encoding and decoding of a signal with a kind of temporal waveform (the so-called optical signature code), allows the selection of a desired signal. Different information bits can share nonerror-producing overlaps of time and wavelength. Simultaneous multiple access can thus be achieved without a complex network protocol to coordinate data transfer among the communicating nodes.

The performance of TDM systems is limited by the time-serial nature of the technology. Each receiver should operate at the total bit rate of the system. The allocation of dedicated time slots does not allow TDM to take advantage of statistical multiplexing gain, which is significant when the data traffic is bursty. In spite of the fact that TDM technologies are well matured and developed at up to some tens of Gbit/s, another problem for the ultrahigh-speed TDM system is the upper limit of electronic circuit operation speed. Although the basic concepts of ultrahigh-speed TDM have been proposed, optical logics still need to be developed. In the WDM system, the available optical bandwidth is divided into fixed wavelength channels that are used concurrently by different channel signals. The problem using WDM is granularity of wavelength. It is limited in that it can only handle traffic on an optical channel of the wavelength path. This may waste the wavelength resources. One of the main applications of WDM systems will be large-capacity long-haul transmission systems because of the relative ease of the transmission technology. OCDMA offers an interesting alternative because neither time management nor frequency management of all nodes is necessary. OCDMA can operate asynchronously and does not suffer from packet collisions; therefore, very low latencies can be achieved. In contrast to TDM and WDM, where the maximum transmission capacity is determined by the total number of channel slots, OCDMA allows flexible network design because the signal quality depends on the number of active channels. Each multiplexing format has its own merits and application area.

In the first half of this chapter, applications in OTDM/WDM are discussed. Some experimental demonstrations for OTDM/WDM networks are reviewed: wavelength-band generation in Section 32.2, tunable wavelength conversion in Section 32.3, multiplexing format conversion in Section 32.4, and wavelength-band conversion in Section 32.5. In the latter half of this chapter, applications in OCDM/WDM are discussed. Some experimental demonstrations for OCDM/WDM networks are reviewed: highly spectral efficient OCDM/WDM transmission in Section 32.6 and transparent virtual optical code/wavelength network in Section 32.7. Finally, discussions are concluded in Section 32.8.

32.2 Wavelength-Band Generation

This section describes a simple configuration of a frequency-standardized simultaneous wavelength-band generation method in carrier-suppressed return-to-zero (CS-RZ) format.[1] One of the critical issues in wavelength division multiplexed (WDM) networks is the linear and nonlinear crosstalks from adjacent channels. The CS-RZ format is one of the promising signal formats that exhibits lower spectrum bandwidth compared to the conventional RZ format and good tolerance against nonlinear effects.[2,3] Another merit of the CS-RZ format is a reduction of interchannel interference in the time domain. Because the relative carrier phase of an adjacent channel in the CS-RZ format is shifted by π in the time domain, interchannel interference due to pulse broadening is reduced.[2,3] The proposed scheme uses a single supercontinuum (SC) source,[4] which is directly pumped by an optically multiplexed CS-RZ signal. Transmission of simultaneously generated 3.24 Tbit/s (81 WDM × 40 Gbit/s) CS-RZ over a 80 km dispersion compensated link has been experimentally demonstrated using tellurite-based erbium-doped fiber amplifiers (T-EDFAs) with a 66 nm continuous signal band.

We generate a 40 Gbit/s CS-RZ by using an optical multiplexer integrated as a planar lightwave circuit (PLC), as shown in Fig. 32.1(a). A 10 Gbit/s RZ signal is optically time-delayed multiplexed into a 40 Gbit/s signal. The optical carrier phase of each delayed adjacent pulse is shifted by π using optical phase shifters. As a result, a 40 Gbit/s CS-RZ format multiplexing is obtained.

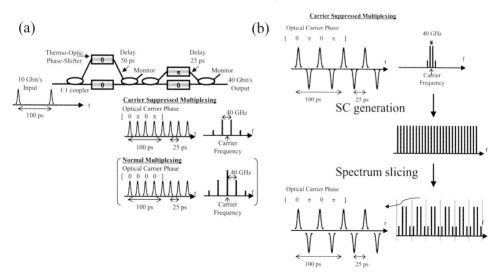

Figure 32.1 Operation principle of (a) CS-RZ generation in the optical domain and (b) simultaneous wavelength-band generation using SC generation and spectrum.

Simultaneous multiwavelength 40 Gbit/s CS-RZ multiplications are performed by SC generation[4] directly pumped by a 40 Gbit/s CS-RZ signal. By spectrum slicing of SC using an arrayed waveguide grating (AWG), simultaneous multiwavelength SC-RZ can be generated as shown in Fig. 32.1(b).[5] The merit of this method is a great ease of WDM channel spacing control. The channel spacing is strictly locked by the microwave mode-locking frequency of the source laser. Thus, the frequency-standardized wavelength band of DWDM in SC-RZ format is generated by a simplified method.

Figure 32.2 shows the experimental setup for simultaneous generation and transmission of a 3.24 Tbit/s (81 WDM × 40 Gbit/s) wavelength band in CS-RZ format. A 10 GHz, 1.5 ps pulse train from the mode-locked laser diode (MLLD) at 1530.33 nm was modulated with a 10 Gbit/s data and optically multiplexed into 40 Gbit/s CS-RZ format by using an optical multiplexer with phase shifter as shown in Fig. 32.3(a). After being amplified, the multiplexed signal was launched into the SC fiber (SCF).[4] The generated 40 Gbit/s CS-RZ induced SC signal was spectrum sliced and recombined by AWGs with a 100 GHz channel spacing to generate multiwavelength 40 Gbit/s CS-RZ signals. T-EDFAs were used for amplification of the continuous signal band in the C and L bands.[6,7] The transmission lines were two pairs of a single-mode dispersion fiber (SMF) and a reversed dispersion fiber (RDF). The total length was 80 km. Signals were wavelength demultiplexed by 100 GHz spacing, 81 channels of AWG (channel 1: 1535.04 nm—channel 81: 1600.60 nm). The resulting WDM DEMUX 40 Gbit/s CS-RZ signal was optically TDM demultiplexed into 10 Gbit/s by using a Symmetric Mach-Zehnder all-optical switch.[8]

Figure 32.3(a) shows the optical spectra of SC at the output of SCF (upper trace), signals before transmission (middle trace), and after transmission and amplified by the T-EDFA (lower trace). Figures 32.3(b)–(d), respectively, show the measured optical spectra after 80 km transmission and WDM demultiplexing of channels 1, 41, and 81. It is clearly shown in the frequency domain that optical carriers were suppressed even after 80 km transmission. Figures 32.3(e)–(g), respectively, show the measured eye diagrams of WDM for channels 1, 41, and 81. Eye diagrams in each WDM channel indicate good eye opening. For all measured channels, the bit error rates (BERs) were less than 1×10^{-9}.

32.3 Tunable Wavelength Conversion

A wideband tunable wavelength conversion of 10 Gbit/s RZ signals is demonstrated by optical time gating of a supercontinuum light source.[9,10] Figure 32.4(a) shows the operational principle of the tunable wavelength conversion by spectrum slicing. By controlling the center wavelength of spectrum slicing, the center wavelength of the spectrum-sliced supercontinuum pulse can be tuned. However, the spectrum-slicing method only generates the RZ pulses at a converted wavelength. Figure 32.4(b) shows the operational principle of the time-gating method. Because supercontinuum pulses generated by nonlinear propagation in the normal dispersion fiber are rectangular pulses with linear upchirping, by shifting the position of the time gating, the center wavelength of time-gated supercontinuum pulse can be tuned. In the data sequence mapping, when data "1" or "0" opens or closes, respectively, the time-gate window, the time-gated supercontinuum pulses possess the same data sequence as the time-gating data sequence. When 10 Gbit/s RZ data sequences are used for controlling the time-gating ON/OFF window of the optical time gating, the time-gated supercontinuum pulses become the wavelength-converted 10 Gbit/s RZ data sequences, since the center wavelength of the time-gated supercontinuum is different, corresponding to its time-gating position as shown in Fig. 32.4(b). As a result, the time-gating method shows the added function, which is data mapping, compared with spectral slicing. In addition, contrast to the conventional wavelength conversion method, such as four-wave mixing-based wavelength conversion, the wavelength tuning can be done only by shifting the time position

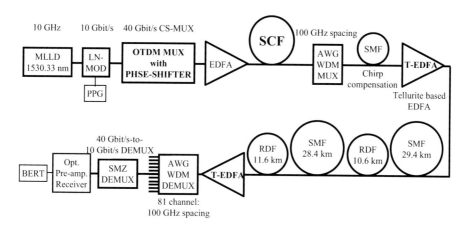

Figure 32.2 Experimental setup.

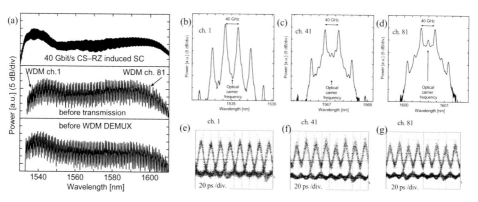

Figure 32.3 (a) Measured optical spectra of SC at the output of the SCF (upper trace), signals before transmission (middle trace), and signals after transmission and amplified by a T-EDFA (lower trace). Measured optical spectra after 80 km transmission and WDM DEMUX of (b) channel 1, (c) channel 41, and (d) channel 81. Measured eye diagrams of (e) channel1, (f) channel 41, and (e) channel 81.

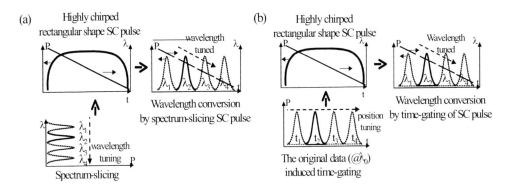

Figure 32.4 Operational principle of tunable wavelength conversion by (a) spectrum slicing and (b) time gating of highly chirped rectangular shape SC pulse.

of the control pulse, resulting in a simple operation. Another advantage of the time-gating method is that the wavelength can be tuned by controlling the pump power of the SCF. Because the amount of frequency chirping depends on the pump pulse power of the SCF, the wavelength tuning range and the center wavelength of the time-gated supercontinuum pulse can be tuned simply by changing the pump pulse power to SCF.

Figure 32.5 shows the experimental setup for tunable wavelength conversion by optical time gating of the supercontinuum. To generate highly linear chirped, wide-width, and rectangular shaped supercontinuum pulses, a 10 GHz repetition rate, 1.5 ps width pulse trains at 1550.0 nm from the MLLD is launched into a 2 km long SCF.[4] A 3.5 km dispersion shifted fiber having normal dispersion is used for further time stretching. For the pump data sequence to the saturable absorber (SA), 10 Gbit/s, 1560 nm, 5.0 ps, mode-locked fiber laser (MLFL) pulses are used. An optical time-gating device is a semiconductor SA with a bias voltage of −1.7 V.[10] The 10 GHz supercontinuum pulses are optical time gated in the SA, which is pumped by 10 Gbit/s RZ data signals. The SA opens and closes the time gates to the supercontinuum pulse when the high-power data "1" and no-power data "0" are used as pump pulses, respectively. The time position of the optical time gating is aligned by an optical delay line. The measured optical

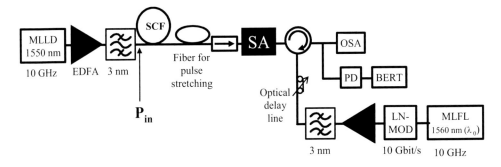

Figure 32.5 Experimental setup.

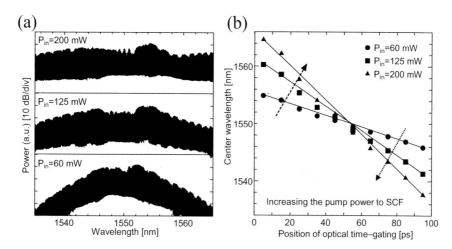

Figure 32.6 (a) Measured SC spectra when the input powers to SCF (P_{in}) are 60 mW, 125 mW, and 200 mW; (b) the center wavelength of the time-gated SC pulses versus time.

spectra are shown in Fig. 32.6(a). When the input powers to the SCF (Pin) are 60 mW, 125 mW, and 200 mW, the 3 dB spectrum widths are 10.1 nm, 20.0 nm, and 30.2 nm, respectively. In each case, pulse shapes are almost rectangular and pulse widths of the full width at half maximum are 100 ps, 103 ps, and 105 ps, and the time-band width products are 131.3, 257.5, and 396.4, respectively.

These results show that the generated supercontinuum pulse is highly chirped and the amount of chirping can be easily controlled by changing the input power to the SCF. Figure 32.6(b) shows the center wavelength of the time gated supercontinuum pulse plotted versus time-gating position. By shifting the time-gating position, the converted wavelength can be precisely tuned almost linearly. Furthermore, by changing the input power to SCF, resulting in supercontinuum spectrum width change, the tuning range can be easily controlled. The entire tuning ranges are 9.0 nm, 18.9 nm, and 27.1 nm for input powers of 60 mW, 125 mW, and 200 mW, respectively. The measured BERs after wavelength conversion were less than 10^{-9}. Power penalties of about 4 dB at a BER of 10^{-9} are mainly due to pulse width changes in the wavelength conversion and amplified spontaneous emission noise in the EDFA allocated before SCF, and 7 dB insertion loss of SA in the open time frame.

32.4 Multiplexing Format Conversion

Multiplexing format conversion and reconversion of OTDM and WDM are demonstrated by use of a supercontinuum generation.[11,12] Conversions of 40 Gbit/s OTDM to 4 × 10 Gbit/s WDM to 40 Gbit/s OTDM are experimentally demonstrated. The operation principle of the multiplexing conversion of TDM to WDM to TDM is illustrated in Fig. 32.7. The conversion scheme is based on ultrafast photonic processing both in the time domain and frequency domain; that is, optical time gating along with time shifting in the time domain, and supercontinuum generation followed by spectrum slicing in the frequency domain. A 40 Gbit/s OTDM signal in the time slot T_i (i = 1,2,3,4) itself generates a supercontinuum, yielding a multiwavelength 40 Gbit/s TDM signal in the same time slot as shown in Fig. 32.7(b). The generated supercontinuum is spectrum sliced at WDM channel wavelengths of λ_i (i = 1,2,3,4) as shown in Fig. 32.7(c). Each WDM signal is time shifted so that the time position of T_i at λ_i aligns within the same time frame as shown in Fig. 32.7(c). Finally, they are optically time gated at 10 GHz repetition rate. Thus, all 40 Gbit/s TDM data in the time slot T_i is converted into 4 × 10 Gbit/s WDM data of λ_i in order, as shown in Fig. 32.7(d). As for WDM-to-TDM conversion, each of four WDM signals is time shifted to position a WDM signal of λ_i located at a time frame of T_i, as shown in Fig. 32.7(f). Multiplexed 4 × 10 Gbit/s WDM signals serve to generate another supercontinuum, as shown in Fig. 32.7(g). By spectrum slicing the supercontinuum at the original wavelength of λ_0, 4 × 10 Gbit/s WDM signals are simultaneously reconverted into a 40 Gbit/s TDM signals, as shown in Fig. 32.7(h).

Figure 32.8 shows the experimental setup of 40 Gbit/s TDM-to-WDM-to-TDM conversions.[13] In the OTDM-to-WDM conversion, 40 Gbit/s data at λ_0 (= 1553.9 nm) was generated by four time multiplexings of 10 Gbit/s signals. A supercontinuum spectrum is generated in SCF #1. An AWG having a channel spacing of 350 GHz is used for four WDM (λ_1 = 1549.7 nm; λ_4 = 1558.2 nm) multiplexing/demultiplexings. The time positions of WDM channels at $\lambda_1 - \lambda_4$ are individually aligned using optical delay lines as shown in Fig. 32.7(c). The time gating at the repetition rate of 10 GHz was carried out by using a semiconductor SA, which was driven by a 10 GHz repetition rate, 2 ps width pulse train at 1532 nm. The time gating opened an about 3 ps time window. The optical time-gated signal was WDM demultiplexed using the four-channel, 350 GHz spacing AWG. Thus, 4 × 10 Gbit/s OTDM to WDM was accomplished.

Regarding the WDM-to-TDM reconversion, after the time positions of WDM channels at $\lambda_1 - \lambda_4$ are individually aligned such as that shown in Fig. 32.7(f) using optical delay lines, 4 × 10 Gbit/s WDM signals were multiplexed using a star coupler. A supercontinuum spectrum is generated in SCF #2. The generated multiwavelength 40 Gbit/s supercontinuum was spectrum sliced at λ_0 (= 1553.9 nm) using a 3 nm optical filter to convert into a 40 Gbit/TDM signal. To begin with the OTDM-to-WDM conversion, at the point indicated in Fig. 32.8, the measured optical spectrum and the corresponding temporal waveform of 40 Gbit/s OTDM are shown in Figs. 32.9(a) and (b), respectively. Figures 32.9(c) and (d) show the converted 4 × 10 Gbit/s WDM signals. Next, the WDM-to-TDM reconversion is performed in series and the experimental results are shown in Figs. 32.9(e) and (f). 40 Gbit/s TDM at λ_0 is converted to 10 Gbit/s WDM channels at $\lambda_1 - \lambda_4$ and reconverted to 40 Gbit/s OTDM at λ_0 with clear eye openings. All of the measured BERs of the converted four channel 10 Gbit/s WDM, and reconverted four channel 10 Gbit/s TDM data were less than 10^{-9}.

The power penalty at BER of 10^{-9} were 4.2 dB in the TDM-to-WDM conversion and 4.3–6.7 dB in the WDM-to-TDM reconversion.

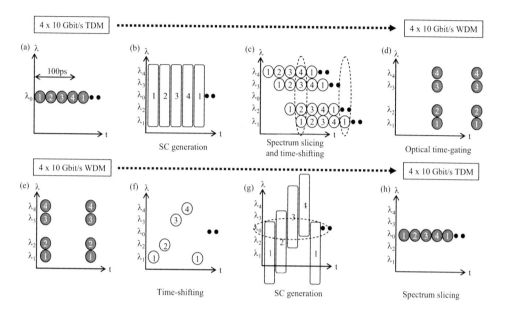

Figure 32.7 Operational principle of conversion and reconversion of OTDM to WDM to OTDM.

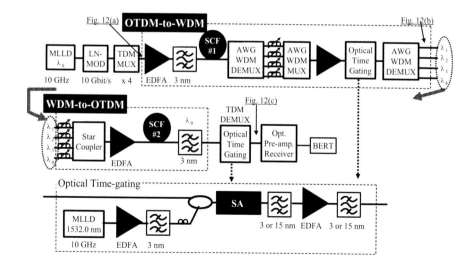

Figure 32.8 Experimental setup.

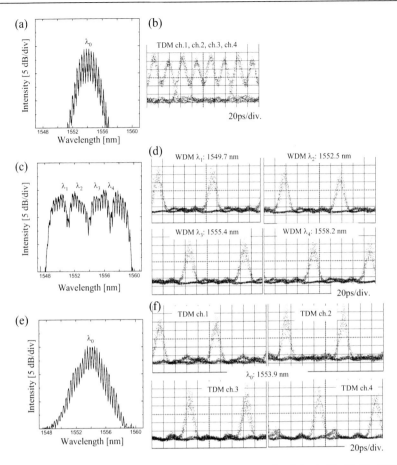

Figure 32.9 (a) Optical spectrum and (b) eye diagram of original 40 Gbit/s TDM; (c) optical spectra and (d) eye diagram of converted 4× 10 Gbit/s WDM; (e) optical spectrum and (f) eye diagram of reconverted 40 Gbit/s TDM.

32.5 Wavelength-Band Conversion

In OTDM/WDM networks, wavelength-band conversions will be key technologies to establish wavelength-band routing.[14] In this section, interwavelength-band conversions of 640 Gbit/s OTDM signals both from C-to-L-band and L-to-C-band followed by 640-to-10 Gbit/s OTDM DEMUX are experimentally demonstrated.[15]

A highly nonlinear dispersion-shifted fiber (HNL-DSF)[16] was used as a wavelength converter. HNL-DSF #1 for wavelength-band conversion has the following characteristics: the zero dispersion wavelength is 1564.8 nm, the dispersion slope 0.032 ps/nm^2/km, the nonlinear coefficient 15 W^{-1} km^{-1}, and the fiber length 100 m. The dependence of the conversion efficiency on the wavelength detuning was measured using two wavelength-tunable CW lasers. The pump wavelength was set to 1565.0 nm. And the pump power P_p was set to 24 dBm and the signal power was −5 dBm. The highest conversion efficiency was −14.6 dB including the insertion loss of 2.1 dB, and the 3 dB bandwidth was 48 nm (1540–1588 nm).

For ultrafast time division demultiplexing using a nonlinear optical loop mirror (NOLM), the ultrafast NOLM should be shortened and the wavelength dispersion and the dispersion slope of the NOLM fiber should be as low as possible. The device length of optical fiber–based NOLM is usually several km long, because the nonlinearly of the optical fiber is low. The minimum switching window and the maximum bit rate for demultiplexing using a NOLM is determined by the pulse width of the control pulse and the walk-off between the signal and control pulses in the NOLM. By the use of a HNL-DSF, which has a high nonlinear coefficient and a low dispersion slope, the walk-off problem can be overcome.[17] HNL-DSF #2 used as the NOLM has the following characteristics: the zero dispersion wavelength is 1561.1 nm, a dispersion slope 0.032 ps/nm²/km, nonlinear coefficient 15 W^{-1} km^{-1}, and fiber length 100 m. In the following experiments, for DEMUX of the C-band (centered at 1550 nm) 640 Gbit/s OTDM signal, the center wavelength of the control pulse was set to 1580 nm. As for the L-band 640 Gbit/s DEMUX, the same wavelength allocation was used; that is, the signal and control wavelengths were 1580 nm and 1550 nm, respectively. In both cases, the walk-off between the signal and control pulses was 100 fs.

Figure 32.10 shows the experimental setup of C-to-L-wavelength-band conversion of a 640 Gbit/s OTDM signal. A 10 GHz repetition rate, 1.5 ps pulse train from a MLLD at 1550 nm was compressed to a pulse width of 700 fs. It was split in two and one was used for a 640 Gbit/s C-band OTDM signal and the other was used for a control signal of wavelength-band converted L-band 640 Gbit/s OTDM signal DEMUX. A compressed pulse train was modulated at 10 Gbit/s and it was optically multiplexed to 640 Gbit/s using an optical time-

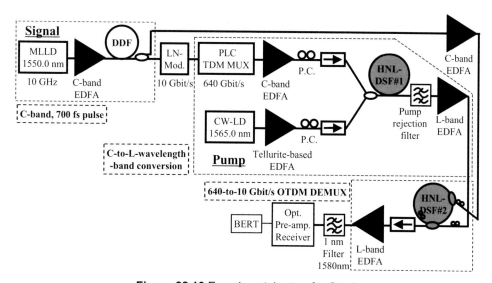

Figure 32.10 Experimental setup for C to L

delayed multiplexer fabricated on a PLC. The pump wavelength λ_p was set to 1565.0 nm, and the pump power P_p was set to 24 dBm. The signal and pump were coupled into HNL-DSF #1 to generate FWM. A wavelength-band-converted 640 Gbit/s signal at 1580 nm was extracted by a rejecting pump wave and amplified using an L-band EDFA. It was optically DEMUX to 10 Gbit/s in a NOLM composed of HNL-DSF #2 controlled by a 10 GHz, 700 fs C-band pulse train.

Figure 32.11(a) shows the optical spectrum measured at the output of HNL-DSF #1. A 640 Gbit/s C-band OTDM signal was wavelength-band converted to an L band centered at 1580 nm. Figure 32.11(b) shows the streak camera traces of the wavelength-band-converted 640 Gbit/s signal and DEMUX 10 Gbit/s signal. These results show that 640 Gbit/s OTDM signal was wavelength-band converted without significant pulse width broadening. The measured BERs of DEMUX wavelength-band-converted signals were less than 10^{-9}.

Figure 32.12 shows the experimental setup of L-to-C wavelength-band conversion of 640 Gbit/s OTDM signal. For the generation of the L-band 1.5 ps pulse train, SC was generated.[8] The SC spectrum was pumped by a 10 GHz, 1.5 ps, 1560 nm, MLLD pulse train and spectrum sliced at 1580 nm using a 3 nm filter. The wavelength-converted pulse train was compressed to a pulse width of 780 fs. After 10 Gbit/s data modulation and 640 Gbit/s optically multiplexing, it was combined with the CW pump and launched into HNL-DSF #1 to generate FWM. A wavelength-band-converted 640 Gbit/s signal at 1550 nm was extracted by rejecting pump wave and amplified using a C-band EDFA. It was optically DEMUX to 10 Gbit/s in a NOLM composed of HNL-DSF #2 controlled by a 10 GHz, 780 fs L-band pulse train.

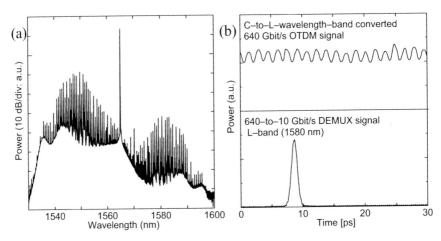

Figure 32.11 (a) Optical spectrum at the output of HNL-DSF #1, (b) streak camera traces of the wavelength-band converted 640 Gbit/s signal (upper trace) and DEMUX 10 Gbit/s signal (lower trace).

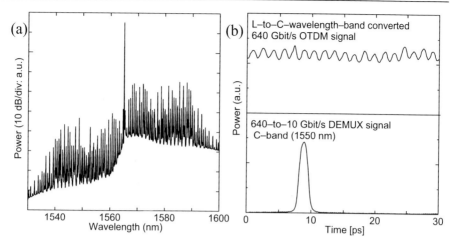

Figure 32.12 Experimental setup for L to C.

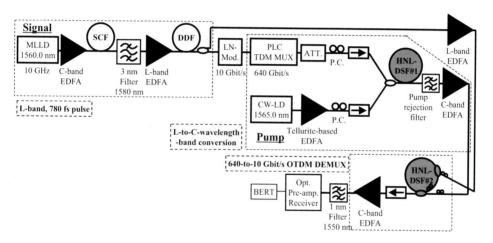

Figure 32.13 (a) Optical spectrum at the output of HNL-DSF #1, (b) streak camera traces of the wavelength-band converted 640 Gbit/s signal (upper trace) and DEMUX 10 Gbit/s signal.

Figure 32.13(a) shows the optical spectrum measured at the output of HNL-DSF #1. A 640 Gbit/s L-band OTDM signal was wavelength converted to a C-band centered at 1550 nm. Figure 32.13(b) shows the streak camera traces of the wavelength-band-converted 640 Gbit/s signal and DEMUX 10 Gbit/s signal. These results show that L-band 640 Gbit/s OTDM signal was also wavelength-band converted without significant pulse width broadening. The measured BERs of DEMUX wavelength converted signals were less than 10^{-9}.

32.6 OCDM Applied for Improved Spectral Efficiency

The efficient utilization of the bandwidth of an optical fiber is a major issue in the design of ultrahigh-speed photonic networks. Therefore, spectral efficiency is becoming a key issue for the full utilization of limited wavelength resources. For the present, the two primary techniques for multiplexing data signals onto the channel of a single fiber are currently OTDM and WDM. OCDM is an alternative method.[18,19] A proper choice of optical codes allows signals from all connected network nodes to be carried without interference between signals. Simultaneous multiple access can thus be achieved without a complex network protocol to coordinate data transfer among the communicating nodes. Therefore, OCDM can provide certain real advantages when applied to photonic networks, due to its unique combination of qualities such as asynchronous transmission, a potential of communication security, soft capacity on demand, and high degree of scalability. As a result, the applications of OCDM range from point-to-point transmission,[20,21] multiple access,[22,23] an optical path network,[24,25] and label switching routing.[26,27]

OCDM is one of the promising multiplexing techniques for ultimate spectral efficiency. A 1.6 bit/s/Hz spectral efficiency OCDM/WDM transmission is demonstrated by applying quaternary phase shift keying (QPSK) optical encoding/decoding accompanied by ultrafast optical time gating and optical hard thresholding for interference noise suppression. As a result, 6.4 Tbit/s OCDM/WDM (4 OCDM × 40 WDM × 40 Gbit/s) transmission using only the C-band wavelength region is experimentally demonstrated.

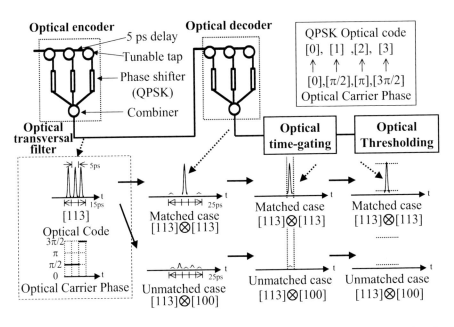

Figure 32.14 Principle operation of QPSK code OCDM with optical time gating and optical hard thresholding.

32.6.1 Key technologies

QPSK coding is applied to upgrade the spectral efficiency in a OCDM/WDM link.[21] Compared to binary phase shift keying (BPSK) optical coding,[20] QPSK optical coding is known to provide a large number of codes with more desirable cross-correlation characteristics. As shown in Fig. 32.14, time-spread QPSK pulse codes are used as optical codes and optical transversal filters are used as optical encoders and decoders. The QPSK-encoded signal is time despread by the decoder. As a result, the QPSK-decoded output shows the correlation waveform; that is, the matched filtering response in the time domain. When the receiver's optical code is matched with the transmitter's optical code, an autocorrelation waveform having a sharp peak at the center is observed. On the other hand, in the case of unmatched codes, a cross-correlation waveform is formed. The transversal filter consists of tunable taps, 5 ps delay lines, programmable quaternary optical phase shifters, and a combiner, all of which are monolithically integrated as a planar lightwave circuit. An optical code of three-chip QPSK pulse code sequence with a chip interval of 5 ps is generated. The impulse response of the optical encoding has a frequency periodicity of 200 GHz, because the chip interval of the optical encoder is 5 ps.[8] As a result, simultaneous multiwavelength encoding having a WDM channel spacing of 200 GHz can be achieved using a single optical encoder when a broadband coherent optical pulse such as an SC pulse is used as a chip pulse.[20,21]

Besides QPSK optical encoding/decoding, ultrafast optical hard thresholding along with optical time gating is applied for high spectral efficiency. In order to achieve the strict sense of high process gain, optical time gating must be introduced to extract only the mainlobe of autocorrelation. As a result, the interference noises arising from interference codes existing outside the time frame of the mainlobe are rejected.[18] In addition, by the introduction of hard thresholding in the optical domain, the interference noise existing inside the time frame of the mainlobe also greatly reduces. Optical time gating was performed at 10 GHz repetition rate by using a 100 m HNLF based NOLM. Optical time gating performs rejecting the interference noise outside the mainlobe of the autocorrelation. For ultrafast operation, a short length of HNLF was used. The control pulse used for 10 GHz optical time gating was 1.5 ps pulse trains and the optical time-gating window ranged from 1.5 ps to 1.8 ps for all WDM channels.[21]

Optical hard thresholding was achieved by the utilization of the nonlinear transmission response of the second NOLM. The NOLM acts as a pulse shaper by setting the proper threshold level, in the sense that it reflects lower-intensity signals but only transmits the higher-intensity signal by limiting its intensity to an appropriate level. By adjusting the input signal power, interference noise inside the autocorrelation mainlobe after optical time gating was suppressed and both signal "0" and "1" level power variations were greatly reduced. For ultrahigh-speed operation, the device length and group delay must be shortened. By use of a short length of 50 m HNLF as a hard thresholding NOLM, the group delay in all WDM channels wavelength range (1533–1564 nm) was < 200 fs, resulting in ultrahigh-speed operation.[21]

32.6.2 Experiments and discussions

Figure 32.15 shows the experimental setup of 6.4 Tbit/s OCDM/WDM (4 OCDM × 40 WDM × 40 Gbit/s) transmission. A 10 GHz, 1.5 ps pulse train of the mode-locked laser diode (MLLD) at 1532 nm was 10 Gbit/s modulated and optically multiplexed to 40 Gbit/s. After amplified to an average power of 0.48 W, it was launched into an SCF.[4] The generated 40 Gbit/s SC signal was linearly polarized and split into eight, and each served as the light source for simultaneous multiwavelength optical encoding using a QPSK optical encoder. Two groups of four different optical encoded signals were generated. One group of which is changed to orthogonal polarized and both groups of four OCDM signals were multiplexed to orthogonal polarized multiwavelength 2 × 4 OCDM × 40 Gbit/s signals. Each group had the WDM channel spacing of 200 GHz, which corresponds to 1.6 bit/s/Hz spectral efficiency. The transmission line was composed of two spans of reversed dispersion fiber (RDF) and single-mode dispersion fiber (SMF) pairs. Each span was 40 km and the total length was 80 km. The average zero dispersion wavelength was 1546.59 nm and the dispersion slope was 0.0087 ps/nm/km/nm. After 80 km transmission, it was split into two and each was WDM DEMUX by 20 channels arrayed wavelength grating (AWG) having the channel spacing of 200 GHz. The pass band wavelengths of two AWGs were separated by 100 GHz (WDM channel 1, 1532.68 nm—channel 40, 1563.86 nm). After WDM DEMUX, frequency chirping was compensated and polarization was demultiplexed. Then, it was decoded by optical decoder and optical time gated at 10 GHz repetition rate by using a NOLM to suppress the

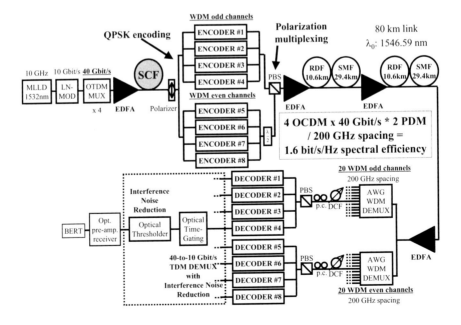

Figure 32.15 Setup of 1.6 bit/s/Hz, 6.4 Tbit/s

interference noise along with 40 Gbit/s to 10 Gbit/s demultiplexing. Optical thresholding was achieved by the utilization of the nonlinear transmission response of the second NOLM. The NOLM acts as a pulse shaper by setting the proper threshold level, in the sense that it reflects lower-intensity signals but only transmits the higher-intensity signal by limiting its intensity to an appropriate level. By adjusting the input signal power, interference noise inside the autocorrelation mainlobe after optical time gating was suppressed and both signal "0" and "1" level power variations were greatly reduced. For the ultrahigh-speed operation, the device length and group delay must be shortened. By use of a short length of 50 m HNLF as a thresholding NOLM, the group delay in all WDM channels wavelength range (1533–1564 nm) was less than 200 fs, resulting in ultrahigh-speed operation.

Figures 32.16(a)–(e), respectively, show the optical spectrum of SC at the output of SCF, multiwavelength 2 × 4 OCDM signals before transmission, after transmission, odd WDM channels, and even WDM channels after WDM and polarization DEMUX. Figure 32.17(a) shows the eye diagrams of WDM channel 40 after optical decoding. The interference noise of the other three unmatched codes severely distorted the signal-to-noise ratio. As shown in Fig. 32.17(b), by optical time gating of the mainlobe of the matched correlation waveforms, interference noise outside the time-gate window was greatly reduced. In addition to this, as shown in Fig. 32.17(c), by introducing optical hard thresholding, interference noise inside the time-gate window was greatly reduced in both signal "0" and "1" levels, resulting in a clear eye opening. Figure 32.18 shows the measured BERs of decoded 4 OCDM × 40 WDM data signals. For all measured 160 channels × 40 Gbit/s signals, the BERs were less than 1×10^{-9}.

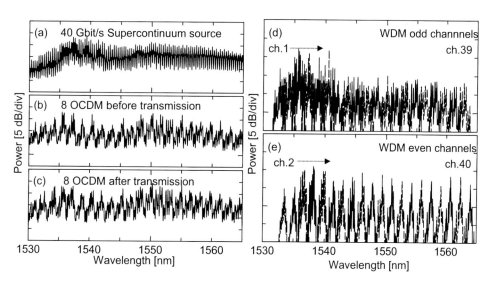

Figure 32.16 (a) SC spectrum, (b) spectra of eight OCDM before transmission, and (c) after transmission. Spectra of (d) odd WDM channels, and (e) even WDM channels after WDM and polarization DEMUX.

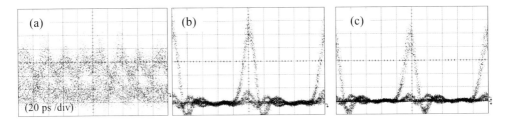

Figure 32.17 Eye diagrams of WDM channel 2 (a) after decoding, (b) after optical time gating, and (c) after optical hard thresholding.

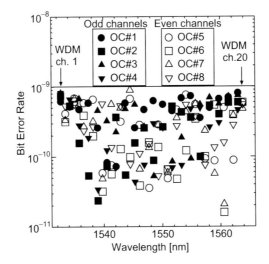

Figure 32.18 Measured BER performances.

32.7 Transparent Virtual Optical Code/Wavelength Path Network

OCDM would be favorably applied not only to multiple access networks, but also to optical path networks. All optical nodes are linked by an optical path such as WP (wavelength path) and VWP (virtual wavelength path). They enhance notonly the transmission capacity but also the cross-connect node throughput cost effectively by capitalizing on the optical routing scheme. The optical code path (OCP), defined as the logical path determined by the optical code (OC), has been proposed within a concept of OCDM networks[24] as shown in Fig. 32.18(a). OCDM can be effectively overlaid onto existing WDM path networks. The introduction of OCs provides soft capacity of networks and saves the network resources. In future hybrid OCDM/WDM networks, flexible OC and wavelength conversion will be key technologies to establish the optical path as shown in Fig. 32.18(b).

A simultaneous OC and wavelength convertible node with optical 3R (retiming, reshaping, regenerating) is experimentally demonstrated by SC generation in the operational bandwidth of 8.05 THz. OC and wavelength convertible virtual optical code/wavelength path (VOCP/VWP) network of a total link length of 180 km with four network nodes is also experimentally demonstrated.[20]

32.7.1 Virtual optical code/wavelength path (VOCP/VWP) network

In OCDM/WDM path networks, there are two approaches to OC and wavelength path assignment: without and with the OC and wavelength conversions. As shown in Fig. 32.19(a), in a wavelength path (WP) network, a wavelength is assigned along the whole optical path; that is, the optical path is identified by a wavelength. Similarly, in OCP as shown in Fig. 32.19(b), an OC is assigned along the entire optical path. In both cases, to establish six optical paths, they require six wavelengths and six optical codes, respectively. On the contrary, optical path provisioning based on OC and wavelength conversion is referred to as a VOCP/VWP. In the latter case, as shown in Fig. 32.19(c), the OC and wavelength are allocated link by link. For example, to establish optical path #3, OC2 at $\lambda 2$ should be converted to OC1 at $\lambda 1$ in node B. In the VOCP/VWP case, to establish the same six optical paths as in Figs. 32.19(a) and (b), there needs to be two wavelengths and optical codes. As a result, the introduction of VOCP/VWP potentially solves the OC and wavelength path assignment problems, which may limit the network and optical path expansion. To maintain the scalability and reconfigurability of OCDM/WDM path networks, simultaneous OC and wavelength conversions are the key technologies, which is analogous to the role of wavelength conversion for VWP networks and OC conversion for VOCP networks.

32.7.2 Experiments and discussions

Figure 32.20(a) shows the experimental setup of a VOCP/VWP path network. For simplicity, data transports of node A to node C in optical path 1 and of node A to node E in optical paths 2 and 3 in Fig. 32.19(b) were demonstrated. Three types of OC and wavelength conversions must be done at node B, i.e., $\lambda 1$-OC1 to $\lambda 1$-OC2 (OC conversion), $\lambda 1$-OC2 to $\lambda 2$-OC2 (wavelength conversion), and $\lambda 2$-

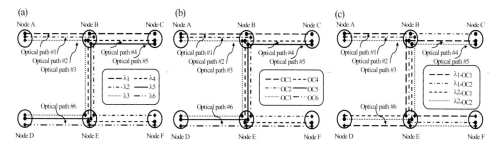

Figure 32.19 Optical paths in (a) WP network, (b) OCP network, and (c) VOCP/VWP network.

OC2 to λ1–OC1 (OC and wavelength conversions) in optical paths 1, 2, and 3, respectively. Each network node was linked by nonzero dispersion-shifted fiber with a dispersion compensation fiber. The total lengths were 180 km, i.e., node A to node B of 80 km, node B to node C of 50 km, and node B to node E of 50 km. The average zero dispersion wavelength was 1550.1 nm and dispersion slope was 0.017 ps/nm/km/nm.

At node A, λ1-OC1, λ1-OC2, and λ2–OC2 were generated and multiplexed. 10 GHz, 1.5 ps pulse trains from the MLLDs at 1549.7 nm (λ1) and 1552.5 nm (λ2) were 10 Gbit/s modulated. Optical transversal filters are used as optical encoders/decoders, each of which consists of eight variable tapped delay lines with phase shifters. The taps were tuned so that they split into an equiamplitude eight-pulse sequence. Each delay line had a 5 ps delay. The carrier phase of each tapped pulse was binary changed by 0 or π. Consequently, an optical code of 8-chip BPSK pulse sequence with a chip interval of 5 ps was generated. In this experiment, OC1 of [00000000] and OC2 of [0π0ππ0π0] were used.

As shown in Fig. 32.20(b), at node B, received signals were decoded, 3R regenerated, wavelength converted by SC generation, and OC converted. First, they were wavelength demultiplexed using an arrayed waveguide grating (AWG) having channel spacing of 350 GHz, the FWHM channel width of 284 GHz, and channel number of 24. And they were decoded by optical decoders. Decoded signals were divided into two, one of which led to injection-locked MLLD for 10 GHz clock recovery. The notable feature of the clock recovery was that waveform distortion and noise accumulation are reset (retiming and reshaping) during this process. By clock recovery, high SNR, negligible excess timing jitter, and coherent optical pulses were generated at λc of 1555 nm. Recovered clock pulses were divided and one of which was used for pump pulses of semiconductor SA time gate. The time window opened while the pump pulse saturated the absorber and its duration was 10 ps when a bias voltage was set to −1.7 V. By optical time gating of the decoded signal, sidelobes of autocorrelation were rejected and interference noise was greatly reduced. After time gating, decoded signals were, in turn, used for pump pulses of the second SA time gate to gate the clock pulse train. By controlling the optical time gate ON/OFF using decoded signals, the data coding could be transferred to newly generated clear optical pulses (regeneration). Thus, all-optical 3R was obtained. For wavelength conversion, SC was produced with pumping by 3R regenerated signals at λc. The SCF was a dispersion-flattened normal dispersion fiber.[4] After spectrum slicing using AWG at the wavelength to be converted, they were optical encoded with the OC to be converted. Consequently, simultaneous OC and wavelength conversion with optical 3R was obtained. As shown in Fig. 32.20(c), after subsequent 50 km transmission, OC and wavelength-converted signals were detected at node C or node E. As was done in node B, after wavelength demultiplexed and decoded, optical time gating was done with the recovered 10 GHz optical clock to reduce interference noise. The bit error rate was measured after detecting by using an optical preamplified receiver.

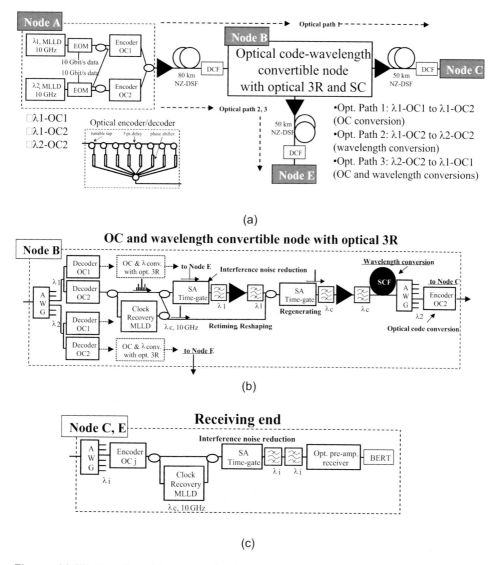

Figure 32.20 Experimental setup of (a) VOCP/VWP network by optical code and wavelength conversion with four network nodes, (b) ultrawideband OC and wavelength conversion by SC generation with optical 3R at Node B, and (c) optical time-gating detection for interference noise

A measured eye diagram of decoded λ1-OC2 after 80 km transmission at node B is shown in Fig. 32.21(a). A measured eye diagram of recovered clock induced optical time-gated signals are shown in Fig. 32.21(b). Compared to Fig. 32.21(a), sidelobe and interference noise were greatly reduced. Figure 32.21(c) shows the optical 3R regenerated decoded signal at λc. As is obviously shown by comparison with Fig. 32.21(b), the SNR was increased and timing jitter was reduced by optical 3R. Figure 32.21(d) shows the optical spectra of 24 WDM

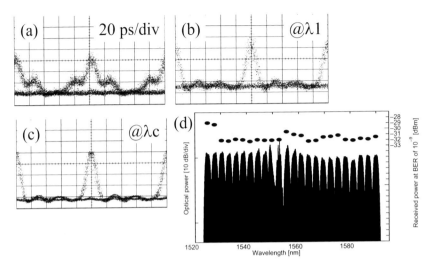

Figure 32.21 (a) Eye diagram of decoded λ1-OC2 at Node B, (b) eye diagram after optical time-gating for interference noise reduction at λ1, (c) eye diagram of decoded signal at λ c after optical 3R and (d) optical spectrum after spectrum-sliced SC pumped by optical 3R decoded signal and received power at BER of 10^{-9} for each wavelength.

channels, which is spectrum-sliced SC pumped by optical 3R decoded signal, which ranged from 1524.9 nm to 1590.0 nm, which corresponds to 8.05 THz bandwidth. Received power at the BER of 10^{-9} for each of the 24 WDM channels are also shown in Fig. 32.21(d). These results show that 24 error-free channels having the total bandwidth of 8.05 THz wavelength and OC conversions were successfully demonstrated using this method.

Figure 32.22 shows the measured BERs of back-to-back, after 80 km transmission, after OC, and wavelength conversion with optical 3R at node B, and after subsequent 50 km transmission at node C or node E. The comparison of BERs after 80 km transmission and BERs after OC and wavelength conversions, almost penalty free, regarding power, OC and wavelength conversions were successfully demonstrated by introducing optical time-gating detection followed by optical 3R. In the viewpoint of a VOCP/VWP path network, OC and wavelength convertible signal transport in the total link length of 180 km with four network nodes were also successfully demonstrated.

32.8 Conclusion

Ultrafast photonic processing is key technology for ultrafast photonic networks. Applications of photonic processing in both OTDM/WDM and OCDM/WDM networks are overviewed. The proposed schemes based on ultrafast photonic processing are to play crucial roles in photonic networks.

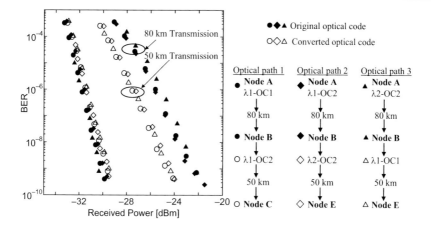

Figure 32.22 Measured BERs: back-to-back, after 80 km transmission, after OC and wavelength conversion with optical 3R, and after subsequent 50 km transmission.

Acknowledgments

The author thanks Prof. Kenichi Kitayama of Osaka University for the collaborations on OCDMA projects and useful discussions, Prof. Takeshi Ozeki of Sophia University for the collaborations on TDM/WDM projects and useful discussions, and Prof. Erich P. Ippen of Massachusetts Institute of Technology for his valuable suggestions and encouragements.

References

1. Sotobayashi, H., Konishi, A., Chujo, W., and Ozeki, T., "Wavelength-band generation and transmission of 3.24-Tbit/s (81-channel WDM40-Gbit/s) carrier-suppressed return-to-zero format by use of a single supercontinuum source for frequency standardization," *OSA J. Opt. Soc. Am. B*, **19**, 2803–2809, 2002.

2. Miyamoto, Y., Hirano, A., Sano, A., Toba, H., Murata,K., and Mitomi, O., "320 bit/s (8 x 40 Gbit/s) WDM transmission over 376-km zero-dispersion-flattened line with 120-km repeater spacing using carrier-suppressed return-to-zero pulse format," Tech. Digest of Optical Amplifiers and Their Applications (OAA '99), ODP4, pp. PdP4-1–PdP4-4, 1999.

3. Kobayashi, Y., Kinjo, K., Ishida, K., Sugihara, T., Kajiya, S., Suzuki, N., and Shimizu, K., "A comparison among pure-RZ, CS-RZ and SSB-RZ format, in 1 Tbit/s (50 x 20 Gbit/s, 0.4 nm spacing) WDM transmission over 4,000 km," Tech. Digest of 26th European Conference on Optical Communication (ECOC 2000), PDP 1.7, 2000.

4. Sotobayashi, H., and Kitayama, K., "325 nm bandwidth supercontinuum generation at 10 Gbit/s using dispersion-flattened and non-decreasing normal dispersion fibre with pulse compression technique," *IEE Electron. Lett.*, **34**, 1336–1337, 1998.

5. Sotobayashi, H., and Kitayama, K., "Observation of phase conservation in multi-wavelength BPSK pulse sequence generation at 10 Gbit/s using spectrum-sliced supercontinuum in an optical fiber," *OSA Opt. Lett.*, **24**(24), 1820–1822, 1999.

6. Yamada, M., Mori, A., Kobayashi, K., Ono, H., Kanamori, T., Oikawa, K., Nishida, Y., Ohishi, Y., "Gain-flattened tellurite-based EDFA with a flat amplification bandwidth of 76 nm," *IEEE Photon. Technol. Lett.*, vol. **10**, 1244–1246, 1998.

7. Makino, T., Sotobayashi, H., and Chujo, W., "1.5 Tbit/s (75 x 20 Gbit/s) DWDM transmission using Er3+-doped tellurite fiber amplifiers with 63 nm continuous signal band," *IEE Electron. Lett.*, **27**(17), 1555–1557, 2002.

8. Nakamura, S., Ueno, Y., Tajima, K., Sasaki, J., Sugimoto, T., Kato, T., Shimoda, T., Itoh, M., Hatakeyama, H., Tamanuki, T., and Sasaki, T., "Demultiplexing of 168-Gb/s data pulses with a hybrid-integrated symmetric Mach-Zehnder all-optical switch," *IEEE Photon. Technol. Lett.*, **12**, 425–427, 2000.

9. Sotobayashi, H., Chujo, W., and Ozeki, T., "Wideband tunable wavelength conversion of 10 Gbit/s RZ signals by optical time-gating of highly chirped rectangular shape supercontinuum light source," *OSA Opt. Lett.*, **26**(17), 1314–1316, 2001).

10. Sotobayashi, H., Chujo, W., and Ozeki, T., "80 Gbit/s simultaneous photonic demultiplexing based on OTDM-to-WDM conversion by four-wave mixing with a supercontinuum light source," *IEE Electron. Lett.*, **37**, 640–642, 2001.

11. Kurita, H., Ogura, I., and Yokoyama, H., "Ultrafast all-optical signal processing with mode-locked semiconductor lasers," *IEICE Trans. Electron.*, **E81-C**, 129–139, 1998.

12. Sotobayashi, H., Kitayama, K., and Chujo, W., "Photonic gateway: TDM-to-WDM-to-TDM conversion and reconversion at 40 Gbit/s (4 channels 10 Gbits/s)," *OSA J. Opt. Soc. Am. B*, **19**, 2810–2816, 2002.

13. Sotobayashi, H., Chujo,W., and Ozeki, T., "Bi-directional photonic conversion between 4x10 Gbit/s OTDM and WDM by optical time-gating wavelength interchange," Optical Fiber Communication Conference (OFC 2001), WM5, WM5-1–WM5-3, Anaheim, March, 2001.

14. Harada, K., Shimizu, K., Kudou, T., and Ozeki, T., "Hierarchical optical path cross-connect systems for large scale WDM networks," *Proc. Optical Fiber Communication Conference* (OFC 1999), **2**, WM55, 356–358, San Diego, February, 1999.

15. Sotobayashi, H., Chujo, W., and Ozeki, T., "Inter-wavelength-band conversions and demultiplexings of 640 Gbit/s OTDM signals," Optical Fiber Communication Conference (OFC 2002), WM2, Anaheim, March, 2002.

16. Aso, O., Arai, S., Yagi, T., Tadakuma, M., Suzuki, Y., and Namiki, S., "Efficient FWM based broadband wavelength conversion using a short high-nonlinearity fiber," *IEICE Transactions on Electronics*, **E83-C**(6), 816-823, 2000.

17. Sotobayashi, H., Sawaguchi, C., Koyamada, Y., and Chujo, W., "Ultrafast walk-off free nonlinear optical loop mirror by a simplified configuration for 320 Gbit/s TDM signal demultiplexing," *Opt. Lett.*, **27**(17), 1555-1557, 2002.

18. Kitayama, K., Sotobayashi, H., and Wada, N., "Optical code division multiplexing (OCDM) and its application to photonic networks," *IEICE Trans. Fundam. Electron., Commun. Comput. Sci.*, **E82-A**, 2616–2626, 1999.

19. Sotobayashi, H., Chujo, W., and Kitayama, K., "Optical code division multiplexing (OCDM) and its application for Peta-bit/s photonic network," *Informat. Sci.*, **149**, 171–182, 2003.

20. Sotobayashi, H., Chujo, W., and Kitayama, K., "1.52 Tbit/s OCDM/WDM (4 OCDM x 19 WDM x 20 Gbit/s) transmission experiment," *IEE Electron. Lett.*, **37**, 700–701, 2001.

21. Sotobayashi, H., Chujo, W., and Kitayama, K., "1.6-b/s/Hz 6.4-Tb/s QPSK-OCDM/WDM (4 OCDM x 40 WDM x 40 Gb/s) transmission experiment using optical hard thresholding," *IEE Photon. Technol. Lett.*, **14**, 555–557, 2002.

22. Sotobayashi, H., and Kitayama, K., "10 Gb/s OCDM/WDM multiple access using spectrum-sliced supercontinuum BPSK pulse code sequences," Optical Amplifiers and their Applications (OAA '99), PD7, pp. Pdp7-1–Pdp7-3, Nara, June, 1999.

23. Kitayama, K., and Murata, M., "Photonic access node using optical code-based label processing and its applications to optical data networking," *IEEE/OSA J. Lightwave Technol.*, **19**, 1401–1415, 2001.

24. Kitayama, K., "Code division multiplexing lightwave networks based upon optical code conversion," *IEEE J. Select. Areas Commun.*, **16**, 1309–1319, 1998.

25. Sotobayashi, H., Chujo, W., and Kitayama, K., "Transparent virtual optical code/wavelength path network," *IEEE J. Sel. Top. Quant. Electron.*, **8**, 699–704, 2002.

26. Sotobayashi, H., and Kitayama, K., "Optical code based label swapping for photonic routing," *IEICE Trans. Commun.*, **E83-B**, 2341–2347, 2000.

27. Kitayama, K., Wada, N., and Sotobayashi, H., "Architectural considerations for photonic IP router based upon optical code correlation," *IEEE/OSA J. Lightwave Technol.*, **18**, 1834–1844, 2000.

28. Sotobayashi, H., and Kitayama, K., "Transfer response measurements of a programmable bipolar optical transversal filter by using the ASE noise of an EDFA," *IEEE Photon. Technol. Lett.*, **11**, 871–873, 1999.

Index

Λ configuration, 537
10Gbps, 589
100Tbps, 597
1D lattice potentials, 559
2D lattice potentials, 559
2R regeneration, 662
3D
    DCT transform, 216
    imaging, 210
    lattice potentials, 560
    visual communication, 222
3R regeneration, 664
4-f lens system, 275, 278, 335, 337

A
ABCD-law, 48, 49
a priori knowledge, 515
absorption coefficient, 325
absorption spectra, 331
acceptor, 295
acoustic modes, 490
acousto-optic (AO), 386
ACT (interferometric autocorrelation trace), 389
actinic radiation, 298
active cable, 594
active components, 583
add-drop multiplexers, 619
ADM (angular division multiplexing), 272–283
air ionization, 279
analytic continuation, 231
anamorphic, 51
angle of view, 215
angular multiplexing, 335
angular selectivity, 291
angular width, 50
antennas, 159
antibunching, 572
anticorrelation dip, 442
anti-Stokes scattering, 497, 509
antisymmetric fractional Fourier transform, 7, 8, 23
antiunitary operation, 524
artificial intelligence, 251
artificial neural networks, 252

astigmatic elliptic dark hollow beams, 59
atmospheric turbulence, 217
atomic ensemble, 536, 548
atom lattices, 539
atom-light interface, 536
Azobenzene, 303
Azo-group photoisomerization, 304
AWG (arrayed waveguide grating), 601

B
backaction cooling, 496
backplane link, 584
bandgap, 144
band-limited function, 237
Bayesian inversion, 518
beam quality, 130
beam splitter, 119, 470
Bell
    basis, 459
    inequality, 460
    states, 451, 460
    state measurement (BSM), 467, 468
BER (bit error rate), 288
Bessel beam, 62, 70
Bessel-Gauss, 67
bidirectional system, 334
    Fourier optical, 337
bidirectional volume holographic read/write head, 334
bilinear relationship, 48
binary phase element, 126
binary phase shift keying (BPSK), 700
binder, 307
biphotons, 438
birefringence, 304, 386
bismuth fiber, 663, 674
bit error rate (BER), 665
bit oriented, 318
black box, 39
"blind" deblurring, 217
Bloch
    band, 541
    mode, 138, 143
    sphere, 524
block matching, 220
board-level, 104
board-level interconnects, 91

Bose-Hubbard Hamiltonian, 540, 561
Bose-Hubbard model, 561
boundary condition, 608
"bra-ket" notation, 157
Bragg
   gratings, 671
   mirror, 146, 148
   selectivity, 330
   matched beam, 334
Bragg's condition, 292
Brownian stochastic force, 493

C

camera fill factor, 2213
canonical integral transform (CT),
   2–4, 9, 11, 12, 16, 21
C-band, 610
CCD/CMOS electronic sensor arrays, 211
central wavelength, 612
CF. See first-order correlation function
chain, 418
chalcogenide
   fiber, 666
   glass, 661
channel waveguide, 611
chaos, 253
chaotic dynamics, 253
characteristic equation, 49
characteristic length scale, 572
charge
   generators, 294
   transport, 294, 296
chirped fiber Bragg gratings, 384
chromatic dispersion, 400, 614
chromatically dispersive medium, 383
chromophore concentrations, 297
chromophores, 295
cloned, 459
CNOT, 544
cognitive science, 251
coherence
   function, 28
   length, 464
coherent
   addition, 129
   code division multiple access
      encoding/decoding, 383
   light, 34
   particle hole admixtures, 567
coincidence
   circuit, 442
   counting rate, 442
   distribution, 452
cold atoms, 534, 535
cold-damping, 490

collective enhancement, 536
collective excitation, 546, 547
collective mode, 538
collective spin, 539
Collins, 41
collisional phase, 543, 545
colloidal solution, 307
color-encoding grating, 272, 280
comb filter, 611
combiner, 613
combining efficiency, 120, 125
common frequency, 123
common path scheme, 230
communications mode, 154, 155, 157
compensating lens, 121
complementary metal oxide semiconductor
   (CMOS), 634, 637
complete tomography, 517
completely positive (CP) map, 513
complex elliptic variables, 67
composite hologram, 273
computational complexity, 213
concentration gradient, 305
conduction band, 145
confinement, 142
continuous medium, 39
continuous-variable, 489
controllable patterns, 367
controlled collisions, 542, 544
convolution, 11
cooling, 490
   regime, 509
   techniques, 494
corner rounding, 609
correlated particles, 459
cosine
   law, 211
   waveforms, 362
counterpart modes, 130
coupling, 636
covalent bonds, 300
covariance matrix, 172
creative thinking, 252
cross-correlation, 373
cross phase modulation (XPM), 664, 668
cross-spectral density, 29, 172
crosstalk, 288, 592, 607
cubic spectral phase, 374, 384
cumulative exposure energy, 327
cumulative grating strength, 327
curvature, 34, 50
curvature matrix, 41
CW laser, 371
cylindrical microlenses, 106

D

dark hollow beam, 57
data coding, 411
data-recording rate, 322
data storage applications, 286
DCF (dispersion compensating fiber), 384
DCT domain filtering, 216
DDF (dispersion-decreasing fiber), 362, 372
dealiasing, 220
deblurring, 216
De Bruijn, 37
deep proton writing, 90
degree of coherence, 175, 177
degree of polarization, 173, 174
(de)multiplexers, 602
denoising, 216
dense integration, 582
density operators, 515
deterministic entangling gate for photons, 534
deterministic phase component, 225
dichromated gelatines, 306
dielectric stacks, 154
differential
  microscopes, 230
  response, 237
diffraction, 156
  efficiency, 287, 291, 302, 608
  limit, 159, 168
  limited spot, 235
  orders, 608
  resolution, 241
diffuse light scattering, 225
digital image processing, 210
dip effect, 443
dipole potential, 556
direction, 50
directional intensity, 29, 31
directional power spectrum, 29
directivity pattern, 225
discrete sinc interpolation, 220
dispersion, 167, 406, 602
dispersion time/frequency map, 399
dispersive spreading of short pulses, 444
dispersive structures, 154
distinguishability, 175
distinguishing information, 451
distortion optical beam, 288
dithering, 605
donor, 295
doped photopolymer, 326
double-well case, 564
Dove prism, 83
downconversion efficiency, 462
DPSS (diode-pumped solid-state), 319

drop-on-demand printing, 93
"dual" gratings, 300
dual-wavelength method, 292
DuPont, 301
Dycril, 298
dye concentration, 308
dynamic
  high, 287
  range, 302, 320, 322, 327, 386

E

ECC (error correction codes), 288
echelle grating, 601
EDFA (erbium-doped fiber amplifier), 389
effective damping, 510
effective index, 139, 606
effective mechanical susceptibility, 506
effective time aperture, 395
effective width, 32
eigenfunctions, 36
eigenmodes, 58
eigenvalues, 36, 47
eigenvectors, 47
elastic pixelwise registration, 220
electrical fixing, 291
electrical I/O, 596
electromagnetic theory, 230
electron beam lithography, 92
electro-optic (EO), 386
electro-optic effect, linear 289
electro-optical photoconductor, 289
elliptic
  Bessel beam, 62
  cylindrical coordinates, 63
  Hermite-Gaussian beam, 60
  Laguerre-Gaussian beam, 62
elliptical
  Gaussian beam, 59
  Hermite-Gaussian beam, 59
embossing, 94
empirical Wiener filtering, 216
encryption, 21
energy efficient optical transceivers, 596
energy-time, 463
entangled state, 519
entanglement, 459, 505, 534
entanglement-based quantum communication, 457
entanglement swapping, 467, 472, 476
entangling gate for photons, 544
EPR, 460
equation of field dynamics, 406
equipartition, 502

erbium-doped fiber amplifiers (EDFAs), 689
etching rate, 100
étendue, 623
evanescently coupled laser, 651
exactly at the same time, 469
excitation spectrum, 568
experimental demonstration, 329
extraocular photoreception, 216
extrapolation interval, 242
eye diagram, 678

F
Fabry-Perot, 138
   resonator, 140, 154
factorization condition, 194, 204
Faraday rotation, 539
FDTD (finite difference time domain), 620
feedback
   bandwidth, 495
   cooling, 494
   gain function, 495
Feigenbaum scenario, 261
femto second, 273
   order, 276
ferroelectric crystal, 289
few-cycle, 405
FFT (fast Fourier transform), 396
fiber, 583
   alignment, 107
   Bragg gratings, 383
   chip coupling, 605
   to the home, 107, 610
figure of merit, 391, 661, 670
filter function, 367
filtering solution, 239
first-order correlation function (CF), 440
first-order interference, 449
first-order optical system, 2
flame hydrolysis (FH), 605
flip-chip interconnects, 585
focal length, 610
focused ion lithography, 92
focusing, 415
Fourier
   holography, 253
   method, 381
   transform AWGs, 623
   transform waveguide array, 611
   transformer, 42
four-wave mixing (FWM), 462, 687
fractional Fourier transform, 7, 19
fractional Fourier transformer, 42
Franson interferometer, 464
free space, 42
free-spaced grating-based shaper, 361

free spectral range (FSR), 604
frequency
   correlation, 447
   entanglement, 447
   doubling, 419
   instability, 368
   modulation, 409
frequency-dependent delay, 374
frequency-dependent phase, 451
frequency-uncorrelated state, 447
Fresnel
   integral, 383
   reflection, 620
   transform, 5, 9
front-side emissions, 590
functionalized systems, 296
fused silica fiber, 408

G
Galilei-type telescope, 337
gate fidelity, 550
Gaussian, 126, 128
   beam, 67, 119
   light, 35
   Schell-model light, 35
   state, 505
Gell-Mann matrices, 178
generalized chirp, 4, 12, 20
generalized convolution, 15–17
generalized radiance, 30
geometric phase, 23
geometric-optical approximation, 43
geometric-optical systems, 46
geometrical optics, 31
geometrical vector flux, 32
Ge-on-SOI, 647
ghost imaging, 425
Gibbs phenomenon, 240
glass waveguides, 605
GRIN, 601, 620
ground state, 500
ground-state cooling, 502
ground states of the Bose-Hubbard
   Hamiltonian, 563
group delay, 614
group dispersion, 418
group index, 139
group-of-lines, 360
group velocity, 138, 165
group velocity difference, 446
group velocity dispersion (GVD), 443
guest-host systems, 296
gyrator, 42
gyrator transform, 7, 8

H

Hanbury-Brown and Twiss correlations, 571
Hanbury-Brown–Twiss intensity-correlation interferometer, 424
harmonic confinement, 557
heating rate, 502
helical laser modes, 76
heliotropism, 216
Helmholtz, 44
Helmholtz equation, 63
Hermite function, 34
Hermite-Gaussian, 14, 21-23, 79, 80
Hermitian function, 29
heterodyne frequency, 232
high index contrast, 616
high refractive index species (HRIS), 309
high repetition rate, 363
high-resolution, 359
higher-order dispersion, 384
higher-order intensity correlation functions, 448
highly coherent CW sources, 472
highly coherent photon-pair source, 474
highly nonlinear fiber, 661, 674
Hilbert-Schmidt distance, 173, 176, 178, 179, 182
Hilbert space, 515
hollow elliptical Gaussian beam, 61
hologram, 79, 83, 84, 154, 168, 223
    composite, 277
    erasure, 292
    macroholograms, 223
    programmed diffuser, 225
    stereoholograms, 224
    synthetic, 223
holographic
    associative memory, 252
    characterization, 325
    data storage, 297, 318, 328
    drives, 286
    medium, 288
    recording materials, 310
holography, 251
homogeneous medium, 45
Hong-Ou-Mandel, 475
horizontal parallax, 224
host matrix, 307
human visual system, 223
hybrid laser, 651
hybrid spread functions, 39
hybridization, 646, 651
hyperpolarizability, 295

I

image
    formation theory, 230, 233
    superresolution, 220
Imaging, 209
imaging system
    optical, 211
    resolving power, 211
imposed, 126
Ince-Gaussian beams, 68
incoherent light, 33
indistinguishable paths, 449
induction period, 300
information processing, 382
informational entropy, 38
informational optics, 209
informational systems, 210
InGaAsP/InP, 605
inhibitor, 299
inhomogeneous medium, 44
injection molding, 94
inorganic crystals, 319
inorganic materials, 293
InPhase Technologies, 303
inseparable field structure, 418
instantaneous bandwidth, 396
integrated photodiodes, 593
intensity distributions, 50
intensity waveform, 389
interaction matrix element, 562, 563
interference, 601
    order, 613
    spectrum, 420
interferogram, 624
interferometric autocorrelation trace. *See* ACT
interferometric combiner, 119
interpolation of sparse data, 220
intra-MCM, 109
intrinsic degrees of coherence, 195, 205
invariance, 194
invariance properties, 190
invariants, 49
ion implantation, 648
ion irradiation, 96
ionic charge distribution, 290
Irgacure 784–doped PMMA, 331
isotropic, 51

J

joint distribution, 447

K

Kerr effect, 643
Kerr nonlinearities, 659, 674

L

Lagrange multipliers, 523
Laguerre-Gaussian, 15, 20, 22, 23, 126, 77, 79, 81
Laguerre-Gaussian beam, 61, 70
Lambertian source, 33
laser cooling rate, 500
laser direct writing, 93
laser photoablation, 93
latent image, 306
lateral superresolution, 231
LCD microdisplay, 337
leaky modes, 145
length difference, 122
lens, 42
    transform, 6
lenses, 211
LIGA, 94, 99
linear
    canonical integral transformation, 40
    integral equation, 238, 243
    interference, 410, 413
    phase, 388
    polarizability, 295
    predictor, 253
linearization, 494
linearized QLEs, 494
line-by-line
    control, 360
    regime, 374
    shaping, 368
line spread function, 234
Liouville approximation, 43, 44
liquid crystal display, 70
liquid crystals, 303
lithium niobate, 289
lithium tantalite, 289
Littrow, 625
local
    frequency spectrum, 30
    measurements, 460
    oscillator, 393
    reflection coefficient, 230, 234
LOCOS, 636
logarithmic negativity, 491, 505
long-distance observation systems, 217
longitudinal spectral decomposition, 383, 400
loss, waveguide propagation, 605
losses, 120, 130
low-shrinkage photopolymers, 320, 333
LPF (low-pass filter), 393
Lucent, 302
Luneburg, 40

M

$M\#$ (M number), 321, 327, 389
Mach-Zehnder, 103
    array spectrometer, 625
    interferometer, 232, 591, 641
magnifier, 42
Mamyshev regenerator, 665
mapping of collective excitations, 545
Maréchal, 103
mask pixelation, 609
material fabrication, 324
material shrinkage, 320
Mathieu
    beams, 63, 64
    functions, 64
    Gauss, 67
    Gauss beam, 66, 67
maximum likelihood, 514, 518
maximum singular value, 182
mean thermal excitation number, 499
mechanical quality factor Q, 500, 502
mechanical susceptibility, 495
median, 220
mesogenic, 303
    composites, 297
metastability, 297
Michelson interferometer, 398
microcavity, 137
microholes, 100
microintegration, 333
microlens, 102, 593
micromechanical oscillator, 491
micromirror, 100, 510
micropillars, 138
microprism, 100, 109
microscope scan, 230
microsolder, 595
microspectrometer, 622
microwave
    signal generation, 383
    spectrum analysis, 383
    spectrum analyzer, 392, 397
mirror, 490
ML (maximum likelihood) estimates, 211, 213
modal
    reflectivity, 139, 143
    volume, 137

Index

mode
    constellation localization, 214
    general localization, 214
    Imaging, 214
    profile mismatch, 142, 143
    splitting, 155
mode-locked fiber laser, 362
modified Iwasawa decomposition, 4
modulation depth, 390
modulator, 45
modules, 592
Moiré-Talbot effect, 624
molecular chains, 96
molecular mass, 97
moment matrix, 32
moments, 47
monochromatic image, 281
monochromator, 600
monomer, 299
Monte Carlo, 110
Montgomery self-imaging, 352
Moore's law, 216, 634
Mott insulator, 541
moving potentials, 543
Moyal, 32
MT ferrule, 593
multimode, 126, 129, 131
multiorder polarization mode dispersion (PMD), 386
multipartite optomechanical entanglement, 503
multiple-quantum well modulators, 589
multiple scattering, 154, 159
multiple-well case, 565
multiplexing, 290, 693
    optical code division (OCDM), 688
    optical time division (OTDM), 688
    wavelength division (WDM), 272–283, 601, 688
multiterabit, 596
multiwavelength lasers, 619
mutual intensity, 36
mutually phase conjugate, 337

N

nanophotonic, 153
nanowire waveguides, 671
Nelder-Mead simplex algorithm, 523
neutral atoms, 535
noise, 429
noise grating, 288
nonclassical light, 437
nondifferential response, 234, 243
nonlinear
    effect, 295
    interaction, 410, 413
    length, 668
    materials, 658
    optics, 461
    polarization rotation, 515, 526
nonlinearity, 406
nonperiodic structures, 155
non-phase-matched, 660
nonresonant electronic
nonstationarity of the pump, 451

O

OAM (orbital angular momentum), 77, 78, 79, 81, 83, 84
OAPTG (optical arbitrary pulse train generation), 367
O-AWG (optical arbitrary waveform generation), 359, 360, 363, 371
object localization, 244
oligomer, 299
OLS (smart) sensors, 211
optical
    aberrations, 334
    axis, 212
    brightness, 131
    carrier, 390
    cavity, 491
    data storage, 310
    dipole force, 556
    fibers, 601
    information processing, 2, 15
    interconnect, 90, 582
    I/O, 586
    lattices, 535, 556
    length, 123
    links, 583
    nonlinearity, 297
    potential, 540
    power monitor, 589, 610
    profile, 233
    pulse shaping, 383
    read/write head, 333
    read/write setup, 336
    reflectometry, 397
    regeneration, 663
    sidebands, 510
    signal processing, 343, 400
    spectrum analyzer, 395
    tapped delay lines, 350
    transmission, 325
opticsless radiation sensors, 211
optoelectronic devices, 596
optoelectronic-VLSI, 587
optomechanical, 490
optomechanical coupling, 494

orbital angular momentum, 14, 22, 50, 65, 66
orbital Poincaré sphere, 21, 22
order one, 194
organic materials, 320
organic photorefractive materials, 293
ormosils, 308
orthonormality relation, 37
orthosymplectic mode, 14, 21
out-of-plane coupling component, 104
overall beam quality, 50
overall degree of coherence, 38

P

π (pi) phase shift, 370
packaging, 586
page oriented, 318, 338
pairs of photons, 461
parallax, 222
parallel optical interconnect, 592
paraxial wave equation, 63
parity, 546
Parseval theorem, 10
partially coherent light, 28
passband, 605
    flattening, 615
pattern recognition, 18
Pauli matrices, 172, 517
PCM (phase-conjugate mirror), 334
peak-to-peak variations, 366
PECVD (plasma-enhanced chemical vapor deposition), 605
Pendellösung
    effect, 310
    fringes, 310
penetration length, 138
periodic modulation, 296
phase component, pseudo-random, 225
phase conjugate, 335
phase distributions, 129
phase error, 609, 614
phase filtering, 367
phase locking, 122
phase-matching, 140, 445, 447, 660, 678
phase-only filtering, 365
phase singularity, 77, 83
phase space, 30
PhC
    cavities, 141
    membranes, 140
    waveguide, 142
photoablation, 107
photochemical reaction, 324
photodiodes, 590

photo-elastic effect, 607
photolithography, 91, 92
photon-correlation imaging, 424, 426
photon-counting detectors, 442
photonic components, 584
photonic crystal cavities, 138
photonic integrated circuits, 658
photonic quantum information interface, 478
photonics-on-VLSI, 595
photons, 533, 534
Photopolymerizable materials, 297
photopolymers, 320
photorefractive
    effect, 289, 293, 295
        polymeric material, 293
        polymers, 295
physical mechanism, 324
PIFSO (planar-integrated free-space optics), 333, 336
planar integration, 338
planar waveguide, 601
plane wave, 34
PMCW, 371
PMD. *See* multiorder polarization mode dispersion
PMMA, 95
Pockels effect, 643
Poincaré sphere, 80, 173
point filtering, 234
point photodetector, 232
point source, 34
point spread function, 39, 103
polarization, 462
    compensator, 607
    dependent loss (PDL), 608
    dependent wavelength shift, 607
    entanglement, 446
    optimization, 197, 198
polarized lights, mixing, 200
Polaroid, 302
polaron, 293
polycondensation, 306
polymer waveguides, 105
polymerization, 298
    ionic, 298
    initiator, 298
porous
    glass, 306
    matrix, 305
position, 50
position measurements, 495
positional intensity, 31
positional power spectrum, 29
power monitoring, 645

power spectrum, 29
power transfer curve, 666, 673
power transfer function, 664
PPLN waveguides, 462
PQ:PMMA [phenanthrenequinone-doped poly(methyl methacrylate)], 305, 323
PQ:PMMA disk, 328
prediction, 252
principle of maximum entropy, 517
printed circuit board, 90
probabilistic process, 465
processes tomography, 519
projection, 549
prolate spheroidal functions, 157
propagation constants, 606
proton beam, 98
proton fluence, 97, 99
prototyping technology, 90
pulse repetition rate, 121
pulse shapers, 387
pulse shaping, 359
pulsed holography, 272
pulsed laser deposition, 662, 676
pulsed pump sources, 444
pupil function, 237
Purcell factor, 137
Pythagorean theorem, 185

Q

Q factor, 137, 669, 671
quadratic-phase signal, 34
quadratic spectral phase, 383
quadrature fluctuations, 494
quadrature variances, 501
quantum
    channel, 514
    complementarity, 184
    efficiency, 294
    gate, 533
    gate protocol, 546
    imaging, 424
    information, 489
    interference, 443, 449
    Langevin equation, 491, 492
    limit, 490
    lithography, 429
    mechanical waves, 167
    memory, 537
    microscopy, 429
    noise correlations, 570
    nondemolition interaction, 537
    observable, 462
    relays, 466, 477
    teleportation, 467, 470

quantum-state transfer, 505
quasi-discrete, 420
quasi-discrete supercontinuum, 406, 411
quasi-homogeneous light, 33
quasi-optimal regularization, 243
qubit, 458
qubit channel, 522
quorum, 517

R

radiant emittance, 32
radiant intensity, 31
radiation losses, 141, 144
radiation pressure, 491
radiation pressure noise, 498
Raman gain, 662
Raman spectroscopy, 383
ray concept, 31
ray invariants, 45
ray-spread function, 40
ray tracing, 105
ray transformation matrix, 3, 41
RDM (recording-plane division multiplexing), 271–283
reactive ion etching (RIE), 608
receivers, 589
reconstruction error, 242, 244, 245
recycling, 147
refractive index, 290
    modulation, 287
regularization operator, 245
regularization procedure, 245
relative phase, 120
reliability, 592
Reoxan, 304
resolving power, 215, 603
resonator, 154
response localization, 244
RF-AWGs (radio-frequency arbitrary waveform generations), 364
RF spectrum analyzer, 366
RF tone, 395
ring resonators, 643
robustness, 507
ROM (read-only memory), 297
Ronchi gratings, 280
rotationally symmetric medium, 44
rotator, 42
rotator transform, 5, 23
roughness, 101
Routh-Hurwitz, 509
Rowland circle, 601
RRM (repetition-rate multiplication), 365

S

samples, 237
sampling functions, 243
sampling grid, 220
sampling interval, 244
satellite pulses, 389
scalability, 123, 125
scalar Helmholtz equation, 63
scaling transform, 5
scattering, 288
scattering rate, 558
Schmidt decomposition, 516
Schwarz, 38
SDHM (scanning differential heterodyne microscope), 232
SDW. *See* spectral decomposition wave
second-order Glauber's CF, 441
second-order interference. *See* two-photon interference
second-order statistical properties, 190
self-action, 408
self-cooling, 490, 496
self-focusing, 418
self-imaging effect, 348
self organization, 124
self-phase modulation, 409, 664, 668
semiconductor optical amplifiers, 659
seminorm, 174, 176, 179
sensing, 601
sensitivity, 293, 302, 320, 327
    of the material, 287
sensitizing dye, 298
sensor noise, 213
sensor SNRs, 214
sequential coherent addition, 125
Shannon sampling theorem, 237
sharpness, 51
shearer, 42
shift multiplexing, 303, 329, 335
shrinkage, 302
    coefficient, 322, 325
    tolerance, 322
sideband-mirror, 506
sideband-sideband, 506
sideband-sideband-mirror, 503
sidebands, 510
sidewall roughness, 619
sidewalls, 608
SiGe, 646
sign gate, 544
signal and idler photons, 440
Silica gels, 306
silica-on-silicon, 605
silicon, 619, 635, 662
    based optoelectronics, 590
    micromachining, 107
    oxynitride, 618
    on-insulator, 605, 619, 634
    photonic, 595
silver halide, 306
similarity transformation, 47
single-mode optical fibers, 337
single-photon superconducting bolometers, 474
singlet state, 451
singular value decomposition, 157, 161
"skin" vision, 216
slab waveguide, 606
SLM, 84
slow light, 143, 154, 165
slow mode, 143
sol-gel, 306
soliton number, 668
SoS, 593
sound waves, 167
space charge field, 293
spatial correlations, 571
    Fourier transform, 29
    frequencies, 234
    light modulators, 79
    modulation, 289
    neighborhood, 216
    selectivity, 211, 212
    width, 50
spatially stationary light, 33
spatiotemporal, 414
spectral
    broadening., 665
    decomposition wave (SDW), 382
    modulation, 420
    phase, 374
    resolution, 361
    width, 416
spectrum extrapolation, 242
spherical OLS sensor, 214
spherical wave, 34, 41
spontaneous parametric downconversion (SPDC), 427, 461
spot-size converters, 620
square-law detection, 393
stability condition, 502, 509
state-dependent optical potentials, 558
state-dependent transport, 541, 542, 551
static athermalization, 617
stereolithography, 94
stigmatization, 604
stochastic media, 46

Stokes
  parameters, 173, 178, 180
  scattering, 497, 509
  sideband, 503
stopping power, 96
storage capacity, 330
storage of information, 286
straggling, 97
stress balancing, 616
stress engineering, 620
strong suppression, 363
subpixels, 220
subwavelength grating (SWG), 620
sum frequency generation, 479
superfluid, 541
superfluid to Mott insulator-transition, 566
super-homodyne receiver, 393
superposition, 458
superprism, 154
superresolution, 221, 243
  coefficient, 242, 244
surface flatness, 100
surface relief gratings, 304
SwissCom, 471
symmetric fractional Fourier transform, 7, 17
symmetry, 51
symplectic, 41
symplectic Gaussian light, 36
synthesize, 363

T
Talbot effect, 348, 365, 624
Talbot imaging, 13
Talbot's bands, 349
tapered mirrors, 146
Tapestry 300r, 303
Taylor expansion, 383
TBWPs (time bandwidth products), 382
TE polarization, 672
telecommunications, 634
temperature dependence, 617
temporal coherence properties, 190
temporal delay, 393
temporal optical processing, 343
theory of diffraction, 211
thermal fixing, 290
thermal lensing, 121
thermal noise, 498
time and frequency domains, 452
time apertures, 391
time-bin, 458, 463
time-bin entanglement, 465
time phase, 559

time resolution, 448
time-separated pump pulses, 449
time-variant, 382, 386
Ti:sapphire, 408
titania nanoparticles, 309
TM polarized, 672
tomography, 514
total energy, 31
total harmonic confinement, 561
total internal reflection, 104
total radiant flux, 32
TPA. *See* two-photon absorption
tracking marks, 329
transport equation, 43
transverse electric (TE), 606
transverse magnetic (TM), 606
trapping frequencies, 561
tripartite entanglement, 510
tunnel matrix element, 562
twist, 35, 49, 50
two-beam interference, 172
two-dimensional polarization, 172
two indexes, 179
two-photon absorption (TPA), 389, 662, 668
two-photon amplitude dispersive spreading, 452
two-photon light, 438
  correlation properties, 452
  interference, 449
  second-order CF, 448
  spectrum of, 440
two-photon spectral amplitude, 439, 443
  asymmetry of, 446
two-qubit gate, 535, 544
two-wavelength storage, 291
type I and type II two-photon light, 439
type II SPDC two-photon amplitude, structure of, 451

U
$U(2) \times U(2)$ invariance, 181
$U(2)$ invariance, 174, 175, 176
$U(3)$ invariance, 180
$U(4)$ invariance, 183
ultrafast
  event, 274
  pulses, 382
  recording, 272
ultra-short reach interconnects, 583
uncertainty principle, 395
uncertainty relations, 37
unidirectional system, 333
unital maps, 522
unitary transformation, 470
universal-NOT gate, 515, 524

unpolarized, 178
    light, 174, 180, 182
    wave, 179
UWB (ultra-wideband), 365

V
valence band, 144
Van Cittert–Zernike, 33
van Cittert-Zernike theorem, 424
van der Waals distances, 300
vectorial wave fields, 46
vertical-cavity surface-emitting laser (VCSEL), 588
vertical tapering, 616
violation, 460
visibility, 174, 183, 446
visibility of interference fringes, 192
VLSI electronics, 584
volume holography, 319
von Neumann entropy, 516
Vycor, 305

W
walk-off, 675
wave equation, 164
wave-spread function, 39
waveform synthesis, 381
waveform synthesizer, 397
waveguide, 139, 583, 671
    array, 611
    birefringence, 607
    gratings, 672
wavelength, 603
    conversion, 674
    splitters, 154
wavelength-to-space-division multiplexing, 587
WDM (wavelength division multiplexing). *See* multiplexing, wavelength division
wearout, 590
Wiener filter, 239
Wigner distribution, 16
wireless communications, 159, 168
WORM (write-once read-many), 297

X
x-ray lithography or LIGA, 93

Y
Young interferometer, 177